BOLYAI SOCIETY
MATHEMATICAL STUDIES

21

BOLYAI SOCIETY MATHEMATICAL STUDIES

1. **Combinatorics, Paul Erdős is Eighty, Vol. 1**
   D. Miklós, V.T. Sós, T. Szőnyi (Eds.)
2. **Combinatorics, Paul Erdős is Eighty, Vol. 2**
   D. Miklós, V.T. Sós, T. Szőnyi (Eds.)
3. **Extremal Problems for Finite Sets**
   P. Frankl, Z. Füredi, G. Katona, D. Miklós (Eds.)
4. **Topology with Applications**
   A. Császár (Ed.)
5. **Approximation Theory and Function Series**
   P. Vértesi, L. Leindler, Sz. Révész, J. Szabados, V. Totik (Eds.)
6. **Intuitive Geometry**
   I. Bárány, K. Böröczky (Eds.)
7. **Graph Theory and Combinatorial Biology**
   L. Lovász, A. Gyárfás, G. Katona, A. Recski (Eds.)
8. **Low Dimensional Topology**
   K. Böröczky, Jr., W. Neumann, A. Stipsicz (Eds.)
9. **Random Walks**
   P. Révész, B. Tóth (Eds.)
10. **Contemporary Combinatorics**
    B. Bollobás (Ed.)
11. **Paul Erdős and His Mathematics I+II**
    G. Halász, L. Lovász, M. Simonovits, V. T. Sós (Eds.)
12. **Higher Dimensional Varieties and Rational Points**
    K. Böröczky, Jr., J. Kollár, T. Szamuely (Eds.)
13. **Surgery on Contact 3-Manifolds and Stein Surfaces**
    B. Ozbagci, A. I. Stipsicz
14. **A Panorama of Hungarian Mathematics in the Twentieth Century, Vol. 1**
    J. Horváth (Ed.)
15. **More Sets, Graphs and Numbers**
    E. Győri, G. Katona, L. Lovász (Eds.)
16. **Entropy, Search, Complexity**
    I. Csiszár, G. Katona, G. Tardos (Eds.)
17. **Horizons of Combinatorics**
    E. Győri, G. Katona, L. Lovász (Eds.)
18. **Handbook of Large-Scale Random Networks**
    B. Bollobás, R. Kozma, D. Miklós (Eds.)
19. **Building Bridges**
    M. Grötschel, G. Katona (Eds.)
20. **Fete of Combinatorics and Computer Science**
    G. Katona, A. Schrijver, T. Szőnyi (Eds.)

Imre Bárány
József Solymosi
(Eds.)

An Irregular Mind

Szemerédi is 70

Springer

JÁNOS BOLYAI MATHEMATICAL SOCIETY

Imre Bárány
Alfréd Rényi Institute of Mathematics
Hungarian Academy of Sciences
Reáltanoda u. 13-15
Budapest 1053
Hungary
e-mail: barany@renyi.hu

József Solymosi
University of British Columbia
Department of Mathematics
Mathematics Road 1984
V6T 1Z2 Vancouver British Columbia
Canada
e-mail: solymosi@math.ubc.ca

Managing Editor:
Gábor Sági
Alfréd Rényi Institute of Mathematics
Hungarian Academy of Sciences
Reáltanoda u. 13-15
Budapest 1053
Hungary
e-mail: sagi@renyi.hu

Mathematics Subject Classification (2000): 11B25, 11B30, 11B75, 05C65, 05C35, 05D10, 05C80, 05D40, and 68R10

ISBN 978-3-662-50641-7 Springer Berlin Heidelberg New York
ISBN 978-3-642-14444-8 (eBook)
DOI 10.1007/978-3-642-14444-8

Springer is a part of Springer Science+Business Media
springer.com

Softcover reprint of the hardcover 1st edition 2010

Cover design: WMXDesign GmbH, Heidelberg

Printed on acid-free paper 44/3142/db – 5 4 3 2 1 0

Contents

FOREWORD

Endre Szemerédi is a mathematician with truely exceptional research power. His influence on today's mathematics is enormous. He solved several fundamental problems that had been raised decades earlier. Many of his results have generated research for the future, and have laid the foundation of new directions in mathematics. Some of his main achievements were born prematurely, their full power and significance became evident only decades later. Although Szemerédi's research interest is combinatiorics, number theory and computer science, his influence on other fields of mathematics, ergodic theory and analysis for instance, is remarkable.

Yet as a mathematician, Szemerédi started out late. He attended medical school for a year, and worked in a factory before studying mathematics. Paul Erdős soon discovered his extraordinary talents and expected great things from him.

Szemerédi lived up to these expectations by proving several fundamental results of tremendous importance. We only mention two of them in this short foreword. Szemerédi was quite young when he proved a central conjecture of Erdős and Turán from the 1930s stating that every sequence of integers with positive density contains arbitrarily long arithmetic progressions. While the significance of this result in combinatorial number theory is obvious, it has led to a new branch of Ramsey theory (called Szemerédi type Ramsey theorems) and of ergodic theory (through the work of Fürstenberg and Katznelson). The recent burst of interest in additive number theory has attracted several outstanding mathematicians and produced spectacular results. But the starting point and also a major tool in almost all of these contributions is Szemerédi's Theorem.

One of the key elements in Szemerédi's solution is a lemma, now called Szemerédi's Regularity Lemma, which is of independent interest, and has an influence that cannot be overestimated. This lemma asserts that every graph can be partitioned into equal parts, whose number only depends on an error bound, so that the bipartite graph between any two such parts is "essentially random" (with a small number of exceptional parts). This statement is counterintuitive since the graph is completely deterministic, and not random. It shows that the randomness is everywhere and inevitably present. It is because of the genius of Szemerédi that the mathematical community (and humankind) has had the opportunity to discover, appreciate, and put to use this ubiquitous and unavoidable presence of randomness.

Szemerédi has an "irregular mind", his brain is wired differently than for most mathematicians. Many of us admire his unique way of thinking, his extraordinary vision. His coauthors often mention that Szemerédi sees things differently, that he is able to find the hidden structure, or able to create one, out of thin air. His insistence that such a structure would work has often proved decisive.

This volume is a celebration of Szemerédi's achievements and personality, on the occasion of his seventieth birthday. It exemplifies his extraordinary vision and unique way of thinking. A number of colleagues and friends, all top authorities in their fields, have contributed their latest research papers to this volume. The topics include extensions and applications of the regularity lemma, the existence of k-term arithmetic progressions in various subsets of the integers, extremal problems in hypergraph theory, and random graphs. All of them are beautiful, Szemerédi type mathematics. It also contains published accounts of the first two, very original and highly successful Polymath projects, one led by Tim Gowers and the other by Terry Tao, and a short and lovely article by András Hajnal on his early encounters with Szemerédi. It is a great shame that Erdős was not able to write one. We finish this foreword by quoting from Hajnal's article:

"Endre has grown to be a great mathematician, one of the best our country has ever given to the world. That is what this conference and this book is about."

Budapest
June 2010

Imre Bárány
József Solymosi

List of Publications of Endre Szemerédi

[1] J. Komlós, A. Sárközy and E. Szemerédi, On sums of powers of complex numbers, *Mat. Lapok,* **15** (1964), 337–347.

[2] A. Sárközy and E. Szemerédi, Über ein Problem von Erdős und Moser, *Acta Arith.,* **11** (1965), 205–208.

[3] A. Sárközy and E. Szemerédi, On the sequence of squares, *Mat. Lapok,* **16** (1965), 76–85.

[4] P. Erdős, A. Sárközy and E. Szemerédi, On divisibility properties of sequences of integers, *Studia Sci. Math. Hungar.,* **1** (1966), 431–435.

[5] P. Erdős, A. Sárközy and E. Szemerédi, On the solvability of the equations $[a_i, a_j] = a_r$ and $(a_i', a_j') = a_r'$ in sequences of positive density, *J. Math. Anal. Appl.,* **15** (1965), 60–64.

[6] P. Erdős, A. Sárközy and E. Szemerédi, On the divisibility properties of sequences of integers. I, *Acta Arith,* **11** (1966), 411–418.

[7] P. Erdős, A. Sárközy and E. Szemerédi, The solvability of certain equations in the dense sequences of integers, *Dokl. Akad. Nauk SSSR,* **167** (1967), 541–544.

[8] P. Erdős, A. Sárközy and E. Szemerédi, On an extremal problem concerning primitive sequences, *J. London Math. Soc.,* **42** (1967), 484–488.

[9] P. Erdős, A. Sárközy and E. Szemerédi, On a theorem of Behrend, *J. Austral. Math. Soc.,* **7** (1967), 9–16.

[10] G. Katona and E. Szemerédi, On a problem of graph theory, *Studia Sci. Math. Hungar,* **2** (1967), 23–28.

[11] P. Erdős, A. Sárközy and E. Szemerédi, On the divisibility properties of sequences of integers. II, *Acta Arith.,* **14** (1967/68), 1–12.

[12] P. Erdős, A. Sárközy and E. Szemerédi, On the solvability of certain equations in sequences of positive upper logarithmic density, *J. London Math. Soc.,* **43** (1968), 71–78.

[13] P. Erdős and E. Szemerédi, On a problem of P. Erdős and S. Stein, *Acta Arith.,* **15** (1968), 85–90.

[14] P. Erdős, A. Sárközy and E. Szemerédi, On some extremal properties of sequences of integers, *Ann. Univ. Sci. Budapest. Eötvös Sect. Math.,* **12** (1964).

[15] P. Erdős, A. Sárközy and E. Szemerédi, Über Folgen ganzer Zahlen, in: *Number Theory and Analysis (Papers in Honor of Edmund Landau)*, 77–86. Plenum, New York, 1969.

[16] E. Szemerédi, On sets of integers containing no four elements in arithmetic progression, *Acta Math. Acad. Sci. Hungar.*, **20** (1969), 89–104.

[17] P. Erdős, A. Sárközy and E. Szemerédi, On divisibility properties of sequences of integers, in: *Number Theory (Colloq., János Bolyai Math. Soc., Debrecen, 1968)*, 35–49. North-Holland, Amsterdam, 1970.

[18] A. Hajnal and E. Szemerédi, Proof of a conjecture of P. Erdős, in: *Combinatorial Theory and its Applications, II (Proc. Colloq., Balatonfüred, 1969)*, 601–623. North-Holland, Amsterdam, 1970.

[19] A. Sárközy and E. Szemerédi, On intersections of subsets of finite sets, *Mat. Lapok*, **21** (1970), 269–278.

[20] E. Szemerédi, On a conjecture of Erdős and Heilbronn, *Acta Arith.*, **17** (1970), 227–229.

[21] E. Szemerédi, On a problem of P. Erdős, in: *Combinatorial Theory and its Applications, III (Proc. Colloq., Balatonfüred, 1969)*, 1051–1053. North-Holland, Amsterdam, 1970.

[22] E. Szemerédi, On sets of integers containing no four elements in arithmetic progression, in: *Number Theory (Colloq., János Bolyai Math. Soc., Debrecen, 1968)*, 197–204. North-Holland, Amsterdam, 1970.

[23] E. Szemerédi, On a problem of W. Schmidt, *Studia Sci. Math. Hungar.*, **6** (1971), 287–288.

[24] P. Erdős and A. Szemerédi, On a Ramsey type theorem, *Period. Math. Hungar.*, **2** (1972), 295–299. Collection of articles dedicated to the memory of Alfréd Rényi, I.

[25] A. Hajnal, E. C. Milner and E. Szemerédi, A cure for the telephone disease, *Canad. Math. Bull.*, **15** (1972), 447–450.

[26] J. Komlós, M. Sulyok and E. Szemerédi, A lemma of combinatorial number theory, *Mat. Lapok*, **23** (1973), 103–108.

[27] E. Szemerédi and G. Petruska. On a combinatorial problem. I, *Studia Sci. Math. Hungar.*, **7** (1972), 363–374.

[28] E. Szemerédi, On graphs containing no complete subgraph with 4 vertices, *Mat. Lapok*, **23** (1973), 113–116.

[29] P. Erdős and E. Szemerédi, On the number of solutions of $m = \sum_{i=1}^{k} \chi_i^k$, in: *Analytic Number Theory (Proc. Sympos. Pure Math., Vol. XXIV, St. Louis Univ., St. Louis, MO., 1972)*, 83–90. Amer. Math. Soc., Providence, R.I., 1973.

[30] E. Szemerédi, On a problem of P. Erdős, *Mathematatika*, **29** (1973), 226–228.

[31] E. Szemerédi, On the difference of consecutive terms of sequences defined by divisibility properties. II, *Acta Arith.*, **23** (1973), 359–361.

[32] E. Szemerédi, On a problem of Davenport and Schinzel, *Acta Arith.*, **25** (1973/74), 213–224.

[33] M. Ajtai and E. Szemerédi, Sets of lattice points that form no squares, *Stud. Sci. Math. Hungar.*, **9** (1975), 9–11.

[34] E. Szemerédi, The number of squares in an arithmetic progression, *Studia Sci. Math. Hungar.*, **9** (1975), 417–422.

[35] B. Bollobás, P. Erdős and E. Szemerédi, On complete subgraphs of r-chromatic graphs, *Discrete Math.*, **13** (1975), 97–107.

[36] S. L. G. Choi, P. Erdős and E. Szemerédi, Some additive and multiplicative problems in number theory, *Acta Arith.*, **27** (1975), 37–50. Collection of articles in memory of Jurij Vladimirovich Linnik.

[37] S. L. G. Choi, J. Komlós and E. Szemerédi, On sum-free subsequences, *Trans. Amer. Math. Soc.*, **211** (1975), 307–313.

[38] S. Józsa and E. Szemerédi, The number of unit distance on the plane, in: *Infinite and Finite Sets (Colloq., Keszthely, 1973; dedicated to P. Erdős on his 60th birthday), Vol. II*, 939–950. Colloq. Math. Soc. Janos Bolyai, Vol. 10. North-Holland, Amsterdam, 1975.

[39] J. Komlós, M. Sulyok and E. Szemerédi, Linear problems in combinatorial number theory, *Acta Math. Acad. Sci. Hungar.*, **26** (1975), 113–121.

[40] J. Komlós and E. Szemerédi, Hamilton cycles in random graphs, in: *Infinite and Finite Sets (Colloq., Keszthely, 1973; dedicated to P. Erdős on his 60th birthday), Vol. II*, 1003–1010. Colloq. Math. Soc. János Bolyai, Vol. 10. North-Holland, Amsterdam, 1975.

[41] E. Szemerédi, On sets of integers containing no k elements in arithmetic progression, in: *Proceedings of the International Congress of Mathematicians (Vancouver, B. C., 1974), Vol. 2*, 503–505. Canad. Math. Congress, Montreal, Que., 1975.

[42] E. Szemerédi, On sets of integers containing no k elements in arithmetic progression, *Acta Arith.*, **27** (1975), 199–245. Collection of articles in memory of Jurij Vladimirovich Linnik.

[43] E. Erdős and E. Szemerédi, On a problem of Graham, *Publ. Math. Debrecen*, **23** (1976), 123–127.

[44] P. Erdős, R. L. Graham and E. Szemerédi, On sparse graphs with dense long paths, in: *Computers and Mathematics with Applications*, 365–369. Pergamon, Oxford, 1976.

[45] P. Erdős and E. Szemerédi, On multiplicative representations of integers, *J. Austral. Math. Soc. Ser. A*, **21** (1976), 418–427.

[46] L. J. Guibas and E. Szemerédi, The analysis of double hashing (extended abstract), in: *Eighth Annual ACM Symposium on Theory of Computing (Hershey, Pa., 1976)*, 187–191. Assoc. Comput. Mach., New York, 1976.

[47] E. Szemerédi, On a problem of P. Erdős, *J. Number Theory*, **8** (1979), 264–270.

[48] M. Ajtai, J. Komlós and E. Szemerédi, There is no fast single hashing algorithm, *Inform. Process. Lett.*, **7** (1979), 270–273.

[49] B. Bollobás, P. Erdős, M. Simonovits and E. Szemerédi, Extremal graphs without large forbidden subgraphs, *Ann. Discrete Math.*, **3** (1978), 29–41. Advances in graph theory (Cambridge Combinatorial Conf., Trinity Coll., Cambridge, 1977).

[50] P. Erdős and E. Szemerédi, Combinatorial properties of systems of sets, *J. Combinatorial Theory Ser. A*, **23** (1978), 308–313.

[51] R. L. Graham and E. Szemerédi, On subgraph number independence in trees, *J. Combinatorial Theory Ser. B*, **24** (1978), 213–222.

[52] L. J. Guibas and E. Szemerédi, The analysis of double hashing, *J. Comput. System Sci.*, **16** (1978), 226–274.

[53] I. Z. Ruzsa and E. Szemerédi, Triple systems with no six points carrying three triangles, in: *Combinatorics (Proc. Fifth Hungarian Colloq., Keszthely, 1976), Vol. II*, volume 18 of *Colloq. Math. Soc. János Bolyai*, 939–945. North-Holland, Amsterdam, 1978.

[54] E. Szemerédi, Regular partitions of graphs, in: *Problèmes Combinatoires et Théorie des Graphes (Colloq. Internat. CNRS, Univ. Orsay, Orsay, 1976)*, volume 260 of *Colloq. Internat. CNRS*, 399–401. CNRS, Paris, 1978.

[55] M. Ajtai, J. Komlós and E. Szemerédi, Topological complete subgraphs in random graphs, *Studia Sci. Math. Hungar.*, **14** (1979), 293–297.

[56] M. Ajtai, J. Komlós and E. Szemerédi, A note on Ramsey numbers, *J. Combin. Theory Ser. A*, **29** (1980), 354–360.

[57] P. Erdős, A. Sárközy and E. Szemerédi, On some extremal properties of sequences of integers. II, *Publ. Math. Debrecen*, **27** (1980), 117–125.

[58] P. Erdős and E. Szemerédi, Remarks on a problem of the *American Mathematical Monthly*, *Mat. Lapok*, **28** (1980), 121–124.

[59] A. Gyárfás, E. Szemerédi and Zs. Tuza, Induced subtrees in graphs of large chromatic number, *Discrete Math.*, **30** (1980), 235–244.

[60] J. Komlós, M. Sulyok and E. Szemerédi, Second largest component in a random graph, *Studia Sci. Math. Hungar.*, **15** (1980), 391–395.

[61] M. Ajtai, P. Erdős, J. Komlós and E. Szemerédi, On Turán's theorem for sparse graphs, *Combinatorica*, **1** (1981), 313–317.

[62] M. Ajtai, J. Komlós and E. Szemerédi, A dense infinite Sidon sequence, *European J. Combin.*, **2** (1981), 1–11.

[63] M. Ajtai, J. Komlós and E. Szemerédi, The longest path in a random graph, *Combinatorica,* **1** (1980), 1–12.

[64] V. Chvátal and E. Szemerédi, On the Erdős–Stone theorem, *J. London Math. Soc. (2),* **23** (1981), 207–214.

[65] J. Komlós, J. Pintz and E. Szemerédi, On Heilbronn's triangle problem, *J. London Math. Soc. (2),* **24** (1981), 385–396.

[66] M. Ajtai, V. Chvátal, M. M. Newborn and E. Szemerédi, Crossing-free subgraphs, in: *Theory and Practice of Combinatorics,* volume 60 of *North-Holland Math. Stud.*, 9–12. North-Holland, Amsterdam, 1982.

[67] M. Ajtai, J. Komlós, J. Pintz, J. Spencer and E. Szemerédi, Extremal uncrowded hypergraphs, *J. Combin. Theory Ser. A,* **32** (1982), 321–335.

[68] M. Ajtai, J. Komlós, V. Rödl and E. Szemerédi, On coverings of random graphs, *Comment. Math. Univ. Carolin.*, **23** (1982), 193–198.

[69] M. Ajtai, J. Komlós and E. Szemerédi, Largest random component of a k-cube, *Combinatorica,* **2** (1982), 1–7.

[70] V. Chvátal and E. Szemerédi, On an extremal problem concerning intervals, *European J. Combin.*, **3** (1982), 215–217.

[71] P. Erdős, A. Hajnal and E. Szemerédi, On almost bipartite large chromatic graphs, in: *Theory and Practice of Combinatorics,* volume 60 of *North-Holland Math. Stud.*, 117–123. North-Holland, Amsterdam, 1982.

[72] M. L. Fredman, J. Komlós and E. Szemerédi, Storing a sparse table with $O(1)$ worst case access time, in: *23rd Annual Symposium on Foundations of Computer Science (Chicago, Ill., 1982),* 165–169. IEEE, New York, 1982.

[73] J. Komlós, J. Pintz and E. Szemerédi, A lower bound for Heilbronn's problem, *J. London Math. Soc. (2),* **2** (1982), 13–24.

[74] E. Szemerédi and Zs. Tuza. Upper bound for transversals of tripartite hypergraphs, *Period. Math. Hungar.*, **13** (1982), 321–323.

[75] M. Ajtai, J. Komlós and E. Szemerédi, Sorting in $c \log n$ parallel steps, *Combinatorica,* **3** (1983), 1–19.

[76] C. Chvatál, V. Rödl, E. Szemerédi and W. T. Trotter, Jr., The Ramsey number of a graph with bounded maximum degree, *J. Combin. Theory Ser. B,* **34** (1983), 239–243.

[77] V. Chvátal and E. Szemerédi, Notes on the Erdős–Stone theorem, in: *Combinatorial Mathematics (Marseille-Luminy, 1981),* volume 75 of *North-Holland Math. Stud.*, 183–190. North-Holland, Amsterdam, 1983.

[78] V. Chvátal and E. Szemerédi, Short cycles in directed graphs, *J. Combin. Theory Ser. B,* **35** (1983), 323–327.

[79] P. Erdős, A. Hajnal, V. T. Sós and E. Szemerédi, More results on Ramsey–Turán type problems, *Combinatorica,* **3** (1983), 69–81.

[80] P. Erdős and E. Szemerédi, On sums and products of integers, in: *Studies in Pure Mathematics,* 213–218. Birkhäuser, Basel, 1983.

[81] J. Komlós and E. Szemerédi, Limit distribution for the existence of Hamiltonian cycles in a random graph, *Discrete Math.,* **43** (1983), 55–63.

[82] E. Szemerédi and W. T. Trotter, Jr. A combinatorial distinction between the Euclidean and projective planes, *European J. Combin.,* **4** (1983), 385–394.

[83] M. Ajtai, J. Komlós and E. Szemerédi, An $O(n \log n)$ sorting network, *Proceedings 15th ACM STOC, Boston, M (1983),* 1–9.

[84] E. Szemerédi and W. T. Trotter, Jr. Extremal problems in discrete geometry, *Combinatorica,* **3** (1983), 381–392.

[85] E. Szemerédi and W. T. Trotter, Jr. Recent progress in extremal problems in discrete geometry, in: *Graphs and Other Combinatorial Topics (Prague, 1982),* volume 59 of *Teubner-Texte Math.,* 316–319. Teubner, Leipzig, 1983.

[86] W. Paul, N. Pippenger, W. Trotter and E. Szemerédi, On determinism versus non-determinism and related problems, in: *Proceedings 24th IEEE FOCS, Tucson, AZ (1983),* 429–438.

[87] M. L. Fredman, J. Komlós and E. Szemerédi, Storing a sparse table with $O(1)$ worst case access time, *J. Assoc. Comput. Mach.,* **31** (1984), 538–544.

[88] A. Gyárfás, J. Komlós and E. Szemerédi, On the distribution of cycle lengths in graphs, *J. Graph Theory,* **8** (1984), 441–462.

[89] H. A. Kierstead, E. Szemerédi and W. T. Trotter, Jr. On coloring graphs with locally small chromatic number, *Combinatorica,* **4** (1984), 183–185.

[90] J. Komlós, J. Pintz and E. Szemerédi, On a problem of Erdős and Straus, in: *Topics in Classical Number Theory, Vol. I, II (Budapest, 1981),* volume 34 of *Colloq. Math. Soc. János Bolyai,* 927–960. North-Holland, Amsterdam, 1984.

[91] L. Babai and E. Szemerédi, On the complexity of matrix group problems, in: *Proceedings 25th IEEE FOCS, Singer Island, FL (1984),* 229–241.

[92] J. Spencer, E. Szemerédi and W. Trotter, Jr. Unit distances in the Euclidean plane, in: *Graph Theory and Combinatorics (Cambridge, 1983),* 293–303. Academic Press, London, 1984.

[93] M. Ajtai, J. Komlós and E. Szemerédi, First occurrence of Hamilton cycles in random graphs, in: *Cycles in Graphs (Burnaby, B.C., 1982),* volume 115 of *North-Holland Math. Stud.,* 173–178. North-Holland, Amsterdam, 1985.

[94] A. Gyárfás, H. J. Prömel, E. Szemerédi and B. Voigt. On the sum of the reciprocals of cycle lengths in sparse graphs, *Combinatorica,* **5** (1985), 41–52.

[95] M. Ajtai, L. Babai, P. Hajnal, J. Komlós, P. Pudlák, V. Rödl, Gy. Turán and E. Szemerédi, Two lower bounds on branching programs, in: *Proceedings 18th ACM STOC, Berkeley, CA (1986)*, 30–38.

[96] L. Babai, P. Hajnal, E. Szemerédi and Gy. Turán, A lower bound for read-once-only branching programs, *J. Comput. System Sci.*, **35** (1987), 153–162.

[97] W. Maass, G. Schnitger and E. Szemerédi, Two tapes are better than one for off-line Turing machines, in: *Proceedings 19th ACM STOC, New York, NY (1987)*, 94–101.

[98] M. Ajtai, J. Komlós and E. Szemerédi, Deterministic simulation in LOGSPACE, in: *Proceedings 19th ACM STOC, New York City, New York (1987)*, 132–140.

[99] V. Chvátal and E. Szemerédi, Many hard examples for resolution, *J. Assoc. Comput. Mach.*, **35** (1988), 759–768.

[100] R. Cole, J. Salowe, W. L. Steiger and E. Szemerédi, Optimal slope selection, in: *Automata, Languages and Programming (Tampere, 1988)*, volume 317 of *Lecture Notes in Comput. Sci.*, 133–146. Springer, Berlin, 1988.

[101] J. Pintz, W. L. Steiger and E. Szemerédi, On sets of natural numbers whose difference set contains no squares, *J. London Math. Soc. (2)*, **37** (1988), 219–231.

[102] P. Ragde, W. Steiger, E. Szemerédi and A. Wigderson, The parallel complexity of element distinctness is $\Omega\left(\sqrt{\log n}\right)$, *SIAM J. Discrete Math.*, **1** (1988), 399–410.

[103] M. Ajtai, D. Karabeg, J. Komlós and E. Szemerédi, Sorting in average time $o(\log n)$, *SIAM J. Discrete Math.*, **2** (1989), 285–292.

[104] M. Ajtai, J. Komlós, W. L. Steiger and E. Szemerédi, Optimal parallel selection has complexity $O(\log\log n)$, *J. Comput. System Sci.*, **38** (1989), 125–133. 18th Annual ACM Symposium on Theory of Computing (Berkeley, CA, 1986).

[105] R. Cole, J. S. Salowe, W. L. Steiger and E. Szemerédi, An optimal-time algorithm for slope selection, *SIAM J. Comput.*, **18** (1989), 792–810.

[106] J. Pintz, W. Steiger and E. Szemerédi, Two infinite sets of primes with fast primality tests, *Proceedings 20th ACM STOC (1988)*, 504–509.

[107] Z. Galil, R. Kannan and E. Szemerédi, On 3-pushdown graphs with large separators, *Combinatorica*, **9** (1989), 9–19.

[108] Z. Galil, R. Kannan and E. Szemerédi, On nontrivial separators for k-page graphs and simulations by nondeterministic one-tape Turing machines, *J. Comput. System Sci.*, **38** (1989), 134–149. 18th Annual ACM Symposium on Theory of Computing (Berkeley, CA, 1986).

[109] M. Geréb-Graus, R. Paturi and E. Szemerédi, There are no p-complete families of symmetric Boolean functions, *Inform. Process. Lett.*, **30** (1989), 47–49.

[110] J. Pintz, W. L. Steiger and E. Szemerédi, Infinite sets of primes with fast primality tests and quick generation of large primes, *Math. Comp.*, **53** (1989), 399–406.

[111] M. Ajtai, H. Iwaniec, J. Komlós, J. Pintz and E. Szemerédi, Construction of a thin set with small Fourier coefficients, *Bull. London Math. Soc.*, **2** (1980), 583–590.

[112] M. Ajtai, J. Komlós and E. Szemerédi, Generating expanders from two permutations, in: *A Tribute to Paul Erdős*, 1–12. Cambridge Univ. Press, Cambridge, 1990.

[113] L. Babai, P. Pudlák, V. Rödl and E. Szemerédi, Lower bounds to the complexity of symmetric Boolean functions, *Theoret. Comput. Sci.*, **74** (1990), 313–323.

[114] P. Hajnal and E. Szemerédi, Brooks coloring in parallel, *SIAM J. Discrete Math.*, **3** (1990), 74–80.

[115] E. Szemerédi, Integer sets containing no arithmetic progressions, *Acta Math. Hungar.*, **56** (1990), 155–158.

[116] F. R. K. Chung, E. Szemerédi and W. T. Trotter, The number of different distances determined by a set of points in the Euclidean plane, *Discrete Comput. Geom.*, **7** (1992), 1–11.

[117] N. Nissan, A. Wigderson and E. Szemerédi, Undirected connectivity in $O\left(\log^{1.5} n\right)$ space, in: *Proceedings 33rd IEEE FOCS, Pittsburgh, PA (1992)*, 24–29.

[118] M. Ajtai, J. Komlós, E. Szemerédi, Halvers and expanders, in: *Proceedings 33rd IEEE FOCS, Pittsburgh, PA (1992)*, 686–692.

[119] J. Pach, W. Steiger and E. Szemerédi, An upper bound on the number of planar k-sets, *Discrete Comput. Geom.*, **7** (1992), 109–123.

[120] M. Ajtai, N. Alon, J. Bruck, R. Cypher, C. T. Ho, M. Naor and E. Szemerédi, Fault tolerant graphs, perfect hash functions and disjoint paths, in: *Proceeding 33rd IEEE FOCS, Pittsburgh, PA (1992)*, 693–702.

[121] P. Erdős, A. Hajnal, M. Simonovits, V. T. Sós and E. Szemerédi, Turán–Ramsey theorems and simple asymptotically extremal structures, *Combinatorica*, **13** (1993), 31–56.

[122] W. Maass, G. Schnitger, E. Szemerédi and Gy. Turán, Two tapes versus one for off-line Turing machines, *Comput. Complexity*, **3** (1993), 392–401.

[123] A. Razborov, E. Szemerédi and A. Wigderson, Constructing small sets that are uniform in arithmetic progressions, *Combin. Probab. Comput.*, **2** (1993), 513–518.

[124] A. Balog, J. Pelikán, J. Pintz and E. Szemerédi, Difference sets without κth powers, *Acta Math. Hungar.*, **65** (1994), 165–187.

[125] A. Balog and E. Szemerédi, A statistical theorem of set addition, *Combinatorica,* **14** (1994), 263–268.

[126] P. Erdős, A. Hajnal, M. Simonovits, V. T. Sós and E. Szemerédi, Turán–Ramsey theorems and K_p-independence numbers, *Combin. Probab. Comput.*, **3** (1994), 297–325.

[127] J. Komlós and E. Szemerédi, Topological cliques in graphs, *Combin. Probab. Comput.*, **3** (1994), 247–256.

[128] A. Sárközy and E. Szemerédi, On a problem in additive number theory, *Acta Math. Hungar.*, **64** (1994), 237–245.

[129] M. Ajtai, J. Komlós and E. Szemerédi, On a conjecture of Loebl, in: *Graph Theory, Combinatorics, and Algorithms, Vol. 1, 2 (Kalamazoo, MI, 1992),* Wiley–Intersci. Publ., 1135–1146. Wiley, New York, 1995.

[130] J. Kahn, J. Komlós and E. Szemerédi, On the probability that a random ± 1-matrix is singular, *J. Amer. Math. Soc.*, **8** (1995), 223–240.

[131] J. Komlós, G. N. Sárközy and E. Szemerédi, Proof of a packing conjecture of Bollobás, *Combin. Probab. Comput.*, **4** (1995), 241–255.

[132] L. Pyber, V. Rödl and E. Szemerédi, Dense graphs without 3-regular subgraphs, *J. Combin. Theory Ser. B,* **63** (1995), 41–54.

[133] J. Komlós, G. N. Sárközy and E. Szemerédi, On the square of a Hamiltonian cycle in dense graphs, in: *Proceedings of the Seventh International Conference on Random Structures and Algorithms (Atlanta, GA, 1995),* volume 9, 193–211, 1996.

[134] J. Komlós and E. Szemerédi, Topological cliques in graphs. II, *Combin. Probab. Comput.*, **5** (1996), 79–90.

[135] P. Erdős, A. Hajnal, M. Simonovits, V. T. Sós and E. Szemerédi, Turán–Ramsey theorems and K_p-independence numbers, in: *Combinatorics, Geometry and Probability (Cambridge, 1993),* 253–281. Cambridge Univ. Press, Cambridge, 1997.

[136] J. Komlós, G. N. Sárközy and E. Szemerédi, Blow-up lemma, *Combinatorica,* **17** (1997), 109–123.

[137] J. Komlós and E. Szemerédi, Topological cliques in graphs, in: *Combinatorics, Geometry and Probability (Cambridge, 1993),* 439–448. Cambridge Univ. Press, Cambridge, 1997.

[138] J. Komlós, Y. Ma and E. Szemerédi, Matching nuts and bolts in $O(n \log n)$ time, *SIAM J. Discrete Math.*, **11** (1998), 347–372.

[139] J. Komlós, G. N. Sárközy and E. Szemerédi, An algorithmic version of the blow-up lemma, *Random Structures Algorithms,* **12** (1998), 297–312.

[140] J. Komlós, G. N. Sárközy and E. Szemerédi, On the Pósa–Seymour conjecture, *J. Graph Theory,* **29** (1998), 167–176.

[141] J. Komlós, G. N. Sárközy and E. Szemerédi, Proof of the Seymour conjecture for large graphs, *Ann. Comb.*, **2** (1998), 43–60.

[142] T. Łuczak, V. Rödl and E. Szemerédi, Partitioning two-coloured complete graphs into two monochromatic cycles, *Combin. Probab. Comput.*, **7** (1998), 423–436.

[143] N. Alon, M. Capalbo, Y. Kohayakawa, V. Rödl, A. Ruciński and E. Szemerédi, Universality and tolerance (extended abstract), in: *41st Annual Symposium on Foundations of Computer Science (Redondo Beach, CA, 2000),* 14–21. IEEE Comput. Soc. Press, Los Alamitos, CA, 2000.

[144] J. Pelikán, J. Pintz and E. Szemerédi, On the running time of the Adleman–Pomerance–Rumely primality test, *Publ. Math. Debrecen,* **56** (2000), 523–534. Dedicated to Professor Kálmán Győry on the occasion of his 60th birthday.

[145] V. Rödl and E. Szemerédi, On size Ramsey numbers of graphs with bounded degree, *Combinatorica,* **20** (2000), 257–262.

[146] N. Alon, M. Capalbo, Y. Kohayakawa, V. Rödl, A. Ruciński and E. Szemerédi, Near-optimum universal graphs for graphs with bounded degrees (extended abstract), in: *Approximation, Randomization and Combinatorial Optimization (Berkeley, CA, 2001),* volume 2129 of *Lecture Notes in Comput. Sci.*, 170–180. Springer, Berlin, 2001.

[147] J. Komlós, G. N. Sárközy and E. Szemerédi, Proof of the Alon–Yuster conjecture, *Discrete Math.*, **235** (2001), 255–269. Combinatorics (Prague, 1998).

[148] J. Komlós, G. N. Sárközy and E. Szemerédi, Spanning trees in dense graphs, *Combin. Probab. Comput.*, **10** (2001), 397–416.

[149] B. Bollobás and E. Szemerédi, Girth of sparse graphs, *J. Graph Theory,* **39** (2002), 194–200.

[150] A. Khalfalah, S. Lodha and E. Szemerédi, Tight bound for the density of sequence of integers the sum of no two of which is a perfect square, *Discrete Math.*, **256** (2002), 243–255.

[151] J. Komlós, A. Shokoufandeh, M. Simonovits and E. Szemerédi, The regularity lemma and its applications in graph theory, in: *Theoretical Aspects of Computer Science (Tehran, 2000),* volume 2292 of *Lecture Notes in Comput. Sci.*, 84–112. Springer, Berlin, 2002.

[152] B. Csaba, A. Shokoufandeh and E. Szemerédi, Proof of a conjecture of Bollobás and Eldridge for graphs of maximum degree three, *Combinatorica,* **23** (2003), 35–72. Paul Erdős and his mathematics (Budapest, 1999).

[153] G. N. Sárközy, S. M. Selkow and E. Szemerédi, On the number of Hamiltonian cycles in Dirac graphs, *Discrete Math.*, **265** (2003), 237–250.

[154] B. Sudakov, E. Szemerédi and V. H. Vu, On a question of Erdős and Moser, *Duke Math. J.*, **129** (2005), 129–155.

[155] E. Szemerédi and V. H. Vu, Long arithmetic progressions in sum-sets and the number of x-sum-free sets, *Proc. London Math. Soc. (3)*, **90** (2005), 273–296.

[156] A. Gyárfás, M. Ruszinkó, G. N. Sárközy and E. Szemerédi, An improved bound for the monochromatic cycle partition number, *J. Combin. Theory Ser. B*, **96** (2006), 855–873.

[157] A. Gyárfás, M. Ruszinkó, G. N. Sárközy and E. Szemerédi, One-sided coverings of colored complete bipartite graphs, in: *Topics in Discrete Mathematics,* volume 26 of *Algorithms Combin.*, 133–144. Springer, Berlin, 2006.

[158] A. Khalfalah and E. Szemerédi, On the number of monochromatic solutions of $x + y = z^2$, *Combin. Probab. Comput.*, **15** (2006), 213–227.

[159] J. Polcyn, V. Rödl, A. Ruciński and E. Szemerédi, Short paths in quasi-random triple systems with sparse underlying graphs, *J. Combin. Theory Ser. B*, **96** (2006), 584–607.

[160] V. Rödl, A. Ruciński and E. Szemerédi, A Dirac-type theorem for 3-uniform hypergraphs, *Combin. Probab. Comput.*, **15** (2006), 229–251.

[161] V. Rödl, A. Ruciński and E. Szemerédi, Perfect matchings in uniform hypergraphs with large minimum degree, *European J. Combin.*, **27** (2006), 1333–1349.

[162] E. Szemerédi and V. Vu, Long arithmetic progressions in sumsets: thresholds and bounds, *J. Amer. Math. Soc.*, **19** (2006), 119–169 (electronic).

[163] E. Szemerédi and V. H. Vu, Finite and infinite arithmetic progressions in sumsets, *Ann. of Math. (2)*, **163** (2006), 1–35.

[164] A. Gyárfás, M. Ruszinkó, G. N. Sárközy and E. Szemerédi, Three-color Ramsey numbers for paths, *Combinatorica,* **27** (2007), 35–69.

[165] A. Gyárfás, M. Ruszinkó, G. N. Sárközy and E. Szemerédi, Tripartite Ramsey numbers for paths, *J. Graph Theory,* **55** (2007), 164–174.

[166] E. Szemerédi, An old new proof of Roth's theorem, in: *Additive Combinatorics,* volume 43 of *CRM Proc. Lecture Notes,* 51–54. Amer. Math. Soc., Providence, RI, 2007.

[167] A. Gyárfás, M. Ruszinkó, G. N. Sárközy and E. Szemerédi, Corrigendum: "Three-color Ramsey numbers for paths" [*Combinatorica,* **27** (2007), no. 1, 35–69], *Combinatorica,* **28** (2008), 499–502.

[168] A. Gyárfás, G. N. Sárközy and E. Szemerédi, The Ramsey number of diamond-matchings and loose cycles in hypergraphs, *Electron. J. Combin.*, **15** (2008), Research Paper 126.

[169] R. Martin and E. Szemerédi, Quadripartite version of the Hajnal–Szemerédi theorem, *Discrete Math.*, **308** (2008), 4337–4360.

[170] H. H. Nguyen, E. Szemerédi and V. H. Vu, Subset sums modulo a prime, *Acta Arith.*, **131** (2008), 303–316.

[171] V. Rödl, A. Ruciński, M. Schacht and E. Szemerédi, A note on perfect matchings in uniform hypergraphs with large minimum collective degree, *Comment. Math. Univ. Carolin.*, **49** (2008), 633–636.

[172] V. Rödl, A. Ruciński and E. Szemerédi, An approximate Dirac-type theorem for k-uniform hypergraphs, *Combinatorica*, **28** (2008), 229–260.

[173] A. Gyárfás, G. N. Sárközy and E. Szemerédi, Stability of the path–path Ramsey number, *Discrete Math.*, **309** (2009), 4590–4595.

[174] V. Rödl, A. Ruciński and E. Szemerédi, Perfect matchings in large uniform hypergraphs with large minimum collective degree, *J. Combin. Theory Ser. A*, **116** (2009), 613–636.

[175] B. Csaba, I. Levitt, J. Nagy-György and E. Szemerédi, Tight bounds for embedding bounded degreee trees, in: *Bolyai Society Mathematical Studies, X. (2010)*, Fete of Combinatorics, 1-44.

[176] I. Levitt, G. N. Sárközy and E. Szemerédi, How to avoid using the Regularity Lemma; Pósa's Conjecture revisited, *Discrete Math.*, **310** (2010), 630–641.

[177] A. Gyárfás, G. N. Sárközy and E. Szemerédi, Long monochromatic Berge cycles in colored 4-unform hypergraphs, *Graphs and Combinatorics*, **26** (2010), 71–76.

[178] A. Gyárfás, G. N. Sárközy and E. Szemerédi, Monochromatic Hamiltonian 3-tight Berge cycles in 2-colored 4-uniform hypergraphs, *Journal of Graph Theory.*, **63** (2010), 288–299.

[179] A. Gyárfás, G. N. Sárközy and E. Szemerédi, Monochromatic matchings in the shadow graph of almost complete hypergraphs, *Annals of Combinatorics*, **14** (2010), 245–249.

BOLYAI SOCIETY
MATHEMATICAL STUDIES, 21

An Irregular Mind
(Szemerédi is 70)
pp. 21–37.

Universality, Tolerance, Chaos and Order

NOGA ALON*

Dedicated to Endre Szemerédi, for his 70th birthday

What is the minimum possible number of edges in a graph that contains a copy of every graph on n vertices with maximum degree a most k ? This question, as well as several related variants, received a considerable amount of attention during the last decade. In this short survey we describe the known results focusing on the main ideas in the proofs, discuss the remaining open problems, and mention a recent application in the investigation of the complexity of subgraph containment problems.

1. Introduction

For a family $\mathcal{H}$ of graphs, a graph G is *$\mathcal{H}$-universal* if it contains a copy of any $H \in \mathcal{H}$. The construction of sparse universal graphs for various families arises in the study of VLSI circuit design. See, for example, [13] and [21] for applications motivating the study of universal graphs with a small number of edges for various families of graphs. There is an extensive literature on universal graphs. In particular, universal graphs for forests have been studied in [12], [19], [20], [25], and universal graphs for planar graphs and other related families have been investigated in [3], [11], [12], [15], [16], [35].

Universal graphs for general bounded-degree graphs have also been considered extensively. For positive integers $k > 2$ and n, let $\mathcal{H}(k, n)$ denote

*Research supported in part by an ERC Advanced grant and by a USA-Israeli BSF grant.

the family of all graphs on n vertices with maximum degree at most k. Various deterministic and randomized constructions of sparse $\mathcal{H}(k,n)$-universal graphs have been found by several researchers, including constructions that satisfy certain fault-tolerance properties, as well as constructions of sparse Ramsey graphs for the problem, namely, sparse graphs for which every two-edge coloring contains a monochromatic $\mathcal{H}(k,n)$-universal graph.

In this survey we discuss the main constructions, including some of those that are not optimal, focusing on the methods used, that apply several interesting ideas. These combine probabilistic techniques with results about graph coloring, tools from matching theory and properties of high girth expanders, as well as sparse versions of the regularity lemma of Szemerédi.

Note that a simple counting argument mentioned in [6] shows that any $\mathcal{H}(k,n)$-universal graph must contain at least $\Omega(n^{2-2/k})$ edges, and a construction showing this is tight is given in [5]. Nevertheless we believe that many of the previous, sub-optimal constructions are interesting in their own right. The study of the Ramsey type question mentioned above is more complicated, and the known estimates for this question are not tight.

2. The Strong Chromatic Number and Universal Graphs

Even the fact that there are $\mathcal{H}(k,n)$-universal graphs with at most $O\big(n^{2-\varepsilon_k}\big)$ edges, for some $\varepsilon_k > 0$ is not obvious. The first construction given in [6] establishes this fact. It is based on the notion of the strong chromatic number of a graph and provides an extremely simple construction of $\mathcal{H}(k,n)$-universal graphs with at most $O\big(n^{2-c/k\log k}\big)$ edges. The construction is in fact so simple that for any n which is a power of $3k-1$, say, $n = (3k-1)^s$, it is a graph $G = G(k,s)$ that can be described in one (short) sentence, as follows. The vertices are all vectors of length s over the alphabet $\{1,2,\dots,3k-1\}$, and two are adjacent if and only if they differ in all coordinates.

Let H be a graph with $|V(H)| = n$. If t divides n we say that H is *strongly t-colorable* if for any partition of $V(H)$ into pairwise disjoint sets V_i, each of cardinality t precisely, there is a proper t-vertex coloring of H in which each color class intersects each V_i in exactly one vertex. If t does not divide n, we say that H is strongly t-colorable if the graph obtained from H by adding to it $t\lceil n/t\rceil - n$ isolated vertices is strongly t-colorable. The *strong chromatic number* of H is the minimum t such that H is strongly t-colorable.

The notion of strong chromatic number is studied in [2], where it shown that the strong chromatic number of any graph with maximum degree k is at most bk, for some (large) absolute constant b. The constant has been improved substantially by Haxell [27], who showed that the estimate bk above can be replaced by $3k-1$.

Given a graph H on $n = (3k-1)^s$ vertices and maximum degree at most k, we have to show it is a subgraph of $G(k,s)$. Partition the vertices of H arbitrarily into sets of size $3k-1$, and, using the fact that the strong chromatic number of H is at most $3k-1$, find a proper $3k-1$-coloring c_1 of it in which each set is multicolored. This provides a partition of the vertices of G into $3k-1$ independent sets of equal size. Partition each of them into new sets of size $3k-1$ each, and find a proper $3k-1$ coloring c_2 in which each of these new sets is multicolored. We now have an ordered pair of colors $\big(c_1(v), c_2(v)\big)$ for each vertex v, all $(3k-1)^2$ color classes are of equal size, and the colors of any pair of adjacent vertices differ in both coordinates. Continuing in this manner s steps, and then mapping the vertex v of H to the vertex $\big(c_1(v), c_2(v), \dots, c_s(v)\big)$ of $G(k,s)$, provides the required embedding of H as a spanning subgraph of $G(k,s)$. The construction for general n is similar, see [6] for more details.

A related construction is given in [3]. Instead of using the notion of the strong chromatic number of a graph, it is based on the fact that if H is an arbitrary graph on n vertices with maximum degree at most k, and $V_1, V_2, \dots, V_m$ is an arbitrary partition of the set of its vertices into pairwise disjoint sets, each of size at least $\frac{c_k}{\varepsilon^2}\log n$, then there are two disjoint independent sets of H, each containing at least a fraction of $\left(\frac{1}{k+1} - \varepsilon\right)$ of each V_i. This is proved by ordering the vertices of H randomly along a line, defining one independent set to be the set of all vertices that appear before all their neighbors, and the other to be the set of all vertices that appear after all their neighbors. A simple probabilistic argument given in [3] (which conveniently applies the Hajnal Szemerédi Theorem [28]) shows that the desired result holds with positive probability. This can now be used in a recursive way that resembles the one in the construction based on strong coloring to construct relatively sparse universal graphs for $\mathcal{H}(k,n)$.

3. Random Universal Fault Tolerant Graphs

It is not surprising that random graphs with appropriate number of vertices and edge-density are $\mathcal{H}(k,n)$-universal with high probability. This is proved in [6]. Let $G(m,p)$ denote, as usual, the random graph on m labelled vertices in which each pair of distinct vertices forms an edge, randomly and independently, with probability p. We say that $G(m,p)$ satisfies a property *asymptotically almost surely*, or *a.a.s.* for short, if the probability it satisfies it tends to 1 as m tends to infinity.

Theorem 3.1 [6]. *For every $\varepsilon > 0$ there exists a positive constant $c = c(\varepsilon)$ such that, for every $k > 2$, the random graph $G(\lceil(1+\varepsilon)n\rceil, p)$ with $p = cn^{-1/k}(\log n)^{1/k}$ is* a.a.s. *$\mathcal{H}(k,n)$-universal. Consequently, for $n > n_0(k)$ there is an $\mathcal{H}(k,n)$-universal graph G with $\lceil(1+\varepsilon)n\rceil$ vertices and at most $(1+\varepsilon)^2cn^{2-1/k}(\log n)^{1/k}$ edges.*

It turns out that if we restrict our attention to bipartite graphs with maximum degree k, then random graphs satisfy, a.a.s., a stronger property. Let $\mathcal{H}(k,n,n)$ denote the set of all bipartite graphs with n vertices in each color class and maximum degree at most k. For a real number α, where $0 < \alpha < 1$, we say that a graph G is *α-fault-tolerant* with respect to a family of graphs $\mathcal{H}$, if every subgraph of G with at least a $1-\alpha$ fraction of the edges of G is $\mathcal{H}$-universal. Note that restricting to bipartite graphs is unavoidable here, as for any graph G, there is a bipartite subgraph G' of G with at least half the edges of G.

Theorem 3.2 [6]. *For every $k > 2$ and $0 < \alpha < 1$ there exist constants $c > 0$ and $C > 0$ such that* a.a.s. *the random graph $G(Cn,p)$ is α-fault-tolerant with respect to $\mathcal{H}(k,n,n)$, where $p = c(\log n/n)^{1/2k}$. Consequently, for $n > n_0(k)$ there is a graph G with $O(n)$ vertices and at most $O(n^{2-1/2k}(\log n)^{1/2k})$ edges, which is α-fault-tolerant with respect to $\mathcal{H}(k,n,n)$.*

It has been shown in [8] (see also [37] for a related result) that, given any *fixed*, particular $H \in \mathcal{H}(k,n)$, the graph H is *a.a.s.* a subgraph of $G(n,p)$, for $p = cn^{-\frac{1}{k}}\log^{1/k} n$, where c is a sufficiently large constant independent of n. By a simple averaging argument, this implies that $G(n,p)$ *a.a.s.* contains **almost every** $H \in \mathcal{H}(k,n)$ as a subgraph. This, however, does not suffice to show that a random graph $G = G((1+\varepsilon)n, p)$ *a.a.s.* contains **every**

$H \in \mathcal{H}(k,n)$ as a subgraph for a fixed $\varepsilon > 0$, as stated in Theorem 3.1. To prove this statement, one first shows that the random graph G satisfies a.a.s. certain properties concerning the number and distribution of sets of common neighbors of arbitrary sets of vertices of size at most k. It is then possible to apply Hall's theorem and show that any graph that satisfies these properties is $\mathcal{H}(k,n)$-universal. See [6] for more details.

The proof of Theorem 3.2 is more complicated. It is based on a combination of a sparse version of the regularity lemma with a hypergraph packing result proved in [36] and several additional ideas. A related problem regarding the construction of sparse fault tolerant graphs is discussed in [1].

4. Universal Graphs and Products of Expanders

A different approach for constructing sparse $\mathcal{H}(k,n)$-universal graphs is described in [4], [5], following an initial construction given in [7]. The first result gives such universal graphs with exactly n vertices.

Theorem 4.1 [4]. *For every $k > 2$ there exists an (explicitly constructible) $\mathcal{H}(k,n)$-universal graph T with n vertices and at most $c(k)n^{2-2/k}\log^{4/k} n$ edges, for some constant $c(k)$.*

The graphs in the second result have more vertices, but have an optimal number of edges, up to a constant factor.

Theorem 4.2 [5]. *For every $k > 2$ there exist positive constants $c_1 = c_1(k)$ and $c_2 = c_2(k)$ so that for every n there is an (explicitly constructible) $\mathcal{H}(k,n)$-universal graph G with at most $c_1 n$ vertices and at most $c_2 n^{2-2/k}$ edges.*

The construction in the two results above are similar, but the proofs of universality are different. In particular, unlike the proof in [5], the proof that the construction of [4] is $\mathcal{H}(k,n)-$universal has the intriguing property that it is probabilistic (although the construction is explicit). We proceed with a description of the construction in [5].

Let $k > 2$ be an integer and put $m = 20n^{1/k}$. Let F be a constant degree high girth expander on m vertices. Specifically, we assume that F is an (m,d,λ)-graph, where d is an appropriate absolute constant. This means that F is d-regular and all its eigenvalues but the largest have absolute value at most λ. It is convenient to assume that F is Ramanujan, that is, $\lambda \leq 2\sqrt{d-1}$. We also assume that the girth of F is at least $\frac{2}{3}\log m/\log(d-1)$. Explicit constructions of such high girth expanders, for every $d = p+1$, where p is a prime congruent to 1 modulo 4, have been given in [31], [32]. Let $G = G_{k,n}$ be the graph whose vertex set is $V(G) = \left(V(F)\right)^k$, where two vertices $(x_1, x_2, \ldots, x_k)$ and $(y_1, y_2, \ldots, y_k)$ are adjacent iff there exist at least two indices i such that x_i and y_i are within distance 4 in F. Note that G has $m^k = O(n)$ vertices and $O\left(nm^{k-2}\right) = O\left(n^{2-2/k}\right)$ edges.

Theorem 4.2 follows by showing that the graph $G_{k,n}$ is $\mathcal{H}(k,n)$-universal. This is done by establishing a graph decomposition result, and by combining it with some properties of high girth expanders. A sketch of the argument follows.

4.1. A graph-decomposition result

A *homomorphism* from a graph Z to a graph T is a mapping of the vertices of Z to those of T such that adjacent vertices in Z are mapped to adjacent ones in T. Note that there is an injective homomorphism from Z to T iff Z is a subgraph of T.

The *k-th power* T^k of a graph $T = \left(V(T), E(T)\right)$ is the graph whose vertices are the vertices of T, and two are adjacent iff the distance between them in T is at most k. Let $P = P_n$ denote the path on n vertices, that is, the graph whose set of vertices is $[n] = \{1, 2, \ldots, n\}$, where i, j are connected iff $|i-j| = 1$.

An *augmentation* of a graph $T = (V, E)$ is any graph obtained from T by choosing an arbitrary (possibly empty) subset $U \subset V$, adding a new set U' of $|U|$ vertices, and adding a matching between U and U'. Thus, an augmentation of T is obtained from it by connecting new vertices of degree 1 to some of its vertices.

Call a graph *thin* if its maximum degree is at most 3 and each connected component of it is either an augmentation of a path or of a cycle, or a graph with at most two vertices of degree 3. It is easy to check that every thin graph H on n vertices is a (spanning) subgraph of the forth power of the path P_n, that is, there is a bijective homomorphism from each such H to P_n^4.

Theorem 4.3 [4]. *Let $k \geq 2$ be an integer, and let H be an arbitrary graph of maximum degree at most k. Then there are k spanning subgraphs*

$H_1, H_2, \ldots, H_k$ of H such that each H_i is thin, and every edge of H lies in precisely two graphs H_i. ∎

The assertion of the theorem for even values of k is an immediate consequence of Petersen's Theorem (c.f., e.g., [39]). The proof for odd values of k requires some work based on techniques from Matching Theory.

4.2. A sketch of the universality of $G_{k,n}$

To prove Theorem 4.2 we have to show that every graph $H \in \mathcal{H}(k,n)$ is a subgraph of $G = G_{k,n}$.

Given such an $H = (V, E)$, let $H_1, H_2, \ldots, H_k$ be as in Theorem 4.3, and note that as all of them are spanning subgraphs of H, the set of vertices of each of them is V. As each H_i is thin, there are injective homomorphisms $g_i : V \mapsto [n]$ from H_i to P_n^4. The main part of the proof is to show that there are homomorphisms $f_i : [n] \mapsto V(F)$ from the path P_n to the expander F, such that the mapping $f : V(H) \mapsto V(G)$ given by $f(v) = \big(f_1(g_1(v)), f_2(g_2(v), \ldots, f_k(g_k(v))\big)$ forms an injective homomorphism from H to G, thus implying that H is a subgraph of G. To do so, we define each f_i as a homomorphism from the path P_n to F, given by a non-backtracking walk. Since the girth of F exceeds 4, this ensures that each composition $f_i(g_i(\cdot))$ is a homomorphism from H_i to the forth power F^4 of F. By the definition of G, this implies that f is indeed a homomorphism from H to G. Indeed, for any pair u, v of adjacent vertices of H there are two indices i such that u, v are adjacent in H_i, as each edge of H is covered by two of the graphs H_i. For each such index i, $g_i(u)$ and $g_i(v)$ are distinct and within distance 4 in P, implying that $f_i(g_i(u))$ and $f_i(g_i(v))$ are distinct and within distance 4 in F, that is, they are adjacent in F^4. Hence $f(u)$ and $f(v)$ are adjacent in G, and f is a homomorphism, as needed.

The crucial part of the proof is to show that the homomorphisms f_i can be defined so that f is injective. This is done by a careful analysis, based on the spectral properties of the expander F. The full details can be found in [5].

5. A Ramsey Type Problem

Theodore Motzkin is credited with the observation that complete disorder, or total chaos, is impossible. This sentence captures the essence of Ramsey Theory. Indeed, Ramsey theory implies that in great generality, every sufficiently large system must contain a substantial ordered sub-system. The quantitative version of this statement for graphs of bounded degree has been considered some 25 years ago by Chvátal, Rödl, Szemerédi and Trotter in [22]. Their main result asserts that the Ramsey number of any graph H on n vertices and maximum degree k is at most $O(n)$. That is, for any fixed k there exists a constant c so that for any graph H on n vertices with maximum degree k, any two coloring of the edges of the complete graph on cn vertices contains a monochromatic copy of H. In a recent paper of Kohayakawa, Rödl, Schacht and Szemerédi [30] it is shown that the complete graph can be replaced by a sparser graph, with only $O\big(n^{2-1/k}\log^{1/k} n\big)$ edges. In fact, a random graph with cn vertices and $Cn^{2-1/k}\log^{1/k} n$ edges satisfies this property with high probability, where c and C are appropriate constants. Moreover, this random graph satisfies, a.a.s., the above Ramsey-type property for all such graphs H simultaneously. Indeed, any two coloring of its edges contains a monochromatic $\mathcal{H}(k,n)$-universal graph. Note that as described in the previous sections, the minimum possible number of edges of any $\mathcal{H}(k,n)$-universal graph is $\Theta(n^{2-2/k})$. The random graph considered here has a somewhat larger number of edges, but satisfies a much stronger condition.

The proof described in [30] is a delicate application of the regularity method, adapted to an appropriate sparse setting. The regularity method, which was initially based on the regularity lemma of Szemerédi proved in [38], turned out to be one of the most powerful tools in Extremal Graph Theory, with applications in other areas including Combinatorial Number Theory and theoretical Computer Science. The initial applications in Graph Theory considered only dense graphs, but it later turned out that sparse versions can be useful as well. The main ingredient in the proof of [30] is an embedding lemma, that enables one to embed bounded degree graphs of linear order in graphs with sufficiently strong pseudo-random properties. A useful phenomenon here is the fact that regularity is typically inherited at a scale that is much finer than the scale at which it is assumed. The detailed proof can be found in [30].

6. Balanced Homomorphisms and Subgraph Containment Problems

The Color Coding technique, introduced in [10], supplies a method for deciding if a given input graph G on n vertices contains a copy of a prescribed graph H with t vertices and treewidth w, in time $2^{O(t)}n^{O(w)}$. This means that the H-subgraph problem for graphs H with bounded treewidth is fixed-parameter tractable when the parameter is the size of the graph H. See [23] for the definition of fixed-parameter tractability, and [34] for the definition of treewidth. It is more convenient to consider a somewhat better understood problem, which we call here the colored H-subgraph problem. The input to this problem is a graph G whose vertices are colored by the numbers $\{1, 2, \ldots, h\}$ that represent the h vertices of H, and the objective is to decide whether or not there is a copy of H in G, in which the vertex playing the role of $i \in V(H)$ is colored i.

The work of Marx [33], improving earlier ideas of Grohe [26] shows that in fact, for every graph H, the treewidth of H essentially captures the complexity of this problem. More precisely, this means that if the Exponential Time Hypothesis of [29] holds, that is, 3-SAT on m variables cannot be solved in time $2^{o(m)}$, then there is no algorithm that solves the colored H-subgraph problem on an n vertex graph in time $n^{o(w/\log w)}$, where H is a fixed graph and $w = w(H)$ is its treewidth. Note that, as usual, the little-o notation here means that formally one has to consider an infinite family of graphs H, and the term $o(w/\log w)$ is a quantity whose ratio to $w/\log w$ tends to zero as w tends to infinity. We will, however, apply here and in what follows a slight abuse of notation, and use the o terminology even when discussing a fixed graph H, having the formal interpretation in mind. Note also that it has been proved already in [17], [18] that under the Exponential Time Hypothesis there is no algorithm that solves the K_w-subgraph containment problem for a clique of size w on an input graph on n vertices in time $n^{o(w)}$, and the novelty in the results of [26] and [33] is to show that the treewidth is the crucial parameter capturing the complexity of the problem for any graph H, and not only for cliques.

A (rough) sketch of the proof in [33] is the following. Given a 3-SAT formula with m variables and a linear number of clauses (which is known to be as difficult as the general case, see [29]), represent it by a graph F with $O(m)$ edges. A function mapping each vertex of F to a connected subset of H is called an *embedding of depth* d (of F into H) if the endpoints of each edge of F are mapped to sets that are within distance 1 or 0 in H, and the inverse image of every vertex of H is of size at most d.

The crucial step in the proof is to use the fact that the treewidth of H is w in order to show that F (and in fact any graph with $O(m)$ edges) has an embedding of depth at most $O(m \log w/w)$ into H.

Next, construct a colored graph G by replacing each vertex i of H by an independent set of size $2^{O(m \log w/w)}$, representing all possible assignments to the variables of the formula mapped to this vertex by the above embedding. All vertices of this set are assigned the color i. The edges of G can now be defined in such a way that each satisfying assignment will correspond to a colored copy of H in G, and vice versa.

If we can now solve the colored H-subgraph problem for G in time $n^{o(w/\log w)}$, where $n = |V(H)| 2^{O(m \log w/w)}$ is the number of vertices of G, we will be able to solve the satisfiability instance in time $2^{o(m)}$, contradicting the Exponential Time Hypothesis.

The main combinatorial part of the argument above is the proof that if the treewidth of H is w, then any graph with m edges can be embedded in it in a balanced way as described above. A natural problem, raised in [33], is whether the $\log w$ term in this embedding result can be omitted; this will make the result tight, up to a constant factor. It turns out that some of the techniques discussed in the present paper can be used to settle this embedding question, show that the logarithmic term is needed, and prove several interesting facts about balanced embeddings of the above type, which supply, in particular, a large class of graphs H for which the colored H-subgraph problem on an n vertex input graph cannot be solved in time $n^{o(|V(H)|)}$ assuming the Exponential Time Hypothesis. These results will appear in [9], here we merely include a brief outline.

The first result proved in [9] is the following.

Proposition 6.1 [9]. *For every fixed integer $k > 2$, real $1/4 > \varepsilon > 0$, integer $w > w_0(\varepsilon, k)$ and for every even $m > m_0(w)$ the following holds. Let $F = (V, E)$ be a random k-regular graph on m vertices. Then a.a.s., for every coloring of the vertices of F by w colors, so that each color appears at most $\frac{m}{w^{1-\varepsilon}}$ times, and for any choice of a set S at most $w^{2-\frac{2}{k}-3\varepsilon}$ pairs of colors, there are at least εm edges of F whose endpoints are not colored by one of the pairs in S.*

This is proved by estimating the number of k-regular graphs F on m vertices for which there is a set S as above and a coloring with less than εm

edges whose endpoints are not colored by a pair in S. The estimate obtained shows that this number is much smaller than the total number of k-regular graphs on m vertices. Note that the exponent $2-2/k$ which appears here (up to the additive error 3ε), is the same exponent that appears in the minimum possible number of edges of an $\mathcal{H}(k,n)$-universal graph. Indeed it turns out that the corresponding problems are closely related.

The above proposition implies that the $\log w$ term in the embedding result of Marx [33] mentioned above is indeed needed, as stated in the next corollary. This settles a problem raised in [33].

Corollary 6.2 [9]. *Let H be a 3-regular graph with w vertices. Then, for all even $m > m_0(w)$, there exists a 3-regular graph F on m vertices so that any embedding of F into H is of depth at least $\Omega\big(\frac{m \log w}{w}\big)$.*

Note that since the above applies to a 3-regular expander H, whose treewidth is $\Theta(w)$, this shows that the $\log w$-term is needed in the embedding result of [33].

Here is a sketch of the proof of the corollary. Take $k=3$, $\varepsilon = \frac{1}{100}$ and a sufficiently large w in Proposition 6.1. Assuming the assertion of the Corollary does not hold, let F be a random cubic graph on m vertices satisfying the assertion of the proposition. Fix an embedding of the required type of F in H in which the maximum size of the inverse image of a vertex of H is of size smaller than $\frac{\varepsilon^2 m \log w}{3w}$. Then there are less than $\frac{\varepsilon m}{3}$ vertices of F that are mapped onto sets of size at least $\varepsilon \log w$, and the total number of edges they touch is less than εm. Let V' denote the set of all vertices of F mapped to sets of size at least $\varepsilon \log w$, and let E' denote the set of all edges they touch.

For each vertex v of F choose an arbitrary vertex of H in the connected subgraph to which it is mapped, and let this vertex be the color of v. This defines a coloring of the vertices of F by w colors (corresponding to the vertices of H), and no color appears more than $\frac{\varepsilon^2 m \log w}{3w} < \frac{m}{w^{1-\varepsilon}}$ times. Let S be the set of all pairs of colors x, y (= pairs of vertices x, y of H) so that the distance in H between x and y does not exceed $2\varepsilon \log w = 0.02 \log w$. Since H is 3-regular, $|S| \le O\big(w \cdot 2^{2\varepsilon \log w}\big) \le O\big(w^{1.02}\big) < w^{4/3-3/100}$. It follows that there must be at least εm edges of F whose endpoints are not colored by a pair of colors in S. As $|E'| < \varepsilon m$ there is such an edge uv that does not belong to E', that is, it does not touch a vertex of V'. But this means that both u and v are mapped onto sets of size at most $\varepsilon \log m$, and hence the properties of our embedding imply that the distance between their colors in H is at most $2\varepsilon \log m$, contradicting the fact that this pair of colors does not belong to S. This completes the proof of the corollary.

The value $2 - 2/k$ (up to the 3ε additive error) in the exponent in Proposition 6.1 is tight in a strong sense. Indeed, if the set S in the proposition is allowed to contain $\Theta(w^{2-2/k})$ pairs, then every k-regular graph has a coloring of the required type in which the endpoints of every edge are colored by a pair in S. Moreover, there is always such a coloring with nearly equal color classes (and even exactly equal, if the number of vertices of F is divisible by w), and such a coloring can be obtained by a homomorphism into an appropriate graph with w vertices.

To state the precise result we need a few definitions. Call a homomorphism f from a graph F to a graph H *nearly balanced* if for every two vertices u, v of H, the ratio between $|f^{-1}(u)|$ and $|f^{-1}(v)|$ is at most 1.1 and at least $\frac{1}{1.1}$. The homomorphism is called *perfectly balanced* if all quantities $|f^{-1}(u)|$ are exactly equal. Note that this means that the graph F is a spanning subgraph of the $|V(F)|/|V(H)|$-blowup of H, that is, the graph obtained from H by replacing each of its vertices by an independent set of size $|V(F)|/|V(H)|$, and each of its edges by a complete bipartite graph between the corresponding sets.

Theorem 6.3 [9]. *Let T be an arbitrary regular connected graph. Let H be the graph whose vertex set is $V(T)^k$ in which two vertices are connected iff in at least two coordinates they are within distance 4 in T. Let w denote the number of vertices of H. Then, for every k-regular graph F with $m > m_0(w)$ vertices, there is a nearly balanced homomorphism of F into H.*

The proof is similar to that given in [4], and is based on the decomposition result described in Section 4 and the fact that the random walk on T converges to a uniform distribution. Starting with a bounded degree T, and combining the construction above with a bounded degree expander on all vertices of H, as done in [7], we can obtain many explicit constructions of graphs H on w vertices with $O(w^{2-2/k})$ edges, so that every k-regular graph whose number of vertices $n \gg w$ is divisible by w admits a perfectly balanced homomorphism into H. Thus, the appropriate blow-ups of the graphs H are $\mathcal{H}(k,n)$-universal (their number of edges is much bigger than the minimum possible, but they have a very special structure).

By the results of [33] and their proofs, the construction in Theorem 6.3 (for $k = 3$) also provides many examples of graphs H with w vertices and maximum degree $O(w^{1/3})$ so that, assuming the Exponential Time Hypothesis of [29], the colored H-subgraph problem on an n vertex graph cannot be solved in time $n^{o(w)}$.

7. Concluding Remarks and Open Problems

- As mentioned in Section 5, it is shown in [30] that there is a graph G with $O(n^{2-1/k} \log^{1/k} n)$ edges so that every two-edge coloring of it contains a monochromatic copy of an $\mathcal{H}(k,n)$-universal graph. The only lower bound known for the minimum possible number of edges of such a graph is $\Omega(n^{2-2/k})$, namely, the minimum possible number of edges of an $\mathcal{H}(k,n)$-universal graph. The problem of closing the gap between the upper and lower bound, raised in [30], seems difficult. Another interesting problem is that of finding an explicit construction of a graph G as above.

- The $\mathcal{H}(k,n)$-universal graph constructed in [5] has an optimal number of edges up to a constant factor, but its number of vertices is (much) bigger than n. By combining it with an appropriate expander, as done in [7], one can reduce the number of vertices to $(1+\varepsilon)n$, for any fixed $\varepsilon > 0$, increasing the number of edges only by a constant factor (depending on ε). It remains open to decide if there are $\mathcal{H}(k,n)$-universal graphs with n vertices and $O_k(n^{2-2/k})$ edges. Note that the construction in [4] provides $\mathcal{H}(k,n)$-universal graphs with n vertices, but their number of edges exceeds that of the graphs constructed in [5] by a logarithmic factor.

- The results of Grohe [26] and Marx [33], described in Section 6 apply to general binary Constraint Satisfaction Problems (CSPs, for short), showing that if the naturally defined graph corresponding to a general binary CSP has treewidth w, then, assuming the Exponential Time Hypothesis of [29], there is no algorithm that solves the problem in time $d^{o(w/\log w)}$, where d is the size of the domain of each variable of the CSP problem. This is tight, up to the $\log w$ term in the exponent, and the results of [9] discussed in Section 6 imply that the method in [33] does not suffice to close this $\log w$ gap.

- In [9] it is shown that for every fixed $\delta > 0$ there are families of graphs H on w vertices with maximum degree at most w^{δ}, so that the colored

H-subgraph problem on an input graph on n vertices cannot be solved in time $n^{o(w)}$, assuming the Exponential Time Hypothesis. It will be interesting to decide if there are sparser examples H with the same property. In particular, if H is a cubic expander, or a random cubic graph on w vertices, it is not clear if the colored H-subgraph problem on an n-vertex input graph can be solved in time $n^{o(w)}$.

- In Proposition 6.1 it is shown that almost every k-regular graph F on m vertices does not admit a vertex coloring by w colors so that the number of pairs of colors appearing in the endpoints of edges of F is smaller than $w^{2-2/k-3\varepsilon}$. It will be interesting to find an explicit graph F with this property, for some fixed small value of ε, say $\varepsilon = 1/100$.

- Corollary 6.2 implies that the $\log w$-term in the embedding result of [33] cannot be omitted. It is still plausible to suspect that the $\log w$-term can be omitted in the result about the complexity of the colored H-subgraph problem, but the proof of this statement, if true, will require a different argument.

- The problem of determining or estimating the minimum possible number of *vertices* of *induced*-universal graphs for bounded degree graphs has also been considered by various authors. Butler [14] showed that for every even k there is a graph G on $O(n^{k/2})$ vertices that contains every $H \in \mathcal{H}(k,n)$ as an induced subgraph. This is tight up to a constant factor. For odd values of k the situation is more complicated. The construction of Butler gives an induced $\mathcal{H}(k,n)$-universal graph with $O(n^{\lceil k/2 \rceil})$ vertices, and this has been improved in [24] to $O(n^{\lceil k/2 \rceil - 1/k} \log^{2+2/k} n)$ by applying the construction in [4]. The methods in [5] can in fact be used to get a tight bound of $O(n^{k/2})$ for odd values of k as well. We omit the details.

Acknowledgment. I would like to thank Dániel Marx for helpful comments.

References

[1] M. Ajtai, N. Alon, J. Bruck, R. Cypher, C. T. Ho, M. Naor and E. Szemerédi, *Fault tolerant graphs, perfect hash functions and disjoint paths,* Proc. 33rd IEEE FOCS, Pittsburgh, IEEE (1992), 693–702.

[2] N. Alon, The strong chromatic number of a graph, *Random Structures and Algorithms,* **3** (1992), 1–7.

[3] N. Alon and V. Asodi, Sparse universal graphs, *Journal of Computational and Applied Mathematics,* **142** (2002), 1–11.

[4] N. Alon and M. Capalbo, Sparse universal graphs for bounded degree graphs, *Random Structures and Algorithms,* **31** (2007), 123–133.

[5] N. Alon and M. Capalbo, *Optimal universal graphs with deterministic embedding,* Proc. of the Nineteenth Annual ACM-SIAM SODA (2008), 373–378.

[6] N. Alon, M. Capalbo, Y. Kohayakawa, V. Rödl, A. Ruciński and E. Szemerédi, *Universality and tolerance,* Proceedings of the 41st IEEE FOCS (2000), 14–21.

[7] N. Alon, M. Capalbo, Y. Kohayakawa, V. Rödl, A. Ruciński and E. Szemerédi, Near-optimum universal graphs for graphs with bounded degrees, *RANDOM-APPROX 2001,* 170–180.

[8] N. Alon and Z. Füredi, Spanning subgraphs of random graphs, *Graphs and Combinatorics,* **8** (1992), 91–94.

[9] N. Alon and D. Marx, *Sparse balanced partitions and the complexity of subgraph problems,* in preparation.

[10] N. Alon, R. Yuster and U. Zwick, Color-coding, *J. ACM,* **42** (1995), 844–856.

[11] L. Babai, F. R. K. Chung, P. Erdős, R. L. Graham and J. Spencer, On graphs which contain all sparse graphs, *Ann. Discrete Math.,* **12** (1982), 21–26.

[12] S. N. Bhatt, F. Chung, F. T. Leighton and A. Rosenberg, Universal graphs for bounded-degree trees and planar graphs, *SIAM J. Disc. Math.,* **2** (1989), 145–155.

[13] S. N. Bhatt and C. E. Leiserson, How to assemble tree machines, in: *Advances in Computing Research,* F. Preparata, ed., 1984.

[14] S. Butler, Induced-universal graphs for graphs with bounded maximum degree, *Graphs and Combinatorics,* **25** (2009), 461–468.

[15] M. Capalbo, A small universal graph for bounded-degree planar graphs, *Proc. SODA,* 1999, 150–154.

[16] M. Capalbo and S. R. Kosaraju, Small universal graphs, *Proc. STOC,* 1999, 741–749.

[17] J. Chen, B. Chor, M. Fellows, X. Huang, D. Juedes, I. Kanj and G. Xia, *Tight Lower Bounds for Certain Parameterized NP-hard Problems,* Proc. 19th Annual IEEE Conference on Computational Complexity, 2004, 150–160.

[18] J. Chen, X. Huang, I. Kanj, and G. Xia, *Linear FPT reductions and computational lower bounds,* Proc. 36th Annual ACM STOC, ACM, New York (2004), 212–221.

[19] F. R. K. Chung and R. L. Graham, On universal graphs for spanning trees, *Proc. London Math. Soc.,* **27** (1983), 203–211.

[20] F. R. K. Chung, R. L. Graham and N. Pippenger, *On graphs which contain all small trees II,* Proc. 1976 Hungarian Colloquium on Combinatorics (1978), 213–223.

[21] F. R. K. Chung, A. L. Rosenberg and L. Snyder, Perfect storage representations for families of data structures, *SIAM J. Alg. Disc. Methods,* **4** (1983), 548–565.

[22] V. Chvátal, V. Rödl, E. Szemerédi and W. T. Trotter, The Ramsey number of a graph with a bounded maximum degree, *J. Combin. Theory Ser. B.* (1983), 239–243.

[23] R.G. Downey and M.R. Fellows, *Parameterized Complexity,* Springer-Verlag, 1999.

[24] L. Esperet, A. Labourel and P. Ochem, On induced-universal graphs for the class of bounded-degree graphs, *Inf. Process. Lett.,* **108(5)** (2008), 255–260.

[25] J. Friedman and N. Pippenger, Expanding graphs contain all small trees, *Combinatorica,* **7** (1987), 71–76.

[26] M. Grohe, The complexity of homomorphism and constraint satisfaction problems seen from the other side, *J. ACM,* **54** (2007).

[27] P. E. Haxell, On the strong chromatic number, Combinatorics, *Probability and Computing,* **13** (2004), 857–865.

[28] A. Hajnal and E. Szemerédi, Proof of a conjecture of Erdős, in: *Combinatorial Theory and its Applications,* Vol. II (P. Erdős, A. Rényi and V. T. Sós, eds.), Colloq. Math Soc. J. Bolyai 4, North Holland, Amsterdam, 1970, 601–623.

[29] R. Impagliazzo, R. Paturi and F. Zane, Which problems have strongly exponential complexity? *J. Comput. Syst. Sci.,* **63** (2001), 512–530.

[30] Y. Kohayakawa, V. Rödl, M. Schacht and E. Szemerédi, *Sparse partition universal graphs for graphs of bounded degree,* to appear.

[31] A. Lubotzky, R. Phillips and P. Sarnak, Ramanujan Graphs, *Combinatorica,* **8** (1988), 261–277.

[32] G. A. Margulis, Explicit group-theoretical constructions of combinatorial schemes and their application to the design of expanders and superconcentrators, *Problems of Information Transmission,* **24** (1988), 39–46.

[33] D. Marx, *Can you beat treewidth?* Proc. 48th Annual IEEE FOCS, (2007), 169–179. Also: Theory of Computing, to appear.

[34] N. Robertson and P. Seymour. Graph minors. II. Algorithmic aspects of tree-width, *Journal of Algorithms,* **7** (1986), 309–322.

[35] V. Rödl, A note on universal graphs, *Ars. Combin.,* **11** (1981), 225–229.

[36] V. Rödl, A. Ruciński and A. Taraz, Hypergraph packing and graph embedding, *Combinatorics, Probability and Computing,* **8** (1999) 363–376.

[37] A. Ruciński, Matching and covering the vertices of a random graph by copies of a given graph, *Discrete Math.,* **105** (1992), 185–197.

[38] E. Szemerédi, Regular partitions of graphs, in: *Proc. Colloque Inter. CNRS* (J. C. Bermond, J. C. Fournier, M. Las Vergnas and D. Sotteau, eds.), 1978, 399–401.

[39] D. P. West, *Introduction to Graph Theory,* Prentice-Hall, 1996.

Noga Alon

Schools of Mathematics and Computer Science
Raymond and Beverly Sackler Faculty of Exact Sciences
Tel Aviv University
Tel Aviv 69978
Israel

e-mail: `ogaa@tau.ac.il`

BOLYAI SOCIETY
MATHEMATICAL STUDIES, 21

An Irregular Mind
(Szemerédi is 70)
pp. 39–129.

Super-Uniformity of the Typical Billiard Path

JÓZSEF BECK

Dedicated to the 70th birthday of Endre Szemerédi

1. What is Super-Uniformity?

1. Introduction. Uniform distribution—also called equidistribution—is a truly fundamental concept. The famous Weyl's Criterion (about uniform distribution) is perhaps the most frequently quoted result in whole mathematics. This volume celebrates the 70th birthday of Endre Szemerédi: the famous Szemerédi's Regularity Lemma is also about uniform distribution. It states, *very* roughly speaking, that every dense graph can be decomposed into a few "regular" parts, where the parts exhibit some kind of local edge-equidistribution. This paper is basically about the quantitative aspects of the continuous Kronecker–Weyl equidistribution theorem.

Let's begin with some of the most natural questions about uniform distribution. What are the most uniformly distributed n-element point sets in the unit interval $[0,1)$? What are the most uniformly distributed infinite sequences in the unit interval $[0,1)$? What are the most uniformly distributed (continuous) curves in the unit square $[0,1)^2$? Our objective is to study these at first sight innocent-looking questions.

We can briefly summarize the results as follows. The case of point sets and sequences is not too surprising. If we test uniformity with "nice" functions (including the characteristic functions of "nice" sets), then the most uniformly distributed n-element point set in the unit interval is the equidistant set $\{0, 1/n, 2/n, \dots, (n-1)/n\}$ (and its shifted copies modulo 1), and the error in approximating the integral with the Riemann sum is $O(n^{-1})$ or less.

If we test uniformity with *very* large classes of functions, such as all continuous (or Riemann, or Lebesgue integrable) functions, then the most uniform n-element sets are (1) the *random* n-sets ("Monte Carlo method"), and the just as good, but much more regular (2) shifted Kronecker sequences $k\alpha + \beta \pmod 1$, $1 \leq k \leq n$ in the unit interval, and, in higher dimensions, the vector version $k\mathbf{a} + \mathbf{b} \pmod 1$, $1 \leq k \leq n$ in the unit cube (or square) $[0,1)^d$, where $\mathbf{a}$ and $\mathbf{b}$ are randomly chosen points in $[0,1)^d$. (The Kronecker sequence $k\mathbf{a} + \mathbf{b} \pmod 1$, $k = 1,2,3,\ldots$ is motivated by the well-known Kronecker–Weyl equidistribution theorem.) For these very large classes of functions the error in approximating the integral with the Riemann sum is $O(n^{-1/2})$ (which is the same as the standard deviation of the symmetric random walk of n steps ± 1). This error $O(n^{-1/2})$, that I call the "random size error", cannot be improved in general.

The case of curves is very different and very surprising. The most uniformly distributed continuous curves in the unit square—or in any rectangle—are the *typical billiard paths,* and unexpectedly it makes very little difference that we test uniformity with very nice sets, say with axis-parallel subsquares, or with extremely complicated sets, such as the class of all Lebesgue measurable subsets—the error is basically the same, and it is "as small as possible".

The goal of this paper is to clarify these vague statements.

2. Super-uniformity of the typical billiard path in the unit square. I show that the typical billiard path in a square (or rectangle) is extremely uniform far beyond "common sense expectation". What most experts would consider a "common sense expectation" is the square-root size error (that is often called the "random error"); what we prove is the much smaller *square-root logarithm*(!). As a byproduct, we obtain that the set of typical billiard paths represents the family of most uniformly distributed curves in the square. Theorem 1 below is a quantitative justification.

But before formulating Theorem 1, first we have to clarify the intuitively plausible concept of "billiard path", and also explain what a " typical" billiard path, or "$1-\varepsilon$ part of all possible billiard paths" means. A "billiard path" means that a point-mass (representing a billiard ball) moves freely along a straight line inside a square-shaped table ("pool") with unit speed until it hits the boundary (i.e., one of the four sides of the unit square). The reflection off the boundary is elastic, meaning the familiar law of reflection: the angle of incidence equals the angle of reflection. After the reflection, the point continues its linear motion with the new velocity (but of course the speed remains the same) until it hits the boundary again, and so forth (we ignore friction, air resistance, etc.). The initial condition, i.e., the starting point of the billiard path and the initial direction, uniquely determine an infinite piecewise linear billiard path $\mathbf{x}(t) = \left(x_1(t), x_2(t)\right)$, $0 < t < \infty$ in the unit square. The law of reflection implies that there are at most four different directions along the billiard path (the initial direction is preserved modulo $\pi/2$, which is one-fourth of the whole angle 2π). Because of the unit speed, "arclength" and "time" are the same.

The vague term of "typical billard path" can be made precise very easily: we just have to define a measure on the set of all initial conditions of the billiard paths. The initial condition consists of a starting point $\mathbf{y} \in [0,1)^2$ and an initial direction (angle) $\theta \in [0, 2\pi)$; therefore, the corresponding measure is simply the product of the two-dimensional Lebesgue measure on the unit square and the normalized one-dimensional Lebesgue measure. This way the somewhat vague term "$1 - \varepsilon$ part of all billiard paths" in Theorem 1 will become perfectly precise.

Next we explain the well-known geometric trick of *unfolding* the billiard path inside the unit square to a straight line in the entire plane. The idea is very simple and elegant: we keep reflecting the square itself in the respective side and unfold the piecewise linear billiard path to a straight line. Two (straight) lines in the plane correspond to the same billiard path if and only if they differ by a translation through an integral vector where both coordinates are even, i.e., where the vector is from the lattice $2\mathbb{Z} + 2\mathbb{Z}$. In other words, the problem of the distribution of a billird path in the unit square is equivalent to the distribution of the corresponding torus-line in the 2×2 square.

As far as I know, the first appearance of the geometric trick of unfolding is in a paper of D. König and A. Szücs from 1913, and it became widely known after Hardy and Wright included it in their famous book on number theory. König and Szücs used the trick of unfolding (combined with the Kronecker–Weyl theorem) to prove the following elegant property of the billiard path in a square (or any rectangle): if the slope of the initial direction is rational, then the billiard path is periodic, and if the slope of the initial direction is irrational, then the billiard path is dense, and what is more, it is uniformly distributed in the unit square. Notice that the analog statement for torus-lines is the famous Kronecker–Weyl theorem.

About the notation: throughout $\log x$ and $\log_2 x$ stand for the natural (i.e., base e) and the binary (i.e., base 2) logarithms (I don't use ln at all). We are now ready to formulate our first main result.

Theorem 1. *Let A be an arbitrary Lebesgue measurable subset of the unit square $[0,1)^2$ with two-dimensional Lebesgue measure* area(A), *and let $T > 100$ be an arbitrarily large (but fixed) real number. Let $\mathbf{x}(t) = \big(x_1(t), x_2(t)\big)$, $0 \le t \le T$ be a billiard path of length T (= time) in the unit square, and let $A(T)$ denote the time the billiard path spends in subset A:*

$$A(T) = \text{measure}\,\big\{t \in [0,T] \,:\, \mathbf{x}(t) \in A\big\}.$$

Let $0 < \varepsilon < 1/2$ be arbitrary. Then for $1-\varepsilon$ part of all billiard paths of length T in the square, the curve-discrepancy $\big|A(T) - T\cdot\text{area}\,(A)\big|$ is estimated from above by

(1.1)
$$\big|A(T) - T\cdot\text{area}\,(A)\big| < \frac{10}{\varepsilon}\sqrt{\text{area}\,(A)\big(1-\text{area}\,(A)\big)}\cdot\sqrt{\log_2 T}\cdot\log_2\log_2 T.$$

Remarks. (1) Note that Theorem 1 remains true if "billiard path" is replaced with "torus line". Indeed, in view of the trick of *unfolding*, the two concepts are basically the same. The same applies for the rest of the results about billiard paths (see, e.g., Theorems 2 and 3 below).

(2) It is astonishing that, given an arbitrarily complicated subset $A \subset [0,1)^2$, the curve-discrepancy in (1.1) is just *square-root logarithmic*, that is, "almost constant"; in other words, the "ugliness" of A plays no role in (1.1). Note that (1.1) is nearly sharp in the following strong sense. Even if Theorem 1 is restricted to the simplest family of subsets, say, the family of axes-parallel subsquares, then constant size curve-discrepancy $O(1)$ in (1.1) is still unavoidable, we cannot have $o(1)$ instead (to be explained below). This shows that in Theorem 1 the complexity of the test set $A \subset [0,1)^2$ is basically irrelevant. We can summarize the message as follows: *the most uniformly distributed curves in the unit square are the typical billiard paths.* (Perhaps the greatest surprise here is that such a vague question has a definite quantitative answer.)

(3) As I promised, I explain the (basically trivial) fact that even for the narrow class of axes-parallel subsquares we must have constant curve-discrepancy $O(1)$ in (1.1). Consider the two subsquares, $A_1 = [0,1/3]^2$ and $A_2 = [2/3,1]^2$; the distance between them is $\sqrt{2}/3$. Let $\mathbf{x}(t)$ be an arbitrary continuous curve in the unit square; we always assume that the arclength of every segment $\mathbf{x}(t)$, $T_1 < t < T_2$ is exactly $T_2 - T_1$ (meaning: t is the time and a point-mass moves along the curve with unit speed). For any real number $\tau > 0$ write

$$A_i(\tau) = \text{measure}\left\{t \in [0,\tau] \, : \, \mathbf{x}(t) \in A_i\right\}, \quad i = 1, 2,$$

where A_i, $i = 1, 2$ are the two subsquares mentioned above. We show that the following four curve-discrepancies:

(1.2)
$$\left|A_i(T) - T \cdot \text{area}\,(A_i)\right|, \quad \left|A_i(T + c) - (T + c) \cdot \text{area}\,(A_i)\right|, \quad i = 1, 2,$$

where $c = \sqrt{2}/3$ is the distance between the two given subsquares A_1 and A_2 (computed for the same curve!), cannot be all $o(1)$. Indeed, the middle segment $\mathbf{x}(t)$, $T < t < T+c$ of the curve cannot visit both subsquares (because the arclength is exactly the distance between A_1 and A_2); consequently, at least one of the four curve-discrepancies in (1.2) must be

$$\geq \frac{1}{2} c \cdot \text{area}\,(A_i) = \frac{1}{2} \cdot \frac{\sqrt{2}}{3} \cdot \frac{1}{9} = \frac{\sqrt{2}}{54}.$$

This trivial argument shows that in Theorem 1 we cannot hope for curve-discrepancy $o(1)$ in (1.1) even for the simplest families of subsets.

(4) As I already said above, we can make the vague term "$1 - \varepsilon$ part of all billard paths" in Theorem 1 precise by using the product measure on the set of all initial conditions of the billiard paths. Since the initial condition consists of a starting point $\mathbf{y} \in [0, 1)^2$ and an initial direction (angle) $\theta \in [0, 2\pi)$, the natural measure here is simply the product of the two-dimensional Lebesgue measure on the unit square and the normalized one-dimensional Lebesgue measure.

(5) By slightly modifying the proof of Theorem 1, we can prove a version with logarithmic curve-discrepancy that holds for *almost every* initial condition. (As usual, *almost every* means that the exception has Lebesgue measure zero.) For the precise statement of this version, see Proposition 4.1 at the end of Section 4.

(6) Theorem 1 gives an interesting new insight to the general question of *discrete versus continuous*. The Kronecker–Weyl equidistribution theorem has two—basically equivalent—forms: a discrete form and a continuous form (see e.g. [9] or [4]).

Kronecker–Weyl Theorem (discrete). *Let $d \geq 1$ be any integer, and let $\mathbf{a} = (\alpha_1, \ldots, \alpha_d)$ be an arbitrary d-dimensional vector with real coordinates. The sequence $n\mathbf{a}$ (mod 1), $n = 1, 2, 3, \ldots$ is uniformly distributed in the unit cube $[0,1)^d$, meaning*

$$\lim_{N\to\infty} \frac{1}{N} \sum_{\substack{1\leq n\leq N: \\ n\mathbf{a}\in R \pmod 1}} 1 = \text{volume}\,(R)$$

for any axes-parallel box $R = I_1 \times \cdots \times I_d \subset [0,1)^d$ (i.e., Cartesian product of intervals) if and only if $1, \alpha_1, \ldots, \alpha_d$ are linearly independent over the rationals.

Kronecker–Weyl Theorem (continuous). *We have*

$$\lim_{T\to\infty} \frac{1}{T} \text{measure}\left\{0 < t < T : t\mathbf{a} \in R \pmod 1\right\} = \text{volume}\,(R)$$

for any axes-parallel box $R = I_1 \times \cdots \times I_d \subset [0,1)^d$ if and only if $\alpha_1, \ldots, \alpha_d$ are linearly independent over the rationals.

This means that the discrete sequence

$$n\mathbf{a} \pmod 1, \quad n = 1, 2, 3, \ldots \tag{1.3}$$

and the continuous torus line passing through the origin

$$t\mathbf{a} \pmod 1, \quad 0 < t < \infty \tag{1.4}$$

have the same equidistribution property—at least from a *qualitative* viewpoint. The surprising message of Theorem 1 is that, in spite of this similarity, the *quantitative* aspects of (1.3) and (1.4) are very different. Indeed, Theorem 1 states that a typical billiard path—which is just a general torus line $t\mathbf{a} + \mathbf{b}$ (mod 1) $0 < t < \infty$ via unfolding—estimates the area of an arbitrary but fixed (measurable) subset $A \subset [0,1]^2$ with "error" $O\left(\sqrt{\log T}\right)$, where T is the length of the time-interval. On the other hand, a typical discrete sequence of the form (which is the discrete version of the general torus line $t\mathbf{a} + \mathbf{b}$ in dimension $d = 1$)

$$n\alpha + \beta \pmod 1, \quad n = 1, 2, \ldots, N$$

cannot estimate the one-dimensional Lebesgue measure of an arbitrary but fixed $A \subset [0,1]$ with "error" $o\left(\sqrt{N}\right)$. This is the message of the following result.

Proposition 1.1. *For every integer $N \geq 1$, there is a measurable subset $A = A_N \subset [0,1]$ (in fact, A is a finite union of intervals) such that, for the majority of the pairs $(\alpha, \beta) \in [0,1)^2$,*

$$\left| \sum_{\substack{1 \leq n \leq N: \\ n\alpha+\beta \in A \pmod 1}} 1 - N \cdot \text{measure}\,(A) \right| > \frac{\sqrt{N}}{8}. \tag{1.5}$$

It is well possible that Proposition 1.1 is folklore, but I couldn't find it in the literature.

Note without proof that Proposition 1.1 is best possible: the error term $\sqrt{N}$ in (1.5) cannot be replaced by any larger function of N.

Comparing Theorem 1 to Proposition 1.1, we see a huge difference between the sizes of the "errors"

$$\sqrt{\log T} \quad \text{and} \quad \sqrt{N}$$

(for simplicity I ignored the negligible iterated logarithmic factor of T). This shows that the quantitative aspects of (1.3) and (1.4) are dramatically different, and explains my choice of word *super-uniformity*.

3. Next I formulate the extension of Theorem 1 to higher dimensions $d \geq 3$.

Theorem 2. *Let A be an arbitrary measurable subset of the unit cube $[0,1)^d$ with Lebesgue measure $\text{vol}\,(A)$, and let $T > 100$ be an arbitrarily large (but fixed) real number. Let $\mathbf{x}(t) = \big(x_1(t), \ldots, x_d(t)\big)$, $0 \leq t \leq T$ be a billiard path of length T (= time) in the unit cube, and let $A(T)$ denote the time the billiard path spends in subset A:*

$$A(T) = \text{measure}\left\{ t \in [0,T] \,:\, \mathbf{x}(t) \in A \right\}.$$

Let $0 < \varepsilon < 1/2$ be arbitrary. Then for $1 - \varepsilon$ part of all billiard paths of length T in the cube,

$$\big| A(T) - T \cdot \text{vol}\,(A) \big| \tag{1.6}$$

$$< \frac{c_0(d)}{\varepsilon} \sqrt{\text{vol}\,(A)\big(1 - \text{vol}\,(A)\big)} \cdot T^{\frac{1}{2} - \frac{1}{2(d-1)}} \cdot (\log T)^{\frac{1}{2(d-1)}} (\log\log T)^{\frac{1}{d-1}},$$

where $c_0(d)$ is a constant depending only on the dimension d (i.e., independent of A and T).

For *convex* sets the upper bound in (1.6) can be improved.

Theorem 3. *Assume that in Theorem 2 we make the extra condition that A is a convex subset of the unit cube $[0,1)^d$ $(d \geq 3)$. Then we have the following improvement of (1.6): given any $0 < \varepsilon < 1/2$, for $1-\varepsilon$ part of all billiard paths of length T in the cube,*

$$\text{(1.7)} \qquad \big|A(T) - T \cdot \mathrm{vol}\,(A)\big|$$

$$< \frac{c_0'(d)}{\varepsilon}\sqrt{\mathrm{vol}\,(A)\big(1-\mathrm{vol}\,(A)\big)} \cdot T^{\frac{1}{2}-\frac{1}{d-1}} \cdot (\log T)^{\frac{1}{d-1}} (\log\log T)^{\frac{2}{d-1}}.$$

For example, if $d = 3$ then (1.7) gives the square-root logarithmic order of magnitude

$$\text{(1.8)} \qquad \sqrt{\log T} \cdot \log\log T,$$

which is shockingly small ("almost constant") similarly to (1.1). Note that, if $d = 3$ then (1.6) gives

$$\text{(1.9)} \qquad T^{1/4} \cdot (\log T)^{\frac{1}{4}} (\log\log T)^{\frac{1}{2}}.$$

If $d = 4$ then (1.7) gives

$$\text{(1.10)} \qquad T^{1/6} \cdot (\log T)^{1/3} (\log\log T)^{\frac{2}{3}},$$

and (1.6) gives

$$\text{(1.11)} \qquad T^{1/3} \cdot (\log T)^{\frac{1}{6}} (\log\log T)^{\frac{1}{3}}.$$

Note that both Theorem 2 and Theorem 3 are best possible apart from logarithmic factors. The proof of the sharpness of Theorem 2 is very similar to the proof of Proposition 1.1—see the Concluding Remarks at the end of Section 8. The sharpness of Theorem 3 follows from Theorem 4 below.

The proofs of Theorems 2 and 3 are very similar in the following sense. For any $0 < \varepsilon < 1/2$, we construct a "large" subset Ω_ε of the space of initial conditions $\subset [0,1)^d \times S^{d-1}$ ("large" means at least $1-\varepsilon$ part) such that, for every measurable $A \subset [0,1)^d$,

(1.12)

$$\int_{\Omega_\varepsilon} \big(A(T;\omega) - T\,\mathrm{vol}\,(A)\big)^2\, d\omega \leq \frac{\mathrm{const}}{\varepsilon} \cdot T^{1-\frac{1}{d-1}} \cdot (\log T)^{\frac{1}{d-1}} (\log\log T)^{\frac{2}{d-1}},$$

where $\omega = (\mathbf{y}, \mathbf{b})$, $\mathbf{y} \in [0,1)^d$, $\mathbf{b} \in S^{d-1}$ denotes an arbitrary initial condition, and if $A \subset [0,1)^d$ is convex, we have,

(1.13)
$$\int_{\Omega_\varepsilon} \big(A(T;\omega) - T\,\mathrm{vol}\,(A)\big)^2\, d\omega \le \frac{\mathrm{const}}{\varepsilon} \cdot T^{1-\frac{2}{d-1}} \cdot (\log T)^{\frac{2}{d-1}} (\log\log T)^{\frac{4}{d-1}}.$$

Notice that (1.12) implies Theorem 2 and (1.13) implies Theorem 3 (via Chebyshev's inequality).

We can prove that (1.13) is basically best possible: the power of T is sharp. What is more, the lower bound is attained for the simplest convex shapes such as the balls (= solid spheres).

Theorem 4. *Let $\{\mathcal{C}_\omega : \omega \in \Omega\}$ be an arbitrary family of continuous curves, parametrized with a finite number of parameters, in the unit torus $[0,1)^d$ with common arclength T. Here "finite number of parameters" means that Ω is a bounded measurable subset of some finite-dimensional euclidean space. For any measurable subset of the unit torus $A \subset [0,1)^d$, write*

$$A(T;\omega) = \mathrm{arclength}\,(\mathcal{C}_\omega \cap A).$$

There is a d-dimensional ball A_0 of radius $\le 1/2$ in the unit torus $[0,1)^d$ such that

$$\int_{\Omega} \big(A_0(T;\omega) - T\,\mathrm{vol}\,(A_0)\big)^2\, d\omega > T^{1-\frac{2}{d-1}}. \tag{1.14}$$

I supply the proofs of Theorems 1, 2, 3, 4 and Proposition 1.1 in Sections 4, 5, 6, 7 and 8, respectively.

We can view Theorems 1-4 and Proposition 1.1 as the starting point of a new direction in the study of the classical subject of uniform distribution. There are many natural questions here, which are worth while pursuing. I just mention a few. Estimating the volume and integral with point sampling, which one is better: regular sampling or random sampling (= Monte Carlo method)? How about "line sampling"? Can we "beat" the Monte Carlo method? I will address some of these exciting questions in the next two sections. This gives me an opportunity to analyze some well-known classical results from a new viewpoint.

2. Can we Beat the Monte Carlo Method? (I)

Regular sampling vs. random sampling. As far as I know, there are two general approaches to build a theory of *uniform distribution.* I can summarize them in a nutshell as follows.

Approach A: a large family of test functions and one single sample;

Approach B: a large family of samples and one single test function.

The theory of "iregularities of distribution" (or "discrepancy theory") and numerical integration are examples of Approach A; the Monte Carlo method and the so-called ergodic theorems are examples of Approach B. I briefly elaborate on this.

Approach A: (1) irregularities of distribution (i.e., discrepancy theory). A typical question (or rather a bunch of questions) in this theory goes as follows. What are the most uniformly distributed N-element point sets (say) in the d-dimensional unit cube $[0,1]^d$ with respect to natural "geometric" families of sets, such as (1) all axes-parallel boxes, (2) all rotated boxes, (3) all balls, (4) all convex subsets, and so on. These questions led to dozens of interesting theorems; I just briefly mention two. I start with a theorem of W. M. Schmidt, proved in 1972, which goes as follows. Given any set of n points in the unit square, there is always an axes-parallel rectangle with discrepancy at least $c_1 \cdot \log n$, where $c_1 > 0$ is an absolute constant independent of n. In other words, there is an axes-parallel rectangle R such that the number of points in R differs from the expected number $n \cdot \text{area}(R)$ by at least $c_1 \cdot \log n$. Here the order of magnitude $\log n$ is the best possible: there is a set of n points in the unit square such that for every axes-parallel rectangle R, the number of points in R differs from the expected number $n \cdot \text{area}(R)$ by less than $c_2 \cdot \log n$, where $c_2 > 0$ is an absolute constant independent of n. We can say, therefore, that Schmidt's theorem is sharp apart from a constant factor (to find the value of the exact constant remains unsolved).

Unfortunately, to find the analog of Schmidt's theorem in higher dimensions remains a notorious open problem.

However, if we switch to the L_2-norm, then the problem of axes-parallel boxes is solved (again apart from a constant factor) by a theorem of K. F.

Roth. In dimension $d \geq 2$, the L_2-norm of the discrepancy is at least constant times $(\log n)^{d/2}$ for any set of n points in $[0,1]^d$. The order of magnitude $(\log n)^{d/2}$ in Roth's L_2-norm result is known to be the best possible.

We have a large literature of similar results for many other shapes (e.g. balls); see e.g. the books by Beck and Chen [1] and Matousek [6]. We will return to "irregularities of distribution with respect to balls" in the proof of Theorem 4; see Section 7.

The theory of numerical integration (Simpson's rule, etc.) also belongs to Approach A. Here I give a little bit longer discussion of this classical subject (see e.g. [3]).

Approach A: (2) classical averages (or quadratures) in numerical integration. We begin with the simplest possible question about the Riemann integral: let $f(x)$, $0 \leq x \leq 1$ be a continuous real function defined on the unit interval, and we want to determine the definite integral $\int_0^1 f(x)\,dx$. If the antiderivative $F(x)$ of $f(x)$ is readily available and sufficiently simple, then the fundamental theorem of integral calculus (Newton–Leibniz law) provides a simple answer:

$$\int_0^1 f(x)\,dx = F(1) - F(0), \quad \text{where} \quad F'(x) = f(x). \tag{2.1}$$

Unfortunately, in most cases the antiderivative is not an elementary function, and recipe (2.1) turns out to be impractical. For example, it is impossible to evaluate the important integrals $\int_a^b e^{-x^2/2}\,dx$ (related to the normal distribution) and $\int_a^b \sqrt{1+x^3}\,dx$ (where a and b are real numbers) exactly. Then we have no choice but use numerical integration. Every undergraduate student knows that the definite integral is the limit of the Riemann sums

$$\sum_{k=0}^{n-1} f(t_k)(x_{k+1} - x_k), \tag{2.2}$$

where $0 = x_0 < x_1 < x_2 < \cdots < x_n = 1$, $t_k \in [x_k, x_{k+1}]$ for $k = 0, 1, 2, \ldots, n-1$, and the maximum gap $\max_k(x_{k+1} - x_k) \to 0$ as $n \to \infty$. The simplest Riemann sums are

$$\frac{1}{n}\sum_{k=0}^{n-1} f(k/n) \quad \text{and} \quad \frac{1}{n}\sum_{k=1}^{n} f(k/n), \tag{2.3}$$

and each approximates the definite integral with error $O(n^{-1})$, provided the first derivative of $f(x)$ is bounded on the unit interval.

If we know more about the smoothness of $f(x)$, say, it has the second, third and the fourth derivative, then we have averages that behave much better than (2.3). Recall that (2.3) is often called the "rectangular rule"; we also have the "midpoint rule":

$$\frac{1}{n}\sum_{k=0}^{n-1} f\left(\frac{k+\frac{1}{2}}{n}\right), \tag{2.4}$$

the "trapezoidal rule":

$$\frac{1}{2n}\left(f(0)+2f\left(\frac{1}{n}\right)+2f\left(\frac{2}{n}\right)+2f\left(\frac{3}{n}\right)+\cdots+2f\left(\frac{n-1}{n}\right)+f(1)\right), \tag{2.5}$$

and the "Simpson's rule":

$$\frac{f(0)+4f(\frac{1}{n})+2f(\frac{2}{n})+4f(\frac{3}{n})+2f(\frac{4}{n})+\cdots}{3n} + \frac{2f(1-\frac{2}{n})+4f(1-\frac{1}{n})+f(1)}{3n}. \tag{2.6}$$

If we assume that the second derivative of $f(x)$ is bounded on the unit interval, then the midpoint, the trapezoidal and the Simpson's rules all approximate the definite integral with error $O(n^{-2})$.

If the function has a third and fourth derivative, then the order of the error is still $O(n^{-2})$ for the midpoint and trapezoidal rules, but for Simpson's formula (2.6) the error is $O(n^{-3})$ and $O(n^{-4})$, respectively. That is, the Simpson's rule has the most benefit from the smoothness of the function (at least up to the fourth derivative; after that more smoothness does not help).

These results are so standard that they are mentioned—usually without proof—in most undergraduate Calculus textbooks.

Note that the list is not complete: there are other classical Riemann sums that I did not mention above; for example, the so-called Newton's three-eights rule

$$\frac{3\big(f(0)+3f(\frac{1}{3n})+3f(\frac{2}{3n})+2f(\frac{3}{3n})+3f(\frac{4}{3n})+3f(\frac{5}{3n})}{8n} + \frac{2f(\frac{6}{3n})+\cdots+f(1)\big)}{8n}. \tag{2.7}$$

Also, it is customary to work with the dyadic numbers

$$\frac{1}{2^k}\sum_{j=0}^{2^k-1} f\left(\frac{j}{2^k}\right), \quad k=1,2,3,\ldots \tag{2.8}$$

as a particularly simple set of division points.

Let's return to the simplest Riemann sum (2.3): it has an unlimited benefit from the smoothness of the function if $f(x)$ is *periodic* with period one. Indeed, already mild smoothness guarantees that the function is represented by its Fourier series:

$$f(x)=\sum_{r\in\mathbb{Z}} a_r e^{2\pi \mathrm{i} rx},$$

where of course $\mathrm{i}=\sqrt{-1}$ and

$$a_r=\int_0^1 f(x)e^{-2\pi \mathrm{i} rx}\,dx.$$

Thus we have

$$\frac{1}{n}\sum_{k=0}^{n-1} f(k/n)=\sum_{r\in\mathbb{Z}} a_r\frac{1}{n}\sum_{k=0}^{n-1} e^{2\pi \mathrm{i} rk/n} \tag{2.9}$$

$$=\sum_{r\in\mathbb{Z}} a_r\frac{e^{2\pi \mathrm{i} r}-1}{n(e^{2\pi \mathrm{i} r/n}-1)}$$

$$=a_0+a_n+a_{-n}+a_{2n}+a_{-2n}+a_{3n}+a_{-3n}+\cdots$$

$$=\int_0^1 f(x)\,dx+a_n+a_{-n}+a_{2n}+a_{-2n}+a_{3n}+a_{-3n}+\cdots.$$

If f is t-times differentable with bounded derivative, then using integration by parts repeatedly, we obtain

$$a_m=\int_0^1 f(x)e^{-2\pi \mathrm{i} mx}\,dx=O\big(|m|^{-t}\big). \tag{2.10}$$

Combining (2.9) and (2.10) with $m=jn$ ($j=\pm1,\pm2,\pm3,\ldots$), we have

$$\frac{1}{n}\sum_{k=0}^{n-1} f(k/n)-\int_0^1 f(x)\,dx=O\big(n^{-t}\big)\sum_{j=1}^{\infty}\frac{1}{j^t}=O\big(n^{-t}\big), \tag{2.11}$$

where t (≥ 2) is the degree of smoothness of the periodic function. In other words, the smoother the function, the better the approximation of the definite integral; formula (2.11) gives a quantitative description of the benefit from the smoothness.

Unfortunately for non-smooth functions, say, continuous but nowhere differentiable $f(x)$, we know nothing about the speed of convergence of (2.3) (or (2.4), or (2.5), or (2.6))—these averages become unreliable. For non-smooth integrable functions in one variable, and in general for most multivariable functions it is customary to use a completely different approach: the so-called Monte-Carlo method, or the method of random sampling. The Monte Carlo method is perhaps the most well-known example of Approach B.

Approach B: (1) Monte Carlo method (i.e., random sampling). The Monte Carlo method was invented in the late 1940s in Los Alamos (a useful byproduct of making the first hydrogen bomb), and the basic idea is probabilistic. Let $X_1, X_2, X_3, \ldots, X_n$ be independent, identically distributed random variables having the uniform distribution in the unit interval. The expectation of each random variable $f(X_k)$ is $= \int_0^1 f(x)\,dx$, so the average

$$\frac{1}{n}\sum_{k=1}^{n} f(X_k), \tag{2.12}$$

a random variable itself, approximates the definite integral with the standard random square-root size error $O\big(n^{-1/2}\big)$ with probability close to one. This is a consequence of the central limit theorem. More precisely,

$$\begin{aligned} \text{Variance}\, f(X_k) &= \mathbf{E}\Big(f(X_k) - \int_0^1 f(u)\,du\Big)^2 \\ &= \int_0^1 \Big(f(z) - \int_0^1 f(u)\,du\Big)^2 dz, \end{aligned} \tag{2.13}$$

where $\mathbf{E}$ stands for the expectation, and so

$$\text{Variance}\,\frac{1}{n}\sum_{k=1}^{n} f(X_k) = \frac{1}{n}\int_0^1 \Big(f(z) - \int_0^1 f(u)\,du\Big)^2 dz. \tag{2.14}$$

Thus by Chebyshev's inequality

$$\text{(2.15)} \qquad \Pr\left[\left|\frac{1}{n}\sum_{k=1}^{n} f(X_k) - \int_0^1 f(u)\,du\right| \geq \lambda\frac{\sigma(f)}{\sqrt{n}}\right] \leq \frac{1}{\lambda^2}$$

holds for any positive real number λ, where $\sigma(f)$ denotes the positive constant

$$\sigma(f) = \left(\int_0^1 \left(f(z) - \int_0^1 f(u)\,du\right)^2 dz\right)^{1/2}$$

depending only on f (of course we always apply (2.15) with a $\lambda > 1$). Note that (2.14) is equivalent to a statement about an n-dimensional multiple integral as follows:

(2.16)

$$\int_{x_1=0}^1 \cdots \int_{x_n=0}^1 \left(\frac{1}{n}(f(x_1) + \cdots + f(x_n)) - \int_0^1 f(u)\,du\right)^2 dx_1 \cdots dx_n$$
$$= \frac{1}{n}\int_0^1 \left(f(z) - \int_0^1 f(u)\,du\right)^2 dz.$$

Identity (2.16) can be proved by a simple application of *orthogonality:* if $j \neq k$ then we have

$$\text{(2.17)} \quad \int_{x_j=0}^1 \int_{x_k=0}^1 \left(f(x_j) - \int_0^1 f(u)\,du\right)\left(f(x_k) - \int_0^1 f(u)\,du\right) dx_k\,dx_j$$

$$= \int_{x_j=0}^1 \left(f(x_j) - \int_0^1 f(u)\,du\right) dx_j \cdot \int_{x_k=0}^1 \left(f(x_k) - \int_0^1 f(u)\,du\right) dx_k = 0,$$

that is, to prove (2.14), we don't really need the full power of independence.

Next we study the advantages and the disadvantages of the Monte Carlo method. A key advantage of the Monte Carlo method is that the approximation error $O(n^{-1/2})$ is independent of the "ugliness" of the function. Also, the Monte Carlo method is equally suitable for the computation of the more general Lebesgue integral. (Note that averages (2.3)–(2.8) do not work in general for the Lebesgue integral, since changing the value of the function in a set of measure zero, the value of the integral should remain the same.)

A second advantage of the Monte Carlo method is that the dimension is totally irrelevant: we can replace the unit interval with the unit cube of any dimension, and the approximation error remains the same $O(n^{-1/2})$, where the implicit constant is independent of the dimension.

On the other hand, an obvious disadvantage of the Monte Carlo method is that it provides only a relatively slow convergence rate of $O(n^{-1/2})$, and it remains the same even if the function is very smooth. That is, no additional benefit is derived from the smoothness of the function. This is the complete opposite of the behavior of averages (2.3)–(2.7), which share the property that more regular functions have a faster convergence rate (the best example is (2.11) for t-times differentiable periodic functions, where t is a large integer).

Another difficulty of the Monte Carlo method is how to generate independent random samples concretely. The practitioners always use *pseudo-random* numbers instead of truly random samples, which raises the question of reliability.

A third difficulty of the Monte Carlo method is that it provides only a probabilistic error bound; we cannot be hundred percent sure that the expected error bound $O(n^{-1/2})$ is actually achieved in a concrete computation. (One way to overcome this difficulty is to repeat the whole computation with independent samples. With sufficiently many repetitions, we can be at least 99 percent sure.)

To summarize, the following points are the advantages and disadvantages of the Monte Carlo method for numerical integration.

Advantages:

(a) the function can be arbitrarily "ugly";

(b) the method works for the more general Lebesgue integral;

(c) the dimension is irrelevant.

Disadvantages:

(i) the rate of convergence $O(n^{-1/2})$ is slow;

(ii) it is not adaptive: there is no benefit from the smoothness of the function: the approximation error is always the same bell curve ("central limit theorem");

(iii) generating random numbers is not easy;

(iv) the approximation error is probabilistic.

Let's say a few words about (iii). The basic idea of generating pseudorandom numbers is to use exponential sequences. This is not too surprising, since independence is equivalent to the product rule, and *exponential* means iterated product.

For example, consider the concrete exponential sequence $(3/2)^n$, $n = 1, 2, 3, \ldots$ modulo one. There is an overwhelming numerical evidence that this sequence is uniformly distributed in the unit interval, and what is more, it exhibits all kind of characteristic properties of "randomness". Unfortunately, we cannot prove anything for 3/2. Note that in the 1930s Koksma could prove that, replacing 3/2 with $\alpha > 1$, the sequence α^n, $n = 1, 2, 3, \ldots$ modulo one is uniformly distributed for *almost all* $\alpha > 1$ (and this "law of large numbers" was extended later to more delicate results such as the law of the iterated logarithm and the central limit theorem), but we don't know a single explicit $\alpha > 1$ featuring uniform distribution.

By far the most important pseudorandom number generator is the "linear congruential method", introduced by Lehman, which results an exponential sequence. I briefly outline it as follows. First we have to choose four integers: (1) $1 \leq m$, called the modulus, (2) $2 \leq A < m$, called the multiplier, (3) $1 \leq C < m$, called the increment, (4) $1 \leq Y_0 < m$, called the starting value, and consider the sequence ($n \geq 0$)

$$Y_{n+1} \equiv A \cdot Y_n + C \pmod{m}, \tag{2.18}$$

which gives integers (remainders modulo m) between 0 and $m - 1$. To have an efficient pseudorandom generator, the implementation of (2.18) should follow a short list of practical advice:

(i) the modulus m should be large, at least 2^{30};

(ii) the starting value Y_0 may be chosen arbitrarily;

(iii) if m is a power of 2, i.e., if a binary computer is being used, pick the multiplier A so that $A \equiv 5 \pmod{8}$, and the binary digits of A should not have a simple, regular pattern;

(iv) the value of multiplier A is critical, but the value of C is almost irrelevant; we may choose $C = 1$.

This completes our nutshell summary of the Monte Carlo method. The so-called *ergodic theorems* is another rich source of examples for Approach B.

Approach B: (2) ergodic theorems. These are general results that address an important question that arose in statistical mechanics: to give a condition under which the time average (meaning the limit of some kind of Riemann sums) and the space-average (meaning a Lebesgue integral) of a "dynamical system" agree. The first ergodic theorem was proved by J. von Neumann in 1931: it is about the case of the convergence in the L_2-norm. A few weeks later G. D. Birkhoff proved his famous individual ergodic theorem about the harder case of pointwise convergence.

A "dynamical system" means a probability space $(\Omega, \mathcal{A}, \mu)$ and a measure-preserving transformation $T : \Omega \to \Omega$ (i.e., Ω is a non-empty set, $\mathcal{A}$ is a σ-algebra in Ω, μ is a measure on $\mathcal{A}$, where of course μ is σ-additive, $\mu(\Omega) = 1$, T is $\mathcal{A}$-measurable, and $\mu\big(T^{-1}(A)\big) = \mu(A)$ for all $A \in \mathcal{A}$). Here comes a key definition: a transformation $T : \Omega \to \Omega$ is called *ergodic* if every T-invariant set $A \in \mathcal{A}$ (i.e., $T^{-1}(A) = A$) has the property that either $\mu(A) = 0$ or $\mu(A^c) = 0$ (where A^c denotes the complement of A).

Unfortunately, it is not easy to decide whether or not a given transformation $T : \Omega \to \Omega$ is ergodic. Here are two important examples of ergodic transformations on the unit interval: $\Omega = [0,1)$, $\mathcal{A}$ is the family of Lebesgue measurable sets, μ is the usual (one-dimensional) Lebesgue measure.

Example 1 ("irrational rotation"). $T(x) = x + \alpha \pmod 1$, where α is any fixed irrational number.

Example 2 ("doubling map"). $T(x) = 2x \pmod 1$.

Note that Example 2 is a non-invertible measure-preserving (ergodic) transformation.

The third example is a generalization of Example 1 to the d-dimensional unit torus $\Omega = [0,1)^d$ (again the σ-algebra is the family of Lebesgue measurable sets, and μ is the d-dimensional Lebesgue measure/volume).

Example 3 ("Kronecker shift"). $T(\mathbf{x}) = \mathbf{x} + \mathbf{u} \pmod 1$ where $1, u_1, u_2, \ldots, u_d$ are linearly independent over the rationals (of course $u_1, u_2, \ldots, u_d$ are the coordinates of $\mathbf{u}$).

Now we are ready to formulate

Birkhoff's ergodic theorem. *Let $(\Omega, \mathcal{A}, \mu)$ be a probability space, and let T be a measure-preserving transformation on $(\Omega, \mathcal{A}, \mu)$. If T is ergodic, then for every integrable function $f \in L_1 = L_1(\Omega, \mathcal{A}, \mu)$,*

$$\lim_{n\to\infty} \frac{1}{n} \sum_{k=0}^{n-1} f\big(T^k(x)\big) = \int_\Omega f\, d\mu$$

for almost every $x \in \Omega$.

The fundamental handicap of the ergodic theorem is that it does not say anything about the speed of convergence; it does not give any explicit error term. In this sense Birkhoff's ergodic theorem is a "soft" result.

We want to apply Birkhoff's theorem in the special case of the Kronecker shift $T(\mathbf{z}) = \mathbf{z} + \mathbf{x} \pmod 1$. First note that, by the Kronecker–Weyl equidistribution theorem,

$$\lim_{n\to\infty} \frac{1}{n} \sum_{j=0}^{n-1} f(j\mathbf{x}) = \int_{I^d} f(\mathbf{u})\, d\mathbf{u} \tag{2.19}$$

holds for all Riemann integrable functions f on $I^d = [0,1)^d$ (we extend f periodically over $\mathbb{R}^d$) if and only if $1, x_1, \ldots, x_d$ are linearly independent over the rationals. It follows that, under the same condition (i.e., linear independence), the translation $\mathbf{z} \to \mathbf{z} + \mathbf{x} \pmod 1$ by the vector $\mathbf{x} = (x_1, \ldots, x_d)$ is an ergodic transformation. So by Birkhoff's ergodic theorem, we can extend (2.19) from the class of Riemann integrable functions to the much larger class of Lebesgue integrable functions as follows. Let f be a Lebesgue integrable function on I^d, and asuume that $1, x_1, \ldots, x_d$ are linearly independent over the rationals, then

$$\lim_{n\to\infty} \frac{1}{n} \sum_{j=0}^{n-1} f(\mathbf{y} + j\mathbf{x}) = \int_{I^d} f(\mathbf{u})\, d\mathbf{u} \tag{2.20}$$

holds for almost all $\mathbf{y}$.

Regular sampling with square-root error: Proposition 2.1. Since we applied Birkhoff's ergodic theorem, we don't know anything about the speed of convergence; in this sense (2.20) is a "soft" result. However, it is quite easy to prove a "hard" result here: a result supplying an explicit error term, which can be interpreted as a "hard analog" of (2.20); see Proposition

2.1(d) below. The price that we pay is that the "hard" result works only for almost all $\mathbf{x}$, meaning that, for a given explicit $\mathbf{x}$ we *cannot* predict that the "hard" result applies or not.

Proposition 2.1. (a) *Suppose that $f(x)$ is a Lebesgue square-integrable function on the unit interval $0 \le x < 1$; for notational simplicity we extend f over the entire real line $\mathbb{R}$ periodically with period one. Then for any natural number $n \ge 1$,*

$$\int_0^1 \int_0^1 \left(\frac{1}{n} \sum_{j=0}^{n-1} f(y + jx) - \int_0^1 f(u)\, du \right)^2 dx\, dy = \frac{\sigma^2(f)}{n}, \tag{2.21}$$

where

$$\sigma^2(f) = \int_0^1 \left(f(z) - \int_0^1 f(u)\, du \right)^2 dz \tag{2.22}$$

is a constant depending only on f.

(b) *We have a straightforward multi-dimensional analog of (a) as follows: Suppose that $f(\mathbf{x}) = f(x_1, \ldots, x_d)$ is a Lebesgue square-integrable function defined in the d-dimensional unit cube $I^d = [0,1)^d$; for notational simplicity we extend f over the entire d-dimensional space $\mathbb{R}^d$ periodically with period one. Then for any natural number $n \ge 1$,*

$$\int_{I^d} \int_{I^d} \left(\frac{1}{n} \sum_{j=0}^{n-1} f(\mathbf{y} + j\mathbf{x}) - \int_{I^d} f(\mathbf{u})\, d\mathbf{u} \right)^2 d\mathbf{x}\, d\mathbf{y} = \frac{\sigma^2(f)}{n}, \tag{2.23}$$

where of course

$$\sigma^2(f) = \int_{I^d} \left(f(\mathbf{z}) - \int_{I^d} f(\mathbf{u})\, d\mathbf{u} \right)^2 d\mathbf{z}, \tag{2.24}$$

just like in (2.22).

(c) *Applying Chebyshev's inequality, we obtain the following straightforward corollary: if X and Y are two independent and uniformly distributed random variables in the unit cube I^d, then*

$$\Pr\left[\left| \frac{1}{n} \sum_{j=0}^{n-1} f(Y + jX) - \int_{I^d} f(\mathbf{u})\, d\mathbf{u} \right| \ge \lambda \frac{\sigma(f)}{\sqrt{n}} \right] \le \frac{1}{\lambda^2} \tag{2.25}$$

holds for any positive real number λ, *where* $\sigma(f)$ *denotes the positive square root of* $\sigma^2(f)$, *defined in* (2.24).

(d) *Finally, we can switch from fixed* n *to all* n*'s: Let* $f(\mathbf{x}) = f(x_1, \ldots, x_d)$ *be a Lebesgue square-integrable function defined in the* d*-dimensional unit cube* $I^d = [0,1)^d$; *for notational simplicity we extend* f *over the entire* d*-dimensional space* $\mathbb{R}^d$ *periodically with period one. For every* $\mu > 1$ *and* $1 > \varepsilon > 0$, *there is a measurable subset* $W = W(\mu) \subset I^d \times I^d$ *with* measure $(W) \geq 1 - \frac{1}{\mu}$ *("2d-dimensional Lebesgue measure") such that the inequality*

(2.26)

$$\left|\frac{1}{n}\sum_{j=0}^{n-1} f(\mathbf{y} + j\mathbf{x}) - \int_{I^d} f(\mathbf{u})\, d\mathbf{u}\right| \leq \frac{\sigma(f)\cdot\sqrt{2\mu}\cdot\left(\log_2 n + (1/\varepsilon)^{\varepsilon} + 1\right)^{2+\varepsilon}}{\sqrt{n}}$$

holds for all $(\mathbf{x}, \mathbf{y}) \in W$ *and all integers* $n \geq 1$. *Here again* $\sigma(f)$ *denotes the positive square root of* $\sigma^2(f)$, *defined in* (2.24).

Remarks. Comparing Proposition 2.1(a) and (d), we notice a loss of factor $O\left(\log^{2+\varepsilon} n\right)$, which is basically negligible compared to the dominant factor of $\sqrt{n}$ in the denominator.

What makes this simple result interesting is that the "ugliness" of the function f is basically irrelevant: there is only a mild indirect dependence of the error term on f via the "standard deviation" constant $\sigma(f)$. For example, if $f(x)$ is bounded, say, $0 \leq f \leq 1$, then of course $0 \leq \sigma(f) \leq 1$, so (2.22) clearly holds with the choice $\sigma = 1$, totally independently of the "ugliness" of f (needless to say, a Lebesgue measurable function can be *very ugly*).

In fact, our proof gives the following generalization of Proposition 2.1. Let $k_0 = 0 < k_1 < k_2 < k_3 < \cdots$ be an arbitrary increasing sequence of positive integers, and consider the sequence

$$Y, Y + k_1 X, Y + k_2 X, Y + k_3 X, \ldots, Y + k_{n-1} X \pmod 1,$$

where X and Y are two independent and uniformly distributed random variables in the unit interval. Then we have the following generalization of (2.21):

$$\int_0^1 \int_0^1 \left(\frac{1}{n}\sum_{j=0}^{n-1} f(y + k_j x) - \int_0^1 f(u)\, du\right)^2 dx\, dy = \frac{\sigma^2(f)}{n}. \tag{2.27}$$

For a proof; see Section 8.

Proposition 2.1 is a simple but elegant result with a short proof. I am convinced that this important result is folklore. To my greatest surprise, I couldn't find it in the literature; did I overlook something obvious?

Notice that Proposition 2.1(a)–(b) is the analog of (2.14) and Proposition 2.1(c) is the analog of (2.15) in the Monte Carlo method. The *regular sampling*

$$\mathbf{y} + j\mathbf{x} \pmod 1, \quad j = 0, 1, 2, \dots$$

in Proposition 2.1 (I call it *regular,* since it is an arithmetic progression modulo one) is clearly more practical than the Monte Carlo method: we just need two random vectors (namely $\mathbf{y}$ and $\mathbf{x}$) instead of n random vectors, where n is usually large like $n = 10^6$. This makes a big difference, especially because the reliability of the pseudonumber generators remains a theoretically unsettled issue. In view of Proposition 2.1, we can say that *the regular sampling is just as good as the random sampling.*

A discrete analog of Proposition 2.1. It is interesting to point out that the discrete analog of Proposition 2.1(a) is false, and the reason is a simple congruence (e.g., parity) argument. Indeed, consider the set S of odd integers $n = 2k + 1$ in $1 \le n = 2k + 1 \le N$; if N is even then S has density exactly 1/2. The most natural discrete analog is to estimate the size of a subset $S \subset \{1, 2, \dots, N\}$ with the intersection of a residue class, i.e. a long arithmetic progression. Notice that about 1/4 of the arithmetic progressions in $1, 2, 3, \dots, N$ has an even starting point and an even gap, and these arithmetic progressions, consisting of even integers only, do not even intersect our large set S of density 1/2. This kills the possibility of a straightforward discrete analog of Proposition 2.1(a).

Nevertheless, we can prove some kind of an analog of Proposition 2.1(a) as follows (we lose a logarithmic factor).

Proposition 2.2. *Let S be an arbitrary subset of the discrete interval $\{1, 2, \dots, N\}$, and let $T \ge 1000$ be an arbitrary integer with $N > 100T^2$. Then there is an integer m in the range $\frac{N}{2T} < m < \frac{N}{T}$ such that*

$$\frac{1}{m} \sum_{r=1}^{m} \left(\frac{|S|}{m} - \sum_{\substack{n \in S: \\ n \equiv r \pmod m}} 1 \right)^2 < c_1 T \log T, \tag{2.28}$$

where c_1 is an absolute constant (say, $c_1 = 400$ is a good choice).

Note that, the proof supplies an efficient algorithm to find such a "good" integer m (it is not true in general that the majority of the integers m in the range $\frac{N}{2T} < m < \frac{N}{T}$ are "good").

As a corollary, we immediately obtain that, there is a residue class $r \pmod{m}$ such that

$$\left| \frac{|S|}{m} - \sum_{\substack{n \in S: \\ n \equiv r \pmod{m}}} 1 \right| < \sqrt{c_1 T \log T}. \tag{2.29}$$

Since $\frac{N}{m} \approx T$, $\sqrt{T \log T}$ is basically the square-root error (apart from the negligible factor of $\sqrt{\log T}$). Can we remove the logarithmic factors in (2.28) and (2.29)? I don't know; perhaps the reader can answer this question.

Proposition 2.2 is basically best possible in the following sense. Let $N > M$ be positive integers satisfying $N/2 > M > \log N$. Then, for the majority of the 2^N subsets $S \subset \{1, 2, \ldots, N\}$,

$$\frac{1}{m} \sum_{r=0}^{m-1} \left(\frac{|S|}{m} - \sum_{\substack{n \in S: \\ n \equiv r \pmod{m}}} 1 \right)^2 > c_2 \frac{N}{M} \tag{2.30}$$

holds for every integer m in the range $M \le m < 2M$; here $c_2 > 0$ is a positive absolute constant.

The proof of Proposition 2.2 is based on a deep tool: the so-called Large Sieve (an important Fourier technique developed for applications in analytic number theory, mainly for the distribution of the primes); see Section 10. We also need the fairly deep fact that there are "many primes" (namely about $x/\log x$ for $p \le x$). The proof of (2.30), on the other hand, is quite easy: it is a routine application of the so-called "probabilistic method"— I leave the details to the reader.

3. Can we Beat the Monte Carlo Method? (II)

1. Regular sampling is adaptive. I recall the message of Proposition 2.1: the regular sampling by a typical arithmetic progression modulo one—i.e., a typical shifted Kronecker sequence—is just as good as the Monte Carlo method. What is more, testing by a regular sampling benefits from the "nicety" of the test function (unlike the Monte Carlo method, which is "blind": it cannot take advantage of "nice" functions: the approximation error is always the very same bell curve).

Identity (2.27) is very general: it works for arbitrary subsequences $k_0 = 0 < k_1 < k_2 < k_3 < \cdots$ of the set of positive integers. We can have all kinds of sequences: the simplest linear sequence (1) $k_j = j$, $j = 1, 2, 3, \ldots$, the quadratic sequence (2) $k_j = j^2$, the cubic sequence (3) $k_j = j^3$, the fourth powers (4) $k_j = j^4$, or the much more rapidly increasing exponential sequences, such as (5) $k_j = 2^j$, the gap sequence (6) satisfying

$$\frac{k_{j+1}}{k_j} \geq 1 + \lambda \quad \text{for every } j \text{ with some fixed constant} \quad \lambda > 0,$$

the factorial sequence (7) $k_j = j!$, and so on.

This leads to the following natural question: Which set

$$y, y + k_1 x, y + k_2 x, y + k_3 x, \ldots, y + k_{n-1} x \pmod 1 \tag{3.1}$$

of division points is the "best" way to approximate the integral? In other words, which sequence $k_0 = 0 < k_1 < k_2 < k_3 < \cdots$ supplies the best approximation in the form

$$\frac{1}{n} \sum_{j=0}^{n-1} f(y + k_j x) \approx \int_0^1 f(u)\, du? \tag{3.2}$$

(for simplicity we study the one-dimensional case, and we always extend function f over the entire real line periodically with period one). We know that the L_2-norm (as $0 < x < 1$ and $0 < y < 1$ independently run through the unit interval) does *not* distinguish among the sequences $k_0 = 0 < k_1 < k_2 < k_3 < \cdots$ (of distinct positive integers): in view of (2.27) the square-integral of the error is always the same

$$\frac{1}{n} \int_0^1 \left(f(z) - \int_0^1 f(u)\, du \right)^2 dz. \tag{3.3}$$

(3.3) tells us that the L_2-norm of the error is the same for "nice" or "ugly" functions, but, if we switch to the L_1-norm or to the "typical error", then the simplest linear sequence $k_j = j$ exhibits a definite advantage: the approximation

$$\frac{1}{n} \sum_{j=0}^{n-1} f(y + jx) \approx \int_0^1 f(u)\, du \tag{3.4}$$

is *adaptive:* it does benefit from the smoothness of the function f.

Switching from the L_2-norm to the L_1-norm. The best illustration is to study the L_1-norm for smooth f. By using the Fourier series of f, we have

$$\left|\frac{1}{n}\sum_{j=0}^{n-1} f(y+kx) - \int_0^1 f(u)\,du\right|$$

$$= \left|\sum_{\substack{r\in\mathbb{Z}:\\ r\neq 0}} a_r \frac{1}{n}\left(\sum_{j=0}^{n-1} e^{2\pi irkx}\right)\cdot e^{2\pi iry}\right|$$

$$\leq \frac{1}{n}\sum_{\substack{r\in\mathbb{Z}:\\ r\neq 0}} |a_r|\cdot\left|\frac{e^{2\pi irnx}-1}{e^{2\pi irx}-1}\right|.$$

Clearly

$$\left|\frac{e^{2\pi irnx}-1}{e^{2\pi irx}-1}\right| \leq \min\left\{n, \frac{2}{|e^{2\pi irx}-1|}\right\}.$$

Since

$$\left|e^{2\pi irx}-1\right| = 2\left|\sin(\pi rx)\right| = 2\left|\sin\left(\pi\|rx\|\right)\right| \geq 4\|rx\|,$$

where I use $\|z\|$ to denote the distance of a real number z from the nearest integer, we have

$$\left|\frac{1}{n}\sum_{k=0}^{n-1} f(y+kx) - \int_0^1 f(u)\,du\right| \tag{3.5}$$

$$\leq \frac{1}{n}\sum_{\substack{r\in\mathbb{Z}:\\ r\neq 0}} |a_r| \min\left\{n, \frac{1}{2\|rx\|}\right\}.$$

Since f is smooth, its Fourier coefficient satisfies the bound ($r \neq 0$):

$$|a_r| \leq \frac{c}{r^2}, \quad \text{where} c = \max\, |f''|. \tag{3.6}$$

On the other hand, for any integer $r \neq 0$,

$$\int_0^1 \min\left\{n, \frac{1}{2\|rx\|}\right\} dx = \int_0^1 \min\left\{n, \frac{1}{2\|x\|}\right\} dx$$

$$= 2\int_0^{1/2} \min\left\{n, \frac{1}{2x}\right\} dx = 2\int_0^{\frac{1}{2n}} n\,dx + \int_{\frac{1}{2n}}^{1/2} \frac{1}{x}\,dx$$

$$= 1 + \log\left(\frac{1}{2}\right) - \log\left(\frac{1}{2n}\right) = 1 + \log n.$$

Summarizing, we have

$$\text{(3.7)} \qquad L_1\text{-norm} = \int_0^1 \int_0^1 \left| \frac{1}{n}\sum_{k=0}^{n-1} f(y+kx) - \int_0^1 f(u)\,du \right| dx\,dy$$

$$\leq \frac{2\max|f''|}{n}\left(\sum_{r=1}^{\infty}\frac{1}{r^2}\right)(1+\log n) = \frac{\pi^2\max|f''|}{3}\cdot\frac{1+\log n}{n}.$$

This proves that for smooth f the L_1-norm is $O(\log n/n)$, that is, the L_1-norm is much smaller than the L_2-norm $\frac{\text{const}}{\sqrt{n}}$ (which, by the way, is independent of the smoothness of f). It follows that the "typical error" is $O(\log n/n)$. But more is true: we can cancel the logarithmic factor in the numerator. We can prove that, for smooth functions the "typical error" is $O(1/n)$; see Proposition 3.1 below.

In sharp contrast to the adaptiveness of the linear sequence, for an (at least) exponential sequence—say, $k_j = 2^j$—the approximation

$$\text{(3.8)} \qquad \frac{1}{n}\sum_{j=0}^{n-1} f(y+2^j x) \approx \int_0^1 f(u)\,du,$$

behaves exactly like the Monte Carlo method; (3.8) does *not* benefit from the smoothness of the function f. This is the message of a large amount of research in uniform distribution initiated by Koksma and Erdős in the 1930–40s; see e.g. the book [4].

It seems very likely that the polynomial sequences, such as $k_j = j^2$ or j^3 or j^4, all behave similarly to the Monte Carlo method in the sense that there is no benefit from the smoothness of the function f, but, unfortunately, there is no technology available to prove this natural guess.

It seems to me that the linear sequence $k_j = j$ is the only one which is adaptive.

The following result is basically an improvement to (3.7): it is another illustration of the key fact that the approximation by a regular sampling

$$\frac{1}{n}\sum_{j=0}^{n-1} f(y+jx) \approx \int_0^1 f(u)\,du$$

does benefit from the smoothness of the function f.

Proposition 3.1. *Suppose that a function f is 2-times differentable with bounded derivative on the unit interval $0 \le x < 1$; for notational simplicity we extend f over the entire real line $\mathbb{R}$ periodically with period one. Then for any natural number $n \ge 1$,*

$$\left|\frac{1}{n}\sum_{j=0}^{n-1} f(y+jx) - \int_0^1 f(u)\,du\right| = O(1/n) \tag{3.9}$$

holds for the majority of the pairs (x, y) in the unit square. In fact, more is true: the value of the shift y is irrelevant in the sense that, (3.9) holds for the majority of x in $0 < x < 1$ and for all $0 \le y < 1$.

The point here is that the error term $O(1/n)$ in (3.9) is much smaller than the error term $O\big(1/\sqrt{n}\big)$ in Proposition 2.1(a).

It is well possible that Proposition 3.1 is folklore; nevertheless, I couldn't find it in the literature. I include a proof in Section 11.

2. A surprising way to "beat" the Monte Carlo method: switching from point samples to curves, surfaces, and so on. The message of Propositions 2.1 and 3.1 in a nutshell goes as follows. Regular sampling is just as good as random sampling (= Monte Carlo method): in the general case they both have the same square-root error. In fact, regular sampling is somewhat better in the sense that it *does* benefit from "nice" functions (and sets).

Theorem 1 (see Section 1) represents a spectacular "win" over the Monte Carlo method: the usual square-root error drops to the much, much smaller square-root logarithmic order (which is *almost* constant), but of course there is a slight cheating here: the point sample is replaced by a continuous curve (namely a typical billiard path) and the test set has to be 2-dimensional (namely, a subset of the unit square and what we want is the area). I refer to this as a "dimension one increase". Next I show that, by making an additional "dimension one increase"—namely, if we approximate the volume of a 3-dimensional set in the unit cube with some "natural" surface (mod 1)—then the square-root logarithmic error drops to constant error! (Needless to say, it is best possible.) The "natural" surface what we actually use is a flat tilted parallelogram in the 3-space.

Theorem 5. *Let A be an arbitrary Lebesgue measurable subset of the unit cube $[0,1)^3$ with 3-dimensional Lebesgue measure $\mathrm{vol}\,(A)$, and let $T_1 > 100$, $T_2 > 100$ be two arbitrarily large (but fixed) real numbers. Let $\mathbf{x}(t_1,t_2) = \big(x_1(t_1,t_2), x_2(t_1,t_2), x_3(t_1,t_2)\big)$ (mod 1) denote the following torus-parallelogram:*

$$x_1(t_1,t_2) = \beta_{1,1}t_1 + \beta_{1,2}t_2 + y_1, \quad x_2(t_1,t_2) = \beta_{2,1}t_1 + \beta_{2,2}t_2 + y_2, \tag{3.10}$$

$$\text{and} \quad x_3(t_1,t_2) = \beta_{3,1}t_1 + \beta_{3,2}t_2 + y_3,$$

where $0 \le t_1 \le T_1$, $0 \le t_2 \le T_2$, and $\mathbf{b}_1 = (\beta_{1,1}, \beta_{2,1}, \beta_{3,1})$, $\mathbf{b}_2 = (\beta_{1,2}, \beta_{2,2}, \beta_{3,2})$ are unit vectors: $|\mathbf{b}_1|^2 = \beta_{1,1}^2 + \beta_{2,1}^2 + \beta_{3,1}^2 = 1$, $|\mathbf{b}_2|^2 = \beta_{1,2}^2 + \beta_{2,2}^2 + \beta_{3,2}^2 = 1$. The quadruple $\big((T_1,T_2);\mathbf{y};\mathbf{b}_1,\mathbf{b}_2\big)$ (where $\mathbf{y} = (y_1,y_2,y_3)$) uniquely determines the torus-parallelogram (3.10); we call the triple $(\mathbf{y},\mathbf{b}_1,\mathbf{b}_2)$ an initial condition. *The total area $A(T_1,T_2) = A(T_1,T_2;\mathbf{y};\mathbf{b}_1,\mathbf{b}_2)$ of the intersection of the torus-parallelogram (3.10) with the given subset A equals*

$$A(T_1,T_2) = A(T_1,T_2;\mathbf{y};\mathbf{b}_1,\mathbf{b}_2) \tag{3.11}$$

$$= |\mathbf{b}_1 \times \mathbf{b}_2| \cdot \mathrm{area}\,\big\{(t_1,t_2) \in [0,T_1] \times [0,T_2] \,:\, \mathbf{x}(t_1,t_2) \in A \text{ (mod 1)}\big\},$$

where $\mathbf{b}_1 \times \mathbf{b}_2$ denotes the usual cross product of 3-dimensional vectors.

Let $0 < \varepsilon < 1/2$ be arbitrary. Then for $1-\varepsilon$ part of all torus parallelograms of parameters T_1, T_2 (see (3.10)), $A(T_1,T_2;\mathbf{y};\mathbf{b}_1,\mathbf{b}_2)$ differs from the "expectation" (see (3.11)) $|\mathbf{b}_1 \times \mathbf{b}_2|T_1T_2 \cdot \mathrm{vol}\,(A)$ by at most a constant: we have

$$\big|A(T_1,T_2;\mathbf{y};\mathbf{b}_1,\mathbf{b}_2) - |\mathbf{b}_1 \times \mathbf{b}_2|T_1T_2 \cdot \mathrm{vol}\,(A)\big| \tag{3.12}$$

$$< \frac{1000}{\varepsilon} \cdot \log^2\left(\frac{1}{\varepsilon}\right) \cdot \sqrt{\mathrm{vol}\,(A)\big(1-\mathrm{vol}\,(A)\big)}\,.$$

Remarks. (1) Theorem 5 is another super-uniformity type result, even sharper than Theorem 1. Again the "ugliness" of A plays absolutely no role; see (3.12). Given an arbitrarily complicated subset $A \subset [0,1)^3$, the typical surface-discrepancy (see (3.12)) is less than an *absolute constant.*

On the other hand, if Theorem 5 is restricted to extremely simple subsets, say to the narrow family of axes-parallel subcubes, then constant surface-discrepancy $O(1)$ in (3.12) is still unavoidable (to be explained below). Shortly speaking, the complexity of the test set $A \subset [0,1)^3$ in Theorem 5 is irrelevant.

(2) Next I briefly explain the trivial fact that even for the narrow class of axes-parallel subcubes we must have constant surface-discrepancy $O(1)$ in (3.12). The following argument is a straightforward adaptation of the argument in part **(3)** of the Remarks after Theorem 1. Consider the two subcubes $A_1 = [0,1/3]^3$ and $A_2 = [2/3,1]^3$; the distance between them is $\sqrt{3}/3$. We show that the following eight surface-discrepancies cannot be all very small like $o(1)$:

$$\left| A_i(T_1,T_2;\mathbf{y};\mathbf{b}_1,\mathbf{b}_2) - T_1T_2|\mathbf{b}_1 \times \mathbf{b}_2| \cdot \mathrm{vol}\,(A)\right|$$

where $i = 1,2$ and (T_1,T_2) runs through the four pairs (T,T) and $(T+c,T)$ and $(T,T+c)$ and $(T+c,T+c)$; here $c > 0$ is a sufficiently small positive constant. Indeed, the following linear combination of the "large" rectangles

$$\begin{aligned}(3.13)\qquad &[0,T+c]\times[0,T+c] - [0,T+c]\times[0,T]\\ &- [0,T]\times[0,T+c] + [0,T]\times[0,T]\end{aligned}$$

is a "small" constant size square $[T,T+c]\times[T,T+c]$, and in terms of the torus parallelogram (3.10), the equation $(t_1,t_2) \in [T,T+c]\times[T,T+c]$ gives a "small" constant size torus parallelogram of diameter less than $\sqrt{3}/3$ (i.e., the diameter is less than the distance between the two subcubes A_1 and A_2) if $c > 0$ is a small enough absolute constant. Therefore, because of the small diameter, this "small" constant size torus parallelogram (see (3.13)) cannot intersect both A_1 and A_2, which yields a constant size surface-discrepancy. This proves that in Theorem 5 we cannot hope for a surface-discrepancy $o(1)$ in (3.12) (i.e., we cannot hope for a surface-discrepancy tending to zero as T_1, T_2 tend to infinity), not even for the simplest families of subsets, such as the family of all axes-parallel subcubes.

(3) In Theorem 5 we test the uniformity of a k-dimensional (finite) surface with respect to a (measurable) subset $A \subset [0,1]^d$, where $k = 2$ and $d = 3$ (Theorems 1 and 2 correspond to the cases $k = 1$ and $d \geq 2$). Consider now the general case with arbitrary $1 \leq k < d$. By a straightforward adaptation of our proof technique, one can show that the corresponding upper bound for the discrepancy is $O(1)$ if $k > d/2$, and basically

$$(3.14)\qquad T^{\frac{1}{2}-\frac{k}{2(d-k)}}$$

if $k \leq d/2$, where T denotes the total (k-dimensional) surface area (arclength for $k = 1$). Here "basically" means that I ignored the logarithmic factors. Similarly to Theorem 2, the exponent $\frac{1}{2} - \frac{k}{2(d-k)}$ in (3.14) is best possible (the same argument works how we prove the sharpness of Theorem 2).

3. Summary. Let me briefly summarize the message of Sections 1–3. If we test volume or integral with a point sample, then in the general case—meaning "ugly" sets and functions—we cannot beat the Monte Carlo method: the square-root error is unavoidable.

However, if we replace the point sample with an appropriate "curve sample", then in low dimensions we can substantially beat the Monte Carlo method in the sense that the error is much less than square-root; see e.g. Theorem 1, where the "curve sample" is a typical billiard path in the unit square. The bad news is that Theorems 2 and 3 are basically sharp, so in "large" dimensions we eventually go back to the square-root error again.

Similarly, if we go one dimension higher, and switch from curves to surfaces, then in low dimensions we can beat the Monte Carlo method in the sense that the error is much less than square-root; see e.g. Theorem 5, where the "surface sample" is a typical flat tilted parallelogram in the 3-space, taken modulo one. In "large" dimensions, however, we eventually go back to the square-root error again.

The message of Theorems 1–3 is that line sampling is much more efficient than the traditional point sampling, at least in low dimensions! This motivates the following new way of computing, or rather approximating, the volume of low-dimensional convex sets (= solids). Note that convex sets are special in the sense that the intersection with a (straight) line is a line segment, so the length is determined by the two endpoints.

How to compute the volume of low-dimensional convex sets? For simplicity, assume that A is a convex subset of the unit cube $[0,1]^3$; we want to approximate the volume $\text{vol}(A)$. First we extend the subset $A \subset [0,1]^3$ *periodically* over the whole 3-space (of course, the period is 1 in each of the three coordinate directions). Then we choose a "typical" straight line segment of length n in the 3-space (n is "large"). It means, more precisely, that first we choose a starting point $\mathbf{y}$ in (say) the unit cube $[0,1]^3$, and then we choose a direction $\mathbf{b}$, represented by a point on the unit sphere S^2. The starting point $\mathbf{y} \in [0,1]^3$, the direction $\mathbf{b} \in S^2$, and the length n uniquely determine a straight line segment $L(n) = L(\mathbf{y}; \mathbf{b}; n)$ in the ordinary

3-space. We consider the intersection of the line segment $L(n) = L(\mathbf{y}; \mathbf{b}; n)$ with the periodic extension of $A \subset [0,1]^3$ over the whole 3-space: it consists of $O(n)$ "pieces", where each "piece" is a line segment itself. For notational simplicity, let total $\left[L(n) \cap A\right]$ denote the total length of the $O(n)$ "pieces". Theorem 3 for $d = 3$ (see (1.8)) says that, for the overwhelming majority of the initial conditions (initial condition=starting point $\mathbf{y} \in [0,1]^3$ plus direction $\mathbf{b} \in S^2$), total $\left[L(n) \cap A\right]$ is shockingly close to the "expected value" $n \cdot \text{vol}(A)$: the discrepancy

$$\left|\text{total}\left[L(n) \cap A\right] - n \cdot \text{vol}(A)\right|$$

is "square-root logarithmic" (i.e., almost independent of n), and the "ugliness" of the convex set $A \subset [0,1]^3$ is irrelevant.

The obvious benefit of working with a convex set $A \subset [0,1]^3$ is that the total length of the $O(n)$ "pieces" total $\left[L(n) \cap A\right]$ can be easily computed. Indeed, each piece is a line segment itself, so we just need to know the coordinates of the two endpoints: the distance comes from a straightforward application of Pythagorean theorem. Finally, we just add up the $O(n)$ distances. It is relatively easy, therefore, to determine the exact value of total $\left[L(n) \cap A\right]$, which happens to be shockingly close to n times the volume vol (A). Dividing by n, we obtain a very good approximation of the volume of the convex set $A \subset [0,1]^3$.

It is an important technical question how to actually determine the two endpoints of a "piece" (= short line segment). The answer heavily depends on the way the convex set $A \subset [0,1]^3$ is described. A typical way to describe a complicated convex set is to represent it as the intersection of a few simpler convex sets. As an illustration, consider the case when $A \subset [0,1]^3$ is the intersection of a ball B, a cube C, and a tetrahedron D : $A = B \cap C \cap D$. Of course, the volumes vol (B), vol (C), vol (D) are easy to be computed, but the volume of the intersection vol $(B \cap C \cap D)$ = vol (A) has nothing to do with the simplicity of vol (B), vol (C), vol (D); the computation of vol $(B \cap C \cap D)$ is hard! But here comes the advantage of the line sampling: the endpoints of a "piece" (= intersection of A with a line) are either on B, or C, or D, and the intersection of a straight line with a ball (or cube, or tetrahedron) is a trivial calculation. This gives a practical solution for the actual computation of the endpoints of the "pieces" for very large classes of convex sets.

I just outlined the 3-dimensional case, but this procedure clearly works in other low dimensions such as, say, 4 and 5. Also, this approach seems very promising even if the set is not convex. What we really need is that the intersection of the line segment $L(n)$ and the periodic extension of A consists of "not too many pieces (= short line-segments)".

Note that the 2-dimensional case (= area computation) is exceptionally simple: then we have a shortcut. Indeed, "area under the curve" is just the definite integral of a function in one variable. This leads us to the classical field of numerical integration and the classical quadrature formulas (e.g., midpoint rule, trapezoidal rule, Simpson's rule).

Unfortunately, in higher dimensions—i.e., for functions of at least two variables—the classical quadrature formulas all break down, mainly because there is no natural analog of the "equidistant set in the unit interval" $0, 1/n, 2/n, \ldots, (n-1)/n$ in higher dimensions. This is why in higher dimensions the only practical solution is the Monte Carlo method ("random sampling"). What we demonstrate here is that, in the classical subject of volume computation, a novel way of regular sampling—namely, line sampling—can beat the Monte Carlo method!

The rest of the paper is devoted to the proofs of Theorems 1–4, Propositions 1.1, 2.1–2, 3.1 and Theorem 5 in this order.

4. Super-Uniformity: Proof of Theorem 1

In view of the trick of *unfolding* the billiard path to a straight line in the plane (explained in Section 1), it suffices to deal with torus-lines (of course we shrink the corresponding 2×2 square to the unit square). Let $A \subset I^2 = [0,1)^2$ be an arbitrary Lebesgue measurable subset (via unfolding it corresponds to the union of four copies of the given subset A in Theorem 1), and consider the Fourier series of the 0-1 valued characteristic function χ_A of the set A:

$$\chi_A(\mathbf{u}) = \sum_{\mathbf{r}\in\mathbb{Z}^2} a_{\mathbf{r}} e^{2\pi i \mathbf{r}\cdot\mathbf{u}} \quad \text{with} \quad a_{\mathbf{r}} = \int_A e^{-2\pi i \mathbf{r}\cdot\mathbf{y}}\, d\mathbf{y}, \tag{4.1}$$

where $\mathbf{r}\cdot\mathbf{u} = r_1u_1 + r_2u_2$ denotes the standard inner product of vectors. Clearly $a_{\mathbf{0}} = \text{area}\,(A)$ (= Lebesgue measure), and by Parseval's formula,

$$\sum_{\substack{\mathbf{r}\in\mathbb{Z}^2:\\ \mathbf{r}\neq\mathbf{0}}} |a_{\mathbf{r}}|^2 = \text{area}\,(A) - \text{area}^2(A). \tag{4.2}$$

Consider the torus-line $\mathbf{x}(t) = \big(x_1(t), x_2(t)\big)$ (mod 1), where

$$(4.3) \qquad x_1(t) = \beta_1 t + y_1, \quad x_2(t) = \beta_2 t + y_2 \quad \text{and} \quad \beta_1^2 + \beta_2^2 = 1.$$

The length of the straight line segment $\mathbf{x}(t)$, $0 < t < T$ is clearly T, i.e., time=arclength. The pair $\big(\mathbf{y}, (\beta_1, \beta_2)\big)$ describes the starting point $\mathbf{y} \in [0,1)^2$ and the angle (by the point (β_1, β_2) on the unit circle) of the torus-line $\mathbf{x}(t)$. The total time $A(T) = A\big(T; \mathbf{y}, (\beta_1, \beta_2)\big)$ that the torus-line $\mathbf{x}(t)$ (defined in (4.3)) spends in subset A during $0 < t < T$ equals

(4.4)
$$A(T) = A\big(T; \mathbf{y}, (\beta_1, \beta_2)\big) = \text{measure}\big\{t \in [0,T] : \mathbf{x}(t) \in A \text{ (mod 1)}\big\}$$

$$= \int_0^T \chi_A\big(\mathbf{x}(t)\big)\, dt = \int_0^T \sum_{\mathbf{r} \in \mathbb{Z}^2} a_{\mathbf{r}} e^{2\pi i \mathbf{r} \cdot \mathbf{x}(t)}\, dt$$

$$= \sum_{\mathbf{r} \in \mathbb{Z}^2} a_{\mathbf{r}} \int_0^T e^{2\pi i \mathbf{r} \cdot \mathbf{x}(t)}\, dt = \sum_{\mathbf{r} \in \mathbb{Z}^2} a_{\mathbf{r}} e^{2\pi i \mathbf{r} \cdot \mathbf{y}} \int_0^T e^{2\pi i (\beta_1 r_1 + \beta_2 r_2) t}\, dt$$

$$= a_{\mathbf{0}} T + \sum_{\substack{\mathbf{r} \in \mathbb{Z}^2: \\ \mathbf{r} \neq \mathbf{0}}} a_{\mathbf{r}} e^{2\pi i \mathbf{r} \cdot \mathbf{y}} \cdot \frac{e^{2\pi i (\beta_1 r_1 + \beta_2 r_2) T} - 1}{2\pi i (\beta_1 r_1 + \beta_2 r_2)}.$$

Since $a_{\mathbf{0}} = \text{area}\,(A)$ (= Lebesgue measure), by (4.4) we have

$$(4.5) \qquad \text{discrepancy} = A\big(T; \mathbf{y}, (\beta_1, \beta_2)\big) - T \cdot \text{area}\,(A)$$

$$= \sum_{\substack{\mathbf{r} \in \mathbb{Z}^2: \\ \mathbf{r} \neq \mathbf{0}}} a_{\mathbf{r}} \cdot \frac{e^{2\pi i (\beta_1 r_1 + \beta_2 r_2) T} - 1}{2\pi i (\beta_1 r_1 + \beta_2 r_2)} \cdot e^{2\pi i \mathbf{r} \cdot \mathbf{y}}.$$

Fix any point (β_1, β_2) on the unit circle $\beta_1^2 + \beta_2^2 = 1$, and run the starting point $\mathbf{y}$ through the unit square; then by Parseval's formula ($I^2 = [0,1]^2$)

$$(4.6) \qquad \int_{I^2} \Big(A\big(T; \mathbf{y}, (\beta_1, \beta_2)\big) - T \cdot \text{area}\,(A)\Big)^2 d\mathbf{y}$$

$$= \sum_{\substack{\mathbf{r} \in \mathbb{Z}^2: \\ \mathbf{r} \neq \mathbf{0}}} |a_{\mathbf{r}}|^2 \cdot \left| \frac{e^{2\pi i (\beta_1 r_1 + \beta_2 r_2) T} - 1}{2\pi (\beta_1 r_1 + \beta_2 r_2)} \right|^2.$$

A technical note. The argument in (4.4) and (4.6) was rather informal: for example, we changed the order of infinite summation and integration, but didn't say anything about under what condition can we really do that. Note, for example, that for an "ugly" measurable set A the Fourier series of the characteristic function χ_A can be divergent in many points, but this kind of technical nuisance is totally irrelevant for us: what we care about is the Parseval formula. It is well-known that the Parseval formula characterizes the L_2 space, meaning the class of functions for which the Lebesgue square integral $\int_0^1 f^2(x)\,dx$ exists (and finite); see the Riesz–Fisher theorem. The point is that the characteristic function $f = \chi_A$ clearly belongs to L_2, and we can *safely work* in the L_2 space ("Lebesgue square-integrable"). The precise proof of this is a standard technical argument that I very briefly recall here. If $f = g$ is a sufficiently smooth function—say, twice differentiable with continuous derivative—then the Fourier series is absolutely convergent, and so every manipulation that we carried out (such as, changing the order of summation and integration) is perfectly justified and legitimate. The last step is to approximate a Lebesgue square-integrable function f with a sequence of smooth functions g_k, $k = 1, 2, 3, \ldots$: we want the approximation error $\big|f(x) - g_k(x)\big|$ to be arbitrarily small (as $k \to \infty$) except on a set of x's with small Lebesgue measure (the measure tends to zero as $k \to \infty$). This last step is totally routine in the theory of the Lebesgue integral.

Let's return to (4.6). We study the last factor; we use the obvious upper bound

$$\left|\frac{e^{2\pi i(\beta_1 r_1+\beta_2 r_2)T} - 1}{2\pi(\beta_1 r_1 + \beta_2 r_2)}\right| \le \min\left\{\frac{1}{\pi|\beta_1 r_1 + \beta_2 r_2|}, T\right\}. \tag{4.7}$$

Key Definition (discarding the bad directions). Let $0 < \varepsilon < 1/2$; we say that a point (β_1, β_2) on the unit circle $\beta_1^2 + \beta_2^2 = 1$ is ε-bad if there exists an $\mathbf{r} \in \mathbb{Z}^2$ such that

$$|\beta_1 r_1 + \beta_2 r_2| \le \frac{\varepsilon}{40|\mathbf{r}| \cdot \log_2|\mathbf{r}| \cdot \big(\log_2\log_2|\mathbf{r}|\big)^2} \tag{4.8}$$

for some $|\mathbf{r}| \ge 8$ or

$$|\beta_1 r_1 + \beta_2 r_2| \le \frac{\varepsilon}{40|\mathbf{r}|} \tag{4.9}$$

for some $1 \le |\mathbf{r}| < 8$, where $|\mathbf{r}| = \sqrt{r_1^2 + r_2^2}$.

Note that the complicated denominator in (4.8) is motivated by the fact that the numerical series

$$\sum_{n=3}^{\infty} \frac{1}{n(\log n)^2} \tag{4.10}$$

is very close to the border of convergence-divergence: the slightly larger series

$$\sum_{n=3}^{\infty} \frac{1}{n \log n}$$

is already divergent, but (4.10) is still convergent (see (4.15) below). Of course we could replace the exponent 2 in (4.10) with $1 + \varepsilon$, but the gain would be negligible.

Let $\mathcal{B}$ denote the set of ε-bad points (β_1, β_2) on the unit circle $\beta_1^2+\beta_2^2 = 1$. Next I show that the "bad" set $\mathcal{B}$ forms a small minority: the measure of $\mathcal{B}$ is negligible compared to the circumference 2π of the unit circle. (Note in advance that at the end of the proof we will throw out all initial conditions having ε-bad angles.)

Lemma 4.1. *The set $\mathcal{B}$ of ε-bad points (see the Key Definition) is small in the sense that*

$$\frac{\text{measure}\,(\mathcal{B})}{2\pi} < \frac{\varepsilon}{2}. \tag{4.11}$$

Proof. Notice that $\beta_1 r_1 + \beta_2 r_2$ is a dot product of two vectors, so the absolute value $|\beta_1 r_1 + \beta_2 r_2|$ equals $|\mathbf{r}| \sin\theta$, where θ is the angle between the unit vector (β_1, β_2) and the vector $(-r_2, r_1)$ perpendicular to $\mathbf{r} = (r_1, r_2)$. Therefore, given any $\mathbf{r} \in \mathbb{Z}^2$ with $|\mathbf{r}| \geq 8$, inequality (4.8) defines two short diametrically opposite arcs on the unit circle $\beta_1^2+\beta_2^2 = 1$ with total arclength

$$4 \arcsin\left(\frac{\varepsilon}{40|\mathbf{r}|^2 \cdot \log_2|\mathbf{r}| \cdot \left(\log_2 \log_2 |\mathbf{r}|\right)^2}\right),$$

where of course arcsin is the inverse of sin. Similarly, given any $\mathbf{r} \in \mathbb{Z}^2$ with $1 \leq |\mathbf{r}| < 8$, inequality (4.9) defines two short diametrically opposite arcs on the unit circle $\beta_1^2 + \beta_2^2 = 1$ with total arclength

$$4 \arcsin\left(\frac{\varepsilon}{40|\mathbf{r}|^2}\right).$$

It follows that

$$\text{measure}\,(\mathcal{B}) < \sum_{\substack{\mathbf{r}\in\mathbb{Z}^2:\\ 1\le|\mathbf{r}|<8}} 4\arcsin\left(\frac{\varepsilon}{40|\mathbf{r}|^2}\right) \tag{4.12}$$

$$+\sum_{\substack{\mathbf{r}\in\mathbb{Z}^2:\\ |\mathbf{r}|\ge 8}} 4\arcsin\left(\frac{\varepsilon}{40|\mathbf{r}|^2\cdot\log_2|\mathbf{r}|\cdot\left(\log_2\log_2|\mathbf{r}|\right)^2}\right).$$

By using the trivial inequality

$$\arcsin(x) < x + x^2 \quad \text{for} \quad 0 < x < 1, \tag{4.13}$$

we can easily estimate the sums in (4.12). We begin with the auxiliary sum

$$\sum\nolimits_1 = \sum_{\substack{\mathbf{r}\in\mathbb{Z}^2:\\ |\mathbf{r}|\ge 8}} \frac{1}{|\mathbf{r}|^2\cdot\log_2|\mathbf{r}|\cdot\left(\log_2\log_2|\mathbf{r}|\right)^2}. \tag{4.14}$$

We estimate (4.14) by applying a standard power-of-two decomposition:

$$\sum\nolimits_1 = \sum_{k=3}^{\infty}\sum_{2^k\le|\mathbf{r}|<2^{k+1}} \frac{1}{|\mathbf{r}|^2\cdot\log_2|\mathbf{r}|\cdot\left(\log_2\log_2|\mathbf{r}|\right)^2} \tag{4.15}$$

$$< \sum_{k=3}^{\infty} \pi 4^{k+1}\cdot\frac{1}{4^k\cdot k\cdot(\log_2 k)^2}$$

$$= 4\pi\sum_{k=3}^{\infty}\frac{1}{k\cdot(\log_2 k)^2}.$$

Note that in (4.15) we used the trivial fact that the number of lattice points in the annulus $2^k \le |\mathbf{r}| < 2^{k+1}$ is less than the area of the big circle $\pi\cdot 4^{k+1}$.

Returning to (4.15), we can estimate the infinite series with the corresponding definite integral:

$$\sum_{k=3}^{\infty}\frac{1}{k\cdot(\log_2 k)^2} < \int_2^{\infty}\frac{dx}{x(\log_2 x)^2} = \log 2,$$

and using this in (4.15), we have

$$\sum\nolimits_1 < 4\pi\log 2. \tag{4.16}$$

Similarly,

$$\sum\nolimits_2 = \sum_{\substack{\mathbf{r}\in\mathbb{Z}^2:\\ |\mathbf{r}|\geq 8}} \frac{1}{|\mathbf{r}|^4 \cdot \left(\log_2 |\mathbf{r}|\right)^2 \cdot \left(\log_2 \log_2 |\mathbf{r}|\right)^4} \tag{4.17}$$

$$= \sum_{k=3}^{\infty} \sum_{2^k \leq |\mathbf{r}| < 2^{k+1}} \frac{1}{|\mathbf{r}|^4 \cdot \left(\log_2 |\mathbf{r}|\right)^2 \cdot \left(\log_2 \log_2 |\mathbf{r}|\right)^4}$$

$$< \sum_{k=3}^{\infty} \pi 4^{k+1} \cdot \frac{1}{16^k \cdot k^2 \cdot (\log_2 k)^4} < \frac{\pi}{100}.$$

We also need the simple numerical facts

$$\sum\nolimits_4 = \sum_{\substack{\mathbf{r}\in\mathbb{Z}^2:\\ 1\leq|\mathbf{r}|<8}} \frac{1}{|\mathbf{r}|^4} < \sum\nolimits_3 = \sum_{\substack{\mathbf{r}\in\mathbb{Z}^2:\\ 1\leq|\mathbf{r}|<8}} \frac{1}{|\mathbf{r}|^2} < 6\pi. \tag{4.18}$$

Combining (4.12)–(4.18), we have

$$\frac{\text{measure}\,(\mathcal{B})}{2\pi} < \frac{\varepsilon}{20\pi}\sum\nolimits_1 + \frac{\varepsilon^2}{800\pi}\sum\nolimits_2 + \frac{\varepsilon}{20\pi}\sum\nolimits_3 + \frac{\varepsilon^2}{800\pi}\sum\nolimits_4 < \frac{\varepsilon}{2},$$

completing the proof of Lemma 4.1. ■

Let $\mathcal{G}$ denote the complement of the "bad" set $\mathcal{B}$, that is, $\mathcal{G}$ is the set of points (β_1, β_2) on the unit circle $\beta_1^2 + \beta_2^2 = 1$ which are *not* ε-bad (see the Key Definition; I used letter $\mathcal{G}$ to indicate "good"). We want to give an upper bound to the integral

$$\int_{\mathcal{G}} \left(\int_{I^2} \left(A(T; \mathbf{y}, (\beta_1, \beta_2)) - T \cdot \text{area}\,(A)\right)^2 d\mathbf{y} \right) ds, \tag{4.19}$$

where in the outer integral of (4.19) "ds" indicates integration with respect to the arclength (since $\mathcal{G}$ is a "large" subset of the unit circle). We prove the following result.

Lemma 4.2. *We have*

$$\int_{\mathcal{G}} \left(\int_{I^2} \left(A(T; \mathbf{y}, (\beta_1, \beta_2)) - T \cdot \text{area}\,(A)\right)^2 d\mathbf{y} \right) ds$$

$$\leq \text{area}\,(A)\left(1 - \text{area}\,(A)\right) \cdot \frac{2688}{\pi^2} \cdot \frac{1}{\varepsilon} \log_2 T \cdot (\log_2 \log_2 T)^2.$$

Proof. By using (4.6)–(4.7), we have

$$\text{integral (4.19)} \le \sum_{\substack{\mathbf{r}\in\mathbb{Z}^2:\\ \mathbf{r}\ne\mathbf{0}}} |a_{\mathbf{r}}|^2 \cdot \int_{\mathcal{G}} \min\left\{\frac{1}{\pi^2(\beta_1 r_1+\beta_2 r_2)^2}, T^2\right\} ds. \tag{4.20}$$

If $(\beta_1,\beta_2)\in\mathcal{G}$ then by definition (see (4.8)–(4.9))

$$|\beta_1 r_1+\beta_2 r_2| > \frac{\varepsilon}{40|\mathbf{r}|\cdot\log_2|\mathbf{r}|\cdot\left(\log_2\log_2|\mathbf{r}|\right)^2} \tag{4.21}$$

for all $|\mathbf{r}|\ge 8$ and

$$|\beta_1 r_1+\beta_2 r_2| > \frac{\varepsilon}{40|\mathbf{r}|} \tag{4.22}$$

for all $1\le|\mathbf{r}|<8$. Let $\mathbf{r}\in\mathbb{Z}^2$ with $|\mathbf{r}|=\sqrt{r_1^2+r_2^2}\ge 8$ be arbitrary but fixed; to estimate the integral at the end of (4.20), we apply a standard power-of-two decomposition of the set

$$\mathcal{G}(\mathbf{r})=\left\{(\beta_1,\beta_2)\,:\,\beta_1^2+\beta_2^2=1,\ \text{(4.21) holds}\right\}\supset\mathcal{G} \tag{4.23}$$

as follows: let ℓ be an arbitrary integer in the range

$$0\le\ell\le L(\mathbf{r})=\log_2\left(\frac{40}{\varepsilon}|\mathbf{r}|\cdot\log_2|\mathbf{r}|\cdot\left(\log_2\log_2|\mathbf{r}|\right)^2\right), \tag{4.24}$$

and write

$$\mathcal{G}_\ell(\mathbf{r})=\left\{(\beta_1,\beta_2)\,:\,\beta_1^2+\beta_2^2=1,\ 2^{-\ell-1}<|\beta_1 r_1+\beta_2 r_2|\le 2^{-\ell}\right\}. \tag{4.25}$$

Finally, write

$$\mathcal{G}_{-1}(\mathbf{r})=\left\{(\beta_1,\beta_2)\,:\,\beta_1^2+\beta_2^2=1,\ |\beta_1 r_1+\beta_2 r_2|>1\right\}, \tag{4.26}$$

and so we have the disjoint decomposition

$$\mathcal{G}(\mathbf{r})=\bigcup_{-1\le\ell\le L(\mathbf{r})}\mathcal{G}_\ell(\mathbf{r})\supset\mathcal{G}. \tag{4.27}$$

For every $\ell\ge 0$ we have the estimation

$$\text{measure}\left(\mathcal{G}_\ell(\mathbf{r})\right)\le 4\arcsin\left(\frac{1}{|\mathbf{r}|2^\ell}\right)\le 4\left(\frac{1}{|\mathbf{r}|\cdot 2^\ell}+\frac{1}{|\mathbf{r}|^2\cdot 4^\ell}\right), \tag{4.28}$$

where (4.28) is just an easy adaptation of the argument at the beginning of the proof of Lemma 4.1.

Motivated by (4.20) and (4.27), we need to estimate the sum

$$\sum_{\substack{\mathbf{r}\in\mathbb{Z}^2:\\ |\mathbf{r}|\geq 8}} |a_{\mathbf{r}}|^2 \cdot \int_{\mathcal{G}(\mathbf{r})} \min\left\{\frac{1}{\pi^2(\beta_1 r_1+\beta_2 r_2)^2}, T^2\right\} ds \tag{4.29}$$

$$= \sum_{\substack{\mathbf{r}\in\mathbb{Z}^2:\\ |\mathbf{r}|\geq 8}} |a_{\mathbf{r}}|^2 \sum_{\ell=-1}^{L(\mathbf{r})} \int_{\mathcal{G}_\ell(\mathbf{r})} \min\left\{\frac{1}{\pi^2(\beta_1 r_1+\beta_2 r_2)^2}, T^2\right\} ds$$

$$\leq \sum_{\substack{\mathbf{r}\in\mathbb{Z}^2:\\ |\mathbf{r}|\geq 8}} |a_{\mathbf{r}}|^2 \sum_{\ell=-1}^{L(\mathbf{r})} \text{measure}\left(\mathcal{G}_\ell(\mathbf{r})\right) \cdot \min\left\{\frac{4^{\ell+1}}{\pi^2}, T^2\right\}$$

$$\leq \sum_{\substack{\mathbf{r}\in\mathbb{Z}^2:\\ |\mathbf{r}|\geq 8}} |a_{\mathbf{r}}|^2 \left(\sum_{\ell=0}^{L(\mathbf{r})} 4\left(\frac{1}{|\mathbf{r}|\cdot 2^\ell}+\frac{1}{|\mathbf{r}|^2\cdot 4^\ell}\right)\min\left\{\frac{4^{\ell+1}}{\pi^2}, T^2\right\} + 2\pi\cdot\frac{1}{\pi^2}\right),$$

where in the last step we used (4.28); the last term in (4.29) is a trivial bound for the special case $\ell = -1$ in the summation; and finally, $L(\mathbf{r})$ is defined in (4.24).

To estimate (4.29), we need some rather long but totally routine calculations. For any $\mathbf{r} \in \mathbb{Z}^2$ with $|\mathbf{r}| \geq 8$, we have

$$\sum_{\ell=0}^{L(\mathbf{r})} \left(\frac{1}{|\mathbf{r}|\cdot 2^\ell}+\frac{1}{|\mathbf{r}|^2\cdot 4^\ell}\right)\min\left\{\frac{4^{\ell+1}}{\pi^2}, T^2\right\} \tag{4.30}$$

$$= \frac{1}{|\mathbf{r}|}\sum_{\ell=0}^{L(\mathbf{r})} \min\left\{\frac{2^{\ell+2}}{\pi^2}, \frac{T^2}{2^\ell}\right\} + \frac{1}{|\mathbf{r}|^2}\sum_{\ell=0}^{L(\mathbf{r})} \min\left\{\frac{4}{\pi^2}, \frac{T^2}{4^\ell}\right\}$$

$$\leq \frac{1}{|\mathbf{r}|}\sum_{\substack{0\leq\ell\leq L(\mathbf{r}):\\ 2^{\ell+1}\leq\pi T}} \frac{2^{\ell+2}}{\pi^2} + \frac{T^2}{|\mathbf{r}|}\sum_{\substack{0\leq\ell\leq L(\mathbf{r}):\\ 2^{\ell+1}>\pi T}} 2^{-\ell} + \frac{\log_2 T}{|\mathbf{r}|^2}$$

$$\leq \frac{1}{|\mathbf{r}|}\cdot\frac{8}{\pi^2}\min\left\{2^{L(\mathbf{r})}, \pi T/2\right\} + \frac{T^2}{|\mathbf{r}|}\cdot 2\chi^*\left(2^{L(\mathbf{r})}\geq \pi T/2\right)\cdot\frac{2}{\pi T} + \frac{\log_2 T}{|\mathbf{r}|^2},$$

where χ^* is the following 0-1 valued characteristic function: $\chi^*\big(2^{L(\mathbf{r})} \geq \pi T/2\big) = 1$ if $2^{L(\mathbf{r})} \geq \pi T/2$ and $\chi^*\big(2^{L(\mathbf{r})} \geq \pi T/2\big) = 0$ if $2^{L(\mathbf{r})} < \pi T/2$.

By (4.24), if $2^{L(\mathbf{r})} < \pi T/2$ then

$$\frac{1}{|\mathbf{r}|} \cdot \frac{8}{\pi^2} \min\left\{2^{L(\mathbf{r})}, \pi T/2\right\} + \frac{T^2}{|\mathbf{r}|} \cdot 2\chi^*\big(2^{L(\mathbf{r})} \geq \pi T/2\big) \cdot \frac{2}{\pi T}$$
$$= \frac{8}{\pi^2} \cdot \frac{40}{\varepsilon} \cdot \log_2 |\mathbf{r}| \cdot \big(\log_2 \log_2 |\mathbf{r}|\big)^2, \tag{4.31}$$

and if $2^{L(\mathbf{r})} \geq \pi T/2$ then

$$\frac{1}{|\mathbf{r}|} \cdot \frac{8}{\pi^2} \min\left\{2^{L(\mathbf{r})}, \pi T/2\right\} + \frac{T^2}{|\mathbf{r}|} \cdot 2\chi^*\big(2^{L(\mathbf{r})} \geq \pi T/2\big) \cdot \frac{2}{\pi T}$$
$$= \frac{4T}{\pi|\mathbf{r}|} + \frac{4T}{\pi|\mathbf{r}|} = \frac{8T}{\pi|\mathbf{r}|}. \tag{4.32}$$

If $2^{L(\mathbf{r})} < \pi T/2$ and $|\mathbf{r}| \geq 8$ then of course

$$\log_2 |\mathbf{r}| < L(\mathbf{r}) < \log_2(\pi T),$$

and so the last term in (4.31) can be estimated from above as follows:

$$\frac{8}{\pi^2} \cdot \frac{40}{\varepsilon} \cdot \log_2 |\mathbf{r}| \cdot \big(\log_2 \log_2 |\mathbf{r}|\big)^2 < \frac{8}{\pi^2} \cdot \frac{40}{\varepsilon} \log_2(\pi T) \cdot \big(\log_2 \log_2(\pi T)\big)^2. \tag{4.33}$$

On the other hand, if we have the equality

$$\pi T/2 = 2^{L(\mathbf{r})} = \frac{40}{\varepsilon}|\mathbf{r}| \cdot \log_2 |\mathbf{r}| \cdot \big(\log_2 \log_2 |\mathbf{r}|\big)^2 \quad \text{and} \quad |\mathbf{r}| \geq 8, \tag{4.34}$$

then clearly

$$\frac{T}{|\mathbf{r}|} = \frac{2}{\pi} \cdot \frac{40}{\varepsilon} \cdot \log_2 |\mathbf{r}| \cdot \big(\log_2 \log_2 |\mathbf{r}|\big)^2 \leq \frac{2}{\pi} \cdot \frac{40}{\varepsilon} \cdot \log_2 T \cdot (\log_2 \log_2 T)^2, \tag{4.35}$$

and (4.35) remains true if we go beyond the equality (4.34) to the whole range $2^{L(\mathbf{r})} \geq \pi T/2$.

Summarizing, by (4.30)–(4.35) for any $\mathbf{r} \in \mathbb{Z}^2$ with $|\mathbf{r}| \geq 8$ we have

$$\sum_{\ell=0}^{L(\mathbf{r})} \left(\frac{1}{|\mathbf{r}| \cdot 2^\ell} + \frac{1}{|\mathbf{r}|^2 \cdot 4^\ell} \right) \min\left\{ \frac{4^{\ell+1}}{\pi^2}, T^2 \right\} \leq \frac{16}{\pi^2} \cdot \frac{41}{\varepsilon} \log_2 T \cdot (\log_2 \log_2 T)^2, \tag{4.36}$$

Applying (4.36) in (4.29), we obtain

$$\sum_{\substack{\mathbf{r}\in\mathbb{Z}^2:\\ |\mathbf{r}|\geq 8}} |a_\mathbf{r}|^2 \cdot \int_{\mathcal{G}(\mathbf{r})} \min\left\{ \frac{1}{\pi^2(\beta_1 r_1 + \beta_2 r_2)^2}, T^2 \right\} ds \leq \sum_{\substack{\mathbf{r}\in\mathbb{Z}^2:\\ |\mathbf{r}|\geq 8}} |a_\mathbf{r}|^2 \cdot \frac{64}{\pi^2} \cdot \frac{42}{\varepsilon} \log_2 T \cdot (\log_2 \log_2 T)^2. \tag{4.37}$$

Similarly,

$$\sum_{\substack{\mathbf{r}\in\mathbb{Z}^2:\\ 1\leq|\mathbf{r}|< 8}} |a_\mathbf{r}|^2 \cdot \int_{\mathcal{G}(\mathbf{r})} \min\left\{ \frac{1}{\pi^2(\beta_1 r_1 + \beta_2 r_2)^2}, T^2 \right\} ds \leq \sum_{\substack{\mathbf{r}\in\mathbb{Z}^2:\\ 1\leq|\mathbf{r}|< 8}} |a_\mathbf{r}|^2 \cdot \frac{64}{\pi^2} \cdot \frac{42}{\varepsilon}. \tag{4.38}$$

Returning to (4.19)–(4.27), and using (4.37)–(4.38),

$$\begin{aligned} &\frac{1}{2\pi} \int_{\mathcal{G}} \left(\int_{I^2} \left(A(T; \mathbf{y}, (\beta_1, \beta_2)) - T \cdot \text{area}\,(A) \right)^2 d\mathbf{y} \right) ds \\ &\leq \sum_{\substack{\mathbf{r}\in\mathbb{Z}^2:\\ |\mathbf{r}|\geq 8}} |a_\mathbf{r}|^2 \cdot \frac{1344}{\pi^3} \cdot \frac{1}{\varepsilon} \cdot \log_2 T \cdot (\log_2 \log_2 T)^2 + \sum_{\substack{\mathbf{r}\in\mathbb{Z}^2:\\ 1\leq|\mathbf{r}|< 8}} |a_\mathbf{r}|^2 \cdot \frac{1344}{\pi^3} \cdot \frac{1}{\varepsilon} \\ &\leq \sum_{\substack{\mathbf{r}\in\mathbb{Z}^2:\\ \mathbf{r}\neq\mathbf{0}}} |a_\mathbf{r}|^2 \cdot \frac{1344}{\pi^3} \cdot \frac{1}{\varepsilon} \log_2 T \cdot (\log_2 \log_2 T)^2 \\ &= \text{area}\,(A)\big(1 - \text{area}\,(A)\big) \cdot \frac{1344}{\pi^3} \cdot \frac{1}{\varepsilon} \log_2 T \cdot (\log_2 \log_2 T)^2, \end{aligned} \tag{4.39}$$

where in the last step we used (4.2). (4.39) gives Lemma 4.2. ■

Now we are ready to finish the proof of Theorem 1: we just throw out the "bad" initial conditions and apply Chebyshev's well-known inequality. First a definition: for any $\lambda > 0$ let

$$\Omega(\lambda) = \Big\{ \big(\mathbf{y}, (\beta_1, \beta_2)\big) \ : \mathbf{y} \in [0,1)^2, \ (\beta_1, \beta_2) \in \mathcal{G}, \tag{4.40}$$

$$\big|A\big(T; \mathbf{y}, (\beta_1, \beta_2)\big) - T \cdot \text{area}\,(A)\big| \geq \lambda \Big\}.$$

Combining (4.39)–(4.40) with Chebyshev's inequality,

$$\frac{1}{2\pi}\,\text{measure}\,\big(\Omega(\lambda)\big) \tag{4.41}$$

$$\leq \text{area}\,(A)\big(1 - \text{area}\,(A)\big) \cdot \frac{1344}{\pi^3} \cdot \frac{1}{\varepsilon} \log_2 T \cdot (\log_2 \log_2 T)^2 \cdot \lambda^{-2},$$

where "measure" stands for the 3-dimensional Lebesgue measure.

By making the choice

$$\lambda = \lambda_0 = \frac{10\sqrt{\text{area}\,(A)\big(1 - \text{area}\,(A)\big)}}{\varepsilon} \sqrt{\log_2 T} \cdot \log_2 \log_2 T \tag{4.42}$$

in (4.41), we conclude

$$\frac{1}{2\pi}\,\text{measure}\,\big(\Omega(\lambda_0)\big) \leq \frac{\varepsilon}{2}. \tag{4.43}$$

If we throw out the set of initial conditions (starting point and angle) $\big(\mathbf{y}, (\beta_1, \beta_2)\big)$ contained in $\Omega(\lambda_0)$, and also throw out those initial conditions $\big(\mathbf{y}, (\beta_1, \beta_2)\big)$ for which the angle (β_1, β_2) is ε-bad (i.e., $(\beta_1, \beta_2) \in \mathcal{B}$), then by (4.43) and Lemma 4.1 the total loss is $\leq \varepsilon/2 + \varepsilon/2 = \varepsilon$. Combining this fact with (4.42)–(4.43), Theorem 1 follows. ■

By slightly modifying the proof above, we can prove the following logarithmic discrepancy type result that holds for *almost every* initial condition (in the usual measure theoretic sense).

Proposition 4.1. *Given any measurable subset $A \subset [0,1]^2$, for almost every initial condition $\big(\mathbf{y}, (\beta_1, \beta_2)\big)$, the torus-line with starting point $\mathbf{y} \in [0,1]^2$ and initial direction (β_1, β_2) with $\beta_1^2 + \beta_2^2 = 1$ is extremely uniform in A for the overwhelming majority of the stopping times T in the following precise sense: for any $\delta > 0$,*

$$\big|A\big(T; \mathbf{y}, (\beta_1, \beta_2)\big) - T \cdot \text{area}\,(A)\big| < (\log T)^{1+\delta} \tag{4.44}$$

holds for all $0 < T < \infty$, *except a set* $T \in \mathcal{E}$ *of density zero* ($\mathcal{E}$ *for "exceptional"*), *meaning*

$$\lim_{N\to\infty} \frac{\text{measure}\left(\mathcal{E} \cap [0,N]\right)}{N} = 0.$$

To prove Proposition 4.1, we use the following diophantine fact: for *almost every* point (β_1, β_2), $\beta_1^2 + \beta_2^2 = 1$ on the unit circle,

$$|\beta_1 r_1 + \beta_2 r_2| > \frac{1}{|\mathbf{r}| \cdot \log_2 |\mathbf{r}| \cdot \left(\log_2 \log_2 |\mathbf{r}|\right)^2} \tag{4.45}$$

for all sufficiently "large" lattice point $\mathbf{r} = (r_1, r_2) \in \mathbb{Z}^2$, i.e., for all $|\mathbf{r}| \geq C_0(\beta_1, \beta_2)$ (where $C_0(\beta_1, \beta_2)$ is some constant depending only on (β_1, β_2)). Note that (4.45) can be easily proved by repeating the proof of Lemma 4.1, and combining it with the Borel–Cantelli lemma.

In view of the diophantine fact (4.45), for any $\varepsilon > 0$ there is a measurable subset $\mathcal{G}(\varepsilon)$ of the unit circle with

$$\frac{\text{measure}\left(\mathcal{G}(\varepsilon)\right)}{2\pi} > 1 - \varepsilon, \tag{4.46}$$

and there is a finite constant $C_1(\varepsilon)$ such that,

$$\sum_{\substack{\mathbf{r}\in\mathbb{Z}^2:\\ \mathbf{r}\neq\mathbf{0}}} |a_{\mathbf{r}}|^2 \cdot \int_{\mathcal{G}(\varepsilon)} \min\left\{\frac{1}{\pi^2(\beta_1 r_1 + \beta_2 r_2)^2}, T^2\right\} ds \tag{4.47}$$

$$\leq C_1(\varepsilon) + 100 \log_2 T \cdot (\log_2 \log_2 T)^2$$

holds for *all* $T > 100$.

Note that the proof of (4.47) is a straightforward adaptation of the proof of Lemma 4.2, combined with (4.45).

By (4.19)–(4.20) and (4.47) (where $I^2 = [0,1]^2$),

$$2^{-n} \int_{T=2^n}^{2^{n+1}} \left(\int_{\mathcal{G}(\varepsilon)} \int_{I^2} \left(A(T; \mathbf{y}, (\beta_1, \beta_2)) - T \cdot \text{area}\,(A)\right)^2 d\mathbf{y}\, ds \right) dT \tag{4.48}$$

$$\leq C_1(\varepsilon) + 100(n+1) \cdot \left(\log_2(n+1)\right)^2$$

for all integers $n \geq 7$.

Now let $\eta > 0$ be arbitrarily small but fixed, and define the set

$$\Omega(n) = \Big\{ \big(\mathbf{y}, (\beta_1, \beta_2)\big) \in [0,1]^2 \times \mathcal{G}(\varepsilon) : \tag{4.49}$$

$$\big|A\big(T; \mathbf{y}, (\beta_1, \beta_2)\big) - T \cdot \text{area}\,(A)\big| \geq n^{1+\delta} \text{ holds for a set of } T's$$

$$\text{in } 2^n \leq T \leq 2^{n+1} \text{ having measure } > \eta 2^n \Big\}.$$

Comparing (4.49) with (4.48), we obtain

$$\text{measure}\,\big(\Omega(n)\big) < \frac{C_1(\varepsilon) + 100(n+1) \cdot \big(\log_2(n+1)\big)^2}{\eta \cdot n^{2+2\delta}}, \tag{4.50}$$

which is less than $n^{-1-\delta}$ if n is large enough.

Since the series $\sum_n n^{-1-\delta}$ is convergent, combining (4.50) with the Borel–Cantelli lemma, we conclude that, for *almost every* initial condition in

$$\big(\mathbf{y}, (\beta_1, \beta_2)\big) \in [0,1]^2 \times \mathcal{G}(\varepsilon),$$

the upper bound

$$\big|A\big(T; \mathbf{y}, (\beta_1, \beta_2)\big) - T \cdot \text{area}\,(A)\big| < (\log T)^{1+\delta} \tag{4.51}$$

holds for all $0 < T < \infty$, except a set $T \in \mathcal{E}(\eta)$ of upper density $< 2\eta$, meaning

$$\limsup_{N \to \infty} \frac{\text{measure}\,\big(\mathcal{E}(\eta) \cap [0,N]\big)}{N} < 2\eta. \tag{4.52}$$

Since the parameters $\varepsilon > 0$ and $\eta > 0$ can be arbitrarily small, (4.46), (4.51) and (4.52) imply (4.44), completing the proof of Proposition 4.1.

Finally, note that, I don't know how to upgrade Proposition 4.1 to a stronger variant with logarithmic discrepancy that holds for *all* large T (not just for the overwhelming majority of T's).

5. Proof of Theorem 2

We generalize the proof of Theorem 1 in higher dimensions $d \geq 3$. It is convenient, therefore, to involve more linear algebra. The following lemma is a consequence of the well-known fact that, roughly speaking, "determinant = volume".

Lemma 5.1. *The n-dimensional region $\mathcal{R}$ defined by n equations of the type*

$$\left|b_{j,1}y_1 + b_{j,2}y_2 + \cdots + b_{j,n}y_n\right| \leq c_j, \quad 1 \leq j \leq n$$

has (n-dimensional) volume

$$V = \frac{2^n c_1 c_2 \cdots c_n}{\left|\det\left(b_{j,k}\right)\right|}.$$

Here we assume that the determinant $\det(b_{j,k})$ is nonzero (which, by the way, implies that $\mathcal{R}$ is bounded).

Proof. Write $z_j = \sum_{i=1}^{n} b_{j,i}y_i$. The volume of the box $|z_j| \leq c_j$, $1 \leq j \leq n$ is clearly $(2c_1)\cdot(2c_2)\cdots(2c_n)$, and so, if $(\beta_{i,j})$ denotes the inverse matrix to $(b_{j,k})$, we have

$$y_i = \sum_{j=1}^{n} \beta_{i,j}z_j, \quad 1 \leq i \leq n,$$

and

$$V = \left|\det\left(\beta_{i,j}\right)\right| \cdot 2^n c_1 c_2 \cdots c_n = \frac{2^n c_1 c_2 \cdots c_n}{\left|\det\left(b_{j,k}\right)\right|},$$

completing the proof of the lemma. ■

Let $d \geq 3$ denote the dimension. Similarly to the 2-dimensional case, it suffices to deal with torus-lines ("unfolding the billiard path"). Let $A \subset I^d = [0,1)^d$ be an arbitrary measurable subset, and again consider the Fourier series of the characteristic function χ_A of the set A:

$$\chi_A(\mathbf{u}) = \sum_{\mathbf{r}\in\mathbb{Z}^d} a_{\mathbf{r}} e^{2\pi i \mathbf{r}\cdot\mathbf{u}} \quad \text{with} \quad a_{\mathbf{r}} = \int_A e^{-2\pi i \mathbf{r}\cdot\mathbf{w}}\, d\mathbf{w}, \tag{5.1}$$

where $\mathbf{r}\cdot\mathbf{u}$ denotes the inner product. Clearly $a_{\mathbf{0}} = \mathrm{vol}\,(A)$ (= the Lebesgue measure of A), and by Parseval's formula,

$$\sum_{\substack{\mathbf{r}\in\mathbb{Z}^d:\\ \mathbf{r}\neq\mathbf{0}}} |a_{\mathbf{r}}|^2 = \mathrm{vol}\,(A) - \mathrm{vol}^2(A). \tag{5.2}$$

Consider the torus-line $\mathbf{x}(t) = \big(x_1(t), \ldots, x_d(t)\big)$ (mod 1), where

$$x_j(t) = \beta_j t + y_j, \quad 1 \le j \le d \quad \text{and} \quad \beta_1^2 + \cdots + \beta_d^2 = 1. \tag{5.3}$$

Again the length of the straight line segment $\mathbf{x}(t)$, $0 < t < T$ is T. The pair $(\mathbf{y}, \mathbf{b})$ with $\mathbf{y} = (y_1, \ldots, y_d) \in [0,1)^d$ and $\mathbf{b} = (\beta_1, \ldots, \beta_d) \in S^{d-1}$ (where S^{d-1} is the unit sphere in the d-space) describes the starting point $\mathbf{y}$ and the direction $\mathbf{b}$ of the torus-line $\mathbf{x}(t)$—we may call $(\mathbf{y}, \mathbf{b})$ the initial condition. The total time $A(T) = A(T; \mathbf{y}, \mathbf{b})$ that the torus-line $\mathbf{x}(t)$ (defined in (5.3)) spends in subset A during $0 < t < T$ equals (perfect analog of (4.4))

$$\begin{aligned} A(T) &= A(T; \mathbf{y}, \mathbf{b}) \\ &= a_0 T + \sum_{\substack{\mathbf{r} \in \mathbb{Z}^d: \\ \mathbf{r} \neq \mathbf{0}}} a_{\mathbf{r}} e^{2\pi i \mathbf{r}\cdot\mathbf{y}} \cdot \frac{e^{2\pi i \mathbf{b}\cdot\mathbf{r} T} - 1}{2\pi i \mathbf{b}\cdot\mathbf{r}}. \end{aligned} \tag{5.4}$$

Since $a_0 = \operatorname{vol}(A)$, by (5.4) we have

$$\begin{aligned} \text{discrepancy} &= A(T; \mathbf{y}, \mathbf{b}) - T \cdot \operatorname{vol}(A) \\ &= \sum_{\substack{\mathbf{r} \in \mathbb{Z}^d: \\ \mathbf{r} \neq \mathbf{0}}} a_{\mathbf{r}} \cdot \frac{e^{2\pi i \mathbf{b}\cdot\mathbf{r} T} - 1}{2\pi i \mathbf{b}\cdot\mathbf{r}} \cdot e^{2\pi i \mathbf{r}\cdot\mathbf{y}}. \end{aligned} \tag{5.5}$$

Fix any point $\mathbf{b} = (\beta_1, \ldots, \beta_d)$ on the unit sphere $\beta_1^2 + \cdots + \beta_d^2 = 1$, and run the starting point $\mathbf{y}$ through the unit cube $I^d = [0,1]^d$; then by Parseval's formula we obtain the analog of (4.6):

$$\begin{aligned} &\int_{I^d} \big(A(T; \mathbf{y}, \mathbf{b}) - T \cdot \operatorname{vol}(A)\big)^2 \, d\mathbf{y} \\ &= \sum_{\substack{\mathbf{r} \in \mathbb{Z}^d: \\ \mathbf{r} \neq \mathbf{0}}} |a_{\mathbf{r}}|^2 \cdot \left| \frac{e^{2\pi i \mathbf{b}\cdot\mathbf{r} T} - 1}{2\pi \mathbf{b}\cdot\mathbf{r}} \right|^2. \end{aligned} \tag{5.6}$$

Also, we need the analog of (4.7): for the last factor in (5.6) we have the obvious upper bound

$$\left| \frac{e^{2\pi i \mathbf{b}\cdot\mathbf{r} T} - 1}{2\pi \mathbf{b}\cdot\mathbf{r}} \right| \le \min\left\{ \frac{1}{\pi |\mathbf{b}\cdot\mathbf{r}|}, T \right\}. \tag{5.7}$$

Key Definition (discarding the bad vectors). Let $0 < \varepsilon < 1/2$; we say that a point $\mathbf{b} = (\beta_1, \ldots, \beta_d)$ on the unit sphere $S^{d-1} : \beta_1^2 + \cdots + \beta_d^2 = 1$ is ε-bad if there exists an $\mathbf{r} = (r_1, \ldots, r_d) \in \mathbb{Z}^d \setminus \mathbf{0}$ such that

$$|\mathbf{b} \cdot \mathbf{r}| \leq \frac{\varepsilon}{|\mathbf{r}|^{d-1} \log(1 + |\mathbf{r}|)\big(\log\log(2 + |\mathbf{r}|)\big)^2}. \tag{5.8}$$

Otherwise we just say that $\mathbf{b} \in S^{d-1}$ is ε-good.

Notice the similarity between (4.8) and (5.8).

Let $\mathcal{B}_d$ denote the set of ε-bad points $\mathbf{b} = (\beta_1, \ldots, \beta_d)$ on the unit sphere $S^{d-1} : \beta_1^2 + \cdots + \beta_d^2 = 1$. Next I show that the "bad" set $\mathcal{B}_d$ forms a small minority in the following sense: the measure (= surface area) of $\mathcal{B}_d$ is negligible compared to the surface area of the unit sphere S^{d-1}, assuming ε is sufficiently small depending only on the dimension d. (At the end of the proof we will throw out all initial conditions having ε-bad directions.) The following lemma is an analog of Lemma 4.1 (measure=surface area).

Lemma 5.2. *The set $\mathcal{B}_d$ of ε-bad points on the unit sphere S^{d-1} is small in the sense that*

$$\frac{\text{measure}(\mathcal{B})}{\text{measure}(S^{d-1})} < \varepsilon \cdot c_1(d), \tag{5.9}$$

where $c_1(d)$ is some constant depending only on d.

Proof. Let $\mathbf{r} = (r_1, \ldots, r_d) \in \mathbb{Z}^d \setminus \mathbf{0}$ be arbitrary but fixed. For notational simplicity, assume that the last coordinate is the largest: $|r_d| = \max_{1 \leq j \leq d} |r_j|$. Consider the d-dimensional region defined by the d equations

$$|u_j| \leq 1, \quad 1 \leq j \leq d-1 \quad \text{and} \tag{5.10}$$

$$|r_1 u_1 + r_2 u_2 + \cdots + r_d u_d| \leq \frac{\varepsilon}{|\mathbf{r}|^{d-1} \log(1 + |\mathbf{r}|)\big(\log\log(2 + |\mathbf{r}|)\big)^2}.$$

We apply Lemma 5.1: since the determinant is r_d, the volume $V = V(5.10)$ of region (5.10) equals

$$V = V(5.10) = \frac{\varepsilon 2^d}{|\det| \cdot |\mathbf{r}|^{d-1} \log(1 + |\mathbf{r}|)\big(\log\log(2 + |\mathbf{r}|)\big)^2} \tag{5.11}$$

$$= \frac{\varepsilon 2^d}{|r_d| \cdot |\mathbf{r}|^{d-1} \log\left(1+|\mathbf{r}|\right)\left(\log\log\left(2+|\mathbf{r}|\right)\right)^2}$$

$$= \frac{\varepsilon\sqrt{d}2^d}{|\mathbf{r}|^d \log\left(1+|\mathbf{r}|\right)\left(\log\log\left(2+|\mathbf{r}|\right)\right)^2},$$

since

$$|\mathbf{r}| = \left(r_1^2 + \cdots + r_d^2\right)^{1/2} \leq \sqrt{d} \max_{1\leq i\leq d} |r_i| = \sqrt{d}|r_d|.$$

If $\mathbf{b} = (\beta_1, \ldots, \beta_d) \in S^{d-1}$ satisfies (5.8), then the whole line segment joining to the origin satisfies (5.10), i.e., (5.10) holds with $\mathbf{u} = \lambda\mathbf{b}$ for all $0 < \lambda \leq 1$. Combining this fact with (5.11), we have that, for any fixed $\mathbf{r} \in \mathbb{Z}^d \setminus \mathbf{0}$, the set of $\mathbf{b} \in S^{d-1}$ satisfying (5.8) has measure (= surface area)

$$\leq \frac{\varepsilon \cdot c_2(d)}{|\mathbf{r}|^d \log\left(1+|\mathbf{r}|\right)\left(\log\log\left(2+|\mathbf{r}|\right)\right)^2}, \tag{5.12}$$

where, as usual, $c_2(d)$ denotes a constant depending only on d (we don't make any effort to find the optimal values of these constants). The last step is to take summation over all $\mathbf{r} \in \mathbb{Z}^d \setminus \mathbf{0}$. A routine power-of-two decomposition yields that the sum is finite:

$$\sum_{\mathbf{r}\in\mathbb{Z}^d\setminus\mathbf{0}} \frac{1}{|\mathbf{r}|^d \log\left(1+|\mathbf{r}|\right)\left(\log\log\left(2+|\mathbf{r}|\right)\right)^2} \tag{5.13}$$

$$= \sum_{k=0}^{\infty} \sum_{\substack{\mathbf{r}\in\mathbb{Z}^d: \\ 2^k\leq|\mathbf{r}|<2^{k+1}}} \frac{1}{|\mathbf{r}|^d \log\left(1+|\mathbf{r}|\right)\left(\log\log\left(2+|\mathbf{r}|\right)\right)^2}$$

$$\leq \sum_{k=0}^{\infty} \frac{c'(d)(2^k)^d}{(2^k)^d \cdot (1+k) \cdot \left(\log(2+k)\right)^2} = c_3(d) < \infty.$$

Finally, notice that (5.12)–(5.13) imply Lemma 5.2. ■

Let $\mathcal{G}_d$ denote the complement of the "bad" set $\mathcal{B}_d$, that is, $\mathcal{G}_d$ is the set of ε-good points $\mathbf{b} = (\beta_1, \ldots, \beta_d) \in S^{d-1}$ on the unit sphere $\beta_1^2 + \cdots + \beta_d^2 = 1$. We want to give an upper bound to the integral

$$\frac{1}{\text{measure}\,(S^{d-1})} \int_{\mathcal{G}_d} \left(\int_{I^d} \left(A(T;\mathbf{y},\mathbf{b}) - T \cdot \text{vol}\,(A)\right)^2 d\mathbf{y} \right) dS, \tag{5.14}$$

where in the outer integral of (5.14) "dS" indicates integration with respect to the surface area (since $\mathcal{G}$ is a "large" subset of the unit sphere S^{d-1}).

By using (5.6)–(5.7), we have

(5.15)
$$\text{integral (5.14)} \le \sum_{\substack{\mathbf{r}\in\mathbb{Z}^d:\\ \mathbf{r}\neq\mathbf{0}}} |a_{\mathbf{r}}|^2 \cdot \frac{1}{\text{measure}\,(S^{d-1})} \int_{\mathcal{G}_d} \min\left\{\frac{1}{\pi^2(\mathbf{b}\cdot\mathbf{r})^2}, T^2\right\} dS.$$

If $\mathbf{b} \in \mathcal{G}_d$ then by definition (see (5.8))

$$|\mathbf{b}\cdot\mathbf{r}| > \frac{\varepsilon}{|\mathbf{r}|^{d-1}\log(1+|\mathbf{r}|)\left(\log\log(2+|\mathbf{r}|)\right)^2} \tag{5.16}$$

for all $\mathbf{r} \in \mathbb{Z}^d \setminus \mathbf{0}$.

Now let $\mathbf{r} \in \mathbb{Z}^d \setminus \mathbf{0}$ be arbitrary but fixed. To estimate the integral at the end of (5.15), we apply a standard power-of-two decomposition of the set

$$\mathcal{G}_d(\mathbf{r}) = \left\{\mathbf{b} = (\beta_1,\ldots,\beta_d) : \beta_1^2+\cdots+\beta_d^2 = 1,\ (5.16)\text{ holds}\right\} \supset \mathcal{G}_d \tag{5.17}$$

as follows: let ℓ be an arbitrary integer in the range

$$0 \le \ell \le L_d(\mathbf{r}) = \log_2\left(\frac{|\mathbf{r}|^{d-1}\log(1+|\mathbf{r}|)\left(\log\log(2+|\mathbf{r}|)\right)^2}{\varepsilon}\right), \tag{5.18}$$

and write

(5.19)
$$\mathcal{G}_{d,\ell}(\mathbf{r}) = \left\{\mathbf{b} = (\beta_1,\ldots,\beta_d) : \beta_1^2+\cdots+\beta_d^2 = 1,\ 2^{-\ell-1} < |\mathbf{b}\cdot\mathbf{r}| \le 2^{-\ell}\right\}.$$

Finally, write

$$\mathcal{G}_{d,-1}(\mathbf{r}) = \left\{\mathbf{b} = (\beta_1,\ldots,\beta_d) : \beta_1^2+\cdots+\beta_d^2 = 1,\ |\mathbf{b}\cdot\mathbf{r}| > 1\right\}. \tag{5.20}$$

Thus we have the disjoint decomposition

$$\mathcal{G}_d(\mathbf{r}) = \bigcup_{-1\le\ell\le L_d(\mathbf{r})} \mathcal{G}_{d,\ell}(\mathbf{r}) \supset \mathcal{G}_d. \tag{5.21}$$

By repeating the proof of Lemma 5.2, for every $\ell \ge 0$ we obtain the following upper bound for the surface area of $\mathcal{G}_{d,\ell}(\mathbf{r})$:

$$\text{measure}\left(\mathcal{G}_{d,\ell}(\mathbf{r})\right) \le \frac{c_4(d)}{|\mathbf{r}|\cdot 2^{\ell}}, \tag{5.22}$$

where, as usual, $c_4(d)$ is some finite constant depending only on d.

Motivated by (5.15) and (5.21), we need to estimate the sum

$$\sum_{\mathbf{r}\in\mathbb{Z}^d\setminus\mathbf{0}} |a_{\mathbf{r}}|^2 \cdot \int_{\mathcal{G}_d(\mathbf{r})} \min\left\{\frac{1}{\pi^2(\mathbf{b}\cdot\mathbf{r})^2}, T^2\right\} dS \tag{5.23}$$

$$= \sum_{\mathbf{r}\in\mathbb{Z}^d\setminus\mathbf{0}} |a_{\mathbf{r}}|^2 \sum_{\ell=-1}^{L_d(\mathbf{r})} \int_{\mathcal{G}_{d,\ell}(\mathbf{r})} \min\left\{\frac{1}{\pi^2(\mathbf{b}\cdot\mathbf{r})^2}, T^2\right\} dS$$

$$\leq \sum_{\mathbf{r}\in\mathbb{Z}^d\setminus\mathbf{0}} |a_{\mathbf{r}}|^2 \sum_{\ell=-1}^{L_d(\mathbf{r})} \text{measure}\left(\mathcal{G}_{d,\ell}(\mathbf{r})\right) \cdot \min\left\{\frac{4^{\ell+1}}{\pi^2}, T^2\right\}$$

$$\leq \sum_{\mathbf{r}\in\mathbb{Z}^d\setminus\mathbf{0}} |a_{\mathbf{r}}|^2 \sum_{\ell=-1}^{L_d(\mathbf{r})} \frac{c_4(d)}{|\mathbf{r}|\cdot 2^\ell} \cdot \min\left\{\frac{4^{\ell+1}}{\pi^2}, T^2\right\},$$

where in the last step we used (5.22), and $L_d(\mathbf{r})$ is defined in (5.18).

We have

$$\sum_{\ell=-1}^{L_d(\mathbf{r})} \frac{1}{2^\ell} \cdot \min\left\{\frac{4^{\ell+1}}{\pi^2}, T^2\right\} \leq \sum_{\ell=-1}^{L_d(\mathbf{r})} \min\left\{2^\ell, \frac{T^2}{2^\ell}\right\} \tag{5.24}$$

$$= \sum_{-1\leq\ell\leq\min\{L_d(\mathbf{r}),\log_2 T\}} 2^\ell + \sum_{\min\{L_d(\mathbf{r}),\log_2 T\}<\ell\leq L_d(\mathbf{r})} \frac{T^2}{2^\ell}$$

$$\leq 4\min\left\{2^{L_d(\mathbf{r})}, T\right\} \leq \frac{4}{\varepsilon}\min\left\{|\mathbf{r}|^{d-1}\log\left(1+|\mathbf{r}|\right)\left(\log\log\left(2+|\mathbf{r}|\right)\right)^2, T\right\},$$

where in the last step we applied (5.18).

Using (5.24) in (5.23), we have

$$\sum_{\mathbf{r}\in\mathbb{Z}^d\setminus\mathbf{0}} |a_{\mathbf{r}}|^2 \cdot \int_{\mathcal{G}_d(\mathbf{r})} \min\left\{\frac{1}{\pi^2(\mathbf{b}\cdot\mathbf{r})^2}, T^2\right\} dS \tag{5.25}$$

$$\leq \frac{c_5(d)}{\varepsilon} \sum_{\mathbf{r}\in\mathbb{Z}^d\setminus\mathbf{0}} |a_{\mathbf{r}}|^2 \cdot \min\left\{|\mathbf{r}|^{d-2}\log\left(1+|\mathbf{r}|\right)\left(\log\log\left(2+|\mathbf{r}|\right)\right)^2, \frac{T}{|\mathbf{r}|}\right\}.$$

To estimate (5.25), we consider first those $\mathbf{r} \in \mathbb{Z}^d \setminus \mathbf{0}$ which satisfy the inequality

$$|\mathbf{r}|^{d-1} \log\left(1+|\mathbf{r}|\right)\left(\log\log\left(2+|\mathbf{r}|\right)\right)^2 \le T. \tag{5.26}$$

Under the hypothesis of (5.26), we can estimate a critical term in (5.25) as follows:

$$\begin{aligned} &|\mathbf{r}|^{d-2} \log\left(1+|\mathbf{r}|\right)\left(\log\log\left(2+|\mathbf{r}|\right)\right)^2 \\ &\le c_6(d)\Big(|\mathbf{r}|^{d-1} \log\left(1+|\mathbf{r}|\right)\left(\log\log\left(2+|\mathbf{r}|\right)\right)^2\Big)^{\frac{d-2}{d-1}} \\ &\quad\cdot \left(\log\left(1+|\mathbf{r}|\right)\right)^{\frac{1}{d-1}}\left(\log\log\left(2+|\mathbf{r}|\right)\right)^{\frac{2}{d-1}} \\ &\le c_6(d)T^{\frac{d-2}{d-1}} \cdot \left(\log\left(1+|\mathbf{r}|\right)\right)^{\frac{1}{d-1}}\left(\log\log\left(2+|\mathbf{r}|\right)\right)^{\frac{2}{d-1}}. \end{aligned} \tag{5.27}$$

On the other hand, if (5.26) fails, then we have

$$|\mathbf{r}| \ge \frac{c_7(d)T^{\frac{1}{d-1}}}{(\log T)^{\frac{1}{d-1}}(\log\log T)^{\frac{2}{d-1}}},$$

implying

$$\frac{T}{|\mathbf{r}|} \le c_8(d)T^{\frac{d-2}{d-1}} \cdot (\log T)^{\frac{1}{d-1}}(\log\log T)^{\frac{2}{d-1}}. \tag{5.28}$$

Now we return to (5.25): by applying (5.27) if (5.26) holds, and applying (5.28) if (5.26) fails, we have

$$\begin{aligned} &\sum_{\mathbf{r}\in\mathbb{Z}^d\setminus\mathbf{0}} |a_{\mathbf{r}}|^2 \cdot \min\left\{|\mathbf{r}|^{d-2} \log\left(1+|\mathbf{r}|\right)\left(\log\log\left(2+|\mathbf{r}|\right)\right)^2, \frac{T}{|\mathbf{r}|}\right\} \\ &\le c_9(d)T^{\frac{d-2}{d-1}} \cdot (\log T)^{\frac{1}{d-1}}(\log\log T)^{\frac{2}{d-1}} \sum_{\mathbf{r}\in\mathbb{Z}^d\setminus\mathbf{0}} |a_{\mathbf{r}}|^2. \end{aligned} \tag{5.29}$$

Combining (5.23), (5.25) and (5.29), we obtain

$$\sum_{\mathbf{r}\in\mathbb{Z}^d\setminus\mathbf{0}} |a_{\mathbf{r}}|^2 \cdot \int_{\mathcal{G}_d(\mathbf{r})} \min\left\{\frac{1}{\pi^2(\mathbf{b}\cdot\mathbf{r})^2}, T^2\right\} dS \tag{5.30}$$

$$\le \frac{c_{10}(d)}{\varepsilon} T^{\frac{d-2}{d-1}} \cdot (\log T)^{\frac{1}{d-1}} (\log\log T)^{\frac{2}{d-1}} \sum_{\mathbf{r}\in\mathbb{Z}^d\setminus\mathbf{0}} |a_{\mathbf{r}}|^2.$$

Summarizing, by (5.2), (5.15), (5.21) and (5.30),

$$\text{integral (5.14)} \tag{5.31}$$

$$\le \frac{c_{11}(d)}{\varepsilon} T^{\frac{d-2}{d-1}} \cdot (\log T)^{\frac{1}{d-1}} (\log\log T)^{\frac{2}{d-1}} \cdot \mathrm{vol}\,(A)\big(1 - \mathrm{vol}\,(A)\big).$$

Notice that (5.31) is the multi-dimensional generalization of (4.39): the special case in (5.31) gives back (4.39).

The rest of the proof of Theorem 2 is exactly the same as that of Theorem 1 (a combination of (5.31) with the Chebyshev's inequality). ■

6. Proof of Theorem 3

For convex $A \subset [0,1]^d$ the Fourier coefficients $a_{\mathbf{r}}$ of the characteristic function χ_A of A (see (5.1)) have the following "decreasing" property.

Lemma 6.1. *If $A \subset [0,1]^d$ is convex, then the Fourier coefficients $a_{\mathbf{r}}$ of the characteristic function of A satisfy*

$$\sum_{\substack{\mathbf{r}\in\mathbb{Z}^d:\\ M\le|\mathbf{r}|<2M}} |a_{\mathbf{r}}|^2 < \frac{c'(d)}{M} \quad \text{for all} \quad M \ge 1, \tag{6.1}$$

where $c'(d)$ is a constant depending only on the dimension.

Applying Lemma 6.1 in (5.25), we have

$$\sum_{\mathbf{r}\in\mathbb{Z}^d\setminus\mathbf{0}} |a_{\mathbf{r}}|^2 \cdot \int_{\mathcal{G}_d(\mathbf{r})} \min\left\{\frac{1}{\pi^2(\mathbf{b}\cdot\mathbf{r})^2}, T^2\right\} dS \tag{6.2}$$

$$\le \frac{c_5(d)}{\varepsilon} \sum_{m=0}^{\infty} \sum_{\substack{\mathbf{r}\in\mathbb{Z}^d:\\ 2^m\le|\mathbf{r}|<2^{m+1}}} |a_{\mathbf{r}}|^2$$

$$\cdot \min\left\{|\mathbf{r}|^{d-2}\log\left(1+|\mathbf{r}|\right)\left(\log\log\left(2+|\mathbf{r}|\right)\right)^2, \frac{T}{|\mathbf{r}|}\right\}$$

$$\leq \frac{c_{12}(d)}{\varepsilon} \sum_{m=0}^{\infty} \min \left\{ 2^{(d-3)m} \cdot (1+m)\big(\log(2+m)\big)^2, T2^{-2m} \right\}$$

$$\leq \frac{c_{13}(d)}{\varepsilon} T^{\frac{d-3}{d-1}} \cdot (\log T)^{\frac{2}{d-1}} (\log\log T)^{\frac{4}{d-1}},$$

where the last step is just a routine calculation. Now we can finish the proof of Theorem 3 the same way as we did for Theorem 2 above, with the minor modification that we use (6.2) instead of (5.30).

It remains to prove the lemma.

Proof of Lemma 6.1. It is a standard argument in Fourier analysis: we use an appropriate "kernel", estimate the corresponding "convolution", and apply Parseval's formula. The details go as follows. Let $B(\rho; \mathbf{u})$ denote the d-dimensional ball of radius ρ centered at $\mathbf{u} \in \mathbb{R}^d$. Let $\chi_{B,\rho}$ denote the characteristic function of the ball $B(\rho; \mathbf{0})$ centered at the origin:

$$\chi_{B,\rho}(\mathbf{w}) = \begin{cases} 1, & \text{if } |\mathbf{w}| \leq \rho; \\ 0, & \text{otherwise.} \end{cases}$$

Let $m \geq 1$ be an arbitrary integer, and let $0 < \rho_1 < \rho_2 < 1/2$. Consider the convolution

$$g = g_{\rho_1,\rho_2} = \chi_A * h, \tag{6.3}$$

where h (= "kernel") is the following difference:

$$h = \chi_{B,\rho_2} - \left(\frac{\rho_2}{\rho_1}\right)^d \chi_{B,\rho_1}. \tag{6.4}$$

Notice that $g(\mathbf{w})$ has the following simple geometric meaning:

$$g(\mathbf{w}) = g_{\rho_1,\rho_2}(\mathbf{w}) \tag{6.5}$$

$$= \text{volume}\big(A \cap B(\rho_2; \mathbf{w})\big) - \left(\frac{\rho_2}{\rho_1}\right)^d \text{volume}\big(A \cap B(\rho_1; \mathbf{w})\big).$$

We are going to use the following three properties of "kernel" h:

$$\int_{[-\rho_2,\rho_2)^d} h(\mathbf{w})\, d\mathbf{w} = 0, \tag{6.6a}$$

(6.6b) $$h(\mathbf{w}) = 0 \quad \text{if} \quad |\mathbf{w}| > \rho_2,$$

(6.6c) $$\left|h(\mathbf{w})\right| \leq \left(\frac{\rho_2}{\rho_1}\right)^d \quad \text{if} \quad |\mathbf{w}| \leq \rho_2.$$

We extend $g = g_{\rho_1,\rho_2}$ (see (6.3)) from the unit size cube $[-1/2, 1/2)^d$ to the whole space $\mathbb{R}^d$ periodically, with period one. Since $g = g_{\rho_1,\rho_2}$ is defined as a convolution, the Fourier coefficients $\widehat{g}(\mathbf{r})$, $\mathbf{r} \in \mathbb{Z}^d$ of g have the following product form:

(6.7) $$g(\mathbf{w}) = g_{\rho_1,\rho_2}(\mathbf{w}) = \sum_{\mathbf{r}\in\mathbb{Z}^d} \widehat{g}(\mathbf{r}) e^{2\pi i \mathbf{r}\cdot\mathbf{w}}$$

with

(6.8) $$\widehat{g}(\mathbf{r}) = a_{\mathbf{r}} \left(b_{\mathbf{r}}(\rho_2) - \left(\frac{\rho_2}{\rho_1}\right)^d b_{\mathbf{r}}(\rho_1) \right),$$

where $a_{\mathbf{r}}$, $b_{\mathbf{r}}(\rho_1)$ and $b_{\mathbf{r}}(\rho_2)$ are the Fourier coefficients of the characteristic functions χ_A, χ_{B,ρ_1} and χ_{B,ρ_2} in this order. Combining Parseval's formula with (6.3)–(6.4) and (6.7)–(6.8), we have

(6.9) $$\int_{[0,1)^d} g^2_{\rho_1,\rho_2}(\mathbf{w})\, d\mathbf{w} = \sum_{\mathbf{r}\in\mathbb{Z}^d} |a_{\mathbf{r}}|^2 \cdot \left| b_{\mathbf{r}}(\rho_2) - \left(\frac{\rho_2}{\rho_1}\right)^d b_{\mathbf{r}}(\rho_1)\right|^2.$$

It is well-known that the Fourier coefficients $b_{\mathbf{r}}(\rho)$, $\mathbf{r} \in \mathbb{Z}^d$ of the (characteristic function $\chi_{B,\rho}$ of the) d-dimensional ball $B(\rho;\mathbf{0})$ (of radius ρ) can be expressed in terms of the classical Bessel function $J_{d/2}$ as follows. By definition, for any $\mathbf{r} \in \mathbb{Z}^d \setminus \mathbf{0}$ we have

(6.10) $$b_{\mathbf{r}}(\rho) = \int_{|\mathbf{u}|\leq\rho} e^{-2\pi \mathbf{r}\cdot\mathbf{u}}\, d\mathbf{u} = c_{14}(d) \int_{-\rho}^{\rho} e^{-2\pi i r y} (\rho^2 - y^2)^{\frac{d-1}{2}}\, dy$$

$$= c_{14}(d)\rho^d \int_{-1}^{1} \cos(2\pi r\rho z)(1 - z^2)^{\frac{d-1}{2}}\, dz,$$

where $c_{14}(d)$ is the volume of the $(d-1)$-dimensional unit ball, and $r = |\mathbf{r}| = \sqrt{r_1^2 + \cdots + r_d^2}$. The classical Bessel function $J_\nu(x)$ (see e.g. (3) on

p. 48 of Watson [8]) has, for $\nu > -1/2$, the integral representation (called Poisson integral)

$$J_\nu(x) = \frac{1}{\sqrt{\pi}\Gamma\left(\nu+\frac{1}{2}\right)}\left(\frac{x}{2}\right)^\nu \int_{-1}^{1} \cos(xz)(1-z^2)^{\nu-\frac{1}{2}}\,dz, \tag{6.11}$$

where $\Gamma(z)$ is the well-known Euler gamma-function ("generalization of $n!$"). Hence, by (6.10) and (6.11),

$$b_{\mathbf{r}}(\rho) = c_{15}(d)\left(\frac{\rho}{r}\right)^{d/2} J_{d/2}(2\pi r\rho). \tag{6.12}$$

The Bessel function has the following asymptotic formula ("Hankel's expansion"; see e.g. (1) on p. 199 of Watson [8]),

$$J_\nu(x) = \left(\frac{2}{\pi x}\right)^{1/2}\cos\left(x - \frac{(2\nu+1)\pi}{4}\right) + O\left(x^{-3/2}\right), \tag{6.13}$$

where the implicit constant in the O-notation depends only on ν. Therefore, by (6.12) and (6.13),

$$b_{\mathbf{r}}(\rho) = c_{16}(d)\frac{\rho^{\frac{d-1}{2}}}{r^{\frac{d+1}{2}}}\cos\left(2\pi r\rho - \frac{(d+1)\pi}{4}\right) + O\left(\rho^{\frac{d-3}{2}}\cdot r^{-\frac{d+3}{2}}\right), \tag{6.14}$$

where $c_{16}(d) > 0$ and the implicit constant in the O-notation depends only on the dimension d (≥ 3).

Let

$$0 < \lambda < \mu < \frac{1}{2}, \quad \rho_1 = \frac{\lambda}{m}, \quad \rho_2 = \frac{\mu}{m}. \tag{6.15}$$

Then by (6.14),

$$b_{\mathbf{r}}(\rho_2) - \left(\frac{\rho_2}{\rho_1}\right)^d b_{\mathbf{r}}(\rho_1) \tag{6.16}$$

$$= \frac{c_{16}(d)}{m^{\frac{d-1}{2}} r^{\frac{d+1}{2}}}\cdot \mu^{\frac{d-1}{2}}\cos\left(2\pi\mu r/m - \frac{(d+1)\pi}{4}\right)\left(1+O\left((\mu r/m)^{-1}\right)\right)$$

$$- \frac{c_{16}(d)}{m^{\frac{d-1}{2}} r^{\frac{d+1}{2}}}\cdot \frac{\mu^d}{\lambda^{\frac{d-1}{2}}}\cos\left(2\pi\lambda r/m - \frac{(d+1)\pi}{4}\right)\left(1+O\left((\lambda r/m)^{-1}\right)\right).$$

We want to guarantee the following "good approximation":

$$b_{\mathbf{r}}(\rho_2) - \left(\frac{\rho_2}{\rho_1}\right)^d b_{\mathbf{r}}(\rho_1) \approx -\frac{c_{16}(d)}{m^{\frac{d-1}{2}} r^{\frac{d+1}{2}}} \cdot \frac{\mu^d}{\lambda^{\frac{d-1}{2}}}. \tag{6.17}$$

To enforce (6.17), we simply satisfy the following three requirements (see (6.15)):

(6.18a) $\frac{r}{m} = \frac{|\mathbf{r}|}{m}$ is a "large" constant;

(6.18b) we choose $\mu = \frac{1}{3}$, and λ is "much smaller" than $\mu = \frac{1}{3}$;

(6.18c)
$\lambda r/m - \frac{d+1}{8}$ is "close" to an integer: (say) the distance is $< 1/10$.

This way, by choosing any sufficiently small constant value for λ, there are constants $0 < c_{17} = c_{17}(d, \lambda) < c_{18} = c_{18}(d, \lambda)$ (where $c_{17} < c_{18}$ are "large", and their difference is "small") such that

$$\frac{1}{2} < \frac{b_{\mathbf{r}}(\rho_2) - \left(\frac{\rho_2}{\rho_1}\right)^d b_{\mathbf{r}}(\rho_1)}{-\frac{c_{16}(d)}{m^{\frac{d-1}{2}} r^{\frac{d+1}{2}}} \cdot \frac{\mu^d}{\lambda^{\frac{d-1}{2}}}} < \frac{3}{2} \tag{6.19}$$

holds for all $\mathbf{r} \in \mathbb{Z}^d$ with $c_{17}m \le r = |\mathbf{r}| < c_{18}m$.

By using (6.19) in the right-hand side of (6.9), we have

$$\sum_{\mathbf{r}\in\mathbb{Z}^d} |a_{\mathbf{r}}|^2 \cdot \left| b_{\mathbf{r}}(\rho_2) - \left(\frac{\rho_2}{\rho_1}\right)^d b_{\mathbf{r}}(\rho_1) \right|^2 \ge \frac{c_{19}(d)}{m^{2d}} \sum_{\substack{\mathbf{r}\in\mathbb{Z}^d: \\ c_{17}m \le |\mathbf{r}| < c_{18}m}} |a_{\mathbf{r}}|^2 \tag{6.20}$$

with some positive absolute constant $c_{19}(d) > 0$.

Next we study the the left-hand side of (6.9):

$$\int_{[0,1)^d} g^2_{\rho_1,\rho_2}(\mathbf{w})\, d\mathbf{w},$$

where (see (6.5) and (6.15))

$$g(\mathbf{w}) = g_{\rho_1,\rho_2}(\mathbf{w}) \tag{6.21}$$

$$= \text{volume}\left(A \cap B(\rho_2; \mathbf{w})\right) - \left(\frac{\rho_2}{\rho_1}\right)^d \text{volume}\left(A \cap B(\rho_1; \mathbf{w})\right),$$

and

$$0 < \lambda < \frac{1}{3}, \quad \rho_1 = \frac{\lambda}{m}, \quad \rho_2 = \frac{1}{3m}.$$

If the boundary ∂A of the convex set $A \subset [0,1]^d$ does *not* intersect the ball $B(\rho_2; \mathbf{w})$, then by (6.21), $g_{\rho_1,\rho_2}(\mathbf{w}) = 0$. The ball $B(\rho_2; \mathbf{w})$ is clearly contained in a (small) cube of size $\frac{1}{m} \times \frac{1}{m} \times \cdots \times \frac{1}{m}$. Motivated by this fact, we divide the unit cube $[0,1]^d$ into m^d congruent subcubes of size $\frac{1}{m} \times \frac{1}{m} \times \cdots \times \frac{1}{m}$: I call them $\frac{1}{m}$-cubes. We need the following

Lemma 6.2. *Let $m \geq 1$ be any integer and $A \subset [0,1]^d$ be any convex set. Then the boundary ∂A of A intersects at most $c_{20}(d)m^{d-1}$ $\frac{1}{m}$-cubes.*

Combining (6.6), (6.21) and Lemma 6.2, we obtain the upper bound

$$\int_{[0,1)^d} g^2_{\rho_1,\rho_2}(\mathbf{w})\, d\mathbf{w} \leq \frac{c_{20}(d)m^{d-1}}{m^d} \cdot \frac{c_{21}(d)}{m^{2d}}. \tag{6.22}$$

Summarizing, by (6.9), (6.20) and (6.22) we have

$$\sum_{\substack{\mathbf{r}\in\mathbb{Z}^d: \\ c_{17}m \leq |\mathbf{r}| < c_{18}m}} |a_{\mathbf{r}}|^2 \leq \frac{c_{22}(d)}{m} \tag{6.23}$$

for any integer $m \geq 1$. Notice that (6.23) implies Lemma 6.1.

Finally, we include a

Proof of Lemma 6.2. The statement of the lemma is obvious for "nice" convex sets, such as the balls (= solid spheres). Unfortunately, a general convex set can be very different from a ball, so we need to involve a little technical trick: we introduce a "nice" auxiliary set (see $A^+[1/m]$ below).

For simplicity, assume that A is not too close to the unit cube $[0,1]^d$: we assume that the distance of A and the complement $\mathbb{R}^d \setminus [0,1]^d$ of the unit cube is $> 1/m$. (This restriction is basically irrelevant, since there are only $c'(d)m^{d-1}$ $\frac{1}{m}$-cubes touching the boundary of the unit cube.) Let $A^+[1/m]$ denote the $\frac{1}{m}$-neighborhood of A:

$$A^+[1/m] = \left\{\mathbf{x} \in \mathbb{R}^d : \text{distance}\,(\mathbf{x}, A) \leq \frac{1}{m}\right\}.$$

We introduce $A^+[1/m]$ because it is "nice" in the following sense. If A intersects a $\frac{1}{m}$-cube Q, then the boundary $\partial A^+[1/m]$ of $A^+[1/m]$ intersects at least one of the $3^d - 1$ neighbors of Q in a relatively "large piece": the surface area of the intersection is at least $c''(d)m^{-d+1}$ with some constant $c''(d) > 0$ (depending only on the dimension), i.e., the surface area of the intersection is at least constant times the surface area of Q. (Note that a general convex set A is not necessarily "nice").

If A is convex then $A^+[1/m]$ is also convex, and we have $A \subset A^+[1/m] \subset [0,1]^d$. Also, we use the well-known general fact that, if $A_1 \subset A_2$ are compact convex sets in $\mathbb{R}^d$, then

$$\text{SurfaceArea}\,(\partial A_1) \le \text{SurfaceArea}\,(\partial A_2).$$

Summarizing, we have

$$\begin{aligned} &\text{number of } \frac{1}{m}\text{-cubes intersecting } \partial A \\ &\le c_{23}(d)\left(\text{number of } \frac{1}{m}\text{-cubes intersecting } \partial A^+[1/m]\right) \\ &\le c_{24}(d)\frac{\text{SurfaceArea}\left([0,1]^d\right)}{c_{25}(d)m^{-d+1}} = c_{26}(d)m^{d-1}, \end{aligned}$$

and Lemma 6.2 follows. ∎

This completes the proof of Theorem 3. ∎

7. Proof of Theorem 4

The basic idea is to "amplify the trivial error". We proceed in three steps.

Step 1: What is the "trivial error"?

I begin with the "trivial error" in the discrete case (i.e., ordinary discrepancy), which is easier to understand than the "trivial error" in the continuous case (i.e., curve-discrepancy). Suppose that we have N points in the unit cube $[0,1]^d$ (N is "large"). Let m be an integer such that $m^d \approx 2N$. We divide $[0,1]^d$ into m^d congruent subcubes of size $\frac{1}{m} \times \frac{1}{m} \times \cdots \times \frac{1}{m}$: I call them $\frac{1}{m}$-cubes. Since $m^d \approx 2N$, at least half of the $\frac{1}{m}$-cubes are empty: do not contain any of the N given points. Note that, in the case of "perfect equidistribution", every $\frac{1}{m}$-cube should contain $Nm^{-d} \approx \frac{1}{2}$ points, but of course this is impossible. In fact, for every $\frac{1}{m}$-cube we have an "inevitable discrepancy": $Nm^{-d} - 0 \approx \frac{1}{2}$ if the $\frac{1}{m}$-cube is empty, and $\geq 1 - Nm^{-d} - 0 \approx \frac{1}{2}$ if the $\frac{1}{m}$-cube is non-empty. This "inevitable discrepancy" is what I call the "trivial error" (a consequence of the trivial fact that there is no integer between 0 and 1).

Next assume that $\mathcal{C}$ is a continuous curve in the unit cube $[0,1]^d$; suppose that arclength$(\mathcal{C}) = T$ is "large". We divide $\mathcal{C}$ into n parts of equal arclength: let $P_j \in \mathcal{C}$, $0 \leq j \leq n$ be the division points, that is, P_0 is the starting point of the curve (we fix an orientation), p_n is the endpoint, and the arclength of the part of $\mathcal{C}$ between P_j and P_{j+1} is T/n $(0 \leq j < n)$. Let m be an integer such that $m^d \approx 2(n+1)$. Then at least half of the $\frac{1}{m}$-cubes do not contain any of the $n+1$ division points P_j, $0 \leq j \leq n$. We divide every one of these empty $\frac{1}{m}$-cubes into 3^d congruent subcubes of side length $\frac{1}{3m}$ each (i.e., $\frac{1}{3m}$-cubes). If $T/n \leq \frac{1}{3m}$, then the $\frac{1}{3m}$-cube in the middle of every division-point-free $\frac{1}{m}$-cube is "curve-free": does not intersect the given continuous curve $\mathcal{C}$. That is, if

$$(7.1) \qquad m^d \approx 2(n+1) \quad \text{and} \quad \frac{T}{n} \leq \frac{1}{3m},$$

then there are at least $m^d/2$ $\mathcal{C}$-free $\frac{1}{3m}$-cubes.

Note that, in the case of "perfect equidistribution", every $\frac{1}{3m}$-cube should intersect the given curve $\mathcal{C}$ in total arclength $T(3m)^{-d}$. Consequently, every $\mathcal{C}$-free $\frac{1}{3m}$-cube has curve-discrepancy $T(3m)^{-d} - 0$. Since a constant percentage of the $\frac{1}{3m}$-cubes are $\mathcal{C}$-free, the "small curve-discrepancy" $T(3m)^{-d} - 0$ can be interpreted as a continuous analog of the "trivial error" in the discrete case above.

Let's return to (7.1): in view of the inequality

$$(7.2) \qquad T \leq \frac{n}{3m} \approx \frac{n+1}{3m} \approx \frac{m^d}{6m} = \frac{1}{6} m^{d-1},$$

by choosing (say)

$$(7.3) \qquad m^{d-1} \approx 12T,$$

we can easily guarantee (7.1).

Step 2: Amplifying the "trivial error"

Let $0 < \rho < 1/2$, and let $B(\rho;\mathbf{u})$ denote the d-dimensional closed ball with radius ρ centered at $\mathbf{u} \in \mathbb{R}^d$. Let χ_ρ denote the characteristic function of $B(\rho;\mathbf{0})$:

$$\chi_\rho(\mathbf{w}) = \begin{cases} 1, & \text{if } |\mathbf{w}| \le \rho; \\ 0, & \text{otherwise.} \end{cases}$$

We extend χ_ρ from the unit size cube $[-1/2, 1/2)^d$ to the whole space $\mathbb{R}^d$ periodically, with period one.

First we discuss the amplification of the "trivial error" in the discrete case. Let $\mathcal{P} = \{P_1, \ldots, P_N\}$ be an N-element point set in the unit cube $I^d = [0,1)^d$. We introduce two measures in I^d. For any $A \subset I^d$, let

$$Z_{\mathcal{P}}(A) = \sum_{P_j \in A} 1.$$

In other words, $Z_{\mathcal{P}}$ denotes the counting measure generated by the given N-element point set $\mathcal{P}$. Next, let λ denote the d-dimensional Lebesgue measure restricted to the unit cube I^d : $\lambda(A) = \operatorname{vol}(A)$ for all measurable $A \subset I^d$. Consider now the function

$$F_\rho = F_{\rho,\mathcal{P}} = \chi_\rho * (dZ_{\mathcal{P}} - N d\lambda),$$

where $*$ denotes the convolution operation. More explicitly,

$$\begin{aligned} F_\rho(\mathbf{x}) = F_{\rho,\mathcal{P}}(\mathbf{x}) &= \int_{I^d} \chi_\rho(\mathbf{x}-\mathbf{y}) \left(dZ_{\mathcal{P}}(\mathbf{y}) - N d\lambda(\mathbf{y})\right) \\ &= \sum_{P_j \in B(\rho;\mathbf{x}) \pmod 1} 1 - N \operatorname{vol}\left(B(\rho;\mathbf{0})\right). \end{aligned}$$

In other words, $F_\rho(\mathbf{x}) = F_{\rho,\mathcal{P}}(\mathbf{x})$ represents the torus discrepancy of the translated ball $B(\rho;\mathbf{x}) = B(\rho;\mathbf{0}) + \mathbf{x}$ (with respect to the given point set $\mathcal{P}$). Since F_ρ is a convolution, the Fourier coefficients $\widehat{F_\rho}(\mathbf{r})$, $\mathbf{r} \in \mathbb{Z}^d$ of F_ρ have the following product form:

$$F_\rho(\mathbf{x}) = F_{\rho,\mathcal{P}}(\mathbf{x}) = \sum_{\mathbf{r} \in \mathbb{Z}^d} \widehat{F_\rho}(\mathbf{r}) e^{2\pi i \mathbf{r}\cdot\mathbf{x}}$$

with

$$\widehat{F_\rho}(\mathbf{r}) = \widehat{\chi_\rho}(\mathbf{r}) \cdot \left(\widehat{dZ_{\mathcal{P}} - N d\lambda}\right)(\mathbf{r}).$$

By Parseval's formula,

$$\int_{I^d} F_\rho^2(\mathbf{x})\,d\mathbf{x} = \sum_{\mathbf{r}\in\mathbb{Z}^d} \left|\widehat{\chi_\rho}(\mathbf{r})\right|^2 \cdot \left|\left(d\widehat{Z_{\mathcal{P}} - N}d\lambda\right)(\mathbf{r})\right|^2. \tag{7.4}$$

Next we need the so-called

Amplification Lemma. *Let $0 < \delta < 1/2$ be arbitrary, then*

$$\frac{2\int_0^{1/2} \left|\widehat{\chi_\rho}(\mathbf{r})\right|^2 d\rho}{\frac{1}{\delta}\int_0^{\delta} \left|\widehat{\chi_\rho}(\mathbf{r})\right|^2 d\rho} \geq c_1(d)\cdot\delta^{1-d} \tag{7.5}$$

for all $\mathbf{r} \in \mathbb{Z}^d$, where the constant $c_1(d) > 0$ depends only on the dimension (i.e., independent of δ and $\mathbf{r}$).

Since $\widehat{\chi_\rho}(\mathbf{r})$ can be expressed in terms of the Bessel function $J_{d/2}$, and the Bessel function has a well-known asymptotic formula—see (6.10)–(6.14)—the proof of the Amplification Lemma is just a straightforward computation. (For the details I refer the reader to the literature to be mentioned below.)

Next we combine the "trivial error", (7.4) and the amplification Lemma. By (7.4) and (7.5), for any $0 < \delta < 1/2$ we have,

(7.6)

$$2\int_0^{1/2}\int_{I^d} F_\rho^2(\mathbf{x})\,d\mathbf{x}\,d\rho = \sum_{\mathbf{r}\in\mathbb{Z}^d} 2\int_0^{1/2} \left|\widehat{\chi_\rho}(\mathbf{r})\right|^2 d\rho \cdot \left|\left(d\widehat{Z_{\mathcal{P}} - N}d\lambda\right)(\mathbf{r})\right|^2$$

$$\geq c_1(d)\cdot\delta^{1-d}\sum_{\mathbf{r}\in\mathbb{Z}^d}\frac{1}{\delta}\int_0^{\delta}\left|\widehat{\chi_\rho}(\mathbf{r})\right|^2 d\rho\cdot\left|\left(d\widehat{Z_{\mathcal{P}} - N}d\lambda\right)(\mathbf{r})\right|^2$$

$$= c_1(d)\cdot\delta^{1-d}\cdot\frac{1}{\delta}\int_0^{\delta}\int_{I^d} F_\rho^2(\mathbf{x})\,d\mathbf{x}\,d\rho.$$

Let ρ_0 denote the radius such that the volume of the d-dimensional ball of radius ρ_0 is $\frac{1}{2N}$ (i.e., the expected number of points $P_j \in \mathcal{P}$ in the ball $B(\rho_0;\mathbf{0})$ (mod 1) is 1/2). Clearly $\rho_0 = c_2(d)N^{-1/d}$. Integrating the "trivial error", we have

$$\frac{1}{\rho_0}\int_0^{\rho_0}\int_{I^d} F_\rho^2(\mathbf{x})\,d\mathbf{x}\,d\rho \geq \frac{1}{4}. \tag{7.7}$$

Using (7.7) in (7.6) with $\delta = \rho_0$, we have,

$$(7.8) \qquad 2\int_0^{1/2}\int_{I^d} F_\rho^2(\mathbf{x})\, d\mathbf{x}\, d\rho \geq c_1(d)\cdot \rho_0^{1-d}\cdot \frac{1}{4} = c_3(d)N^{\frac{d-1}{d}}.$$

As an easy consequence of (7.8), we obtain the following statement. Given any N-element point set $\mathcal{P} = \{P_1, \ldots, P_N\}$ in the unit cube $I^d = [0,1)^d$, there is a ball of radius $\leq 1/2$ with torus discrepancy $\geq c_4(d)N^{\frac{1}{2}-\frac{1}{2d}}$, where the constant factor $c_4(d) > 0$ depends only on the dimension. This is an old result in "irregularities of distribution" (see the beginning of Section 2) that was proved independently by me and H. Montgomery (see the book [1] and the paper [7]; note that a slightly weaker result was first proved by W. M. Schmidt). This proof is due to me; it is a special case of the so-called "Fourier transform method" that I have developed in a series of papers in the 1980s (a good source is Chapter 6 in [1], or Section 7.1 in [6]).

It is easy to adapt this amplification technique for the continuous case, where the N-element point set $\mathcal{P} = \{P_1, \ldots, P_N\}$ is replaced with a continuous curve $\mathcal{C}$ of arclength T (we assume that T is "large"). Instead of working with the counting measure $Z_\mathcal{P}$ generated by the given N-element point set $\mathcal{P}$, we switch to the following "arclength measure": for any measurable $A \subset I^d$, let

$$Z_\mathcal{C}(A) = \text{arclength}\,(\mathcal{C}\cap A).$$

We keep $\lambda : \lambda(A) = \text{vol}\,(A)$ for all measurable $A \subset I^d$.

Consider now the function

$$G_\rho = G_{\rho,\mathcal{C}} = \chi_\rho * (dZ_\mathcal{C} - Td\lambda),$$

where $*$ denotes the convolution operation. More explicitly,

$$(7.9) \qquad G_{\rho,\mathcal{C}}(\mathbf{x}) = \text{arclength}\left(\mathcal{C}\cap B(\rho;\mathbf{x}) \pmod 1\right) - T\,\text{vol}\left(B(\rho;\mathbf{0})\right).$$

In other words, $G_\rho(\mathbf{x}) = G_{\rho,\mathcal{C}}(\mathbf{x})$ represents the torus curve-discrepancy of the translated ball $B(\rho;\mathbf{x}) = B(\rho;\mathbf{0})+\mathbf{x}$ (with respect to the given curve $\mathcal{C}$).

Again applying the Amplification Lemma, we obtain the following analog of (7.6): for any $0 < \delta < 1/2$ we have,

$$(7.10) \qquad 2\int_0^{1/2}\int_{I^d} G_\rho^2(\mathbf{x})\, d\mathbf{x}\, d\rho \geq c_1(d)\cdot \delta^{1-d}\cdot \frac{1}{\delta}\int_0^{\delta}\int_{I^d} G_\rho^2(\mathbf{x})\, d\mathbf{x}\, d\rho.$$

Next we apply the "trivial error" in the continuous case. Motivated by (7.1)–(7.3), we choose

$$\delta_0 = c_5(d)T^{-\frac{1}{d-1}}. \tag{7.11}$$

The "continuous trivial error" is

$$\geq c_6(d)T\delta_0^d = c_7(d)T^{-\frac{1}{d-1}}. \tag{7.12}$$

Using (7.12) in (7.10) with $\delta = \delta_0$ (see (7.11)), we have,

$$2\int_0^{1/2}\int_{I^d} G_\rho^2(\mathbf{x})\,d\mathbf{x}\,d\rho \geq c_8(d)\cdot\delta_0^{1-d}\cdot\left(T^{-\frac{1}{d-1}}\right)^2 = c_9(d)T^{\frac{d-3}{d-1}}. \tag{7.13}$$

As an easy consequence of (7.13), we obtain the following statement. Given any continuous curve $\mathcal{C}$ in the unit torus $I^d = [0,1)^d$ with arclength T, there is a ball of radius $\leq 1/2$ with torus curve-discrepancy $\geq c_{10}(d)T^{\frac{1}{2}-\frac{1}{d-1}}$, where the constant factor $c_{10}(d) > 0$ depends only on the dimension. This result, with this proof, is due to Drmota [5]. For more about "continuous uniform distribution"; see Section 2.3 in [4].

Step 3: Completing the proof of Theorem 4

I recall that Theorem 4 is about an arbitrary family of continuous curves $\{\mathcal{C}_\omega : \omega \in \Omega\}$, parametrized with a finite number of parameters, in the unit torus $[0,1)^d$ with common arclength T. (Ω is a bounded measurable subset of some finite-dimensional euclidean space.) For any measurable subset of the unit torus $A \subset [0,1)^d$, we write

$$A(T;\omega) = \text{arclength}\,(\mathcal{C}_\omega \cap A).$$

Let's return to (7.13), which is about a fixed curve $\mathcal{C}$. We integrate (7.13) over the family $\mathcal{C} = \mathcal{C}_\omega : \omega \in \Omega$, and obtain the lower bound,

$$\frac{1}{\text{measure}\,(\Omega)}\int_\Omega\left(2\int_0^{1/2}\int_{I^d} G_{\rho,\omega}^2(\mathbf{x})\,d\mathbf{x}\,d\rho\right)d\omega \geq c_{11}(d)T^{\frac{d-3}{d-1}}. \tag{7.14}$$

Therefore, we must have a radius $0 < \rho_0 \leq 1/2$ and a center $\mathbf{x}_0 \in I^d$ such that

$$\frac{1}{\text{measure}\,(\Omega)}\int_\Omega G_{\rho_0,\omega}^2(\mathbf{x}_0)\,d\omega \geq c_{11}(d)T^{\frac{d-3}{d-1}}. \tag{7.15}$$

Since $G_{\rho_0,\omega}(\mathbf{x}_0)$ is the torus curve-discrepancy of the ball $B(\rho_0, \mathbf{x}_0)$ with respect to the curve $\mathcal{C}_\omega$ (where $\omega \in \Omega$), Theorem 4 follows. ∎

8. Proof of Proposition 1.1

We combine the ideas of *trivial error* and *random construction.* Let $\mathcal{X} = \{x_1, x_2, \ldots, x_N\}$ be an arbitrary set of N points in the unit interval $[0,1]$. Let r denote the (unique) integer satisfying $2N \le 2^r < 4N$, and divide the unit interval into 2^r congruent subintervals

$$(8.1) \qquad \big[(i-1)2^{-r}, i2^{-r}\big), \quad i = 1, 2, 3, \ldots, 2^r.$$

Then of course at least half of the subintervals (8.1) do not contain any element of $\mathcal{X}$—I call them empty subintervals. I refer to the simple fact that at least half of the subintervals (8.1) are empty as the *trivial error.*

The *random construction* simply means that we take the union of a random subset of intervals (8.1). More precisely, let $Y_1, Y_2, \ldots, Y_{2^r}$ be a sequence of 2^r independent and indentically distributed Bernoulli variables with $\Pr[Y_i = 0] = \Pr[Y_i = 1] = 1/2$ (one may call it a sequence of Heads and Tails). Independence means that the underlying probability space Ω is a cartesian product:

$$\Omega = \big\{\omega = (\delta_1, \ldots, \delta_{2^r}) : \delta_i \in \{0,1\},\ i = 1, 2, \ldots, 2^r\big\},$$

and $Y_i(\omega)$ is simply the ith coordinate δ_i of ω. Note that $|\Omega| = 2^{2^r}$.

Every $\omega \in \Omega$ defines a measurable subset $B(\omega)$ (in fact, a union of some intervals (8.1)) of the unit interval in the following natural way:

$$(8.2) \qquad B(\omega) = \bigcup_{i : Y_i(\omega) = 1} \big[(i-1)2^{-r}, i2^{-r}\big).$$

Now let $\mathcal{X} = \{x_1, x_2, \ldots, x_N\} \subset [0,1]$ be arbitrary. Since $2N \le 2^r < 4N$, at least half of the subintervals (8.1) are empty ("trivial error"): let $I = I(\mathcal{X})$ denote the set of indices i such that the interval $\big[(i-1)2^{-r}, i2^{-r}\big)$ contains no element of $\mathcal{X}$. Clearly $\big|I(\mathcal{X})\big| \ge 2^r - N \ge 2^{r-1} \ge N$.

Consider the sum of independent (and identically distributed) random variables

$$\sum_{i \in I(\mathcal{X})} Y_i$$

with expectation (denoted by $\mathbf{E}$) and variance

$$\mathbf{E}\Big(\sum_{i \in I(\mathcal{X})} Y_i\Big) = \frac{1}{2}\big|I(\mathcal{X})\big| \quad \text{and} \quad \text{Variance}\Big(\sum_{i \in I(\mathcal{X})} Y_i\Big) = \frac{1}{4}\big|I(\mathcal{X})\big|.$$

The central limit theorem says that, for any fixed real number $\lambda > 0$,

$$\text{(8.3)} \qquad \Pr\left[\sum_{i\in I(\mathcal{X})} Y_i - \frac{1}{2}\big|I(\mathcal{X})\big| > \frac{\lambda}{2}\sqrt{\big|I(\mathcal{X})\big|}\,\right] = \frac{1}{\sqrt{2\pi}}\int_\lambda^\infty e^{-u^2/2}\,du + O\big(N^{-1/2}\big).$$

(Here "Pr" means equiprobability in the discrete probability space Ω.)

For any (measurable) subset $A \subset [0,1]$, consider the "error"

$$\text{error}\,(\mathcal{X}, A) = |\mathcal{X} \cap A| - |\mathcal{X}|\,\text{measure}\,(A).$$

Let

$$J(\mathcal{X}) = \{1, 2, \ldots, 2^r\} \setminus I(\mathcal{X})$$

$$= \Big\{i : \big[(i-1)2^{-r}, i2^{-r}\big) \text{ contains no element of } \mathcal{X}\Big\},$$

and $\big|J(\mathcal{X})\big| = \ell$. Then there are 2^ℓ possible vectors of the form

$$\mathbf{v} = \big(Y_j(\omega) : j \in J(\mathcal{X})\big) \in \{0,1\}^\ell,$$

where the elements $j \in J(\mathcal{X})$ are written in increasing order.

Let $\mathbf{v} =\in \{0,1\}^\ell$ be arbitrary but fixed, and consider the subset

$$\Omega_{\mathbf{v}} = \Omega_{\mathbf{v}}(\mathcal{X}) = \Big\{\omega \in \Omega : \big(Y_j(\omega) : j \in J(\mathcal{X})\big) = \mathbf{v}\Big\}$$

of the probability space Ω. We study the fluctuation of the "error"

$$\text{error}\,(\mathcal{X}, A) = |\mathcal{X} \cap A| - |\mathcal{X}|\,\text{measure}\,(A)$$

as A runs through the sets $B(\omega)$, $\omega \in \Omega_{\mathbf{v}}$.

By definition, the cardinality $\big|\mathcal{X} \cap B(\omega)\big|$ is independent of the choice $\omega \in \Omega_{\mathbf{v}}(\mathcal{X})$, and clearly

$$|\mathcal{X}|\,\text{measure}\,\big(B(\omega)\big) = \frac{N}{2^r}\sum_{i\in I(\mathcal{X})} Y_i.$$

Thus for any $\omega \in \Omega_{\mathbf{v}}(\mathcal{X})$,

$$\text{error}\,(\mathcal{X}, \omega) = \text{error}\,\big(\mathcal{X}, B(\omega)\big) = \big|\mathcal{X} \cap B(\omega)\big| - |\mathcal{X}|\,\text{measure}\,\big(B(\omega)\big)$$

$$= C(\mathcal{X}, \mathbf{v}) - \frac{N}{2^r}\bigg(\sum_{i\in I(\mathcal{X})} Y_i - \frac{1}{2}\big|I(\mathcal{X})\big|\bigg),$$

where $C(\mathcal{X}, \mathbf{v})$ is a constant depending only on $\mathcal{X}$ and $\mathbf{v}$.

It follows from (8.3) that, independently of the value of the constant $C(\mathcal{X}, \mathbf{v})$, we can estimate the following conditional probability:

$$\begin{aligned}(8.4)\qquad &\Pr\left[\left|\text{error}\,(\mathcal{X},\omega)\right| \leq \frac{1}{10}\sqrt{\left|I(\mathcal{X})\right|} \mid \omega \in \Omega_{\mathbf{v}}(\mathcal{X})\right] \\ &\leq \Pr\left[\left|\text{error}\,(\mathcal{X},\omega)\right| \leq \frac{1}{10}\sqrt{2^r} \mid \omega \in \Omega_{\mathbf{v}}(\mathcal{X})\right] \leq \frac{1}{3},\end{aligned}$$

where we used the trivial upper bound $\left|I(\mathcal{X})\right| \leq 2^r$ and $\Pr[A_1 \mid A_2]$ denotes the probability of event A_1 under the condition A_2. Since (8.4) is independent of the choice of the vector $\mathbf{v}$, we can switch to the unconditional probability, and obtain

$$(8.5)\qquad \Pr\left[\left|\text{error}\,(\mathcal{X},\omega)\right| \leq \frac{1}{10}\sqrt{2^r}\right] \leq \frac{1}{3}.$$

Next we restrict ourselves to the special cases $\mathcal{X} = \mathcal{X}(\alpha, \beta; N)$, where $0 < \alpha < 1$, $0 < \beta < 1$, and $\mathcal{X}(\alpha, \beta; N)$ denotes the N-term arithmetic progression $\beta+\alpha, \beta+2\alpha, \ldots, \beta+N\alpha$ modulo one (i.e., we take the fractional parts). By (8.5) we have

(8.6)
$$\frac{1}{3} \geq \int_0^1 \int_0^1 \frac{1}{|\Omega|}\left|\left\{\omega \in \Omega \,:\, \left|\text{error}\,\big(\mathcal{X}(\alpha,\beta;N),\omega\big)\right| \leq \frac{1}{10}\sqrt{2^r}\right\}\right| \,d\alpha\, d\beta.$$

Since the minimum is less or equal to the average, (8.6) implies the existence of an $\omega_0 \in \Omega$ such that

$$(8.7)\qquad \text{area}\left\{(\alpha,\beta) \in [0,1]^2 \,:\, \left|\text{error}\,\big(\mathcal{X}(\alpha,\beta;N),\omega_0\big)\right| \leq \frac{1}{10}\sqrt{2^r}\right\} \leq \frac{1}{3}.$$

By (8.7), the set $B = B(\omega_0)$ (a finite union of intervals, see (8.2)) has the property that for the majority of the pairs $(\alpha, \beta) \in [0,1]^2$,

$$\left|\sum_{\substack{1\leq n\leq N:\\ n\alpha+\beta\in B \pmod 1}} 1 - N \cdot \text{measure}\,(B)\right| > \frac{1}{10}\sqrt{2^r} \geq \frac{1}{10}\sqrt{2N} > \frac{\sqrt{N}}{8}.$$

This proves Proposition 1.1. ■

Concluding Remarks: explaining why Theorem 2 is best possible. A slight modification of the proof of Proposition 1.1 yields that Theorem 2 is best possible apart from the logarithmic factor. Again we combine the *random construction* with the *trivial error*, but this time it is the trivial error of a continues curve of arclength T (instead of an N-element point set). For simplicity we just consider the special case $d = 3$ (the same argument works in higher dimensions). We divide the unit cube $[0,1]^3$ into congruent subcubes of side length $cT^{-1/2}$, where $c > 0$ is some positive absolute constant to be specified later (we assume that $c^{-1}T^{1/2}$ is an integer). There are $c^{-3}T^{3/2}$ such small subcubes. The corresponding "trivial error" goes as follows: given an arbitrary continuous curve $\mathcal{C}$ of arclength T in $[0,1]^3$, at least half of the $c^{-3}T^{3/2}$ small subcubes are disjoint from $\mathcal{C}$ (assuming the absolute constant $c > 0$ is small enough). Indeed, assume that $\mathcal{C}$ intersects exactly M small subcubes. Note that every small subcube has (at most) $3^3 - 1 = 26$ neighbors, and the distance between any two non-neighboring small subcubes is $\geq cT^{-1/2}$. Applying the standard greedy algorithm, we can find at least $M/27$ among the M intersecting small subcubes such that no two are neighbors. Since the distance between non-neighbors is $\geq cT^{-1/2}$, the arclength of $\mathcal{C}$ is at least $(M/27) \cdot cT^{-1/2}$, implying the inequality

$$T \geq \frac{M}{27} \cdot cT^{-1/2},$$

or equivalently,

$$M \leq \frac{27T^{3/2}}{c}. \tag{8.8}$$

Since the total number of small subcubes is $c^{-3}T^{3/2}$, by choosing $c = 1/8$, we can guarantee that

$$\frac{27T^{3/2}}{c} < \frac{1}{2}c^{-3}T^{3/2}. \tag{8.9}$$

Comparing (8.8) and (8.9), we conclude that, the number M of $\mathcal{C}$-intersecting small subcubes is less than half of the total. This proves our statement about the "trivial error" of continuous curves.

The corresponding *random construction* gives a standard deviation in the range of

$$\sqrt{\text{total}} = \sqrt{T^{3/2}} = T^{3/4}.$$

Each small subcube has volume in the range of $T^{-3/2}$, and we have to multiply up with T (length of the curve). Summarizing, the analog random construction gives an error in the range

$$T^{3/4} \cdot T^{-3/2} \cdot T = T^{1/4},$$

which is exactly the main factor $T^{\frac{1}{2}-\frac{1}{2(d-1)}}$ with $d = 3$ in (1.6).

The same argument shows that the main factor $T^{\frac{1}{2}-\frac{1}{2(d-1)}}$ is sharp for any $d \geq 3$, proving that Theorem 2 is basically best possible.

What is more, the same argument shows that, in the general case of Theorem 5, the main factor $T^{\frac{1}{2}-\frac{k}{2(d-k)}}$ is best possible (see (3.14)).

9. Proof of Proposition 2.1

We begin with the proof of part **(a)**. I cannot help, but give two short proofs. In fact, both proofs give the stronger result (2.27).

First Proof of (2.27)**:** using pairwise independence. We just repeat the proof of (2.14). We know from (2.17) that orthogonality suffices. What we have in case (a) is the stronger property of *pairwise independence.* Notice that pairwise independence follows from the following *simple fact.* Let $0 \leq k < \ell$ be arbitrary integers; then given any pair $0 \leq z_1 < 1$, $0 \leq z_2 < 1$ of real numbers, there exist(s) exactly $\ell - k$ pair(s) (x, y) of real numbers in $0 \leq x < 1$, $0 \leq y < 1$ such that

$$kx + y \equiv z_1 \pmod 1 \quad \text{and} \quad \ell x + y \equiv z_2 \pmod 1. \tag{9.1}$$

Notice that $\ell - k$ is exactly the determinant of the matrix of the linear equations in (9.1).

To prove the *simple fact,* we just solve (9.1) for x and y: taking the difference of the two equations, we have $(\ell - k)x \equiv z_2 - z_1 \pmod 1$, which gives

$$x = \frac{z_2 - z_1}{\ell - k} + \frac{\text{integer}}{\ell - k},$$

and there are exactly $\ell - k$ numbers $x = x_j$, $j = 1, 2, \ldots, \ell - k$ of this type in the interval $0 \leq x < 1$. For each one of these numbers $x = x_j$, $j = 1, 2, \ldots, \ell - k$ there is a uniquely determined $0 \leq y = y_j < 1$ satisfying $kx_j + y_j \equiv z_1 \pmod 1$, completing the proof. ∎

Second Proof of (2.27): using Parseval's formula. To prove (2.27), we extend the real function $f(x)$, originally defined only in the unit interval $0 \le x < 1$, to the whole real line periodically, with period one. We work with the Fourier series

$$f(x) = \sum_{r\in\mathbb{Z}} a_r e^{2\pi i r x} \quad \text{where} \quad a_r = \int_0^1 f(y) e^{-2\pi i r y}\, dy. \tag{9.2}$$

To evaluate the left hand side of (2.27), we apply Parseval's formula several times:

$$\int_0^1 \int_0^1 \left(\frac{1}{n} \sum_{j=0}^{n-1} f(y + k_j x) - \int_0^1 f(u)\, du \right)^2 dx\, dy \tag{9.3}$$

$$= \int_0^1 \left(\int_0^1 \left| \sum_{\substack{r\in\mathbb{Z}:\\ r\neq 0}} a_r \frac{1}{n} \left(\sum_{j=0}^{n-1} e^{2\pi i r k_j x} \right) \cdot e^{2\pi i r y} \right|^2 dy \right) dx$$

$$= \frac{1}{n^2} \int_0^1 \sum_{\substack{r\in\mathbb{Z}:\\ r\neq 0}} |a_r|^2 \cdot \left| \sum_{j=0}^{n-1} e^{2\pi i r k_j x} \right|^2 dx$$

$$= \frac{1}{n^2} \sum_{\substack{r\in\mathbb{Z}:\\ r\neq 0}} |a_r|^2 \int_0^1 \left| \sum_{j=0}^{n-1} e^{2\pi i r k_j x} \right|^2 dx$$

is the first application of Parseval's formula. The second application is the equality

$$\int_0^1 \left| \sum_{j=0}^{n-1} e^{2\pi i r k_j x} \right|^2 dx = n,$$

which holds for any $r \neq 0$, and using it in (9.3), we have

$$\int_0^1 \int_0^1 \left(\frac{1}{n} \sum_{j=0}^{n-1} f(y + k_j x) - \int_0^1 f(u)\, du \right)^2 dx\, dy \tag{9.4}$$

$$= \frac{1}{n^2} \sum_{\substack{r\in\mathbb{Z}:\\ r\neq 0}} |a_r|^2 \cdot n = \frac{1}{n} \sum_{\substack{r\in\mathbb{Z}:\\ r\neq 0}} |a_r|^2.$$

By (9.2) and applying Parseval's formula one more time,

$$(9.5)\qquad \sum_{\substack{r\in\mathbb{Z}:\\ r\neq 0}} |a_r|^2 = \int_0^1 \left(f(z)-a_0\right)^2 dz = \int_0^1 \left(f(z) - \int_0^1 f(u)\,du\right)^2 dz.$$

Combining (9.4) and (9.5), formula (2.27) follows. ■

To prove **(b)**, we just note that formula (2.27) has a straightforward generalization in higher dimensions. In fact, both proofs work. For example, the Fourier series approach clearly works in any dimension: we just have to replace numbers with vectors.

Case **(c)** is a corollary of **(b)**: a trivial application of Chebyshev's inequality.

The proof of **(d)** is a routine adaptation of the "dyadic technique" in orthogonal series, and goes as follows. By using **(b)**, for any Lebesgue square-integrable function f and for any pair $\ell \geq 0$, $m \geq 0$ of integers we have

(9.6)

$$\int_{I^d}\int_{I^d} \sum_{j=0}^{2^\ell-1} \left(\sum_{k=j2^m}^{(j+1)2^m-1} f(\mathbf{y}+j\mathbf{x}) - 2^m \int_{I^d} f(\mathbf{u})\,d\mathbf{u}\right)^2 d\mathbf{y}\,d\mathbf{x} = \sigma^2(f)2^{\ell+m}.$$

Let $1 > \varepsilon > 0$ be arbitrarily small but fixed. Then

$$\sum_{r=1}^{\infty} \frac{1}{r^{1+\varepsilon}} < \infty,$$

so there is a threshold $r_0 = r_0(\varepsilon)$ such that

(9.7)

$$\sum_{r\geq r_0}^{\infty} \frac{1}{r^{1+\varepsilon}} \leq 1, \quad \text{for example,} \quad \mathrm{r}_0 = r_0(\varepsilon) = \left(\frac{1}{\varepsilon}\right)^{\varepsilon} + 1 \quad \text{is a good choice.}$$

By (9.6) and Chebyshev's inequality, for any $\mu > 1$ we have that the probability of the event

$$\sum_{j=0}^{2^\ell-1} \left(\sum_{k=j2^m}^{(j+1)2^m-1} f(Y+jX) - 2^m \int_{I^d} f(\mathbf{u})\,d\mathbf{u}\right)^2$$

$$\geq \sigma^2(f)\cdot \mu(\ell+r_0)^{1+\varepsilon}(m+r_0)^{1+\varepsilon}2^{\ell+m}$$

is less or equal to

$$\frac{1}{\mu(\ell + r_0)^{1+\varepsilon}(m + r_0)^{1+\varepsilon}}, \tag{9.8}$$

where X and Y are two independent and uniformly distributed random variables in the d-dimensional unit cube I^d.

Applying (9.7) we have

$$\sum_{\ell \geq 0} \sum_{m \geq 0} \frac{1}{(\ell + r_0)^{1+\varepsilon}(m + r_0)^{1+\varepsilon}} \leq \left(\sum_{r \geq 0} \frac{1}{(r + r_0)^{1+\varepsilon}} \right)^2 \leq 1,$$

so by (9.8) for every $\mu > 1$ there is a measurable subset $W = W(\mu) \subset I^d \times I^d$ with measure$(W) \geq 1 - \frac{1}{\mu}$ ("$2d$-dimensional Lebesgue measure") such that the inequality

$$\sum_{j=0}^{2^\ell - 1} \left(\sum_{k=j2^m}^{(j+1)2^m - 1} f(\mathbf{y} + j\mathbf{x}) - 2^m \int_{I^d} f(\mathbf{u})\, d\mathbf{u} \right)^2 < \sigma^2(f)\mu(\ell + r_0)^{1+\varepsilon}(m + r_0)^{1+\varepsilon} 2^{\ell+m} \tag{9.9}$$

holds for all pairs $(\mathbf{x}, \mathbf{y}) \in W$ and for all integers $\ell \geq 0$, $m \geq 0$.

Now we are ready to involve the binary decomposition of an arbitrary integer $n \geq 1$:

$$n = 2^{m_1} + 2^{m_2} + \cdots + 2^{m_t} \tag{9.10}$$

where $m_1 > m_2 > \cdots > m_t \geq 0$ and $1 \leq t = t(n) < 1 + \log_2 n$ are integers ("binary logarithm"). For notational simplicity, write

$$S(j; m) = \sum_{k=j2^m}^{(j+1)2^m - 1} f(\mathbf{y} + j\mathbf{x}) - 2^m \int_{I^d} f(\mathbf{u})\, d\mathbf{u}. \tag{9.11}$$

By (9.10), (9.11), and the Cauchy–Schwartz inequality,

$$\left| \frac{1}{n} \sum_{k=0}^{n-1} f(\mathbf{y} + j\mathbf{x}) - \int_{I^d} f(\mathbf{u})\, d\mathbf{u} \right| \tag{9.12}$$

$$= \frac{1}{n} \left| S(0; m_1) + S(2^{m_1 - m_2}; m_2) + S(2^{m_1 - m_3} + 2^{m_2 - m_3}; m_3) + \cdots \right.$$

$$+ S(2^{m_1-m_t} + \cdots + 2^{m_{t-1}-m_t}; m_t)\big|$$

$$\leq \frac{1}{n}\sqrt{t(n)}\big(S^2(0;m_1) + S^2(2^{m_1-m_2};m_2) + S^2(2^{m_1-m_3} + 2^{m_2-m_3};m_3) + \cdots$$

$$+ S^2(2^{m_1-m_t} + \cdots + 2^{m_{t-1}-m_t}; m_t)\big)^{1/2},$$

where of course here S^2 means the square of the sum S (and not the unit sphere). By (9.9), for all $(\mathbf{x},\mathbf{y}) \in W = W(\mu)$ and for all integers $\ell \geq 0$, $m \geq 0$,

$$\sum_{j=0}^{2^\ell-1} S^2(j;m) < \sigma^2(f)\mu(\ell + r_0)^{1+\varepsilon}(m + r_0)^{1+\varepsilon}2^{\ell+m}.$$

Thus we have

$$S^2(2^{m_1-m_2};m_2) \leq \sum_{j=0}^{2^{1+m_1-m_2}-1} S^2(j;m_2)$$

$$< \sigma^2(f)\mu(1 + m_1 - m_2 + r_0)^{1+\varepsilon}(m_2 + r_0)^{1+\varepsilon}2^{m_1+1}.$$

Similarly,

$$S^2(2^{m_1-m_3} + 2^{m_2-m_3};m_3) \leq \sum_{j=0}^{2^{1+m_1-m_3}-1} S^2(j;m_3)$$

$$< \sigma^2(f)\mu(1 + m_1 - m_3 + r_0)^{1+\varepsilon}(m_3 + r_0)^{1+\varepsilon}2^{m_1+1},$$

and so on. Using these inequalities in (9.12), we obtain

$$\left|\frac{1}{n}\sum_{k=0}^{n-1} f(\mathbf{y} + j\mathbf{x}) - \int_{I^d} f(\mathbf{u})\,d\mathbf{u}\right| \tag{9.13}$$

$$\leq \frac{1}{n}\sqrt{t(n)}\sigma(f)\sqrt{\mu}2^{(m_1+1)/2} \cdot \sqrt{\sum\nolimits^*}$$

where

$$\sum\nolimits^* = r_0^{1+\varepsilon}(m_1 + r_0)^{1+\varepsilon} + (1 + m_1 - m_2 + r_0)^{1+\varepsilon}(m_2 + r_0)^{1+\varepsilon}$$

$$+ (1 + m_1 - m_3 + r_0)^{1+\varepsilon}(m_3 + r_0)^{1+\varepsilon} + \cdots .$$

Clearly

$$0 \le \sum\nolimits^{*} \le t(n)(m_1 + r_0)^{2+2\varepsilon},$$

so returning to (9.13), we have

$$\left|\frac{1}{n}\sum_{k=0}^{n-1} f(\mathbf{y}+j\mathbf{x}) - \int_{I^d} f(\mathbf{u})\,d\mathbf{u}\right| \tag{9.14}$$

$$\le \frac{1}{n}\sqrt{t(n)}\sigma(f)\sqrt{\mu}2^{(m_1+1)/2}\cdot\sqrt{t(n)}(m_1+r_0)^{1+\varepsilon}.$$

Finally, using the inequality $1 \le t(n) \le 1 + \log_2 n$ and the explicit value of $r_0 = r_0(\varepsilon)$ (see (9.7)) in (9.14), we have

$$\left|\frac{1}{n}\sum_{k=0}^{n-1} f(\mathbf{y}+j\mathbf{x}) - \int_{I^d} f(\mathbf{u})\,d\mathbf{u}\right| \tag{9.15}$$

$$\le \sigma(f)\sqrt{2\mu}\frac{1+\log_2 n}{\sqrt{n}}\left(\log_2 n + (1/\varepsilon)^{\varepsilon} + 1\right)^{1+\varepsilon}.$$

Since (9.15) implies (2.26), the proof of Proposition 2.1 is complete. ■

10. Proof of Proposition 2.2

We combine the so-called Large Sieve with a descent process. The Large Sieve, a sophisticated second moment argument, is one of the most important tools in analytic number theory. Let $\mathcal{A}$ be a set of $A = |\mathcal{A}|$ integers in the interval $1 \le n \le N$, and let $A(q,h)$ denote the number of the elements of $\mathcal{A}$ which are congruent to h (mod q). Clearly

$$\sum_{h=1}^{q} A(q,h) = A = |\mathcal{A}|.$$

The main goal of the Large Sieve is to estimate the "variance"

$$V(q) = V_{\mathcal{A}}(q) = \sum_{h=1}^{q}\left(A(q,h) - \frac{A}{q}\right)^2$$

from above. We have the following remarkable average result (see e.g. Section 27 in Davenport's well-known book [2]).

Large Sieve Inequality. *If $\mathcal{A} \subset \{1,2,\ldots,N\}$ with $|\mathcal{A}| = A$, then*

$$\sum_{p\le Q} p \sum_{h=1}^{p} \left(A(p,h) - \frac{A}{p} \right)^2 \le (N + 2Q^2)A, \tag{10.1}$$

where p denotes the primes.

Remarks. The Large Sieve Inequality is not too useful in the range where Q is much smaller than $\sqrt{N}$.

Next I explain why (10.1) is (nearly) optimal. Suppose that $\sqrt{N} < Q < N^{1-\varepsilon}$ and the set $\mathcal{A}$ has cardinality $|\mathcal{A}| = A = \text{const} \cdot N$. In the range $Q/2 \le p \le Q$ there are const $\cdot\, Q/\log Q$ primes. It follows from (10.1) that, for some prime p_0 in $Q/2 \le p_0 \le Q$ and for some remainder h_0 in $1 \le h_0 \le p_0$,

$$\left| A(p_0,h_0) - \frac{1}{p_0} A \right| \le \text{const} \cdot \sqrt{\frac{N}{p_0}} \cdot \sqrt{\log p_0}.$$

If $\mathcal{A}$ is a "random subset" of $\{1,2,\ldots,N\}$ (say, we toss a fair coin N times and include the corresponding integer if the outcome is Heads) then the standard deviation is const $\cdot \sqrt{\frac{N}{p_0}}$. This means that (10.1) is optimal apart from the logarithmic factor $\sqrt{\log p_0}$.

If p_0 is close to N then the logarithmic factor $\sqrt{\log p_0}$ is not negligible compared to $\sqrt{\frac{N}{p_0}}$. This is the reason why need an iteration process—called the process of descent—instead of making one giant leap.

Let's return to the proof of Proposition 2.2. Let $\mathcal{A} \subset \{1,2,\ldots,N\}$ be arbitrary, and write $|\mathcal{A}| = A$. The first step in our descent process goes as follows. Let $\sqrt{N} < M < N^{3/4}$, then by (10.1) we have

$$\sum_{M\le p<2M} p \sum_{h=1}^{p} \left(\frac{A}{p} - A(p,h) \right)^2 \le 9M^2 A. \tag{10.2}$$

Let $\mathcal{P}_M$ denotes the set of primes p between $M \le p < 2M$. By the prime number theorem

$$|\mathcal{P}_M| = \big(1 + o(1)\big) \frac{M}{\log M} > \frac{3M}{4\log M} \tag{10.3}$$

if M is large enough. We can rewrite (10.2) as follows:

$$\frac{1}{|\mathcal{P}_M|}\sum_{p\in\mathcal{P}_M}\frac{1}{p}\sum_{h=1}^{p}\left(\frac{A-pA(p,h)}{N}\right)^2\le\frac{9M^2A}{N^2|\mathcal{P}_M|}. \tag{10.4}$$

Since the minimum is less or equal to the average, (10.4) implies the existence of a prime $p_0\in\mathcal{P}_M$ such that

$$\frac{1}{p_0}\sum_{h=1}^{p_0}\left(\frac{A-p_0A(p_0,h)}{N}\right)^2\le\frac{9M^2A}{N^2|\mathcal{P}_M|}, \tag{10.5}$$

completing step one in the descent process.

Next we switch from the set $\{1,2,\ldots,N\}$ of consecutive integers to any arithmetic progression $n\equiv h_0 \pmod{p_0}$ of gap p_0 restricted to our underlying interval $1\le n\le N$. Let $N(p_0,h_0)$ denote the number of elements of this finite arithmetic progression; clearly $(1\le h_0\le p_0)$

$$\left|N(p_0,h_0)-\frac{N}{p_0}\right|<1. \tag{10.6}$$

Here comes the second step in the descent process. Write $N_1=N/p_0$. We use the trivial (but important) fact that an arithmetic progression of an arithmetic progression is also an arithmetic progression. Again we apply the Large Sieve. Let $\sqrt{N_1}<M_1<N_1^{3/4}$, then by (10.1) for every $h_0=1,2,\ldots,p_0$ we have

$$\sum_{M_1\le p<2M_1}p\sum_{h=1}^{p}\left(\frac{A(p_0,h_0)}{p}-A(p_0p,h_0+hp_0)\right)^2\le 9M_1^2A(p_0,h_0). \tag{10.7}$$

I recall that $\mathcal{P}_{M_1}$ denotes the set of primes p between $M_1\le p<2M_1$. We can rewrite (10.7) as follows:

$$\frac{1}{|\mathcal{P}_{M_1}|}\sum_{p\in\mathcal{P}_{M_1}}\frac{1}{p}\sum_{h=1}^{p}\left(\frac{A(p_0,h_0)-pA(p_0p,h_0+hp_0)}{N_1}\right)^2\le\frac{9M_1^2A(p_0,h_0)}{N_1^2|\mathcal{P}_{M_1}|}. \tag{10.8}$$

Since the minimum is less or equal to the average, (10.8) implies the existence of a prime $p_1\in\mathcal{P}_{M_1}$ such that

$$\frac{1}{p_1}\sum_{h=1}^{p_1}\frac{1}{p_0}\sum_{h_0=1}^{p_0}\left(\frac{A(p_0,h_0)-p_1A(p_0p_1,h_0+hp_0)}{N_1}\right)^2\le\frac{9M_1^2A}{p_0N_1^2|\mathcal{P}_{M_1}|}, \tag{10.9}$$

completing step two in the descent process.

Next we switch to the arithmetic progression $n \equiv h_0 + h_1 p_0 \pmod{p_0 p_1}$ restricted to our underlying interval $1 \le n \le N$. Let $N(p_0p_1, h_0 + h_1 p_0)$ denote the number of elements of this finite arithmetic progression; clearly $(1 \le h_0 \le p_0, 1 \le h_1 \le p_1)$

$$\left| N(p_0p_1, h_0 + h_1p_0) - \frac{N}{p_0p_1} \right| < 1. \tag{10.10}$$

Here comes the third step in the descent process. Write $N_2 = N/p_0p_1$. Let $\sqrt{N_2} < M_2 < N_2^{3/4}$, then by (10.1) for every $1 \le h_0 \le p_0$, $1 \le h_1 \le p_1$ we have the analog of (10.8):

(10.11)

$$\frac{1}{|\mathcal{P}_{M_2}|} \sum_{p \in \mathcal{P}_{M_2}} \frac{1}{p} \sum_{h=1}^{p} \left(\frac{A(p_0p_1, h_0 + h_1p_0) - pA(p_0p_1p, h_0 + h_1p_0 + hp_0p_1)}{N_2} \right)^2$$
$$\le \frac{9M_2^2 A(p_0p_1, h_0 + h_1p_0)}{N_2^2 |\mathcal{P}_{M_2}|}.$$

Since the minimum is less or equal to the average, (10.11) implies the existence of a prime $p_2 \in \mathcal{P}_{M_2}$ such that

$$\frac{1}{p_2} \sum_{h=1}^{p_2} \frac{1}{p_0p_1} \tag{10.12}$$
$$\times \sum_{h_0=1}^{p_0} \sum_{h_1=1}^{p_1} \left(\frac{A(p_0p_1, h_0 + h_1p_0) - p_2A(p_0p_1p_2, h_0 + h_1p_0 + hp_0p_1)}{N_2} \right)^2$$
$$\le \frac{9M_2^2 A}{p_0p_1 N_2^2 |\mathcal{P}_{M_2}|},$$

completing step three in the descent process.

We keep doing this. Since by hypothesis $N > 100T^2$, we can easily find a sequence $\sqrt{N} < M = M_0 < N^{3/4}$, $p_0 \in \mathcal{P}_M$, $N_1 = N/p_0$, $\sqrt{N_1} < M_1 < N_1^{3/4}$, $p_1 \in \mathcal{P}_{M_1}$, $N_2 = N_1/p_1$, $\sqrt{N_2} < M_2 < N_2^{3/4}$, $p_2 \in \mathcal{P}_{M_2}, \ldots$ of parameters such that, at the end, the whole product $m = p_0p_1p_2 \cdots p_\ell$ (where ℓ is the last one) falls into the required range

$$\frac{N}{2T} < m = p_0p_1p_2 \cdots p_\ell < \frac{N}{T}. \tag{10.13}$$

Consider the sum

$$\text{(10.14)} \qquad \frac{A - p_0 A(p_0, h_0)}{N} + \frac{A(p_0, h_0) - p_1 A(p_0 p_1, h_0 + h_1 p_0)}{N_1}$$
$$+ \frac{A(p_0 p_1, h_0 + h_1 p_0) - p_2 A(p_0 p_1 p_2, h_0 + h_1 p_0 + h_2 p_0 p_1)}{N_2} + \cdots$$
$$+ \left(\frac{A(p_0 p_1 \cdots p_{\ell-1}, h_0 + h_1 p_0 + \cdots + h_{\ell-1} p_0 p_1 \cdots p_{\ell-2})}{N_\ell} \right.$$
$$\left. - \frac{p_\ell A(p_0 p_1 \cdots p_\ell, h_0 + h_1 p_0 + \cdots + h_\ell p_0 p_1 \cdots p_{\ell-1})}{N_\ell} \right).$$

Since $N_1 = N/p_0$, $N_2 = N_1/p_1$, $N_3 = N_2/p_2$, and so on, (10.14) is a telescoping sum, which simplifies to

(10.15)
$$\text{sum (10.14)} = \frac{A - p_0 p_1 \cdots p_\ell A(p_0 p_1 \cdots p_\ell, h_0 + h_1 p_0 + \cdots + h_\ell p_0 p_1 \cdots p_{\ell-1})}{N}$$
$$= \frac{A - mA(m, r)}{N},$$

where

$$m = p_0 p_1 \cdots p_\ell \quad \text{and} \quad r = h_0 + h_1 p_0 + \cdots + h_\ell p_0 p_1 \cdots p_{\ell-1}.$$

In view of (10.14)–(10.15), we need the following simple lemma about sequences of real numbers $C(h_0), C(h_0, h_1), \ldots$.

Lemma 10.1. *Assume that*

$$\frac{1}{p_0} \sum_{h_0=1}^{p_0} C^2(h_0) \le \varepsilon_0, \quad \frac{1}{p_0 p_1} \sum_{h_0=1}^{p_0} \sum_{h_1=1}^{p_1} C^2(h_0, h_1) \le \varepsilon_1, \quad \ldots$$
$$\frac{1}{p_0 p_1 \cdots p_\ell} \sum_{h_0=1}^{p_0} \cdots \sum_{h_\ell=1}^{p_\ell} C^2(h_0, \ldots, h_\ell) \le \varepsilon_\ell.$$

Then

(10.16)
$$\frac{1}{p_0 p_1 \cdots p_\ell} \sum_{h_0=1}^{p_0} \cdots \sum_{h_\ell=1}^{p_\ell} \left(C(h_0) + C(h_0, h_1) + \cdots + C(h_0, h_1, \ldots, h_\ell) \right)^2$$
$$\le 2 \sum_{i=0}^{\ell} (\ell + 1 - i)^2 \varepsilon_i.$$

Proof. We use the Cauchy–Schwarz inequality:

$$(10.17)\qquad \big(C(h_0)+C(h_0,h_1)+\cdots+C(h_0,h_1,\ldots,h_\ell)\big)^2$$

$$=\Big(\frac{1}{\ell+1}\cdot C(h_0)(\ell+1)+\frac{1}{\ell}\cdot C(h_0,h_1)\ell+\frac{1}{\ell-1}\cdot C(h_0,h_1,h_2)(\ell-1)$$

$$+\cdots+1\cdot C(h_0,h_1,\ldots,h_\ell)\Big)^2\le\Big(\sum_{j=1}^{\ell+1}\frac{1}{j^2}\Big)$$

$$\cdot\big((\ell+1)^2C^2(h_0)+\ell^2C^2(h_0,h_1)+(\ell-1)^2C^2(h_0,h_1,h_2)+\cdots$$

$$+C^2(h_0,\ldots,h_\ell)\big).$$

Taking average over all $1\le h_0\le p_0, 1\le h_1\le p_1,\ldots,1\le h_\ell\le p_\ell$ in (10.17), the lemma follows. ■

We want to apply Lemma 10.1 with

$$(10.18)\qquad C(h_0)=\frac{A-p_0A(p_0,h_0)}{N},$$

$$(10.19)\qquad C(h_0,h_1)=\frac{A(p_0,h_0)-p_1A(p_0p_1,h_0+h_1p_0)}{N_1},$$

(10.20)
$$C(h_0,h_1,h_2)=\frac{A(p_0p_1,h_0+h_1p_0)-p_2A(p_0p_1p_2,h_0+h_1p_0+h_2p_0p_1)}{N_2},$$

and so on, ending with

(10.21)
$$C(h_0,h_1,\ldots,h_\ell)=\Big(\frac{A(p_0p_1\cdots p_{\ell-1},h_0+h_1p_0+\cdots+h_{\ell-1}p_0p_1\cdots p_{\ell-2})}{N_\ell}$$

$$-\frac{p_\ell A(p_0p_1\cdots p_\ell,h_0+h_1p_0+\cdots+h_\ell p_0p_1\cdots p_{\ell-1})}{N_\ell}\Big).$$

In view of (10.5), (10.9), (10.12), we can choose ε_0, ε_1, ε_2 as follows:

$$\frac{9M^2A}{N^2|\mathcal{P}_M|}\le\frac{9M^2}{N|\mathcal{P}_M|}=\varepsilon_0,$$

$$\frac{9M_1^2A}{p_0N_1 \cdot N_1|\mathcal{P}_{M_1}|} = \frac{9M_1^2A}{N \cdot N_1|\mathcal{P}_{M_1}|} \leq \frac{9M_1^2}{N_1|\mathcal{P}_{M_1}|} = \varepsilon_1,$$

$$\frac{9M_2^2A}{p_0p_1N_2 \cdot N_2|\mathcal{P}_{M_2}|} = \frac{9M_2^2A}{N \cdot N_2|\mathcal{P}_{M_2}|} \leq \frac{9M_2^2}{N_2|\mathcal{P}_{M_2}|} = \varepsilon_2,$$

and in general,

$$\varepsilon_j = \frac{9M_j^2}{N_j|\mathcal{P}_{M_j}|}, \qquad 0 \leq j \leq \ell. \tag{10.22}$$

Combining (10.22) with (10.3), we have

$$\varepsilon_j < \frac{9M_j^2}{N_j \frac{3M_j}{4\log M_j}} = \frac{12M_j \log M_j}{N_j}. \tag{10.23}$$

Combining Lemma 10.1 with (10.14)–(10.16) and (10.18)–(10.23), we have with $m = p_0p_1 \cdots p_\ell$,

$$\frac{1}{m}\sum_{r=1}^{m}\left(\frac{A - mA(m,r)}{N}\right)^2 \leq 2\sum_{i=0}^{\ell}(\ell+1-i)^2\varepsilon_i \tag{10.24}$$

$$< 24\sum_{i=0}^{\ell}(\ell+1-i)^2\frac{M_j \log M_j}{N_j}.$$

Since $\sqrt{N_j} < M_j < N_j^{3/4}$ and $N_j < \sqrt{N_{j-1}}$ $(0 \leq j \leq \ell)$, the sequence $\varepsilon_\ell > \varepsilon_{\ell-1} > \varepsilon_{\ell-2} > \varepsilon_{\ell-3} > \cdots$ decreases *extremely rapidly*. In particular, an easy calculation gives that

$$\sum_{i=0}^{\ell}(\ell+1-i)^2\frac{M_j \log M_j}{N_j} < \sum_{i=0}^{\ell}2^{i-\ell}\frac{M_\ell \log M_\ell}{N_\ell} < 2\frac{M_\ell \log M_\ell}{N_\ell}. \tag{10.25}$$

Using (10.25) in (10.24), we obtain

$$\frac{1}{m}\sum_{r=1}^{m}\left(\frac{A - mA(m,r)}{N}\right)^2 < 50\frac{M_\ell \log M_\ell}{N_\ell}, \tag{10.26}$$

where (see (10.13))

$$\frac{N}{2T} < m = p_0p_1p_2 \cdots p_\ell < \frac{N}{T}, \tag{10.27}$$

$$N_\ell = \frac{N}{p_0p_1 \cdots p_{\ell-1}} = \frac{p_\ell N}{m}, \quad M_\ell \leq p_\ell < 2M_\ell.$$

Multiplying (10.26) with $(N/m)^2$, and using (10.27), we have

$$\text{(10.28)} \qquad \frac{1}{m}\sum_{r=1}^{m}\left(\frac{A}{m}-A(m,r)\right)^2 < 50\frac{M_\ell \log M_\ell}{N_\ell}\cdot\left(\frac{N}{m}\right)^2$$

$$\leq 50\frac{p_\ell \log p_\ell}{\frac{p_\ell N}{m}}\cdot\left(\frac{N}{m}\right)^2 = 50\log p_\ell \cdot \frac{N}{m} < 50\log p_\ell \cdot 2T.$$

Since $N_\ell = p_\ell N/m < 2p_\ell T$ (see (10.27)), we have

$$p_\ell < 2M_\ell < 2N_\ell^{3/4} < 2(2p_\ell T)^{3/4} < 4p_\ell^{3/4}T^{3/4},$$

implying $p_\ell^{1/4} < 4T^{3/4}$, and so

$$\text{(10.29)} \qquad \log p_\ell < 3\log T + 4\log 4.$$

Using (10.29) in (10.28), we conclude that

$$\text{(10.30)} \qquad \frac{1}{m}\sum_{r=1}^{m}\left(\frac{A}{m}-A(m,r)\right)^2 < 100(3\log T + 4\log 4)T.$$

Since (10.30) implies (2.28), the proof is complete. ∎

Concluding Remark. It is easy to modify the argument above to obtain an efficient algorithm supplying a "good" $m = p_0 p_1 \cdots p_\ell$. All what we have to do is to replace *minimum* with *typical* (= randomly chosen). More precisely, instead of finding the prime $p = p_0 \in \mathcal{P}_M$ in (10.4) that attains the minimum, which is hard(!), we just choose a typical $p = p_0 \in \mathcal{P}_M$. And similarly, we just choose a typical $p = p_1 \in \mathcal{P}_{M_1}$, choose a typical $p = p_2 \in \mathcal{P}_{M_2}$, and so on. The Large Sieve guarantees that, replacing *minimum* with *typical* leads only to a negligable constant factor loss in the value of c_1 in (2.28).

11. Proof of Proposition 3.1

By (3.5)–(3.6) we have

$$\left|\frac{1}{n}\sum_{k=0}^{n-1} f(y+kx) - \int_0^1 f(u)\,du\right| \tag{11.1}$$

$$\leq \frac{1}{n}\sum_{\substack{r\in\mathbb{Z}:\\ r\neq 0}} |a_r|\frac{1}{2\|rx\|}$$

$$\leq \frac{c}{n}\sum_{r=1}^{\infty}\frac{1}{r^2\|rx\|} \quad \text{where} \quad c=\max|f''|.$$

The critical part of the proof is to show that, for the majority of the real numbers x in $0<x<1/2$, we have

$$\sum_{r=1}^{\infty}\frac{1}{r^2\|rx\|}=O(1). \tag{11.2}$$

To prove (11.2), we apply standard arguments from diophantine approximation. We start with a power-of-two type decomposition:

$$\sum_{r=1}^{\infty}\frac{1}{r^2\|rx\|}=\sum_{k=0}^{\infty}\left(\sum_{2^k\leq r<2^{k+1}}\frac{1}{r^2\|rx\|}\right), \tag{11.3}$$

and focus on the binary block $2^k\leq r<2^{k+1}$. Let

$$\delta_k=\delta_k(x)=\min_{1\leq r<2^{k+1}}\|rx\|, \tag{11.4}$$

we can clearly assume that $\delta_k>0$.

For every integer $\ell\geq 1$ define the set of integers

$$A_\ell=A_\ell(k)=\left\{r\in\mathbb{Z}\,:\,1\leq r<2^{k+1},\ \|rx\|<2^\ell\delta_k\right\}.$$

The Pigeonhole Principle guarantees that the set A_ℓ cannot contain more than $2^{\ell+1}$ integers. Indeed, otherwise there exist two different elements $r_1,r_2\in A_\ell$ with (say) $r_1<r_2$ such that $\big\|(r_2-r_1)x\big\|<\delta_k$, a contradiction.

Thus we have ($\log_2$ denotes the binary logarithm)

$$\sum_{2^k \le r < 2^{k+1}} \frac{1}{\|rx\|} \le \sum_{\substack{\ell \ge 1: \\ 2^\ell \le \frac{1}{\delta_k}}} 2^{\ell+1} \cdot \frac{1}{2^{\ell-1}\delta_k} \tag{11.5}$$

$$= \frac{4}{\delta_k} \sum_{\substack{\ell \ge 1: \\ 2^\ell \le \frac{1}{\delta_k}}} 1 = \frac{4}{\delta_k} \log_2\left(\frac{1}{\delta_k}\right).$$

Returning to (11.3), by (11.5) we have

$$\sum_{2^k \le r < 2^{k+1}} \frac{1}{r^2\|rx\|} \le 4^{-k} \cdot \frac{4}{\delta_k} \log_2\left(\frac{1}{\delta_k}\right) = 4^{-k+1} \cdot \frac{1}{\delta_k} \log_2\left(\frac{1}{\delta_k}\right). \tag{11.6}$$

The equation in (11.4) means that, there exists an integer multiple rx of x such that $1 \le r < 2^{k+1}$ and rx =integer$\pm\delta_k$, or equivalently,

$$x = \frac{\text{integer}}{r} \pm \frac{\delta_k}{r} \quad \text{for some integer} \quad 1 \le r < 2^{k+1}. \tag{11.7}$$

The message of (11.7) is that x can be well approximated by a rational number. This gives the idea to restrict our attention only to the x's which have no good rational approximation. We make this intuition precise by the following definition. For every rational number q/r, where $0 \le q < r$, consider the interval

$$I(q;r) = I_\lambda(q;r) = \left[\frac{q}{r} - \frac{1}{\lambda r^{5/2}}, \frac{q}{r} + \frac{1}{\lambda r^{5/2}}\right], \tag{11.8}$$

where the value of the constant parameter $\lambda > 1$ will be specified later, and form the union set

$$B = B(\lambda) = \bigcup_{r \ge 1} \bigcup_{0 \le q < r} I(q;r). \tag{11.9}$$

It is easy to estimate the Lebesgue measure ("length") of the set $B = B(\lambda)$ from above:

$$\text{measure}\,(B) \le \sum_{r \ge 1} \sum_{0 \le q < r} \frac{2}{\lambda r^{5/2}} = \sum_{r \ge 1} \frac{2}{\lambda r^{3/2}} < \frac{6}{\lambda}. \tag{11.10}$$

The set B is interpreted as the set of "bad" x's, so we consider the complement set with respect to the unit interval:

$$A = A(\lambda) = [0,1) \setminus B(\lambda). \tag{11.11}$$

Comparing (11.7)–(11.11), we see that, for any $x \in A$,

$$\delta_k = \delta_k(x) > \frac{1}{\lambda 2^{3(k+1)/2}}. \tag{11.12}$$

Applying (11.12) in (11.6)

$$\sum_{2^k \le r < 2^{k+1}} \frac{1}{r^2 \|rx\|} \le 4^{-k+1} \cdot \lambda 2^{3(k+1)/2} \log_2\left(\lambda 2^{3(k+1)/2}\right)$$

$$= 8\sqrt{2}\lambda 2^{-k/2} \cdot \left(\log_2 \lambda + \frac{3(k+1)}{2}\right).$$

Returning to (11.3), we obtain that

$$\sum_{r=1}^{\infty} \frac{1}{r^2\|rx\|} \le \sum_{k=0}^{\infty} \frac{8\sqrt{2}\lambda\left(\log_2 \lambda + \frac{3(k+1)}{2}\right)}{2^{k/2}} \tag{11.13}$$

$$= \frac{8\sqrt{2}}{1-\frac{1}{\sqrt{2}}}\lambda \log_2 \lambda + \frac{24}{\left(1-\frac{1}{\sqrt{2}}\right)^2}\lambda$$

holds for any $x \in A$, defined by (11.11), where measure$(A) > 1 - \frac{6}{\lambda}$.

The last step is to combine (11.1) and (11.13). Choosing λ as a large positive constant, Proposition 3.1 follows from (11.1) and (11.13). ∎

12. More on Super-Uniformity: Proof of Theorem 5

Let $A \subset I^3 = [0,1)^3$ be an arbitrary Lebesgue measurable subset, and, as usual, we consider the Fourier series of the characteristic function χ_A of the set A:

$$\chi_A(\mathbf{u}) = \sum_{\mathbf{r} \in \mathbb{Z}^3} a_{\mathbf{r}} e^{2\pi i \mathbf{r}\cdot\mathbf{u}} \quad \text{with} \quad a_{\mathbf{r}} = \int_A e^{-2\pi i \mathbf{r}\cdot\mathbf{y}}\, d\mathbf{y}, \tag{12.1}$$

where $\mathbf{r} \cdot \mathbf{u} = r_1u_1 + r_2u_2 + r_3u_3$ denotes the inner product. Clearly $a_0 = \mathrm{vol}\,(A)$ (3-dimensional Lebesgue measure), and by Parseval's formula,

$$\sum_{\substack{\mathbf{r}\in\mathbb{Z}^3:\\ \mathbf{r}\neq\mathbf{0}}} |a_{\mathbf{r}}|^2 = \mathrm{vol}\,(A) - \mathrm{vol}^2(A). \tag{12.2}$$

Consider the torus-parallelogram $\mathbf{x}(t_1,t_2) = \big(x_1(t_1,t_2), x_2(t_1,t_2), x_3(t_1,t_2)\big)$ (mod 1) where

$$x_1(t_1,t_2) = \beta_{1,1}t_1 + \beta_{1,2}t_2 + y_1, \quad x_2(t_1,t_2) = \beta_{2,1}t_1 + \beta_{2,2}t_2 + y_2, \tag{12.3}$$
$$\text{and} \quad x_3(t_1,t_2) = \beta_{3,1}t_1 + \beta_{3,2}t_2 + y_3,$$

where $0 \le t_1 \le T_1$, $0 \le t_2 \le T_2$, $\mathbf{b}_1 = (\beta_{1,1}, \beta_{2,1}, \beta_{3,1})$ and $\mathbf{b}_2 = (\beta_{1,2}, \beta_{2,2}, \beta_{3,2})$ satisfy $|\mathbf{b}_1|^2 = \beta_{1,1}^2 + \beta_{2,1}^2 + \beta_{3,1}^2 = 1$, $|\mathbf{b}_2|^2 = \beta_{1,2}^2 + \beta_{2,2}^2 + \beta_{3,2}^2 = 1$. The quadruple $\big((T_1,T_2);\mathbf{y};\mathbf{b}_1,\mathbf{b}_2\big)$ (where $\mathbf{y} = (y_1,y_2,y_3)$) uniquely determines the torus-parallelogram (12.3); we call the triple $(\mathbf{y},\mathbf{b}_1,\mathbf{b}_2)$ an *initial condition.* As usual, $\mathbf{b}_1 \times \mathbf{b}_2$ denotes the cross product of the two vectors $\mathbf{b}_1$ and $\mathbf{b}_2$. The magnitude of a cross product is the area of a parallelogram, and so the total area $A(T_1,T_2) = A(T_1,T_2;\mathbf{y};\mathbf{b}_1,\mathbf{b}_2)$ of the intersection of the torus-parallelogram (12.3) with the given subset A equals

$$A(T_1,T_2) = A(T_1,T_2;\mathbf{y};\mathbf{b}_1,\mathbf{b}_2) \tag{12.4}$$
$$= |\mathbf{b}_1 \times \mathbf{b}_2| \cdot \mathrm{area}\,\big\{(t_1,t_2) \in [0,T_1]\times[0,T_2] \,:\, \mathbf{x}(t_1,t_2) \in A \text{ (mod 1)}\big\}$$
$$= |\mathbf{b}_1 \times \mathbf{b}_2| \cdot \int_0^{T_1}\int_0^{T_2} \chi_A\big(\mathbf{x}(t_1,t_2)\big)\, dt_1\, dt_2$$
$$= |\mathbf{b}_1 \times \mathbf{b}_2| \int_0^{T_1}\int_0^{T_2} \sum_{\mathbf{r}\in\mathbb{Z}^3} a_{\mathbf{r}} e^{2\pi i \mathbf{r}\cdot\mathbf{x}(t_1,t_2)}\, dt_1\, dt_2$$
$$= |\mathbf{b}_1 \times \mathbf{b}_2| \sum_{\mathbf{r}\in\mathbb{Z}^3} a_{\mathbf{r}} \int_0^{T_1}\int_0^{T_2} e^{2\pi i \mathbf{r}\cdot\mathbf{x}(t_1,t_2)}\, dt_1\, dt_2$$
$$= |\mathbf{b}_1 \times \mathbf{b}_2| \sum_{\mathbf{r}\in\mathbb{Z}^3} a_{\mathbf{r}} e^{2\pi i \mathbf{r}\cdot\mathbf{y}} \int_0^{T_1} e^{2\pi i(\beta_{1,1}r_1+\beta_{2,1}r_2+\beta_{3,1}r_3)t_1}\, dt_1$$
$$\cdot \int_0^{T_2} e^{2\pi i(\beta_{1,2}r_1+\beta_{2,2}r_2+\beta_{3,2}r_3)t_2}\, dt_2$$

$$= a_0 T_1 T_2 |\mathbf{b}_1 \times \mathbf{b}_2|$$

$$+ |\mathbf{b}_1 \times \mathbf{b}_2| \sum_{\substack{\mathbf{r}\in\mathbb{Z}^3:\\ \mathbf{r}\neq\mathbf{0}}} a_{\mathbf{r}} e^{2\pi i \mathbf{r}\cdot\mathbf{y}} \cdot \frac{e^{2\pi i(\beta_{1,1}r_1+\beta_{2,1}r_2+\beta_{3,1}r_3)T_1}-1}{2\pi i(\beta_{1,1}r_1+\beta_{2,1}r_2+\beta_{3,1}r_3)}$$

$$\cdot \frac{e^{2\pi i(\beta_{1,2}r_1+\beta_{2,2}r_2+\beta_{3,2}r_3)T_2}-1}{2\pi i(\beta_{1,2}r_1+\beta_{2,2}r_2+\beta_{3,2}r_3)}.$$

Since $a_0 = \mathrm{vol}\,(A)$ (= the Lebesgue measure of A), by (12.4) we have

$$\text{(12.5)} \qquad \text{discrepancy} = A(T_1, T_2; \mathbf{y}; \mathbf{b}_1, \mathbf{b}_2) - T_1T_2|\mathbf{b}_1 \times \mathbf{b}_2| \cdot \mathrm{vol}\,(A)$$

$$= |\mathbf{b}_1 \times \mathbf{b}_2| \sum_{\substack{\mathbf{r}\in\mathbb{Z}^3:\\ \mathbf{r}\neq\mathbf{0}}} a_{\mathbf{r}} \cdot \frac{e^{2\pi i\mathbf{b}_1\cdot\mathbf{r}T_1}-1}{2\pi i\mathbf{b}_1\cdot\mathbf{r}} \cdot \frac{e^{2\pi i\mathbf{b}_2\cdot\mathbf{r}T_2}-1}{2\pi i\mathbf{b}_2\cdot\mathbf{r}} \cdot e^{2\pi i\mathbf{r}\cdot\mathbf{y}}.$$

Fix any two points $\mathbf{b}_1$ and $\mathbf{b}_2$ on the unit sphere S^2 (i.e., two unit vectors), and run the third parameter $\mathbf{y}$ through the unit cube; then by (12.5) and Parseval's formula ($I^3 = [0,1]^3$)

$$\text{(12.6)} \qquad \int_{I^3} \left(A(T_1, T_2; \mathbf{y}; \mathbf{b}_1, \mathbf{b}_2) - T_1T_2|\mathbf{b}_1 \times \mathbf{b}_2| \cdot \mathrm{vol}\,(A)\right)^2 d\mathbf{y}$$

$$= |\mathbf{b}_1 \times \mathbf{b}_2|^2 \sum_{\substack{\mathbf{r}\in\mathbb{Z}^3:\\ \mathbf{r}\neq\mathbf{0}}} |a_{\mathbf{r}}|^2 \cdot \left|\frac{e^{2\pi i\mathbf{b}_1\cdot\mathbf{r}T_1}-1}{2\pi i\mathbf{b}_1\cdot\mathbf{r}}\right|^2 \cdot \left|\frac{e^{2\pi i\mathbf{b}_2\cdot\mathbf{r}T_2}-1}{2\pi i\mathbf{b}_2\cdot\mathbf{r}}\right|^2.$$

Let's study the last two factors in (12.6): we have the upper bound

$$\left|\frac{e^{2\pi i\mathbf{b}\cdot\mathbf{r}T}-1}{2\pi i\mathbf{b}\cdot\mathbf{r}}\right| \le \min\left\{\frac{1}{\pi|\mathbf{b}\cdot\mathbf{r}|}, T\right\}$$

that we already used in (4.7), but here it suffices to apply the simpler inequality

$$\text{(12.7)} \qquad \left|\frac{e^{2\pi i\mathbf{b}\cdot\mathbf{r}T}-1}{2\pi i\mathbf{b}\cdot\mathbf{r}}\right| \le \frac{1}{\pi|\mathbf{b}\cdot\mathbf{r}|}.$$

Key Definition: discarding bad pairs of vectors. Let $0 < \gamma < 1/10$; we say that a pair $(\mathbf{b}_1, \mathbf{b}_2) \in S^2 \times S^2$ of 3-dimensional vectors of unit length is γ-bad if there exists an $\mathbf{r} \in \mathbb{Z}^3 \setminus \mathbf{0}$ such that

$$\text{(12.8)} \qquad |\mathbf{b}_1 \cdot \mathbf{r}| \cdot |\mathbf{b}_2 \cdot \mathbf{r}| \cdot |\mathbf{r}|^{3/2} \le \gamma,$$

where of course $|\mathbf{r}| = \sqrt{r_1^2 + r_2^2 + r_3^2}$.

Let $\mathcal{B}(\gamma)$ denote the set of all γ-bad pairs $(\mathbf{b}_1, \mathbf{b}_2) \in S^2 \times S^2$. Next I show that the "bad" set $\mathcal{B}(\gamma)$ forms a small minority: the measure of $\mathcal{B}(\gamma)$ is negligible compared to the measure $(4\pi)^2$ of the Cartesian product $S^2 \times S^2$ of two unit spheres. (Note in advance that at the end, we will specify the value of γ in terms of the given ε, and we will throw out all initial conditions containing a γ-bad pair $(\mathbf{b}_1, \mathbf{b}_2)$.)

Lemma 12.1. *The set $\mathcal{B}(\gamma)$ of γ-bad pairs $(\mathbf{b}_1, \mathbf{b}_2) \in S^2 \times S^2$ (see the Key Definition) is small in the sense that*

$$\frac{\text{measure}\left(\mathcal{B}(\gamma)\right)}{(4\pi)^2} < 100\gamma \log\left(\frac{1}{\gamma}\right). \tag{12.9}$$

Proof. I recall the following well-known fact about the surface area of some spherical regions: for any $0 < \eta < 1$ and any 3-dimensional vector $\mathbf{r} \neq \mathbf{0}$,

$$\text{SurfaceArea}\left(\left\{\mathbf{b} \in S^2 \,:\, |\mathbf{r}\cdot\mathbf{b}| \le \eta|\mathbf{r}|\right\}\right) = 4\pi \cdot \eta. \tag{12.10}$$

Let's return to (12.8); note that

$$|\mathbf{b}_1\cdot\mathbf{r}| \cdot |\mathbf{b}_2\cdot\mathbf{r}| \cdot |\mathbf{r}|^{3/2} \le \gamma \iff \frac{|\mathbf{b}_1\cdot\mathbf{r}|}{|\mathbf{r}|} \le \min\left\{\frac{\gamma}{|\mathbf{b}_2\cdot\mathbf{r}|\cdot|\mathbf{r}|^{5/2}}, 1\right\},$$

and combining this with (12.10), for a fixed $\mathbf{r} \in \mathbb{Z}^3 \setminus \mathbf{0}$ we have

$$\begin{aligned}
&(4\pi)^{-2}\,\text{measure}\left\{(\mathbf{b}_1, \mathbf{b}_2) \in S^2 \times S^2 \,:\, |\mathbf{b}_1\cdot\mathbf{r}| \cdot |\mathbf{b}_2\cdot\mathbf{r}| \cdot |\mathbf{r}|^{3/2} \le \gamma\right\} \\
&= \int_{\eta=0}^{1} \min\left\{\frac{\gamma}{\eta r^{7/2}}, 1\right\} d\eta \\
&= \int_{\eta=0}^{\gamma r^{-7/2}} \min\left\{\frac{\gamma}{\eta r^{7/2}}, 1\right\} d\eta + \int_{\eta=\gamma r^{-7/2}}^{1} \min\left\{\frac{\gamma}{\eta r^{7/2}}, 1\right\} d\eta \\
&= \gamma r^{-7/2} + \frac{\gamma}{r^{7/2}} \int_{\eta=\gamma r^{-7/2}}^{1} \frac{d\eta}{\eta} \\
&= \frac{\frac{7}{2}\gamma \log r + \gamma \log\left(\frac{1}{\gamma}\right) + \gamma}{r^{7/2}},
\end{aligned}$$

where $|\mathbf{r}| = r$ and $|\mathbf{b}_2 \cdot \mathbf{r}| = \eta r$. It follows that

$$\text{(12.11)} \qquad \frac{\text{measure}\left(\mathcal{B}(\gamma)\right)}{(4\pi)^2} < \sum_{\mathbf{r}\in\mathbb{Z}^3\setminus\mathbf{0}} \frac{\frac{7}{2}\gamma\log|\mathbf{r}| + \gamma\log\left(\frac{1}{\gamma}\right) + \gamma}{|\mathbf{r}|^{7/2}}.$$

The infinite series on the right-hand side of (12.11) is clearly convergent, since the number of lattice points $\mathbf{r} \in \mathbb{Z}^3$ in a big ball of radius R centered at the origin is about the volume $\frac{4}{3}\pi \cdot R^3$, on the other hand, $7/2 > 3$, so $R^{7/2}$ is *much* larger than R^3 (and $\log R$ is relatively negligible) if $R \to \infty$. A routine calculation/estimation in (12.11) gives the crude upper bound

$$\text{(12.12)} \qquad \frac{\text{measure}\left(\mathcal{B}(\gamma)\right)}{(4\pi)^2} < 100\gamma\log\left(\frac{1}{\gamma}\right),$$

which completes the proof of Lemma 12.1. ∎

Let $\mathcal{G}(\gamma)$ denote the complement of the "bad" set $\mathcal{B}(\gamma)$, that is, $\mathcal{G}(\gamma)$ is the set of pairs $(\mathbf{b}_1, \mathbf{b}_2) \in S^2 \times S^2$ which are *not* γ-bad (see the Key Definition; of course $\mathcal{G}$ indicates "good"). We want to give an upper bound to the integral (see (12.5)–(12.7))

$$\text{(12.13)} \qquad \int_{\mathcal{G}}\int_{I^3} (\text{discrepancy})^2\, d\mathbf{y}\, d\mathbf{b}_1\, d\mathbf{b}_2$$

$$= \int_{\mathcal{G}}\int_{I^3} \left(A(T_1, T_2; \mathbf{y}; \mathbf{b}_1, \mathbf{b}_2) - T_1T_2|\mathbf{b}_1 \times \mathbf{b}_2| \cdot \text{vol}\,(A)\right)^2 d\mathbf{y}\, d\mathbf{b}_1\, d\mathbf{b}_2$$

$$\leq \sum_{\substack{\mathbf{r}\in\mathbb{Z}^3:\\ \mathbf{r}\neq\mathbf{0}}} |a_{\mathbf{r}}|^2 \int_{\mathcal{G}} \frac{1}{\pi^4|\mathbf{b}_1 \cdot \mathbf{r}|^2|\mathbf{b}_2 \cdot \mathbf{r}|^2}\, d\mathbf{b}_1\, d\mathbf{b}_2$$

$$= \sum_{\substack{\mathbf{r}\in\mathbb{Z}^3:\\ \mathbf{r}\neq\mathbf{0}}} \frac{|a_{\mathbf{r}}|^2}{\left(\pi|\mathbf{r}|\right)^4} \int_{\mathcal{G}} \frac{1}{\left|\mathbf{b}_1 \cdot \mathbf{e}(\mathbf{r})\right|^2\left|\mathbf{b}_2 \cdot \mathbf{e}(\mathbf{r})\right|^2}\, d\mathbf{b}_1\, d\mathbf{b}_2,$$

where $\mathbf{e}(\mathbf{r}) = \mathbf{r}/|\mathbf{r}|$ is a unit vector. We prove the following result.

Lemma 12.2. *We have*

$$\int_{\mathcal{G}}\int_{I^3} (\text{discrepancy})^2\, d\mathbf{y}\, d\mathbf{b}_1\, d\mathbf{b}_2$$

$$\leq (4\pi)^2 \frac{\left(10\log_2\left(\frac{1}{\gamma}\right)\right)^2}{6\gamma} \cdot \text{vol}\,(A)\left(1 - \text{vol}\,(A)\right).$$

Proof. First note that if $(\mathbf{b}_1, \mathbf{b}_2) \in \mathcal{G}$ then by definition (see (12.8))

$$|\mathbf{b}_1 \cdot \mathbf{e}(\mathbf{r})| \cdot |\mathbf{b}_2 \cdot \mathbf{e}(\mathbf{r})| \cdot |\mathbf{r}|^{7/2} > \gamma. \tag{12.14}$$

To estimate the integral at the end of (12.13), we apply a standard power-of-two decomposition of the set

$$\mathcal{G}(\mathbf{r}) = \{ (\mathbf{b}_1, \mathbf{b}_2) \in S^2 \times S^2 : \ (12.14) \text{ holds} \} \supset \mathcal{G} \tag{12.15}$$

as follows: let $\ell_1 \geq 0$ and $\ell_2 \geq 0$ be arbitrary nonnegative integers satisfying

$$\ell_1 + \ell_2 \leq L(\mathbf{r}) = \log_2 \left(\frac{1}{\gamma} |\mathbf{r}|^{7/2} \right), \tag{12.16}$$

and write

$$\begin{aligned} &\mathcal{G}_{\ell_1,\ell_2}(\mathbf{r}) \\ &= \{ (\mathbf{b}_1, \mathbf{b}_2) \in S^2 \times S^2 : 2^{-\ell_1 - 1} < |\mathbf{b}_1 \cdot \mathbf{e}(\mathbf{r})| \leq 2^{-\ell_1}, \\ &\qquad 2^{-\ell_2 - 1} < |\mathbf{b}_2 \cdot \mathbf{e}(\mathbf{r})| \leq 2^{-\ell_2} \}. \end{aligned} \tag{12.17}$$

Thus we have the disjoint decomposition (see (12.14)–(12.17))

$$\mathcal{G}(\mathbf{r}) = \bigcup_{\substack{\ell_1 \geq 0, \ \ell_1 \geq 0: \\ \ell_1 + \ell_2 \leq L(\mathbf{r})}} \mathcal{G}_{\ell_1,\ell_2}(\mathbf{r}) \supset \mathcal{G}. \tag{12.18}$$

Note that (12.10) implies the upper bound

$$\frac{\text{measure}\left(\mathcal{G}_{\ell_1,\ell_2}(\mathbf{r}) \right)}{(4\pi)^2} \leq 2^{-\ell_1} \cdot 2^{-\ell_1} = 2^{-(\ell_1 + \ell_2)}. \tag{12.19}$$

By (12.13)–(12.19) we have

$$\begin{aligned} &(4\pi)^{-2} \int_{\mathcal{G}} \int_{I^3} (\text{discrepancy})^2 \, d\mathbf{y} \, d\mathbf{b}_1 \, d\mathbf{b}_2 \\ &= (4\pi)^{-2} \int_{\mathcal{G}} \int_{I^3} \left(A(T_1, T_2; \mathbf{y}; \mathbf{b}_1, \mathbf{b}_2) - T_1 T_2 |\mathbf{b}_1 \times \mathbf{b}_2| \cdot \text{vol}\,(A) \right)^2 d\mathbf{y} \, d\mathbf{b}_1 \, d\mathbf{b}_2 \\ &\leq \sum_{\substack{\mathbf{r} \in \mathbb{Z}^3: \\ \mathbf{r} \neq \mathbf{0}}} \frac{|a_{\mathbf{r}}|^2}{\left(\pi |\mathbf{r}| \right)^4} \sum_{\substack{\ell_1 \geq 0, \ \ell_1 \geq 0: \\ \ell_1 + \ell_2 \leq L(\mathbf{r})}} (4\pi)^{-2} \int_{\mathcal{G}_{\ell_1,\ell_2}(\mathbf{r})} \frac{1}{|\mathbf{b}_1 \cdot \mathbf{e}(\mathbf{r})|^2 |\mathbf{b}_2 \cdot \mathbf{e}(\mathbf{r})|^2} \, d\mathbf{b}_1 \, d\mathbf{b}_2 \end{aligned} \tag{12.20}$$

$$\leq \sum_{\substack{\mathbf{r}\in\mathbb{Z}^3:\\ \mathbf{r}\neq\mathbf{0}}} \frac{|a_{\mathbf{r}}|^2}{\left(\pi|\mathbf{r}|\right)^4} \sum_{\substack{\ell_1\geq 0,\ \ell_1\geq 0:\\ \ell_1+\ell_2\leq L(\mathbf{r})}} (4\pi)^{-2} \,\mathrm{measure}\left(\mathcal{G}_{\ell_1,\ell_2}(\mathbf{r})\right) 4^{\ell_1+1}\cdot 4^{\ell_2+1}$$

$$\leq \sum_{\substack{\mathbf{r}\in\mathbb{Z}^3:\\ \mathbf{r}\neq\mathbf{0}}} \frac{|a_{\mathbf{r}}|^2}{\left(\pi|\mathbf{r}|\right)^4} \sum_{\substack{\ell_1\geq 0,\ \ell_1\geq 0:\\ \ell_1+\ell_2\leq L(\mathbf{r})}} 2^{-(\ell_1+\ell_2)}\cdot 4^{\ell_1+1}\cdot 4^{\ell_2+1}$$

$$\leq \frac{1}{6} \sum_{\substack{\mathbf{r}\in\mathbb{Z}^3:\\ \mathbf{r}\neq\mathbf{0}}} \frac{|a_{\mathbf{r}}|^2}{|\mathbf{r}|^4} \sum_{\substack{\ell_1\geq 0,\ \ell_1\geq 0:\\ \ell_1+\ell_2\leq L(\mathbf{r})}} 2^{\ell_1+\ell_2}.$$

By (12.16) we have

$$\sum_{\substack{\ell_1\geq 0,\ \ell_1\geq 0:\\ \ell_1+\ell_2\leq L(\mathbf{r})}} 2^{\ell_1+\ell_2} \leq \left(L(\mathbf{r})\right)^2\cdot 2^{L(\mathbf{r})} \tag{12.21}$$

$$= \left(\log_2\left(\frac{1}{\gamma}|\mathbf{r}|^{7/2}\right)\right)^2\cdot\frac{1}{\gamma}|\mathbf{r}|^{7/2}.$$

Simple estimation gives the upper bound

$$\max_{\substack{\mathbf{r}\in\mathbb{Z}^3:\\ \mathbf{r}\neq\mathbf{0}}} \frac{\left(\log_2\left(\frac{1}{\gamma}|\mathbf{r}|^{7/2}\right)\right)^2\cdot\frac{1}{\gamma}|\mathbf{r}|^{7/2}}{|\mathbf{r}|^4} \tag{12.22}$$

$$= \max_{\substack{\mathbf{r}\in\mathbb{Z}^3:\\ \mathbf{r}\neq\mathbf{0}}} \frac{\left(\log_2\left(\frac{1}{\gamma}|\mathbf{r}|^{7/2}\right)\right)^2}{\gamma\sqrt{|\mathbf{r}|}} \leq \frac{\left(10\log_2\left(\frac{1}{\gamma}\right)\right)^2}{\gamma}.$$

Combining (12.20)–(12.22) we have

$$(4\pi)^{-2}\int_{\mathcal{G}}\int_{I^3} (\mathrm{discrepancy})^2\, d\mathbf{y}\, d\mathbf{b}_1\, d\mathbf{b}_2 \tag{12.23}$$

$$= (4\pi)^{-2}\int_{\mathcal{G}}\int_{I^3} \left(A(T_1,T_2;\mathbf{y};\mathbf{b}_1,\mathbf{b}_2) - T_1T_2|\mathbf{b}_1\times\mathbf{b}_2|\cdot\mathrm{vol}\,(A)\right)^2 d\mathbf{y}\, d\mathbf{b}_1\, d\mathbf{b}_2$$

$$\leq \frac{\left(10\log_2\left(\frac{1}{\gamma}\right)\right)^2}{6\gamma} \sum_{\substack{\mathbf{r}\in\mathbb{Z}^3:\\ \mathbf{r}\neq\mathbf{0}}} |a_{\mathbf{r}}|^2 = \frac{\left(10\log_2\left(\frac{1}{\gamma}\right)\right)^2}{6\gamma}\cdot\mathrm{vol}\,(A)\big(1-\mathrm{vol}\,(A)\big),$$

where in the last step we used (12.2). (12.23) gives Lemma 12.2. ■

The main point here is that the upper bound in (12.23) is a constant independent of the values of T_1 and T_2.

Now we are ready to finish the proof of Theorem 5 in the usual way: we just throw out the "bad" initial conditions and apply Chebyshev's inequality. First a definition: for any $\lambda > 0$ let

$$\Omega(\lambda) = \Big\{(\mathbf{y}, \mathbf{b}_1, \mathbf{b}_2) : \mathbf{y} \in [0,1)^3,\ (\mathbf{b}_1, \mathbf{b}_2) \in \mathcal{G},$$
$$\big|A(T_1, T_2; \mathbf{y}; \mathbf{b}_1, \mathbf{b}_2) - T_1 T_2 |\mathbf{b}_1 \times \mathbf{b}_2| \cdot \mathrm{vol}\,(A)\big| \geq \lambda\Big\}. \tag{12.24}$$

Combining (12.23)–(12.24) with Chebyshev's inequality,

$$\frac{1}{(4\pi)^2}\,\mathrm{measure}\,\big(\Omega(\lambda)\big) \leq \mathrm{vol}\,(A)\big(1 - \mathrm{vol}\,(A)\big) \cdot \frac{\big(10 \log_2\big(\frac{1}{\gamma}\big)\big)^2}{6\gamma} \cdot \lambda^{-2}, \tag{12.25}$$

where "measure" stands for the usual product measure.

By making the choice

$$\lambda = \lambda_0 = \frac{\sqrt{\mathrm{vol}\,(A)\big(1 - \mathrm{vol}\,(A)\big)}}{\gamma} \sqrt{\log_2\left(\frac{1}{\gamma}\right)} \tag{12.26}$$

in (12.25), we conclude

$$\frac{1}{(4\pi)^2}\,\mathrm{measure}\,\big(\Omega(\lambda_0)\big) \leq \frac{100}{6}\gamma \log_2\left(\frac{1}{\gamma}\right). \tag{12.27}$$

Now let

$$100\gamma \log\left(\frac{1}{\gamma}\right) = \frac{\varepsilon}{2}. \tag{12.28}$$

If we throw out the set of initial conditions $(\mathbf{y}, \mathbf{b}_1, \mathbf{b}_2)$ contained in $\Omega(\lambda_0)$, and also throw out those initial conditions $(\mathbf{y}, \mathbf{b}_1, \mathbf{b}_2)$ for which the "tail" $(\mathbf{b}_1, \mathbf{b}_2)$ is γ-bad (i.e., $(\mathbf{b}_1, \mathbf{b}_2) \in \mathcal{B}(\gamma)$), then by (12.27)–(12.28) and Lemma 12.1 the total loss is

$$100\gamma \log\left(\frac{1}{\gamma}\right) + \frac{100}{6}\gamma \log_2\left(\frac{1}{\gamma}\right) \leq \varepsilon/2 + \varepsilon/2 = \varepsilon. \tag{12.29}$$

Also, the choice (12.28) implies that

$$\lambda = \lambda_0 = \frac{\sqrt{\mathrm{vol}\,(A)\big(1 - \mathrm{vol}\,(A)\big)}}{\gamma}\sqrt{\log_2\left(\frac{1}{\gamma}\right)} \tag{12.30}$$

$$< \frac{1000}{\varepsilon} \cdot \log^2\left(\frac{1}{\varepsilon}\right) \cdot \sqrt{\mathrm{vol}\,(A)\big(1 - \mathrm{vol}\,(A)\big)}\,.$$

Combining (12.29) with (12.24), (12.26), (12.30), Theorem 5 follows. ■

References

[1] J. Beck and W. W. L. Chen, *Irregularities of Distribution,* Cambridge Tracts in Math., 89, Cambridge University Press, 1987.

[2] H. Davenport, *Multiplicative Number Theory,* Second Edition, Graduate Texts in Math., 74, Springer-Verlag, 1980.

[3] D. J. Davis and P. Rabinowitz, *Methods of Numerical Integration,* Academic Press, 1984.

[4] M. Drmota and R. F. Tichy, *Sequences, Discrepancies and Applications,* Lecture Notes in Math., 1651, Springer, 1997.

[5] M. Drmota, Irregularities of continuous distributions, *Ann. Inst. Fourier,* **39** (1989), 501–527.

[6] J. Matousek, *Geometric Discrepancy: An Illustrated Guide,* Algorithms and Combinatorics, 18, Springer, 1999.

[7] H. L. Montgomery, On irregularities of distribution. In *Congress of Number Theory* (Zarautz, 1984), pp. 11–27. Univ. del Pais Vasco, Bilbao, 1989.

[8] G. N. Watson, *A Treatise on the Theory of Bessel Functions,* Cambridge University Press, 1958.

[9] H. Weyl, Über die Gleichverteilung von Zahlen mod Eins, *Math. Ann.,* **77** (1916), 313–352.

József Beck

Mathematics Department
Busch Campus
Hill Center
Rutgers University
New Brunswick
NJ 08903 USA

e-mail: jbeck@math.rutgers.edu

BOLYAI SOCIETY
MATHEMATICAL STUDIES, 21

An Irregular Mind
(Szemerédi is 70)
pp. 131–217.

Percolation on Self-Dual Polygon Configurations

BÉLA BOLLOBÁS* and OLIVER RIORDAN

Recently, Scullard and Ziff noticed that a broad class of planar percolation models are self-dual under a simple condition that, in a parametrized version of such a model, reduces to a single equation. They state that the solution of the resulting equation gives the critical point. However, just as in the classical case of bond percolation on the square lattice, self-duality is simply the starting point: the mathematical difficulty is precisely showing that self-duality implies criticality. Here we do so for a generalization of the models considered by Scullard and Ziff. In these models, the states of the bonds need not be independent; furthermore, increasing events need not be positively correlated, so new techniques are needed in the analysis. The main new ingredients are a generalization of Harris's Lemma to products of partially ordered sets, and a new proof of a type of Russo–Seymour–Welsh Lemma with minimal symmetry assumptions.

1. Introduction

In 1963 Sykes and Essam [30] noticed that, in independent bond percolation, a star with bond probabilities p_1, p_2 and p_3 may be replaced by a triangle with bond probabilities r_1, r_2 and r_3, provided the p_i and r_i satisfy certain equations; in particular, a star in which each bond has probability $p_0 = 1-2\sin(\pi/18)$ of being open may be replaced by a triangle with bond probabilities $1-p_0$. Sykes and Essam went on to use this star-triangle transformation to predict that p_0 and $1-p_0$ should be the critical probabilities for bond percolation on the hexagonal and triangular lattices, respectively. In 1981 Wierman [31] gave a rigorous proof of this result with the aid of a

*Research supported in part by NSF grants DMS-0906634, CNS-0721983 and CCF-0728928, and ARO grant W911NF-06-1-0076.

Russo–Seymour–Welsh-type theorem. In 1982 Kesten [17] extended Wierman's theorem to describe the 'critical surface' of weighted bond percolation on the triangular lattice, although the details were worked out only in 1999 by Grimmett [13]. Later, in 2008, it was shown [7] that the sum of the critical probabilities of a centrally symmetric planar lattice and its dual is 1; more generally, an analogous statement allowing for percolation with different probabilities for different bonds was proved, giving the Kesten–Grimmett theorem as an easy corollary.

In 1984 Wierman [32] used the general star-triangle transformation to determine the exact critical probability for bond percolation (with equal bond probabilities) on a lattice obtained from the square lattice by adding some diagonals. Using a simpler transformation, Suding and Ziff [29] deduced the critical probability for site percolation on the extended Kagomé lattice from Wierman's result for the hexagonal lattice.

Recently, extending work of Scullard [24] and Chayes and Lei [10], Ziff [35] and Ziff and Scullard [36] proposed a simple criterion predicting the value of the critical probability for a wide variety of percolation models in the plane. In addition to the usual independent site and/or bond percolation models on a number of lattices, these models include cases which can be seen as bond percolation with local dependencies between the states of certain bonds.

The predictions of Scullard and Ziff are similar in nature to those of Sykes and Essam [30] mentioned above: having shown that for a certain probability p_0 the percolation model is 'self-dual', they state that this probability p_0 is therefore critical. Mathematically, there is a folklore 'conjecture' (with, as far as we are aware, no precise formulation) stating that any 'reasonable' self-dual planar model is critical. This conjecture is still wide open. It is well known to hold in certain special cases, in particular for site or bond percolation on lattices with certain symmetries, such as reflection in a line (see Kesten [17]), or rotational symmetry of any order, as shown in [7]. More generally, as remarked in [7], it can be shown for site or bond percolation on any lattice by combining results of Sheffield [27, Corollary 9.4.6], Aizenman, Kesten and Newman [1] and Menshikov [21].

Chayes and Lei [10] independently described a special case of the Scullard–Ziff criterion (as well as a generalization to random cluster models), and gave a sketch proof of criticality under an extra assumption. In a recent preprint, Wierman and Ziff [33] proved criticality in certain special cases, using known results on self-dual planar lattices.

In this paper we shall prove that the Scullard–Ziff criterion does indeed give the critical point for a wide variety of planar percolation models. In the original papers in the physics literature, the exact scope of applicability of the criterion is not entirely clear. In this paper we shall define *precisely* a general class of models that are self-dual in the appropriate sense, and use new methods to show that the self-dual point is indeed critical in *all* cases.

Although full definitions will be given only in the next section, let us illustrate some simple special cases of our main result, starting with one very concrete (but rather specific) example, and then turning to a more general family.

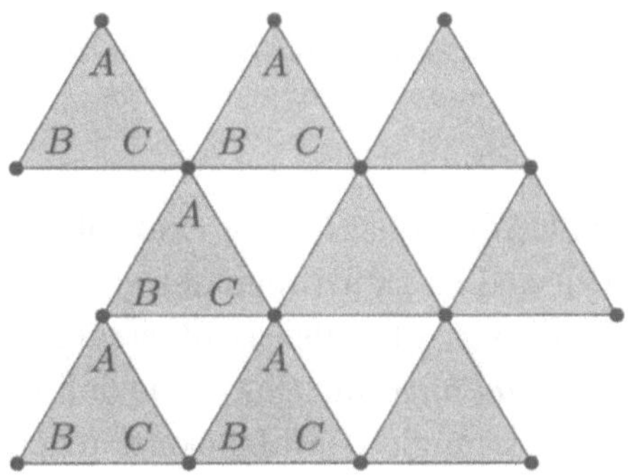

Fig. 1.1. The triangular lattice, with alternate triangles shaded.

Example 1.1. Consider the usual triangular lattice shown in Figure 1.1. Given a parameter $0 < p < 1$, we initially select each bond (i.e., edge) independently with probability p. But then bonds within a shaded triangle 'compete': if there are two bonds, the first in the clockwise order 'wins', and the other is deleted. However, if all three bonds are present there is a standoff, and all three remain. This results in a configuration of bonds such that, within each shaded triangle, with probability $(1-p)^3$ no bonds are present, with probability p^3 all three are present, and otherwise exactly one bond is present, with each of these cases having probability $\frac{1}{3}\left(1-p^3-(1-p)^3\right) = p(1-p)$. The configurations in different triangles are of course independent.

Applying our main results to this particular model we shall see that (with probability 1) the remaining bonds form a graph containing an infinite component if and only if $p > 1/2$; in other words, the model *percolates* if and only if $p > 1/2$. Note for later that considering a single shaded triangle, the probability that the vertices A and B are connected within this triangle is $p(1-p)+p^3$, as is the probability that B and C are connected. The probability that both events hold is p^3, since this happens if and only if all three bonds are selected.

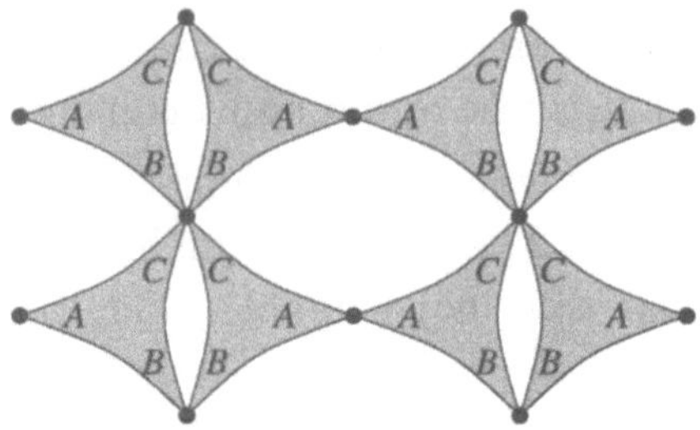

Fig. 1.2. A lattice of labelled triangles with no axes of symmetry, and no rotational symmetries.

Example 1.2. More generally, consider, for example, either the usual triangular lattice as above (viewed as an arrangement of shaded triangles), or the lattice of triangles a small part of which is illustrated in Figure 1.2. These are both examples of *self-dual hyperlattices*, to be defined in the next section. Suppose that each shaded triangle contains some mechanism that connects certain subsets of its vertices, with these processes independent in different triangles; we have given one example above, but the mechanism is irrelevant, only the final connection probabilities matter. Suppose that in each triangle the probability that all vertices are connected (inside the triangle) is p_{ABC}, that none are connected is $p_\emptyset$, that A and B are connected to each other but not to C is p_{AB}, and so on, with these probabilities the same for all triangles. Then, except possibly in the degenerate case $p_{ABC} = p_\emptyset = 0$, there is (with probability 1) an infinite connected component if and only if $p_{ABC} > p_\emptyset$.

Even the very special cases described above are outside the scope of existing results, for several reasons. Firstly, in general they do not correspond to independent bond or site percolation on any lattice (as required in [33]). This is certainly the case when increasing events are not positively correlated; see the discussion in Section 2. Such correlation is absent in Example 1.1 when p is equal to or close to $1/2$, since $p^3 < \left(p(1-p)+p^3\right)^2$ when $p = 1/2$. Positive correlation is also required for the arguments in [10]. Secondly, in Example 1.2 there is no rotational or reflectional symmetry, as required in [33] and in [10], so existing Russo–Seymour–Welsh-type results do not apply.

Our results show that self-duality implies criticality in a context that is much broader than the Scullard–Ziff criterion; an example is given in Figure 2.7.

Mathematically, the main interest of this paper is perhaps in the development of new tools needed to analyze the general model, including new proofs of analogues of the Russo–Seymour–Welsh Lemma, and a generalization of Harris's Lemma. The rest of the paper is organized as follows. In Section 2 we describe the model we shall study, and state our main results. In Section 3 we present a generalization of Harris's Lemma to products of posets. In Section 4 we prove various technical results about the model. The heart of the paper is Section 5, where we prove a new RSW-type lemma; the proof is rather involved, so we first illustrate the ideas in a simpler setting (bond percolation on $\mathbb{Z}^2$) in Subsection 5.1. In Section 6 we show how to apply this lemma using only the minimal symmetry guaranteed by self-duality. Finally, in Section 7 we show that (as in [4]), using a suitable sharp-threshold result, it is but a small step from rectangle crossings to the final results. In Section 8 we conclude with a brief discussion of the behaviour of the model in the critical case.

2. The Model and Results

The starting point of our investigation is an embedding of a suitable hypergraph in the plane, as described implicitly by Ziff and Scullard [36] and explicitly by Wierman and Ziff [33]. In these papers the hypergraphs considered are 3-uniform, but much of the time there is no need for this restriction. Since the concept of a plane hypergraph seems fundamental, we give several equivalent definitions. To avoid irrelevant difficulties we always assume piecewise linearity of all curves.

By a *polygon* we mean a piecewise-linear closed curve P in the plane that, if it touches itself at all, does so only externally at some vertices. (To be pedantic, P is the image of some regular n-gon P' under a piecewise linear map defined on the closed domain bounded by P' that is injective except possibly at the vertices of P'.) Such a polygon surrounds (more precisely winds around) a simply connected open set that we call its *interior*.

By a *plane hypergraph* $\mathcal{H}$ we mean a set of points of $\mathbb{R}^2$, the *vertices*, together with a set of polygons, the *hyperedges*, with the following properties:

(i) any bounded subset of the plane contains only finitely many vertices, and meets only finitely many hyperedges,

(ii) the interiors of the hyperedges are disjoint,

(iii) each hyperedge is incident with at least one vertex, and

(iv) hyperedges meet themselves or each other only at vertices.

Note that we allow a hyperedge to meet the same vertex several times, as in Figure 2.1, simply because there turns out to be no reason not to. Property (i) ensures that each vertex meets only finitely many hyperedges, and vice versa. In this paper, all plane hypergraphs we consider are connected, in the natural sense.

Fig. 2.1. On the left is part of a plane hypergraph $\mathcal{H}$: the shaded regions are the (interiors of the) polygons corresponding to hyperedges. The same picture may be seen as a proper 2-colouring of the faces of a plane graph. On the right is the corresponding 3-coloured cubic map. Note that not all edges are drawn as straight lines (see the top right of the figure).

Two plane hypergraphs are *isomorphic* if there is a homeomorphism from the plane to itself mapping one into the other, in the obvious sense. Of course, given a plane hypergraph (or indeed an isomorphism class of plane hypergraphs) there is a corresponding abstract hypergraph; more precisely, a (multi-)hyper(-multi-)graph, where we have included 'multi' twice to indicate that two or more hyperedges may be incident with the same set of vertices, and a single hyperedge may be incident with a vertex more than once. However, we shall work directly with the drawings throughout.

Plane hypergraphs are natural generalizations of plane (multi)graphs. Indeed, we may think of a plane graph as a representation of an abstract graph, with the vertices represented by points, and the edges by connected sets meeting only at the vertices. The edges of a graph are naturally represented by minimal sets connecting the corresponding vertices, i.e., simple curves, but in the hypergraph case polygons are more natural, so we use them even for hyperedges with only two vertices (or indeed, one vertex). When we come to percolation in a moment, the idea is that instead of each edge in a graph randomly either connecting its vertices or not, independently of the other edges, each hyperedge will randomly connect some subsets of its vertices, independently of the other hyperedges.

It is easy to check that (connected, as always in this paper) plane hypergraphs $\mathcal{H}$ correspond exactly to *shaded* locally finite connected plane multigraphs G, where the faces of G are properly coloured grey and white, so that every edge borders faces of different colours, with any unbounded faces white. (At this stage there may be one or more unbounded faces; in the bulk of the paper all faces will be bounded.) Indeed, vertices correspond to vertices, and the hyperedges of $\mathcal{H}$ are simply the grey faces of G, as on the left in Figure 2.1. Of course, a plane graph G with at most one unbounded face has an appropriate shading if and only if every degree is even, and then it has either one or two shadings, depending on whether or not it has an unbounded face.

Note that if e is a hyperedge incident with $|e|$ vertices (counting multiplicity), then, as a polygon, e is made up of $|e|$ segments joining vertices, corresponding to the edges of G. We cannot in general draw these segments as straight lines. For example, $\mathcal{H}$ may contain triples uvw and uvw', say; furthermore, there may be further hyperedges inside the region bounded by the two curves joining u and v associated to these triples.

By a *face* of a plane hypergraph $\mathcal{H}$ (defined as above) we mean a component of what is left of the plane after removing all hyperedges and their interiors, i.e., a white face of the corresponding graph G. For our next few definitions (in particular that of duality) to make sense, it is convenient to insist that each face of $\mathcal{H}$ has finitely many edges in its boundary. This is equivalent to imposing the condition that if $\mathcal{H}$ is infinite, then it has no unbounded faces. (Of course, if $\mathcal{H}$ is finite, then it necessarily has exactly one unbounded face.) When it comes to percolation, we naturally consider only infinite $\mathcal{H}$.

There is yet another way of defining plane hypergraphs, which will turn out to be much more convenient to work with, but is at first sight perhaps less natural (at least for percolation). Given a (connected, as usual) plane hypergraph $\mathcal{H}$ and the corresponding graph G as above, replace each vertex

v of $\mathcal{H}$ by a *black* $2d$-gon, where $d = d(v)$ is the degree of v, each face by a *white* $2d$-gon, where d is the number of edges (of G) bounding the face, and each hyperedge e incident with d vertices (counted with multiplicity) by a *grey* $2d$-gon. In this way we obtain a cubic planar map M in which the faces are properly coloured black, white and grey; see Figure 2.1. (As before, we cannot necessarily draw the edges of the polygons as straight lines.)

The reverse transformation is even simpler: starting from a cubic map M (i.e., a connected locally finite 3-regular plane graph in which each face has finitely many edges in its boundary) in which the faces are properly coloured black, white and grey, we simply contract each black face to a point to form a vertex of $\mathcal{H}$, and take the grey faces to form the hyperedges. In what follows we shall refer to such a coloured map M as a *map hypergraph* (or simply a hypergraph, when there is no danger of confusion), and denote it also by $\mathcal{H}$.

Note that when $\mathcal{H}$ is a graph (i.e., $\mathcal{H}$ is 2-uniform), then the corresponding map is the one considered in Chapter 3 of [6] (see Figures 2 and 4, for example). There, the 4-gon corresponding to an edge is coloured black or white according to whether the edge is open or closed; here, the $2|e|$-gon corresponding to a hyperedge e is grey for now, but will be coloured with a mixture of black and white later.

By a *plane hyperlattice* $\mathcal{H}$ we shall mean an infinite connected plane hypergraph (defined in any of the three ways above) with a lattice $\mathcal{L}$ of translational symmetries, i.e., such that there are linearly independent vectors $\mathbf{a}$ and $\mathbf{b}$ with the property that translation of the plane through either vector maps the drawing into itself in the obvious sense, corresponding to an isomorphism of the underlying hypergraph. Throughout, we view $\mathcal{L} = \{m\mathbf{a} + n\mathbf{b} : m, n \in \mathbb{Z}\}$ as a subset of $\mathbb{R}^2$. More formally, we define a *plane hyperlattice* to be a pair $(\mathcal{H}, \mathcal{L})$ as above, since in what follows $\mathcal{L}$ need not be the full lattice of translational symmetries of $\mathcal{H}$; in spite of this, we usually omit $\mathcal{L}$ from the notation. Naturally, when we consider isomorphisms of plane hyperlattices, these are required to preserve the corresponding lattices of symmetries. More precisely, a homeomorphism $S : \mathbb{R}^2 \to \mathbb{R}^2$ is an isomorphism from the plane hyperlattice $(\mathcal{H}, \mathcal{L})$ to $(\mathcal{H}', \mathcal{L}')$ if it corresponds to a plane hypergraph isomorphism and satisfies $S(x + \ell) = S(x) + T(\ell)$ for all $x \in \mathbb{R}^2$ and $\ell \in \mathcal{L}$, where T is a linear map with $T(\mathcal{L}) = \mathcal{L}'$.

In the context of percolation, the natural notion of the dual of a plane hypergraph $\mathcal{H}$ turns out to be the plane hypergraph $\mathcal{H}^*$ defined as follows. Take a vertex of $\mathcal{H}^*$ inside each face of $\mathcal{H}$. To obtain the hyperedges of $\mathcal{H}^*$, replace each hyperedge e of $\mathcal{H}$ by the *dual hyperedge* e^* joining the vertices corresponding to the faces that e meets, as in Figure 2.2. In the special case

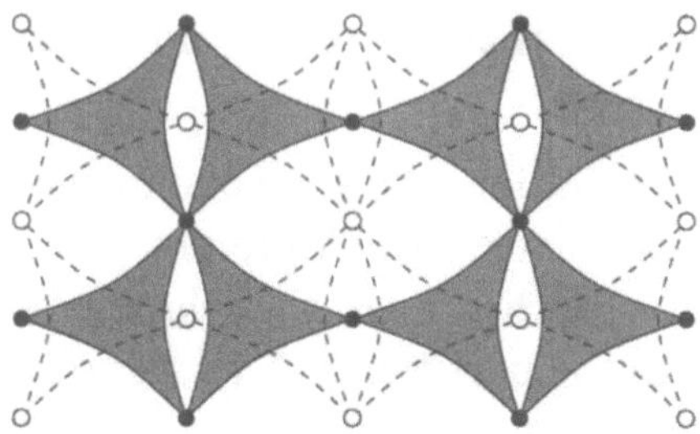

Fig. 2.2. Part of a 3-uniform plane hyperlattice (filled circles and shaded triangles) and its dual (open circles and dashed triangles). The hyperedges of the dual are the concave dashed triangles.

of 2-uniform hypergraphs, i.e., graphs, this is the usual notion of planar duality. In the 3-uniform case considered (with mild additional restrictions) by Wierman and Ziff [33], the notion of duality just defined coincides with theirs; the description of Ziff and Scullard [36] seems to be ambiguous.

In terms of the shaded graph G, the operation of taking the dual is rather complicated – it is *not* simply reversing the shading. However, in the 3-coloured map formulation, it is very simple to construct the dual: simply exchange black and white. Indeed, one can think of a hyperedge and its dual together as corresponding to a grey face of the map M (for example, in Figure 2.2 one can take the intersection of e and e^* as the grey face); the vertices of $\mathcal{H}$ correspond to the black faces of M, and the vertices of $\mathcal{H}^*$ to the white faces of M.

Of course, choosing the drawing appropriately, we may take the dual $\mathcal{H}^*$ of a plane hyperlattice $\mathcal{H}$ to be a plane hyperlattice, and the dual of $\mathcal{H}^*$ to be $\mathcal{H}$. A plane hyperlattice $\mathcal{H}$ is *self-dual* if $\mathcal{H}^*$ is isomorphic to $\mathcal{H}$; examples are shown in Figures 2.2, 2.5 and 2.7.

If P is a polygon in the plane, then by a *non-crossing partition* of its vertex set $V(P)$ we mean a partition π such that no two distinct parts of π contain interlaced pairs of vertices: if x, y, z, w are four distinct vertices appearing in this cyclic order around P, a non-crossing partition π cannot contain two parts one of which includes x and z, and the other y and w. Equivalently, a partition π of $V(P)$ is a non-crossing partition if and only if it may be realized by constructing disjoint (path-)connected subsets S_i of P (which we take to include its interior) so that each part of π is the intersection of some S_i with $V(P)$.

The *dual* π^* of a non-crossing partition π is the non-crossing partition of the *edges* of P in which two edges e and f are in the same part if and only if they are not interlaced with two vertices x and y in a common part of π. Here interlaced means that the edges and vertices occur in the cyclic order e, x, f, y or its reverse.

Turning finally to percolation, the *state* of a hyperedge e will simply be a non-crossing partition of its vertices. (More precisely, of the incidences of e with its vertex set, so if e touches itself, the relevant vertex appears multiple times in the groundset of the partition.) A *configuration* ω is an assignment of a state to each hyperedge of the hypergraph $\mathcal{H}$ under consideration. We think of the state of a hyperedge e as describing connections within e. In particular, by an *open path* in a configuration ω we mean a sequence $v_0e_1v_1e_2\dots e_\ell v_\ell$ such that, for each i, the partition of the vertices of e_i has a part containing both v_{i-1} and v_i. Two vertices are *connected* in ω if they are joined by an open path, and the *open clusters* of ω are the maximal connected sets of vertices.

Finally, a *hyperlattice percolation model* consists of a plane hyperlattice $(\mathcal{H}, \mathcal{L})$ together with a probability measure on configurations on $\mathcal{H}$ such that the states of different hyperedges are independent, and the measure is preserved by the action of $\mathcal{L}$. In other words, if e' is a translate of e under an element of $\mathcal{L}$, then corresponding states in e' and e have the same probabilities. Note that for a single hyperedge e, all probability distributions on the set of non-crossing partitions associated to e are allowed.

As usual, the sigma-field of measurable events is the one generated by cylindrical sets, i.e., by events depending only on the states of a finite set of hyperedges. In fact, except when defining percolation, throughout this paper we can work with large enough finite regions of $\mathcal{H}$, so there are no issues of measurability.

The *dual* ω^* of a configuration ω on $\mathcal{H}$ is the configuration on $\mathcal{H}^*$ in which the state of e^* is the dual of the state of e (noting that vertices of e^* correspond to edges of the polygon e). It is not hard to check that finite open clusters in ω are surrounded by open cycles in ω^* and *vice versa*; this is most easily seen in the colouring formulation described at the start of Section 4.

Given a plane hyperlattice $(\mathcal{H}, \mathcal{L})$, suppressing $\mathcal{L}$ in the notation as usual, the hyperlattice percolation models on $\mathcal{H}$ may be parametrized as follows. First pick one representative e_i of each orbit of the action of $\mathcal{L}$ on the hyperedges. Then for each non-crossing partition π of the vertices of e_i, choose a probability $p_{i,\pi}$, subject only to $\sum_\pi p_{i,\pi} = 1$ for each i. We call such a vector $\mathbf{p} = (p_{i,\pi})_{i,\pi}$ a *probability vector* (for $\mathcal{H}$), and write $\mathcal{H}(\mathbf{p})$ for the corresponding percolation model. The *dual vector* $\mathbf{p}^*$ assigns the probability $p_{i,\pi}$ to the partition π^* of e_i^*, so $\mathcal{H}^*(\mathbf{p}^*)$ is a hyperlattice percolation model on $\mathcal{H}^*$.

The hyperlattice percolation model $\mathcal{H}(\mathbf{p})$ is *self-dual* if $\mathcal{H}(\mathbf{p})$ and $\mathcal{H}^*(\mathbf{p}^*)$ are isomorphic, i.e., if there is an isomorphism from $\mathcal{H}$ to $\mathcal{H}^*$ such if $e \in E(\mathcal{H})$ and $f \in E(\mathcal{H}^*)$ correspond under the isomorphism, then each partition π of f has the same probability in $\mathcal{H}^*(\mathbf{p}^*)$ as the corresponding partition of e does in $\mathcal{H}(\mathbf{p})$. Our aim is to show that self-dual hyperlattice models are 'critical', but first we must define what critical means.

The set of partitions of a (here finite) set S forms a poset $\mathcal{P}$ in a natural way: we have $\pi \preccurlyeq \pi'$ if any two elements in the same part of π are in the same part of π', i.e., the parts of π' are unions of those of π, i.e., if π *refines* π'.

Given a hyperedge e_i as above, let $\mathcal{P} = \mathcal{P}_{e_i}$ be the poset formed by the *non-crossing* partitions of the vertices of e_i. An *upset* $\mathcal{U}$ in $\mathcal{P}$ is a subset of $\mathcal{P}$ such that if $\pi \in \mathcal{U}$ and $\pi \prec \pi'$ then $\pi' \in \mathcal{U}$. Given an upset $\mathcal{U}$ in $\mathcal{P}_{e_i}$ and a probability vector $\mathbf{p}$, let $p_i(\mathcal{U}) = \sum_{\pi \in \mathcal{U}} p_{i,\pi}$ denote the probability that the state of e_i is in $\mathcal{U}$. Given two probability vectors $\mathbf{p}$ and $\mathbf{q}$, we say that $\mathbf{q}$ *dominates* $\mathbf{p}$ if $q_i(\mathcal{U}) \geq p_i(\mathcal{U})$ for each i and each upset $\mathcal{U} \subset \mathcal{P}_{e_i}$. We say that $\mathbf{q}$ *strictly dominates* $\mathbf{p}$, and write $\mathbf{q} \succ \mathbf{p}$, if $q_i(\mathcal{U}) > p_i(\mathcal{U})$ for each i and each non-trivial upset $\mathcal{U} \subset \mathcal{P}_{e_i}$, i.e., for all upsets apart from $\mathcal{U} = \emptyset$ and $\mathcal{U} = \mathcal{P}_{e_i}$. Note that we can have $\mathbf{p} \neq \mathbf{q}$ such that $\mathbf{q}$ dominates $\mathbf{p}$ but does not strictly dominate it.

Hall's theorem implies that $\mathbf{q}$ dominates $\mathbf{p}$ if and only if $\mathbf{q}$ can be obtained from $\mathbf{p}$ by moving 'probability mass' from elements $p_{i,\pi}$ to elements $p_{i,\pi'}$ with $\pi \prec \pi'$. In the case of strict domination, we can assume that, for each i, a non-zero mass is moved from each π to each $\pi' \succ \pi$.

A percolation model $\mathcal{H}(\mathbf{p})$ *percolates* if the probability that the open cluster containing any given vertex is infinite is positive. As usual, this is equivalent to the existence with probability 1 of an infinite open cluster. The model $\mathcal{H}(\mathbf{p})$ is *critical* if two conditions hold: for any $\mathbf{q} \succ \mathbf{p}$ the model $\mathcal{H}(\mathbf{q})$ percolates, and for any $\mathbf{q} \prec \mathbf{p}$ the model $\mathcal{H}(\mathbf{q})$ does not percolate.

We say that the model $\mathcal{H}(\mathbf{p})$ exhibits *exponential decay* (of the volume) if there is a constant $\alpha > 0$ such that for any fixed vertex v the probability that the open cluster containing v contains at least n vertices is at most $e^{-\alpha n}$ for all $n \geq 2$. Our main result is the following.

Theorem 2.1. *Let $\mathcal{H}(\mathbf{p})$ be a self-dual hyperlattice percolation model. Then for any $\mathbf{q} \succ \mathbf{p}$ the model $\mathcal{H}(\mathbf{q})$ percolates, and for any $\mathbf{q} \prec \mathbf{p}$ the model $\mathcal{H}(\mathbf{q})$ exhibits exponential decay. In particular, $\mathcal{H}(\mathbf{p})$ is critical.*

Although this is far from the main point, a very special case is that the critical probability for bond percolation on any self-dual planar lattice is 1/2. Note that the condition $\mathbf{q} \succ \mathbf{p}$ is stronger than what one could hope for, namely a similar result with this condition replaced by $\mathbf{q}$ dominating $\mathbf{p}$ and $\mathbf{q} \neq \mathbf{p}$. However, one would then need to rule out degenerate special cases (corresponding to increasing the probability of bonds that are 'dead ends' in a bond percolation model, for example). Also, Theorem 2.1 with the present conditions is strong enough for the main application, Corollary 2.3 below.

It turns out that in proving Theorem 2.1, we do not require an exact isomorphism between $\mathcal{H}(\mathbf{p})$ and $\mathcal{H}^*(\mathbf{p}^*)$. We call two hyperlattice percolation models $\mathcal{H}_1(\mathbf{p}_1)$ and $\mathcal{H}_2(\mathbf{p}_2)$ *equivalent* if they can be coupled so that, for some constant C, for every open path P in either model there is an open path P' in the other model at Hausdorff distance at most C from P. Roughly speaking, the typical reason for two models to be equivalent is that they can be viewed as different ways of realizing a single underlying model.

For example, consider a plane triangulation G with a lattice $\mathcal{L}$ of translational symmetries, such as the triangular lattice. Then there are two natural ways to form a hyperlattice from G, illustrated (for a more complicated lattice) in Figure 2.3. In the first, there is a vertex for each edge of G, in the second, a vertex for each face of G. In either case, there is a hyperedge for each vertex v of G; this hyperedge is incident to all vertices corresponding to edges or faces of G that v is incident to. Within each hyperedge, assign probability p to the partition in which all vertices are in a single part, and $1-p$ to that in which every vertex is in a separate part. Let us call a hyperedge *open* if we select the partition into one part, and *closed* otherwise. Then the resulting models $\mathcal{H}(p)$ and $\mathcal{H}'(p)$ are equivalent using the natural coupling, i.e., the coupling in which the hyperedges in the two models corresponding to a vertex v of G have the same state, open or closed. Indeed, in either model an open path of length more than 1 consists of a sequence of open hyperedges with consecutive ones sharing (hyperlattice) vertices. But two hyperedges share a vertex in $\mathcal{H}$ if and only if the corresponding vertices of G are joined by an edge, and share a vertex in $\mathcal{H}'$ if and only if the corresponding vertices of G are in a common face. In a triangulation, two vertices are in a common face if and only if they are joined by an edge.

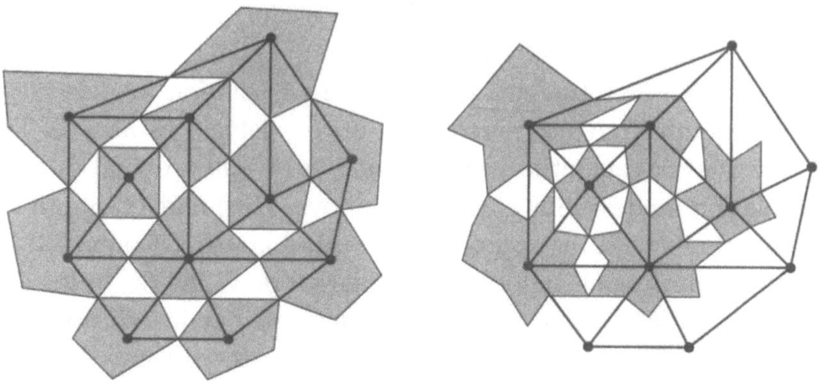

Fig. 2.3. The filled circles and thick lines show part of a planar triangulation G, which is assumed to have a lattice of symmetries not visible on this scale. On the left is part of the hyperlattice $\mathcal{H}$ formed from G by taking a vertex for each edge, and a hyperedge for each vertex. On the right is part of the hyperlattice $\mathcal{H}'$ formed by taking a vertex for each face of G, and a hyperedge for each vertex. Note that $\mathcal{H}$ and $\mathcal{H}'$ are dual as hyperlattices.

Emphasizing the lattice in the notation, for a change, we say that a hyperlattice percolation model $(\mathcal{H}(\mathbf{p}), \mathcal{L})$ is *approximately self-dual* if there is a model $(\mathcal{H}'(\mathbf{p}'), \mathcal{L})$ equivalent to $(\mathcal{H}^*(\mathbf{p}^*), \mathcal{L})$ such that $(\mathcal{H}'(\mathbf{p}'), \mathcal{L})$ and $(\mathcal{H}(\mathbf{p}), \mathcal{L})$ are isomorphic as plane hyperlattices, with the corresponding linear map T an isometry of the plane. The last restriction is a technicality: in the case of (exact) self-duality we did not impose it, but as we shall see in Lemma 6.7, any isomorphism witnessing self-duality has this property (after a suitable affine transformation). This is presumably also true for approximate self-duality, but as the condition will (we believe) self-evidently hold in any applications, we do not bother checking this. For our proofs, approximate self-duality is (apart from one technicality) just as good as self-duality, so we obtain the following strengthening of Theorem 2.1. In this result 'malleability' is a technical condition we shall introduce later (see Definitions 5.8 and 5.9); in the 3-uniform case, all hyperlattice percolation models are malleable. Also, any 'site percolation' model, where only the partition into singletons and that into a single part occur, is malleable.

Theorem 2.2. *Let $\mathcal{H}(\mathbf{p})$ be a malleable approximately self-dual hyperlattice percolation model. Then for any $\mathbf{q} \succ \mathbf{p}$ the model $\mathcal{H}(\mathbf{q})$ percolates, and for any $\mathbf{q} \prec \mathbf{p}$ the model $\mathcal{H}(\mathbf{q})$ exhibits exponential decay. In particular, $\mathcal{H}(\mathbf{p})$ is critical.*

To indicate that this extension may be useful, consider a plane triangulation G with a lattice of translational symmetries, and a real number $0 < p < 1$. Consider the two hyperlattice percolation models $\mathcal{H}(p)$ and $\mathcal{H}'(p)$ corresponding to site percolation on G, defined as above. Then $\mathcal{H}$ and $\mathcal{H}'$ are dual to each other, so $\mathcal{H}(p)$ and $\mathcal{H}'(1-p)$ are dual as hyperlattice percolation models. Since, as noted above, $\mathcal{H}(p)$ and $\mathcal{H}'(p)$ are equivalent, we see that $\mathcal{H}(1/2)$ is approximately self-dual, so Theorem 2.2 implies that the critical probability for site percolation on G is $1/2$. In itself this is not new (see the discussion in the introduction), but it indicates that the models to which Theorem 2.2 applies include ones with a site percolation 'flavour'.

We have just seen that site percolation on a triangulation with a lattice of symmetries, which for $p = 1/2$ is easily seen to be self-dual in an appropriate sense, may be transformed to a hyperlattice percolation model that is only approximately self-dual. In this case, the duality is clearer in the site percolation formulation than the hyperlattice one. Unsurprisingly, there are also cases where the reverse holds. Indeed, for any plane hyperlattice $\mathcal{H}$, selecting only the partitions in which all vertices in a given edge are connected or none are, we obtain a site percolation model on an (in general non-planar) graph. For example, taking $\mathcal{H}$ as in Figure 1.2, one obtains the non-planar graph in Figure 2.4. Theorem 2.1 shows that if $\mathcal{H}$ is self-dual,

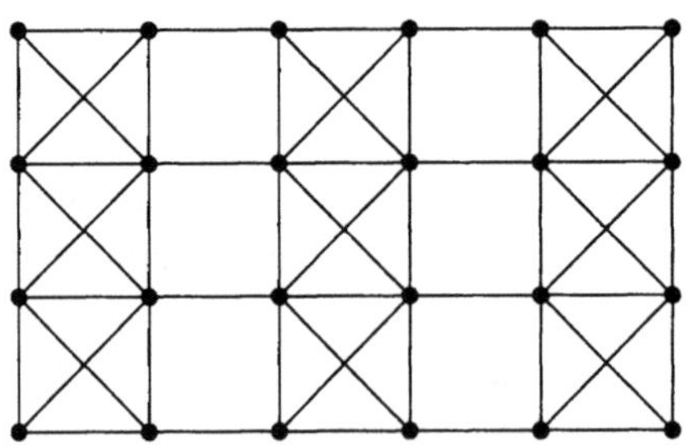

Fig. 2.4. A non-planar lattice which is 'self-dual' for site percolation.

then the critical probability for site percolation on the resulting graph is $1/2$.

Returning to the original motivation, the key observation of Ziff [35] and Ziff and Scullard [36] (present also in the original papers of Scullard [24] and Chayes and Lei [10] in the special case where $\mathcal{H}$ is the 'triangular hyperlattice' T shown in Figure 2.5) is that if $\mathcal{H}$ is 3-uniform, then the condition for self-duality becomes very simple, at least if one takes the same connection probabilities in all triangles. Unfortunately, this involves some further definitions, to allow for non-symmetric cases.

By a *labelled plane hyperlattice* we mean a plane hyperlattice in which the vertices around each hyperedge are labelled $1, 2, \dots, k$ in a way that is consistent with either the clockwise or anti-clockwise cyclic order within each hyperedge, and is globally consistent with the lattice $\mathcal{L}$ of translational symmetries. When $k = 3$ or $k = 4$ we often use letters $A, B, \dots,$ to denote the labels rather than numbers. In the 3-uniform case, assigning labels as above amounts to designating the vertices of a hyperedge e as its A-, B- and C-vertex in any order, as in [33]. The simplest example of a labelled plane hyperlattice is the *labelled triangular hyperlattice* shown in Figure 2.5. Another example is illustrated in Figure 2.6.

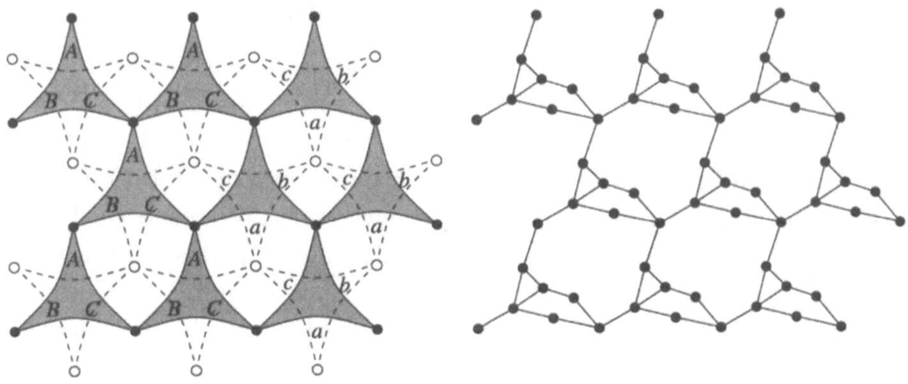

Fig. 2.5. The shaded triangles depict the *labelled triangular hyperlattice* T, i.e., the labelled 3-uniform hyperlattice obtained by taking alternate triangles in the triangular lattice, and labelling them consistently; capital letters denote the labelling of T. The dashed lines and lower case letters depict the dual T^*. Note that T is self-dual: there is a rotation through π mapping T to T^*. The figure on the right shows a lattice L obtained by substituting a 'generator' into each hyperedge of T.

Note that the A-vertex of one hyperedge may be the B-vertex of another hyperedge (or indeed, of the same hyperedge if it touches itself). Given a k-uniform labelled plane hyperlattice and a probability vector $\mathbf{p}$ consisting of probabilities p_π, $\pi \in \mathcal{P}_{\{1,2,\dots,k\}}$, that sum to 1, there is a corresponding hyperlattice percolation model $\mathcal{H}(\mathbf{p})$: for every hyperedge e, we assign p_π as the probability that the vertices of e are partitioned according to π, with the groundset $\{1, 2, \dots, k\}$ of π corresponding to the vertices of e in a manner indicated by the labelling.

Specializing to the 3-uniform case, the dual e^* of an edge e inherits a labelling from e: take the first vertex to be the one opposite the first vertex

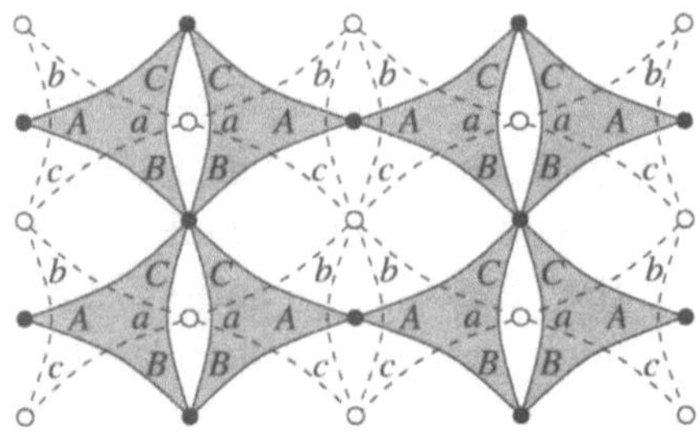

Fig. 2.6. A labelled 3-uniform hyperlattice with only translational symmetries that is isomorphic to its dual by reflection in a horizontal line. For visual clarity, the labelling of the dual is indicated with lower case letters.

of e, and so on, so a labelled 3-uniform plane hyperlattice $\mathcal{H}$ has a dual $\mathcal{H}^*$ that is again a labelled 3-uniform plane hyperlattice.

If $AcBaCb$ is a triangle with vertices A, B, C and edges a, b and c, then there are five possible partitions of the vertex set, all of which are non-crossing. These are represented by Wierman and Ziff [33] as $A|B|C$, $AB|C$, $AC|B$, $BC|A$, and ABC; we use the more compact notation $\emptyset$, AB, AC, BC and ABC. The duals of these partitions are, respectively, abc, ab, ac, bc and $\emptyset$. In this setting, a probability vector $\mathbf{p}$ is simply a vector $\mathbf{p} = (p_\emptyset, p_{AB}, p_{AC}, p_{BC}, p_{ABC})$ of non-negative reals summing to 1, and the dual of $\mathcal{H}(\mathbf{p})$ is simply $\mathcal{H}^*(\mathbf{p}^*)$, where $\mathbf{p}^*$ is formed from $\mathbf{p}$ by interchanging $p_\emptyset$ and p_{ABC}.

The observation of Scullard and Ziff mentioned earlier may be formulated as follows: if the labelled plane hyperlattice $\mathcal{H}$ is self-dual, then the model $\mathcal{H}(\mathbf{p})$ is self-dual if and only if $p_\emptyset = p_{ABC}$. The key point is that the other three partitions are all self-dual, as long as the dual triangle is labelled in the appropriate way, so only two entries in the probability vector, namely $p_\emptyset$ and p_{ABC}, change when we pass to the dual. [In the papers [35, 36], the formulation of duality is not totally clear. Wierman and Ziff [33] clearly formulate the notion of self-duality for unlabelled plane hyperlattices, and state that one can consider any labelling with lattice structure, but this seems to be an oversight; one needs the labelled hyperlattice to be self-dual *as a labelled hyperlattice,* which is not always the case.]

If the probabilities in $\mathbf{p}$ are appropriate functions of a single parameter p, the condition $p_\emptyset = p_{ABC}$ allows one to determine the critical point of the model; in general it gives the critical surface. As remarked earlier, Scullard and Ziff do not discuss whether self-duality in fact implies criticality; that it does is shown by Theorem 2.1.

Corollary 2.3. *Let $\mathcal{H}$ be a labelled 3-uniform plane hyperlattice that is isomorphic (as a labelled plane hyperlattice) to its dual, and let* $\mathbf{p}$ *be a probability vector* $(p_\emptyset, p_{AB}, p_{AC}, p_{BC}, p_{ABC})$. *Then* $\mathcal{H}(\mathbf{p})$ *percolates if* $p_{ABC} > p_\emptyset$, *and exhibits exponential decay if* $p_{ABC} < p_\emptyset$.

Proof. Let $\mathbf{q}$ be the probability vector $(p, p_{AB}, p_{AC}, p_{BC}, p)$, where $p = (p_\emptyset + p_{ABC})/2$, and note that $\mathbf{q}^* = \mathbf{q}$. Then the dual of the hyperlattice percolation model $\mathcal{H}(\mathbf{q})$ is $\mathcal{H}^*(\mathbf{q}^*) = \mathcal{H}^*(\mathbf{q})$, which is isomorphic to $\mathcal{H}(\mathbf{q})$ by the assumption on $\mathcal{H}$. Thus $\mathcal{H}(\mathbf{q})$ is a self-dual hyperlattice percolation model, and so is critical by Theorem 2.1. If $p_{ABC} > p_\emptyset$, then $\mathbf{p} \succ \mathbf{q}$, while if $p_{ABC} < p_\emptyset$, then $\mathbf{p} \prec \mathbf{q}$, so the result follows from Theorem 2.1. ∎

When $p_{ABC} = p_\emptyset$, the model $\mathcal{H}(\mathbf{p})$ may or may not percolate. A (degenerate) example that percolates is given by taking $p_{AB} = 1$ and all other probabilities zero in the triangular hyperlattice shown in Figure 2.5. An example that does not is given by taking connection probabilities in the same hyperlattice corresponding to critical bond percolation on the triangular lattice. As we shall see in Section 8, in non-degenerate models there is no percolation at the self-dual point.

As we have seen, Corollary 2.3 follows from Theorem 2.1 simply by restricting the parametrization of the percolation model, using the same partition probabilities for all triangles, rather than allowing different ones for each orbit under the action of the lattice of symmetries. In other words, we took all triangles to be of the same *type*. Of course, one can restrict the model in other ways, considering two or more types of triangle, or one type of triangle and one type of 4-gon, etc. Since any results obtained in this way are simply special cases of Theorem 2.1 we omit the details; the case of a single type of triangle is of special importance, since (for self-dual $\mathcal{H}$), self-duality reduces to a single equation, so one obtains the entire critical surface, rather than a lower dimensional subset of it.

For example, consider the self-dual plane hyperlattice $\mathcal{H}$ shown in Figure 2.7. Depending on the parameters we choose, $\mathcal{H}(\mathbf{p})$ can be self-dual via several different maps S from the plane to itself. Letting e_1 denote the hyperedge $ABCD$ and e_2 $xyzw$, there is a translation mapping e_1 into e_2^* and e_2^* into a hyperedge congruent to e_1. The model $\mathcal{H}(\mathbf{p})$ is self-dual under this translation if and only if the following equations hold: $p_{1,\emptyset} = p_{2,xyzw}$, $p_{1,AB} = p_{2,xyz}$, $p_{1,ABC} = p_{2,yz}$, $p_{1,AC} = p_{2,xw|yz}$, $p_{1,AB|CD} = p_{2,xz}$, and $p_{1,ABCD} = p_{2,\emptyset}$, together with the images of these equations under rotation. One natural way to satisfy these equations is simply to take the same probability $p = 1/14$ for each of the 14 non-crossing partitions in each square;

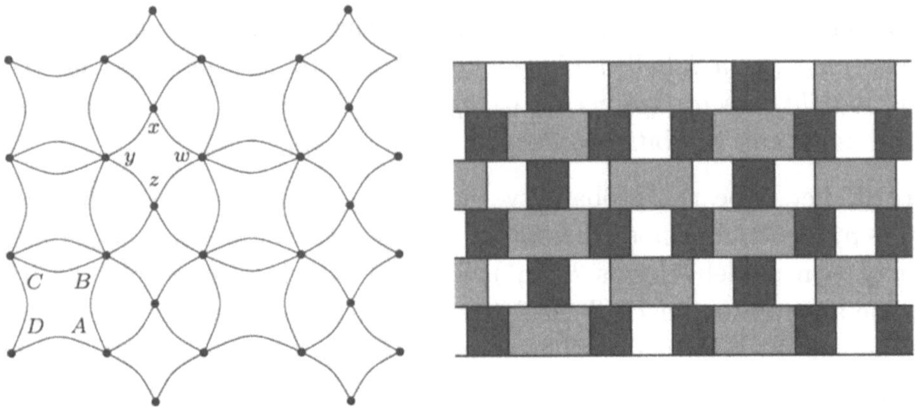

Fig. 2.7. Part of a self-dual 4-uniform hyperlattice $\mathcal{H}$, shown both as a plane hypergraph and as the corresponding 3-coloured cubic map.

taking probability $1/14$ for all partitions other than that into a single part or into four singletons, and probabilities p and $1/7-p$ for these two partitions, it follows that the percolation threshold in this model is at $p = 1/14$.

Before turning to the proof of Theorem 2.1, let us make some remarks. Firstly, an important special class of hyperlattice percolation models consists of those corresponding to bond or site percolation on (planar) lattices. By a *k-generator* we mean a finite graph G with k distinguished vertices; when $k = 3$, we denote these vertices by A, B and C. In a *planar generator* we insist that G is planar, and that the distinguished vertices lie in a common face, which we may take to be the outer face. Suppose that each bond (edge) of G is assigned a probability. Then, taking the bonds open independently with these probabilities, for each partition π of the distinguished vertices, let p_π be the probability that precisely those vertices in the same part of π are connected by open paths in G. Replacing each hyperedge in a labelled k-uniform plane hyperlattice $\mathcal{H}$ by a copy of G one obtains a (planar, if G is planar) lattice L. Taking all bonds open independently with the appropriate probabilities, the resulting bond percolation model is equivalent to $\mathcal{H}(\mathbf{p})$ in an obvious sense. In this case we say that $\mathcal{H}(\mathbf{p})$ is *bond realizable.* We may define *site realizability* analogously; this time, in each generator we insist that the distinguished vertices are always open. The definitions extend to general plane hyperlattices using one generator for each equivalence class of hyperedges.

The key observation of Ziff [35] and Ziff and Scullard [36] is that, for any self-dual labelled 3-uniform plane hyperlattice $\mathcal{H}$ and any generator

G, no matter how complicated, the equation $\mathbf{p} = \mathbf{p}^*$ reduces to a *single* polynomial equation in the probabilities associated to the bonds or sites of G, so, assuming self-duality implies criticality, this equation gives the entire critical surface in the inhomogeneous case. Moreover, if we take the same probability for each bond or site, this polynomial equation gives the critical probability for bond or site percolation on the lattice generated. Indeed, this was how Scullard [24] found the critical point for the 'martini lattice'.

As an example, this method predicts the critical point for bond percolation on the lattice L shown on the right in Figure 2.5, as the root of a certain polynomial of degree 9. Corollary 2.3 proves that this predicted value is indeed critical. To see the power of the method, consider the original proof of Wierman [31] that the critical probability for bond percolation on the triangular lattice is the root $2\sin(\pi/18)$ of the equation $p^3 - 3p + 1 = 0$. This proof relied on the star-triangle transformation, and an apparent coincidence between the various connection probabilities associated to a star and to a triangle. Scullard and Ziff's method gives an argument (which becomes a proof using the results of Wierman and Ziff [33] or Corollary 2.3) that does not rely on this – one simply considers a triangle and writes down the equation that the probability $p^3 + 3p^2(1-p)$ that all vertices are connected is equal to the probability $(1-p)^3$ that none are. The fact that one need not consider the dual lattice is key for examples such as the lattice shown on the right in Figure 2.5, which is not simply related to its dual.

We should emphasize that having observed that a certain percolation model is self-dual, one is very far from proving criticality. The classical example is bond percolation on the square lattice with $p = 1/2$. This model is obviously self-dual; proving the conjectured criticality was one of the key open problems in the early days of percolation theory, finally settled after 20 years by Kesten [16]. Since then, criticality at the self-dual point has been proved for a number of other models, but these remain the exceptions. In the hyperlattice context, Wierman and Ziff [33] proved a result of this type using existing results on planar lattices; for this reason they consider only a subclass of bond-realizable models, with planar generators and certain symmetries. This does not include examples such as that in Figure 2.5, which has no symmetries other than translations. However, as outlined earlier, one can easily adapt their method to such lattices, as long as the generator is planar. Chayes and Lei [10] proved a result of this class for triangular hyperlattices; they considered only the symmetric case, and imposed an additional condition on the parameters to ensure positive correlation (see below). They sketched an argument using standard techniques for independent percolation on planar lattices that they claimed extends to this case.

The real significance of Theorem 2.1 is that it applies to models that are not bond or site realizable. In such cases, standard results (such as Menshikov's Theorem [21], for example) do not apply, and there seems to be no simple way to adapt the arguments of Chayes and Lei or Wierman and Ziff. Indeed, considerable work will be needed to prove lemmas corresponding to, for example, Harris's Lemma and the Russo–Seymour–Welsh Lemma. Let us note that there *are* hyperlattice percolation models that are not bond or site realizable. Indeed, considering a single generator G and the corresponding probabilities $\mathbf{p}$, the event that two given vertices are connected in G is an upset in terms of the states of the individual bonds or sites. Thus Harris's Lemma implies that if the vector $\mathbf{p}$ is realizable, then for any two upsets $\mathcal{U}_1, \mathcal{U}_2$ in $\mathcal{P}$ we have $\mathbb{P}(\mathcal{U}_1 \cap \mathcal{U}_2) \geq \mathbb{P}(\mathcal{U}_1)\mathbb{P}(\mathcal{U}_2)$. For example, we must have $p_{ABC} \geq (p_{ABC} + p_{AB})(p_{ABC} + p_{BC})$. (In the 3-uniform case, Chayes and Lei [10] showed that a necessary and sufficient condition for such positive correlation is that $p_{ABC}p_\emptyset \geq p_{AB}(p_{BC} + p_{AC})$, and the equations obtained from this by permuting A, B and C, all hold.) Of course it is trivial to find a probability vector $\mathbf{p}$ for which this does not hold; an example is given in the introduction.

The applicability of Corollary 2.3 is also not limited to models that are bond or site realizable. Indeed Scullard and Ziff [24, 35, 36] noted that their duality observation does not require the model to be bond or site realizable. As in the original paper of Scullard [24], one may think of any model $\mathcal{H}(\mathbf{p})$ as bond percolation on a suitable (planar) lattice, where the states of the bonds within a generator may be dependent, although those in different generators must be independent. For example, we may take each 3-generator to be a triangle, and declare that with probability $p_{i,\emptyset}$ none of the edges are open, with probability $p_{i,ABC}$ all three are, with probability $p_{i,AB}$ the edge AB is open and the other edges are closed, and so on. Alternatively, we can take the generator to be a star. However, there is no need to think of bonds at all; for percolation, the only relevant property of the configuration within a triangle is which of the vertices A, B and C the configuration connects to which others, so it is natural to take this (random) partition of the vertices as the fundamental object of study. For the mathematical work (deducing criticality from self-duality), the details of the hyperlattice turn out to be mostly irrelevant; this is why we consider general plane hyperlattices in the rest of the paper.

Finally, let us note that extensions of the Scullard–Ziff criterion to the random cluster model and Potts model have been described by Chayes and Lei [10] and Wu [34]; establishing criticality at the self-dual point remains an open problem in these cases. Returning to hyperlattice percolation, in a few very special cases results have been proved that go further than determining the critical point. For example, Sedlock and Wierman [25] established the equality of the critical exponents between certain pairs of models, and Chayes and Lei [11] extended Smirnov's conformal invariance result [28] to what is essentially a very restricted case of the present model. In this paper we shall not consider such extensions; rather we shall prove that self-duality does imply criticality, in the full generality of hyperlattice percolation.

3. A Generalization of Harris's Lemma

Given posets $\mathcal{P}_1, \ldots, \mathcal{P}_n$, their *product* is the poset $\mathcal{P}_1 \times \cdots \times \mathcal{P}_n$ where each element $\mathbf{x}$ is a list $(x_1, \ldots, x_n)$ with x_i an element of $\mathcal{P}_i$, with $\mathbf{x} \preccurlyeq \mathbf{y}$ if and only if $x_i \preccurlyeq y_i$ for $i = 1, \ldots, n$. If $\mathbb{P}$ is a product probability measure on a product of posets, then with a slight abuse of notation we write $\mathbb{P}$ for any of the corresponding marginal measures.

In the later sections of this paper we shall make repeated use of the following generalization of Harris's Lemma [14] to products of posets. We only need the case where all $\mathcal{P}_i$ are equal and finite, but since the proof gives a little more, we state the result more generally. As usual, a *greatest element* in a poset $\mathcal{P}$ means an element y such that $x \preccurlyeq y$ for all $x \in \mathcal{P}$. Of course, if a greatest element exists, then it is unique.

Lemma 3.1. *Let $p > 0$. There is a constant $C = C(p) > 0$ such that if $\mathbb{P}$ is a product measure on a product $\mathcal{P} = \mathcal{P}_1 \times \cdots \times \mathcal{P}_n$ in which each factor $\mathcal{P}_i$ is a poset with a greatest element whose probability is at least p, then for any two upsets A and B in $\mathcal{P}$ we have*

$$\mathbb{P}(A \cap B) \geq \big(\mathbb{P}(A)\mathbb{P}(B)\big)^C.$$

Proof. We shall prove the result with $C = \lceil 2/p \rceil$.

As usual, we use induction on n. When $n = 0$, the set $\mathcal{P}$ contains only a single element, and the inequality is trivial. (It is also not hard to verify directly for $n = 1$.) Suppose then that $n \geq 1$, and that the result holds for smaller n. Suppose for notational convenience that $\mathcal{P}_n$ is finite, and list its elements as $x_0, x_1, \ldots, x_k$, with x_0 the greatest element. Let p_i denote the probability of element i in $\mathcal{P}_n$, so our assumption is that $p_0 \geq p$.

Given a set $S \subset \mathcal{P}$, let

$$S_i = \{y \in \mathcal{P}_1 \times \cdots \times \mathcal{P}_{n-1} : (y, x_i) \in S\} \subset \mathcal{P}_1 \times \cdots \times \mathcal{P}_{n-1}$$

denote the ith *slice* of S (with respect to the last factor in the product). Clearly, if S is an upset, then so is S_i, so the induction hypothesis gives $\mathbb{P}\big((A \cap B)_i\big) \geq \big(\mathbb{P}(A_i)\mathbb{P}(B_i)\big)^C$ for each i. Also, the upset conditions give $A_i \subset A_0$ and $B_i \subset B_0$ for $i > 0$. Since $\mathbb{P}$ is a product measure, we have $\mathbb{P}(S) = \sum_i p_i \mathbb{P}(S_i)$ for any S. Using these observations, it suffices to show that

$$\sum_{i=0}^{k} p_i (a_i b_i)^C \geq (ab)^C \tag{1}$$

holds whenever the non-negative real numbers p_i, a_i and b_i satisfy the following conditions: $p_0, \ldots, p_k$ sum to 1, $p_0 \geq p$, $a_0 = \max_i a_i$, $b_0 = \max_i b_i$, $a = \sum p_i a_i$, and $b = \sum p_i b_i$.

In proving (1) we may assume that $a, b > 0$. Dividing the a_i through by a and the b_i by b, we may assume that $a = b = 1$. Let $\alpha_i = a_i - 1$ and $\beta_i = b_i - 1$. Since $\sum_i p_i \alpha_i = 0$, we have $\alpha_0 = \max_i \alpha_i \geq 0$. Also,

$$\sum_{i : \alpha_i < 0} -p_i \alpha_i = \sum_{i : \alpha_i > 0} p_i \alpha_i \leq \sum_i p_i \alpha_0 = \alpha_0, \tag{2}$$

and similarly for the β_i. Our aim is to prove that $\sum_i p_i (1 + \alpha_i)^C (1 + \beta_i)^C \geq 1$. Recalling that $\sum_i p_i (\alpha_i + \beta_i) = 0$, this is equivalent to showing that

$$\Delta = \sum_i p_i \big((1 + \alpha_i)^C (1 + \beta_i)^C - 1 - C(\alpha_i + \beta_i)\big) \geq 0. \tag{3}$$

Since $\alpha_0, \beta_0 \geq 0$, we have $(1 + \alpha_0)^C (1 + \beta_0)^C \geq 1 + C\alpha_0 + C\beta_0 + C^2 \alpha_0 \beta_0$, so the contribution to the sum Δ from the $i = 0$ term is at least $p_0 C^2 \alpha_0 \beta_0$.

Turning to the remaining terms, since $(1 + x)^n \geq 1 + nx$ if $x \geq -1$ and n is a positive integer, we have

$$(1 + \alpha_i)^C (1 + \beta_i)^C = (1 + \alpha_i + \beta_i + \alpha_i \beta_i)^C \geq 1 + C(\alpha_i + \beta_i + \alpha_i \beta_i).$$

If α_i and β_i have the same sign, then the contribution of the ith summand to (3) is nonnegative. If $\alpha_i < 0$ and $\beta_i > 0$, then the negative of the contribution of the ith summand to (3) is at most $p_i C|\alpha_i|\beta_i \le p_i C|\alpha_i|\beta_0$. By (2), the negative of the sum of the contribution of all such terms is at most

$$C\beta_0 \sum_{i:\alpha_i<0} p_i(-\alpha_i) \le C\beta_0\alpha_0.$$

The same bound holds for terms with $\alpha_i > 0$ and $\beta_i < 0$, so we conclude that the sum in (3) satisfies

$$\Delta \ge p_0 C^2 \alpha_0\beta_0 - 2C\alpha_0\beta_0 = (p_0 C - 2)C\alpha_0\beta_0.$$

Since $p_0 C \ge 2$ by our choice of C, this establishes the inequality, and hence the lemma. ∎

Note that we have not attempted to optimize the value of C above. Indeed, for p_0 small, the proof above goes through with C only slightly larger than $1/p_0$, noting that the contribution to Δ from $i = 0$ is at least $p_0\big(C(C-1)\big(\alpha_0^2+\beta_0^2\big)/2 + C^2\alpha_0\beta_0\big) \ge p_0\big(C(C-1)+C^2\big)\alpha_0\beta_0$.

Of course, Harris's Lemma itself does not apply in this setting, i.e., one cannot simply take $C = 1$. Indeed, considering the upsets $\{x_0, x_1\}$ and $\{x_0, x_2\}$ in the poset on $\{x_0, x_1, x_2\}$ in which x_0 is greatest and x_1 and x_2 are incomparable, with $\mathbb{P}(x_0) = p_0$ and $\mathbb{P}(x_1) = \mathbb{P}(x_2) = (1-p_0)/2$, we may have $\mathbb{P}(A) = \mathbb{P}(B) = (1+p_0)/2$ and $\mathbb{P}(A\cap B) = p_0$. For p_0 small, this shows that we need the exponent C to be at least a constant times $\log(1/p_0)$.

Since the form of the bound obtained will be irrelevant in our remaining arguments, let us state as a corollary a weaker, more abstract version of the result.

Lemma 3.2. *Let $\mathcal{P}$ be a finite poset with a greatest element x_0 and let $\mathbb{P}$ be a probability measure on $\mathcal{P}$ with $\mathbb{P}(x_0) > 0$. There is a function $F = F_{\mathcal{P},\mathbb{P}}$ from $(0,1]^2$ to $(0,1]$ that is strictly increasing in each argument such that, for any $n \ge 1$ and any upsets A and B in $\mathcal{P}^n$ with $\mathbb{P}(A), \mathbb{P}(B) > 0$, we have*

$$\mathbb{P}(A\cap B) \ge F\big(\mathbb{P}(A), \mathbb{P}(B)\big).$$

Proof. Immediate from Lemma 3.1. ∎

As usual, the extension to infinite products is immediate by approximating with the finite case.

Note that while the form of the function F is irrelevant, it is natural to look for an F of the form $F(a,b) = (ab)^C$. Indeed, given upsets A_i and B_i in $\mathcal{P}^{n_i}$ then, considering the product upsets $A_1 \times A_2$ and $B_1 \times B_2$ in $\mathcal{P}^{n_1+n_2}$, one sees that the optimal F satisfies $F(a_1a_2, b_1b_2) \le F(a_1,b_1)F(a_2,b_2)$. Of course the optimal F cannot be precisely $F(a,b) = (ab)^C$, since we certainly need $C > 1$, and then the bound is not tight if $a = 1$, for example.

3.1. High probability unions of upsets

In many applications of Harris's Lemma in percolation, the exact form of the bound is not important, so the weaker conclusion of the more generally applicable Lemma 3.2 may be used instead of Harris's bound. We give one example that we shall use later: a form of the 'square-root' trick, showing that if the union of a fixed number of upsets has high enough probability, then one of the upsets has high probability.

Corollary 3.3. *Let $\mathcal{P}$ be a finite poset with a least element x_0 and let $\mathbb{P}$ be a probability measure on $\mathcal{P}$ with $\mathbb{P}(x_0) > 0$. Given $\varepsilon > 0$ and a positive integer k there is a $\delta = \delta(\mathcal{P}, \mathbb{P}, k, \varepsilon) > 0$ such that, for any $n \ge 1$, if $A_1, \ldots, A_k$ are upsets in $\mathcal{P}^n$ with $\mathbb{P}\big(\bigcup A_i\big) \ge 1 - \delta$, then $\mathbb{P}(A_i) \ge 1 - \varepsilon$ for some i.*

Note that x_0 is a *least* element here, not a greatest one.

Proof. Set $\delta_1 = \varepsilon$. For $j \ge 2$ let $\delta_j = F(\delta_{j-1}, \varepsilon)$, where F is the function given by Lemma 3.2 applied to the reverse of $\mathcal{P}$, and set $\delta = \delta_k/2$.

If $\mathbb{P}(A_i) < 1 - \varepsilon$ for each i, then the downsets A_i^c each have probability at least ε. Viewing these downsets as upsets in the reversed poset, it follows by Lemma 3.2 and induction on j that $\mathbb{P}\big(A_1^c \cap \cdots \cap A_j^c\big) \ge \delta_j$. Thus $\mathbb{P}\big(\bigcup_{i=1}^k A_i\big) \le 1 - \delta_k < 1 - \delta$, a contradiction. ∎

We shall also need a related result, stating that if we have a union of upsets which is extremely likely to hold, then it is very likely that *many* of the individual upsets hold, as long as we rule out the trivial case that the union is extremely likely because one of the individual upsets is itself extremely likely.

Lemma 3.4. *Let $\mathcal{P}$ be a poset with a least element x_0, and let $\mathbb{P}$ be a probability measure on $\mathcal{P}$ with $\mathbb{P}(x_0) > 0$. Given an integer $N > 0$ and a real number $\varepsilon > 0$, there exists a $\delta = \delta\big(\mathbb{P}(x_0), N, \varepsilon\big) > 0$ such that, for any n and any collection $A_1, \ldots, A_m$ of upsets in $\mathcal{P}^n$ with $\mathbb{P}(A_i) \leq 1 - \varepsilon$ for all i and $\mathbb{P}\big(\bigcup A_i\big) \geq 1 - \delta$, the probability that at least N of the events A_i hold is at least $1 - \varepsilon$.*

Proof. Let F be the function appearing in Lemma 3.2 applied to the reverse of the poset $\mathcal{P}$. Set $\varepsilon' = F(\varepsilon/N, \varepsilon)$ and $\delta_1 = \varepsilon/N$. Inductively define δ_k by $\delta_k = F(\varepsilon', \delta_{k-1})$ for $k \geq 2$.

We claim that, for any $k \geq 1$, if $A_1, \ldots, A_m$ is any collection of upsets in any power $\mathcal{P}^n$ of $\mathcal{P}$ with $\mathbb{P}\big(\bigcup A_i\big) \geq 1 - \delta_k$, then we can find disjoint index sets $I_1, I_2, \ldots, I_k$ such that for each $1 \leq j \leq k$ we have $\mathbb{P}\big(\bigcup_{i \in I_j} A_i\big) \geq 1 - \varepsilon/N$. The result then follows by setting $\delta = \delta_N$ and $k = N$: the claim tells us that with probability at least $1 - N\varepsilon/N = 1 - \varepsilon$, for every j at least one of the events $\{A_i : i \in I_j\}$ holds, so at least N of the A_i hold.

For $k = 1$ the claim is trivial, taking $I_1 = \{1, 2, \ldots, m\}$.

Suppose then that $k \geq 2$ and that the claim holds when we replace k by $k - 1$. Let f_i be the probability that none of $A_1, \ldots, A_i$ holds. For any k we have $\delta_k \leq \delta_1 = \varepsilon/N$, so $f_m \leq \delta_k \leq \varepsilon/N$, and

$$i_1 = \min\{i : f_i \leq \varepsilon/N\}$$

is defined. Setting $I_1 = \{1, 2, \ldots, i_1\}$, note that the event $\bigcup_{j \in I_1} A_i$ has probability at least $1 - \varepsilon/N$.

Consider the downsets $\bigcap_{1 \leq i \leq i_1 - 1} A_i^{\mathrm{c}}$ and $A_{i_1}^{\mathrm{c}}$. Applying Lemma 3.2 to these events, seen as upsets in the reversed poset, we have $f_{i_1} \geq F\big(f_{i_1 - 1}, 1 - \mathbb{P}(A_{i_1})\big)$. Since F is increasing, using the definition of i_1 and our assumption on $\mathbb{P}(A_i)$, it follows that $f_{i_1} \geq F(\varepsilon/N, \varepsilon) = \varepsilon'$. Let $\mathcal{D}_1 = \bigcap_{1 \leq i \leq i_1} A_i^{\mathrm{c}}$ and $\mathcal{D}_2 = \bigcap_{i_1 + 1 \leq i \leq m} A_i^{\mathrm{c}}$. Applying Lemma 3.2 to $\mathcal{D}_1$ and $\mathcal{D}_2$, we have

$$\delta_k \geq \mathbb{P}(\mathcal{D}_1 \cap \mathcal{D}_2) \geq F\big(\mathbb{P}(\mathcal{D}_1), \mathbb{P}(\mathcal{D}_2)\big) \geq F\big(f_{i_1}, \mathbb{P}(\mathcal{D}_2)\big).$$

Since F is strictly increasing, $f_{i_1} \geq \varepsilon'$, and $\delta_k = F(\varepsilon', \delta_{k-1})$, it follows that

$$F(\varepsilon', \delta_{k-1}) = \delta_k \geq F\big(\varepsilon', \mathbb{P}(\mathcal{D}_2)\big),$$

so $\mathbb{P}(\mathcal{D}_2) \leq \delta_{k-1}$, and the union of $A_{i_1+1}, \ldots, A_m$ has probability at least $1 - \delta_{k-1}$. Applying the induction hypotheses to this set of events gives us $I_2, \ldots, I_m$ with the required properties, completing the proof. ∎

4. Colourings, Hypergraphs and Crossings

Our next aim is to prove a form of rectangle-crossing lemma loosely analogous to the Russo–Seymour–Welsh Lemma [22, 26], but applicable in the hyperlattice percolation context. Naturally, this involves considering 'open crossings of rectangles'. As in [3, 5], for example (see also [6]), to make this precise and clean we shall work instead with 'black crossings' in a suitable black/white colouring of the faces of a cubic map (i.e., 3-regular plane graph). We assume throughout that our maps are *well-behaved*, meaning that the edges are drawn as piecewise-linear curves, every face is bounded, and any bounded subset of the plane contains only finitely many vertices and meets only finitely many edges.

Recall that a plane hyperlattice $\mathcal{H}$ may be thought of as a 3-coloured cubic map, where the faces are properly coloured black, white and grey, with a lattice $\mathcal{L}$ of translational symmetries; as usual we view $\mathcal{L}$ as a subset of $\mathbb{R}^2$. From now on this is our default viewpoint when considering any plane hyperlattice. A *colouring* C of $\mathcal{H}$ is a 2-coloured cubic map obtained as follows: first subdivide each grey face of $\mathcal{H}$ into one or more *subfaces*, in such a way that the resulting map is still cubic. Then colour each subface black or white, as in Figure 4.1, for example. The resulting colouring C is an

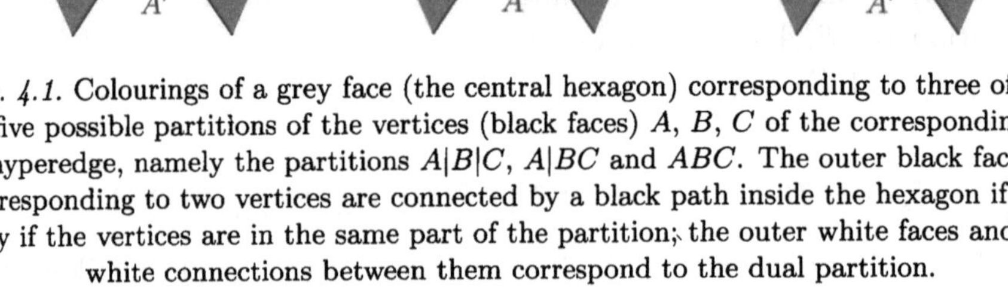

Fig. 4.1. Colourings of a grey face (the central hexagon) corresponding to three of the five possible partitions of the vertices (black faces) A, B, C of the corresponding hyperedge, namely the partitions $A|B|C$, $A|BC$ and ABC. The outer black faces corresponding to two vertices are connected by a black path inside the hexagon if and only if the vertices are in the same part of the partition; the outer white faces and the white connections between them correspond to the dual partition.

(improper, of course) black/white colouring of the faces of a (well-behaved) cubic map. It will be convenient to declare that points in the boundary of a face have the colour of that face, so some points are both black *and* white. A *black path* in C is then simply a (piecewise-linear) path in the plane every point of which is black; such a path corresponds to a sequence of black faces in which consecutive faces share a point and thus (since the map is cubic) an edge. White paths are defined similarly.

Recall that a grey face F_e of $\mathcal{H}$ corresponds to a hyperedge e. Also, since $\mathcal{H}$ is properly 3-coloured, F_e is surrounded alternately by black and white faces, corresponding to vertices and dual vertices. In $\mathcal{C}$, certain pairs of vertices incident with e are connected by black paths within F_e; this generates a partition π of the vertices of e, which is easily seen to be non-crossing. Thus $\mathcal{C}$ corresponds to a configuration ω on $\mathcal{H}$, with open paths in ω corresponding to black paths in $\mathcal{C}$ and vice versa. (Of course, each open path is represented by many 'nearby' black paths). Crucially, *white* paths within $F_{e^*} = F_e$ induce the dual partition π^* of the dual vertices incident with e^*, so the *negative* of $\mathcal{C}$, obtained by interchanging black and white throughout, corresponds to the dual configuration ω^* on $\mathcal{H}^*$.

By a *black cluster* in $\mathcal{C}$ we mean a maximal connected black subset of the plane. A *white cycle* is a white path that starts and ends at the same point. Since $\mathcal{C}$ is a black/white colouring of a cubic planar map, it is easy to see that a black cluster is surrounded by a white cycle if and only if it is finite. This is a precise form of the duality property relating ω and ω^* mentioned in Section 2.

Remark 4.1. It will be convenient later to assume that all partitions corresponding to e are realized by colourings of a single subdivision of F_e into subfaces, as in Figure 4.1. This can be achieved for all non-crossing partitions of hyperedges with any number of vertices, as illustrated in Figure 4.2.

Note that there are many possible colourings $\mathcal{C}$ corresponding to a given configuration ω on $\mathcal{H}$: even if we fix the division of each grey face into subfaces (as we shall), there may be many colourings of the subfaces giving the same partition of the vertices. For much of the rest of the paper, we shall take the random colouring $\mathcal{C}$ as the fundamental object of study, rather than the random configuration ω.

More formally, an *independent lattice colouring* $\mathcal{C}$, or simply a *colouring*, is a colouring obtained from a plane hyperlattice $(\mathcal{H}, \mathcal{L})$ by subdividing each grey face in a deterministic manner, and then colouring the resulting subfaces randomly black and white, in such a way that the colourings inside different grey faces are independent, and translations through elements of the lattice $\mathcal{L}$ preserve the distribution of $\mathcal{C}$. From the remarks above, any plane lattice percolation model $\mathcal{H}(\mathbf{p})$ can be realized by an independent lattice colouring $\mathcal{C}$ associated to $\mathcal{H}$.

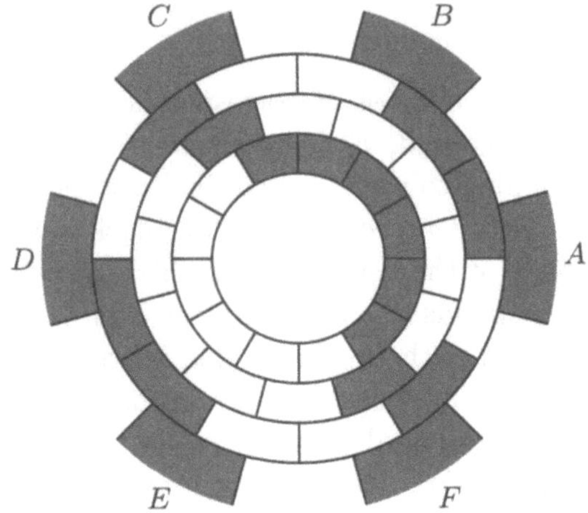

Fig. 4.2. A subdivision of the $2n$-gon corresponding to a hyperedge e with $n = 6$ vertices, and a colouring corresponding to the partition $AB|CF|DE$. In general, it suffices to take $2k - 1$ rings of subfaces of the type shown to realize any non-crossing partition, where $k \leq n/2$ is the maximum 'nesting depth' of a non-crossing partition of n objects.

We assume throughout that our colourings $\mathcal{C}$ are *non-degenerate,* meaning that within any grey face, the all-black and all-white colourings have positive probability. Later on we shall have to impose some additional conditions for faces corresponding to hyperedges with more than three vertices.

Note that the state space Ω underlying the random colouring $\mathcal{C}$ may be viewed as a product of one poset $\mathcal{P}_F$ for each grey face F of $\mathcal{H}$: in the partial order, we have $c_1 \preccurlyeq c_2$ if every subface that is black in c_1 is black in c_2. Picking a finite set $F_1, \ldots, F_n$ of faces representing the orbits of $E(\mathcal{H})$ (the set of grey faces) under the action of $\mathcal{L}$, from lattice invariance we may regard Ω as a power of the poset $\mathcal{P} = \mathcal{P}_{F_1} \times \cdots \times \mathcal{P}_{F_n}$. From independence, the probability measure associated to $\mathcal{C}$ is then a product probability measure on Ω. The non-degeneracy condition implies that $\mathcal{P}$ has a greatest element (all subfaces of each F_i black) and a least element (all white), and that each has positive probability.

The event that a given path is black, or that a black path exists with certain properties, is an upset in Ω, in the sense of Section 3. Thus Lemma 3.2 applies to two such events. Similarly, considering the reverse poset, Lemma 3.2 applies to two events each defined by the existence of a white path with certain properties. This is the reason for the non-degeneracy assumption.

Note that we are always considering two coloured maps: $\mathcal{H}$ (which is deterministic) and $\mathcal{C}$. To avoid ambiguity, we say that a point or face is *$\mathcal{H}$-black* if it is black in $\mathcal{H}$, and *$\mathcal{C}$-black* if it is black in $\mathcal{C}$, and similarly for other colours. By default, *black* or *white* refers to $\mathcal{C}$, while *grey* necessarily refers to $\mathcal{H}$.

The lattice structure ensures that the faces of the hyperlattice $\mathcal{H}$ (seen as a map, as usual) cannot be too wild.

Lemma 4.2. *Let $(\mathcal{H}, \mathcal{L})$ be a plane hyperlattice, viewed as a cubic map. There are finitely many faces $F_1, \ldots, F_N$ of $\mathcal{H}$ such that for any face F, there is an element of $\mathcal{L}$ such that the corresponding translation of $\mathbb{R}^2$ induces an isomorphism of $\mathcal{H}$ mapping F to one of $F_1, \ldots, F_N$. Furthermore, there is a constant d_0 such that* (i) *every face has diameter at most d_0 and* (ii) *every point of $\mathbb{R}^2$ is within distance d_0 of an element of $\mathcal{L}$, and for each $r > 0$ there is a constant N_r such that any disk of radius r meets at most N_r faces.*

Proof. Let D be a fundamental domain of $\mathcal{L}$. Since D is bounded, by the definition of a plane hypergraph D contains finitely many vertices of $\mathcal{H}$, and meets finitely many hyperedges. It follows that D meets only finitely many faces of $\mathcal{H}$, viewed now as a cubic map. Hence there is a finite set $F_1, \ldots, F_N$ of faces, all meeting D, containing one representative of each orbit of the action of $\mathcal{L}$ on the faces of $\mathcal{H}$. The remaining statements follow easily, taking d_0 to be the larger of $\max_i \operatorname{diam}(F_i)$ and $\operatorname{diam}(D)$. ∎

The parameter $d_0 = d_0(\mathcal{H})$ appearing in Lemma 4.2 will be used throughout this and the next section. For example, we say that a rectangle is *large* if all its sides have length at least $100d_0$. In what follows, to avoid trivialities such as a rectangle having a black crossing with probability 1, we only ever consider large rectangles.

Remark 4.3. Let us remark briefly on the numerical constants appearing in this paper. In many places, rather than argue that some constants exist with certain properties, we simply give numerical values that work, such as 100 (here) or the less natural constants 0.1, 1.1, 8, etc appearing later. Of course the precise values are not important.

Given an angle θ, by a *θ-aligned rectangle* we mean a rectangle $R \subset \mathbb{R}^2$ such that one pair of sides makes an angle θ to the x-axis, measured in the positive sense from the x-axis. We refer to these sides as *horizontal* and the other sides as *vertical.* Thus, after rotating R clockwise through an angle θ, the horizontal sides become horizontal in the usual sense. Whenever we speak of a rectangle R, we have an angle θ in mind and assume that R is θ-aligned. Note that the same geometric rectangle is θ-aligned for two values of θ differing by $\pi/2$.

By the *width* and *height* of a rectangle, we mean the length of the horizontal and vertical sides, respectively; which is which depends on whether we view R as θ-aligned or $(\theta + \pi/2)$-aligned.

We always assume that our rectangles R are in *general position* with respect to our colouring $\mathcal{C}$, meaning that each vertex of R lies in the interior of a face of the colouring, no vertices of the colouring are on the boundary of R, and the edges of the colouring can only cross the edges of R transversely.

Given a rectangle R, by a *black horizontal crossing* of R we mean a $\mathcal{C}$-black path within R starting at some point on one vertical side of R and ending at some point on the other vertical side. White vertical crossings are defined similarly, and so on. We write $H_{\mathrm{b}}(R)$ for the event that R has a black horizontal crossing (in the random colouring $\mathcal{C}$), and $V_{\mathrm{w}}(R)$ for the event that it has a white vertical crossing, and so on. For a proof of the following 'obvious' lemma concerning (well-behaved) 2-coloured maps see [6, Ch. 8, Lemma 12].

Lemma 4.4. *Given any well-behaved colouring $\mathcal{C}$ of the plane and any rectangle R in general position with respect to $\mathcal{C}$, exactly one of the events $H_{\mathrm{b}}(R)$ and $V_{\mathrm{w}}(R)$ holds.* ■

This lemma, together with self-duality, will be the starting point for our Russo–Seymour–Welsh-type argument. This argument will be rather involved. There are various technical complications arising from the generality of plane hyperlattices; we deal with most of these in the rest of this section. In the next section we turn to the core of the argument, where the complications are mostly due to the lack of symmetry.

For the rest of the section we consider a given non-degenerate independent lattice colouring $\mathcal{C}$, associated to a plane hyperlattice $\mathcal{H}$.

4.1. How crossing probabilities vary

Let $h(R) = \mathbb{P}\big(H_{\mathrm{b}}(R)\big)$ be the probability that R has a black horizontal crossing, and let $v(R) = \mathbb{P}\big(V_{\mathrm{b}}(R)\big)$. Note that if we switch from viewing a given geometric rectangle R as θ-aligned to viewing it as $(\theta + \pi/2)$-aligned, then $h(R)$ and $v(R)$ swap. Our next aim is to show that $h(R)$ and $v(R)$ do not change too much if we move the edges of R slightly. This is not very surprising, but giving full details in the present generality requires a little work. We start with a technical lemma.

We say that a path P in the plane is *potentially black* with respect to a hyperlattice $\mathcal{H}$ if no point of P is $\mathcal{H}$-white, so P corresponds to a sequence of black and grey faces of $\mathcal{H}$. In other words, P is potentially black if and only if there is a positive probability that P is actually black in the random colouring $\mathcal{C}$. We write d_{H} for the Hausdorff distance between subsets of $\mathbb{R}^2$. Let $d_0 = d_0(\mathcal{H})$ be the constant given by Lemma 4.2, so every face of $\mathcal{H}$ has diameter at most d_0.

Lemma 4.5. *Let $\mathcal{H}$ be a hyperlattice and P a piecewise-linear path. Then there is a potentially black path P' with $d_{\mathrm{H}}(P, P') \leq 2d_0$.*

Proof. Recall that in the 3-coloured map $\mathcal{H}$, no two white faces are adjacent. Whenever P passes through a white face, simply take a detour around (or just outside) this face. Similarly, if P starts or ends in a white face, modify P to start/end just outside this face. ∎

Lemma 4.6. *Let $\mathcal{C}$ be a non-degenerate independent lattice colouring. Then there exists a constant $c > 0$, depending only on (the distribution of) $\mathcal{C}$, such that, for any large rectangle R, if R_+ is a rectangle formed by moving one vertical side of R outwards by a distance of at most 1, then $h(R) \geq h(R_+) \geq c\,h(R)$.*

Proof. Note first that any black horizontal crossing of R_+ contains a black horizontal crossing of R, so $h(R) \geq h(R_+)$. Also, we may assume without loss of generality that we move a side of R outwards by distance *exactly* 1 to obtain R_+.

In the arguments that follow, various constants appear that depend on $\mathcal{C}$. However, they will depend only on (a) the quantity $d_0 = d_0(\mathcal{H})$, where $\mathcal{H}$ is the hyperlattice underlying $\mathcal{C}$, and (b) the minimum probability of the all-black state in a grey face. These are invariant under rotation and translation, so, rotating and translating R *and* $\mathcal{C}$, without loss of generality we may assume that $\theta = 0$, so horizontal means horizontal in the usual sense, and that $R = [0, a] \times [0, b]$ and $R_+ = [0, a+1] \times [0, b]$, even though $h(R)$ itself varies as R is rotated and/or translated with $\mathcal{C}$ fixed.

We may assume that $a, b \geq 100d_0$. Applying Lemma 4.5 to the line segment L from $(a - 3d_0, -3d_0)$ to $(a - 3d_0, b + 3d_0)$, and truncating the resulting path P' when it last hits the bottom side of R and first hits the

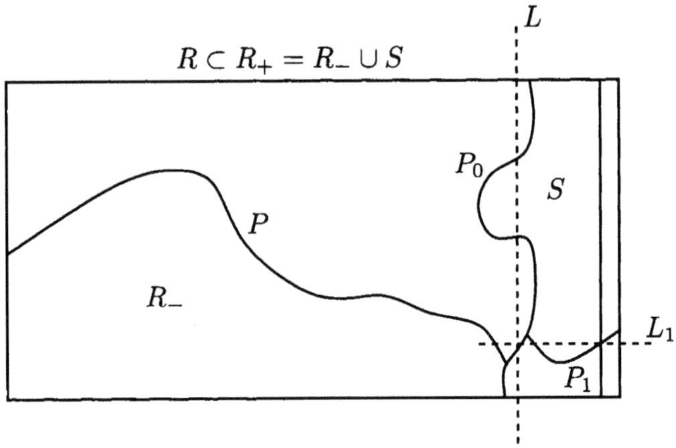

Fig. 4.3. A rectangle R_+ slightly extending a rectangle R, divided into R_- and S by a potentially black path P_0 that crosses R from top to bottom near its right-hand side. The path P starts on the left-hand side of R and ends on P_0. Finally, P_1 starts on P_0 and ends on the right of R_+. All paths lie inside R_+.

top side, we find a potentially black path P_0 crossing R from top to bottom, where all points have x-coordinate between $a - 5d_0$ and $a - d_0$. Let S be the 'strip' consisting of those points of R_+ to the right of P_0, and let R_- denote the rest of R_+, so $R_- \subset R$ is a 'distorted rectangle' whose right-hand side is potentially black. Let E_0 be the event that R_- has a black horizontal crossing, noting that if $H_b(R)$ holds then so does E_0, so $\mathbb{P}(E_0) \geq h(R)$.

Let $\mathcal{F}_-$ be the set of grey faces meeting R_- but not S, and let $\mathcal{F}_S$ be the set of grey faces meeting S, so E_0 depends on the colourings of the faces in $\mathcal{F}_- \cup \mathcal{F}_S$. Let E_0' be the event that E_0 would hold after recolouring (in $\mathcal{C}$) all faces in $\mathcal{F}_S$ to black; thus E_0' is the event that there is a path P crossing R_- from left to right, every point of which is either $\mathcal{C}$-black or in a face in $\mathcal{F}_S$. If E_0 holds then so does E_0', so $\mathbb{P}(E_0') \geq \mathbb{P}(E_0) \geq h(R)$.

Now E_0' depends only on the states of the faces in $\mathcal{F}_-$. Let us condition on these states, assuming that E_0' holds. Our aim is to show that the conditional probability that $H_b(R_+)$ holds is not too small. To do this we shall first modify P in a certain way (if needed), and then extend P, obtaining a path P_+ crossing R_+ from left to right in which every point is either $\mathcal{C}$-black or in a face in $\mathcal{F}_S$, in such a way that the set of faces in $\mathcal{F}_S$ that P_+ meets has size $O(1)$. Then we reveal the (as yet unexamined) states of these faces. Since there are $O(1)$ of them, with probability bounded away from zero they are all entirely black in $\mathcal{C}$.

We start with the modification of P. Let us call a face *bad* if it meets both R_- and S, but does not lie entirely in $R_+ = R_- \cup S$, and *good* otherwise. Since all faces have diameter at most d_0, any bad face must contain a point within distance $10d_0$ of either $(a, 0)$ or (a, b). In particular, there are $O(1)$ bad faces. Suppose P meets one or more good faces in $\mathcal{F}_S$. Tracing P from the left, stop the first time it meets such a face F. Since F is connected and meets S, we can continue within this face to a point of S. Since F is contained in R_+, in doing so we do not go outside R_+, but we do leave R_-, so we cross the right-hand side of R_-. Stop when this happens.

After this modification, P has the properties above (all points $\mathcal{C}$-black or in faces in $\mathcal{F}_S$), and it meets at most one good face in $\mathcal{F}_S$, and thus (since there are $O(1)$ bad faces in total) $O(1)$ faces in $\mathcal{F}_S$.

Let (x, y) be the right-hand end of P, so (x, y) lies on P_0. Pick $y' \in [3d_0, b-3d_0]$ with $|y-y'| \leq 3d_0$. Let L_1 be the line-segment from $(a-7d_0, y')$ to $(a+1+2d_0, y')$. Apply Lemma 4.5 to L_1 to obtain a potentially black path P_1'. Then P_1' starts inside R_-, ends outside R_+, and cannot cross the lines $y = 0$ and $y = b$, so it contains a sub-path P_1 within S crossing S from left to right. The left-hand end of P_1 is within distance $10d_0$ of (x, y). To construct our final path P_+, trace P from left to right, run along P_0 from (x, y) to the first end of P_1, and then trace P_1. This path crosses R_+ from left to right. Furthermore, any point of $P_+ \setminus P$ is within distance $20d_0$ of (x, y), so $P_+ \setminus P$ meets $O(1)$ grey faces. Now P_+ is potentially black. Any point of P_+ in a grey face in $\mathcal{F}_-$ is necessarily a point of P, and so is $\mathcal{C}$-black by the properties of P. Finally, P_+ meets $O(1)$ grey faces in $\mathcal{F}_S$. With (conditional) probability bounded away from 0 the latter faces are all entirely $\mathcal{C}$-black, and then P_+ is a black path, so $H_{\mathrm{b}}(R_+)$ holds. ■

When $h(R)$ is very close to 1, Lemma 4.6 is not very informative; it does not rule out $h(R)$ dropping from 1 to 1/100, say, as R is extended a tiny bit horizontally. In this case the probability of a white vertical crossing would increase from 0 to 99/100. Note that this vertical crossing probability *can* change by a large ratio: if the white colouring is subcritical, with a constant and b large, the white vertical crossing probability is approximately of the form $e^{-c_a b}$, where c_a is a positive constant depending on a. Increasing a by 1 decreases c_a to a new constant value, which can change $e^{-c_a b}$ by an arbitrarily large ratio.

However, it is still true that the white vertical crossing probability cannot jump from very small to fairly large. We phrase the result in terms of black horizontal crossings as above.

Lemma 4.7. *Let $\mathcal{C}$ be a non-degenerate independent lattice colouring. Given $\varepsilon > 0$ there is a $\delta > 0$ such that for any large rectangle R and any rectangle R_+ formed by moving one vertical side of R outwards by a distance of at most 1, if $h(R) \geq 1 - \delta$ then $h(R_+) \geq 1 - \varepsilon$.*

Proof. The proof is an extension of that of Lemma 4.6; we define the path P_0 splitting R_+ into R_- and S as before, and let $\mathcal{F}_-$ be the set of grey faces meeting R_- but not S, and $\mathcal{F}_S$ the set of grey faces meeting S. As before, let E_0' be the event that R_- has a horizontal crossing every point of which is $\mathcal{C}$-black or in a grey face in $\mathcal{F}_S$, recalling that $\mathbb{P}(E_0') \geq h(R)$.

Let $f_1, \ldots, f_N$ list all grey or $\mathcal{H}$-black faces of $\mathcal{H}$ meeting S. Let E_i be the event that R_- contains a $\mathcal{C}$-black path P starting on the left-hand side of R_-, ending at a boundary point of f_i, and meeting no other f_j, nor the interior of f_i. Note that E_i depends only on the states of grey faces in $\mathcal{F}_-$, not those in $\mathcal{F}_S$. If E_0' holds then, truncating a path P witnessing this event the first time P meets any f_i, we see that one of the E_i must hold. Hence,

$$\mathbb{P}\Big(\bigcup E_i\Big) \geq \mathbb{P}(E_0') \geq h(R). \tag{4}$$

We claim that any given face f of $\mathcal{H}$ is surrounded by a 'ring' of white and grey faces of $\mathcal{H}$ with bounded size such that if all grey faces in the ring happen to be coloured white in $\mathcal{C}$, then no $\mathcal{C}$-black path starting outside the ring can end at a point of f. Here the bound depends only on $\mathcal{H}$, not the face chosen. Indeed, if f is $\mathcal{H}$-black, we simply take the faces neighbouring f to form the ring. If f is white or grey, we may simply take all white or grey faces of $\mathcal{H}$ within distance $2d_0$ of f as our ring: since no black face of $\mathcal{H}$ touches any other black face, any $\mathcal{C}$-black path to f from distance more than $2d_0$ must meet a grey face in our ring.

If every grey face in the ring just described about f_i happens to be coloured entirely white in $\mathcal{C}$, then no black path ends at any point of f_i, and E_i does not hold. It follows that for some constant $\varepsilon_1 > 0$ we have $\mathbb{P}(E_i) \leq 1 - \varepsilon_1$ for all i.

Let N_0 be the maximum number of faces meeting any disk of radius $100d_0$, and let N_1 be the number of *bad* grey or $\mathcal{H}$-black faces, i.e., grey or black faces meeting S, R_-, and the exterior of R_+. Note that N_0 and N_1 are bounded by constants as before. Let M be a huge constant to be chosen in a moment.

By Lemma 3.4 (applied with $\min\{\varepsilon/2, \varepsilon_1\}$ in place of ε), our assumption $h(R) \geq 1 - \delta$, and (4), if we choose δ small enough, then with probability at least $1-\varepsilon/2$ at least $K = N_1 + MN_0$ of the events $E_1, \ldots, E_N$ hold. Let us condition on the states of all faces in $\mathcal{F}_-$, assuming that at least K of the E_i hold. It suffices to show that the conditional probability that $H_b(R_+)$ holds is then at least $1-\varepsilon/2$. As before, we use the fact that we have not yet looked at the faces in $\mathcal{F}_S$.

Since there are at most N_1 bad faces among the f_i, there is a set I_0 of size at least MN_0 such that for every $i \in I_0$ the event E_i holds and f_i is good. Using the greedy algorithm, we may pick a subset $I \subset I_0$ of size at least M such that for distinct $i, j \in I_0$ the faces f_i and f_j are at distance at least $30d_0$. For $i \in I_0$, let P_i be a path witnessing E_i. Our aim is to complete the proof as before, but now showing that each P_i has a not-too-small chance of being extendable to cross R_+, and that these events (that the particular extensions we look for are present) are independent, so with high probability at least one holds. The details are essentially as before: since f_i is good, we may extend P_i within the face f_i to meet our right-hand side P_0. Then we find an extension P_i^+ of P_i as before, remaining within distance $10d_0$ of the end of P_i. The extensions meet disjoint sets of faces, so we are done. ■

Together, Lemmas 4.6, 4.7 and 4.4 show that no crossing probability changes 'too much' when a rectangle is moved slightly. This statement needs a little interpretation: we could in principle obtain explicit bounds in Lemmas 4.6 and 4.7. However, these turn out to be irrelevant. In the end, all we care about is whether certain probabilities tend to 0 or tend to 1 as some parameter (the area of the rectangles we consider) tends to infinity. It will thus be convenient to 're-scale' all probabilities by an increasing function $\varphi : (0,1) \to \mathbb{R}$ with $\varphi(x) \to -\infty$ as $x \to 0$ and $\varphi(x) \to \infty$ as $x \to 1$, in such a way that the maximum change in a probability p 'allowed' by our lemmas corresponds to a change in $\varphi(p)$ of at most 2, say.

To make this precise, let c_b be the constant given by Lemma 4.6; recall that this does not depend on the orientation of R. Let c_w be the corresponding constant with black and white exchanged, and let $c_0 = \min\{c_b, c_w\}$. Similarly, given $\varepsilon > 0$, let $\delta_0(\varepsilon) = \min\{\delta_b, \delta_w\}$, where $\delta_b = \delta_b(\varepsilon)$ is given by Lemma 4.7, and δ_w by Lemma 4.7 with black and white exchanged.

Define a sequence $(\varepsilon_n)_{n\geq 0}$ inductively by setting $\varepsilon_0 = 1/2$ and $\varepsilon_{n+1} = \min\left\{c_0\varepsilon_n, \delta_0(\varepsilon_n)\right\}$. Set $\pi_n = 1-\varepsilon_n$ for $n \geq 0$ and $\pi_n = \varepsilon_{-n}$ for $n \leq 0$. Consider the *scaling function* $\varphi : (0,1) \to \mathbb{R}$ defined as follows: set $\varphi(\pi_n) = n$ for all $n \in \mathbb{Z}$, and interpolate linearly between these points. Note that

$\varphi(1-p) = -\varphi(p)$. This function (or rather its inverse) is illustrated in Figure 4.4.

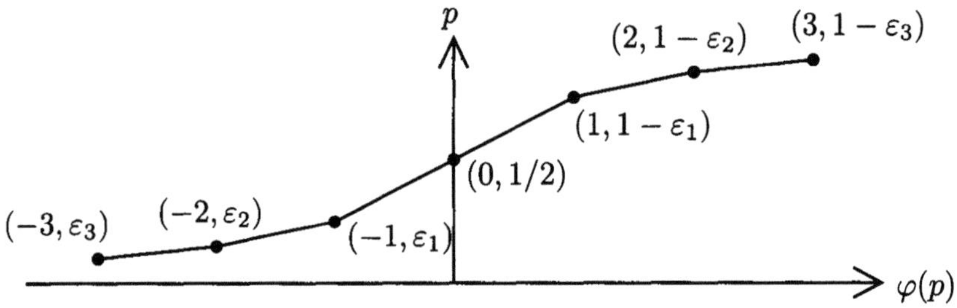

Fig. 4.4. The inverse of the scaling function $p \mapsto \varphi(p)$.

Recall that we call a rectangle *large* if both sides have length at least $100d_0$.

Lemma 4.8. *Let R_1 and R_2 be two large rectangles such that one of R_1 and R_2 is obtained from the other by moving one side outwards by a distance between 0 and 1. Let $f(R)$ be any of the four functions $\mathbb{P}(H_b(R))$, $\mathbb{P}(H_w(R))$, $\mathbb{P}(V_b(R))$ and $\mathbb{P}(V_w(R))$. Then $|\varphi(f(R_1)) - \varphi(f(R_2))| \le 2$.*

Proof. Without loss of generality, R_2 extends R_1 horizontally. Using Lemma 4.4 and the fact that $|\varphi(1-p_1)-\varphi(1-p_2)| = |-\varphi(p_1)+\varphi(p_2)| = |\varphi(p_1) - \varphi(p_2)|$, we may assume that we are considering horizontal crossings. Exchanging colours, we may assume that $f(R) = \mathbb{P}(H_b(R)) = h(R)$. Let $p_j = f(R_j) = h(R_j)$, noting that $p_1 \ge p_2$. Let $i = \lfloor \varphi(p_1) \rfloor$, so $\pi_i \le p_1 < \pi_{i+1}$. If $i \le 0$ then by definition of π_i we have $\pi_{i-1} = \varepsilon_{-i+1} \le c_0\varepsilon_{-i} = c_0\pi_i$. By Lemma 4.6 we have $p_2 \ge c_0 p_1$, so $p_2 \ge \pi_{i-1}$ and $\varphi(p_2) \ge i-1$.

On the other hand, if $i \ge 1$ then $\pi_i = 1 - \varepsilon_i \ge 1 - \delta_0(\varepsilon_{i-1})$. Since $p_1 \ge \pi_i$, Lemma 4.7 gives $p_2 \ge 1 - \varepsilon_{i-1} = \pi_{i-1}$. In either case we have $\pi_{i-1} \le p_2 \le p_1 < \pi_{i+1}$, so $i - 1 \le \varphi(p_2) \le \varphi(p_1) < i+1$ and the result follows. ∎

Lemma 4.8 shows that if we measure probabilities in the right way, they don't change too much when we move a rectangle slightly. Since we have a lattice of translational symmetries, this has the following consequence. (Recall that a rectangle is 'large' if its sides have length at least $100d_0$, where $d_0 = d_0(\mathcal{H})$ is the constant from Lemma 4.2.)

Corollary 4.9. *There is a constant C such that if R' is a translate of a large rectangle R, then $|\varphi(h(R)) - \varphi(h(R'))| \le C$.*

Proof. Pick a fundamental domain D of the lattice $\mathcal{L}$ of symmetries. Since our colouring $\mathcal{C}$ is invariant under translations corresponding to elements of $\mathcal{L}$, we may assume that R and R' are related by translation by a vector in D. Since D is bounded, the result follows by applying Lemma 4.8 a bounded number of times. ■

Define $\psi : (0,1) \to (0,1)$ by $\psi(p) = \varphi^{-1}(\varphi(p) - C)$, where C is the constant given by Corollary 4.9. Then ψ is increasing. In fact, although we shall not use this, ψ is strictly increasing, and $\psi(p)$ tends to 0 as $p \to 0$ and to 1 as $p \to 1$. We may rewrite Corollary 4.9 is the following more convenient form.

Corollary 4.10. *Suppose that R and R' are large rectangles with the same orientation, width and height. Then $h(R') \geq \psi\big(h(R)\big)$ and $v(R') \geq \psi\big(v(R)\big)$.* ■

We may also rotate a rectangle slightly without changing the crossing probabilities much.

Corollary 4.11. *Let R be any rectangle of width a and height b, with $a, b \geq 100d_0$ (so R is large), and let R' be obtained from R by rotating it about its centre through an angle $\theta \leq 1/\big(10\max\{a,b\}\big)$. If $f(\cdot)$ denotes any of the four crossing probability functions considered in Lemma 4.8, then $\big|\varphi\big(f(R)\big) - \varphi\big(f(R')\big)\big| \leq 8$.*

Proof. Without loss of generality we may assume that f is the function $f(\cdot) = h(\cdot)$ giving the probability of a black horizontal crossing. Rotating and translating the rectangles and $\mathcal{C}$ together as before (or simply changing coordinates) we may assume that $R = [-a,a] \times [-b,b]$. Let $R'' = [-a-1, a+1] \times [-b+1, b-1]$. Then any horizontal crossing of R'' crosses R' horizontally, so $h(R') \geq h(R'')$. Hence, by Lemma 4.8, $\varphi\big(h(R')\big) \geq \varphi\big(h(R'')\big) \geq \varphi\big(h(R)\big) - 8$.

A similar argument interchanging R and R' gives $\varphi\big(h(R)\big) \geq \varphi\big(h(R')\big) - 8$, so the result follows. ■

A key consequence of the lemma above is that for any given orientation, we may find a large rectangle R with $h(R)$ not too close to 0 or 1, and that when we rotate, we can assume that the dimensions of R vary 'smoothly'. For now we formalize only the first of these statements. Given an angle θ, we write $h_\theta(m,n)$ and $v_\theta(m,n)$ for $\mathbb{P}\big(H_{\mathrm{b}}(R)\big)$ and $\mathbb{P}\big(V_{\mathrm{b}}(R)\big)$, where R is an m-by-n θ-aligned rectangle centred on the origin. (Thus $v_\theta(m,n) = h_{\theta+\pi/2}(n,m)$.)

Lemma 4.12. *Let $\mathcal{C}$ be a non-degenerate independent lattice colouring associated to a hyperlattice $\mathcal{H}$. Given $L \geq 100d_0$, there is a constant A_0 that for any $A \geq A_0$ and any angle θ, there are $m, n \geq L$ with $mn = A$ such that $-4 \leq \varphi\big(h_\theta(m,n)\big) \leq 4$.*

Proof. By Lemma 4.2 there is a constant $N = N(L)$ such that any disk of radius $2L$ meets at most N faces of $\mathcal{H}$. By Lemma 4.5, if R is a rectangle of any orientation with width $m \geq L$ and height $n = L$, then we can find $\lfloor m/(10d_0) \rfloor \geq m/(20d_0)$ potentially white paths P_i crossing R from top to bottom, with these paths separated by distances of at least d_0. Since the events that these paths are white are independent, and each path meets at most N grey faces, the probability that no P_i is white is at most $\exp(-\alpha m)$ for some constant $\alpha > 0$ that does not depend on θ. By Lemma 4.4, if any P_i is white, then $H_{\mathrm{b}}(R)$ does not hold. Taking A large enough (i.e., $A \geq L^2$ and $A \geq L\alpha^{-1}\big|\log\big(\varphi^{-1}(-4)\big)\big|$), we thus have $\varphi\big(h_\theta(A/L, L)\big) \leq -4$; similarly, if A is large enough then $\varphi\big(h_\theta(L, A/L)\big) \geq 4$.

Now consider a rectangle R with area A that varies smoothly between these two extremes, centred always on the origin. As m varies, Lemma 4.8 implies that $\varphi\big(h_\theta(m, A/m)\big)$ cannot jump by more than 8 at any point, and the result follows. ∎

5. A Rectangle-Crossing Lemma

In the original context of independent bond percolation on the square lattice, there are now several different proofs of the key lemma of Russo [22] and Seymour and Welsh [26]; see, for example, [4, 5, 3]. The various proofs extend (with differing degrees of additional complication) to various more general classes of percolation model. However, as far as we are aware, none of the published proofs can be made to work in the context of general self-dual hyperlattices – in addition to various technical problems, they all require symmetry assumptions that may not hold here.

In this section we shall prove an analogue of the Russo–Seymour–Welsh Lemma for self-dual hyperlattice percolation. Since this proof is a little involved, we first illustrate the key ideas by writing out the argument for bond percolation on $\mathbb{Z}^2$. This amounts to reproving the original RSW Lemma in a more complicated way than necessary. Even among proofs using our new strategy, we do not aim to present the simplest, but rather one that extends easily to hyperlattices.

5.1. Bond percolation on $\mathbb{Z}^2$

Throughout this subsection we consider independent bond percolation on $\mathbb{Z}^2$ with $p = 1/2$. Thus a *configuration* ω is an assignment of a *state* (*open* or *closed*) to each edge e of $\mathbb{Z}^2$, and $\mathbb{P}$ is the probability measure on the set $\Omega = 2^{E(\mathbb{Z}^2)}$ of configurations in which the states $\omega(e)$ of different bonds are independent and each bond is open with probability 1/2.

All rectangles R we consider will be aligned with the axes and have corners with integer coordinates; a rectangle includes its boundary. An *open horizontal crossing* of R is a path of open bonds in R joining a vertex on the left to one on the right; we write $H(R)$ for the event that R has such a crossing. Similarly, $V(R)$ is the event that R has an open vertical crossing, defined analogously.

As inputs to the argument we shall present, we need two simple lemmas, which *do* make use of the symmetries of $\mathbb{Z}^2$; the main argument will then use only translational symmetries. The first lemma is a standard fact which is an easy consequence of the self-duality of $\mathbb{Z}^2$. It is well known to hold with $c_1 = 1/2$; see [4], for example. We write c_1 rather than 1/2 since the main argument below does not depend on the particular value of c_1, and in the case of hyperlattices, the value of c_1 in the analogous statement will be different.

Lemma 5.1. *There is a constant $c_1 > 0$ such that if S is any square in $\mathbb{Z}^2$ then $\mathbb{P}\big(H(S)\big) = \mathbb{P}\big(V(S)\big) \geq c_1$.* ■

Our second 'input lemma' is the following consequence of Lemma 5.1, whose proof also requires the use of symmetry. Here we can take $c_2 = c_1^2/16$, but again the value of c_2 is irrelevant later.

Lemma 5.2. *Given an n-by-n square S, let $E = E(S)$ be the event that there is an open vertical crossing P of S such that the x-coordinates of the endpoints of P differ by at most $3n/5$. There is a constant $c_2 > 0$ such that $\mathbb{P}\big(E(S)\big) \geq c_2$ for all squares S.*

Proof. Let $c_1 > 0$ be as in Lemma 5.1 and consider $S = [0, n]^2$. We may suppose that $\mathbb{P}(E) < c_1/2$. Let F_1 be the event that S has an open vertical crossing from some point $(x, 0)$ to some point (x', n) with $x' - x > 3n/5$, and F_2 the horizontal mirror image of this event. Now $V(S) = E \cup F_1 \cup F_2$, so we must have $\mathbb{P}(F_i) \geq c_1/4$ for some i. Since $\mathbb{P}(F_1) = \mathbb{P}(F_2)$ by symmetry, it follows that $\mathbb{P}(F_1) = \mathbb{P}(F_2) \geq c_1/4$. But then, by Harris's Lemma, $\mathbb{P}(F_1 \cap F_2) \geq c_1^2/16$. Suppose that F_1 and F_2 hold, and let P_1 and P_2

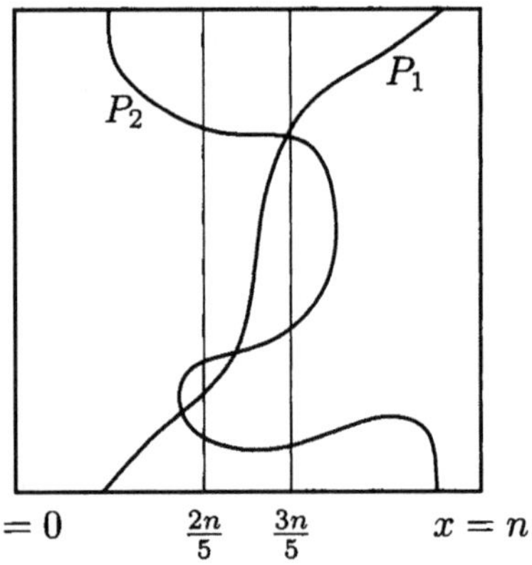

Fig. 5.1. Two open paths crossing a square S vertically, P_1 from bottom-left to top-right, and P_2 from bottom-right to top-left. Their union contains a path P from bottom-left to top-left.

be open paths witnessing these events, with P_i joining $(x_i, 0)$ to (x_i', n); see Figure 5.1. Since $x_1' - x_1 \geq 3n/5$, we have $x_1 \leq 2n/5$ and $x_1' \geq 3n/5$, and similarly $x_2 \geq 3n/5$ and $x_2' \leq 2n/5$. It follows that P_1 and P_2 cross. Hence there is an open path P within S joining $(x_1, 0)$ to (x_2', n). Since $0 \leq x_1, x_2' \leq 2n/5$, this shows that E holds. In conclusion, if $\mathbb{P}(E) < c_1/2$, then $\mathbb{P}(E) \geq c_1^2/16$, so $\mathbb{P}(E) \geq c_1^2/16 > 0$. ∎

Let us write $h(m, n) = \mathbb{P}\big(H(R)\big)$ for the probability that a rectangle R of width m and height n has an open horizontal crossing, and $v(m, n) = \mathbb{P}\big(V(R)\big)$ for the probability that it has an open vertical crossing. Our aim is to prove the following form of the RSW Lemma.

Theorem 5.3. *There is a constant $c > 0$ such that $v(n, 100n) \geq c$ for all n.*

Of course, Theorem 5.3 is a well known result of Russo [22] and Seymour and Welsh [26]. As noted above, our aim in this subsection is to present a (complicated) way of deducing Theorem 5.3 from Lemmas 5.1 and 5.2 using minimal properties of the measure $\mathbb{P}$. In particular, we shall restrict ourselves to properties that extend to general hyperlattice percolation, so the argument will adapt to prove Theorem 5.12 below. For example, we shall use translational symmetry but no other symmetry. We shall make repeated use of Harris's Lemma, and of the geometric fact that open horizontal and vertical crossings of the same rectangle must meet.

We also use one more very important property of the measure $\mathbb{P}$: if a rectangle R has an open horizontal crossing, then it has an *uppermost open horizontal crossing* $UH(R)$ with the property that the event $UH(R) = P_0$ is independent of the states of all bonds below P_0. Indeed, $UH(R)$ may be found by 'exploring R from above'; see [4], for example. Similarly, if $H(R)$ holds then R has a *lowest open horizontal crossing* $LH(R)$, defined analogously, and found by exploring from below.

The proof of Theorem 5.3 that we shall present, although simpler than that of Theorem 5.12, is still somewhat lengthy. We shall start with three lemmas, the first two of which are standard observations.

In what follows, we shall often implicitly assume that n is 'large enough', meaning larger than a suitable constant n_0 depending on the parameters, e.g., ε, that we choose. To avoid clutter, we ignore the rounding of plane coordinates to integers; it is easy to see that the effect of rounding can be handled by adjusting the various constants suitably. (Recall that in this subsection we are not proving new results, merely rehearsing the arguments that we shall use in the next subsection; formally, nothing outside this subsection depends on anything inside it. So we do not feel the need to dot all i's and cross all t's.)

Lemma 5.4. *For any $\varepsilon > 0$ and $c' > 0$ there is a $c > 0$ such that for any n, if $v\big(n, (1+\varepsilon)n\big) > c'$ then $v(n, 100n) > c$.*

Proof. Given $\gamma \geq 1$, let R_1 and R_3 be rectangles of width n and heights γn and $(1+\varepsilon)n$, respectively, overlapping in a square R_2 of side n. If $V(R_1)$, $H(R_2)$ and $V(R_3)$ all hold, then so does $V(R_1 \cup R_3)$; see Figure 5.5 for an illustration of this in a slightly different context. By the original form of Harris's Lemma [14], it follows that

$$v\big(n, (\gamma+\varepsilon)n\big) \geq v(n, \gamma n)h(n, n)v\big(n, (1+\varepsilon)n\big)\,. \tag{5}$$

Since $v(n, n) \geq c_1$ and $h(n, n) \geq c_1$ by Lemma 5.1, applying (5) inductively $\lceil 99/\varepsilon \rceil$ times gives the result. ■

Given two overlapping n-by-n squares S_1 and S_2 such that S_2 is obtained from S_1 by translating it upwards through a distance of at most n, let $T = T(S_1, S_2)$ denote the infinite strip bounded by the vertical lines containing the vertical sides of S_1 and S_2. Let $J(S_1, S_2)$ be the event that S_1 and S_2 have open horizontal crossings P_1 and P_2 that are *joined* within T, meaning that there is an open path P within T joining some point of P_1 to some point of P_2; this includes the case where P_1 and P_2 meet; see Figure 5.2. Note that when $J(S_1, S_2)$ holds, a minimal P lies entirely between P_1 and P_2 in the strip, so we may assume that P is contained in $S_1 \cup S_2$. For later, note also that if there is an open horizontal crossing P_1 of S_1 above an open horizontal crossing P_2 of S_2, then P_1 is in fact contained in S_2, and the crossings P_1 of S_1 and P_1 of S_2 meet, so $J(S_1, S_2)$ holds.

Lemma 5.5. *For any $\varepsilon > 0$ and $c' > 0$ there is a $c > 0$ such that for any n, if there exist n-by-n squares S_1 and S_2 with S_2 obtained by translating S_1 upwards by a distance of εn such that $\mathbb{P}\big(J(S_1, S_2)\big) \geq c'$, then $v(n, 100n) \geq c$.*

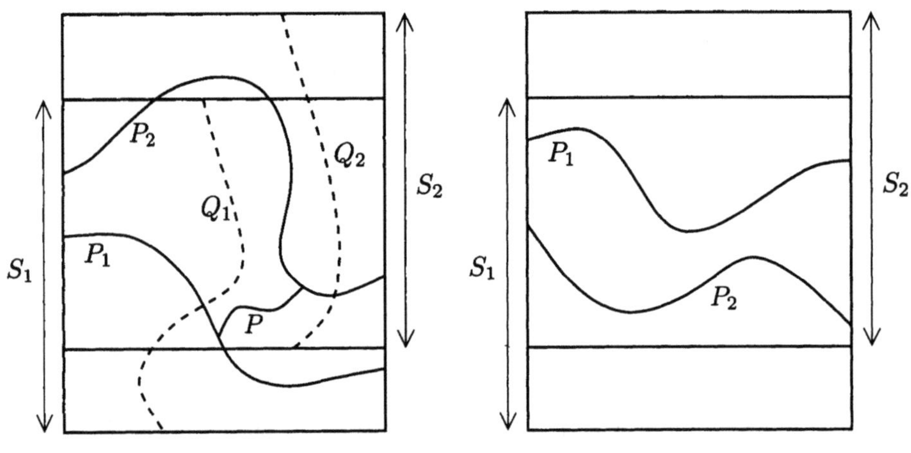

Fig. 5.2. In the left figure, the solid paths illustrate the event $J(S_1, S_2)$. The dashed paths Q_i are vertical crossings of the squares S_i. Since P_i and Q_i meet, the event $V(S_1 \cup S_2)$ holds. The figure on the right shows that if any horizontal crossing of S_1 is above any horizontal crossing of S_2, then $J(S_1, S_2)$ holds – either crossing crosses both squares.

Proof. The event $J = J(S_1, S_2)$ is increasing. Hence, by Harris's Lemma and Lemma 5.1, the event $E = J \cap V(S_1) \cap V(S_2)$ has probability at least $c_1^2 c'$.

Using the fact that horizontal and vertical crossings of the same square must meet, it is easy to see that whenever E holds, so does $V(S_1 \cup S_2)$; see Figure 5.2. Hence $v\big(n, (1+\varepsilon)n\big) = \mathbb{P}\big(V(S_1 \cup S_2)\big) \geq \mathbb{P}(E) \geq c_1^2 c'$, and the result follows by applying Lemma 5.4 with $c_1^2 c'$ in place of c'. ∎

Our next lemma is less run-of-the-mill. Given $\varepsilon > 0$ and two n-by-n squares S_1 and S_2 with S_2 obtained by translating S_1 upwards by a distance of $\varepsilon n/10$, define the strip $T = T(S_1, S_2)$ as above, and let $G_\varepsilon(S_1, S_2)$ be the event that $H(S_1)$ and $H(S_2)$ hold, the path $P_1 = LH(S_1)$ is below $P_2 = UH(S_2)$ in T, and the area of T between P_1 and P_2 is at most εn^2. In other words, the lowest open horizontal crossing of the lower square and the highest of the higher square do not meet, but they are 'close together', in the sense that the area between them is small.

Lemma 5.6. *For any $0 < \varepsilon < 1/10$ there are constants $c_3 > 0$ and $c > 0$ such that for any n, either there exist n-by-n squares S_1 and S_2 as above with $\mathbb{P}\big(G_\varepsilon(S_1, S_2)\big) \geq c_3$, or $v(n, 100n) > c$.*

Proof. Set $N = 2\lceil 2/\varepsilon \rceil$, and, for $i = 0, 1, \dots, N$, let $S_i = [0, n] \times [i\varepsilon n/10, n + i\varepsilon n/10]$, so S_{i+1} is obtained by translating S_i upwards through a distance $\varepsilon n/10$.

Let $H = H(S_0) \cap \ldots \cap H(S_N)$. Since each $H(S_i)$ has probability at least c_1, by Harris's Lemma there is some $c' > 0$ such that $\mathbb{P}(H) \geq c'$. (We may take $c' = c_1^{N+1}$, but we prefer to be less specific, with an eye to the hyperlattice case.) Set $c_3 = c'/(2N)$. If for some i the event $J(S_i, S_{i+1})$ has probability at least c_3, then we are done by Lemma 5.5. Let $J = \bigcup_{i=0}^{N-1} J(S_i, S_{i+1})$. Then we may assume that $\mathbb{P}(J) \leq Nc_3 \leq c'/2$. Hence $\mathbb{P}(H \setminus J) \geq c'/2$.

We claim that if $H \setminus J$ holds, then so does $F = \bigcup_{i=0}^{N-1} G_\varepsilon(S_i, S_{i+1})$. Assuming the claim, the result follows, since for some i we have

$$\mathbb{P}\big(G_\varepsilon(S_i, S_{i+1})\big) \geq N^{-1}\mathbb{P}(H \setminus J) \geq c'/(2N) = c_3.$$

Suppose then that $H \setminus J$ holds, and let P_i^- and P_i^+ be the lowest and highest open horizontal crossings of S_i. Note that P_i^- and P_i^+ may meet, but P_i^- lies (weakly) below P_i^+. Since J does not hold, for $i = 0, \dots, N-1$, the path P_i^+ is strictly below P_{i+1}^-; it follows that P_i^- is strictly below P_{i+1}^-. For $i = 0, \dots, N-2$, let A_i be the region in T between P_i^- and P_{i+2}^-. Then the A_i are disjoint. Since there are $\lfloor N/2 \rfloor \geq 2/\varepsilon$ such regions A_i, and their union is contained in a rectangle of width n and height $n + N\varepsilon n/10 \leq n + (6/\varepsilon)\varepsilon n/10 < 2n$, it follows that some A_i has area at most εn^2. Since the region between P_i^- and P_{i+1}^+ is contained in A_i, it follows that $G_\varepsilon(S_i, S_{i+1})$ holds, as required. ∎

Clearly, in the present context the events $G_\varepsilon(S_i, S_{i+1})$ considered above all have the same probability due to translational symmetry; with an eye to the general case, we avoided using this fact.

We now turn to the key idea, which is a rather involved way of generating a configuration. Given a configuration ω and a vector $v \in \mathbb{Z}^2$, let ω^v be obtained by translating ω through the vector v. Thus the state of a bond e in ω^v is the state of its translate $e - v$ in ω. Let us say that a random configuration ω has the *standard distribution* if it has the distribution corresponding to $\mathbb{P}$, so bonds are open independently in ω and each is open with probability $1/2$. From translation invariance, if ω is random with the standard distribution and v is constant, then ω^v has the standard distribution. This conclusion also holds if v is random, as long as ω and v are independent.

Suppose we have some algorithm $\mathbb{A}$ whose input is a configuration ω, and that $\mathbb{A}$ examines the states of bonds one by one, with the next bond to be examined depending on the results of previous examinations, but not on the states of any other bonds. We assume that $\mathbb{A}$ terminates after a finite number of steps, and write $\mathcal{S} = \mathcal{S}_{\mathbb{A}}(\omega)$ for the set of bonds examined by $\mathbb{A}$ when $\mathbb{A}$ is run on the configuration ω. Let ω_1 and ω_2 be independent standard configurations. Define a new configuration ω by running $\mathbb{A}$ on ω_1, setting $\omega(e) = \omega_1(e)$ if $e \in \mathcal{S}_{\mathbb{A}}(\omega_1)$ and $\omega(e) = \omega_2(e)$ otherwise. Then it is easy to check that ω has the standard distribution: we can think of tossing coins corresponding to ω_1 to determine the states of all bonds, looking at certain bonds (corresponding to $\mathcal{S}$), and then retossing the coins we have not yet looked at.

Given an algorithm $\mathbb{A}$ as above, and a corresponding map $\mathcal{S}_{\mathbb{A}}$ from the set Ω of all configurations to the set of subsets of $E(\mathbb{Z}^2)$, define a map $f_{\mathbb{A}}$ from $\Omega \times \Omega \times \mathbb{Z}^2$ to Ω by

$$\big(f_{\mathbb{A}}(\omega_1, \omega_2, v)\big)(e) = \begin{cases} \omega_1(e) & \text{if } e \in \mathcal{S}_{\mathbb{A}}(\omega_1), \\ \omega_2(e - v) & \text{if } e \notin \mathcal{S}_{\mathbb{A}}(\omega_1). \end{cases} \tag{6}$$

Combining the two observations above, we see that if ω_1, ω_2 and v are independent and ω_1 and ω_2 have the standard distribution, then $\omega = f_{\mathbb{A}}(\omega_1, \omega_2, v)$ does too.

Proof of Theorem 5.3. Recall that our task is to show that the probability $v(n, 100n)$ that an n-by-$100n$ rectangle has an open vertical crossing is bounded away from zero, using Lemmas 5.1 and 5.2 as 'inputs', and otherwise making no use of reflectional or rotational symmetry.

With an eye to later generalizations, suppose that α, β_1, β_2 and η are positive constants satisfying

$$\alpha, \beta_1 \leq 1/3, \quad \beta_2 \leq 2, \quad \text{and} \quad \eta > 3\beta_1. \tag{7}$$

Let $E_v = E_v(n)$ be the event that the rectangle $[0, \beta_1 n] \times [0, \beta_2 n]$ contains an open path from some point (x, y) to some point (x', y') with $y' \geq y + \alpha n$ and $|x - x'| \leq (1 - \eta)|y - y'|$. (Here 'v' stands for 'vertical': the overall orientation of the path is significantly closer to vertical than to horizontal.) Taking $\alpha = \beta_1 = \beta_2 = 1/100$ and $\eta = 1/10$, Lemma 5.2 tells us that for all (large enough) n, we have $\mathbb{P}(E_v(n)) \geq c_2 > 0$. In the rest of the proof we assume only that our various constants satisfy (7) and that, for these constants, $\mathbb{P}(E_v)$ is bounded away from 0.

Pick $\gamma > 0$ such that

$$\gamma \leq 1/3 \quad \text{and} \quad \eta \geq 3\beta_1 + 3\gamma, \tag{8}$$

and choose $\varepsilon > 0$ such that $\varepsilon < \gamma$ and

$$\varepsilon < \gamma^2 \alpha / 10. \tag{9}$$

Let c_3 and c be the constants given by Lemma 5.6. For any n, by Lemma 5.6, either $v(n, 100n) \geq c$, in which case we are done, or there are squares S_1 and S_2 with S_2 obtained by translating S_1 upwards by a distance of $\varepsilon n/10$ such that

$$\mathbb{P}\big(G_\varepsilon(S_1, S_2)\big) \geq c_3. \tag{10}$$

We may assume that the second case holds. By translational symmetry, we may assume that $S_1 = [0, n]^2$ and $S_2 = [0, n] \times \big[\varepsilon n/10, (1 + \varepsilon/10)n\big]$.

Recall that $J = J(S_1, S_2)$ is the event that there are open horizontal crossings of S_1 and S_2 that meet, or are connected by an open path lying within the strip

$$T = \big\{(x, y) : 0 \leq x \leq n\big\} \subset \mathbb{R}^2$$

generated by $S_1 \cup S_2$. Also, $G_\varepsilon = G_\varepsilon(S_1, S_2)$ is the event that $H(S_1)$ and $H(S_2)$ hold, the path $P_1 = LH(S_1)$ is below $P_2 = UH(S_2)$ in T, and the area of T between P_1 and P_2 is at most εn^2. We shall show that

$$\mathbb{P}(J \cap G_\varepsilon) \geq c' \tag{11}$$

for some constant c' depending on the various constants we have chosen so far, but not on n. Then $\mathbb{P}(J) \geq c'$ and so, applying Lemma 5.5 and reducing c if necessary, we have $v(n, 100n) \geq c$, as required.

Consider the following algorithm $\mathbb{A}$ for testing whether G_ε holds: explore S_1 from below (as in [4]) to find its lowest horizontal crossing P_1, if $H(S_1)$ holds. Similarly, explore S_2 from above to find its uppermost horizontal crossing P_2, if $H(S_2)$ holds; then, from the positions of P_1 and P_2, decide whether G_ε holds. Define $f_{\mathbb{A}}$ as in (6) above.

Let ω_1, ω_2 and X be independent, where the ω_i are random configurations (with the standard distribution) and X is uniformly random on

$$[-5n, 5n-1]^2 = \{-5n, -5n-1, \ldots, 5n-1\}^2 \subset \mathbb{Z}^2,$$

and let $\omega = f_{\mathbb{A}}(\omega_1, \omega_2, X)$, so ω has the standard distribution. The reason for the slightly incongruous notation is that the core of our argument will involve conditioning on ω_1 and ω_2, but keeping X random.

To establish (11), we first examine ω to check whether $\omega \in G_\varepsilon$. This depends only on the states of bonds examined by the algorithm $\mathbb{A}$ above; by the definition of $f_{\mathbb{A}}$, the state of such bonds in ω is the same as in ω_1. Thus $\omega \in G_\varepsilon$ if and only if $\omega_1 \in G_\varepsilon$. When this event holds, let P_1 and P_2 be the paths defined above, and write A for the part of the strip T on or below P_1, B for the part of T on or above P_2, and G for the 'gap' between P_1 and P_2, i.e., the rest of T; see Figure 5.3. Note that all bonds whose interiors lie within G have their states in $\omega = f_{\mathbb{A}}(\omega_1, \omega_2, X)$ given by ω_2^X. Suppose that ω_2 contains an open path P from a point u to a point v. Then ω_2^X contains the open path $P+X$, the translate of P through X, joining $u+X$ to $v+X$. If $u+X \in A$, $v+X \in B$, and $P+X$ remains within the strip $0 \leq x \leq n$, then the minimal subpath P' of $P+X$ meeting A and B contains bonds only in G, so this path P' is present in ω and joins P_1 and P_2, and $\omega \in J$.

Let E be the event that $\omega_1 \in G_\varepsilon$ and $\omega_2 \in E_{\mathrm{v}}$. Since ω_1 and ω_2 are independent, we have $\mathbb{P}(E) = \mathbb{P}(G_\varepsilon)\mathbb{P}(E_{\mathrm{v}}) > c_3 c_2$. For the rest of the proof we condition on ω_1 and ω_2, so the only remaining randomness is in the choice of X. We assume that E holds; we shall show that for any ω_1 and ω_2 such that E holds, the conditional probability that J holds satisfies

$$\mathbb{P}(J \mid \omega_1, \omega_2) \geq c_4 = \gamma^2 \alpha / 400. \tag{12}$$

Then we have $\mathbb{P}(J) \geq c_4 \mathbb{P}(E) \geq c_4 c_3 c_2 > 0$, establishing (11). It remains only to prove (12).

Let us choose an open path P in the configuration ω_2 witnessing $\omega_2 \in E_{\mathrm{v}}$; thus P lies within $[0, \beta_1 n] \times [0, \beta_2 n]$ and joins some point (x, y) to some (x', y') with $y' - y \geq \alpha n$ and $|x' - x| \leq (1 - \eta)|y - y'|$. Let v be the vector $(x' - x, y' - y)$. Let $I = \lceil (1 + 2\gamma)n/(y' - y) \rceil$, and note that

$$I \leq \lceil (1 + 2\gamma)/\alpha \rceil \leq 2/\alpha, \tag{13}$$

since $y' - y \geq \alpha n$ and $\alpha, \gamma \leq 1/3$. For $0 \leq i \leq I$, set $w_i = iv$; we think of w_i as an *offset*, for reasons that will hopefully become clear.

Let P^* be the ('virtual', in the sense that it is not known to be open in any configuration we are considering) path formed by starting at the origin and concatenating I copies of P. Thus P^* is the union of the paths $P_1^*, \ldots, P_I^*$, where each P_i^* is the translate of P joining the point w_{i-1} to w_i; see Figure 5.3.

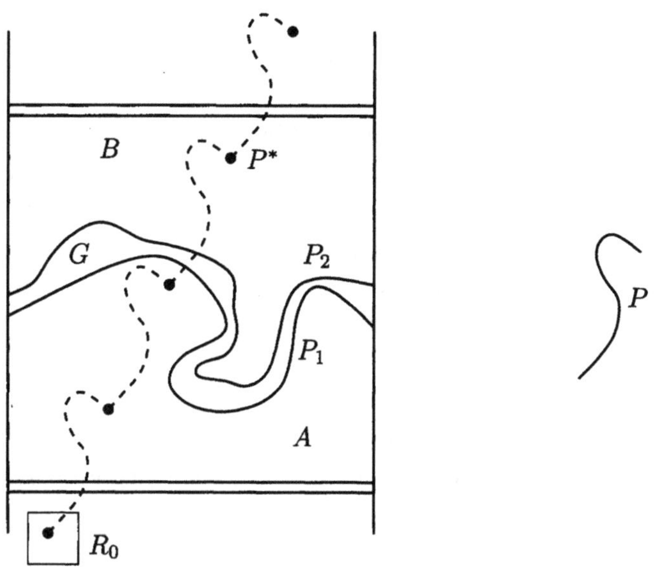

Fig. 5.3. The main part of the figure shows the lowest crossing P_1 of S_1 and highest crossing P_2 of S_2, dividing the strip T into the regions A, G and B. In the final configuration ω, bonds in $A \cup B$ have their states given by ω_1; those in G are given by a random translate of ω_2. On the right, an open path P in ω_2 is shown; on the left we see a 'virtual' chain P^* of translates of P, starting somewhere in the square R_0. Since G is small, for most starting points in R_0, at least one copy of P in P^* crosses from A to B. It follows that with not-too-small probability, the *random* translate of P present in ω_2^X crosses from A to B, giving an open path in ω joining P_1 to P_2.

If $x' \geq x$, let v_0 be the point $\big((\beta_1 + \gamma/2)n, -\gamma n/2\big)$; otherwise, set $v_0 = \big(n - (\beta_1 + \gamma/2)n, -\gamma n/2\big)$. We claim that if a translate $\widetilde{P}^*$ of P^* starts at a point within the square R_0 of side γn centred at v_0, then this translate lies entirely within the strip T. Indeed, in the case $x' \geq x$, the x-coordinate of any point of $\widetilde{P}^*$ is within $(\beta_1 + \gamma/2)n$ of the x-coordinate of some point $v_0 + w_i$, and hence is at least 0 and at most

$$\begin{aligned}(2\beta_1 + \gamma)n + I(x' - x) &\leq (3\beta_1 + \gamma)n + (I - 1)(x' - x) \\ &\leq (3\beta_1 + \gamma)n + (I - 1)(1 - \eta)(y' - y) \\ &\leq \big(3\beta_1 + \gamma + (1 - \eta)(1 + 2\gamma)\big)\, n \leq n,\end{aligned}$$

using the assumption that $\eta \geq 3\beta_1 + 3\gamma$ in the final step. In the case $x' < x$ the argument is similar, subtracting all x-coordinates from n.

Note that R_0 lies entirely below S_1, so, recalling that A and B are the portions of the strip T that lie below P_1 and above P_2, respectively, we have $R_0 \subset A$.

If a translate of P^* starts at a point in R_0, then it ends at a point with y-coordinate at least $-\gamma n + I(y' - y) \geq -\gamma n + (1 + 2\gamma)n = (1 + \gamma)n \geq (1 + \varepsilon)n$, so its upper endpoint is above S_2, and hence lies in B.

For any point $v \in R_0$, we have seen that $v + w_0 = v \in A$, while $v + w_I \in B$. Also, $v + w_i \in T = A \cup G \cup B$ for $0 \leq i \leq I$. Hence, either some point $v + w_i$, $1 \leq i < I$, lies in the 'gap' G, or there is some i such that $v + w_{i-1} \in A$ and $v + w_i \in B$. Let us colour the points of R_0 with $I + 1$ colours, assigning colour 0 in the first case, and colour i in the second (choosing the minimal i if there are several). Let C_i denote the set of points in R_0 assigned colour i.

Now C_0 is a subset of the union of $I - 1$ translates of G. Since G has area at most εn^2, while $I \leq 2/\alpha$, the area of C_0 is thus at most $2\varepsilon n^2/\alpha \leq \gamma^2 n^2/5$, recalling (9). Since R_0 has area $\gamma^2 n^2$, it follows that there is some $i > 0$ for which C_i has area at least $I^{-1}\gamma^2 n^2/2 \geq \gamma^2 \alpha n^2/4$.

Recall that in constructing our random configuration ω as $\omega = f_{\mathbb{A}}(\omega_1, \omega_2, X)$, we shift the configuration ω_2 by a random vector X uniformly distributed on (the integer points in) $[-5n, 5n - 1]^2$. Recall also that the open path P in ω_2 starts at $(x, y) \in [0, \beta_1 n] \times [0, \beta_2 n] \subset [0, 2n]^2$. Consider the set $C_i' = C_i + w_{i-1} - (x, y)$. Then C_i' has the same area as C_i, and certainly lies within $[-5n, 5n - 1]^2$. Hence the probability that X falls in C_i' is at least $(\gamma^2 \alpha n^2/4)/(100 n^2) = \gamma^2 \alpha/400$. But when this happens, the translate $P + X$ of P starts at a point of $C_i + w_{i-1}$. Hence we may think of $P + X$ as the path $P_i^* = P_1^* + w_{i-1}$ in a translate of P^* starting at a point of C_i. From the comments above and the definition of C_i it follows that $P + X$ lies entirely within T, starts in A, and ends in B. As noted earlier, the presence of such a path in ω_2^X guarantees that the configuration $\omega = f_{\mathbb{A}}(\omega_1, \omega_2, X)$ has the property J. Hence the conditional probability that ω belongs to J given ω_1 and ω_2 is at least $\gamma^2\alpha/400$, establishing (12) and completing the proof. ∎

Remark 5.7. The key step of the proof above involved selecting an open path P in ω_2, and then chaining together translates of P to form a path P^* with the following properties: P^* stays well within the strip T, starts well below S_1, and ends well above S_2, where 'well within' means at least a distance $\gamma n/2$ away from the boundary. This elbow room ensures that we can translate P^* through distances of up to $\gamma n/2$ while retaining the properties of starting below S_1, ending above S_2, and remaining within T. Then we randomly shifted the whole path P^*, and randomly chose one of the links in the chain to focus on, thinking of this link as being the final random shift P' of the path P in ω_2. Since it is unlikely that any randomly shifted link starts or ends in the gap G, there is almost always at least one link that crosses from A to B, so the probability that P' does so is bounded away from zero. In order to construct a suitable P^*, we required that P remain within a fairly small region, and that its endpoints be significantly further apart vertically than horizontally.

It is just as easy to start from more than one path in ω_2. Indeed, suppose that there are constants $c > 0$ and C such that with probability at least c the configuration ω_2 contains a set of open paths P_j such that we can chain together at most C paths P_i^*, each of which is a translate of some P_j, to form a path P^* with the properties above. Note that this will hold (for example) whenever ω_2 contains paths P_1 and P_2 each of which stays within some not-too-large region, such that the overall directions of P_1 and P_2 are significantly different, and each is not too short, in the sense that the vector from the start to the end is not too short. Then one can always chain $O(1)$ copies together to produce an approximation to a vertical line. Taking w_{i-1} to be the start of the ith path in the chain as above, the proof goes through essentially unaltered, except that at the very end we set $C_i' = C_i + w_{i-1} - (x_j, y_j)$ if the ith path in our chain is a copy of P_j, where (x_j, y_j) is the starting point of P_j.

5.2. A rectangle-crossing lemma for hyperlattices

Our aim in this subsection is to prove an analogue of Theorem 5.3 in the context of percolation on plane hyperlattices; we have already illustrated the main ideas in a simpler context in the previous subsection.

Recall that a plane hyperlattice $(\mathcal{H}, \mathcal{L})$, originally defined as an embedding of a hypergraph, may also be defined simply as a cubic map whose faces are properly coloured black, white and grey, in a way that is invariant under translations through elements of the lattice $\mathcal{L}$; see Section 2. Recall from Section 4 that an *independent lattice colouring* $\mathcal{C}$ associated to $\mathcal{H}$ is a random black/white-coloured map obtained from $\mathcal{H}$ as follows: First subdivide each grey face into *subfaces* in a deterministic manner, keeping the resulting map cubic. Then recolour these subfaces randomly black and white, with the colourings inside different grey faces independent, such that if one grey face is obtained by translating another through an element of $\mathcal{L}$, then their colourings have the same distribution. Such a random colouring $\mathcal{C}$ gives rise to a hyperlattice percolation model $\mathcal{H}(\mathbf{p})$: simply take the non-crossing partition of a hyperedge e to be the partition of the black faces around the corresponding grey face F_e induced by the (black part of) the colouring of F_e. We say that $\mathcal{C}$ *realizes* the model $\mathcal{H}(\mathbf{p})$.

It turns out that, at one point in the coming argument, we may need to modify the colouring within certain faces. To enable this, we need our random colouring to satisfy a certain technical assumption. Given a colouring χ of a grey face F_e, the *colour components* of χ are the maximal connected monochromatic subsets of F_e.

Definition 5.8. An independent lattice colouring $\mathcal{C}$ is *malleable* if two conditions hold. First, within each grey face F_e, the all-white colouring has positive probability. Second, if χ is a colouring of F_e with positive probability, and χ' is obtained from χ by recolouring a white component black, then χ' has positive probability.

Note that, in a malleable colouring, the all-black colouring of a grey face necessarily has positive probability, so a malleable colouring is non-degenerate (meaning, as before, that within each grey face, the all-black and all-white colourings have positive probability).

In general, recolouring as above may change the partition π corresponding to the colouring χ in many different ways. However, if we restrict the colourings suitably, the situation becomes simpler. Let us call a part of a partition π *non-trivial* if it is not a singleton. A colouring χ of a grey face F_e is *minimal* if its black components are in one-to-one correspondence with the non-trivial parts of π, and its white components with those of the dual partition π^*. Let us say that a part P of π is *adjacent* to a part P' of π^* if, in the original polygon formulation of non-crossing partitions, P contains a vertex v and P' an edge e incident to v. This corresponds to P containing a black vertex adjacent to a white dual vertex in P'. If χ is a minimal colouring associated to a partition π, then recolouring a white component black has the effect of uniting all parts of π adjacent to some non-trivial part of π^*. We call such an operation on a partition a *joining*. In the dual, the operation is simpler: simply split a non-trivial part into singletons.

Given a hyperedge e, by the *top* partition of its vertices we mean the partition into a single part. The *bottom* partition is that into singletons. Recall that a probability vector $\mathbf{p}$ associated to a hyperlattice $(\mathcal{H}, \mathcal{L})$ assigns a probability $p_{i,\pi}$ to each non-crossing partition π of the vertices of a hyperedge e, where i encodes which orbit of the action of $\mathcal{L}$ the hyperedge e belongs to.

A hyperlattice percolation model $\mathcal{H}(\mathbf{p})$ is *non-degenerate* if $p_{i,\pi} > 0$ whenever π is a top or bottom partition.

Definition 5.9. A hyperlattice percolation model $\mathcal{H}(\mathbf{p})$ is *malleable* if it is non-degenerate and, whenever $p_{i,\pi} > 0$ and π' is obtained from π by a joining operation as defined above, then $p_{i,\pi'} > 0$.

Note that if $\mathcal{H}$ is 3-uniform, then any non-degenerate model $\mathcal{H}(\mathbf{p})$ is automatically malleable: any joining operation results in the top partition. Also, any $\mathbf{p}$ assigning positive probability to all top and bottom partitions but to no other partitions gives a malleable model: the unique joining operation that may be performed on a bottom partition yields the corresponding top partition. Thus malleability holds automatically in the 'site percolation' models considered in the discussion surrounding Theorem 2.2.

If $\mathcal{H}(\mathbf{p})$ is self-dual, then malleability is equivalent to its dual formulation, that if $p_{i,\pi} > 0$ and π' is obtained from π by splitting a part into singletons, then $p_{i,\pi'} > 0$.

The next lemma captures the connection between the notions of malleability for probability vectors and for lattice colourings.

Lemma 5.10. *Let $\mathcal{H}(\mathbf{p})$ be a malleable hyperlattice percolation model. Then $\mathcal{H}(\mathbf{p})$ may be realized by a malleable independent lattice colouring.*

Proof. Regard $\mathcal{H}$ as a 3-coloured cubic map, as usual. For each grey face F_e corresponding to hyperedge e with at least 3 vertices, subdivide it into subfaces as in Figure 4.2. (When $|e| = 3$ one can also use the simpler subdivision shown in Figure 4.1.) It is not hard to check that for any non-crossing partition π of the vertices of e, there is at least one black/white colouring of the subfaces of F_e that gives a *minimal* colouring realizing the partition π. If there are N_π such colourings, assign each probability $p_{i,\pi}/N_\pi$, where i is the probability vector entry corresponding to e. Since $\mathcal{H}(\mathbf{p})$ is non-degenerate, the all-black and all-white colourings (which are minimal) receive positive probability. Also, if χ is any colouring of F_e receiving positive probability, then $p_{i,\pi} > 0$ for the corresponding π. If χ' is obtained from χ by recolouring a white component to black, then χ' is minimal, and corresponds to a partition π' obtained from π by a joining operation. Since $\mathcal{H}(\mathbf{p})$ is malleable, we have $p_{i,\pi'} > 0$, so χ' has positive probability.

For a hyperedge e with $|e| = 2$, there is no need to subdivide F_e at all; simply colour F_e black or white with the appropriate (positive) probabilities. Finally, if $|e| = 1$ then the colouring of F_e is irrelevant, so we may colour F_e black with probability $1/2$ and white otherwise. ∎

Of course, a similar but simpler argument shows that any non-degenerate hyperlattice percolation model may be realized by a non-degenerate independent lattice colouring.

As noted in Section 4, a non-degenerate independent lattice colouring $\mathcal{C}$ corresponds to a product probability measure on a power of a certain poset in a natural way; furthermore, the non-degeneracy condition ensures that Lemma 3.2 applies both to this poset and to its reverse. Events defined by the existence of black paths with certain properties are upsets; events defined by the existence of white paths are downsets.

For the rest of this section we fix a malleable independent lattice colouring $\mathcal{C}$. Recall that a black horizontal crossing of a rectangle R is a piecewise-linear path P in the plane joining a point on the left-hand side of R to a point on the right and otherwise lying in the interior of R, such that every point of P is black. White vertical crossings are defined similarly, and so on. In our product probability space, events such as $H_{\mathrm{b}}(R)$ are increasing.

We shall use the following lemma, which applies to all black/white colourings of cubic planar maps, i.e., involves no randomness.

Lemma 5.11. *Let R be a rectangle in $\mathbb{R}^2$ in general position with respect to a colouring $\mathcal{C}$. Then precisely one of the events $H_b(R)$ and $V_w(R)$ holds.*

Proof. As in [6, Ch. 8, Lemma 12], from where Figure 5.4 is adapted, recolour the points outside R as in Figure 5.4, and consider the interfaces between black and white regions. ∎

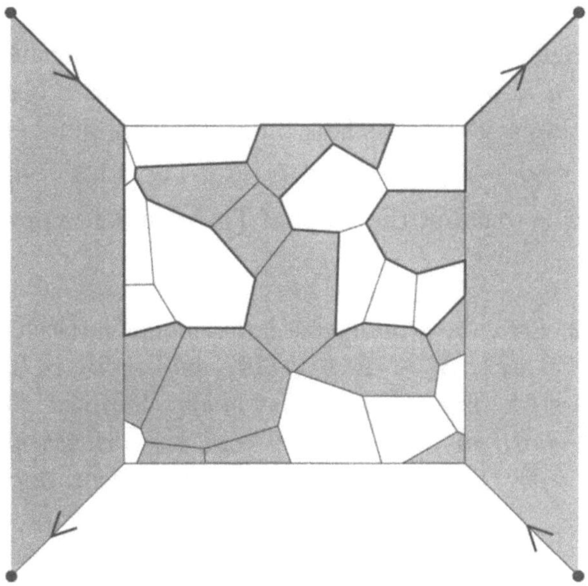

Fig. 5.4. The shading inside the (square) rectangle R is from $\mathcal{C}$. The event $H_b(R)$ holds if and only if the outer black regions are joined by a black path, and $V_w(R)$ holds if and only if the outer white regions are joined by a white path. Tracing the interface between black and white regions shows that one of these events must hold.

We shall need the equivalent of highest and lowest open crossings; these are the highest and lowest *black* crossings of a rectangle, defined below. The definitions require a little care, as the independence properties are not quite what one would like.

Given a rectangle R and our colouring $\mathcal{C}$, shade the outside of the rectangle as in Figure 5.4. Let I^+ be the *interface* shown by the thick black line, starting at the top left corner. Thus if $H_b(R)$ holds, then I^+ leaves the rectangle R from the top right corner. Let I_0^+ be the minimal subpath of I^+ meeting both vertical sides of R. In the figure, points just to the right of I^+ are black. Hence, points just to the right of I_0^+ give a black path within R joining the left to the right. We call this path the *highest black horizontal crossing* of R, and denote it $UH(R)$. Similarly, again assuming $H_b(R)$ holds, the interface I^- starting at the bottom right corner leaves at the bottom left; we define I_0^- to be the minimal subpath joining the vertical sides of R; the points to the right of I_0^- form the *lowest black horizontal crossing* of R, written $LH(R)$. In fact, we can usually work directly with the interfaces I_0^- and I_0^+.

As before, a rectangle R is *large* if all its sides have length at least $100d_0$, where d_0 is the constant from Lemma 4.2.

Given an angle θ, a θ-aligned rectangle R', and 'length scales' m and n, let $E_v(R', m, n, \alpha, \eta)$ be the event that R' contains a black path P such that, after rotating so that R' is horizontal, the endpoints (x, y) and (x', y') of P satisfy $(y' - y)/n \geq \alpha$ and $|x' - x|/m \leq (1 - \eta)(y' - y)/n$.

Our aim now is to adapt the proof of Theorem 5.3 to prove the following result.

Theorem 5.12. *Let C be a malleable independent lattice colouring. Given constants $\alpha, \beta_1 \leq 1/3$, $\beta_2 \leq 2$, $\eta > 3\beta_1$, and $c_1, c_2 > 0$, there exists a constant $c > 0$ such that the following holds. Suppose that R is a large rectangle with width m and height n and any orientation, and R' is a large rectangle with the same orientation, width $\beta_1 m$, and height $\beta_2 n$. If $\mathbb{P}(H_b(R)) \geq c_1$, $\mathbb{P}(V_b(R)) \geq c_1$, and $\mathbb{P}(E_v(R', m, n, \alpha, \eta)) \geq c_2$, then $\mathbb{P}(V_b(R'')) \geq c$ for any rectangle R'' with the same orientation as R, width m and height $100n$.*

Our proof of Theorem 5.12 will follow that of Theorem 5.3 in the previous subsection, rescaling to map the square S_1 considered there onto the rectangle R. Unfortunately, there are various additional complications; for example, we cannot assume that congruent rectangles with the same orientation have the same crossing probabilities. These complications can be dealt with using Corollary 4.10. There will also be some other difficulties.

Remark 5.13. As noted in Section 4, although the probabilities of events such as the existence of various crossings of a rectangle R will depend very much on the orientation of R, and to a lesser extent on its position, all the lower bounds we shall prove will depend on C only via three quantities: the quantity d_0 appearing in Lemma 4.2 (which provides an upper bound on the diameter of a face), the maximum number N of faces meeting any disk of radius 1, and the minimum probability p_0 of a configuration within a face. These three quantities are preserved by rotations and translations, so whenever we consider a single rectangle R, we may rotate and translate R *and* $\mathcal{C}$ together so that R has the form $[0,m] \times [0,n]$.

Let us write $h(R)$ for $\mathbb{P}\big(H_\mathrm{b}(R)\big)$ and $v(R)$ for $\mathbb{P}\big(V_\mathrm{b}(R)\big)$. Also, we write $h(m,n)$ for the probability that $[0,m]\times[0,n]$ has a black horizontal crossing, and $v(m,n)$ for the probability that it has a black vertical crossing. Recall that by Corollary 4.10 there is an increasing function $\psi : (0,1) \to (0,1)$ such that if R and R' are large rectangles with the same orientation, width and height, then $h(R') \ge \psi\big(h(R)\big)$ and $v(R') \ge \psi\big(v(R)\big)$.

Recall that the colouring $\mathcal{C}$ we are considering is malleable and hence non-degenerate, meaning that within any grey face, the all-black and all-white colourings have positive probabilities. As noted earlier, non-degeneracy allows us to apply our Harris-type lemma, Lemma 3.2, to two black-increasing events, or two white-increasing events. Throughout the proof of Theorem 5.12 we write F for the function whose existence is guaranteed by Lemma 3.2, so for any two black-increasing events A and B we have

$$\mathbb{P}(A \cap B) \ge F\big(\mathbb{P}(A), \mathbb{P}(B)\big). \tag{14}$$

The first step in our proof of Theorem 5.12 is the analogue of Lemma 5.4; there is an additional assumption (that $h(m,n)$ and $v(m,n)$ are at least c_1) since we do *not* have the analogue of Lemma 5.1 in this context.

Lemma 5.14. *Let $\mathcal{C}$ be a non-degenerate independent lattice colouring, and let $\varepsilon > 0$, $c_1 > 0$ and $c' > 0$. There is a $c > 0$ such that for any $m, n \ge 100d_0$, if $h(m,n) \ge c_1$, $v(m,n) \ge c_1$, and $v\big(m,(1+\varepsilon)n\big) \ge c'$, then $v(m,100n) \ge c$.*

Proof. Fix ε, c_1 and $c' > 0$. Let $\alpha_0 = c_1$; for $i \ge 0$ let

$$\alpha_{i+1} = F\Big(F\big(\alpha_i, \psi(c_1)\big)\ , \psi(c')\Big),$$

where ψ is the function appearing in Corollary 4.10, and set $c = \alpha_{\lceil 99/\varepsilon \rceil}$. Note that $c > 0$.

Suppose that $m, n \ge 100d_0$, $h(m,n) \ge c_1$, $v(m,n) \ge c_1$, and $v\big(m, (1+\varepsilon)n\big) \ge c'$. We claim that for every $i \ge 0$ we have

$$v\big(m,(1+i\varepsilon)n\big) \ge \alpha_i. \tag{15}$$

Setting $i = \lceil 99/\varepsilon \rceil$, we then have $1 + i\varepsilon \ge 100$, so $v(m,100n) \ge v\big(m, (1+i\varepsilon)n\big) \ge c$, and the result follows.

We prove (15) by induction. For $i = 0$ it is true by assumption. Turning to the induction step, suppose that (15) holds for some i, and consider the rectangles $R_1 = [0,m] \times \big[0,(1+i\varepsilon)n\big]$, $R_2 = [0,m] \times \big[i\varepsilon n,(1+i\varepsilon)n\big]$, and $R_3 = [0,m] \times \big[i\varepsilon n, 1+(i+1)\varepsilon n\big]$, as in Figure 5.5. Let $R_2' = [0,m] \times [0,n]$ and $R_3' = [0,m] \times \big[0,(1+\varepsilon)n\big]$, so R_2' and R_3' are images of R_2 and R_3 under appropriate translations.

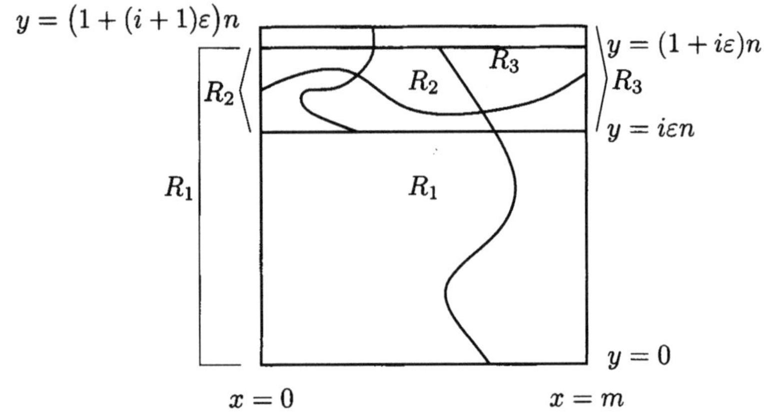

Fig. 5.5. The rectangles R_1 and R_3, with their intersection R_2. Whenever R_1 and R_3 have black vertical crossings and R_2 has a black horizontal crossing, then these crossings can be combined to form a black vertical crossing of $R_1 \cup R_3$.

By the induction hypothesis we have $\mathbb{P}\big(V_{\mathrm{b}}(R_1)\big) = v\big(m,(1+i\varepsilon)n\big) \geq \alpha_i$ and by assumption we have $\mathbb{P}\big(H_{\mathrm{b}}(R_2')\big) = h(m,n) \geq c_1$. Hence, by Corollary 4.10, we have $\mathbb{P}\big(H_{\mathrm{b}}(R_2)\big) \geq \psi(c_1)$. Similarly, by assumption $\mathbb{P}\big(V_{\mathrm{b}}(R_3')\big) = v\big(m,(1+\varepsilon)n\big) \geq c'$, so $\mathbb{P}\big(V_{\mathrm{b}}(R_3)\big) \geq \psi(c')$.

Since $V_{\mathrm{b}}(R_1)$ and $H_{\mathrm{b}}(R_2)$ are increasing events, from (14) it follows that $\mathbb{P}\big(V_{\mathrm{b}}(R_1) \cap H_{\mathrm{b}}(R_2)\big) \geq F\big(\alpha_i,\psi(c_1)\big)$. Applying (14) to the increasing events $V_{\mathrm{b}}(R_1) \cap H_{\mathrm{b}}(R_2)$ and $V_{\mathrm{b}}(R_3)$, it follows that with probability at least α_{i+1}, the events $V_{\mathrm{b}}(R_1)$, $H_{\mathrm{b}}(R_2)$ and $V_{\mathrm{b}}(R_3)$ all hold. Choosing black paths P_1, P_2 and P_3 witnessing these events, P_2 meets both P_1 and P_3, and it follows that $V_{\mathrm{b}}(R_1 \cup R_3)$ holds. Hence

$$v\big(m,\big(1+(i+1)\varepsilon\big)\,n\big) = \mathbb{P}\big(V_{\mathrm{b}}(R_1 \cup R_3)\big) \geq \alpha_{i+1},$$

proving the induction step and so completing the proof of (15). ∎

The next step is the analogue of Lemma 5.5; this concerns crossings of nearby squares (now rectangles) that are joined. As in the previous lemma, we work with rectangles, not squares, and add the assumptions that $h(m,n)$

and $v(m,n)$ are at least some constant c_1 rather than appeal to Lemma 5.1. The definition of the event J is as before, *mutatis mutandis:* specifically, wherever n appears in the definition, it is replaced by m or n depending on whether we are considering an x-coordinate or a y-coordinate.

Lemma 5.15. *Let $\mathcal{C}$ be a non-degenerate independent lattice colouring. For any $\varepsilon > 0$, $c_1 > 0$ and $c' > 0$, there is a $c > 0$ such that for any $m, n \geq 100d_0$ with $h(m,n)$, $v(m,n) \geq c_1$, if there exist m-by-n rectangles R_1 and R_2, with R_2 obtained by translating R_1 upwards by a distance of εn, for which $\mathbb{P}\big(J(R_1,R_2)\big) \geq c'$, then $v(m,100n) \geq c$.*

Proof. Modify the proof of Lemma 5.5 as above: replace each application of Harris's Lemma by an appeal to (14), and use the fact that, under our assumptions, any m-by-n rectangle R has $h(R)$, $v(R) \geq \psi(c_1) > 0$ by Corollary 4.10. ■

Next comes the analogue of Lemma 5.6. For two m-by-n rectangles R_1 and R_2, with R_2 obtained by translating R_1 upwards by a distance of $\varepsilon n/10$, define the strip $T = T(R_1,R_2)$ to be the region between the vertical lines containing the vertical sides of R_1 and R_2, and let $G_\varepsilon(R_1,R_2)$ be the event that $H_b(R_1)$ and $H_b(R_2)$ hold, the path $P_1 = LH(R_1)$ is below $P_2 = UH(R_2)$ in T, and the area of T between P_1 and P_2 is at most εmn.

Lemma 5.16. *Let $\mathcal{C}$ be a non-degenerate independent lattice colouring. For any $\varepsilon > 0$ and $c_1 > 0$ there are constants $c_3 > 0$ and $c > 0$ such that for any $m, n \geq 100d_0$ with $h(m,n)$, $v(m,n) \geq c_1$, either there exist m-by-n rectangles R_1 and R_2 as above with $\mathbb{P}\big(G_\varepsilon(R_1,R_2)\big) \geq c_3$, or $v(m,100n) > c$.*

Proof. Imitate the proof of Lemma 5.6, *mutatis mutandis.* ■

In adapting the main part of the proof of Theorem 5.3 there is only one genuine additional complication: we shall have to work to join up paths in our two configurations ω_1 and ω_2^X. The other changes are mostly in notation.

Recall that our independent lattice colouring $\mathcal{C}$ is obtained by randomly colouring the subdivided grey faces of the 3-coloured planar map $\mathcal{H}$. As before, let Ω denote the set of all configurations, i.e., assignments of states to the grey faces of $\mathcal{H}$. Let $\mathbb{A}$ be an algorithm that examines the states of grey faces in its input configuration one by one, with the next face to be examined determined by the states of the faces examined so far. Assume that $\mathbb{A}$ terminates and write $S_{\mathbb{A}}(\omega)$ for the set of grey faces examined by $\mathbb{A}$

when run on the configuration ω. Recalling that $\mathcal{H}$ is invariant under translations through elements of the lattice $\mathcal{L}$, define a function $f_{\mathbb{A}}$ from $\Omega^2 \times \mathcal{L}$ to Ω by $(\omega_1, \omega_2, \ell) \mapsto \omega$, where $\omega = f_{\mathbb{A}}(\omega_1, \omega_2, \ell)$ is the configuration given by

$$\omega(g) = f_{\mathbb{A}}(\omega_1, \omega_2, \ell)(g) = \begin{cases} \omega_1(g) & \text{if } g \in S_{\mathbb{A}}(\omega_1), \\ \omega_2(g-\ell) & \text{if } g \notin S_{\mathbb{A}}(\omega_1). \end{cases} \tag{16}$$

Here g denotes an arbitrary grey face, and $g - \ell$ the grey face obtained by translating g through the vector $-\ell$. In other words, as before, the state of a grey face in $\omega = f_{\mathbb{A}}(\omega_1, \omega_2, \ell)$ is given either by its state in ω_1 or by its state in the translate ω_2^{ℓ} of ω_2, according to whether or not the algorithm $\mathbb{A}$ examines g when run on the configuration ω_1.

Let $\mathcal{L}_{m,n}$ be the set of points of $\mathcal{L}$ in $[-5m, 5m] \times [-5n, 5n]$, and interpret $\Omega^2 \times \mathcal{L}_{m,n}$ as the product probability space in which the two configurations have the distribution associated to $\mathcal{C}$ and are independent, and the random vector $X \in \mathcal{L}_{m,n}$ is chosen uniformly from $\mathcal{L}_{m,n}$. As before, $\omega = f_{\mathbb{A}}(\omega_1, \omega_2, X)$ has the distribution appropriate for $\mathcal{C}$.

Proof of Theorem 5.12. We follow the proof of Theorem 5.3, concentrating on the differences. In the light of Remark 5.13, rotating the rectangles under consideration together with $\mathcal{C}$, we may assume that R and R' are aligned with the coordinate axes, i.e., are 0-aligned.

First, note that (7) now holds by assumption. As before, choose $\gamma > 0$ such that (8) holds, and then choose $\varepsilon > 0$ such that $\varepsilon < \gamma$ and (9) holds. Let m and n satisfy the assumptions of the theorem. Then, since $h(m,n)$, $v(m,n) \geq c_1$, we can apply Lemma 5.16. Let c_3 and c be the constants given by Lemma 5.16, so either $v(m, 100n) \geq c$, in which case we are done, or there are m-by-n 0-aligned rectangles S_1 and S_2 with S_2 obtained by translating S_1 upwards by a distance of $\varepsilon n/10$ such that

$$\mathbb{P}\big(G_{\varepsilon}(S_1, S_2)\big) \geq c_3.$$

We may assume that the second case holds. Translating the rectangles under consideration together with $\mathcal{C}$, we may assume that $S_1 = [0, m] \times [0, n]$ and $S_2 = [0, m] \times \big[\varepsilon n/10, (1 + \varepsilon/10)n\big]$.

As before, we explore S_1 from below to find its lowest black horizontal crossing P_1, if it exists, and S_2 from above to find its highest black horizontal crossing P_2. More precisely, we let I_1 be the interface I_0^- in S_1 described earlier in the subsection, and I_2 the interface I_0^+ in S_2. Note that orienting

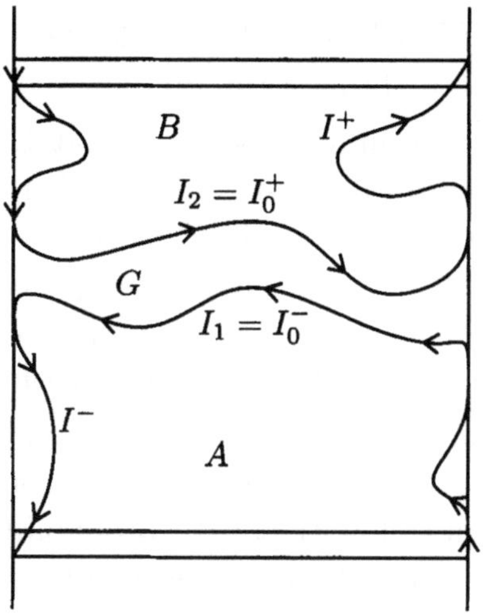

Fig. 5.6. The overlapping congruent rectangles S_1 (below) and S_2 (above), with the interfaces I^- in S_1 and I^+ in S_2 defined as in Figure 5.4. I_0^- and I_0^+ are the minimal subpaths of I^- and I^+ crossing the strip $S_1 \cup S_2$. Note that every point just to the right of $I^\pm$ is black or outside $S_1 \cup S_2$ horizontally; points just to the right of $I_0^\pm$ are always inside $S_1 \cup S_2$ and hence black. Points to the left of $I^\pm$ are white or outside $S_1 \cup S_2$ vertically. The 'gap' G is the region between $I_1 = I_0^-$ and $I_2 = I_0^+$.

these interfaces as in Figure 5.6, the points just to the right of each interface are black while those to the left are either white, or inside T but outside the relevant rectangle S_i vertically, if the interface runs along the top/bottom of S_i. It follows that I_1 and I_2 cannot meet. Our definitions allow us to view I_i as a black path itself, although it is perhaps clearer to think of P_i as running next to I_i. In the present context, the *gap* G is the region between I_1 and I_2; we define A and B to be the regions of our strip $[0, m] \times \mathbb{R}$ below I_1 and above I_2, respectively.

As before, we first test whether $\omega_1 \in G_\varepsilon$, using the algorithm $\mathbb{A}$ implicitly defined above. More precisely, we implement $\mathbb{A}$ by following the full interfaces I^- and I^+ in the relevant rectangles. Note that this only involves 'testing' the state of grey faces f that meet one of these interfaces, where f *meets* I if I passes through f, or along one of the sides of f. It is easy to check that all of I^- lies below $I_0^- = I_1$, and all of I^+ above I_2. Let $Z(I_i)$ denote the union of I_i and all grey faces that meet I_i, which we think of as the *zone of influence* of I_i. Then when G_ε holds, the algorithm $\mathbb{A}$ establishes this by looking only at grey faces in $A \cup B \cup Z(I_1) \cup Z(I_2)$. Defining $\omega = f_{\mathbb{A}}(\omega_1, \omega_2, X)$ as before, it follows that any grey face contained in $G\setminus\big(Z(I_1)\cup Z(I_2)\big)$ has its state in ω given by its state in ω_2^X, the configuration ω_2 translated through the random vector X.

As before, we condition on ω_1, assuming that G_ε holds, which it does with probability bounded away from zero. Furthermore, we condition on ω_2, assuming the existence of a path P with the property described in the definition of $E_{\mathrm{v}}(R', m, n, \alpha, \eta)$. Again, the probability of this event is bounded away from zero, this time by assumption. As before, the only remaining randomness is in the choice of the random translation X.

Recalling that we may take m and n large, apart from one very minor technical issue that we postpone to the end of the proof, trivial modifications to our previous arguments show that with probability bounded away from zero, the random translate $P+X$ of P crosses from A to B, while remaining within the strip T. In fact, adjusting the constants slightly if necessary, we can assume that it does not come within distance d_0 of the edges of T.

Let $E\subset\Omega^2\times\mathcal{L}_{m,n}$ denote the set of triples (ω_1,ω_2,ℓ) such that $\omega_1\in G_\varepsilon$, and ω_2 contains a path P as above whose translate $P+\ell$ meets the interfaces I_1 and I_2 in ω_1. We have shown that $\mathbb{P}(E)$ is bounded away from zero. Unfortunately, unlike in the $\mathbb{Z}^2$ setting, it is not true that if $(\omega_1,\omega_2,\ell)\in E$, then $f_{\mathbb{A}}(\omega_1,\omega_2,\ell)\in J$; the problem is illustrated in Figure 5.7.

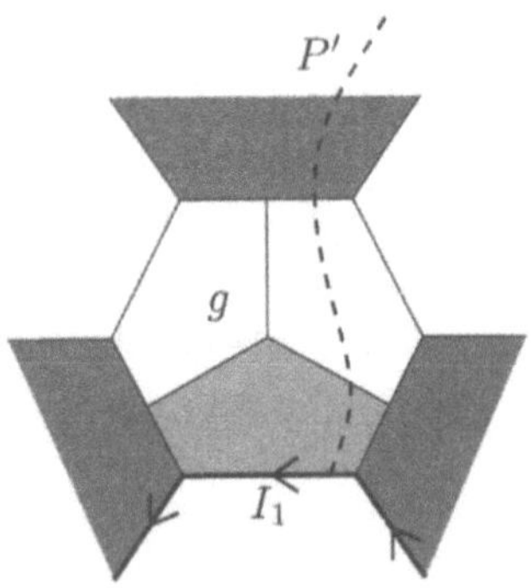

Fig. 5.7. A hexagonal grey face g surrounded by black and white faces of $\mathcal{H}$. In the random colouring associated to ω_1, the internal colouring of g is indicated by the lighter shading. In this case I_1 touches g. The dashed line shows a black path P' in the configuration ω_2^X; since the state of g is read from ω_1 (even though g is in the 'gap'), in the combined configuration ω, the path P' fails to join up with I_1.

To overcome this problem, we adjust the colourings of at most two faces. Let $(\omega_1,\omega_2,\ell)\in E$, and let P' denote a minimal part of the path $P+\ell$ joining I_1 to I_2. Then P' necessarily lies in the gap G. Let P'' denote a minimal subpath of P' joining $Z(I_1)$ to $Z(I_2)$, and let v_1 and v_2 denote its endpoints (which may coincide, if $Z(I_1)$ and $Z(I_2)$ meet); see Figure 5.8. Note that any interior point of P'' lies in $G \setminus \big(Z(I_1) \cup Z(I_2)\big)$, so its colour in $\omega = f_{\mathbb{A}}(\omega_1, \omega_2, \ell)$ is its colour in ω_2^ℓ; since $P'' \subset P + \ell$, such points are therefore black.

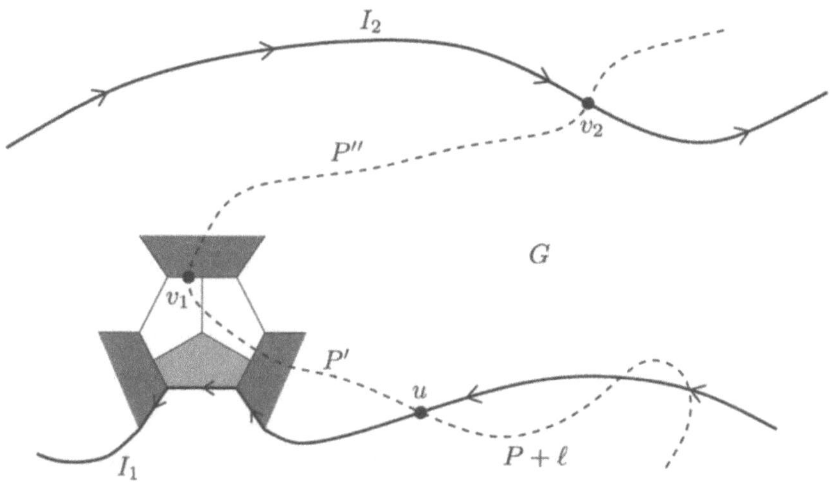

Fig. 5.8. The path $P + \ell$ (dashed line) crossing the gap G, from below I_1 to above I_2. The subpath P' of $P + \ell$, here from u to v_2, is minimal subject to crossing G, and so lies within G. In turn, P'', from v_1 to v_2, is minimal subject to joining $Z(I_1)$ to $Z(I_2)$. In this example, v_2 is on I_2, while v_1 is in a grey face f_1 (the hexagon in the figure) touching I_1; the two white subfaces inside f_1 will be recoloured black.

Now each v_i is either on I_i, or belongs to a grey face f_i which meets I_i. Suppose the latter case holds for $i = 1, 2$. To handle the remaining cases we simply recolour at most one face in the following argument, rather than two. Recalling that $P+\ell$ does not pass within distance d_0 of the edges of T, note that each f_i lies entirely with the strip T. Let ω_1' be the configuration obtained from ω_1 by recolouring all points of $(f_1 \cup f_2) \cap G$ black. As we shall show in a moment, this is a legal configuration. Assuming this for the moment, if we restrict our attention overall to a finite region of the plane, as we may, the ratio $\mathbb{P}(\omega_1')/\mathbb{P}(\omega_1)$ of the probabilities of the individual configurations is bounded below by p_0^2, where $p_0 > 0$ is the minimum probability of any possible state of a grey face.

The key point is that recolouring a set of points within the gap G black does not change the interfaces I_1 and I_2; these interfaces are determined by the full interfaces $I^\pm$ shown in Figure 5.6, which are defined 'locally', and the side of I_i on which the gap lies is already 'locally black'. Thus the algorithm $\mathbb{A}$ examines the states of the same set of faces if run on ω_1' or on ω_1. Recalling that the states of f_1 and f_2 in ω are those in ω_1, it follows that in the configuration $\omega' = f_{\mathbb{A}}(\omega_1', \omega_2, \ell)$ there is a black path joining I_1 to I_2, given by the union of P'' and two short paths from v_i to I_i within $f_i \cap G$. Hence $\omega' \in J$.

Let $g : E \to \Omega^2 \times \mathcal{L}_{m,n}$ denote the map $(\omega_1, \omega_2, \ell) \mapsto (\omega_1', \omega_2, \ell)$, and let $E' = g(E)$. We have shown that $E' \subset f_{\mathbb{A}}^{-1}(J)$. Recalling that $f_{\mathbb{A}}$ is measure preserving, it follows that $\mathbb{P}(J) \geq \mathbb{P}(E')$. Since our recolouring does not change the interfaces I_1 and I_2, given some $(\tilde{\omega}, \omega_2, \ell) \in E'$ known to be the image of some unknown $(\omega_1, \omega_2, \ell)$ under g, we can read off the interfaces I_1 and I_2 (defined in ω_1) by looking at $\tilde{\omega}$. We also know the path P from ω_2. This allows us to determine P' and P'' as defined above, and hence f_1 and f_2. In other words, we know which two (or at most two) faces were recoloured, though not how. It follows that $g^{-1}\big(\{\,(\tilde{\omega}, \omega_2, \ell)\}\big)$ consists of a bounded number of configurations, each of whose probabilities is at most p_0^{-2} times that of $(\tilde{\omega}, \omega_2, \ell)$. Hence there is a constant C such that $\mathbb{P}\big(g^{-1}(E')\big) \leq C\mathbb{P}(E')$. Since $g^{-1}(E') = g^{-1}\big(g(E)\big) \supset E$, it follows that $\mathbb{P}(E') \geq \mathbb{P}(E)/C$. Since E is known to have probability bounded away from zero, the result follows.

It remains to establish that the recolouring is permissible, i.e., to show that if we recolour within a grey face g (either f_1 or f_2 above), the new colouring has positive probability; this is where we use the assumption that $\mathcal{C}$ is malleable (a condition of Theorem 5.12). Now G is bounded by interfaces running between black and white regions, as well as the sides of the strip T; the latter do not meet g. It follows that $g \cap G$ is the union of one or more colour components (maximal connected monochromatic subsets) of g. Our recolouring thus recolours one or more white components within g to black; the definition of malleability ensures that the resulting colouring has positive probability.

Finally, let us comment on the technical issue we overlooked, which is that since the endpoints of P may not differ by a lattice element, we cannot exactly join up translates of P though lattice elements to form P^*. One way to handle this is to find a short (length $O(1)$) black path P' in ω_2 joining appropriate points within a fundamental domain of $\mathcal{L}$, and form P^* by chaining together alternate copies of P and P', as described in Remark 5.7 at the end of Subsection 5.1. Alternatively, simply leave small (length at most $d_0 = O(1)$) gaps between the translates of P making up P^*: when colouring R_0, assign colour 0 to any point v such that in the copy of P^* starting at v, one or more of these small gaps meets G. It remains the case that few points receive colour 0: the set of such points is contained in the union of $I = O(1)$ translates of the d_0-neighbourhood G^{d_0} of G. Since G is made up of faces whose size is bounded below, the area of G^{d_0} is at most a constant times that of G, so the area receiving colour 0 is still $O(\varepsilon n^2)$; choosing ε small enough, the rest of the argument is unchanged. ■

5.3. A stronger rectangle-crossing lemma

Although technically we can do without it, we now present a more convenient version of Theorem 5.12 giving the same conclusion under a weaker assumption. First we need a simple geometric lemma.

By the *displacement* of a path P we mean the Euclidean distance from its start to its endpoint. By the *direction* of P we mean the direction from its start to its endpoint, considered as an angle modulo π. The *angle* between two paths is simply the (unsigned) angle between their directions, taken as a real number between 0 and $\pi/2$.

Lemma 5.17. *Given $\delta > 0$ there exists a constant $C = C(\delta)$ such that whenever P_1 and P_2 are two paths in $[0, 1/10]^2$ with displacement at least δ such that the angle between P_1 and P_2 is at least δ, then we can chain together at most C paths each of which is a translate of P_1 or P_2 to form a path P^* that lies within $[0.1, 0.9] \times [-0.5, 1.5]$, starts below $y = -0.1$, and ends above $y = 1.1$.*

Proof. Let v_i be the vector from the start of P_i to its end, and let $\mathcal{L}$ be the lattice $\{av_1 + bv_2 : a, b \in \mathbb{Z}\}$. Let D be the fundamental domain of $\mathcal{L}$ whose corners are the origin, v_1, v_2 and $v_1 + v_2$. Note that the area of D is bounded below by a constant depending on δ.

The idea is simply to approximate the line-segment L joining the points $(0.5, -0.3)$ and $(0.5, 1.3)$ by a path Q in the graph associated to $\mathcal{L}$ where two lattice points are neighbours if they differ by $\pm v_i$ for some i; then replace each edge by an appropriate translate of P_i. We can find such a path Q so that every point of Q is in the same domain $D + av_1 + bv_2$ as some point of L, and Q starts and ends in the same domains as the start and end of L; to see this, simply apply an affine transformation mapping $\mathcal{L}$ to the usual square grid, and approximate the image of L by an appropriate path.

Since any two points of D have x-coordinates that differ by at most $2/10$ and y-coordinates that differ by at most $2/10$, the path P^* obtained from Q has the required properties. ■

Theorem 5.18. *Let $\mathcal{C}$ be a malleable independent lattice colouring, and let δ, c_1 and c_2 be positive constants. Then there exists a constant $c > 0$ such that the following holds. Suppose that R is a large rectangle with width m, height n and any orientation, and R' is a large rectangle with the same orientation, width $m/10$, and height $n/10$. Let E be the event that R' contains two black paths such that their images under the affine transformation mapping R' to $[0, 1/10]^2$ satisfy the conditions of Lemma 5.17. If $\mathbb{P}(H_b(R)) \geq c_1$, $\mathbb{P}(V_b(R)) \geq c_1$, and $\mathbb{P}(E) \geq c_2$, then $\mathbb{P}(V_b(R'')) \geq c$ for any rectangle R'' with the same orientation as R, width m and height $100n$.*

Proof. The proof is the same as that of Theorem 5.12, except that we construct the 'virtual' path P^* as the union of translates of our black paths P_1 and P_2 in ω_2 given by Lemma 5.17. As before, since we can only translate by elements of the lattice $\mathcal{L}$ of symmetries of $\mathcal{C}$, we may need additional short paths to join up these translates. Each of these additional paths lies within a fundamental domain of $\mathcal{L}$, and we may take them to be translates of appropriate short black paths $P_3, \ldots$ in ω_2. (With extremely high probability, such paths will exist; we only need one of the many domains to have the property that every grey face meeting it is coloured black.)

Arguing as before, after conditioning on ω_1 and ω_2, we find that with probability bounded away from zero our random translate of one of the P_i joins I_1 to I_2, and the rest of the proof is as before. ∎

The precise constants appearing in Lemma 5.17 are not important; the key point is that we can chain together translates of our paths P_i to give a path P^* with the properties discussed in Remark 5.7. After appropriate rescaling, these properties are that P^* crosses a square from bottom to top, with some 'elbow room', i.e., starting well below the bottom and ending well above the top, without coming too close to the vertical sides.

6. Self-Duality and Rectangle Crossings

Our aim in this section is to show that if $\mathcal{C}$ is a malleable independent lattice colouring associated to a self-dual plane hyperlattice $\mathcal{H}$, then an analogue of Theorem 5.3 holds for $\mathcal{C}$. It turns out that, due to the lack of symmetry, we cannot specify in advance the orientation of the rectangles we work with. In fact, we cannot even fix their aspect ratio.

Recall that $h_\theta(m,n)$ and $v_\theta(m,n)$ denote $\mathbb{P}\big(H_{\mathrm{b}}(R)\big)$ and $\mathbb{P}\big(V_{\mathrm{b}}(R)\big)$, where R is an m-by-n θ-aligned rectangle centred on the origin.

Definition 6.1. Let $\rho > 1$. We say that a lattice colouring $\mathcal{C}$ has the 'large rectangles' property LR_ρ if there exists a constant $c > 0$ such that for all L and for all large enough A (depending on L) there are $m, n \geq L$ with $mn = A$ and an angle θ such that $h_\theta(\rho m, n) \geq c$ and $v_\theta(m, \rho n) \geq c$.

In symbols, this large rectangles property with parameter ρ may be stated as follows:

$$\exists c > 0 \; \forall L \; \exists A_0 \; \forall A \geq A_0 \; \exists m, n, \theta :$$

$$m, n \geq L, \; mn = A, \; h_\theta(\rho m, n) \geq c \text{ and } v_\theta(m, \rho n) \geq c.$$

Roughly speaking, the idea is that we can find a rectangle R so that when we extend it by a factor of ρ horizontally or vertically, the probability of a black horizontal or vertical crossing, respectively, is not too small. We can take this rectangle to have any given large area, and can assume that both sides are at least any given length. However, the orientation and aspect ratio cannot be specified in advance.

Recall that $\mathcal{C}$ is non-degenerate if, within each grey face, the all-black and all-white colourings have positive probability. Under this assumption, one can adapt the usual argument from Harris's Lemma to show that for any $\rho_1, \rho_2 > 1$, the property LR_{ρ_1} implies LR_{ρ_2}; the argument is as for Lemma 5.14. In the light of this, the following definition makes sense.

Definition 6.2. A non-degenerate independent lattice colouring $\mathcal{C}$ has the *large rectangles property* if it has the property LR_ρ for some $\rho > 1$, and hence for all $\rho > 1$.

Recall that $\mathcal{C}$ is *malleable* if it is non-degenerate and satisfies certain technical extra conditions; for the full definition see Subsection 5.2. Our aim in this section is to prove the following result; as we shall see in the next section, it is then easy to deduce Theorems 2.1 and 2.2.

Theorem 6.3. *Let $\mathcal{C}$ be a malleable independent lattice colouring realizing an (approximately) self-dual hyperlattice percolation model $\mathcal{H}(\mathbf{p})$. Then $\mathcal{C}$ has the large rectangles property.*

The proof of Theorem 6.3 will require a little preparation. First, let us restate the property slightly. Given an ellipse E with centre x_0, let $2E/3$

denote the ellipse $\{x_0 + 2(x - x_0)/3 : x \in E\}$ obtained by shrinking E by a factor of 3/2, keeping the centre the same. Let E^0 denote the 'annulus' between E and $2E/3$, and let $O_b(E)$ denote the event that there is a closed black path in E^0 surrounding the central hole.

Definition 6.4. A lattice colouring $\mathcal{C}$ has the *large ellipses property* if there exists a constant $c > 0$ such that for all L and for all large enough A (depending on L) there is an ellipse E centred on the origin with area A and with both axes having length at least L such that $\mathbb{P}(O_b(E)) \geq c$.

It is easy to see that the large rectangles property and the large ellipses property are equivalent.

Lemma 6.5. *A non-degenerate independent lattice colouring $\mathcal{C}$ has the large rectangles property if and only if it has the large ellipses property.*

Fig. 6.1. (a) Rectangles in an annulus, and (b) circles crossing two rectangles.

Proof. Suppose first that $\mathcal{C}$ has the large rectangles property; in particular, it has the property LR_{20}.

Consider the arrangement of overlapping rectangles shown in Figure 6.1(a). Taking the shorter side of each rectangle to have length 1, the arrangement is such that the longer side of each rectangle has length at most 20. Also, if each rectangle has a black crossing in the long direction, then $O_b(C)$ holds, where C is the outer circle. Let a denote the area of C, so a is an absolute constant.

We must show the existence of an ellipse E with (large enough) area A and both axes at least L such that $\mathbb{P}(O_b(E)) \geq c_1$, for some constant c_1. Let $d_0 = d_0(\mathcal{C})$ be the constant in Lemma 4.2. The large rectangles property LR_{20} gives us a θ and $m, n \geq \max\{L, 100d_0\}$ with $mn = A/a$ such that $h_\theta(20m, n) \geq c$ and $v_\theta(m, 20n) \geq c$, for some constant c that is independent of A. Consider the image of Figure 6.1(a) under a linear transformation mapping each 1-by-1 square to a θ-aligned m-by-n rectangle. Note that the resulting ellipse E (the image of C) has both axes at least L and has area A. Using Corollary 4.10 to allow for translation, each image rectangle has a black crossing in the relevant direction with probability at least some constant $c' = \psi(c) > 0$. Using Lemma 3.2 (in the form (14)), it follows that $\mathbb{P}(O_b(E))$ is bounded away from zero.

For the reverse implication, we assume the large ellipses property and deduce the property LR_2. As shown in Figure 6.1(b), one can arrange 9 circles C_i of radius 0.9 to 'cross' a 2-by-1 rectangle R_1 and a 1-by-2 rectangle R_2 in such a way that if $O_b(C_i)$ holds for each i, then $H_b(R_1)$ and $V_b(R_2)$ hold. Moreover, this remains true if each circle is translated by a small distance (at most 0.01, say). Given an ellipse E with both axes at least $\max\{100d_0, L\}$ such that $\mathbb{P}(O_b(E)) \geq c$, one can rotate the arrangement in Figure 6.1(b) and then scale it along the directions of the axes of E so that each circle is mapped to a translate E_i of E, and R_1 and R_2 are mapped to θ-aligned $2m$-by-n and m-by-$2n$ rectangles R'_1 and R'_2, for some θ, m and n. Note that $m, n \geq L$, and mn is an absolute constant times the area of E.

Moving each E_i by a distance of at most d_0 (which corresponds to translating C_i through a distance of at most 0.01), we may assume that $E_i = E + \ell_i$ for some $\ell_i \in \mathcal{L}$. Then $\mathbb{P}(O_b(E_i)) = \mathbb{P}(O_b(E)) \geq c$, and if all 9 upsets $O_b(E_i)$ hold, then $H_b(R'_1)$ and $V_b(R'_2)$ hold. Using Lemma 3.2 thus gives a constant lower bound on $h_\theta(2m, n)$ and $v_\theta(m, 2n)$, establishing the property LR_2. ∎

Most of the time, we work with the large rectangles property; the large ellipses property will be convenient to use in Section 8. The equivalence is also useful in that it leads to a quick proof that the large rectangles property is affine-invariant.

Lemma 6.6. *Let T be an invertible linear map from $\mathbb{R}^2$ to $\mathbb{R}^2$ and let $\mathcal{C}$ be a non-degenerate independent lattice colouring. If $\mathcal{C}$ has the large rectangles property, then so does $T(\mathcal{C})$.*

Proof. There is a constant $a = a(T) > 0$ such that if E is an ellipse both of whose axes have length at least L then $T(E)$ is an ellipse both of whose axes have length at least aL. (For example, note that E contains a circle of radius L, and the image of this circle contains a circle of radius aL for some $a > 0$.) Hence the large ellipses property is invariant under T. Applying Lemma 6.5 twice, we see that the large rectangles property is too. ∎

Our next lemma shows that after a linear transformation, a self-dual hyperlattice percolation model must be related to its dual by one of a small number of linear transformations. We consider both the self-dual case (for Theorem 2.1) and the approximately self-dual case (for Theorem 2.2).

Lemma 6.7. *Let $\mathcal{H}(\mathbf{p})$ be an (approximately) self-dual hyperlattice percolation model. Then there is a map $S : \mathbb{R}^2 \to \mathbb{R}^2$ giving an isomorphism between $\mathcal{H}(\mathbf{p})$ and a hyperlattice model equivalent to $\mathcal{H}^*(\mathbf{p}^*)$ such that S can be written in the form $S(x) = T(x)+\Delta(x)$, where T is linear and $|\Delta(x)|$ is bounded. Furthermore, after a linear change of coordinates (if necessary), we may assume that T is either reflection in some line, or rotation through one of the angles 0, $\pi/2$, or π.*

Proof. Suppose first that $\mathcal{H}(\mathbf{p})$ is self-dual. By the definition of self-duality for hyperlattices, there is a homeomorphism S from $\mathbb{R}^2$ to $\mathbb{R}^2$ mapping $\mathcal{H}$ to its dual, and preserving the lattice structure. More precisely, there is a linear map T with $T(\mathcal{L}) = \mathcal{L}$ such that $S(x + \ell) = S(x) + T(\ell)$ whenever $\ell \in \mathcal{L}$. Since $\Delta = S - T$ is continuous and doubly periodic, it is bounded, giving the first statement.

Since $T(\mathcal{L}) = \mathcal{L}$, the map T preserves area, so $T \in GL_2(\mathbb{R}^2)$ with $\det(T) = \pm 1$. From standard results, T is conjugate in $GL_2(\mathbb{R}^2)$ to a map T' that is either a rotation, a shear with matrix $\left(\begin{smallmatrix} 1 & \lambda \\ 0 & 1 \end{smallmatrix}\right)$, or a stretch with matrix $\left(\begin{smallmatrix} \lambda & 0 \\ 0 & \pm 1/\lambda \end{smallmatrix}\right)$. Changing coordinates (or applying a linear transformation to $\mathcal{H}$ and its dual simultaneously), we may assume that $T = T'$.

Now S^2 maps $\mathcal{H}$ into itself, and $S^2(x) = T^2(x) + \Delta_2(x)$ where Δ_2 is bounded. If T is a shear with $\lambda \neq 0$ or a stretch with $|\lambda| \neq 1$, then repeated application of S^2 shows the existence of arbitrarily long edges in $\mathcal{H}$, giving a contradiction. Thus T is either a rotation or a reflection. In the former case, the fact that T maps $\mathcal{L}$ into itself guarantees that the angle of rotation θ is a multiple of either $\pi/3$ or $\pi/2$. Replacing S by the isomorphism S^3 from $\mathcal{H}(\mathbf{p})$ to its dual allows us to reduce the cases $\theta = k\pi/3$, $k \in \mathbb{Z}$, to the cases $\theta = 0$ or $\theta = \pi$, and the case $\theta = 3\pi/2$ to the case $\theta = \pi/2$.

The argument for the approximately self-dual case is similar, except that from the definition of approximate self-duality we may simply assume that T is a rotation or a reflection. ∎

For the rest of the section, in the light of Lemma 6.6, we assume as we may that $\mathcal{H}(\mathbf{p})$ and its dual are related as described in Lemma 6.7. In the case where the map T is a reflection, we take it to be reflection in the x-axis, and call this the *reflection* case. The remaining cases are *rotation* cases; we write θ^* for the angle of rotation passing from $\mathcal{H}$ to $\mathcal{H}^*$, so $\theta^* \in \{0, \pi/2, \pi\}$.

Let $\mathcal{C}^*$ be the *negative* of the colouring $\mathcal{C}$, defined simply by interchanging black and white (both in $\mathcal{H}$, and in the colours of the subfaces of the grey faces of $\mathcal{H}$). Even in the self-dual case, although $\mathcal{C}$ realizes $\mathcal{H}(\mathbf{p})$, which is self-dual via the map $S : \mathbb{R}^2 \to \mathbb{R}^2$, we cannot assume that $\mathcal{C}$ is self-dual in the natural sense. For example, in the dual colouring to that shown in the centre of Figure 4.1, there are two black subfaces within the hexagon. This pattern may not occur in $\mathcal{C}$. However, writing d_0 for the constant given by Lemma 4.2, since $\mathcal{C}$ realizes $\mathcal{H}(\mathbf{p})$, there is a natural coupling of $\mathcal{C}$ and $\mathcal{H}(\mathbf{p})$ such that any black path P in $\mathcal{C}$ is within Hausdorff distance d_0 of an open path in $\mathcal{H}(\mathbf{p})$, and vice versa. Applying this observation also to $\mathcal{C}^*$, which realizes $\mathcal{H}^*(\mathbf{p}^*) = S\big(\mathcal{H}(\mathbf{p})\big)$, it follows that we can couple $\mathcal{C}$ and $\mathcal{C}^*$ so that for any black path P in $\mathcal{C}$ there is a black path in $\mathcal{C}^*$ within distance $O(1)$ of $S(P)$, and vice versa. Recalling that black in $\mathcal{C}^*$ simply means white in $\mathcal{C}$, and using Lemma 6.7, this has the following consequence. Here, φ is the 'probability scaling function' defined before Lemma 4.8.

Lemma 6.8. *Let $\mathcal{C}$ be a non-degenerate independent lattice colouring realizing an (approximately) self-dual hyperlattice percolation model $\mathcal{H}(\mathbf{p})$, and let T be the linear map given by Lemma 6.7. Then there are constants L and C such that, for any angle θ and any θ-aligned m-by-n rectangle R with $m, n \geq L$, we have*

$$\varphi\big(\mathbb{P}\big(H_{\mathrm{b}}(R)\big)\big) - \varphi\big(\mathbb{P}\big(H_{\mathrm{w}}\big(T(R)\big)\big)\big) \in [-C, C]$$

Proof. We write out only the self-dual case; since we in any case allow some 'elbow room' when passing from $\mathcal{H}(\mathbf{p})$ to its dual, there are no additional difficulties in the approximately self-dual case.

Let $L = 100d_0 + 10D$, where $D = \sup\big\{\big|S(x) - T(x)\big|\big\}$, which is finite by Lemma 6.7. Couple $\mathcal{C}$ and $\mathcal{C}_1 = \mathcal{C}^*$ as above. Let $\mathcal{C}_2$ be the colouring obtained from $\mathcal{C}_1$ by interchanging white and black, so $\mathcal{C}_2$ has the same distribution as $\mathcal{C}$.

Let R be a rectangle as described, and let R^+ be obtained by moving the vertical sides of R outwards by a distance D and the horizontal sides inwards by the same distance. Suppose R^+ has a black horizontal crossing in $\mathcal{C}$. Then from the remarks before the lemma, there is a black path in $\mathcal{C}_1$ close to $S(P)$ and hence to $T(P)$, and thus a white path in $\mathcal{C}_2$ close to $T(P)$. But any such path crosses $T(R)$. Hence

$$\mathbb{P}_{\mathcal{C}}\big(H_{\mathrm{w}}\big(T(R)\big)\big) = \mathbb{P}_{\mathcal{C}_2}\big(H_{\mathrm{w}}\big(T(R)\big)\big) \geq \mathbb{P}_{\mathcal{C}}\big(H_{\mathrm{b}}(R^+)\big).$$

Writing $\mathbb{P}$ for $\mathbb{P}_{\mathcal{C}}$ as usual, and using Lemma 4.8, this gives

$$\varphi\big(\mathbb{P}\big(H_{\mathrm{w}}\big(T(R)\big)\big)\big) \geq \varphi\big(\mathbb{P}\big(H_{\mathrm{b}}(R)\big)\big) - O(1).$$

The reverse inequality is proved similarly. ∎

In what follows we assume throughout that our 'large' length L is at least $100d_0$, and is such that Lemma 6.8 applies. Note that rectangles with dimensions at least L are 'large' in the terminology of the previous sections.

Lemma 6.9. *Let $\mathcal{C}$ be a non-degenerate independent lattice colouring realizing an (approximately) self-dual hyperlattice percolation model $\mathcal{H}(\mathbf{p})$. For any L there are constants $A_0 = A_0(\mathcal{C}, L)$ and $C = C(\mathcal{C}, L)$ such that for any $A \geq A_0$, and any θ in the rotation case or $\theta = 0$ in the reflection case, we may find $m, n \geq L$ with $mn = A$ such that $\varphi\big(h_\theta(m,n)\big)$, $\varphi\big(v_\theta(m,n)\big) \in [-C, C]$.*

In other words, both the horizontal and vertical crossing probabilities for some rectangle of the given area and orientation are bounded away from 0 and 1.

Proof. Let us write $h_\theta^*(m,n)$ and $v_\theta^*(m,n)$ for the probabilities that an m-by-n θ-aligned rectangle has a white horizontal crossing or a white vertical crossing, respectively.

Consider first the rotation case with $\theta^* = \pi/2$. In this case we simply set $A_0 = L^2$, and choose $m = n = \sqrt{A}$. Let R be the m-by-m θ-aligned square centred on the origin. Note that $T(R)$ is the same rectangle R, but viewed as $(\theta + \pi/2)$-aligned, so $v_\theta^*(m,m) = \mathbb{P}\big(H_{\mathrm{w}}\big(T(R)\big)\big)$. Hence Lemma 6.8 gives $\varphi\big(h_\theta(m,m)\big) - \varphi\big(v_\theta^*(m,m)\big) = O(1)$. But by Lemma 4.4, we have $h_\theta(m,m) + v_\theta^*(m,m) = 1$, so $\varphi\big(v_\theta^*(m,m)\big) = -\varphi\big(h_\theta(m,m)\big)$. It follows that $\varphi\big(h_\theta(m,m)\big) = O(1)$. Similarly, $\varphi\big(v_\theta(m,m)\big) = O(1)$.

For the remaining cases, for A large enough, Lemma 4.12 gives us $m, n \geq L$ with $mn = A$ and $\varphi\big(h_\theta(m,n)\big) \in [-4, 4]$. This time (noting that

$\theta = 0$ in the reflection case), the map T maps our θ-aligned m-by-n rectangle into itself, and Lemma 6.8 gives $\varphi\big(h_\theta(m,n)\big) - \varphi\big(h_\theta^*(m,n)\big) = O(1)$. Hence $\varphi\big(v_\theta(m,n)\big) = -\varphi\big(h_\theta^*(m,n)\big) = \varphi\big(h_\theta(m,n)\big) + O(1) = O(1)$. ■

We are now ready to prove Theorem 6.3.

Proof of Theorem 6.3. We assume as we may that $L \geq 100d_0$. Throughout we fix an arbitrary $A \geq 10^6 A_0(\mathcal{C}, L)$, where $A_0(\mathcal{C}, L)$ is the constant in Lemma 6.9. All constants c or c_i below will depend only on $\mathcal{C}$, not on A or L. We shall show that for some θ, a and b with $ab = 10A$ we have

(17) $$h_\theta(10a, b),\ v_\theta(a, 10b) \geq c,$$

where $c > 0$ does not depend on A. This will establish that $\mathcal{C}$ has the large rectangles property LR_{10}.

Fix an orientation θ, with $\theta = 0$ in the reflection case, and θ arbitrary otherwise. Suppressing the dependence on θ in the notation, by Lemma 6.9 there are $m, n \geq L$ with $mn = A$ such that

(18) $$h(m,n),\ v(m,n) \geq c_0,$$

where $c_0 > 0$ is a constant depending only on $\mathcal{C}$.

Suppose for the moment that $h(100m, n) \geq c_1$, where c_1 is a positive constant not depending on A. Set $m' = 10m$ and $n' = n$, and consider an m'-by-n' rectangle R and the $m'/10$-by-n' rectangle R' inside it. Note that $h(R) = h(10m, n) \geq h(100m, n) \geq c_1$. Also, since R' is m-by-n, we have $v(R') \geq c_0$. We may thus apply Theorem 5.12 with m and n replaced by m' and n', and $\alpha = \beta_1 = 1/10$, $\beta_2 = 1$, and $\eta = 1/2$. Any black vertical crossing of R' satisfies the conditions for the event E_{v} considered in Theorem 5.12 simply because, seen with respect to the m'-by-n' grid, R' is much taller than wide. Theorem 5.12 thus gives $v(m', 100n') \geq c$, for some constant $c > 0$. But then $h(10m', n') = h(100m, n)$ and $v(m', 10n') \geq v(m', 100n')$ are both at least $\min\{c_0, c\}$, establishing (17) with $a = m' = 10m$ and $b = n' = n$.

A similar argument (using a 'rotated' version of Theorem 5.12) shows that if $v(m, 100n) \geq c_1$, then (17) holds with $a = m$ and $b = 10n$. Thus, in what follows, it suffices to show that one of $h(100m, n)$ and $v(m, 100n)$ is not too small.

Applying Lemma 6.9 again we find $m_1, n_1 \geq L$ with $m_1 n_1 = mn/400 = A/400$ such that
$$h(m_1, n_1),\ v(m_1, n_1) \geq c_0.$$

Let R_1 be the m_1-by-n_1 rectangle centred on the origin, so $\mathbb{P}(H_b(R_1)) \geq c_0$ and $\mathbb{P}(V_b(R_1)) \geq c_0$.

Fix a small constant $\delta > 0$; for example, $\delta = 1/1000$ will do. Suppose first that $n_1/n \geq (1+\delta)m_1/m$, say. Then on the m-by-n scale, the rectangle R_1 is significantly taller than wide, and has area $1/400$. After truncating R_1 vertically if necessary (after which we still have $\mathbb{P}(V_b(R_1)) \geq c_0$), we may apply Theorem 5.12 with R an m-by-n rectangle, with $\beta_1 = m_1/m \leq 1/20$, $\alpha = \beta_2 = \min\{n_1/n, 1/3\} \geq (1+\delta)m_1/m$, and $\eta = \delta/2$. A black vertical crossing of R_1 once again ensures that E_v holds, and it follows that $v(m, 100n) \geq c$ for some constant $c > 0$. As noted above, we are done in this case.

A similar argument applies if $m_1/m \geq (1+\delta)n_1/n$, so we may assume that $(1-\delta)/20 \leq m_1/m,\ n_1/n \leq (1+\delta)/20$.

As before, let F be the function given by Lemma 3.2 applied to the product of partially ordered sets corresponding to the random partitions induced by each edge e of $\mathcal{H}$. Recalling that $h(m_1, n_1),\ v(m_1, n_1) \geq c_0$, set $c_2 = F(\psi(c_0), \psi(c_0))$, where ψ is the function appearing in Corollary 4.10. For any rectangle R, let $X(R)$ denote the event $H_b(R) \cap V_b(R)$ that R has both horizontal and vertical black crossings. For any m_1-by-n_1 rectangle R, we have $\mathbb{P}(H_b(R)), \mathbb{P}(V_b(R)) \geq \psi(c_0)$ by Corollary 4.10, and thus

$$\mathbb{P}(X(R)) \geq c_2. \tag{19}$$

Set $c_3 = F(c_2/4, c_2/4) < c_2/4$, and $c_4 = F(c_3, c_3/2) < c_3/2$.

Let us say that a rectangle R is *useful* if it contains two black paths P_1 and P_2 such that, after scaling by dividing all x-coordinates by m and all y-coordinates by n, the paths satisfy the assumptions of Lemma 5.17. In other words, for each i, the vector v_i from the start of P_i to the end has (after rescaling) length at least δ, and the angle between v_1 and v_2 is at least δ. We write $U(R)$ for the event that R is useful. If there is any rectangle R with width at most $m/10$ and height at most $n/10$ for which $\mathbb{P}(U(R)) \geq c_4$ then, recalling (18), Theorem 5.18 gives $v(m, 100n) \geq c$ for some $c > 0$. Hence we may assume that

$$\mathbb{P}(U(R)) < c_4 < c_3/2 < c_2/8 \tag{20}$$

for any rectangle with these dimensions. Note that $1.5m_1 \leq m/10$ and $1.5n_1 \leq n/10$, so the rectangle R_1 defined earlier satisfies the size restriction above with room to spare.

Suppose that some m_1-by-n_1 rectangle R is coloured in such a way that $X(R) \setminus U(R)$ holds. Then R has a black horizontal crossing P_1 and a black vertical crossing P_2. The angle (after scaling) between these crossings must be very close to 0, so it follows that either both P_1 and P_2 cross R from near the top left to near the bottom right, or both cross from near the bottom left to near the top right. Suppose the former holds. Then, in addition, every black horizontal or vertical crossing stays (after the usual scaling) within distance 5δ of the diagonal; otherwise, such a crossing can be split into two parts (AB and BC in the figure) with an angle of at least δ between them, so $U(R)$ holds; see Figure 6.2.

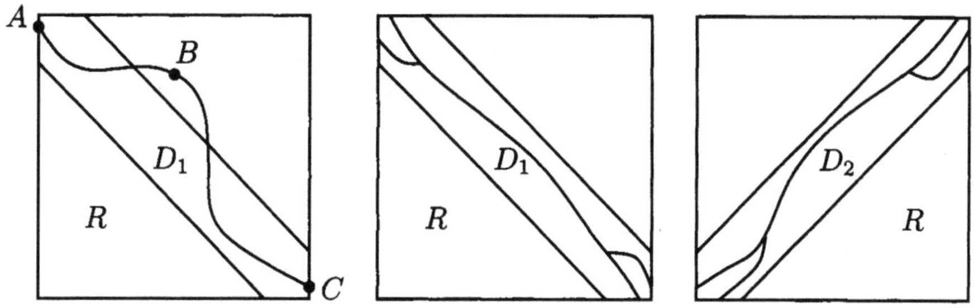

Fig. 6.2. Various possible configurations in an m_1-by-n_1 rectangle R, rescaled by dividing x-coordinates by m and y-coordinates by n. Note that the width and height of R are then between $(1-\delta)/20$ and $(1+\delta)/20$. The strips D_1 and D_2 have width 10δ. The second and third figures illustrate $X_1(R)$ and $X_2(R)$.

Let $D_1 = D_1(R)$ and $D_2 = D_2(R)$ denote the strips of width (after rescaling as above) 5δ about the two diagonals of R. Let $X_i(R)$ denote the event that $D_i(R)$ contains both horizontal and vertical crossings of R; then $X(R) \setminus U(R) \subset X_1(R) \cup X_2(R)$. Hence, for any m_1-by-n_1 rectangle R, for some i we have $\mathbb{P}\big(X_i(R)\big) \geq \big(\mathbb{P}\big(X(R)\big) - \mathbb{P}\big(U(R)\big)\big)/2 \geq c_2/4$, using (19) and (20). We say that R is *of type* i if $\mathbb{P}\big(X_i(R)\big) \geq c_2/4$, so any R is of type 1 or type 2. Furthermore, no R can be of both types: otherwise, using Lemma 3.2 again, we have $\mathbb{P}\big(U(R)\big) \geq \mathbb{P}\big(X_1(R) \cap X_2(R)\big) \geq F(c_2/4, c_2/4) = c_3$, contradicting (20).

At this point we consider two separate cases.

1. *Reflection case.* Let R_1 and R_2 be two m_1-by-n_1 rectangles with R_2 obtained by translating R_1 vertically through a distance $10\delta n \leq n_1/10$. Recalling that our orientation θ is the standard orientation $\theta = 0$ in this case, we may choose R_1 and R_2 so that the rectangle $R_1 \cup R_2$ is centred at the origin; in particular, the x-axis is an axis of symmetry of $R_1 \cup R_2$.

Suppose first that R_1 and R_2 are of opposite types, say with R_i of type i. Then $\mathbb{P}\big(U(R_1 \cup R_2)\big) \geq \mathbb{P}\big(X_1(R_1) \cap X_2(R_2)\big) \geq F(c_2/4, c_2/4) = c_3$, contradicting (20). It follows that R_1 and R_2 must be of the same type, say type 1.

Let E be the event $X_1(R_1) \cap X_1(R_2) \setminus U(R_1 \cup R_2)$, so

$$\mathbb{P}(E) \geq F(c_2/4, c_2/4) - c_3/2 = c_3/2$$

by (20). If E holds, then R_1 contains a black horizontal crossing P_1 in the strip $D_1 = D_1(R_1)$, and R_2 contains a black horizontal crossing P_2 in the strip $D_1' = D_1(R_2)$. It is easy to check that if P_1 and P_2 are joined by a black path in $R_1 \cup R_2$, then $U(R_1 \cup R_2)$ holds. (For example, since we also have vertical crossings of R_1 and R_2, we obtain a vertical crossing of $R_1 \cup R_2$, whose direction is necessarily more than an angle δ away from that of P_1 or P_2.) Hence, whenever E holds, so does the event F, that $D_1 \cup D_1'$ contains a *white* horizontal crossing of $R_1 \cup R_2$. Thus, $\mathbb{P}(F) \geq c_3/2$.

We now apply self-duality as in the proof of Lemma 6.8, recalling that now T is reflection in the x-axis, an axis of symmetry of $R_1 \cup R_2$. Let $\tilde{F}$ be the image of F under the symmetry transformation. Then $\mathbb{P}(\tilde{F}) = \mathbb{P}(F)$. Also, if $\tilde{F}$ holds then there is a black path in $R_1 \cup R_2$ that almost crosses $R_1 \cup R_2$ horizontally, and lies within or close to the mirror image of $D_1 \cup D_1'$. Since $\tilde{F}$ is an upset, by Lemma 3.2 we have $\mathbb{P}(X_1(R_1) \cap X_2(R_2) \cap \tilde{F}) \geq F(c_3, c_3/2) = c_4$. But whenever this event holds, $R_1 \cup R_2$ is clearly useful: the relevant crossings must in fact meet, but we do not even need this, simply the observation that $R_1 \cup R_2$ then contains two longish black paths that, after the usual rescaling, are at almost 90 degrees to each other. Thus $\mathbb{P}\big(U(R_1 \cup R_2)\big) \geq c_4$, contradicting (20). This contradiction completes the proof in the reflection case.

2. *Rotation case.* In this case, Lemma 6.9 applies *regardless of the orientation* θ. The idea is to observe that the argument above shows that for each θ, the θ-aligned m_1-by-n_1 rectangle R_1 centred on the origin is either of type 1 or of type 2. As we rotate, we can assume (as we will shortly show) that R_1 varies continuously. Thus it should not jump from type 1 to type 2. But after rotating by 90 degrees, we return to the same rectangle viewed with a different orientation. Changing orientation in this way interchanges types 1 and 2.

To make this precise, first note that the argument above establishes the following. Let C be the constant in Lemma 6.9, let θ be any orientation, and let m_1 and n_1 satisfy $m_1 n_1 = A/400$ and $m_1, n_1 \geq L$. If R is the

θ-aligned m_1-by-n_1 rectangle centred on the origin, and $\varphi(h_\theta(m_1,n_1))$, $\varphi(v_\theta(m_1,n_1)) \in [-C-8, C+8]$, then R is either of type 1 or of type 2 (and not both, although we shall not use this). To see this, simply use the rectangle R under consideration as R_1 in the argument above: so far we selected an arbitrary rectangle with the properties described by Lemma 6.9;. now we choose a specific one with the same properties, except that we have replaced C by $C+8$, which makes no difference.

Pick an integer $k \geq A/d_0$, and let $\varepsilon = (\pi/2)/k$. We claim that we can construct a finite sequence $R_0, R_1, \ldots, R_N$ of rectangles, all centred on the origin, where R_i is θ_i-aligned and satisfies the conditions above, with $\theta_0 = 0$, $\theta_N = \pi/2$ and R_0 and R_N the same geometric rectangle, such that for $i = 0, 1, \ldots, N-1$, either R_i and R_{i+1} have the same dimensions and $\theta_{i+1} = \theta_i + \varepsilon$, or R_i and R_{i+1} have the same orientation, and their corresponding dimensions differ by at most one.

To establish the claim, simply start with $\theta_0 = 0$ and R_0 of the dimensions given by Lemma 6.9. At each stage, if R_i satisfies $\varphi(h(R_i))$, $\varphi(v(R_i)) \in [-C, C]$, then rotate it through an angle ε to obtain R_{i+1}. Since the largest dimension of R is at most $(A/400)/(100d_0) \leq k/10$, Corollary 4.11 ensures that R_{i+1} has the required properties. Otherwise, by assumption we have $\varphi(h(R_i))$, $\varphi(v(R_i)) \in [-C-8, C+8]$. Applying Lemma 6.9, pick a rectangle R with the same orientation and area for which $\varphi(h(R))$, $\varphi(v(R)) \in [-C, C]$. Since R_i and R have the same orientation, we may pass gradually from one to the other with $h(\cdot)$ increasing and $v(\cdot)$ decreasing or vice versa; the intermediate rectangles thus all satisfy our requirements. Once we reach R, we are ready for the next rotation step.

Since the type of R_N is the opposite of that of R_0, there is some i such that R_i and R_{i+1} have different types; suppose that R_i is of type 1 and R_{i+1} of type 2. Then by Lemma 3.2 we have $\mathbb{P}(X_1(R_i) \cap X_2(R_{i+1})) \geq c_3$. Although the dimensions of our rectangles may change radically as we rotate them, they cannot do so in one step. In particular, after the rescaling above applied to R_i, the paths witnessing $X_1(R_i)$ and $X_2(R_{i+1})$ are close to orthogonal. It follows that if $X_1(R_i) \cap X_2(R_{i+1})$ holds, then R_i is useful, so $\mathbb{P}(U(R_i)) \geq c_3$, contradicting (20) above. ■

7. From Rectangle Crossings to Percolation

In this section we shall deduce Theorems 2.1 and 2.2 from Theorem 6.3; this turns out to be relatively straightforward, adapting the very simple argument for bond percolation on $\mathbb{Z}^2$ described in [4]. First, we get one technical detail out of the way.

Recall that we can represent any hyperlattice percolation model $\mathcal{H}(\mathbf{p})$ by an independent lattice colouring $\mathcal{C}$; our results in the previous section apply only to *malleable* $\mathcal{C}$. In the bulk of this section we shall prove the following variant of Theorem 2.1, differing only in the additional assumption of malleability.

Theorem 7.1. *Let $\mathcal{H}(\mathbf{p})$ be a malleable (approximately) self-dual hyperlattice percolation model. Then for any $\mathbf{q} \succ \mathbf{p}$ the model $\mathcal{H}(\mathbf{q})$ percolates, and for any $\mathbf{q} \prec \mathbf{p}$ the model $\mathcal{H}(\mathbf{q})$ exhibits exponential decay.*

Before proving this result, we note that Theorem 2.1 follows.

Proof of Theorem 2.1. Let $\mathcal{H}(\mathbf{p})$ be self-dual, and let S be an isomorphism from $\mathcal{H}$ to $\mathcal{H}^*$ witnessing this. For this S, the self-duality condition reduces to a set of equations equating certain entries $p_{i,\pi}$ of $\mathbf{p}$. In particular, any entry $p_{i,\pi}$ where π is a partition into singletons is equated with some $p_{j,\pi'}$, where π' is a partition into one part. Recall that we call entries of these two types *bottom* and *top* entries, respectively.

If $\mathbf{q} \succ \mathbf{p}$, then by definition of our partial order, each bottom entry of $\mathbf{p}$ is non-zero. Hence by self-duality each top entry is non-zero. Since corresponding top or bottom entries of $\mathbf{p}$ and $\mathbf{q}$ cannot be equal, it follows that we can adjust $\mathbf{p}$ slightly to find some $\mathbf{p}'$ all of whose entries are strictly positive such that $\mathcal{H}(\mathbf{p}')$ is self-dual, with $\mathbf{p}' \prec \mathbf{q}$. Since $\mathbf{p}'$ is malleable, Theorem 7.1 implies that $\mathcal{H}(\mathbf{q})$ percolates, as required.

The argument that any $\mathbf{q} \prec \mathbf{p}$ exhibits exponential decay proceeds similarly. ■

The argument above shows that in proving Theorem 2.1, we may impose the condition of malleability (which we need in the proof) without loss of generality, so we do not need to assume malleability in the statement of the theorem. Unfortunately, there does not seem to be an obvious analogous argument in the approximately self-dual case: it is not clear how to adjust the probabilities slightly while preserving approximate self-duality. For this reason we simply impose malleability as a condition in Theorem 2.2, so Theorem 2.2 trivially follows from Theorem 7.1.

To deduce Theorem 7.1 from Theorem 6.3, we shall use an analogue for posets of the well-known Friedgut–Kalai sharp-threshold result for symmetric events, Theorem 2.1 of [12], which is itself a consequence of a result of Kahn, Kalai and Linial [15] (see also [8]) concerning the influences of coordinates in a product space. This sharp-threshold result has been applied in many contexts – it was first used to prove criticality (for random Voronoi percolation) in [3]; we shall use the same technique here (see also [4]).

Let $\mathcal{P}$ be a finite poset. Given two probability measures $\mathbb{P}_0$ and $\mathbb{P}_1$ on $\mathcal{P}$, recall that $\mathbb{P}_1$ *strictly dominates* $\mathbb{P}_0$, written $\mathbb{P}_1 \succ \mathbb{P}_0$, if $\mathbb{P}_1(\mathcal{U}) > \mathbb{P}_0(\mathcal{U})$ for every upset $\mathcal{U} \subset \mathcal{P}$, except the trivial upsets $\mathcal{U} = \emptyset, \mathcal{P}$.

As usual, given a (po)set $\mathcal{P}$ and a subset $\mathcal{A}$ of $\mathcal{P}^n$, a coordinate i is *pivotal* for $\mathcal{A}$ in a configuration $\omega \in \mathcal{P}^n$ if changing the ith coordinate of ω can affect whether $\omega \in \mathcal{A}$. Let $\mathcal{A}_i(\omega) \subset \mathcal{P}$ denote the set of values that, when substituted for the ith coordinate of ω, give some $\omega' \in \mathcal{A}$. Thus i is pivotal for $\mathcal{A}$ in ω if and only if $\emptyset \neq \mathcal{A}_i(\omega) \neq \mathcal{P}$. If $\mathcal{A}$ is an upset, then $\mathcal{A}_i(\omega)$ is an upset.

Given $\mathbb{P}_0 \prec \mathbb{P}_1$, for $0 < h < 1$ define $\mathbb{P}_h$ by linear interpolation: $\mathbb{P}_h(x) = h\mathbb{P}_1(x) + (1-h)\mathbb{P}_0(x)$ for all $x \in \mathcal{P}$. Let $c_0 = c_0(\mathbb{P}_0, \mathbb{P}_1)$ be the minimum of $\mathbb{P}_1(\mathcal{U}) - \mathbb{P}_0(\mathcal{U})$ over all non-trivial upsets in $\mathcal{P}$, so $c_0 > 0$ by assumption. Then, for any non-trivial upset $\mathcal{U}$, we have $\frac{\mathrm{d}}{\mathrm{d}h}\mathbb{P}_h(\mathcal{U}) \geq c_0$. Considering partial derivatives in a product with different values for h in each coordinate, one obtains an analogue of the Margulis–Russo formula [20, 23]: if $\mathcal{A} \subset \mathcal{P}^n$ is an upset, then with $c_0 = c(\mathbb{P}_0, \mathbb{P}_1) > 0$ as above, we have

$$\frac{\mathrm{d}}{\mathrm{d}h}\mathbb{P}_h^n(\mathcal{A}) \geq c_0\, \mathbb{E}_h\, N, \tag{21}$$

where $\mathbb{E}_h$ denotes expectation with respect to the product measure $\mathbb{P}_h^n$, and $N = N(\omega)$ is the number of pivotal coordinates for $\mathcal{A}$ in the random configuration ω.

Bourgain, Kahn, Kalai, Katznelson and Linial [8] showed that if X is any probability space, and $\mathcal{A}$ is a subset of X^n, then then there is some coordinate i such that the probability that i is pivotal for $\mathcal{A}$ is at least $ct(1-t)\log n/n$, where $c > 0$ is an absolute constant, and t is the probability of $\mathcal{A}$. As usual, we say that $\mathcal{A}$ is *symmetric* if there is a permutation group acting transitively on $\{1, 2, \ldots, n\}$ whose induced action on X^n preserves $\mathcal{A}$. If $\mathcal{A}$ is symmetric, each coordinate has the same probability of being pivotal, so the expected number of pivotal elements is at least $ct(1-t)\log n$. Using (21) in place of the usual Margulis–Russo formula, one then obtains the following result; we omit the simple calculation, noting that one may take $c_1(\mathbb{P}_0,\mathbb{P}_1) = c\,c_0(\mathbb{P}_0,\mathbb{P}_1)/2$.

Theorem 7.2. *Let $\mathbb{P}_0$ and $\mathbb{P}_1$ be probability measures on a poset $\mathcal{P}$ with $\mathbb{P}_0 \prec \mathbb{P}_1$. There is a constant $c_1(\mathbb{P}_0,\mathbb{P}_1) > 0$ with the following property. Let $0 < \varepsilon < 1/2$, and let $\mathcal{A}$ be a symmetric, increasing event in a power $\mathcal{P}^n$ of $\mathcal{P}$ with $\mathbb{P}_0^n(\mathcal{A}) > \varepsilon$. If*

$$c_1(\mathbb{P}_0,\mathbb{P}_1)\log n \geq \log(1/\varepsilon),$$

then $\mathbb{P}_1^n(\mathcal{A}) > 1-\varepsilon$. ∎

Using Theorem 7.2 in place of the Friedgut–Kalai result, it is very simple to adapt (one of) the simple arguments given in [4] to deduce Theorem 7.1 from Theorem 6.3.

Proof of Theorem 7.1. Let $\mathcal{H}(\mathbf{p})$ be a malleable (approximately) self-dual hyperlattice percolation model, with $\mathcal{L}$ the corresponding lattice of translational symmetries, and let $\mathbf{q} \succ \mathbf{p}$. Note that since $\mathcal{H}(\mathbf{p})$ is non-degenerate, so is $\mathcal{H}(\mathbf{q}')$, where $\mathbf{q}' = (\mathbf{p}+\mathbf{q})/2$. If some bottom entries in $\mathbf{q}$ are zero, we replace $\mathbf{q}$ by $\mathbf{q}'$ in what follows: since $\mathbf{q} \succ \mathbf{q}' \succ \mathbf{p}$, it suffices to prove percolation in $\mathcal{H}(\mathbf{q}')$. Thus we may assume without loss of generality that $\mathcal{H}(\mathbf{q})$ is non-degenerate.

As usual, we wish to work with crossings of rectangles, so it is more convenient to work with independent lattice colourings.

By Lemma 5.10 there is a malleable independent lattice colouring $\mathcal{C}_0$ realizing $\mathcal{H}(\mathbf{p})$. As in Section 4 (before Lemma 4.2), we regard the state space Ω underlying the random colouring $\mathcal{C}_0$ as a product of one poset $\mathcal{P}_F$ for each grey face F of $\mathcal{H}$: in the partial order, we have $c_1 \preccurlyeq c_2$ if every subface that is black in c_1 is black in c_2. Picking a finite set $F_1,\ldots,F_k$ of faces representing the orbits of $E(\mathcal{H})$ (the set of grey faces) under the action of $\mathcal{L}$, from lattice invariance we may regard Ω as a countable power of the poset $\mathcal{P} = \mathcal{P}_{F_1} \times \cdots \times \mathcal{P}_{F_k}$. (As usual, the events we consider in the following arguments will be defined in terms of finite regions of the plane, and so can be viewed as events in a finite power of $\mathcal{P}$.)

From independence, the probability measure associated to $\mathcal{C}_0$ is a power of a probability measure $\mathbb{P}_0$ on $\mathcal{P}$. Furthermore, since $\mathbf{p} \prec \mathbf{q}$, we may choose another measure $\mathbb{P}_1$ on $\mathcal{P}$ with $\mathbb{P}_0 \prec \mathbb{P}_1$ such that the corresponding independent lattice colouring $\mathcal{C}_1$ realizes $\mathcal{H}(\mathbf{q})$. We may and shall assume that $\mathcal{C}_1$ is non-degenerate.

Let $\delta(\mathcal{P}, \mathbb{P}, k, \varepsilon)$ be the function appearing in Corollary 3.3 (our version of the square-root trick), and set

$$\delta = \delta(\mathcal{P}, \mathbb{P}_1, 100, 0.01) > 0. \tag{22}$$

Let $c > 0$ be the constant in the LR_{10} property of $\mathcal{C}_0$; such a constant exists by Theorem 6.3. Let $\varepsilon > 0$ be the minimum of c and δ, and choose N such that

$$c_1(\mathbb{P}_0, \mathbb{P}_1) \log N \geq \log(1/\varepsilon), \tag{23}$$

where $c_1(\mathbb{P}_0, \mathbb{P}_1)$ is defined as in Theorem 7.2. Let $D = F_1 \cup \cdots \cup F_k$, and choose L_0 so that $L_0^2 \geq N \text{area}(D)$, so any region of area at least L_0^2 meets at least N translates of D by elements of $\mathcal{L}$.

Let d_1 be the maximum of the quantity d_0 appearing in Lemma 4.2 and diam(D), the diameter of D. Since $\mathcal{C}_0$ has the large rectangles property LR_{10}, we can find an angle θ and $m, n \geq \max\{L_0, 100d_1\}$ such that $h_\theta(10m, n) \geq c$ and $v_\theta(m, 10n) \geq c$. In other words, there are orthogonal vectors v_1 and v_2 (obtained by rotating $(m, 0)$ and $(0, n)$ through an angle θ) with the following property: the rectangles R_1 with corners $\pm 5v_1 \pm v_2/2$ and R_2 with corners $\pm v_1/2 \pm 5v_2$ are such that the probability that R_i has a 'long' (parallel to the $10v_i$ side) black crossing in $\mathcal{C}_0$ is at least c.

By Lemma 4.2, every point of $\mathbb{R}^2$ is within distance d_1 of some point of $\mathcal{L}$, so we may find $\ell_1, \ell_2 \in \mathcal{L}$ within distance d_1 of $1.1v_1$ and $1.1v_2$, respectively. Let $\widetilde{R}_1$ be the parallelogram with corners 0, $8\ell_1$, $8\ell_1 + \ell_2$ and ℓ_2. Since v_1 and v_2 have length at least $100d_1$, this parallelogram is obtained from a translate of R_1 by first 'distorting it very slightly', and then making it significantly shorter and thicker. It is easy to check that a translate of $\widetilde{R}_1$ through a suitable lattice element has the property that any 'long' crossing of R_1 includes a 'long' crossing of $\widetilde{R}_1$, so, in $\mathcal{C}_0$, the probability that $\widetilde{R}_1$ has a 'long' black crossing is at least c. Define $\widetilde{R}_2$ from R_2 similarly. Since ℓ_i is close to $1.1v_i$ and the vectors v_i are not too short and are orthogonal, the area of $\widetilde{R}_i$ is (crudely) at least $8|v_1||v_2| \geq 8mn \geq 8L_0^2 \geq 8N\text{area}(D)$.

Applying a linear transformation mapping ℓ_1 to $(1, 0)$ and ℓ_2 to $(0, 1)$, we find that in the transformed model $\mathcal{C}_0'$, with lattice of symmetries $\mathcal{L}' \supset \mathbb{Z}^2$, the probability that the rectangle $R_1' = [0, 8] \times [0, 1]$ has a black horizontal crossing is at least c, as is the probability that $R_2' = [0, 1] \times [0, 8]$ has a black vertical crossing. Note that the image D' of our fundamental domain D has diameter at most $1/10$, say, since $|\ell_i| \geq 100d_1 \geq 100\,\mathrm{diam}\,(D)$ and the ℓ_i are close to orthogonal. Also, the area of D' is $\mathrm{area}\,(D) \times \mathrm{area}\,(R_1')/\mathrm{area}\,(\widetilde{R}_1) \leq 1/N$.

Let $\mathbb{T}$ be the torus obtained by taking the quotient of $\mathbb{R}^2$ by the lattice $10\mathbb{Z}^2$ generated by $(10,0)$ and $(0,10)$. Since $10\mathbb{Z}^2 \subset \mathcal{L}'$, we may choose $n = 100/\mathrm{area}\,(D')$ translates of D' by elements of $\mathcal{L}'$ so that their images in $\mathbb{T}$ cover $\mathbb{T}$ exactly once. This allows us to define a natural equivalent of $\mathcal{C}_0'$ on $\mathbb{T}$; the corresponding probability measure may be seen as $\mathbb{P}_0^n$, where $n = 100/\mathrm{area}\,(D') \geq 100N$. Moreover, given a rectangle that does not come 'close' to wrapping around the torus, the events that it has a horizontal black crossing in the plane or in the torus have the same probability.

Let E be the event that some translate of $[0,8] \times [0,1]$ in $\mathbb{T}$ has a black horizontal crossing. Then $\mathbb{P}_0^n(E) \geq c$, and E is a symmetric, increasing event in $\mathbb{P}_0^n$ in the sense of Theorem 7.2. Since $n \geq N$, from our choice (23) of N and Theorem 7.2, we have $\mathbb{P}_1^n(E) \geq 1 - \delta$. As in [4], let $R_1, \ldots, R_{100}$ be translates in $\mathbb{T}$ of the rectangle $[0,6] \times [0,2]$ arranged so that any 8-by-1 rectangle crosses some R_i horizontally. Then we have $\mathbb{P}_1^n\big(\bigcup H_{\mathrm{b}}(R_i)\big) \geq 1 - \delta$, so by Corollary 3.3 and our choice (22) of δ we have $\mathbb{P}_1^n\big(H_{\mathrm{b}}(R_i)\big) \geq 0.99$ for some i, and thus for all i.

Translating back to the plane, we see that in $\mathcal{C}_1'$ (obtained from $\mathcal{C}_1$ by the linear transformation mapping $\mathcal{C}_0$ to $\mathcal{C}_0'$), any 6-by-2 rectangle R with corners at points of the lattice $\mathcal{L}'$ has $\mathbb{P}\big(H_{\mathrm{b}}(R)\big) \geq 0.99$. The same argument shows that any 2-by-6 rectangle R' with lattice point corners has $\mathbb{P}\big(V_{\mathrm{b}}(R')\big) \geq 0.99$. From here it is very easy to prove that percolation occurs, using any of several standard methods; we shall give one example.

Note that if S is a 2-by-2 square then, with $\mathbb{P}$ denoting the probability measure associated to $\mathcal{C}_1'$, we have $\mathbb{P}\big(H_{\mathrm{b}}(S)\big) \geq \mathbb{P}\big(H_{\mathrm{b}}(R)\big) \geq 0.99$ and $\mathbb{P}\big(V_{\mathrm{b}}(S)\big) \geq \mathbb{P}\big(V_{\mathrm{b}}(R')\big) \geq 0.99$. As in [4] (the third version of the proof of Theorem 10 there), let $G(R)$ be the event that $H_{\mathrm{b}}(R)$ holds and each of the two 2-by-2 'end squares' of R has a black vertical crossing, and define $G(R')$ similarly. Then $\mathbb{P}\big(G(R)\big)$, $\mathbb{P}\big(G(R')\big) \geq 1 - 3(1-0.99) = 0.97$. Of course, the bound 0.97 here can be replaced by any constant less than 1, although, as we shall see, 0.97 is more than good enough.

Considering a square grid of 6-by-2 and 2-by-6 rectangles overlapping in 2-by-2 squares as in [4], and taking a bond of $\mathbb{Z}^2$ to be open if $G(R)$ holds for the corresponding rectangle, one obtains a dependent bond percolation measure on $\mathbb{Z}^2$. Given sets S and T of bonds of $\mathbb{Z}^2$ separated by a distance (in the graph $\mathbb{Z}^2$) of at least 1, the corresponding unions of rectangles are disjoint, and are separated in the plane by a distance of at least $2 \geq 1/10$. It follows that the states of the bonds in S are independent of the sates of the bonds in T, i.e., the bond percolation measure is 1-*independent*. It is rather easy to see that any such measure in which each bond is open with high enough probability has an infinite open cluster with probability 1; see, for example, the general domination result of Liggett, Schonmann and Stacey [19]. The best current bound on what 'high enough' means is due to Balister, Bollobás and Walters [2], who showed that all bond probabilities at least 0.8639 will do. Since $\mathbb{P}\big(G(R)\big), \mathbb{P}\big(G(R')\big) \geq 0.97$, we see that with probability 1 there is an infinite open cluster in $\mathbb{Z}^2$. Translating back, the definition of $G(R)$ ensures that we find a corresponding infinite black cluster in $\mathcal{C}'_1$. Since $\mathcal{C}'_1$ is simply a linear image of $\mathcal{C}_1$, it follows that $\mathcal{C}_1$ contains an infinite black cluster with probability 1; hence $\mathcal{H}(\mathbf{q})$ percolates, as required.

It remains to establish exponential decay of the volume in $\mathcal{H}(\mathbf{q})$ for $\mathbf{q} \prec \mathbf{p}$. But first note that with $\mathbf{q} \succ \mathbf{p}$ as above, and with the arbitrary constant 0.99 replaced by a suitable constant $a < 1$, the argument in [5, Section 3] (again using locally-dependent percolation) shows that the dual of the model $\mathcal{C}'_1$ exhibits exponential decay (of the volume). It follows that the dual $\mathcal{H}^*(\mathbf{q}^*)$ of $\mathcal{H}(\mathbf{q})$ exhibits exponential decay.

Given a self-dual model $\mathcal{H}(\mathbf{p})$ and $\mathbf{q} \prec \mathbf{p}$, the model $\mathcal{H}^*(\mathbf{p}^*)$ is self-dual (it is isomorphic to $\mathcal{H}(\mathbf{p})$), and $\mathbf{q}^* \succ \mathbf{p}^*$. Applying the result above to $\mathcal{H}^*(\mathbf{p}^*)$ and $\mathcal{H}^*(\mathbf{q}^*)$, we see that the dual $\mathcal{H}(\mathbf{q})$ of $\mathcal{H}^*(\mathbf{q}^*)$ exhibits exponential decay, as required.

Suppose instead that $\mathcal{H}(\mathbf{p})$ is approximately self-dual. Then by definition there is a model $\mathcal{H}'(\mathbf{p}')$ that is isomorphic to $\mathcal{H}(\mathbf{p})$, such that $\mathcal{H}^*(\mathbf{p}^*)$ and $\mathcal{H}'(\mathbf{p}')$ are equivalent, in the sense that they may be coupled so that for any open path in either model there is a nearby open path in the other model. Recall also that we may take the isomorphism to be given by an isometry of the plane (plus a small 'distortion', if needed). Taking the colouring viewpoint, open paths are simply black paths. Now the notion of equivalence is not obviously preserved under taking duals, i.e., inverting the colouring. However, the condition for approximate self-duality is exactly that for every *white* path in $\mathcal{H}(\mathbf{p})$ (corresponding to a black path in $\mathcal{H}^*(\mathbf{p}^*)$), there is a nearby *black* path in $\mathcal{H}'(\mathbf{p}')$, and vice versa. Since $\mathcal{H}(\mathbf{p})$ and $\mathcal{H}'(\mathbf{p}')$ are isomorphic, $\mathcal{H}'(\mathbf{p}')$ also satisfies this condition, so we may couple $\mathcal{H}'(\mathbf{p}')$ and $\mathcal{H}(\mathbf{p})$ so that for every white path in $\mathcal{H}'(\mathbf{p}')$ there is a nearby black path in $\mathcal{H}(\mathbf{p})$, and vice versa. In other words, approximate self-duality holds after interchanging black and white, i.e., $\mathcal{H}^*(\mathbf{p}^*)$ is approximately self-dual. From this point the argument for exponential decay is as in the self-dual case. ■

8. On the Critical Surface

In the bulk of this paper we have shown that any self-dual hyperlattice percolation model $\mathcal{H}(\mathbf{p})$ is 'critical' in the sense that if $\mathbf{q} \succ \mathbf{p}$ then $\mathcal{H}(\mathbf{q})$ percolates, while if $\mathbf{q} \prec \mathbf{p}$ then $\mathcal{H}(\mathbf{q})$ exhibits exponential decay.

As noted earlier, the model $\mathcal{H}(\mathbf{p})$ itself may or may not percolate. Here we show that, except for degenerate cases, it does not. Furthermore, we show that one has power-law decay of the radius, as expected. Let v_0 be any fixed vertex of $\mathcal{H}$; we write $v_0 \to r$ for the event that there is an open path from v_0 to a vertex at distance at least r from v_0.

Theorem 8.1. *Let $\mathcal{H}(\mathbf{p})$ be a malleable self-dual hyperlattice percolation model. Then there are constants $0 < a_1 < a_2$ and r_0 such that $r^{-a_2} \le \mathbb{P}(v_0 \to r) \le r^{-a_1}$ for all $r \ge r_0$.*

Note that since we argue directly about properties of the self-dual case, we need to impose the technical condition of malleability defined in Definition 5.9. It seems likely that this can be weakened to non-degeneracy; the latter condition is used throughout the proofs in the previous sections, whereas malleability is only used at one point, where the need for it could perhaps be circumvented.

Proof. By Lemma 5.10, $\mathcal{H}(\mathbf{p})$ can be realized by a malleable independent lattice colouring $\mathcal{C}$. By Theorem 6.3, $\mathcal{C}$ has the large rectangles property and hence, by Lemma 6.5, the large ellipses property.

Let S and T be the maps exhibiting self-duality, as described in Lemma 6.7. Let $L = 100 \max\{d_0, d_1\}$, where d_0 is the constant in Lemma 4.2 and d_1 is the bound on $\big|S(x) - T(x)\big|$ from Lemma 6.7.

The large ellipses property tells us that there are constants $c > 0$ and A_0 such that for every $A \ge A_0$ there is an ellipse $E = E(A)$ centred on the origin with area A and with both axes having length at least L such that $\mathbb{P}\big(O_{\mathrm{b}}(E)\big) \ge c$. Let N be an integer such that $(1-c)^N < c$.

For $i \ge 0$ let $E_i = E\big((10N)^i A_0\big)$ be an ellipse as above with area $(10N)^i A_0$. We claim that for each i, $T(E_i)$ fits inside a copy of E_{i+1} scaled by a factor 1/2. To see this, rotate and scale so that the ellipse $T(E_i)$ becomes a circle with diameter 1, and E_{i+1} has horizontal major axis with length a and vertical minor axis with length b. Our claim is exactly that $b \geq 2$. But if not, then $b \leq 2$ so, since $\text{area}\,(E_{i+1}) = 10N\text{area}\,(E_i) = 10N\text{area}\big(T(E_i)\big)$, we have $a \geq 5N$. Writing E^0 for the annulus between an ellipse E and the concentric ellipse $2E/3$, it is not hard to see that when $b \leq 2$ and $a \geq 5N$ one can arrange N disjoint copies of $T(E_i)$ to 'cross' the annulus E^0_{i+1} as in Figure 8.1. Moreover, in the rescaled arrangement one can easily ensure that

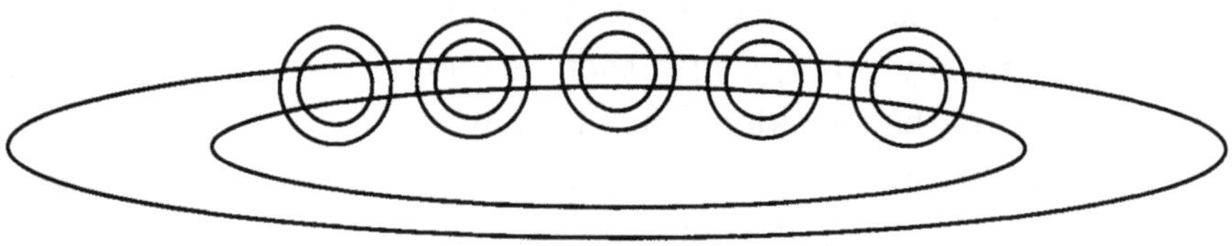

Fig. 8.1. Circles 'crossing' an elliptical annulus. Depending on the height of the ellipse, the circles may or may not intersect the bottom half of the ellipse; this is irrelevant for the argument.

the circles are separated by a distance of at least 1/10, say, and each 'crosses' E^0_{i+1} even after any transformation moving points by a distance of at most 1/10. Since lengths with original scale $\max\{d_0, d_1\}$ have transformed scale at most 1/100, it follows that each image $S(E_i)$ crosses E^0_{i+1}, and the $S(E_i)$ are separated by a distance of at least d_0. But by self-duality each $S(E_i)$ contains a *white* path surrounding its centre with probability at least c, and these events are independent. Whenever one of these white paths is present, $O_{\mathrm{b}}(E_{i+1})$ cannot hold. Hence $\mathbb{P}\big(O_{\mathrm{b}}(E_{i+1})\big) \leq (1-c)^N < c$, a contradiction.

Recalling from Lemma 6.7 that T^2 is either the identity or reflection in the origin, we see that E^0_{i+2} surrounds $T(E^0_{i+1})$ which surrounds $T\big(T(E^0_i)\big) = E^0_i$. Hence the annuli E^0_{2i} are disjoint. Moreover, since $1/2 < 2/3$ and all axes of all E_i have length at least L, the annuli E^0_{2i} are separated by distances of at least d_0, and so meet disjoint sets of faces of $\mathcal{H}$. Now the shorter axis of E_i has length at least $L \geq 1$, so its longer axis has length at most $(10N)^i A_0$. It follows that for large r, any point at distance r from the origin is outside E_k, where $k = \Theta(\log r)$. There can only be a *white* path starting at the origin and ending at least distance r away if none of the events $O_{\mathrm{b}}(E_{2i})$, $2i < k$, holds. Since these events are independent, this has probability at most $(1-c)^{\lfloor k/2 \rfloor} = \exp\big(-\Theta(\log r)\big)$. Passing to the dual, this proves the upper bound on $\mathbb{P}(v_0 \to r)$.

The lower bound is essentially immediate, with 1 as the exponent. Indeed, Lemma 4.4 tells us that if R is an r-by-r square, then either

$\mathbb{P}(H_b(R)) \geq 1/2$ or $\mathbb{P}(V_w(R)) \geq 1/2$. Suppose first that $\mathbb{P}(H_b(R)) \geq 1/2$. Then considering one of the $O(r)$ translates of our reference vertex v_0 within distance $O(1)$ of the left-hand side of R, by the union bound there is a translate of v_0 such that the probability that there is an open path starting at v_0 with length at least $r - O(1)$ is at least $O(1/r)$.

If $\mathbb{P}(V_w(R)) \geq 1/2$ then we apply the same argument in the dual, which is isomorphic to the original percolation model. ∎

It is easy to see that the argument above extends to the approximately self-dual case. The key point is that there is enough 'elbow room' for small distortions of the paths considered not to matter.

It seems very probable that the general conjecture of Aizenman and Langlands, Pouliot and Saint-Aubin [18] concerning conformal invariance of the scaling limit of critical plane percolation will hold for all non-degenerate self-dual hyperlattice percolation models. However, this is likely to be very hard to prove. This conjecture asserts, among other things, that if R is any rectangle, λR denotes its image under a dilation with scale-factor λ, and $H(R)$ denotes the event that R has an open (or here, black) horizontal crossing, then for any fixed R, the limit $\lim_{\lambda\to\infty} \mathbb{P}(H(\lambda R))$ exists and lies strictly between 0 and 1. Moreover, this limit should be given by Cardy's formula [9], after first applying a suitable linear transformation to the model.

This conjecture has been proved by Smirnov [28] for site percolation on the triangular lattice; this is essentially the only case known. For many other models, such as bond percolation on the square lattice, RSW-type theorems give the much weaker result that

$$0 < \liminf_{\lambda\to\infty} \mathbb{P}(H(\lambda R)) \leq \limsup_{\lambda\to\infty} \mathbb{P}(H(\lambda R)) < 1. \tag{24}$$

(This applies just as well to shapes other than rectangles.) Unfortunately, Theorem 6.3, while strong enough to establish criticality, lacks the uniformity needed to prove (24), so we leave this as a conjecture.

Conjecture 8.2. Let $\mathcal{H}(\mathbf{p})$ be a non-degenerate (approximately) self-dual hyperlattice percolation model. Then for any fixed rectangle R, the bounds (24) hold.

As far as we are aware, this conjecture is open even for the simple special case of inhomogeneous bond percolation on the square lattice, where each horizontal bond is open with probability p and each vertical one with probability $1 - p$, with the states of all bonds independent.

Acknowledgements. We are grateful to Robert Ziff for bringing the results of Scullard and himself to our attention: this paper started from an attempt to show that the self-duality they established in certain (quite general) cases does imply criticality. The last section of this paper was added in response to a question asked by Marek Biskup.

References

[1] M. Aizenman, H. Kesten and C. M. Newman, Uniqueness of the infinite cluster and related results in percolation, in: *Percolation Theory and Ergodic Theory of Infinite Particle Systems* (Minneapolis, Minn., 1984–1985), Springer (1987), pp. 13–20.

[2] P. Balister, B. Bollobás and M. Walters, Continuum percolation with steps in the square or the disc, *Random Struct. Alg.,* **26** (2005), 392–403.

[3] B. Bollobás and O. Riordan, The critical probability for random Voronoi percolation in the plane is 1/2, *Probability Theory and Related Fields,* **136** (2006), 417–468.

[4] B. Bollobás and O. M. Riordan, A short proof of the Harris–Kesten Theorem, *Bull. London Math. Soc.,* **38** (2006), 470–484.

[5] B. Bollobás and O. Riordan, Sharp thresholds and percolation in the plane, *Random Struct. Alg.,* **29** (2006), 524–548.

[6] B. Bollobás and O. Riordan, *Percolation,* Cambridge University Press, 2006, x + 323 pp.

[7] B. Bollobás and O. Riordan, Percolation on dual lattices with k-fold symmetry, *Random Struct. Alg.,* **32** (2008), 463–472.

[8] J. Bourgain, J. Kahn, G. Kalai, Y. Katznelson and N. Linial, The influence of variables in product spaces, *Israel J. Math.,* **77** (1992), 55–64.

[9] J. L. Cardy, Critical percolation in finite geometries, *J. Phys. A,* **25** (1992), L201–L206.

[10] L. Chayes and H. K. Lei, Random cluster models on the triangular lattice, *J. Stat. Phys.,* **122** (2006), 647–670.

[11] L. Chayes and H. K. Lei, Cardy's formula for certain models of the bond-triangular type, *Reviews in Math. Physics,* **19** (2007), 511–565.

[12] E. Friedgut and G. Kalai, Every monotone graph property has a sharp threshold, *Proc. Amer. Math. Soc.,* **124** (1996), 2993–3002.

[13] G. Grimmett, *Percolation,* Second edition, Springer, 1999, xiv + 444 pp.

[14] T. E. Harris, A lower bound for the critical probability in a certain percolation process, *Proc. Cam. Philos. Soc.,* **56** (1960), 13–20.

[15] J. Kahn, G. Kalai and N. Linial, The influence of variables on boolean functions, Proc. 29th Annual Symposium on Foundations of Computer Science, 68–80, Computer Society Press, 1988.

[16] H. Kesten, The critical probability of bond percolation on the square lattice equals 1/2, *Comm. Math. Phys.,* **74** (1980), 41–59.

[17] H. Kesten, *Percolation Theory for Mathematicians,* Birkhäuser, 1982, iv + 423 pp.

[18] R. Langlands, P. Pouliot and Y. Saint-Aubin, Conformal invariance in two-dimensional percolation, *Bull. Amer. Math. Soc. (N.S.),* **30** (1994), 1–61.

[19] T. M. Liggett, R. H. Schonmann and A. M. Stacey, Domination by product measures, *Ann. Probab.,* **25** (1997), 71–95.

[20] G. A. Margulis, Probabilistic characteristics of graphs with large connectivity, *Problemy Peredači Informacii,* **10** (1974), 101–108.

[21] M. V. Menshikov, Coincidence of critical points in percolation problems, *Soviet Math. Dokl.,* **33** (1986), 856–859.

[22] L. Russo, A note on percolation, *Z. Wahrsch. Verw. Gebiete,* **43** (1978), 39–48.

[23] L. Russo, On the critical percolation probabilities, *Z. Wahrsch. Verw. Gebiete,* **56** (1981), 229–237.

[24] C. R. Scullard, Exact site percolation thresholds using a site-to-bond transformation and the star-triangle transformation, *Phys. Rev. E,* **73** (2006), 016107 [6 pages]

[25] M. R. A. Sedlock and J. C. Wierman, Equality of bond percolation critical exponents for pairs of dual lattices, *Phys. Rev. E,* **79** (2009), 051119 [10 pages]

[26] P. D. Seymour and D. J. A. Welsh, Percolation probabilities on the square lattice, in: *Advances in Graph Theory* (Cambridge Combinatorial Conf., Trinity College, Cambridge, 1977). *Ann. Discrete Math.,* **3** (1978), pp. 227–245.

[27] S. Sheffield, Random surfaces, *Astérisque,* **304** (2005), vi + 175 pp.

[28] S. Smirnov, Critical percolation in the plane: conformal invariance, Cardy's formula, scaling limits, *Comptes Rendus de l'Académie des Sciences. Série I. Mathématique,* **333** (2001), 239–244.
Expanded version available at `www.math.kth.se/~stas/papers`.

[29] P. N. Suding and R. M. Ziff, Site percolation thresholds for Archimedean lattices, *Phys. Rev. E,* **60** (1999), 275–283.

[30] M. F. Sykes and J. W. Essam, Some exact critical percolation probabilities for bond and site problems in two dimensions, *Physical Review Letters,* **10** (1963), 3–4.

[31] J. C. Wierman, Bond percolation on honeycomb and triangular lattices, *Adv. in Appl. Probab.,* **13** (1981), 298–313.

[32] J. C. Wierman, A bond percolation critical probability determination based on the star-triangle transformation, *J. Phys. A,* **17** (1984), 1525–1530.

[33] J. C. Wierman and R. M. Ziff, Triangle-duality and equality of infinitely many bond percolation thresholds, preprint (2009). `http://arxiv.org/abs/0903.3135v1`

[34] F. Y. Wu, New critical frontiers for the Potts and percolation models, *Physical Review Letters,* **96** (2006), 090602 [4 pages]

[35] R. M. Ziff, Generalized cell–dual-cell transformation and exact thresholds for percolation, *Phys. Rev. E,* **73** (2006), 016134 [6 pages]

[36] R. M. Ziff and C. R. Scullard, Exact bond percolation thresholds in two dimensions, *J. Phys. A: Math. Gen.,* **39** (2006), 15083–15090.

Béla Bollobás

Department of Pure Mathematics and Mathematical Statistics
Wilberforce Road
Cambridge CB3 0WB, UK
and
Department of Mathematical Sciences
University of Memphis
Memphis TN 38152
USA

e-mail:
B.Bollobas@dpmms.cam.ac.uk

Oliver Riordan

Mathematical Institute
University of Oxford
24–29 St. Giles'
Oxford OX1 3LB
UK

e-mail: riordan@maths.ox.ac.uk

BOLYAI SOCIETY
MATHEMATICAL STUDIES, 21

An Irregular Mind
(Szemerédi is 70)
pp. 219–242.

On Exponential Sums in Finite Fields

JEAN BOURGAIN*

0. Introduction

The purpose of this paper is to establish certain multilinear exponential sums in arbitrary finite fields, extending some of the results from [1] for prime fields.

Let us first recall the main result from [1].

Theorem A. *Let $1 > \delta > 0$ and $r \in \mathbb{Z}_+$, $r \geq 2$. There is $\delta' > \left(\frac{\delta}{r}\right)^{Cr}$ such that if p is a sufficiently large prime and $A_1, \dots, A_r \subset \mathbb{F}_p$ satisfy*

$$|A_i| > p^\delta \quad \text{for} \quad 1 \leq i \leq r \tag{0.1}$$

$$\prod_{i=1}^{r} |A_i| > p^{1+\delta}. \tag{0.2}$$

Then we have the exponential sum bound

$$\left| \sum_{x_1 \in A_1, \dots, x_r \in A_r} e_p(x_1 \dots x_r) \right| < p^{-\delta'} |A_1| \dots |A_r|. \tag{0.3}$$

Consider now a field $\mathbb{F}_q$, $q = p^n$. An obvious issue one encounters with a generalization of Theorem A is the presence of non-trivial subfields. More surprisingly perhaps, it turns out that even if $\mathbb{F}_q$ has no large non-trivial subfields, the condition (0.2) still needs to be modified.

Theorem 3 on p. 20 below implies the following statement:

*The author was supported in part by the NSF grant DMS 0808042.

Theorem B. *Let* $0 < \delta, \delta_2 < 1$ *and* $r \in \mathbb{Z}_+$, $r \geq 2$. *Let* $q = p^n$ *be sufficiently large and* $A_1, \dots, A_r \subset \mathbb{F}_q$ *satisfy*

$$|A_i| > q^\delta \quad \text{for} \quad 1 \leq i \leq r \tag{0.4}$$

$$\big| A_i \cap (aG + b)\big| < q^{-\delta_2}|A_i| \quad \text{for} \quad 3 \leq i \leq r, \tag{0.5}$$

whenever $a, b \in \mathbb{F}_q$ *and* G *a proper subfield*

$$|A_1|.|A_2|.\prod_{i=3}^{r} |A_i|^{\frac{1}{2}} > q^{1+\delta}. \tag{0.6}$$

Then, denoting $\psi(x) = e_p(Tr_{\mathbb{F}_q/\mathbb{F}_p} x)$, *we have*

$$\Big| \sum_{x_1 \in A_1, \dots, x_r \in A_r} \psi(x_1 \dots x_r)\Big| < q^{-\delta'}|A_1| \dots |A_r| \tag{0.7}$$

where we may take $\delta' = C^{-\frac{r}{\delta_2}} \left(\frac{\delta}{r}\right)^{Cr}$.

Remarks.

(0.8) Condition (0.5) may in fact be replaced by

$$\big| A_i \cap (aG + b)\big| < |A_i|^{1-\delta_2} \quad (3 \leq i \leq r).$$

It follows in particular that if we fix the characteristic p and let n be prime, we may take $\delta' = \left(c\frac{\delta}{r}\right)^{C_r}$.

(0.9) Assume $|A_1| = \cdots = |A_r| = q^\sigma$. Condition (0.6) becomes then

$$\sigma > \frac{2}{r+2}. \tag{0.10}$$

This condition is in some sense optimal, as seen from the obvious example $\mathbb{F}_q = \mathbb{F}_{p^n} = \mathbb{F}_p[\xi]$, $r = n - 1$, $A_1 = \cdots = A_r = \mathbb{F}_p + \xi\mathbb{F}_p$, $\sigma = \frac{2}{n}$.

Multilinear exponential sums arise naturally if one applies Weyl's differencing scheme to Gauss sums. More precisely, consider $B \subset \mathbb{F}_q$, $r \in \mathbb{Z}_+$, $r \geq 2$ and

$$S = \sum_{x \in B} \psi(x^r) \tag{0.11}$$

with ψ as above.

One obtains (cf. [2], Lemma 3.1)

$$|S|^{2^{r-1}} \leq |B - B|^{2^{r-1}-r} \times \sum_{x_1 \in B-B} \cdots \sum_{x_{r-1} \in B-B} \Big| \sum_{x_r \in B(x_1,\dots,x_{r-1})} \psi(2^r r! x_1 \cdots x_r) \Big| \tag{0.12}$$

where

$$B(x_1, \dots, x_{r-1}) = \bigcap_{\varepsilon_1=0}^{1} \cdots \bigcap_{\varepsilon_{r-1}=0}^{1} (B - \varepsilon_1 x_1 - \cdots - \varepsilon_{r-1} x_{r-1}). \tag{0.13}$$

If $B \subset \mathbb{F}_q$ is a linear subspace over $\mathbb{F}_p$, we derive immediately from Theorem B and the preceding

Theorem C. *Let $r \in \mathbb{Z}_+$, $r \geq 2$, $p > r$ and V a linear subspace of $\mathbb{F}_{p^n}$ over $\mathbb{F}_p$ of dimension*

$$m = \dim V > (1+\delta)\frac{2n}{r+2}. \tag{0.14}$$

where $0 < \delta < 1$.

Assume further that

$$|V \cap aG| < q^{-\delta_2} p^m \tag{0.15}$$

if G is a proper subfield, $a \in \mathbb{F}_{p^n}^*$.

Then (assuming q large enough)

$$\max_{a \in \mathbb{F}_q^*} \Big| \sum_{x \in V} \psi(ax^r) \Big| < q^{-\delta'} p^m \text{ with } \delta' > C^{-\frac{r}{\delta_2}} \left(\frac{\delta}{r}\right)^{Cr}. \tag{0.16}$$

From Remark (0.9), we see that condition (0.14) on $\dim V$ is essentially optimal.

If $e_1, \dots, e_n$ is an (arbitrary) basis of $\mathbb{F}_q$ over $\mathbb{F}_p$, we define a 'box' as a translate of a set

$$B = \{t_1 e_1 + \cdots + t_n e_n \mid 1 \leq t_i \leq H_i,\ 1 \leq i \leq n\} \tag{0.17}$$

where $1 \le H_1, \dots, H_n \le p$. For $H_1 = \cdots = H_n = H$, denote (0.17) by B_H. One easily verifies that if G is a proper subfield of $\mathbb{F}_q$ and $a, b \in \mathbb{F}_q$,

$$\big| B_H \cap (aG + b) \big| \le |B_H|^{\frac{1}{2}}. \tag{0.18}$$

Also, in (0.13), the set $B_H(x_1, \dots, x_{r-1})$ ia a union of at most 2^{nr} boxes (0.17) with $H_i \le H$.

From (0.12), one obtains therefore

$$|S|^{2^{r-1}} \le 2^{nr} |B - B|^{2^{r-1} - r} \times \sum_{x_1, \dots, x_{r-1} \in B - B} \prod_{i=1}^{n} \min \left(H, \left\| \frac{2^r r!}{p} Tr(x_1 \dots x_{r-1} e_i) \right\|^{-1} \right). \tag{0.19}$$

Denote $\varphi = \varphi_H$ the function on $\mathbb{F}_p$

$$\varphi(z) = \min \left(H, \left\| \frac{z}{p} \right\|^{-1} \right), \quad z \in \mathbb{F}_p \tag{0.20}$$

and

$$\hat{\varphi}(t) = \frac{1}{p} \sum_{0 \le z < p} \varphi(z) e_p(-tz). \tag{0.21}$$

Hence

$$\varphi(z) = \sum_{0 \le t \le p} \hat{\varphi}(t) e_p(tz)$$

and

$$cH < \|\hat{\varphi}\|_1 < C(\log p) H \tag{0.22}$$

$$\|\hat{\varphi}\|_2 = \frac{1}{\sqrt{p}} \|\varphi\|_2 \sim c\sqrt{H}. \tag{0.23}$$

Thus

(0.24)

$$(0.19) = 2^{nr} |B - B|^{2^{r-1} - r} \sum_{x_1, \dots, x_{r-1} \in B - B} \prod_{i=1}^{n} \varphi\big(2^r r! (Tr x_1 \dots x_{r-1} e_i)\big)$$

$$= 2^{nr}|B-B|^{2^{r-1}-r} \sum_{x_1\dots x_{r-1}\in B-B} \sum_{x\in\mathbb{F}_q} \alpha(x)\psi(x_1\dots x_{r-1}x)$$

where

$$\alpha(x) = \prod_{i=1}^{n} \hat{\varphi}(t_i) \text{ for } x = t_1e_1 + \dots + t_ne_n$$

satisfies by (0.22), (0.23)

$$(cH)^n < \|\alpha\|_1 < C^n(\log p)^n H^n \tag{0.25}$$

$$\|\alpha\|_2 \sim C^n H^{n/2} \tag{0.26}$$

The double sum in (0.24) is estimated using Theorem 3 (stated on p. 20), taking $\alpha_1 = \frac{\alpha}{\|\alpha\|_1}$ and $\alpha_2 = \dots = \alpha_r = \frac{1_{|B-B|}}{|B-B|}$.

Take $H = p^\sigma$ and σ satisfying

$$\sigma > (1+\delta)\frac{2}{r+2} \tag{0.27}$$

$(0 < \delta < 1)$. Assume $p > p(r,\delta)$. It follows from (0.18) that (8.3) holds with $\delta_2 = \frac{\sigma}{2}$. From (0.25), (0.26) and (0.27), (8.2) and (8.4) hold with $\delta = \min\left(\frac{\delta}{4}, \frac{\sigma}{2}\right)$. From (8.5), we obtain

$$\big|(0.24)\big| < C^{n2^r}|B|^{2^{r-1}-1}\|\alpha\|_1 q^{-\delta'} < C^{n2^r}(\log p)^n|B|^{2^{r-1}} q^{-\delta'} \tag{0.28}$$

with $\delta' > C^{-\frac{r}{\sigma}}\left(\frac{\delta}{r}\right)^{Cr}$.

Hence, we proved

Theorem D. *Let $q = p^n$, $r \in \mathbb{Z}_+$, $r \geq 2, 0 < \delta < 1$ and $p > p(r,\delta)$. Let $H = p^\sigma$, with*

$$\sigma > (1+\delta)\frac{2}{r+2} \tag{0.29}$$

and $B_H \subset \mathbb{F}_q$ the box as defined above. Then, with $\delta' > C^{-\frac{r}{\sigma}}\left(\frac{\delta}{r}\right)^{Cr}$

$$\max_{a\in\mathbb{F}_q^*}\left|\sum_{x\in B_H} \psi(ax^r)\right| < C^n(\log p)^{n2^{-r+1}} q^{-\delta'} H^n. \tag{0.30}$$

Remarks.

1. Both Theorem C and Theorem D remain of course valid if we replace ax^r by an arbitrary polynomial $f(x) = a_r x^r + a_{r-1}x^{r-1} + \cdots + a_0 \in \mathbb{F}_q[X]$ with $a_r \neq 0$, as r-fold Weyl differencing leads to the same multilinear bound (0.12).

2. Theorem D should be compared with Theorem 2 from [2] on incomplete exponential sums in one and several variables (only the 1-variable result, i.e. $s = 1$ in the notation from [2], is of relevance here). In [2], Theorem 2, a nontrivial estimate on $\sum_{x\in B_H} \psi(f(x))$ is obtained, $f(x) \in \mathbb{F}_q[X]$ as above, under the assumption

$$H = p^\sigma, \quad \sigma > \frac{1}{r} \tag{0.31}$$

which is weaker than (0.29) (and optimal). However the result from [2] is not uniform in $n(q = p^n)$, in the sense that it requires $p > p(r, n)$, while (0.30) provides non-trivial bounds for $p > p(r)$ (assuming σ fixed). The method from [2] relies on geometry of numbers and the dependence on n results from dimensional factors in Minkowski's second theorem. Whether (or to what extent) they are avoidable in this particular application seems an interesting question.

The reminder of the paper is organized as follows:

In §1, we establish a 'sum-product' type result in a general finite field $\mathbb{F}_q$, which is the main new underlying ingredient (compared with [1]). The later sections are basically an adjustment from [1] to convert this set-theoretical property (Lemma 1 below) in bounds on convolutions and exponential sums.

1. A Sum-Product Property

The following will be the substitute for Lemma 2 in [1].

Lemma 1. *Let $X, Y \subset \mathbb{F}_q^*$ and assume Y not contained in a proper subfield of $\mathbb{F}_q$.*

There are elements $x_1, x_2, x_3, x_4 \in \pm X$ and $y_1, y_2, y_3, y_4 \in \pm Y \cup \{1\}$ such that for all $X' \subset X$, $Y' \subset Y$

$$|y_1X' + y_2X' + y_3X' + y_4X' + x_1Y' + x_2Y' + x_3Y' + x_4Y'| \tag{1.1}$$

$$\geq \min\left\{\frac{1}{6}|X'|\,|Y'|^{\frac{1}{2}}, q\left(\frac{|X'|\,|Y'|}{|X|\,|Y|}\right)^2\right\}.$$

Remark. Assuming Y not contained in a multiplicative coset of a proper subfield, we may take y_1, y_2, y_3, y_4 above in $\pm Y$.

Proof of Lemma 1. We may clearly assume $|X| > 1$.

Define

$$V = \frac{X - X}{Y - Y} \neq \{0\}$$

and notice that the properties

$$YV \subset V \tag{1.2}$$

$$V + V \subset V \tag{1.3}$$

can not both hold unless

$$V = \mathbb{F}_q. \tag{1.4}$$

Indeed, if (1.2) + (1.3), then V contains any sum of products of elements of Y and hence the field generated by Y, multiplied with V.

If (1.2) fails, there are $y_1, y_2, y_3 \in Y (y_1 \neq y_2)$ and $x_1, x_2 \in X$ such that $\xi = y_3 \frac{x_1 - x_2}{y_1 - y_2} \notin V$.

Hence, if $X' \subset X$, $Y' \subset Y$, $\xi \notin \frac{X' - X'}{Y' - Y'}$ implying

$$\begin{aligned}
|X'|\,|Y'| &= |X' + \xi Y'| \\
&= \big|(y_1 - y_2)X' + y_3(x_1 - x_2)Y'\big| \\
&\leq \big|(y_1 - y_2)X' + y_3X'\big|\,\big|(x_1 - x_2)Y' - X'\big|\,|X'|^{-1}.
\end{aligned}$$

Hence, either

$$\big|(y_1 - y_2)X' + y_3X'\big| \geq |X'|\,|Y'|^{\frac{1}{2}}$$

or

$$\big|X' + (x_2 - x_1)Y'\big| \geq |X'|\,|Y'|^{\frac{1}{2}}$$

and certainly

$$|X' + y_1X' - y_2X' + y_3X' - x_1Y' + x_2Y'| \geq |X'|\,|Y'|^{\frac{1}{2}}. \tag{1.5}$$

If (1.3) fails, there are $x_1, x_2, x_3, x_4 \in X$ and $y_1, y_2, y_3, y_4 \in Y$, $y_1 \neq y_2$, $y_3 \neq y_4$ such that

$$\xi = \frac{x_1 - x_2}{y_1 - y_2} + \frac{x_3 - x_4}{y_3 - y_4}V.$$

Let $X' \subset X$, $Y' \subset Y$ and define

$$Z = \frac{x_1 - x_2}{y_1 - y_2}Y' \cup \frac{x_3 - x_4}{y_3 - y_4}Y'.$$

Write

$$|X' + Z| = K|X'|. \tag{1.6}$$

Applying Corollary 1.5 from [3], we obtain a subset $X'' \subset X'$, $|X''| > \frac{1}{2}|X'|$ and such that

$$|X'' + Z + Z| \leq 4K^2|X'|. \tag{1.7}$$

Hence

$$\begin{aligned} 4K^2|X'| &\geq \left| X'' + \frac{x_1 - x_2}{y_1 - y_2}Y' + \frac{x_3 - x_4}{y_3 - y_4}Y' \right| \\ &\geq |X'' + \xi Y'| \\ &= |X''|\,|Y'| \\ &\geq \frac{1}{2}|X'|\,|Y'| \end{aligned}$$

and

$$K > \frac{1}{\sqrt{8}}|Y'|^{\frac{1}{2}}. \tag{1.8}$$

Returning to (1.6), we showed

$$\big|(y_1 - y_2)X' + (x_1 - x_2)Y'\big| + \big|(y_3 - y_4)X' + (x_3 - x_4)Y'\big| \geq \frac{1}{\sqrt{8}}|X'|\,|Y'|^{\frac{1}{2}}$$

and therefore

$$(1.9)\qquad |y_1X' - y_2X' + y_3X' - y_4X' + x_1Y' - x_2Y' + x_3Y' - x_4Y'| \geq \frac{1}{2\sqrt{8}}|X'|\,|Y'|^{\frac{1}{2}}.$$

Finally, assume (1.4).

Take $\xi \in V$, $\xi = \frac{x_1 - x_2}{y_1 - y_2}$, s.t.

$$(1.10)\qquad \left|\left\{(x, x', y, y') \in X \times X \times Y \times Y \,\middle|\, \xi = \frac{x - x'}{y - y'}\right\}\right| \leq \frac{|X|^2\,|Y|^2}{q}.$$

If $X' \subset X, Y' \subset Y$, we have

$$(1.11)\qquad |X' + \xi Y'| \geq \frac{|X'|^2|Y'|^2}{E_+(X', \xi Y')}$$

where

$$E_+(A, B) = \left|\{\,(a, a', b, b') \in A^2 \times B^2; a + b = a' + b'\}\,\right|$$

is the additive energy.

Clearly

$$E_+(X', \xi Y') \leq |X'|\,|Y'| + (1.10)$$

implying

$$(1.12)\quad \left|(y_1 - y_2)X' + (x_1 - x_2)Y'\right| \geq \frac{1}{2}\min\left(|X'|.|Y'|, q\left(\frac{|X'|\,|Y'|}{|X|\,|Y|}\right)^2\right).$$

Thus (1.1) holds again.

This proves Lemma 1. ■

With Lemma 1 at hand, we may follow the method from [1] almost verbatim (the main steps with details of the modifications will be given). This part of the analysis in [1] does indeed not depend on the primality of the field. Of course, in the applications of Lemma 1, one has to ensure that the set Y under consideration is not contained in a proper subfield.

Recall the following property from additive combinatorics ([1], Lemma 3), which holds in the context of an arbitrary additive group.

Lemma 2. *Let $X_i \subset \mathbb{F}_q$ $(1 \le i \le j)$ and $Y \subset \mathbb{F}_q$. There is $y_0 \in Y$ such that*

$$\left|(Y - y_0) \cap \bigcap_{i=1}^{j} (X_i - X_i)\right| \ge \left(\prod_{i=1}^{j} \frac{|X_i|}{|X_i - Y|}\right)|Y|. \tag{1.13}$$

2. Preliminary Estimates (1)

Recall [1], Lemma 5, which is deduced from the Balog–Szemerédi–Gowers theorem ([1], Prop. 1). Only the additive structure is involved.

Lemma 3. *Let $\alpha : \mathbb{F}_q \to \mathbb{R}_+$ satisfy $\|\alpha\|_1 = \sum_{x\in\mathbb{F}_q} |\alpha(x)| \le 1$. Fix $1 < K < q$. There are the following alternatives.*

Either

$$\|\alpha * \alpha\|_2 < \frac{1}{K}\|\alpha\|_2 \quad (* \text{ denotes additive convolution}) \tag{2.1}$$

or there is a subset $A \subset \mathbb{F}_q$ with the following properties (we ignore multiplicative constants).

$$\frac{2\sigma}{|A|} > \alpha|_A > \frac{\sigma}{|A|} \text{ where } 1 \ge \sigma > (\log q)^{-6} K^{-3}; \tag{2.2}$$

$$\big\|\alpha|_A\big\|_1 > (\log q)^{-6} K^{-3} \tag{2.3}$$

$$\big\|\alpha|_A\big\|_2 > (\log q)^{-4} K^{-2} \|\alpha\|_2 \tag{2.4}$$

$$|A + A| < (\log q)^{76} K^{38} |A|. \tag{2.5}$$

The argument is identical to the prime case.

Iteration of Lemma 3 gives ([1], Lemma 6).

Lemma 4. *Let $\alpha : \mathbb{F}_q \to \mathbb{R}_+$, $\|\alpha\|_1 \le 1$. Fix $1 \le K \le q$.*

Then there is a decomposition (with disjointedly supported components)

$$\alpha = \sum_{j \le J} \alpha_j + \beta \tag{2.6}$$

where each α_j satisfies for some $B_j \subset \mathbb{F}_q$

$$\frac{\sigma_j}{|B_j|} \mathcal{X}_{B_j} < \alpha_j < \frac{2\sigma_j}{|B_j|} \mathcal{X}_{B_j} \text{ with } 1 \ge \sigma_j > (\log q)^{-6} K^{-3} \tag{2.7}$$

and

$$\|\alpha_j\|_1 > (\log q)^{-6} K^{-3}, \tag{2.8}$$

$$\|\alpha_j\|_2 > K^{-2} (\log q)^{-4} \|\alpha\|_2. \tag{2.9}$$

$$|B_j + B_j| < K^{38} (\log q)^{76} |B_j|, \tag{2.10}$$

$$J < (\log q)^6 K^3, \tag{2.11}$$

$$\|\beta * \beta\|_2 < \frac{1}{K} \|\beta\|_2 < \frac{1}{K} \|\alpha\|_2. \tag{2.12}$$

Denote $\psi(x) = e_p(Trx), Tr = Tr_{\mathbb{F}_q/\mathbb{F}_p}$ the additive character of $\mathbb{F}_q$, $q = p^n$.

Lemma 5. *Let $\alpha, \beta, \gamma : \mathbb{F}_q \to \mathbb{R}_+$; $\|\alpha\|_1, \|\beta\|_1, \|\gamma\|_1 \le 1$.*

Take $1 \le K \le q$.

Then

$$(2.1) \qquad |S| = \Big| \sum_{x,y,z} \alpha(x)\beta(y)\gamma(z)\psi(xyz) \Big| \le 3|S_1| \tag{2.13}$$

$$+ 8 \max_{(*)} \frac{1}{|A|.|B|.|C|} \Big| \sum_{x \in A,\ y \in B,\ z \in C} \psi(xyz) \Big| \tag{2.14}$$

where $(*)$ refers to sets $A, B, C \subset \mathbb{F}_q$ such that

$$\frac{2\sigma}{|A|} > \alpha|_A > \frac{\sigma}{|A|} \quad \text{where} \quad 1 \ge \sigma > (\log q)^{-6} K^{-3} \tag{2.15}$$

$$(2.16)\qquad (\log q)^{-12}K^{-6}\|\alpha\|_2^{-2} < |A| < K^4(\log q)^8\|\alpha\|_2^{-2}$$

$$(2.17)\qquad |A+A| < (\log q)^{76}K^{38}|A|$$

and similarly for B, C and

$$(2.18)\qquad S_1 = \sum_{x,y,z} \alpha'(x)\beta'(y)\gamma'(z)\psi(xyz)$$

with $0 \le \alpha' \le \alpha$, $0 \le \beta' \le \beta$, $0 \le \gamma' \le \gamma$ *and*

$$(2.19)\quad \|\alpha' * \alpha'\|_2 < \frac{1}{K}\|\alpha'\|_2 \text{ or } \|\beta' * \beta'\|_2 < \frac{\|\beta'\|_2}{K} \text{ or } \|\gamma' * \gamma'\|_2 < \frac{1}{K}\|\gamma'\|_2.$$

Proof. Apply decomposition from Lemma 4 to each of the factors α, β, γ.

Note that in (2.7), $\sum \sigma_j \le \|\alpha\|_1$. In order to justify the characteristic functions $\mathcal{X}_A$, $\mathcal{X}_B$, $\mathcal{X}_C$ in (2.14), we use the fact that if $\mathcal{X}_\Omega \le f \le 2\mathcal{X}_\Omega$, then f may be recovered as an average of $\pm\mathcal{X}_{\Omega'}$ for subset $\Omega' \subset \Omega$, $|\Omega'| \sim |\Omega|$. ∎

Recall also that by Cauchy–Schwarz, we have

$$|S_1|^2 \le \sum_{y,z} \beta'(y)\gamma'(z)\Big|\sum_x \alpha'(x)\psi(xyz)\Big|^2$$

since $\|\beta'\|_1, \|\gamma'\|_1 \le 1$. Hence

$$(2.20)\qquad |S_1| \le \Big|\sum_{x,y,z}(\alpha' * \alpha')(x)\beta'(y)\gamma'(z)\psi(xyz)\Big|^{1/2}.$$

For $a, b, c > 0$, denote $\zeta(a,b,c)$ the maximum of

$$(2.21)\qquad \Big|\sum \alpha'(x)\beta'(y)\gamma'(z)\psi(xyz)\Big|$$

where $\alpha', \beta', \gamma' : \mathbb{F}_q \to \mathbb{R}_+$ satisfy

(2.22)
$$\|\alpha'\|_1 \le 1,\quad \|\beta'\|_1 \le 1,\quad \|\gamma'\|_1 \le 1 \text{ and } \|\alpha'\|_2 \le a,\quad \|\beta'\|_2 \le b,\quad \|\gamma'\|_2 \le c.$$

Lemma 5 implies then that

$$(2.23)\quad \zeta\big(\|\alpha\|_2, \|\beta\|_2, \|\gamma\|_2\big) \le (2.14) + 3\left\{\zeta\left(\frac{1}{K}\|\alpha\|_2, \|\beta\|_2, \|\gamma\|_2\right) + ''\right\}$$

('' referring to the other 2 terms).

3. Preliminary Estimates (2)

We will use the following construction.

Let

$$S = \sum \alpha(x)\beta(y)\psi(xy)$$

with $0 \leq \alpha, \beta$; $\|\alpha\|_1, \|\beta\|_1 = 1$.

Write

$$|S|^2 \leq \left| \sum (\alpha * \alpha)(x)\beta(y)\psi(xy) \right|,$$

and more generally (denoting $\alpha^{(\ell)}$ the ℓ-fold additive convolution of α)

$$|S|^{2^s} \leq \left| \sum \alpha^{(2^s)}(x)\beta(y)\psi(xy) \right|. \tag{3.1}$$

Fix $s \in \mathbb{Z}_+$ and define $L = L_s$ by

$$\left\|\alpha^{(2^{s+1})}\right\|_2 = \frac{1}{L}\left\|\alpha^{(2^s)}\right\|_2. \tag{3.2}$$

Applying Lemma 3 with α replaced by $\alpha^{(2^s)}$, it follows that there is a subset $A \subset \mathbb{F}_q$ satisfying

$$\left\|\alpha^{(2^s)}|_A\right\|_1 > (\log q)^{-6} L^{-3} \tag{3.3}$$

$$|A| < L^4 (\log q)^8 \left\|\alpha^{(2^s)}\right\|_2^{-2} \tag{3.4}$$

$$|A + A| < L^{38} (\log q)^{76}\, |A|. \tag{3.5}$$

Note that

$$\left\|\alpha^{(2^s)}|_A\right\|_1 \leq \max_{x \in \mathbb{F}} \left\| \alpha|_{x+A} \right\|_1.$$

Replacing A by a translate and denoting $\alpha_1 = \alpha^{(2^s)}$, it follows from (3.1) that

$$|S| \leq \left| \sum \alpha_1(x)\beta(y)\psi(x.y) \right|^{2^{-s}} \tag{3.6}$$

and there exists a set $A \subset \mathbb{F}_q$ such that

$$\left\| \alpha|_A \right\|_1 > (\log q)^{-6} L^{-3} \tag{3.7}$$

$$|A| < (\log q)^8 L^4 \|\alpha_1\|_2^{-2} \tag{3.8}$$

$$|A + A| < (\log q)^{76} L^{38} |A|. \tag{3.9}$$

where

$$L = \frac{\|\alpha_1\|_2}{\|\alpha_1 * \alpha_1\|_2} \tag{3.10}$$

Returning to (2.14), consider

$$S_0 = \frac{1}{|A|\,|B|\,|C|} \sum_{x\in A,\ y\in B,\ z\in C} \psi(xyz) = \frac{1}{|C|} \sum_{\substack{u\in\mathbb{F}_p\\ z\in C}} \eta(u)\psi(uz) \tag{3.11}$$

with η the image measure of $\mathcal{X}_A/|A| \otimes \mathcal{X}_B/|B|$ under the product map $(x, y) \mapsto x.y$.

Apply the considerations above with $\alpha = \eta, \beta = \frac{1}{|C|}\mathcal{X}_C$. Fix $s \in \mathbb{Z}_+$ and let $\eta_1 = \eta^{(2^s)}$.

Hence from (3.6)–(3.10)

$$|S_0| < \left| \frac{1}{|C|} \sum_{\substack{z\in C\\ u\in\mathbb{F}}} \eta_1(u)\ \psi(uz) \right|^{2^{-s}} \tag{3.12}$$

and there is a set $U \subset \mathbb{F}$ satisfying (U plays the role of A in (3.7)–(3.9))

$$\big|\{x \in A,\ y \in B \mid xy \in U\}\big| > (\log q)^{-6} L^{-3} |A|.|B| \tag{3.13}$$

$$|U| < (\log q)^8 L^4 \|\eta_1\|_2^{-2} \tag{3.14}$$

$$|U + U| < (\log q)^{76} L^{38} |U|. \tag{3.15}$$

where

$$L = \frac{\|\eta_1\|_2}{\|\eta_1 * \eta_1\|_2} \tag{3.16}$$

4. Further Assumptions

We make the following further assumptions on $\alpha, \beta, \gamma : \mathbb{F}_q^* \to \mathbb{R}_+$

$$\|\alpha\|_2, \|\beta\|_2, \|\gamma\|_2 < q^{-\delta_0} \tag{4.1}$$

$$\|\alpha\|_2, \|\beta\|_2^{\frac{1}{2}} \|\gamma\|_2 < q^{-\frac{1}{2}-\delta_1} \tag{4.2}$$

$$\max_{\substack{a,b \in \mathbb{F} \\ G \text{ proper subfield}}} \Big\{ \sum_{x \in G} \beta(ax+b) \Big\} < q^{-\delta_2} \tag{4.3}$$

where $\delta_0, \delta_1, \delta_2 > 0$.

In the definition of $\zeta(a, b, c)$ in §2, we make the extra hypothesis that β' satisfies (4.3). Since obviously the left side of (4.3) decreases when β' is replaced by $\beta' * \beta'$ (recall that $\|\beta'\|_1 \leq 1$), inequality (2.23) still holds in this restricted setting.

Let $B \subset \mathbb{F}_q$ be the set corresponding to β from Lemma 5. Thus by (2.15)

$$\frac{2\sigma}{|B|} > \beta|_B > \frac{\sigma}{|B|} \quad \text{where} \quad (\log q)^{-6} K^{-3} < \sigma \leq 1. \tag{4.4}$$

If $B_0 \subset B$, clearly $\big\| \beta|_{B_0} \big\|_1 > (\log q)^{-6} K^{-3} \frac{|B_0|}{|B|}$ by (4.4) and (4.3) implies that B_0 is not contained in a set $aG + b$, G a proper subfield, provided

$$|B_0| > (\log q)^6 K^3 q^{-\delta_2} |B|. \tag{4.5}$$

5. Preliminary Estimates (3)

Returning to (3.11)–(3.16), we establish a lower found on $|U|$ using Lemma 1. From (3.14), this will give a bound on $\|\eta_1\|_2$.

Denote $\mathcal{G} = \big\{ (x, y) \in A \times B \mid x, y \in U \big\}$.

Hence by (3.13)

$$|\mathcal{G}| > (\log q)^{-6} L^{-3} |A|.|B|. \tag{5.1}$$

Define

$$A_0 = \left\{ x \in A \mid |\mathcal{G}(x)| > \frac{1}{2}(\log q)^{-6} L^{-3} |B| \right\}$$

and

$$B_0 = \left\{ y \in B \mid |\mathcal{G}(y)| > \frac{1}{2}(\log q)^{-6} L^{-3} |A| \right\}$$

denoting $\mathcal{G}(x)$ and $\mathcal{G}(y)$ the fibers of $\mathcal{G}$.

Clearly

$$|A_0| > \frac{1}{2}(\log q)^{-6} L^{-3} |A| \tag{5.2}$$

$$|B_0| > \frac{1}{2}(\log q)^{-6} L^{-3} |B|. \tag{5.3}$$

We apply Lemma 1 with $X = A_0; Y = B_0$. In view of (4.5), (5.3), the assumption

$$K.L < \frac{1}{2}(\log q)^{-4} q^{\frac{1}{3}\delta_2} \tag{5.4}$$

ensures that B_0 is not contained in a multiplicative coset of a proper subfield. From Lemma 1 and the related Remark, we obtain $a_1, a_2, a_3, a_4 \in A_0 \cup (-A_0)$ and $b_1, b_2, b_3, b_4 \in B_0 \cup (-B_0)$ such that

$$\begin{aligned} &|b_1 A' + b_2 A' + b_3 A' + b_4 A' + a_1 B' + a_2 B' + a_3 B' + a_4 B'| \\ &\geq \frac{1}{2} \min \left\{ |A'|.|B'|^{\frac{1}{2}},\ q \left(\frac{|A'|.|B'|}{|A|.|B|} \right)^2 \right\} \end{aligned} \tag{5.5}$$

if $A' \subset A_0$, $B' \subset B_0$.

Next we apply Lemma 2.

Take $X_i = \mathcal{G}(b_i) \subset A$ $(1 \leq i \leq 4)$ and $Y = A_0$. From (1.13), there is $A' \subset A_0$ and $a' \in A_0$ s.t.

$$A' - a' \subset \bigcap_{i=1}^{4} (\mathcal{G}(b_i) - \mathcal{G}(b_i)) \tag{5.6}$$

and

$$(5.7)\qquad |A'| \geq \left(\prod_{i=1}^{4} \frac{|\mathcal{G}(b_i)|}{|\mathcal{G}(b_i) - A_0|}\right)|A_0| \geq \left(\prod_{i=1}^{4} \frac{|\mathcal{G}(b_i)|}{|A-A|}\right).|A_0|$$

$$> (\log q)^{-C} K^{-304} L^{-15} |A|$$

by (5.2), (2.17).

By (5.6) and definition of $\mathcal{G}$

$$(5.8)\qquad b_1(A'-a') + b_2(A'-a') + b_3(A'-a') + b_4(A'-a') \subset 4U - 4U.$$

Similarly we obtain $B' \subset B_0$ and $b' \in B_0$ s.t.

$$(5.9)\qquad |B'| > (\log q)^{-C} K^{-304} L^{-15} |B|$$

and

$$(5.10)\qquad a_1(B'-b') + a_2(B'-b') + a_3(B'-b') + a_4(B'-b') \subset 4U - 4U.$$

From (5.5), (5.8), (5.10), it follows

$$(5.11)\qquad |8U - 8U| \geq (\log q)^{-C} \min\left\{K^{-456} L^{-23} |A|.|B|^{\frac{1}{2}},\ qK^{-1208} L^{-60}\right\}.$$

Recalling (3.14), (3.15) and the Plunnecke–Ruzsa inequality, (5.11) implies

$$|U| \geq (\log q)^{-C} \min\left\{K^{-456} L^{-631} |A|.|B|^{\frac{1}{2}},\ qK^{-1208} L^{-668}\right\}$$

and

$$(5.12)\qquad \|\eta_1\|_2 \leq (\log q)^{C} \left(K^{228} L^{318} |A|^{-\frac{1}{2}} |B|^{-\frac{1}{4}} + q^{-\frac{1}{2}} K^{604} L^{336}\right).$$

From (3.16) defining L and (5.12)

$$(5.13)\qquad \|\eta_1 * \eta_1\|_2 \leq (\log q)^{C} K^{\frac{151}{84}} \theta^{\frac{1}{336}} \|\eta_1\|_2^{1-\frac{1}{336}}$$

where $\eta_1 = \eta^{(2^s)}$ and

$$(5.14)\qquad \theta = |A|^{-\frac{1}{2}} |B|^{-\frac{1}{4}} + \frac{1}{\sqrt{q}}.$$

The validity of (5.13) is conditional to (5.4), thus

$$\frac{\left\|\eta^{(2^s)}\right\|_2}{\left\|\eta^{(2^{s+1})}\right\|_2} < (\log q)^{-4} q^{\frac{1}{3}\delta_2} K^{-1}. \tag{5.15}$$

Set

$$K = q^{\delta_3} \text{ where } \delta_3 = 10^{-4} \min(\delta_1, \delta_2). \tag{5.16}$$

From (5.13), (5.15), either

$$\left\|\eta^{(2^{s+1})}\right\|_2 < q^{-\frac{1}{4}\delta_2} \left\|\eta^{(2^s)}\right\|_2 \tag{5.17}$$

or

$$\left\|\eta^{(2^{s+1})}\right\|_2 < K^2 \theta^{\frac{1}{336}} \left\|\eta^{(2^s)}\right\|_2^{1-\frac{1}{336}}. \tag{5.18}$$

Note also that since $\eta \geq 0$, $\|\eta\|_1 = 1$, the sequence $\left\|\eta^{(2^s)}\right\|_2$ is monotonically decreasing in s.

Iterating (5.17)–(5.18) $s_1 = s + \left[\frac{5}{\delta_2}\right]$ times, we obtain

$$\begin{aligned} \left\|\eta^{(2^{s_1})}\right\|_2 &< \frac{1}{q} + K^{672} \theta^{1-\left(\frac{335}{336}\right)^s} \|\eta\|_2^{\left(\frac{335}{336}\right)^s} \\ &< K^{672} q^{\frac{1}{2}\left(\frac{335}{336}\right)^s} \theta. \end{aligned} \tag{5.19}$$

Choose s such that

$$\left(\frac{335}{336}\right)^s < \frac{\delta_1}{100}$$

which is possible for s_1 satisfying

$$2^{s_1} < 32^{(\delta_2^{-1})} \left(\frac{100}{\delta_1}\right)^{300}. \tag{5.20}$$

From (5.16), (5.19), we conclude that

$$\left\|\eta^{(2^{s_1})}\right\|_2 < q^{\frac{\delta_1}{10}} \left(|A|^{-\frac{1}{2}} |B|^{-\frac{1}{4}} + \frac{1}{\sqrt{q}}\right). \tag{5.21}$$

6. Estimation of Trilinear Sums

Return to (3.12) with $\eta_1 = \eta^{(2^{s_1})}$. Estimate using Cauchy–Schwarz

$$\sum_{\substack{u\in\mathbb{F}\\ z\in C}} \eta_1(u)\psi(uz) \le \sum_u \eta_1(u)\Big|\sum_{z\in C}\psi(uz)\Big| \tag{6.1}$$

$$\le \|\eta_1\|_2\Big(\sqrt{q}|C|^{\frac{1}{2}}\Big)$$

and hence by (5.21)

$$\frac{1}{|C|}\Big|\sum_{\substack{u\in\mathbb{F}\\ z\in C}}\eta_1(u)\psi(uz)\Big| < q^{\frac{\delta_1}{10}}\Big(q^{\frac{1}{2}}|A|^{-\frac{1}{2}}|B|^{-\frac{1}{4}}|C|^{-\frac{1}{2}} + |C|^{-\frac{1}{2}}\Big). \tag{6.2}$$

From Lemma 5 and (3.12), (6.2), we obtain the following bound on (2.14).

$$(2.14) \le 8\Big[q^{\frac{\delta_1}{9}}\Big(q^{\frac{1}{2}}\|\alpha\|_2\|\beta\|_2^{\frac{1}{2}}\|\gamma\|_2 + \|\gamma\|_2\Big)\Big]^{\kappa} \tag{6.3}$$

where by (5.20)

$$\kappa > \Big(\frac{1}{32}\Big)^{\frac{1}{\delta_2}}\Big(\frac{\delta_1}{100}\Big)^{300}. \tag{6.4}$$

Note that, since $\gamma(0)=0$, certainly

$$|S| = \Big|\sum_{x,y,z}\alpha(x)\beta(y)\gamma(z)\psi(xyz)\Big| \le \sqrt{q}\|\alpha\|_2\|\beta\|_2$$

and hence, if $\|\gamma\|_2 \ge q^{-\frac{1}{2}\delta_1}$, $|S| < q^{-\frac{1}{2}\delta_1}$ by (4.2). Assuming $\|\gamma\|_2 < q^{-\frac{1}{2}\delta_1}$, (6.3) and (4.2) imply

$$(2.14) < q^{-\frac{\delta_1}{2}\kappa}. \tag{6.5}$$

Hence, we proved (recalling (2.23))

$$\zeta\big(\|\alpha\|_2, \|\beta\|_2, \|\gamma\|_2\big) < q^{-\frac{\delta_1}{2}\kappa} + 3\zeta(a,b,c)^{\frac{1}{2}}, \tag{6.6}$$

where certainly

$$a \le \|\alpha\|_2, \quad b \le \|\beta\|_2, \quad c \le \|\gamma\|_2$$

and

$$abc \leq \frac{1}{K}\|\alpha\|_2\|\beta\|_2\|\gamma\|_2 = q^{-\delta_3}\|\alpha\|_2\|\beta\|_2\|\gamma\|_2. \tag{6.7}$$

Straightforward iteration of (6.6), (6.7), until reaching $abc < \frac{1}{q}$ for which $\zeta(a,b,c) < \sqrt{q}\, q^{-2/3} = q^{-\frac{1}{6}}$, gives by (5.16)

$$\zeta\big(\|\alpha\|_2, \|\beta\|_2, \|\gamma\|_2\big) < q^{-\kappa'} \tag{6.8}$$

where

$$\kappa' > C^{-\frac{1}{\delta_2}}\delta_1^C \tag{6.9}$$

(C some constant).

Hence, we obtain

Theorem 1. *Let $\alpha, \beta, \gamma : \mathbb{F}_q \to \mathbb{R}_+$ and $\delta_0, \delta_1, \delta_2 > 0$ satisfying $\|\alpha\|_1, \|\beta\|_1, \|\gamma\|_1 \leq 1$ and (4.1)–(4.3). Then*

$$\Big|\sum \alpha(x)\beta(y)\gamma(z)\psi(xyz)\Big| < q^{-\kappa'} + 3q^{-\delta_0} \tag{6.10}$$

with

$$\kappa' > C^{-\frac{1}{\delta_2}}\delta_1^C. \tag{6.11}$$

7. Convolution of Product Densities

From Theorem 1, we deduce

Theorem 2. *Let $\alpha, \beta : \mathbb{F} \to \mathbb{R}_+$ satisfying*

$$\|\alpha\|_1, \|\beta\|_1 \leq 1 \tag{7.1}$$

$$\|\alpha\|_2 < q^{-\delta} \tag{7.2}$$

$$\max_{\substack{a,b\in\mathbb{F} \\ G \text{ proper subfield}}} \Big\{\sum_{x\in G} \beta(ax+b)\Big\} < q^{-\delta_2} \tag{7.3}$$

Let η be the image measure of $\alpha \otimes \beta$ under the product map $(x, y) \mapsto x.y$.

There is $k = k(\delta, \delta_2) < \delta^{-C} C^{\frac{1}{\delta_2}}$ such that

$$\|\eta^{(k)}\|_2 < q^{\delta} \|\alpha\|_2 . \|\beta\|_2^{\frac{1}{2}} + q^{\delta - \frac{1}{2}} \tag{7.4}$$

where $\eta^{(k)}$ denotes k-fold (additive) convolution.

Proof. Write

$$\|\eta^{(k)}\|_2^2 = q^{-1} \sum |\hat{\eta}(\xi)|^{2k} < q^{-2} + q^{-1}|Z| \tag{7.5}$$

with

$$Z = \{\xi \in \mathbb{F} \mid |\hat{\eta}(\xi)| > q^{-1/k}\}.$$

Defining $\gamma : \mathbb{F} \to \mathbb{C}$ by

$$\gamma(\xi) = \begin{cases} \dfrac{\overline{\hat{\eta}(\xi)}}{|Z||\hat{\eta}(\xi)|} & \text{if } \xi \in Z \\ 0 & \text{otherwise} \end{cases}$$

we have $\|\gamma\|_1 = 1$, $\|\gamma\|_2 = |Z|^{-\frac{1}{2}}$ and

$$q^{-1/k} \leq \sum_{\xi} \hat{\eta}(\xi)\gamma(\xi) = \sum_{x,y,\xi} \alpha(x)\beta(y)\gamma(\xi)\psi(xy\xi). \tag{7.6}$$

Apply Theorem 1 to (7.6) with $\delta_0 = \min\left(\delta, \frac{\delta_2}{2}, \frac{2}{k}, \frac{2}{k}\right)$ and $\delta_1 = \delta$. Then (4.1) holds, unless

$$\|\gamma\|_2 \geq q^{-\delta_0} \quad \text{hence} \quad |Z| \leq q^{\frac{4}{k}} \tag{7.7}$$

and (4.2), unless

$$|Z| \leq q^{1+2\delta} \|\alpha\|_2^2 \|\beta\|_2. \tag{7.8}$$

Clearly for $k > \frac{1}{\kappa'}, \kappa' = C^{-\frac{1}{\delta_2}} \delta^C$ given by (6.11), (6.10) and (7.6) are contradictory. Therefore, either (7.7) or (7.8) hold, i.e.

$$|Z| \leq q^{\frac{4}{k}} + q^{1+2\delta} \|\alpha\|_2^2 \|\beta\|_2. \tag{7.9}$$

Substitution in (7.5) gives (7.4). ∎

8. The General Case

From Theorem 2, we obtain the multilinear extension of Theorem 1.

Theorem 3. *Let $r \geq 2$ and $\alpha_1, \dots, \alpha_r : \mathbb{F} \to \mathbb{R}_+$ satisfy*

$$\|\alpha_i\|_1 \leq 1 \quad (1 \leq i \leq r) \tag{8.1}$$

$$\|\alpha_i\|_2 < q^{-\delta} \quad (1 \leq i \leq r) \tag{8.2}$$

$$\max_{\substack{a,b\in\mathbb{F}\\ G \text{ proper subfield}}} \alpha_i(aG+b) < q^{-\delta_2} \quad (3 \leq i \leq r) \tag{8.3}$$

$$\|\alpha_1\|_2 . \|\alpha_2\|_2 . \prod_{i=3}^{r} \|\alpha_i\|_2^{\frac{1}{2}} < q^{-\frac{1}{2}-\delta}. \tag{8.4}$$

Then there is the exponential sum bound

$$\left| \sum_{x_1,\dots,x_r\in\mathbb{F}} \prod_{i=1}^{r} \alpha_i(x_i)\psi\left(\prod_{i=1}^{r} x_i\right)\right| < q^{-\delta'} \text{ with } \delta' > C^{-\frac{r}{\delta_2}}\left(\frac{\delta}{r}\right)^{Cr}. \tag{8.5}$$

Proof. By induction on r.

For $r = 2$, there is the obvious bound $q^{\frac{1}{2}}\|\alpha_1\|_2\|\alpha_2\|_2 < q^{-\delta}$.

The case $r = 3$ is given by Theorem 1.

For the inductive step, we will use Theorem 2. Let $r \geq 4$. Denote S the exponential sum on the left of (8.5) and let η be the image density of $\alpha_1 \otimes \alpha_3$ under the product map $(x_1, x_3) \mapsto x_1x_3$. Thus

$$S = \sum_{x,x_2,x_4,\dots,x_r} \eta(x) \prod_{i\neq 1,3} \alpha_i(x_i)\psi\left(x \prod_{i\neq 1,3} x_i\right) \tag{8.6}$$

and estimate using Hölder's inequality

$$\begin{aligned} |S| &\leq \left\{ \sum_{x_2,x_4,\dots,x_r} \prod_{i\neq 1,3} \alpha_i(x_i) \left| \sum_{x\in\mathbb{F}} \eta(x)\psi\left(x. \prod_{i\neq 1,3} x_i\right)\right|^{2k}\right\}^{\frac{1}{2k}} \\ &= \left\{ \sum_{x,x_2,x_4,\dots,x_r} \eta^{(2k)}(x) \cdot \prod_{i\neq 1,3} \alpha_i(x_i) \cdot \psi\left(x. \prod_{i\neq 1,3} x_i\right)\right\}^{\frac{1}{2k}}. \end{aligned} \tag{8.7}$$

Theorem 2 is then applied to bound $\left\|\eta^{(2k)}\right\|_2$. Replacing δ by $\frac{\delta}{2r}$ in Theorem 2, it follows from (7.4)

$$\left\|\eta^{(2k)}\right\|_2 < q^{\frac{\delta}{2r}}\|\alpha_1\|_2\|\alpha_3\|_2^{\frac{1}{2}} + q^{\frac{\delta}{2r}-\frac{1}{2}} \tag{8.8}$$

for

$$k \sim \left(\frac{2r}{\delta}\right)^C C^{\frac{1}{\delta_2}}. \tag{8.9}$$

Hence

$$\left\|\eta^{(2k)}\right\|_2 < q^{-\delta\left(\frac{3}{2}-\frac{1}{2r}\right)} + q^{\frac{\delta}{2r}-\frac{1}{2}} < q^{-\delta}$$

and since $r \geq 4$, from (8.8), (8.4)

$$\left\|\eta^{(2k)}\right\|_2\|\alpha_2\|_2\prod_{i=4}\|\alpha_i\|_2^{\frac{1}{2}} < q^{\frac{\delta}{2r}-\frac{1}{2}-\delta} + q^{\frac{\delta}{2r}-\frac{1}{2}-\frac{r-1}{2}\delta} \leq 2q^{-\frac{1}{2}-\left(1-\frac{1}{2r}\right)\delta}. \tag{8.10}$$

At this point, invoke the induction hypothesis with r replaced by $r-1$ and δ by $\left(1-\frac{1}{2r}\right)\delta$. Recalling (8.7), it follows that

$$|S| < q^{-\delta_r'} \tag{8.11}$$

where

(8.12)

$$\delta_r' = \delta_r'(\delta) = \frac{1}{2k}\delta_{r-1}'\left(\left(1-\frac{1}{2r}\right)\delta\right) \underset{(8.9)}{>} \left(\frac{\delta}{r}\right)^C C^{-\frac{1}{\delta_2}}\delta_{r-1}'\left(\left(1-\frac{1}{2r}\right)\delta\right).$$

It remains to iterate (8.12). ■

Following the argument to derive Theorem 2 from Theorem 1, Theorem 3 implies also

Theorem 4. *Let $\alpha_1, \dots, \alpha_r : \mathbb{F}_q \to \mathbb{R}_+$ satisfy*

$$\|\alpha_i\|_1 \leq 1 \quad (1 \leq i \leq r) \tag{8.13}$$

$$\|\alpha_i\|_2 < q^{-\delta} \quad (1 \leq i \leq r) \tag{8.14}$$

$$\max_{\substack{a,b\in\mathbb{F}\\ G \text{ proper subfield}}} \alpha_i(aG+b) < q^{-\delta_2} \quad (3 \leq i \leq r). \tag{8.15}$$

Denote η *the image density of* $\alpha_1 \otimes \cdots \otimes \alpha_r$ *under the product map* $(x_1, \dots, x_r) \mapsto x_1 \cdots x_r$. *Then*

$$\left\|\eta^{(k)}\right\|_2 < q^{\delta} \|\alpha_1\|_2 . \|\alpha_2\|_2 . \prod_{i=3}^{r} \|\alpha_i\|_2^{\frac{1}{2}} + q^{\delta - \frac{1}{2}} \tag{8.16}$$

provided $k > C^{r/\delta_2} \left(\frac{r}{\delta}\right)^{Cr}$.

REFERENCES

[1] J. Bourgain, Multilinear exponential sums in prime fields under optimal entropy conditions on the sources, *GAFA*, Vol. **18** (2009), 1477–1502.

[2] W. Schmidt, Bounds on exponential sums, *Acta Arith.* (1964).

[3] N. Katz and C. Shen, A slight improvement to Garaev's sum-product estimate, to appear in *Proc. AMS*.

Jean Bourgain
Institute for Advanced Study
Princeton, NJ 08540
USA

e-mail: bourgain@ias.edu

BOLYAI SOCIETY
MATHEMATICAL STUDIES, 21

An Irregular Mind
(Szemerédi is 70)
pp. 243–250.

An Estimate of Incomplete Mixed Character Sums

MEI-CHU CHANG*

Dedicated to Endre Szemerédi for his 70th birthday.

生日快樂![†]

In this note we consider incomplete mixed character sums over a finite field $\mathbb{F}_{p^n}$ of the form $\sum_{x\in B_H}\psi\big(f(x)\big)\chi(x)$, where ψ is an additive character, $f(x)\in\mathbb{F}_{p^n}$ a polynomial, χ a non-trivial multiplicative character and B_H a 'box' of the form $B_H=\left\{\sum_{j=1}^n x_j\omega_j : x_j\in[1,H]\right\}$. (Here $\{\omega_i\}_{i=1}^n$ is an arbitrary basis of $\mathbb{F}_{p^n}$ over $\mathbb{F}_p$.)

If $f(x)=0$ and $n=1$, Burgess' well-known theorem provides a non-trivial estimate under the assumption $H>p^{1/4+\varepsilon}$. A generalization to arbitrary finite fields was obtained in [3], [4] and very recently [9], eventually providing a statement of the same strength as Burgess, in $\mathbb{F}_{p^n}$.

If $n=1$ and $f(x)$ is linear, [7] proved a non-trivial bound assuming $H>p^{1/4+\varepsilon}$. For a general polynomial $f(x)$ the only available result are that of P. Enflo [6] and a comment made by Heath-Brown [8] in the review of [6]. Heath-Brown's estimate (for $n=1$) assumes again that $H>p^{1/4+\varepsilon}$ and comes with a saving of the form $p^{-c(\varepsilon)/2^d}$, where d is the degree of $f(x)$.

Our result below treats the situation of a field $\mathbb{F}_{p^n}$ (relying on Konyagin's bound for the multiplicative energy of a box B_H as described above) and a polynomial $f(x)$ of arbitrary degree d, assuming $H>p^{1/4+\varepsilon}$. We obtain a saving over the trivial bound of the form $p^{-c(n,\varepsilon)/(d+1)^2}$, so that, interestingly, even for $n=1$ the result seems new.

*Research partially financed by the National Science Foundation.

†Happy Birthday!

Notation and Convention

1. $e(\theta) = e^{2\pi i\theta}$, $e_p(\theta) = e\left(\frac{\theta}{p}\right)$
2. When there is no ambiguity, $p^\varepsilon = [p^\varepsilon] \in \mathbb{Z}$.
3. Multiplicative energy

$$E(A,B) = \left|\left\{ (a_1,a_2,b_1,b_2) \in A\times A\times B\times B \,:\, a_1b_1 = a_2b_2\right\}\right|.$$

Let $\omega_1,\dots,\omega_n$ be an arbitrary basis for $\mathbb{F}_{p^n}$ over $\mathbb{F}_p$. Then for any $x \in \mathbb{F}_{p^n}$, there is a unique representation of x in terms of the basis.

$$x = x_1\omega_1 + \cdots + x_n\omega_n.$$

A *box* $B_H \subset \mathbb{F}_{p^n}$ *of size* H is a set such that for each j, the coefficients x_j form an interval.

$$B_H = \left\{\sum_{j=1}^{n} x_j\omega_j \,:\, x_j \in [1,H], \quad \forall j\right\}. \tag{1}$$

Theorem. *Let χ (respectively, ψ) be a non-principal multiplicative (resp. additive) character of $\mathbb{F}_{p^n}$. For a basis $\omega_1,\omega_2,\dots,\omega_n$ of $\mathbb{F}_{p^n}$ over $\mathbb{F}_p$, let B_H be a box as defined in (1) by the basis with*

$$H > p^{\frac{1}{4}+\kappa} \quad \text{for some} \quad \kappa > 0. \tag{2}$$

Then for a polynomial $f \in \mathbb{F}_{p^n}$ of degree d, we have

$$\left|\sum_{x\in B_H} \psi\big(f(x)\big)\,\chi(x)\right| < c(n,\kappa)(d+1)^2 p^{-\delta}|B|,$$

where

$$\delta = \frac{\kappa^2 n}{4(1+2\kappa)\big(2n+(d+1)^2\big)}$$

and $c(n,\kappa)$ is a constant depending on n and κ.

Sketch of Proof. As in [3], [4] and [9], we use Burgess' method [1].

Let $\varepsilon > 0$ be specified later (see (16)) and let $B_{p^{-2\varepsilon}H}$ be a box of size $p^{-2\varepsilon}H$ as defined in (1). For $y \in B_{p^{-2\varepsilon}H}$ and $0 < t < p^{\varepsilon}$, since $yt \in B_{p^{-\varepsilon}H}$, we have

$$\Big|\sum_{x\in B_H} \psi\big(f(x)\big)\chi(x) - \sum_{x\in B_H} \psi\big(f(x+yt)\big)\chi(x+yt)\Big|$$

$$\leq \big|B\setminus(B+yt)\big| + \big|(B+yt)\setminus B\big| < 2np^{-\varepsilon}H^n.$$

Hence

$$(3)\quad \Big|\sum_{x\in B_H} \psi\big(f(x)\big)\chi(x)\Big|$$

$$\leq \frac{1}{p^{\varepsilon}|B_{p^{-2\varepsilon}H}|}\Bigg|\sum_{\substack{x\in B_H,\ y\in B_{p^{-2\varepsilon}H}\\ 0<t<p^{\varepsilon}}} \psi\big(f(x+yt)\big)\chi(x+yt)\Bigg| + O\big(p^{-\varepsilon}H^n\big).$$

An additive character is of this form

$$\psi(z) = e_p(\operatorname{Tr}\xi z), \quad \text{for some} \quad \xi \in \mathbb{F}_{p^n}.$$

Expanding

$$f(x+yt) = a_d(x,y)t^d + a_{d-1}(x,y)t^{d-1} + \cdots + a_0(x,y),$$

and we write

$$(4)\qquad \psi\big(f(x+yt)\big) = e\bigg(\sum_{j=0}^{d} \frac{\operatorname{Tr}\xi a_j(x,y)}{p}\, t^j\bigg).$$

Fix $\varepsilon_1 > 0$ (to be specified later) and partition $[0,1]^{d+1}$ in boxes Q_α of size $p^{-\varepsilon_1}$. There are $p^{\varepsilon_1(d+1)}$ boxes. Partition $B_H \times B_{p^{-2\varepsilon}H}$ according to the boxes Q_α.

$$B_H \times B_{p^{-2\varepsilon}H} = \bigcup_\alpha \Omega_\alpha,$$

where

$$\Omega_\alpha = \bigg\{(x,y) \in B_H \times B_{p^{-2\varepsilon}H} :\ \bigg(\frac{\operatorname{Tr}\xi a_j(x,y)}{p}\bigg)_{1\leq j\leq d+1} \in Q_\alpha \ (\bmod 1)\bigg\}.$$

Hence for $\theta_\alpha = (\theta_{\alpha,1}, \dots, \theta_{\alpha,d+1}) \in Q_\alpha$ and $(x,y) \in \Omega_\alpha$, we have

(5) $$\left|\frac{\operatorname{Tr}\xi a_j(x,y)}{p} - \theta_{\alpha,j}\right| < p^{-\varepsilon_1}, \quad \text{for} \quad j = 1, \dots, d+1.$$

Since $t < p^\varepsilon$, (4) and (5) imply that for $(x,y) \in \Omega_\alpha$,

(6) $$\begin{aligned} &\left|\psi(f(x+yt)) - e\Big(\sum_{j=0}^{d} \theta_{\alpha,j} t^j\Big)\right| \\ &\le 2\pi \sum_j \left|\frac{\operatorname{Tr}\xi a_j(x,y)}{p} - \theta_{\alpha,j}\right| t^j \\ &< 2\pi(d+1)p^{d\varepsilon - \varepsilon_1} \lesssim p^{-\varepsilon}, \end{aligned}$$

for

(7) $$\varepsilon_1 = (d+1)\varepsilon.$$

Therefore, the bound in (3) is bounded by

(8) $$\frac{1}{p^\varepsilon |B_{p^{-2\varepsilon}H}|} \sum_\alpha \sum_{(x,y)\in\Omega_\alpha} \left|\sum_{t=1}^{p^\varepsilon} e\Big(\sum_{j=0}^{d} \theta_{\alpha,j} t^j\Big)\chi(x+yt)\right| + O(p^{-\varepsilon}H^n).$$

For $z \in \mathbb{F}_{p^n}$, denote

(9) $$\mu_\alpha(z) = \left|\left\{(x,y) \in \Omega_\alpha : \frac{x}{y} = z\right\}\right|.$$

The sum in the first term of (8) equals

(10) $$\sum_\alpha \sum_{z\in\mathbb{F}_{p^n}} \mu_\alpha(z) \left|\sum_{t=1}^{p^\varepsilon} e\Big(\sum_{j=0}^{d} \theta_{\alpha,j} t^j\Big)\chi(z+t)\right|.$$

Take $r \in \mathbb{Z}$ specified later. Hölder's inequality bounds (10) by

(11)
$$\underbrace{\Big(\sum_\alpha \sum_{z\in\mathbb{F}_{p^n}} \mu_\alpha(z)^{\frac{2r}{2r-1}}\Big)^{1-\frac{1}{2r}}}_{(A)} \underbrace{\Big(\sum_\alpha \sum_{z\in\mathbb{F}_{p^n}} \left|\sum_{t=1}^{p^\varepsilon} e\Big(\sum_{j=0}^{d} \theta_{\alpha,j} t^j\Big)\chi(z+t)\right|^{2r}\Big)^{\frac{1}{2r}}}_{(B)}.$$

Hölder's inequality also gives

$$(A) \le \Big(\sum_{\alpha,z} \mu_\alpha(z)\Big)^{1-\frac{1}{r}} \Big(\sum_{\alpha,z} \mu_\alpha(z)^2\Big)^{\frac{1}{2r}} \tag{12}$$

$$= \Big(\sum_\alpha |\Omega_\alpha|\Big)^{1-\frac{1}{r}} E\big(B_H, B_{p^{-2\varepsilon}H}\big)^{\frac{1}{2r}}$$

$$\le c(n)\big(p^{-2\varepsilon}H^2\big)^{n(1-\frac{1}{r})}\big(p^{-2n\varepsilon}H^{2n}\big)^{\frac{1}{2r}} \log p.$$

Here the equality follows from the definitions of $\mu_\alpha(z)$ and the multiplicative energy. For the last inequality, we use Konyagin's bound on multiplicative energy [9] and that

$$E(B_H, B_{p^{-2\varepsilon}H}) \le E\big(B_H, B_H\big)^{\frac{1}{2}} E\big(B_{p^{-2\varepsilon}H}, B_{p^{-2\varepsilon}H}\big)^{\frac{1}{2}}.$$

(This is by Cauchy–Schwarz. (See [10] Corollary 2.10.))

To bound (B), we write

$$(B)^{2r} = \sum_\alpha B_\alpha$$

with

$$B_\alpha = \sum_z \left|\sum_{t=1}^{p^\varepsilon} e\Big(\sum_{j=0}^{d} \theta_{\alpha,j} t^j\Big)\chi(z+t)\right|^{2r}.$$

For fixed α, we expand $\big|\sum_{t=1}^{p^\varepsilon} e\big(\sum_{j=0}^{d} \theta_{\alpha,j} t^j\big)\chi(z+t)\big|^{2r}$ and obtain

$$\left|\sum_{t=1}^{p^\varepsilon} e\Big(\sum_{j=0}^{d} \theta_{\alpha,j} t^j\Big)\chi(z+t)\right|^{2r} \tag{13}$$

$$= \sum_{t_1,\dots,t_{2r}} c_\alpha(t_1,\dots,t_{2r})\chi\left(\frac{(z+t_1)\cdots(z+t_r)}{(z+t_{r+1})\cdots(z+t_{2r})}\right)$$

with $\big|c_\alpha(t_1,\dots,t_{2r})\big| = 1$. This gives

$$(B)^{2r} \le (2r)^{2r} p^{\varepsilon_1(d+1)} \sum_{t_1,\dots,t_{2r}<p^\varepsilon} \left|\sum_z \chi\left(\frac{(z+t_1)\cdots(z+t_r)}{(z+t_{r+1})\cdots(z+t_{2r})}\right)\right|$$

$$\le (2r)^{2r} p^{\varepsilon_1(d+1)}\big[p^n p^{r\varepsilon} + p^{\frac{n}{2}} p^{2r\varepsilon}\big]$$

(The last inequality is given by Weil's estimate.)

Therefore,

$$(B) < crp^{\frac{\varepsilon_1(d+1)}{2r}}\left[p^{\frac{n}{2r}+\frac{\varepsilon}{2}} + p^{\frac{n}{4r}+\varepsilon}\right]. \tag{14}$$

Putting (10)–(12) and (14) together, we have the first term of (8) bounded by

$$\frac{c(n)r\log p}{p^{\varepsilon}\left(p^{-2\varepsilon}H\right)^{\frac{n}{r}}}\left(p^{-2\varepsilon}H^2\right)^{n(1-\frac{1}{r})}\left(p^{-2n\varepsilon}H^{2n}\right)^{\frac{1}{2r}}p^{\frac{\varepsilon_1(d+1)}{2r}}\left[p^{\frac{n}{2r}+\frac{\varepsilon}{2}}+p^{\frac{n}{4r}+\varepsilon}\right] \tag{15}$$

$$\leq c(n)r\log p\, H^{n-\frac{n}{r}}p^{\frac{\varepsilon_1(d+1)}{2r}+\frac{\varepsilon n}{r}}\left[p^{\frac{n}{2r}-\frac{\varepsilon}{2}}+p^{\frac{n}{4r}}\right]$$

$$\leq c(n)r\, H^n p^{\frac{\varepsilon}{2r}\left((d+1)^2+2n\right)}\left[\left(\frac{p^{\frac{1}{2}}}{H}\right)^{\frac{n}{r}}p^{-\frac{\varepsilon}{2}}+\left(\frac{p^{\frac{1}{4}}}{H}\right)^{\frac{n}{r}}\right]$$

$$< c(n)r\, H^n p^{\frac{\varepsilon}{2r}\left((d+1)^2+2n\right)}\left[p^{\frac{n}{4r}-\frac{\varepsilon}{2}}+p^{-\kappa\frac{n}{r}}\right].$$

(The last inequality is by our assumption (2).)

Take

$$\varepsilon = \kappa\frac{n}{(d+1)^2+2n} \tag{16}$$

and

$$r = \left\lceil (2\kappa+1)\frac{n}{\varepsilon}\right\rceil = \left\lceil \left((d+1)^2+2n\right)\left(2+\frac{1}{\kappa}\right)\right\rceil. \tag{17}$$

Substituting (16) in the second factor of (15), we obtain $p^{\frac{\kappa n}{2r}}$. Our choice of r implies that $r > (2\kappa+1)\frac{n}{\varepsilon}$ and hence $\frac{\kappa n}{2r}+\frac{n}{4r} < \frac{\varepsilon}{4}$. Therefore, (15) is bounded by

$$c(n,\kappa)(d+1)^2H^n\left(p^{-\frac{\varepsilon}{4}}+p^{-\kappa\frac{n}{2r}}\right) < c(n,\kappa)(d+1)^2H^np^{-\frac{\kappa^2 n}{4(1+2\kappa)((d+1)^2+2n)}}.$$

(The inequality is because $r < 2\left((d+1)^2+2n\right)\left(2+\frac{1}{\kappa}\right)$ by (17).)

Remark. One may estimate the quantity

$$\sum_{t_1,\ldots,t_{2r}} \left| \sum_{\alpha} c_\alpha(t_1,\ldots,t_{2r}) \right|,$$

which essentially equals to

$$\int_{\Pi^{d+1}} \left| \sum_{t=1}^{p^\varepsilon} e\Big(\sum_{j=0}^{d} \theta_j t^j \Big) \right|^{2r} d\theta_0 \cdots d\theta_d$$

and may be estimated using the classical Vinogradov's mean value theorem. This will lead to some further saving of δ that may be significant for specific values of κ and d. In the context of our theorem where we focus on small κ and large d, the improvement turns out to be without interest.

Acknowledgement. The author would like to thank referee for many helpful comments. The author would also like to thank Lih-Chung Wang for technical support.

References

[1] D. A. Burgess, On character sums and primitive roots, *Proc. LMS (3),* **12** (1962), 179–192.

[2] J. Bourgain and M.-C. Chang, On a multilinear character sum of Burgess, *Comptes rendus – Mathematique,* to appear.

[3] M.-C. Chang, On a question of Davenport and Lewis and new character sum bounds in finite fields, *Duke Math. J.,* **145** (2008), No. 3, 409–442.

[4] M.-C. Chang, Burgess inequality in F_{p^2}, *Geom. Funct. Anal,* to appear.

[5] M.-C. Chang, Character Sums in Finite Fields, *AMS Contemporary Mathematics Series,* to appear.

[6] P. Enflo, Some problems in the interface between number theory, harmonic analysis and geometry of Euclidean space, *First International Conference in Abstract Algebra, Quaestiones Math.,* **18** (1995), no. 1–3, 309–323.

[7] J. Friedlander and H. Iwaniec, Estimates for character sums, *Proc. Amer. Math. Soc.,* **119**, No. 2, (1993), 265–372.

[8] D. R. Heath-Brown, *MR1340486 (96h:11079).*

[9] S. V. Konyagin, Estimates of character sums in finite fields, *Matematicheskie Zametki,* to appear, (in Russian).

[10] T. Tao and V. Vu, *Additive Combinatorics,* Cambridge University Press, (2006).

Mei-Chu Chang
Department of Mathematics
University of California
Riverside, CA 92521

e-mail: mcc@math.ucr.edu

BOLYAI SOCIETY
MATHEMATICAL STUDIES, 21

An Irregular Mind
(Szemerédi is 70)
pp. 251–260.

Crossings Between Curves with Many Tangencies

JACOB FOX*, FABRIZIO FRATI, JÁNOS PACH† and ROM PINCHASI

Let $\mathcal{A}$ and $\mathcal{B}$ be two families of two-way infinite x-monotone curves, no three of which pass through the same point. Assume that every curve in $\mathcal{A}$ lies above every curve in $\mathcal{B}$ and that there are m pairs of curves, one from $\mathcal{A}$ and the other from $\mathcal{B}$, that are tangent to each other. Then the number of proper crossings among the members of $\mathcal{A}\cup\mathcal{B}$ is at least $(1/2-o(1))m\ln m$. This bound is almost tight.

1. Introduction

Studying the incidence structure of a family of curves in the plane is a classical theme in combinatorial geometry with many applications in computational geometry. Venn diagrams were introduced in the 19th century to analyze logical relationships between various statements [9, 7]. The incidence structure of non-overlapping circular disks was investigated by Koebe [3], while Erdős [2] raised several questions about tangencies between possibly overlapping congruent disks, including his famous problem on unit distances: How many pairs of points can be at distance one from each other in a set of n points in the plane? In other words, how many tangencies can occur among n unit diameter disks in the plane? These are hard questions, see [5] for a survey.

*Research supported by an NSF Graduate Research Fellowship and a Princeton Centennial Fellowship.

†Research supported by NSF grant CCF-08-30272, by grants from NSA, OTKA, BSF, and SNF.

An equally tantalizing innocent-looking question was asked by Richter and Thomassen [6]. We say that two closed curves γ_1 and γ_2 in the plane *properly cross* if they share at least one point p (called a *crossing point*) such that γ_1 passes from one side to the other side of γ_2 in a small neighborhood of p. We say that two closed curves γ_1 and γ_2 in the plane *touch* or are *tangent to* each other, if they share exactly one point. This point is called the *point of tangency* of the two curves. We say that two closed curves are *intersecting* if they have at least one point in common. A *family* $\mathcal{F}$ of closed curves is *intersecting* if every pair of them is intersecting. The family $\mathcal{F}$ is in *general position* if any *two* of its members share only a finite number of points and no *three* members pass through the same point. According to the Richter–Thomassen conjecture, any intersecting family of n closed curves in general position in the plane determines a total of at least $\big(1 - o(1)\big)\, n^2$ crossing points. This, of course, holds automatically if no two curves of the family touch each other, because then the number of crossing points is at least $2\binom{n}{2}$. Therefore, in order to settle the problem, we have to analyze families of curves with many tangencies.

In this note, we take the first step in this direction by studying the system of tangencies between *two* intersecting families $\mathcal{A}$ and $\mathcal{B}$ of curves in general position, with the property that no curve in $\mathcal{A}$ properly crosses any curve in $\mathcal{B}$ (see Fig. 1). In this case, we are going to prove that, if m denotes the number of pairs of touching curves (α, β) with $\alpha \in \mathcal{A}$ and $\beta \in \mathcal{B}$, the total number of crossing points in $\mathcal{F} = \mathcal{A} \cup \mathcal{B}$ divided by m tends to infinity, as $m \to \infty$. Consequently, if $|\mathcal{F}| = n$ and $m > \varepsilon n^2$ for some $\varepsilon > 0$, then the total number of crossing points in $\mathcal{F}$ is superquadratic in n. In particular, any intersecting family $\mathcal{F}$ of n closed curves in general position that can be partitioned into two subfamilies $\mathcal{A}$ and $\mathcal{B}$ such that every curve in $\mathcal{A}$ is tangent to all curves in $\mathcal{B}$, determine at least constant times $n^2 \log n$ crossing points. This is, of course, a lot more than what is required by the Richter–Thomassen conjecture in this special case.

For aesthetical reasons, we formulate our results for *two-way infinite x-monotone curves,* that is, for graphs γ_f of continuous functions $f : \mathbb{R} \to \mathbb{R}$. For simplicity, in the sequel, we use the term *curve* in this sense. We say that a curve γ_f lies *above* a curve γ_g if $f(x) \geq g(x)$ for all $x \in \mathbb{R}$.

For any family $\mathcal{F}$ of curves in general position, let $\mathrm{CN}(\mathcal{F})$ denote the number of crossing points.

Our main result is the following.

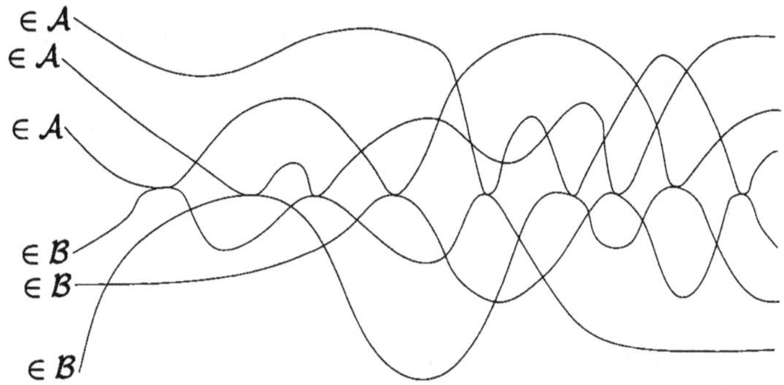

Fig. 1. Two intersecting families $\mathcal{A}$ and $\mathcal{B}$ of curves in general position such that no curve in $\mathcal{A}$ properly crosses any curve in $\mathcal{B}$.

Theorem 1. *Let $\mathcal{A}$ and $\mathcal{B}$ be two families of two-way infinite x-monotone curves such that $\mathcal{A} \cup \mathcal{B}$ is in general position. Assume that every curve in $\mathcal{A}$ lies above every curve in $\mathcal{B}$ and that there are m pairs of curves, one from $\mathcal{A}$ and the other from $\mathcal{B}$, that touch. Then the sum of the numbers of proper crossings among the members of $\mathcal{A}$ and among the members of $\mathcal{B}$ satisfies*

$$\mathrm{CN}(\mathcal{A}) + \mathrm{CN}(\mathcal{B}) \geq \left(\frac{1}{2} - o(1)\right) m \ln m,$$

where the $o(1)$ term goes to 0 as m tends to ∞.

We say that $\mathcal{A}$ and $\mathcal{B}$ *completely touch* if every member of $\mathcal{A}$ touches every member of $\mathcal{B}$.

Theorem 2. *For all positive integers n, there exist two completely touching n-member families $\mathcal{A}$ and $\mathcal{B}$ of two-way infinite x-monotone curves such that $\mathcal{A} \cup \mathcal{B}$ is in general position, every curve in $\mathcal{A}$ lies above every curve in $\mathcal{B}$, and*

$$\mathrm{CN}(\mathcal{A}) + \mathrm{CN}(\mathcal{B}) \leq \left(\frac{3}{4} + o(1)\right) n^2 \log_2 n.$$

Comparing Theorems 1 and 2, we obtain that if $c(n)$ denotes the minimum number of crossing points in the union $\mathcal{A} \cup \mathcal{B}$ of two completely touching n-member families of curves, $\mathcal{A}$ and $\mathcal{B}$, such that all the members of $\mathcal{A}$ are above all the members of $\mathcal{B}$, then we have:

$$(1 - o(1))\, n^2 \ln n \leq c(n) \leq \left(\frac{3}{4} + o(1)\right) n^2 \log_2 n = \left(\frac{3}{4 \ln 2} + o(1)\right) n^2 \ln n.$$

This shows that Theorem 1 is tight up to a multiplicative factor of roughly $\frac{3}{4 \ln 2} \approx 1.082$.

In Sections 2 and 3 of this note, we establish Theorems 1 and 2, respectively. In the final section, we make some concluding remarks. In particular, we formulate a combinatorial result of independent interest on alternations in certain sequences over finite alphabets (Theorem 3), which can also be used to prove Theorem 1.

2. Levels – Proof of Theorem 1

The *lower k-level* of a family $\mathcal{F}$ of curves is the closure of the set of all points that lie on exactly one member of $\mathcal{F}$ and strictly above exactly $k-1$ members (see Fig. 2). Let $\ell_k(\mathcal{F})$ denote the number of all proper crossings among members of $\mathcal{F}$ that lie on the lower k-level of $\mathcal{F}$. Analogously, the *upper k-level* of a family $\mathcal{F}$ of curves is the closure of the set of all points that lie on exactly one member of $\mathcal{F}$ and strictly below exactly $k-1$ members. Let $u_k(\mathcal{F})$ denote the number of all proper crossings among members of $\mathcal{F}$ that lie on the upper k-level of $\mathcal{F}$. Note that each proper crossing among two members of a family $\mathcal{F}$ of curves in general position lies on two consecutive levels, so that we have

$$\sum_{k=1}^{|\mathcal{F}|} \ell_k(\mathcal{F}) = \sum_{k=1}^{|\mathcal{F}|} u_k(\mathcal{F}) = 2\,\mathrm{CN}\,(\mathcal{F}). \tag{1}$$

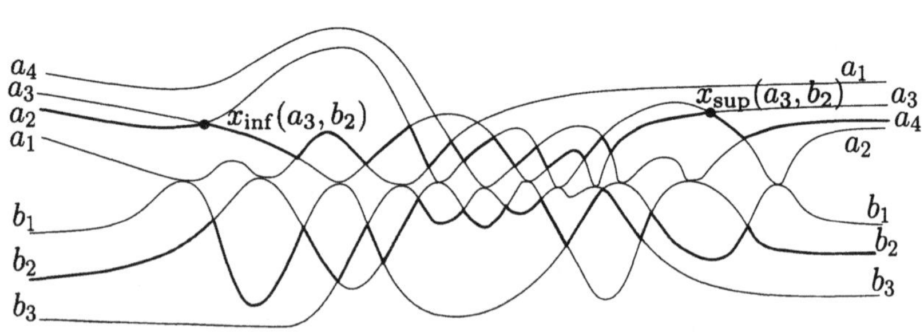

Fig. 2. Two families $\mathcal{A}$ and $\mathcal{B}$ of curves, with $|\mathcal{A}| = 4$ and $|\mathcal{B}| = 3$. The lower 2-level of $\mathcal{A}$ and the upper 2-level of $\mathcal{B}$ are shown by thick lines. Black dots show $x_{\inf}(a_3, b_2)$ and $x_{\sup}(a_3, b_2)$ when $k = 2$.

Theorem 1 can be easily deduced from the following lemma.

Lemma 1. *Let $k > 1$ and $\mathcal{A}$ and $\mathcal{B}$ be two families of two-way infinite x-monotone curves, each of cardinality at least k, such that $\mathcal{A} \cup \mathcal{B}$ is in general* *position. Assume that every curve in $\mathcal{A}$ lies above every curve in $\mathcal{B}$ and that there are m pairs of curves, one from $\mathcal{A}$ and the other from $\mathcal{B}$, that touch. Then, we have*

$$\ell_1(\mathcal{A}) + u_1(\mathcal{B}) \geq m - 1,$$

and

$$\ell_k(\mathcal{A}) + u_k(\mathcal{B}) \geq 2\frac{m}{k} - 4k.$$

Proof. We may assume without loss of generality that all crossing points between members of $\mathcal{A} \cup \mathcal{B}$ have distinct x-coordinates and that all of these values belong to the open interval $0 < x < 1$.

Note that, as x varies between the x-coordinates of two consecutive points at which a member of $\mathcal{A}$ touches a member of $\mathcal{B}$, the lowest curve of $\mathcal{A}$ or the highest curve of $\mathcal{B}$ must change. This yields the inequality

$$\ell_1(\mathcal{A}) + u_1(\mathcal{B}) \geq m - 1.$$

Fix $k > 1$. For any $0 \leq \xi \leq 1$ which is not the x-coordinate of an intersection point, let $\mathcal{A}_k(\xi)$ denote the kth lowest curve in $\mathcal{A}$ at the vertical line $x = \xi$ and let $\mathcal{B}_k(\xi)$ denote the kth highest curve in $\mathcal{B}$ at the vertical line $x = \xi$. Analogously, $\mathcal{A}_{\leq k}(\xi)$ denotes the family consisting of the k lowest curves in $\mathcal{A}$ at the vertical line $x = \xi$ and $\mathcal{B}_{\leq k}(\xi)$ denotes the family consisting of the k highest curves in $\mathcal{B}$ at the vertical line $x = \xi$.

For $\xi = 0$ or $\xi = 1$, the number of pairs $(a, b) \in \mathcal{A}_{\leq k}(\xi) \times \mathcal{B}_{\leq k}(\xi)$ is k^2. A pair $(a, b) \in \mathcal{A} \times \mathcal{B}$ is said to be *internally touching* if a and b touch each other and

$$(a, b) \notin \left(\mathcal{A}_{\leq k}(0) \times \mathcal{B}_{\leq k}(0)\right) \cup \left(\mathcal{A}_{\leq k}(1) \times \mathcal{B}_{\leq k}(1)\right).$$

Let I stand for the number of internally touching pairs (a, b). Clearly, we have $I \geq m - 2k^2$. For any internally touching pair (a, b), let (see Fig. 2)

1. $x_{\text{inf}}(a, b)$ be the infimum of all x-values for which $\mathcal{A}_k(x) = a$ and $b \in \mathcal{B}_{\leq k}(x)$, or $a \in \mathcal{A}_{\leq k}(x)$ and $\mathcal{B}_k(x) = b$, and let

2. $x_{\text{sup}}(a, b)$ be the supremum of all x-values for which $\mathcal{A}_k(x) = a$ and $b \in \mathcal{B}_{\leq k}(x)$, or $a \in \mathcal{A}_{\leq k}(x)$ and $\mathcal{B}_k(x) = b$.

Obviously, we have $x_{\text{inf}}(a, b) < x_{\text{sup}}(a, b)$ as the x-coordinate of the touching point between a and b lies strictly between these two numbers. It is also clear that the numbers $x_{\text{inf}}(a, b)$ and $x_{\text{sup}}(a, b)$ are x-coordinates of crossing points lying on the kth lowest level of $\mathcal{A}$ or on the kth highest level of $\mathcal{B}$.

For any $0 < \xi < 1$, there are at most k internally touching pairs (a, b) with $x_{\inf}(a, b) = \xi$. Indeed, for any $a \in \mathcal{A}$ such that $a = \mathcal{A}_k(\xi + \varepsilon)$, say, for all sufficiently small $\varepsilon > 0$, all curves $b \in \mathcal{B}$ with $x_{\inf}(a, b) = \xi$ must belong to the set $\mathcal{B}_{\leq k}(\xi)$. This is a set of size k. Thus, the number of distinct x-coordinates ξ at which either $\mathcal{A}_k(\xi)$ or $\mathcal{B}_k(\xi)$ changes is at least $2I/k$. That is, we have

$$\ell_k(\mathcal{A}) + u_k(\mathcal{B}) \geq \frac{2I}{k} \geq 2\frac{m - 2k^2}{k} = 2\frac{m}{k} - 4k. \qquad \blacksquare$$

A similar argument was used in [1].

Now we are in a position to establish Theorem 1.

Proof of Theorem 1. Assume without loss of generality that $|\mathcal{A}| \geq |\mathcal{B}|$ and that every curve in $\mathcal{A} \cup \mathcal{B}$ participates in at least one touching pair. This implies that any two members of $\mathcal{A}$ properly cross at least once and any two members of $\mathcal{B}$ properly cross at least once. Hence, we have

$$\mathrm{CN}(\mathcal{A}) + \mathrm{CN}(\mathcal{B}) \geq \binom{|\mathcal{A}|}{2} + \binom{|\mathcal{B}|}{2}.$$

This completes the proof in the special case where $m \leq |\mathcal{A}|^2/\ln|\mathcal{A}|$, because then the term $\binom{|\mathcal{A}|}{2}$ already exceeds the desired lower bound. In particular, since the total number m of touching pairs is at most $|\mathcal{A}|\,|\mathcal{B}|$, we are done if $|\mathcal{B}| \leq |\mathcal{A}|/\ln|\mathcal{A}|$.

From now on, we can assume that

$$m > |\mathcal{A}|^2/\ln|\mathcal{A}|$$

and

$$|\mathcal{A}|/\ln|\mathcal{A}| \leq |\mathcal{B}| \leq |\mathcal{A}|.$$

Let $\varepsilon > 0$ be a very small constant. Set $K = m^{\frac{1}{2}-\varepsilon}$, and add up $\ell_k(\mathcal{A}) + u_k(\mathcal{B})$ for all $1 \leq k \leq K$. Note that we can apply Lemma 1, since the last two inequalities imply that $K \leq |\mathcal{B}|$. In view of (1), we obtain

$$\mathrm{CN}(\mathcal{A}) + \mathrm{CN}(\mathcal{B}) \geq \frac{1}{2}\sum_{k=1}^{K}\left(\ell_k(\mathcal{A}) + u_k(\mathcal{B})\right) \geq \frac{1}{2}\left(m - 1 + \sum_{k=2}^{K}\left(2\frac{m}{k} - 4k\right)\right)$$

$$\geq \frac{1}{2}\left(m - 2K(K+1) + 3 + 2m\sum_{k=2}^{K}\frac{1}{k}\right)$$

$$= \left(\frac{1}{2} - \varepsilon - o(1)\right) m \ln m.$$

Letting $\varepsilon \to 0$, we can conclude that $\mathrm{CN}(\mathcal{A}) + \mathrm{CN}(\mathcal{B})$ is at least $\big(\frac{1}{2} - o(1)\big) m \ln m$, as required. ■

3. Constructive Upper Bound – Proof of Theorem 2

Let $c(n)$ denote the minimum number of crossing points in the union of any two completely touching n-member families of curves $\mathcal{A} \cup \mathcal{B}$, where all members of $\mathcal{A}$ are above all members of $\mathcal{B}$.

We need the following:

Lemma 2. *For any pair of positive integers i and j, we have*

$$c(ij) \leq i^2 c(j) + j^2 c(i).$$

Proof. Let $(\mathcal{A}', \mathcal{B}')$ be a pair of completely touching i-member families of curves with

$$\mathrm{CN}(\mathcal{A}') + \mathrm{CN}(\mathcal{B}') = c(i).$$

Replace each curve $\gamma \in \mathcal{A}' \cup \mathcal{B}'$ by j curves that closely follow γ. For any $\alpha \in \mathcal{A}'$ and for any $\beta \in \mathcal{B}'$, let each of the j curves corresponding to α touch each of the j curves corresponding to β in a small neighborhood of the point where α and β touch each other. This can be achieved by introducing $c(j)$ crossings near each point of tangency between α and β. Apart from the crossings introduced in the neighborhoods of these points, the j new curves corresponding to an "old" curve $\gamma \in \mathcal{A}' \cup \mathcal{B}'$ are disjoint.

Denote the family of ij curves obtained from the members of $\mathcal{A}'$ by $\mathcal{A}$, and the family of ij curves obtained from $\mathcal{B}'$ by $\mathcal{B}$. Since the number of tangencies between $\mathcal{A}'$ and $\mathcal{B}'$ is i^2, there are at most $i^2 c(j)$ crossings among the members of $\mathcal{A} \cup \mathcal{B}$ that occur near these touching points. On the other hand, in a small neighborhood of each crossing between two members of $\mathcal{A}'$ or two members of $\mathcal{B}'$, we create j^2 crossings in $\mathcal{A}$ or in $\mathcal{B}$. Therefore, there are $j^2c(i)$ crossings among members of $\mathcal{A} \cup \mathcal{B}$ that occur near crossings in $\mathcal{A}'$ or $\mathcal{B}'$. In view of the fact that each crossing in $\mathcal{A} \cup \mathcal{B}$ occurs in a small neighborhood of either a touching point or a crossing point in $\mathcal{A}' \cup \mathcal{B}'$, we obtain that $c(ij) \le \mathrm{CN}(\mathcal{A}) + \mathrm{CN}(\mathcal{B}) \le i^2c(j) + j^2c(i)$, as required. ■

Using the fact $c(2) = 3$, by repeated application of Lemma 2 with $j = 2$ and $i = 2, 2^2, \ldots, 2^{k-1}$. we obtain that $c(2^k) \le \frac{3}{4}k4^k$. Starting with a completely touching pair of 2-member families of curves, after $k-1$ iterations we obtain a completely touching pair $(\mathcal{A}, \mathcal{B})$ of 2^k-member families with $m = 2^{2k}$ touching pairs. Thus, there exists a configuration with only $\frac{3}{4}k4^k = \frac{3}{8}m\log_2 m$ crossings, meeting the requirements. This completes the proof of Theorem 2 in the case n is a power of 2. If n is not a power of 2, let k be the least integer such that $2^k > n$. By randomly sampling n curves in $\mathcal{A}$ and n curves in $\mathcal{B}$, we obtain $c(n) \le (n/2^k)2c(2^k)$, and the desired bound follows.

4. Concluding Remarks

The assumption in Theorem 1 that the curves are two-way infinite is not important. If we have a family $\mathcal{F} = \mathcal{A} \cup \mathcal{B}$ of arbitrary x-monotone curves such that, for any pair of curves $\alpha \in \mathcal{A}$, $\beta \in \mathcal{B}$ which can be met by a vertical line, α lies above β, we can make each curve *two-way infinite* without destroying this property, by adding only at most

$$2\binom{|\mathcal{A}|}{2} + 2\binom{|\mathcal{B}|}{2} < |\mathcal{F}|^2$$

crossings.

One can give an alternative proof of Theorem 1 by reducing it to a combinatorial statement about sequences. Let $(x_1, \ldots, x_m)$ be a sequence of m elements taken from a finite alphabet Φ. For any pair of distinct elements $a, b \in \Phi$, define the number of *alternations* of a and b in the sequence, as the largest number t such that there is a subsequence $(x_{i(0)}, x_{i(1)}, \ldots, x_{i(t)})$ of length $t+1$ with $1 \le i(0) < i(1) < \cdots < i(t) \le m$ such that its elements alternate between a and b (or between b and a). That is,

$$x_{i(0)} = x_{i(2)} = \cdots = a, \quad x_{i(1)} = x_{i(3)} = \cdots = b,$$

or

$$x_{i(0)} = x_{i(2)} = \cdots = b, \quad x_{i(1)} = x_{i(3)} = \cdots = a.$$

This number t is denoted by $\text{alt}_{\{a,b\}}(x_1, \ldots, x_n)$.

Define the *alternation number* of the sequence $(x_1, \ldots, x_m)$, as

$$\sum_{\{a,b\} \subseteq \Sigma} \text{alt}_{\{a,b\}}(x_1, \ldots, x_m),$$

where the sum is taken over all unordered pairs $\{a, b\}$ of distinct elements from Φ.

Theorem 1 can also be proved using the following result, which is perhaps of independent interest.

Theorem 3. *Let $(x_1, \ldots, x_m)$ be a sequence of length m over an alphabet Φ. Assume that there exists an absolute constant $c > 0$ such that for all $1 \le z \le m$, every z consecutive elements of the sequence contain at least $c\sqrt{z}$ distinct symbols. Then the alternation number of the sequence $(x_1, \ldots, x_m)$ is at least $dm \log m$, for a suitable constant $c' > 0$, depending only on c.*

Salazar [8] verified the Richter–Thomassen conjecture in the special case when any pair of curves have at most k points in common, for a fixed constant k. The best known general bound is due to Mubayi [4], who proved that any family of n closed curves in general position in the plane determines at least $\left(\frac{4}{5} + o(1)\right)n^2$ intersection points.

References

[1] T. M. Chan, On Levels in Arrangements of Curves, II: A Simple Inequality and Its Consequences, *Discrete & Computational Geometry,* **34(1)** (2005), 11–24.

[2] P. Erdős, On sets of distances of n points, *The American Mathematical Monthly,* **53**, (1946), 248–250.

[3] P. Koebe, Kontaktprobleme der konforman Abbildung, *Berichte über die Verhandlungen d. Sächs. Akademie der Wissenschaften Leipzig,* **88** (1936), 141–164.

[4] D. Mubayi, Intersecting curves in the plane, *Graphs and Combinatorics,* **18(3)** (2002), 583–589.

[5] J. Pach and P. K. Agarwal, *Combinatorial Geometry,* Wiley, New York, 1995.

[6] R. B. Richter and C. Thomassen, Intersection of curves systems and the crossing number of $C_5 \times C_5$, *Discrete & Computational Geometry,* **13**, (1995), 149–159.

[7] F. Ruskey and M. Weston, Venn Diagram Survey, *Electronic Journal of Combinatorics,* **DS#5** (2005).

[8] G. Salazar, On the intersections of systems of curves, *Journal of Combinatorial Theory Series B,* **75** (1999), 56–60.

[9] J. Venn, On the diagrammatic and mechanical representation of propositions and reasonings, *Philosophical Magazine and Journal of Science, Series 5,* **10(59)** (1880).

Jacob Fox
Department of Mathematics
Princeton University
Princeton, NJ
e-mail: jacobfox@math.princeton.edu

Fabrizio Frati
Dipartimento di Informatica e Automazione
Roma Tre University
Italy
e-mail: frati@dia.uniroma3.it

János Pach
EPFL Lausanne
Switzerland
and
Rényi Institute
Budapest
Hungary
e-mail: pach@cims.nyu.edu

Rom Pinchasi
Mathematics Department
Technion –
Israel Institute of Technology
Haifa 32000
Israel
e-mail: room@math.technion.ac.il

BOLYAI SOCIETY
MATHEMATICAL STUDIES, 21

An Irregular Mind
(Szemerédi is 70)
pp. 261–334.

An Arithmetic Regularity Lemma, an Associated Counting Lemma, and Applications

BEN GREEN and TERENCE TAO

To Endre Szemerédi on the occasion of his 70th birthday

Szemerédi's regularity lemma can be viewed as a rough structure theorem for arbitrary dense graphs, decomposing such graphs into a structured piece, a small error, and a uniform piece. We establish an *arithmetic regularity lemma* that similarly decomposes bounded functions $f : [N] \to \mathbb{C}$, into a (well-equidistributed, virtual) s-step nilsequence, an error which is small in L^2 and a further error which is minuscule in the Gowers U^{s+1}-norm, where $s \geqslant 1$ is a parameter. We then establish a complementary *arithmetic counting lemma* that counts arithmetic patterns in the nilsequence component of f.

We provide a number of applications of these lemmas: a proof of Szemerédi's theorem on arithmetic progressions, a proof of a conjecture of Bergelson, Host and Kra, and a generalisation of certain results of Gowers and Wolf.

Our result is dependent on the inverse conjecture for the Gowers U^{s+1} norm, recently established for general s by the authors and T. Ziegler.

1. Introduction

Szemerédi's celebrated *regularity lemma* [48, 49] is a fundamental tool in graph theory; see for instance [36] for a survey of some of its many applications. It is often described as a structure theorem for graphs $G = (V, E)$, but one may also view it as a decomposition for arbitrary functions $f : V \times V \to [0, 1]$. For instance, one can recast the regularity lemma in the following "analytic" form. Define a *growth function* to be any monotone increasing function $\mathcal{F} : \mathbb{R}^+ \to \mathbb{R}^+$ with $\mathcal{F}(M) \geqslant M$ for all M.

Lemma 1.1 (Szemerédi regularity lemma, analytic form). *Let V be a finite vertex set, let $f : V \times V \to [0,1]$ be a function, let $\varepsilon > 0$, and let $\mathcal{F} : \mathbb{R}^+ \to \mathbb{R}^+$ be a growth function. Then there exists a positive integer*[1] *$M = O_{\varepsilon,\mathcal{F}}(1)$ and a decomposition*

$$f = f_{\mathrm{str}} + f_{\mathrm{sml}} + f_{\mathrm{unf}} \tag{1.1}$$

of f into functions $f_{\mathrm{str}}, f_{\mathrm{sml}}, f_{\mathrm{unf}} : V \times V \to [-1,1]$ such that:

(i) *(f_{str} structured) V can be partitioned into M cells $V_1, \ldots, V_M$, such that f_{str} is constant on $V_i \times V_j$ for all i, j with $1 \leqslant i,j \leqslant M$;*

(ii) *(f_{sml} small) The quantity*[2] *$\|f_{\mathrm{sml}}\|_{L^2(V\times V)} := \left(\mathbb{E}_{v,w\in V}\left|f_{\mathrm{sml}}(v,w)\right|^2\right)^{1/2}$ is at most ε.*

(iii) *(f_{unf} very uniform) The box norm $\|f_{\mathrm{unf}}\|_{\square^2(V\times V)}$, defined to be the quantity*

$$\left(\mathbb{E}_{v_1,v_2,w_1,w_2\in V} f_{\mathrm{unf}}(v_1,w_1) f_{\mathrm{unf}}(v_1,w_2) f_{\mathrm{unf}}(v_2,w_1) f_{\mathrm{unf}}(v_2,w_2)\right)^{1/4},$$

is at most $1/\mathcal{F}(M)$.

(iv) *(Nonnegativity) f_{str} and $f_{\mathrm{str}} + f_{\mathrm{sml}}$ take values in $[0,1]$.*

Informally, this regularity lemma decomposes any bounded function into a structured part, a small error, and an extremely uniform error. While this formulation does not, at first sight, look much like the usual regularity lemma, it easily implies that result: see [53]. The idea of formulating the regularity lemma with an arbitrary growth function $\mathcal{F}$ first appears in [1], and is also very useful for generalisations of the regularity lemma to hypergraphs. See, for example, [52]. The bound on M turns out to essentially be an iterated version of the growth function $\mathcal{F}$, with the number of iterations being polynomial in $1/\varepsilon$. In applications, one usually selects the growth function to be exponential in nature, which then makes M essentially tower-exponential in $1/\varepsilon$. See [51, 54] for a general discussion of these sorts of structure theorems and their applications in combinatorics. See also [42] for a related analytical perspective on the regularity lemma.

---

[1]As usual, we use $O(X)$ to denote a quantity bounded in magnitude by CX for some absolute constant X; if we need C to depend on various parameters, we will indicate this by subscripts. Thus for instance $O_{\varepsilon,\mathcal{F}}(1)$ is a quantity bounded in magnitude by some expression $C_{\varepsilon,\mathcal{F}}$ depending on ε, $\mathcal{F}$.

[2]We use here the expectation notation $\mathbb{E}_{a\in A} f(a) := \frac{1}{|A|}\sum_{a\in A} f(a)$ for any finite non-empty set A, where $|A|$ denotes the cardinality of A.

In applications the regularity lemma is often paired with a *counting lemma* that allows one to control various expressions involving the function f. For example, one might consider the expression

$$\mathbb{E}_{u,v,w\in V} f(u,v)f(v,w)f(w,u), \tag{1.2}$$

which counts triangles in V weighted by f. Applying the decomposition (1.1) splits expressions such as (1.2) into multiple terms (in this instance, 27 of them). The key fact, which is a slightly non-trivial application of the Cauchy–Schwarz inequality, is that the terms involving the box-norm-uniform error f_{unf} are negligible if the growth function $\mathcal{F}$ is chosen rapidly enough. The terms involving the small error f_{sml} are somewhat small, but one often has to carefully compare those errors against the main term (which only involves f_{str}) in order to get a non-trivial bound on the final expression (1.2). In particular, one often needs to exploit the positivity of f_{str} and $f_{\mathrm{str}} + f_{\mathrm{sml}}$ to first *localise* expressions such as (1.2) to a small region (such as the portion of a graph between a "good" triple V_i, V_j, V_k of cells in the partition of V associated to f_{str}) before one can obtain a useful estimate.

The graph regularity and counting lemmas can be viewed as the first non-trivial member of a hierarchy of *hypergraph* regularity and counting lemmas, see e.g. [9, 19, 20, 43, 44, 52]. The formulation in [52] is particularly close to the formulation given in Theorem 1.1. These lemmas are suitable for controlling higher order expressions such as

$$\mathbb{E}_{u,v,w,x\in V} f(u,v,w)f(v,w,x)f(w,x,u)f(x,u,v).$$

Our objective in this paper is to introduce an analogous hierarchy of such regularity and counting lemmas (one for each integer $s \geqslant 1$), in *arithmetic* situations. Here, the aim is to decompose a function $f : [N] \to [0,1]$ defined on an arithmetic progression $[N] := \{1,\ldots,N\}$ instead of a graph. One is interested in counting averages such as

$$\mathbb{E}_{n,r\in[N]} f(n)f(n+r)f(n+2r),$$

which counts 3-term arithmetic progressions weighted by f, as well as higher order expressions such as

$$\mathbb{E}_{n,r\in[N]} f(n)f(n+r)f(n+2r)f(n+3r).$$

As it turns out, the former average will be best controlled using the $s = 1$ regularity and counting lemmas, while the latter requires the $s = 2$ versions of these lemmas. In this paper we shall see several examples of these types of applications of the two lemmas.

The arithmetic regularity lemma. We begin with by formulating our regularity lemma. Following the statement we explain the terms used here.

Theorem 1.2 (Arithmetic regularity lemma). *Let $f : [N] \to [0,1]$ be a function, let $s \geqslant 1$ be an integer, let $\varepsilon > 0$, and let $\mathcal{F} : \mathbb{R}^+ \to \mathbb{R}^+$ be a growth function. Then there exists a quantity $M = O_{s,\varepsilon,\mathcal{F}}(1)$ and a decomposition*

$$f = f_{\mathrm{nil}} + f_{\mathrm{sml}} + f_{\mathrm{unf}}$$

of f into functions $f_{\mathrm{nil}}, f_{\mathrm{sml}}, f_{\mathrm{unf}} : [N] \to [-1,1]$ of the following form:

(i) *(f_{nil} structured) f_{nil} is a $(\mathcal{F}(M), N)$-irrational virtual nilsequence of degree $\leqslant s$, complexity $\leqslant M$, and scale N;*
(ii) *(f_{sml} small) f_{sml} has an $L^2[N]$ norm of at most ε;*
(iii) *(f_{unf} very uniform) f_{unf} has a $U^{s+1}[N]$ norm of at most $1/\mathcal{F}(M)$;*
(iv) *(Nonnegativity) f_{nil} and $f_{\mathrm{nil}} + f_{\mathrm{sml}}$ take values in $[0,1]$.*

Remark. This result easily implies the recently proven *inverse conjecture for the Gowers norms* (Theorem 2.1). Conversely, this inverse conjecture, together with the equidistribution theory of nilsequences, will be the main ingredient used to prove Theorem 1.2.

We prove this theorem in §2. We turn now to a discussion of the various concepts used in the above statement. Readers who are interested in applications may skip ahead to the end of the section.

The $L^2[N]$ norm, used to control f_{sml}, is simply

$$\|f\|_{L^2[N]} := \left(\mathbb{E}_{n\in[N]}|f(n)|^2\right)^{1/2}.$$

We turn next to the Gowers uniformity norm $U^{s+1}[N]$, used to control f_{unf}. If $f : G \to \mathbb{C}$ is a function on a finite additive group G, and $k \geqslant 1$ is an integer, then the *Gowers uniformity norm* $\|f\|_{U^k(G)}$ is defined by the formula

$$\|f\|_{U^k(G)} := \left(\mathbb{E}_{x,h_1,\dots,h_k\in G}\Delta_{h_1}\cdots\Delta_{h_k}f(x)\right)^{1/2^k},$$

where $\Delta_h f : G \to \mathbb{C}$ is the multiplicative derivative of f in the direction h, defined by the formula

$$\Delta_h f(x) := f(x+h)\overline{f(x)}.$$

In this paper we will be concerned with functions on $[N]$, which is not quite a group. To define the Gowers norms of a function $f : [N] \to \mathbb{C}$, set $G := \mathbb{Z}/\tilde{N}\mathbb{Z}$ for some integer $\tilde{N} \geqslant 2^k N$, define a function $\tilde{f} : G \to \mathbb{C}$ by $\tilde{f}(x) = f(x)$ for $x = 1, \dots, N$ and $\tilde{f}(x) = 0$ otherwise, and set $\|f\|_{U^k[N]} := \|\tilde{f}\|_{U^k(G)} / \left\|1_{[N]}\right\|_{U^k(G)}$, where $1_{[N]}$ is the indicator function of $[N]$. It is easy to see that this definition is independent of the choice of $\tilde{N}$, and so for definiteness one could take $\tilde{N} := 2^k N$. Henceforth we shall write simply $\|f\|_{U^k}$, rather than $\|f\|_{U^k[N]}$, since all Gowers norms will be on $[N]$. One can show that $\|\cdot\|_{U^k}$ is indeed a norm for any $k \geqslant 2$, though we shall not need this here; see [18]. For further discussion of the Gowers norms and their relevance to counting additive patterns see [18], [27, §5] or [55, §11].

Finally, we turn to the notion of a irrational virtual nilsequence, which is the concept that defines the structural component f_{nil}. This is the most complicated concept, and requires a certain number of preliminary definitions. We first need the notion of a *filtered nilmanifold.* The first two sections of [30] may be consulted for a more detailed discussion.

Definition 1.3 (Filtered nilmanifold). Let $s \geqslant 1$ be an integer. A *filtered nilmanifold* $G/\Gamma = (G/\Gamma, G_\bullet)$ of degree $\leqslant s$ consists of the following data:

(i) A connected, simply-connected nilpotent Lie group G;

(ii) A discrete, cocompact subgroup Γ of G (thus the quotient space G/Γ is a compact manifold, known as a *nilmanifold*);

(iii) A *filtration* $G_\bullet = \left(G_{(i)}\right)_{i=0}^{\infty}$ of closed connected subgroups

$$G = G_{(0)} = G_{(1)} \geqslant G_{(2)} \geqslant \dots$$

of G, which are *rational* in the sense that the subgroups $\Gamma_{(i)} := \Gamma \cap G_{(i)}$ are cocompact in $G_{(i)}$, such that $\left[G_{(i)}, G_{(j)}\right] \subseteq G_{(i+j)}$ for all $i, j \geqslant 0$, and such that $G_{(i)} = \{\mathrm{id}\}$ whenever $i > s$;

(iv) A *Mal'cev basis* $\mathcal{X} = \left(X_1, \dots, X_{\dim(G)}\right)$ adapted to $G_\bullet$, that is to say a basis $X_1, \dots, X_{\dim(G)}$ of the Lie algebra of G that exponentiates to elements of Γ, such that $X_j, \dots, X_{\dim(G)}$ span a Lie algebra ideal for all $j \leqslant i \leqslant \dim(G)$, and $X_{\dim(G) - \dim(G_{(i)}) + 1}, \dots, X_{\dim(G)}$ spans the Lie algebra of $G_{(i)}$ for all $1 \leqslant i \leqslant s$. (For a detailed discussion of this concept, see [30, §2].)

Once a Mal'cev basis has been specified, notions such as the rationality of subgroups may be quantified in terms of it. Furthermore one may use a

Mal'cev basis to define a metric $d_{G/\Gamma}$ on the nilmanifold G/Γ. The results of this paper are rather insensitive to the precise metric that one takes, but one may proceed for example as in [30, Definition 2.2]. We encourage the reader not to think too carefully about the precise definition (or about Mal'cev bases in general), but it is certainly important to have some definite metric in mind so that one can make sense of notions such as that of a *Lipschitz function* on G/Γ.

Observe that every filtered nilmanifold G/Γ comes with a canonical *probability Haar measure* $\mu_{G/\Gamma}$, defined as the unique Borel probability measure on G/Γ that is invariant under the left action of G. We abbreviate $\int_{G/\Gamma} F(x)\, d\mu_{G/\Gamma}(x)$ as $\int_{G/\Gamma} F$.

We will need a quantitative notion of *complexity* for filtered nilmanifolds, though once again, the precise definition is somewhat unimportant.

Definition 1.4 (Complexity). Let $M \geqslant 1$. We say that a filtered nilmanifold $G/\Gamma = (G/\Gamma, G_\bullet)$ has *complexity* $\leqslant M$ if the dimension of G, the degree of $G_\bullet$, and the rationality of the Mal'cev basis $\mathcal{X}$ (cf. [30, Definition 2.4]) are bounded by M.

Heisenberg example. The model example of a degree $\leqslant 2$ filtered nilmanifold is the *Heisenberg nilmanifold*

$$G/\Gamma := \begin{pmatrix} 1 & \mathbb{R} & \mathbb{R} \\ 0 & 1 & \mathbb{R} \\ 0 & 0 & 1 \end{pmatrix} \Big/ \begin{pmatrix} 1 & \mathbb{Z} & \mathbb{Z} \\ 0 & 1 & \mathbb{Z} \\ 0 & 0 & 1 \end{pmatrix}$$

with the lower central series $G_{(0)} = G_{(1)} = G$ and

$$G_{(2)} = [G, G] = \begin{pmatrix} 1 & 0 & \mathbb{R} \\ 0 & 1 & 0 \\ 0 & 0 & 1 \end{pmatrix}$$

with Mal'cev basis $\mathcal{X} = \{X_1, X_2, X_3\}$ consisting of the matrices

$$X_1 = \begin{pmatrix} 0 & 1 & 0 \\ 0 & 0 & 0 \\ 0 & 0 & 0 \end{pmatrix}, \quad X_2 = \begin{pmatrix} 0 & 0 & 0 \\ 0 & 0 & 1 \\ 0 & 0 & 0 \end{pmatrix}, \quad X_3 = \begin{pmatrix} 0 & 0 & 1 \\ 0 & 0 & 0 \\ 0 & 0 & 0 \end{pmatrix}.$$

With the definition of filtered nilmanifold in place, the next thing we need is the idea of a *polynomial sequence*. The basic theory of such sequences was laid out in Leibman [37], and was extended slightly to general filtrations in [30]. An extensive discussion may be found in Section 6 of that paper.

Definition 1.5 (Polynomial sequence). Let $(G/\Gamma, G_\bullet)$ be a filtered nilmanifold, with filtration $G_\bullet = \left(G_{(i)}\right)_{i=0}^{\infty}$. A *(multidimensional) polynomial sequence* adapted to this filtered nilmanifold is a sequence $g : \mathbb{Z}^D \to G$ for some $D \geqslant 1$ with the property that

$$\partial_{h_1} \dots \partial_{h_i} g(n) \in G_{(i)}$$

for all $i \geqslant 0$ and $h_1, \dots, h_i, n \in \mathbb{Z}^D$, where $\partial_h g(n) := g(n+h)g(n)^{-1}$ is the derivative of g with respect to the shift h. The space of all such polynomial sequences will be denoted $\operatorname{poly}(\mathbb{Z}^D, G_\bullet)$. The space of polynomial sequences taking values in Γ will be denoted $\operatorname{poly}(\mathbb{Z}^D, \Gamma_\bullet)$. When $D = 1$, we refer to multidimensional polynomial sequences simply as *polynomial sequences.*

Remark. We will be primarily interested in the one-dimensional case $D = 1$, but will need the higher D case in order to establish the counting lemma, Theorem 1.11.

One of the main reasons why we work with polynomial sequences, instead of just linear sequences such as $n \mapsto g_0 g_1^n$, is that objects of the former type constitute a group.

Theorem 1.6 (Lazard–Leibman). *If $(G/\Gamma, G_\bullet)$ is a filtered nilmanifold and $D \geqslant 1$ is an integer, then* $\operatorname{poly}(\mathbb{Z}^D, G_\bullet)$ *is a group (and* $\operatorname{poly}(\mathbb{Z}^D, \Gamma_\bullet)$ *is a subgroup).*

Proof. See [38] or [30, Proposition 6.2]. ∎

With the concept of a polynomial sequence in hand, it is easy to define a *polynomial orbit.*

Definition 1.7 (Orbits). Let $D, s \geqslant 1$ be integers, and $M, A > 0$ be parameters. A *(multidimensional) polynomial orbit* of degree $\leqslant s$ and complexity $\leqslant M$ is any function[3] $n \mapsto g(n)\Gamma$ from $\mathbb{Z}^D \to G/\Gamma$, where $(G/\Gamma, G_\bullet)$ is a filtered nilmanifold of complexity $\leqslant M$, and $g \in \operatorname{poly}(\mathbb{Z}^D, G_\bullet)$ is a (multidimensional) polynomial sequence.

Using the concept of polynomial orbit, we can define the notion of a (polynomial) nilsequence, as well as a generalisation which we call a *virtual* nilsequence, in analogy with virtually nilpotent groups (groups with a finite index nilpotent subgroup).

[3]Strictly speaking, the orbit is the tuple of data $(G, \Gamma, G/\Gamma, G_\bullet, n \mapsto g(n)\Gamma)$, rather than just the sequence $n \mapsto g(n)\Gamma$, but we shall abuse notation and use the sequence as a metonym for the whole orbit.

Definition 1.8 (Nilsequences). A *(multidimensional, polynomial) nilsequence* of degree $\leqslant s$ and complexity $\leqslant M$ is any function $f : \mathbb{Z}^D \to \mathbb{C}$ of the form $f(n) = F\big(g(n)\Gamma\big)$, where $n \mapsto g(n)\Gamma$ is a polynomial orbit of degree $\leqslant s$ and complexity $\leqslant M$, and $F : G/\Gamma \to \mathbb{C}$ is a function of Lipschitz norm[4] at most M.

Definition 1.9 (Virtual nilsequences). Let $N \geqslant 1$. A *virtual nilsequence* of degree $\leqslant s$ and complexity $\leqslant M$ at scale N is any function $f : [N] \to \mathbb{C}$ of the form $f(n) = F\big(g(n)\Gamma, n(\text{mod } q), n/N\big)$, where $1 \leqslant q \leqslant M$ is an integer, $n \mapsto g(n)\Gamma$ is a polynomial orbit of degree $\leqslant s$ and complexity $\leqslant M$, and $F : G/\Gamma \times \mathbb{Z}/q\mathbb{Z} \times \mathbb{R} \to \mathbb{C}$ is a function of Lipschitz norm at most M. (Here we place a metric on $G/\Gamma \times \mathbb{Z}/q\mathbb{Z} \times \mathbb{R}$ in some arbitrary fashion, e.g. by embedding $\mathbb{Z}/q\mathbb{Z}$ in $\mathbb{R}/\mathbb{Z}$ and taking the direct sum of the metrics on the three factors.)

One concept that featured in Theorem 1.2 remains to be defined: that of an *irrational* orbit. The definition is a little technical (see Definition A.6) and takes some setting up, and so we defer it and the discussion of some motivating examples to Appendix A. Very roughly speaking, an irrational orbit is one whose coefficients are not close to rationals (of bounded height) and for which the filtration $G_\bullet$ is as small as possible. For instance, with a polynomial sequence $P : [N] \to \mathbb{R}/\mathbb{Z}$ of the form $P(n) = \alpha_s n^s + \ldots + \alpha_0$, then (roughly speaking) this sequence would be considered irrational if one takes $G_{(i)} = \mathbb{R}$ for $i \leqslant s$ and $G_{(i)} = \{0\}$ for $i > s$, and if there was no positive integer $q = O(1)$ for which $\|q\alpha_s\|_{\mathbb{R}/\mathbb{Z}} \ll N^{-s}$. Again, we refer the reader to Appendix A for further examples and discussion.

This concludes our attempt to discuss all the concepts involved in the arithmetic regularity lemma, Theorem 1.2; we turn now to a statement and discussion of the counting lemma.

Counting Lemma. In applications of the arithmetic regularity lemma, we will be interested in counting additive patterns such as arithmetic progressions or parallelepipeds. To understand the phenomena properly it is advantageous to work in a somewhat general setting similar to that taken in

[4]The (inhomogeneous) Lipschitz norm $\|F\|_{\mathrm{Lip}}$ of a function $F : X \to \mathbb{C}$ on a metric space $X = (X, d)$ is defined as

$$\|F\|_{\mathrm{Lip}} := \sup_{x \in X} |F(x)| + \sup_{x,y \in X : x \neq y} \frac{|F(x) - F(y)|}{|x - y|}.$$

[22, 23, 24, 31]. In the latter paper one works with a family $\Psi = (\psi_1, \ldots, \psi_t)$ of integer-coefficient linear forms (or equivalently, group homomorphisms) $\psi_1, \ldots, \psi_t : \mathbb{Z}^D \to \mathbb{Z}$, and consider expressions such as

$$\mathbb{E}_{\mathbf{n}\in\mathbb{Z}^D\cap P} f(\psi_1(\mathbf{n})) \ldots f(\psi_t(\mathbf{n})) \tag{1.3}$$

where P is a convex subset of $\mathbb{R}^D$. Thus, for instance, if counting arithmetic progressions, one might use the linear forms

$$\psi_i(n_1, n_2) := n_1 + (i-1)n_2; \quad i = 1, \ldots, k \tag{1.4}$$

whilst for counting parallelepipeds one might instead use the linear forms

$$\psi_{\omega_1,\ldots,\omega_k}(n_0, n_1, \ldots, n_k) := n_0 + \omega_1 n_1 + \ldots + \omega_k n_k; \quad \omega_1, \ldots, \omega_k \in \{0, 1\}. \tag{1.5}$$

In order to understand the contribution to (1.3) coming from the structured part f_{nil} of f, one is soon faced with the question of understanding the equidistribution of the orbit

$$\big(g(\psi_1(\mathbf{n}))\Gamma, \ldots, g(\psi_t(\mathbf{n}))\Gamma\big) \tag{1.6}$$

inside $(G/\Gamma)^t$, where $\mathbf{n} = (n_1, \ldots, n_D)$ ranges over $\mathbb{Z}^D \cap P$. We abbreviate this orbit as $g^\Psi(\mathbf{n})\Gamma^t$, where $g^\Psi : \mathbb{Z}^D \to G^t$ is the polynomial sequence

$$g^\Psi(\mathbf{n}) := \big(g(\psi_1(\mathbf{n})), \ldots, g(\psi_t(\mathbf{n}))\big). \tag{1.7}$$

A very useful model for this question, in which *infinite* orbits were considered in the "linear" case $g(n) = g^n x$, was studied by Leibman [41]. His work leads one to the following definition.

Definition 1.10 (The Leibman group). Let $\Psi = (\psi_1, \ldots, \psi_t)$ be a collection of linear forms $\psi_1, \ldots, \psi_t : \mathbb{Z}^D \to \mathbb{Z}$. For any $i \geqslant 1$, define $\Psi^{[i]}$ to be the linear subspace of $\mathbb{R}^k$ spanned by the vectors $\big(\psi_1^j(\mathbf{n}), \ldots, \psi_t^j(\mathbf{n})\big)$ for $1 \leqslant j \leqslant i$ and $\mathbf{n} \in \mathbb{Z}^D$. Given a filtered nilmanifold $(G/\Gamma, G_\bullet)$, we define the *Leibman group* $G^\Psi \lhd G^t$ to be the Lie subgroup of G^t generated by the elements $g_i^{\vec{v}_i}$ for $i \geqslant 1$, $g_i \in G_{(i)}$, and $\vec{v}_i \in \Psi^{[i]}$, with the convention that[5]

$$g^{(v_1,\ldots,v_t)} := (g^{v_1}, \ldots, g^{v_t})$$

for each $g \in G$. Note that G^Ψ is normal in G^t because $G_{(i)}$ is normal in G. We will show in §3 that G^Ψ is also a rational subgroup of G^t, thus $\Gamma^\Psi := \Gamma^t \cap G^\Psi$ is a discrete cocompact subgroup of G^Ψ.

[5]We define g^v for real v by the formula $g^v := \exp(v \log(g))$, where $\exp : \mathfrak{g} \to G$ is the usual exponential map from the Lie algebra $\mathfrak{g}$ to G (this is a homeomorphism since G is nilpotent, connected, and simply connected).

Example. Two particular instances of this construction correspond to the two lattices (1.4) and (1.5) above. In the case of arithmetic progressions, where Ψ is as in (1.4), the Leibman group G^{Ψ} is sometimes (see, for example, [12]) referred to as the *Hall–Petresco group* $\mathrm{HP}^k(G_\bullet)$ and has the particularly simple alternative description

$$\mathrm{HP}^k(G_\bullet) = G^{\Psi} = \big\{ \big(g(0), \dots, g(k-1)\big) \ :\ g \in \mathrm{poly}\,(G_\bullet) \big\},$$

We will prove this fact in §3. In the case of parallelepipeds, where Ψ is as in (1.5), the Leibman group G^{Ψ} has been referred to as the *Host–Kra cube group* [31] and it too has an alternative description. See [31, Appendix E] for more information: we will not be making use of this particular group here.

Let $g \in \mathrm{poly}\,(\mathbb{Z}, G_\bullet)$ be a polynomial sequence, and let $\Psi = (\psi_1, \dots, \psi_t)$ be a collection of linear forms $\psi_1, \dots, \psi_t : \mathbb{Z}^d \to \mathbb{Z}$. It turns out (see Lemma 3.2) that the sequence g^{Ψ} takes values in G^{Ψ}. More remarkably, the orbit (1.6) is in fact *totally equidistributed* on G^{Ψ}/Γ^{Ψ} if g is sufficiently irrational. It is this result that we refer to as our counting lemma.

Theorem 1.11 (Counting lemma). *Let M, D, t, s be integers with $1 \leqslant D, t, s \leqslant M$, let $(G/\Gamma, G_\bullet)$ be a degree $\leqslant s$ filtered nilmanifold of complexity $\leqslant M$, let $g : \mathbb{Z} \to G$ be an (A, N)-irrational polynomial sequence adapted to $G_\bullet$, let $\Psi = (\psi_1, \dots, \psi_t)$ be a collection of linear forms $\psi_1, \dots, \psi_t : \mathbb{Z}^D \to \mathbb{Z}$ with coefficients of magnitude at most M, and let P be a convex subset of $[-N, N]^D$. Then for any Lipschitz function $F : (G/\Gamma)^t \to \mathbb{C}$ of Lipschitz norm at most M, one has*[6]

$$\sum_{\mathbf{n} \in \mathbb{Z}^D \cap P} F(g^{\Psi}(\mathbf{n})\Gamma^t) = \mathrm{vol}\,(P) \int_{g(0)^{\Delta} G^{\Psi}/\Gamma^{\Psi}} F$$

$$+ \, o_{A \to \infty; M}\big(N^D\big) + o_{N \to \infty; M}\big(N^D\big),$$

where $g(0)^{\Delta} := \big(g(0), \dots, g(0)\big) \in G^t$ and the integral is with respect to the probability Haar measure $\mu_{g(0)^{\Delta} G^{\Psi}/\Gamma^{\Psi}}$ on the coset

$$g(0)^{\Delta} G^{\Psi}/\Gamma^{\Psi},$$

viewed as a subnilmanifold of $(G/\Gamma)^t$, and $\mathrm{vol}\,(P)$ is the Lebesgue measure of P in $\mathbb{R}^D$.

[6] We use $o_{A \to \infty; M}(X)$ to denote a quantity bounded in magnitude by $c_M(A)X$, where $c_M(A) \to 0$ as $A \to \infty$ for fixed M. Similarly for other choices of subscripts.

More generally, whenever $\Lambda \leqslant \mathbb{Z}^D$ is a sublattice of index $[\mathbb{Z}^D : \Lambda] \leqslant M$, and $\mathbf{n}_0 \in \mathbb{Z}^D$, one has

$$\sum_{\mathbf{n}\in(\mathbf{n}_0+\Lambda)\cap P} F(g^{\Psi}(\mathbf{n})\Gamma^t) = \frac{\mathrm{vol}\,(P)}{[\mathbb{Z}^D : \Lambda]} \int_{g(0)^{\triangle}G^{\Psi}/\Gamma^{\Psi}} F$$

$$+ o_{A\to\infty;M}\big(N^D\big) + o_{N\to\infty;M}\big(N^D\big).$$

The counting lemma is, of course, best understood by seeing it in action as we shall do several times later on. The errors $o_{A\to\infty;M}\big(N^D\big)$ and $o_{N\to\infty;M}\big(N^D\big)$ are negligible in most applications, as A will typically be a huge function $\mathcal{F}(M)$ of M, and N can also be taken to be arbitrarily large compared to M.

We remark that one could easily extend the above lemma to control averages of virtual irrational nilsequences, rather than just irrational sequences, by introducing some additional integrations over the local factors $\mathbb{Z}/q\mathbb{Z}$ and $\mathbb{R}$, but this would require even more notation than is currently being used and so we do not describe such an extension here.

Applications. The proofs of the regularity and counting lemmas occupy about half the paper. In the remaining half, we give a number of applications of these results to problems in additive combinatorics. The scheme of the arguments in all of these cases is similar. First, one applies the arithmetic regularity lemma to decompose the relevant function f into structured, small, and (very) uniform components $f = f_{\mathrm{nil}} + f_{\mathrm{sml}} + f_{\mathrm{unf}}$. Very roughly speaking, these are analysed as follows:

(i) f_{nil} is studied using algebraic properties of nilsequences, particularly the counting lemma;

(ii) f_{sml} is shown to be negligible, though often (unfortunately) some additional algebraic input is required to ensure that this error does not conspire to destroy the contribution from f_{nil};

(iii) f_{unf} is easily shown to be negligible using results of "generalised von Neumann" type as discussed in §4.

As we shall see, dealing with the error f_{sml} can cause a certain amount of pain. To show that this error is truly negligible, one often has to prove that patterns guaranteed by f_{nil} (such as arithmetic progressions) do not concentrate on some small set which might be contained in the support of f_{sml}.

We now give specific examples of this paradigm. In §6 we give a "new" proof of Szemerédi's famous theorem on arithmetic progressions. This is hardly exciting nowadays, with at least 16 proofs already in the literature [2, 3, 6, 10, 14, 18, 19, 44, 45, 48, 50, 52, 56] as well as (slightly implicitly) in [4, 35, 58]. However this proof makes the point that for a certain class of problems it suffices to "check the result for nilsequences", and in so doing one really sees the structure of the problem. Just as random and structured graphs are two obvious classes to test conjectures against in graph theory, we would like to raise awareness of nilsequences as potential (and, in certain cases such as this one, the *only*) sources of counterexamples.

The second application, proven in §5, is to establish a conjecture of Bergelson, Host and Kra [4]. Here and in the sequel we use the notation $X \ll_{\alpha,\varepsilon} Y$ or $Y \gg_{\alpha,\varepsilon} X$ synonymously with $X = O_{\alpha,\varepsilon}(Y)$, and similarly for other choice of subscripts.

Theorem 1.12 (Bergelson–Host–Kra conjecture). *Let $k = 1, 2, 3$ or 4, and suppose that $0 < \alpha < 1$ and $\varepsilon > 0$. Then for any $N \geqslant 1$ and any subset $A \subseteq [N]$ of density $|A| \geqslant \alpha N$, one can find $\gg_{\alpha,\varepsilon} N$ values of $d \in [-N, N]$ such that there are at least $(\alpha^k - \varepsilon)N$ k-term arithmetic progressions in A with common difference d.*

Remarks. The claim is trivial for $k = 1$, and follows from an easy averaging argument when $k = 2$. This theorem was established in the case $k = 3$ by the first author in [25]: we give a new proof of this result which may be of independent interest. The case $k = 4$ is new, although a finite field analogue of this result previously appeared in lecture notes of the first author [26] (reporting on joint work). Our proof of the $k = 4$ argument relies on the inverse conjecture for the U^3 norm, proven in [28]. A counterexample example of Ruzsa in the appendix to [4] shows that Theorem 1.12 fails when $k \geqslant 5$. An ergodic counterpart to Theorem 1.12 (which, roughly speaking, replaces a single scale N with a sequence of scales going to infinity and takes a limit), using a related but slightly different argument, was established in [4].

Finally, in §7, we establish a generalisation of a recent result of Gowers and Wolf [22, 23, 24] regarding the "true" complexity of a system of linear forms.

Theorem 1.13. *Let $\Psi = (\psi_1, \dots, \psi_t)$ be a collection of linear forms from $\mathbb{Z}^D \to \mathbb{Z}$, and let $s \geqslant 1$ be an integer such that the polynomials $\psi_1^{s+1}, \dots, \psi_t^{s+1}$ are linearly independent. Then for any function $f : [N] \to \mathbb{C}$*

bounded in magnitude by 1 *(and defined to be zero outside of* $[N]$*) obeying the bound* $\|f\|_{U^{s+1}[N]} \leqslant \delta$ *for some* $\delta > 0$*, one has*

$$\mathbb{E}_{\mathbf{n}\in[N]^D} \prod_{i=1}^{t} f\big(\psi_i(\mathbf{n})\big) = o_{\delta\to 0;s,D,t,\Psi}(1).$$

Remarks. This result was conjectured in [22], where it was shown that the linear independence hypothesis was necessary. The programme in [22, 23, 24] gives an alternate approach to this result that avoids explicit mention of nilsequences, and in particular establishes the counterpart to Theorem 1.13 in finite characteristic; their work also gives a proof of this theorem in the case when the Cauchy–Schwarz complexity of the system (see Theorem 4.1) is at most two, and with better bounds than our result, which is all but ineffective. It is worth mentioning that the arguments in [22, 23, 24] also develop several structural decomposition theorems along the lines of Theorem 1.2, but using the language of (high-rank) locally polynomial phases rather than (irrational) nilsequences.

Relation to previous work. A result closely related to Theorem 1.2 in the case $s = 1$ was proved by Bourgain as long ago as 1989 [7]. In that paper, the decomposition was applied to give a different proof of *Roth's theorem*, that is to say Szemerédi's theorem for 3-term progressions. A different take on this result was supplied by the first author in [25], where the application to the case $k = 3$ of the Bergelson–Host–Kra conjecture was noted. In that same paper a construction of Gowers [16] was modified to show that any application of the arithmetic regularity lemma must lead to awful (tower-type) bounds; the same kind of construction would show that the cases $s \geqslant 2$ of Theorem 1.2 lead to tower-type bounds as well. In[7] [26] the analogue of the case $s = 2$ of Theorem 1.2 in a finite field setting was stated, proved, and used to deduce the finite field analogue of the Bergelson–Host–Kra conjecture in the case $k = 4$. In that same paper the present work was promised (as reference [22]) at "some future juncture". Four years later we have reached that juncture and we apologise for the delay. We note, however, that until the very recent resolution of the inverse conjectures for the Gowers norms [33, 34] many of our results would have been conditional; furthermore, we are heavily dependent on our work [30], which had not been envisaged when the earlier promise was made.

[7]The relevant part of these lecture notes by the first author reported on joint work of the two of us.

In the meantime a greater general understanding of decomposition theorems of this type has developed through the work of Gowers [21], Reingold–Trevisan–Tulsiani–Vadhan [46], and Gowers-Wolf [22, 23, 24]; see also the survey [54] of the second author. While Theorem 1.2 is related to several of these general decomposition theorems, it also relies upon specific structure of nilmanifolds. In any case it seems appropriate, in this volume, to give a proof using the "energy increment argument" pioneered by Szemerédi.

The ergodic theory analogue of Theorem 1.2 is the classification of characteristic factors for the Gowers-Host–Kra seminorms $\| \cdot \|_{U^{s+1}(X)}$ (the ergodic theory counterpart of the Gowers norms) as inverse limits of nilsystems, which was first established by Host and Kra [35]. Roughly speaking, this classification allows one to decompose any bounded non-negative function $f \in L^\infty(X)$ in an (ergodic) measure-preserving system as a sum $f = f_{\mathrm{str}} + f_{\mathrm{sml}} + f_{\mathrm{unf}}$, where $\|f_{\mathrm{unf}}\|_{U^{s+1}(X)} = 0$, f_{sml} is as small as one wishes in the $L^2(X)$ norm, and f_{str} arises from an s-step nilsystem factor of X. This fundamental decomposition has many applications; for instance, in [4] it was used (together with the Furstenberg correspondence principle) to establish an ergodic analogue of Theorem 1.12, in which A is a set of integers rather than a finite subset of $[N]$, with the notion of upper density replacing the notion of cardinality. It appears however that this correspondence principle does not directly yield "single-scale" results such as Theorem 1.12 from the ergodic theory results.

Acknowledgments. BG was, while this work was being carried out, a fellow at the Radcliffe Institute at Harvard. He is very happy to thank the Institute for proving excellent working conditions. TT is supported by a grant from the MacArthur Foundation, by NSF grant DMS-0649473, and by the NSF Waterman award. The authors also thank Tim Gowers, Julia Wolf, and Tamar Ziegler for helpful comments and corrections.

2. Proof of the Arithmetic Regularity Lemma

We now prove Theorem 1.2. The proof proceeds in two main stages. Firstly, we establish a "non-irrational regularity lemma", which establishes a weaker version of Theorem 1.2 in which the structured component f_{nil} is a polynomial nilsequence, but one which is not assumed to be irrational. The main tool here is the *inverse conjecture* GI(s) *for the Gowers norms* [34], combined with the energy incrementation argument that appears in proofs of the graph regularity lemma. In the second stage, we upgrade this weaker regularity lemma to the full regularity lemma by converting the nilsequence to a irrational nilsequence. The main tool here is a dimension reduction argument and a factorisation of nilsequences similar to that appearing in [30].

The non-irrational regularity lemma. We begin the first stage of the argument. As mentioned above, the key ingredient is the following result.

Theorem 2.1 (GI (s)). *Let $s \geqslant 1$, and suppose that $f : [N] \to \mathbb{C}$ is a function bounded in magnitude by 1 such that $\|f\|_{U^{s+1}[N]} \geqslant \delta$ for some $\delta > 0$. Then there is a degree $\leqslant s$ polynomial nilsequence $\psi : \mathbb{Z} \to \mathbb{C}$ of complexity $O_{s,\delta}(1)$ such that $|\langle f, \psi\rangle_{L^2[N]}| \gg_{s,\delta} 1$, where*

$$\langle f, \psi\rangle_{L^2[N]} := \mathbb{E}_{n\in[N]} f(n)\overline{\psi(n)}$$

is the usual inner product.

Remark. The difficulty of this conjecture increases with s. The case $s = 1$ easily follows from classical harmonic analysis. The case $s = 2$ was established by the authors in [28], building upon the breakthrough paper of Gowers [17]. The case $s = 3$ was recently established by the authors and Ziegler in [33], and the general case will appear in the forthcoming paper [34] by the authors and Ziegler.

For technical reasons, it is convenient to replace the notion of a degree $\leqslant s$ polynomial nilsequence by a slightly different concept. The following definition is not required beyond the end of the proof of Proposition 2.7.

Definition 2.2 (s-measurability). Let $\Phi : \mathbb{R}^+ \to \mathbb{R}^+$ be a growth function and $s \geqslant 1$. A subset $E \subseteq [N]$ is said to be *s-measurable* with growth function Φ if for every $M \geqslant 1$, there exists a degree $\leqslant s$ polynomial nilsequence $\psi : \mathbb{Z} \to [0, 1]$ of complexity $\leqslant \Phi(M)$ such that

$$\|\psi - 1_E\|_{L^2[N]} \leqslant 1/M.$$

An example of a 1-measurable function would be a regular Bohr set, as introduced in [8] and discussed further in [28, §2]. We will not need Bohr sets elsewhere in this paper, so we shall not dwell any longer on this example. However the reader will see ideas related to the basic theory of those sets in the proof of Corollary 2.3 below.

We make the simple but crucial observation that if E, F are s-measurable with some growth functions Φ, Φ' respectively, then boolean combinations of E, F such as $E \cap F$, $E \cup F$, or $[N] \backslash E$ are also s-measurable with some growth function depending on Φ, Φ'. Underlying this, of course, is that fact that the product and sum of two nilsequences is also a nilsequence, and hence the set of nilsequences form a kind of algebra (graded by complexity). The role of algebraic structure of this kind was brought to the fore in the work of Gowers [21] cited above.

Theorem 2.1 then implies

Corollary 2.3 (Alternate formulation of GI(s)). *Let $s \geqslant 1$, and suppose that $f : [N] \to [-1,1]$ is such that $\|f\|_{U^{s+1}[N]} \geqslant \delta$ for some $\delta > 0$. Then there exists a growth function $\Phi_{s,\delta}$ depending only on s, δ, and an s-measurable set $E \subset N$ with growth function $\Phi_{s,\delta}$, such that*

$$\left|\mathbb{E}_{n \in [N]} f(n) 1_E(n)\right| \gg_{s,\delta} 1.$$

Proof. We allow implied constants to depend on s, δ. By Theorem 2.1, there exists a degree $\leqslant s$ polynomial nilsequence ψ of complexity $O(1)$ such that

$$\left|\mathbb{E}_{n \in [N]} f(n) \overline{\psi(n)}\right| \gg 1.$$

By taking real and imaginary parts of ψ, and then positive and negative parts, and rescaling, we may assume without loss of generality that ψ takes values in $[0,1]$. By Fubini's theorem, we then have

$$\left|\int_0^1 \mathbb{E}_{n \in [N]} f(n) 1_{E_t}(n)\, dt\right| \gg 1$$

where $E_t := \{n \in [N] : \psi(n) \geqslant t\}$. We thus see that there is a subset $\Omega \subset [0,1]$ of Lebesgue measure $|\Omega| \gg 1$ such that

$$\left|\mathbb{E}_{n \in [N]} f(n) 1_{E_t}(n)\right| \gg 1$$

uniformly for all $t \in \Omega$.

It remains to show that at least one[8] of the E_t is s-measurable with respect to a suitable growth function. For any $t \in \mathbb{R}$, we consider the maximal function

$$M(t) := \sup_{r>0} \frac{1}{2r} \frac{1}{N} \left|\{n \in [N] : |\psi(n) - t| \leqslant r\}\right|.$$

[8]Here we are, in some sense, finding a "regular" nil-Bohr set $\{n \in [N] : \psi(n) \geqslant t\}$, that is to say one rather insensitive to small changes in the value of t. A similar idea also appears in [46, Claim 2.2].

From the Hardy–Littlewood maximal inequality or the Besicovitch covering lemma we have that the set $\{t \in \mathbb{R} : M(t) \geqslant \lambda\}$ has Lebesgue measure $O(1/\lambda)$ for any $\lambda > 0$. Thus, we can find $t \in \Omega$ such that $M(t) = O(1)$. Fixing such a t, we then see that

$$\big|\{n \in [N] : |\psi(n) - t| \leqslant r\}\big| \ll rN$$

for all $r > 0$. As a consequence, for any $r > 0$, one can then approximate 1_{E_t} to within $O(\sqrt{r})$ in $L^2[N]$ norm by a Lipschitz function of ψ with Lipschitz norm $O(1/r)$. This implies that 1_{E_t} is s-measurable with some growth function Φ depending only on s, δ, and the claim follows. ■

We rephrase this fact in terms of conditional expectations. The following definition, like Definition 2.2, will only be needed until the end of the proof of Proposition 2.7.

Definition 2.4 (s-factors). An *s-factor* $\mathcal{B}$ of complexity $\leqslant M$ and growth function Φ is a partition of $[N]$ into at most M sets (or *cells*) $E_1, \dots, E_m$ which are s-measurable of growth function Φ. Given an s-factor $\mathcal{B}$ and a function $f : [N] \to \mathbb{C}$, we define the *conditional expectation* $\mathbb{E}(f|\mathcal{B}) : [N] \to \mathbb{C}$ of f with respect to the s-factor to be the function which equals $\mathbb{E}_{n \in E_j} f(n)$ on each cell of the partition. We define the *index* or *energy* $\mathcal{E}(\mathcal{B})$ of the s-factor $\mathcal{B}$ relative to f to be the quantity $\|\mathbb{E}(f|\mathcal{B})\|^2_{L^2[N]}$.

An s-factor $\mathcal{B}'$ is said to *refine* another $\mathcal{B}$ if every cell of $\mathcal{B}'$ is contained in a cell of $\mathcal{B}$.

Corollary 2.5 (Lack of uniformity implies energy increment). *Let $s \geqslant 1$, let $\mathcal{B}$ be an s-factor of complexity $\leqslant M$ and some growth function Φ, and suppose that $f : [N] \to [0,1]$ is such that $\|f - \mathbb{E}(f|\mathcal{B})\|_{U^{s+1}[N]} \geqslant \delta$ for some $\delta > 0$. Then there exists a refinement $\mathcal{B}'$ of $\mathcal{B}$ of complexity $\leqslant 2M$ and some growth function depending on s, δ, M, Φ, such that*

$$\mathcal{E}(\mathcal{B}') - \mathcal{E}(\mathcal{B}) \gg_{s,\delta} 1.$$

Proof. By Corollary 2.3, we can find an s-measurable set E with a growth function depending on s, δ such that

$$\big|\langle f - \mathbb{E}(f|\mathcal{B}), 1_E\rangle_{L^2[N]}\big| \gg_{s,\delta} 1 \tag{2.1}$$

Now let $\mathcal{B}'$ be the partition generated by $\mathcal{B}$ and E; then $\mathcal{B}'$ clearly has complexity $\leqslant 2M$ and a growth function depending on s, δ, M, Φ. Since 1_E

is measurable with respect to the partition $\mathcal{B}'$ (that is to say it is constant on each cell of this partition), we can rewrite the left-hand side of (2.1) as

$$\left|\left\langle \mathbb{E}(f|\mathcal{B}') - \mathbb{E}(f|\mathcal{B}), 1_E\right\rangle_{L^2[N]}\right|$$

and hence by the Cauchy–Schwarz inequality

$$\left\|\mathbb{E}(f|\mathcal{B}') - \mathbb{E}(f|\mathcal{B})\right\|_{L^2[N]} \gg_{s,\delta} 1.$$

The claim then follows from Pythagoras' theorem. ∎

We can iterate this to obtain a weak regularity lemma, analogous to the weak graph regularity lemma of Frieze and Kannan [15].

Corollary 2.6. *Let $s \geqslant 1$, let $\mathcal{B}$ be an s-factor of complexity $\leqslant M$ and some growth function Φ, let $f : [N] \to [0,1]$, and let $\varepsilon > 0$. Then there exists a refinement $\mathcal{B}'$ of $\mathcal{B}$ of complexity $O_{s,M,\varepsilon}(1)$ and some growth function depending on s, ε, M, Φ, such that*

$$\left\|f - \mathbb{E}(f|\mathcal{B}')\right\|_{U^{s+1}[N]} \leqslant \varepsilon. \tag{2.2}$$

Proof. We define a sequence of successively more refined factors $\mathcal{B}'$, starting with $\mathcal{B}' := \mathcal{B}$. If (2.2) already holds then we are done, so suppose that this is not the case. Then by Corollary 2.5, we can find a refinement $\mathcal{B}''$ of complexity $O_{s,M,\varepsilon}(1)$ and some growth function depending on s, ε, M, Φ whose energy is larger than that of $\mathcal{B}'$ by a factor $\gg_{s,\varepsilon} 1$. On the other hand, the energy clearly ranges between 0 and 1. Thus after replacing $\mathcal{B}'$ with $\mathcal{B}''$ and iterating this algorithm at most $O_{s,\varepsilon}(1)$ times we obtain the claim. ∎

One final iteration then gives the full non-irrational regularity lemma.

Proposition 2.7. *Let $f : [N] \to [0,1]$, let $s \geqslant 1$, let $\varepsilon > 0$, and let $\mathcal{F} : \mathbb{R}^+ \to \mathbb{R}^+$ be a growth function. Then there exists a quantity $M = O_{s,\varepsilon,\mathcal{F}}(1)$ and a decomposition*

$$f = f_{\mathrm{nil}} + f_{\mathrm{sml}} + f_{\mathrm{unf}}$$

of f into functions $f_{\mathrm{nil}}, f_{\mathrm{unf}} : [N] \to [-1,1]$ such that:

(i) *(f_{nil} structured) f_{nil} equals a degree $\leqslant s$ polynomial nilsequence of complexity $\leqslant M$.*

(ii) *(f_{sml} small) $\|f_{\mathrm{sml}}\|_{L^2[N]} \leqslant \varepsilon$.*

(iii) *(f_{unf} very uniform) $\|f_{\mathrm{nil}}\|_{U^{s+1}[N]} \leqslant 1/\mathcal{F}(M)$.*

(iv) *(Nonnegativity) f_{nil} and $f_{\mathrm{nil}} + f_{\mathrm{sml}}$ take values in $[0,1]$.*

Proof. We need a growth function $\tilde{\mathcal{F}} : \mathbb{R}^+ \to \mathbb{R}^+$, somewhat more rapidly growing than $\mathcal{F}$ in manner that depends on $\mathcal{F}$, s, ε. We will specify the exact requirements we have of it later. We then define a sequence $1 = M_0 \leqslant M_1 \leqslant \ldots$ by setting $M_0 := 1$ and $M_{i+1} := \tilde{\mathcal{F}}(M_i)$.

Applying Corollary 2.6 repeatedly, we may find for each $i \geqslant 0$ an s-factor $\mathcal{B}_i$ of complexity $O_{s,M_i}(1)$ and a growth function depending on s, M_i, such that each $\mathcal{B}_i$ refines $\mathcal{B}_{i-1}$, and such that

$$\left\| f - \mathbb{E}(f|\mathcal{B}_i) \right\|_{U^{s+1}[N]} \leqslant 1/M_i$$

for all $i \geqslant 0$.

By Pythagoras' theorem, the energies $\mathcal{E}(\mathcal{B}_i)$ are non-decreasing, and also range between 0 and 1. Thus by the pigeonhole principle, one can find $i = O_\varepsilon(1)$ such that

$$\mathcal{E}(\mathcal{B}_{i+1}) - \mathcal{E}(\mathcal{B}_i) \leqslant \varepsilon^2/4,$$

which by Pythagoras' theorem again is equivalent to

$$\left\| \mathbb{E}(f|\mathcal{B}_{i+1}) - \mathbb{E}(f|\mathcal{B}_i) \right\|_{L^2[N]} \leqslant \varepsilon/2.$$

Meanwhile, as $\mathcal{B}_i$ is an s-factor and f is bounded, we can find a degree $\leqslant s$ polynomial nilsequence $f_{\mathrm{nil}} : [N] \to \mathbb{R}$ of complexity $O_{s,M_i}(1)$ such that

$$\left\| \mathbb{E}(f|\mathcal{B}_i) - f_{\mathrm{nil}} \right\|_{L^2[N]} \leqslant \varepsilon/2.$$

Since $\mathbb{E}(f|\mathcal{B}_i)$ ranges in $[0,1]$, we may retract f_{nil} to $[0,1]$ also (note that this does not increase the complexity of f_{nil}). If we then set $f_{\mathrm{unf}} := f - \mathbb{E}(f|\mathcal{B}_{i+1})$ and $f_{\mathrm{sml}} := \mathbb{E}(f|\mathcal{B}_{i+1}) - f_{\mathrm{nil}}$, we obtain the claim. ■

Remark. The application of the Hardy–Littlewood maximal inequality in the proof of Corollary 2.3 makes for a reasonably tidy argument. A more direct approach would be to carve up $[N]$ into approximate level sets of nilsequences, and then to approximate the projections onto the factors thus defined by nilsequences using the Weierstrass approximation theorem. There are a number of technicalities involved in this approach, chiefly involving the need to choose the approximate level sets randomly. This kind of argument was employed, in a closely related context, in [27, Chapter 7]. One can also use utilise arguments based on the Hahn–Banach theorem instead; see [21], [46], and [22, 23, 24].

Obtaining irrationality. Our task now is to replace the nilsequence f_{nil} appearing in Proposition 2.7 with a highly "irrational" nilsequence as advertised in the statement of our main theorem, Theorem 1.2. It turns out to be sufficient to establish the following claim.

Proposition 2.8. *Let $s, M_0 \geqslant 1$, let $\mathcal{F}$ be a growth function, and let $f : \mathbb{Z} \to [0,1]$ be a degree $\leqslant s$ nilsequence of complexity $\leqslant M_0$. Then there exists an $M = O_{s,M_0,\mathcal{F}}(1)$, such that f (when restricted to $[N]$) is also a $(\mathcal{F}(M), N)$-irrational degree $\leqslant s$ virtual nilsequence of complexity $\leqslant M$ at scale N.*

To establish Theorem 1.2 from this and Proposition 2.7 one first applies the latter result with $\mathcal{F}$ replaced by a much more rapid growth function $\mathcal{F}'$, and then one applies Proposition 2.8 to the structured component f_{nil} obtained in Theorem 2.6.

It remains to prove Proposition 2.8. Let $s, M_0, \mathcal{F}, \psi$ be as in that proposition. By definition, we have $\psi = F_0(g_0(n)\Gamma)$ for some degree $\leqslant s$ filtered nilmanifold $(G/\Gamma, G_\bullet)$ of complexity $\leqslant M_0$, a polynomial sequence $g_0 \in \mathrm{poly}(\mathbb{Z}, G_\bullet)$, and a function $F_0 : G/\Gamma \to \mathbb{C}$ which has a Lipschitz norm of at most M_0. Since ψ takes values in $[0,1]$, we may assume without loss of generality that F_0 is real, and by replacing F_0 with the retraction $\max(\min(F_0, 1), 0)$ to $[0,1]$ if necessary, we may assume that F_0 also takes values in $[0,1]$. Henceforth $(G/\Gamma, G_\bullet)$, g_0, and F_0 are fixed.

Factorisation results. One of the main results of our paper [30] was a decomposition of an arbitrary polynomial nilsequence g on G/Γ into a product[9] $\beta g'\gamma$, where β is "smooth", γ is "rational", and $g'(n)\Gamma$ is equidistributed inside some possibly smaller nilmanifold G'/Γ'. We need a similar result here, but with g' having the somewhat stronger property of being *irrational* that we mentioned in the introduction. The notion of irrationality is discussed in more detail in Appendix A.

We will be also using the notions of *smooth* and *rational* polynomial sequences from [30]. Again, the basic definitions and properties of these concepts are recalled in Appendix A.

Define a *complexity $\leqslant M$ subnilmanifold* of $(G/\Gamma, G_\bullet)$ to be a degree $\leqslant s$ filtered nilmanifold $(G'/\Gamma', G'_\bullet)$ of complexity $\leqslant M$, where each subgroup $G'_{(i)}$ in the filtration $G'_\bullet$ is a rational subgroup of the associated subgroup

[9] In our paper [30] the letter ε was used for a smooth nilsequence, but we use β here to avoid conflict with various uses of ε to denote a small positive real number.

$G_{(i)}$ of complexity $\leqslant M$, $\Gamma' = G' \cap \Gamma$, and each element of the Mal'cev basis of $(G'/\Gamma', G'_\bullet)$ is a rational linear combination of the Mal'cev basis of $(G/\Gamma, G_\bullet)$, where the coefficients all have height $\leqslant M$. We define the *total dimension* of such a nilmanifold to be the quantity $\sum_{i=0}^{s} \dim \left(G'_{(i)}\right)$; this is also the dimension of $\mathrm{poly}(\mathbb{Z}, G_\bullet)$ (thanks to the Taylor series expansion, Lemma A.1).

We make the easy remark that if $(G'/\Gamma', G'_\bullet)$ is a complexity $\leqslant M$ subnilmanifold of $(G/\Gamma, G_\bullet)$ for some $M \geqslant M_0$, and $(G''/\Gamma'', G''_\bullet)$ is a complexity $\leqslant M$ subnilmanifold of $(G'/\Gamma', G'_\bullet)$, then $(G''/\Gamma'', G''_\bullet)$ is a complexity $O_M(1)$ subnilmanifold of $(G/\Gamma, G_\bullet)$.

Our first lemma is very similar in form to [30, Lemma 7.9].

Lemma 2.9 (Initial factorisation). *Let $(G'/\Gamma', G'_\bullet)$ be a complexity $\leqslant M$ subnilmanifold of $(G/\Gamma, G_\bullet)$ for some $M \geqslant M_0$, let $g' \in \mathrm{poly}(\mathbb{Z}, G'_\bullet)$, and let $A > 0$ and $N \geqslant 1$. Then at least one of the following statements hold:*

(Irrationality) g' is (A, N)-irrational in $(G'/\Gamma', G'_\bullet)$.

(Dimension reduction) There exists a factorisation

$$g' = \beta g'' \gamma$$

where $\beta \in \mathrm{poly}(\mathbb{Z}, G'_\bullet)$ is $\left(O_{M,A}(1), N\right)$-smooth, $g'' \in \mathrm{poly}(\mathbb{Z}, G''_\bullet)$ takes values in a subnilmanifold $(G''/\Gamma'', G''_\bullet)$ of $(G'/\Gamma', G'_\bullet)$ of strictly smaller total dimension and of complexity $O_{M,A}(1)$, and $\gamma \in \mathrm{poly}(\mathbb{Z}, G'_\bullet)$ is $O_{M,A}(1)$-rational.

Proof. To make this proof a little more readable, we drop one dash from every expression. Thus g' becomes g, G'' becomes G', and so on. Suppose that g is not (A, N)-irrational. Recall (see Lemma A.1) that g has a Taylor expansion that we may write in the form

$$g(n) = g_0 g_1^{\binom{n}{1}} g_2^{\binom{n}{2}} \cdots g_s^{\binom{n}{s}},$$

where $g_i \in G_{(i)}$ for each i. It follows from Lemma A.7 that for some i, $1 \leqslant i \leqslant s$, we can factorise

$$g_i = \beta_i g_i' \gamma_i,$$

where $g_i' \in G_{(i)}$ lies in the kernel of some horizontal character $\xi_i : G_{(i)} \to \mathbb{R}$ of complexity $O_{A,M}(1)$, $\gamma_i \in G_{(i)}$ is $O_{A,M}(1)$-rational in the sense that $\gamma_i^m \in \Gamma_{(i)}$ for some $m = O_{A,M}(1)$, and $\beta_i \in G_{(i)}$ has distance $O_{A,M}(1/N^i)$ from the origin.

We now divide into two cases, depending on whether $i > 1$ or $i = 1$. First suppose that $i > 1$. Then the Taylor expansion of g reads, with an obvious notation,

$$g(n) = g_{<i}(n)\big(\beta_i g_i' \gamma_i\big)^{\binom{n}{i}} g_{>i}(n).$$

By commutating all the β_is to the left and all the γ_is to the right, and using the group properties of polynomial sequences (Theorem 1.6), one can rewrite this as

$$g(n) = \beta_i^{\binom{n}{i}} g'(n) \gamma_i^{\binom{n}{i}}$$

where

$$g'(n) := g_{<i}(n) g_i'^{\binom{n}{i}} \tilde{g}_{>i}(n)$$

and $\tilde{g}_{>i}(n)$ is another polynomial sequence taking values in $G_{(i+1)}$. Observe that g' is then a polynomial sequence adapted to the subnilmanifold $(G'/\Gamma', G'_\bullet)$, where $G'/\Gamma' = G/\Gamma$ and $G'_{(j)} = G_{(j)}$ for $j \neq i$, but $G'_{(i)} = \ker(\xi_i')$. This is indeed a subnilmanifold, with complexity $O_{A,M}(1)$; note that $\big(G'_{(l)}\big)_{l=0}^{\infty}$ is a filtration, thanks to our insistence in the definition of i-horizontal character (cf. Definition A.6) that $\big[G_{(j)}, G_{(i-j)}\big] \subseteq \ker(\xi_i')$ for all $0 \leqslant j \leqslant i$. Meanwhile, $\beta_i^{\binom{n}{i}}$ is a $\big(O_{A,M}(1), N\big)$-smooth sequence and $\gamma_i^{\binom{n}{i}}$ is a $O_{A,M}(1)$-rational sequence, so we have the desired factorisation in the $i > 1$ case.

When $i = 1$, the above argument does not quite work, because $G'_{(1)}$ would be distinct from $G'_{(0)}$ and would thus not qualify as a filtration. But this can be easily remedied by performing an additional factorisation

$$g_0 = \beta_0 g_0'$$

where $\beta_0 \in G'$ is a distance $O_{A,M}(1)$ from the identity, and g_0' lies in the kernel of ξ_1'. This leads to a factorisation of the form

$$g(n) = \beta_0 \beta_1^n g'(n) \gamma_1^n$$

where

$$g'(n) = g_0' g_1'^n g_{>1}'(n)$$

and $g_{>1}'$ is a polynomial sequence taking values in $G'_{(2)}$. One then argues as before, but now one sets both $G''_{(0)}$ and $G''_{(1)}$ equal to the kernel of ξ_1'. ■

We can iterate the above lemma to obtain the following result, which is analogous to [30, Theorem 1.19]. Apart from dealing with irrationality rather than equidistribution, the following result is somewhat different to that just cited in that one requires an arbitrary (rather than polynomial) growth function, but one does not (of course) need polynomial complexity bounds. A variant of [30, Theorem 1.19] was also given in [33, Theorem 4.2].

Lemma 2.10 (Complete factorisation). *Let $(G/\Gamma, G_\bullet)$ be a degree $\leqslant s$ filtered nilmanifold of complexity $\leqslant M_0$, and let $g \in \mathrm{poly}(\mathbb{Z}, G_\bullet)$. For any growth function $\mathcal{F}'$, we can find a quantity $M_0 \leqslant M \leqslant O_{M,\mathcal{F}'}(1)$ and a factorisation $g = \beta g' \gamma$ where:*

(i) *$\beta \in \mathrm{poly}(\mathbb{Z}, G_\bullet)$ is $\big(O_M(1), N\big)$-smooth;*

(ii) *$g' \in \mathrm{poly}(\mathbb{Z}, G_\bullet)$ is $\big(\mathcal{F}'(M), N\big)$-irrational in a subnilmanifold $(G'/\Gamma', G'_\bullet)$ of $(G/\Gamma, G_\bullet)$ of complexity $O_M(1)$, and*

(iii) *$\gamma \in \mathrm{poly}(\mathbb{Z}, G_\bullet)$ is $O_M(1)$-periodic.*

Proof. We use an iterative argument, setting $\beta = \gamma = \mathrm{id}$, $g' = g$, $M = M_0$, and $(G'/\Gamma', G'_\bullet) = (G/\Gamma, G_\bullet)$ to begin with. In particular, $(G', \Gamma', G'_\bullet)$ is initially a subnilmanifold of $(G/\Gamma, G_\bullet)$ of complexity $O_M(1)$. If g' is $\mathcal{F}'(M)$-equidistributed in $(G'/\Gamma', G'_\bullet)$ then we are done; otherwise, by Lemma 2.9 we may factorise $g' = \beta' g'' \gamma'$ where c' is $\big(O_{\mathcal{F}'(M)}(1), N\big)$-smooth, γ is $O_{\mathcal{F}'(M)}(1)$-periodic, and g'' now takes values in a subnilmanifold $(G''/\Gamma'', G''_\bullet)$ of $(G'/\Gamma', G'_\bullet)$ of complexity $O_{\mathcal{F}'(M)}(1)$ and smaller total dimension than $(G'/\Gamma', G'_\bullet)$. We then replace β by $\beta\beta'$, γ by $\gamma'\gamma$, g' by g'', $(G'/\Gamma', G'_\bullet)$ by $(G''/\Gamma'', G''_\bullet)$, and increase M to a quantity of the form $O_{\mathcal{F}'(M)}(1)$, using Lemma A.4 to conclude that the new β is smooth and the new γ is rational. We then iterate this process. Since the total dimension of $(G/\Gamma, G_\bullet)$ is initially $O_{M_0}(1)$, this process can iterate at most $O_{M_0}(1)$ times, and the claim follows. ∎

With this lemma we can now establish Proposition 2.8 and hence Theorem 1.2. Let $\mathcal{F}'$ be a rapid growth function (depending on $\varepsilon, M_0, \mathcal{F}$) to be chosen later. We apply Lemma 2.10, obtaining some M with $M_0 \leqslant M \leqslant O_{M_0,\mathcal{F}'}(1)$ and a factorisation

$$\psi(n) = F\big(\beta(n) g'(n) \gamma(n) \Gamma\big)$$

with β, g' and γ having the properties described in that lemma.

The sequence γ is $O_M(1)$-rational and so, by Lemma A.4, the orbit $n \mapsto \gamma(n)\Gamma$ is periodic with some period $q = O_M(1)$, and thus $\gamma(n)\Gamma$ depends only on n mod q.

For each n, the rationality of $\gamma(n)$ ensures that $\gamma(n)\Gamma$ intersects Γ in a subgroup of Γ of index $O_M(1)$. Since there are only $O_M(1)$ different possible values of $\gamma(n)\Gamma$, we may thus find a subgroup Γ' of Γ of index $O_M(1)$ such that $\Gamma' \subseteq \gamma(n)\Gamma$ for all n.

We can thus express ψ as a virtual nilsequence

$$\psi(n) = \tilde{F}\big(g'(n)\Gamma', n \bmod q, n/N\big)$$

where $\tilde{F} : G/\Gamma' \times \mathbb{Z}/q\mathbb{Z} \times \mathbb{R}$ is defined by the formula

$$\tilde{F}(x, a, y) := F\big(\beta(Ny)\tilde{x}\gamma(\tilde{a})\Gamma\big)$$

whenever $y \in \frac{1}{N}\mathbb{Z}$ and by Lipschitz extension to all $y \in \mathbb{R}$. where $\tilde{a}$ is any integer with $\tilde{a} = a \bmod q$, and $\tilde{x}$ is any element of G such that $\tilde{x}\Gamma' = x$. One easily verifies that $\tilde{F}$ is well-defined and has a Lipschitz norm of $O_M(1)$. Also, since g' was already $\big(\mathcal{F}(M), N\big)$-irrational in G/Γ, and Γ' has index $O_M(1)$ in Γ, we see that g' is $\big(\gg_M \mathcal{F}(M), N\big)$-irrational in G/Γ'. Proposition 2.8 now follows by replacing M by a suitable quantity of the form $O_M(1)$, and choosing $\mathcal{F}'$ sufficiently rapidly growing depending on $\mathcal{F}$.

3. Proof of the Counting Lemma

The purpose of this section is to prove the counting lemma, Theorem 1.11. We begin by recalling from the introduction the definition of the Leibman group G^Ψ.

Definition 3.1 (The Leibman group). Let $\Psi = (\psi_1, \dots, \psi_t)$ be a collection of linear forms $\psi_1, \dots, \psi_t : \mathbb{Z}^D \to \mathbb{Z}$. For any $i \geqslant 1$, define $\Psi^{[i]}$ to be the linear subspace of $\mathbb{R}^t$ spanned by the vectors $\big(\psi_1^j(\mathbf{n}), \dots, \psi_t^j(\mathbf{n})\big)$ for $1 \leqslant j \leqslant i$ and $\mathbf{n} \in \mathbb{Z}^D$. Given a filtered nilmanifold $(G/\Gamma, G_\bullet)$, we define the *Leibman group* $G^\Psi \lhd G^t$ to be the Lie subgroup of G^t generated by the elements $g_i^{\vec{v}_i}$ for $i \geqslant 1$, $g_i \in G_{(i)}$, and $\vec{v}_i \in \Psi^{[i]}$, with the convention that if $\vec{v} = (v_1, \dots, v_t)$ then

$$g^{\vec{v}} := (g^{v_1}, \dots, g^{v_t}).$$

Now might be a good time to remark explicitly that we have introduced a slightly vulgar convention that we hope will help the reader follow this section and other parts of the paper. Bold font letters such as $\mathbf{n} \in \mathbb{R}^D$ denote D-dimensional vectors, whilst arrows such as $\vec{v} \in \mathbb{R}^t$ denote t-vectors. Occasionally we shall write $m_i := \dim\left(\Psi^{[i]}\right)$.

When reading this section, it might be found helpful to have a running example in mind. We will take as an illustrative example the case $D = 2$, $t = 4$ and $\Psi = (\psi_1, \dots, \psi_4)$, where $\psi_i(\mathbf{n}) = n_1 + in_2$ for $i = 0, 1, 2, 3$. The system Ψ, of course, defines a 4-term arithmetic progression. As we remarked in the introduction the corresponding Leibman group G^Ψ is also known as the *Hall–Petresco group* $\mathrm{HP}^4(G)$. The reader will easily confirm that in this case we have

$$\Psi^{[1]} = \mathbb{R}(1,1,1,1) \oplus \mathbb{R}(0,1,2,3)$$

and

$$\Psi^{[2]} = \mathbb{R}(1,1,1,1) \oplus \mathbb{R}(0,1,2,3) \oplus \mathbb{R}(0,0,1,3)$$

and

$$\Psi^{[3]} = \mathbb{R}(1,1,1,1) \oplus \mathbb{R}(0,1,2,3) \oplus \mathbb{R}(0,0,1,3) + \mathbb{R}(0,0,0,1) = \mathbb{R}^4.$$

Some work must be done before we can describe $G^\Psi = \mathrm{HP}^4(G)$ in a pleasant way. However we can already establish the following lemma, whose statement and proof go some way towards explaining the introduction of the Leibman group.

Lemma 3.2. *Let $\Psi = (\psi_1, \dots, \psi_t)$ be a collection of linear forms $\psi_1, \dots, \psi_t : \mathbb{Z}^D \to \mathbb{Z}$. Suppose that $(G/\Gamma, G_\bullet)$ is a filtered nilmanifold and that $g \in \mathrm{poly}(\mathbb{Z}, G_\bullet)$ is a polynomial sequence. Then the sequence $g^\Psi : \mathbb{Z}^D \to G^t$ defined by $g^\Psi(\mathbf{n}) := \left(g\left(\psi_1(\mathbf{n})\right), \dots, g\left(\psi_t(\mathbf{n})\right)\right)$ takes values in G^Ψ.*

Proof. The sequence $g(n)$ has a (unique) Taylor expansion

$$g(n) = g_0 g_1^{\binom{n}{1}} \cdots g_s^{\binom{n}{s}}$$

with $g_i \in G_{(i)}$ for all i (see Lemma A.1). Substituting in, it follows that

$$g^\Psi(\mathbf{n}) = \prod_{i=0}^{s} g_i^{\left(\binom{\psi_1(\mathbf{n})}{i}, \dots, \binom{\psi_t(\mathbf{n})}{i}\right)},$$

and it is immediate from the definition that each element in this product lies in G^Ψ. ■

The counting lemma, whose proof is the main objective of this section, was stated as Theorem 1.11. Essentially, it states that $g^{\Psi}(\mathbf{n})\Gamma^{\Psi}$ is equidistributed in G^{Ψ}/Γ^{Ψ} as $\mathbf{n}$ ranges over "nice" subsets of "big" lattices, provided that the original sequence g is suitably irrational. We will recall what that means in due course, but our first task is to develop the basic theory of the Leibman group G^{Ψ}. At the moment, for example, we have not established that G^{Ψ} is a connected Lie subgroup of G^t or that G^{Ψ}/Γ^{Ψ} has the structure of a filtered nilmanifold. Nor have we developed tools for calculating inside this group.

Basic facts about the Leibman group and nilmanifold. We can endow $\mathbb{R}^t$ with the structure of a commutative algebra over $\mathbb{R}$ by using the pointwise product

$$\vec{x}.\vec{y} = (x_1y_1, \ldots, x_ty_t)$$

and setting $\vec{1} = (1, \ldots, 1)$ to be the multiplicative identity. With this algebra structure, one can view the spaces $\Psi^{[i]}$ defined in Definition 1.10 as the span of the powers $\Psi(\mathbf{n})^j$ for $\mathbf{n} \in \mathbb{Z}^D$ and $1 \leqslant j \leqslant i$, where we view Ψ as a homomorphism from $\mathbb{Z}^D$ to $\mathbb{Z}^t$. We have the following alternate definition of the $\Psi^{[i]}$.

Lemma 3.3 (Depolarisation). *$\Psi^{[i]}$ is the span of the products*

$$\Psi(\mathbf{n}_1)\ldots\Psi(\mathbf{n}_j),$$

where $1 \leqslant j \leqslant i$ and $\mathbf{n}_1, \ldots, \mathbf{n}_j \in \mathbb{Z}^D$.

Proof. Clearly $\Psi^{[i]}$ is contained in this span. To establish the reverse containment, we observe the elementary depolarisation identity

$$\Psi(\mathbf{n}_1)\ldots\Psi(\mathbf{n}_j) = \frac{(-1)^j}{j!}\sum_{\omega\in\{0,1\}^j}(-1)^{|\omega|}\Psi(\omega_1\mathbf{n}_1 + \ldots + \omega_j\mathbf{n}_j)^j$$

where $\omega = (\omega_1, \ldots, \omega_j)$ and $|\omega| := \omega_1 + \ldots + \omega_j$, and the claim follows. ∎

As an immediate consequence we have

Corollary 3.4 (Filtration property). *For any $i, j \geqslant 0$, we have $\Psi^{[i]} \cdot \Psi^{[j]} \subseteq \Psi^{[i+j]}$.*

Let $(G/\Gamma, G_\bullet)$ be a degree $\leqslant s$ filtered nilmanifold. From Definition 1.10, the Leibman group G^Ψ is the subgroup of G^t generated by the group elements $g_i^{v_i}$ for $i \geqslant 1$, $v_i \in \Psi^{[i]}$, and $g_i \in G_{(i)}$. For any $i_0 \geqslant 1$, let $G^\Psi_{(i_0)}$ be the subgroup of G^Ψ generated by those $g_i^{\vec{v}_i}$ with $i \geqslant i_0$, $\vec{v}_i \in \Psi^{[i]}$, $g_i \in G_{(i)}$, with the convention that $G^\Psi_{(0)} := G^\Psi$.

Lemma 3.5 (Filtration property for $G^\Psi_\bullet$). *$G^\Psi_\bullet := \left(G^\Psi_{(i)}\right)_{i=0}^\infty$ is a filtration on G^Ψ. In other words, the $G^\Psi_{(i)}$ are nested with $\left[G^\Psi_{(i)}, G^\Psi_{(j)}\right] \subset G^\Psi_{(i+j)}$ for all $i, j \geqslant 0$.*

Proof. It suffices to check that if $g_i \in G_{(i)}$, $g_j \in G_{(j)}$, $\vec{v}_i = (v_{i1}, \ldots, v_{it}) \in \Psi^{[i]}$ and $\vec{v}_j = (v_{j1}, \ldots, v_{jt}) \in \Psi^{[j]}$ then $\left[g_i^{\vec{v}_i}, g_j^{\vec{v}_j}\right] \in G^\Psi_{(i+j)}$. But this follows from the Baker–Campbell–Hausdorff formula (see (C.2)), the filtration property of $G_{(i)}$ and Corollary 3.4. ■

The spaces $\Psi^{[i]}$ form a flag

$$0 \leqslant \Psi^{[1]} \leqslant \ldots \leqslant \Psi^{[s]} \leqslant \mathbb{R}^t$$

of subspaces which are rational (i.e. they can be defined over $\mathbb{Q}$). From a greedy algorithm (and clearing denominators) we may thus find a basis $\vec{v}_1, \ldots, \vec{v}_{m_s} \in \Psi^{[s]}$ with the following properties:

(i) (Integrality) $\vec{v}_1, \ldots, \vec{v}_{m_s}$ all lie in $\mathbb{Z}^t$;

(ii) (Partial span) For every $1 \leqslant i \leqslant s$, $\vec{v}_1, \ldots, \vec{v}_{m_i}$ span $\Psi^{[i]}$;

(iii) (Row echelon form) For each $1 \leqslant j \leqslant m_s$, there exists l_j, $1 \leqslant l_j \leqslant t$, such that $\vec{v}_j$ has a non-zero l_j coordinate, but such that $\vec{v}_{j'}$ has a zero l_j coordinate for all $j < j' \leqslant m_s$.

For instance, the basis

$$\vec{v}_1 := (1,1,1,1); \quad \vec{v}_2 := (0,1,2,3); \quad \vec{v}_3 := (0,0,1,3); \quad \vec{v}_4 := (0,0,0,1)$$

we implicitly gave above for our running example is already in this form.

Fix such a basis. For each basis element $\vec{v}_j$, we can define the *degree* $\deg(\vec{v}_j)$ of that element to be the first i for which $j \leqslant m_i$, thus $\deg(\vec{v}_j)$ is an integer between 1 and s, and $\vec{v}_j \in \Psi^{[\deg(\vec{v}_j)]}$.

Observe that an arbitrary element of G^Ψ can be expressed as a product of finitely many elements of the form $g_j^{\vec{v}_j}$ for $0 \leqslant j \leqslant m_s$ and $g_j \in G_{(\deg(\vec{v}_j))}$.

By many applications[10] of the Baker–Campbell–Hausdorff formula (see (C.1)) and Lemma 3.5, we can now express any element of G^Ψ in the form

$$\prod_{j=1}^{m_s} g_j^{\vec{v}_j} \tag{3.1}$$

where $g_j \in G_{(\deg(\vec{v}_j))}$ for all $1 \leqslant j \leqslant m_s$.

Thus, in our running example, we have the explicit description of $G^\Psi = \mathrm{HP}^4(G)$ as

$$\big\{(g_0, g_0g_1, g_0g_1^2g_2, g_0g_1^3g_2^3g_3) \;:\; g_0 \in G_{(0)},\; g_1 \in G_{(1)},\; g_2 \in G_{(2)},\; g_3 \in G_{(3)}\big\}.$$

Note that from results on the Taylor expansion (see Lemma A.1) this group may also be identified as

$$\big\{(g(0), g(1), g(2), g(3)) \;:\; g \in \mathrm{poly}(\mathbb{Z}, G_\bullet)\big\}.$$

The group nature of $\mathrm{HP}^4(G)$ is then easily deduced from Theorem 1.6, but this presentation is somewhat specific to the Hall–Petresco case and we shall not require it further.

From the row-echelon form one can verify inductively that the representation (3.1) is unique (this can be seen clearly by working with the Hall–Petresco example presented above). This gives G^Ψ the structure of a connected, simply connected Lie group, with dimension

$$\dim(G^\Psi) = \sum_{i=1}^{s} \dim\big(G^{(i)}\big)\big(\dim\big(\Psi^{[i]}\big) - \dim\big(\Psi^{[i-1]}\big)\big) \tag{3.2}$$

(with the convention that $\Psi^{[0]}$ is trivial). A similar argument also shows that every element of $G^\Psi_{(i_0)}$ can be expressed uniquely in the form (3.1), where now g_j is constrained to lie in $G_{(\max(\deg(v_j), i_0))}$ rather than $G_{(\deg(v_j))}$. In particular, by reading off the coefficients g_j one at a time, this implies the pleasant identity

$$G^\Psi_{(i)} = G^\Psi \cap \big(G_{(i)}\big)^k. \tag{3.3}$$

[10]Indeed, one uses (C.1) and Lemma 3.5 to extract out and collects all terms with degree $\deg(\vec{v}_j) = 1$, leaving only terms with base g_j in $G_{(2)}$. Then one extracts out those terms with degree 2 (merging them with the $i = 1$ terms as necessary), leaving only terms with base in $G_{(3)}$. Continuing this process gives the desired factorisation.

Remark. From Taylor expansion (see Lemma A.1) we see that the sequence g^Ψ in (1.7) lies in poly $(\mathbb{Z}, G^\Psi_\bullet)$. While we do not directly use this fact here, it may help explain why the filtration $G^\Psi_\bullet$ will plays a prominent role in the proof of the counting lemma that we will shortly come to.

Recall that we normalised the basis vectors $\vec{v}_j \in \mathbb{Z}^t$ to have integer coefficients. As a consequence, we see that if the g_j are in Γ, then the expression (3.1) lies in Γ^k. From this (and many applications of Lemma 3.5) we see that $\Gamma^\Psi_{(i)} := \Gamma^k \cap G^\Psi_{(i)}$ is cocompact in $G^\Psi_{(i)}$ for each i, and so $(G^\Psi/\Gamma^\Psi, G^\Psi_\bullet)$ is a filtered nilmanifold. Furthermore, the same argument shows that the $G^\Psi_{(i)}$ are rational subgroups of G^k and so $(G^\Psi/\Gamma^\Psi, G^\Psi_\bullet)$ is a subnilmanifold of $(G^k/\Gamma^k, G^k_\bullet)$.

The counting lemma: preliminary manœuvres. Now that we have verified that G^Ψ/Γ^Ψ is indeed a nilmanifold, we can begin the proof of Theorem 1.11.

We begin with some easy reductions. First, observe that for fixed M, there are only finitely many possibilities for s, D, t, Ψ, and (up to isomorphism) there are only finitely many possibilities for $(G/\Gamma, G_\bullet)$ and Γ. Thus it will suffice to establish the result for a single choice of s, D, t, Ψ, $(G/\Gamma, G_\bullet)$, with the bounds depending on these quantities. Hence, we fix these quantities and allow all implicit constants to depend on these quantities (thus, in this section, we will not explicitly subscript out $O(1)$ quantities).

Similarly, because the space of Lipschitz functions with Lipschitz norm $O(1)$ is precompact in the uniform topology (by the Arzelá-Ascoli theorem), it suffices to prove the desired bound for each fixed F, as the uniformity in F then follows from an easy approximation argument. Thus we fix F and allow all quantities to depend on F.

Next, we observe that we may normalise $g(0) = \mathrm{id}$. Indeed, we may factorise $g(0) = c_0\gamma_0$ where $d_G(c_0, \mathrm{id}) = O(1)$ and $\gamma_0 \in \Gamma$. Factorising, we obtain

$$g(n) = c_0 g'(n)\gamma_0$$

where $g'(n) := c_0\gamma_0(\gamma_0^{-1}g(n)\gamma_0)$. Note that $g'(0) = \mathrm{id}$ and that Taylor coefficients of g' are given by $g'_i = \gamma_0^{-1}g_i\gamma_0$, and so g' is also (A, N)-irrational. It is then an easy matter to see that Theorem 1.11 for g and F follows from Theorem 1.11 for g' and for the shifted function $F'(x) := F(c_0x)$, which is still Lipschitz with norm $O(1)$.

Note that we may assume that A and N are large, as the claim is trivial otherwise.

Equidistribution in the Leibman group. Let us recall what we are trying to prove. In the counting lemma, Theorem 1.11, our aim is to show that if $g(n)$ is suitably irrational then the orbit $\big(g^{\psi}(\mathbf{n})\big)_{\mathbf{n}\in(\mathbf{n_0}+\Lambda)\cap P}$ is equidistributed on the Leibman nilmanifold G^{Ψ}/Γ^{Ψ}. We shall proceed by contradiction, supposing this orbit is not equidistributed and deducing that $g(n)$ could not have been irrational. The reader should recall the definition of *irrational* in this context: it is given in Definition A.6.

Our main tool will be a mild generalisation of the "multiparameter Leibman criterion", which is [30, Theorem 8.6]. Here is the statement we shall use.

Theorem 3.6. *Suppose that $(G/\Gamma, G_\bullet)$ is a filtered nilmanifold of complexity $\leqslant M$ and that $g \in \mathrm{poly}\,(\mathbb{Z}^D, G_\bullet)$ is a polynomial sequence for some $D \leqslant M$. Suppose that $\Lambda \subseteq \mathbb{Z}^D$ is a lattice of index $\leqslant M$, that $\mathbf{n_0} \in \mathbb{Z}^D$ has magnitude $\leqslant M$, and that $P \subseteq [-N, N]^D$ is a convex body. Suppose that $\delta > 0$, and that*

$$\Big| \sum_{\mathbf{n}\in(\mathbf{n_0}+\Lambda)\cap P} F(g(\mathbf{n})\Gamma) - \frac{\mathrm{vol}\,(P)}{[\mathbb{Z}^D : \Lambda]} \int_{G/\Gamma} F \Big| > \delta N^D \|F\|_{\mathrm{Lip}}$$

for some Lipschitz function $F : G/\Gamma \to \mathbb{C}$. Then there is a nontrivial homomorphism $\eta : G \to \mathbb{R}$ which vanishes on Γ, has complexity $O_M(1)$ and such that

$$\|\eta \circ g\|_{C^\infty([N]^D)} = O_{\delta,M}(1).$$

Remarks. This differs from [30, Theorem 8.6] in several insubstantial ways. On the one hand we have no concern here with the polynomial bounds that were important in that setting. However, we are dealing here with a sublattice $\Lambda \subseteq \mathbb{Z}^D$ rather than $\mathbb{Z}^D$ itself, and with an arbitrary convex body P rather than the box $[N]^D$. This more general result can be deduced from [30, Theorem 8.6] in a somewhat routine, though slightly tedious, manner. We sketch the details in Appendix B. The notation $C^\infty\big([N]^D\big)$ is recalled both in the appendix and later in this section.

Later on, the notation will get a little complicated. Let us, then, first apply Theorem 3.6 to establish the following very simple special case of the counting lemma (it is, of course, the special case in which Ψ consists of the single form $\psi_1(\mathbf{n}) = n_1$).

Lemma 3.7 (Irrational implies equidistributed). *Suppose that $(G/\Gamma, G_\bullet)$ is a filtered nilmanifold of complexity at most M and that $g : \mathbb{Z} \to G$ is an (A, N)-irrational polynomial sequence. Then we have the equidistribution property*

$$\mathbb{E}_{n\in[N]}F\big(g(n)\Gamma\big) = \int_{G/\Gamma} F + O_M\big(A^{-c_M}\|F\|_{\mathrm{Lip}}\big)$$

for all Lipschitz $F : G/\Gamma \to \mathbb{C}$ and some $c_M > 0$.

Proof. Suppose the conclusion is false. Then by[11] Theorem 3.6 there is some continuous homomorphism $\eta : G \to \mathbb{R}$ which vanishes on $[G, G]$ and Γ, has complexity $O_\delta(1)$, and for which $\|\eta \circ g\|_{C^\infty[N]} \leqslant \delta^{-O(1)}$. Recall (cf. [30, Definition 2.7]) what this means: in the Taylor expansion

$$\eta \circ g(n) = \alpha_0 + \alpha_1\binom{n}{1} + \cdots + \alpha_s\binom{n}{s},$$

the jth coefficient α_j satisfies $\|\alpha_j\|_{\mathbb{R}/\mathbb{Z}} \leqslant \delta^{-O(1)}/N^j$ for $j = 1, \ldots, s$. If the sequence g is developed as a Taylor expansion

$$g(n) = g_0 g_1^{\binom{n}{1}} \cdots g_s^{\binom{n}{s}}$$

then we of course have $\alpha_j = \eta(g_j)$. Choose i maximal so that the restriction $\eta|_{G_{(i)}}$ is nontrivial. Then certainly $\big\|\eta(g_i)\big\|_{\mathbb{R}/\mathbb{Z}} \leqslant \delta^{-O(1)}/N^i$. We claim that η is an i-horizontal character in the sense of Definition A.5, a statement which will clearly contradict the supposed (A, N)-irrationality of g if δ is a sufficiently small power of $1/A$. To this end all we need do is confirm that η vanishes on $G_{(i+1)}$, $\Gamma_{(i)}$ and on $\big[G_{(j)}, G_{(i-j)}\big]$ for $0 \leqslant j \leqslant i$. The first of these follows from the maximality of i, whilst the second and third follow immediately from the properties of η stated at the beginning of the proof. ■

Let us turn now to the more notationally intensive general case. Now, we apply Theorem 3.6 to G^Ψ/Γ^Ψ to conclude that there is a non-trivial continuous homomorphism $\eta : G^\Psi \to \mathbb{R}$ which maps Γ^Ψ to $\mathbb{Z}$, has complexity $O_\delta(1)$, and satisfies

$$\|\eta \circ g^\Psi\|_{C^\infty([N]^D)} = O_\delta(1). \tag{3.4}$$

[11] In fact here we only need the rather simpler 1-parameter version, which is [30, Theorem 1.16].

Much as in the proof of Lemma 3.7, what this means is that if $\eta \circ g^{\Psi}(\mathbf{n})$ is developed as a Taylor series in multi-binomial coefficients $\binom{\mathbf{n}}{\mathbf{j}} = \binom{n_1}{j_1} \cdots \binom{n_D}{j_D}$ (see Lemma A.1), the coefficient $\alpha_{\mathbf{j}}$ satisfies $\|\alpha_{\mathbf{j}}\|_{\mathbb{R}/\mathbb{Z}} \ll_\delta N^{-|\mathbf{j}|}$. Our aim is to use this information to contradict the assumption that $g(n)$ is (A, N)-irrational.

Let us once again take i maximal such that $\eta|_{G^{\Psi}_{(i)}}$ is nontrivial. Considering again the Taylor expansion of $g(n)$, we have

$$(\eta \circ g^{\Psi})(\mathbf{n}) = \sum_{j=1}^{i} \eta\Big(g_j^{\binom{\psi_1(\mathbf{n})}{j}}, \dots, g_j^{\binom{\psi_t(\mathbf{n})}{j}}\Big). \tag{3.5}$$

Take the basis $\vec{v}_1, \vec{v}_2, \dots$ for $\Psi^{[i]}$ described earlier. Then, since the vector

$$\left(\binom{\psi_1(\mathbf{n})}{j}, \dots, \binom{\psi_t(\mathbf{n})}{j}\right)$$

lies in $\Psi^{[j]}$, there is an expansion

$$\left(\binom{\psi_1(\mathbf{n})}{j}, \dots, \binom{\psi_t(\mathbf{n})}{j}\right) = P_{j,1}(\mathbf{n})\vec{v}_1 + \dots + P_{j,m_j}(\mathbf{n})\vec{v}_{m_j} \tag{3.6}$$

for $j = 1, \dots, i$, where the $P_{j,k} : \mathbb{Z}^D \to \mathbb{R}$ are polynomials of degree at most j, recalling that $m_j := \dim\left(\Psi^{[j]}\right)$. Comparing with (3.5), we obtain

$$(\eta \circ g^{\Psi})(\mathbf{n}) = \sum_{j=1}^{i} \sum_{k=1}^{m_j} P_{j,k}(\mathbf{n}) \eta(g_j^{\vec{v}_k}). \tag{3.7}$$

We are going to look at the coefficients $\alpha_{\mathbf{i}}$ of (3.7) for the monomial $\mathbf{n}^{\mathbf{i}} := n_1^{i_1} \dots n_D^{i_D}$, where $\mathbf{i} = (i_1, \dots, i_D)$ and $|\mathbf{i}| := |i_1| + \dots + |i_d| = i$. We are assuming that every such coefficient satisfies $\|\alpha_{\mathbf{i}}\|_{\mathbb{R}/\mathbb{Z}} \ll_\delta N^{-i}$. Note also that

$$\alpha_{\mathbf{i}} = \sum_{k=1}^{m_i} (P_{i,k})_{\mathbf{i}} \eta\big(g_i^{\vec{v}_k}\big), \tag{3.8}$$

where $(P_{i,k})_{\mathbf{i}}$ is the $\mathbf{n}^{\mathbf{i}}$ coefficient of $P_{i,k}(\mathbf{n})$; this is because terms of total degree i cannot arise from the terms $j = 1, \dots, i-1$ in the sum on the right hand side of (3.7).

On the other hand by taking $j = i$ in (3.6) we have

$$(3.9)\quad \big(P_{i,1}(\mathbf{n})\big)_{\mathbf{i}}\vec{v}_1 + \cdots + \big(P_{i,m_i}(\mathbf{n})\big)_{\mathbf{i}}\vec{v}_{m_i}$$

$$= \frac{1}{i_1!\ldots i_D!}\big(\psi_1(\mathbf{e}_1)^{i_1}\cdots\psi_1(\mathbf{e}_D)^{i_D},\ldots,\psi_t(\mathbf{e}_1)^{i_1}\cdots\psi_t(\mathbf{e}_D)^{i_D}\big)$$

$$= \frac{1}{i_1!\ldots i_D!}\Psi(\mathbf{e}_1)^{i_1}\cdots\Psi(\mathbf{e}_D)^{i_D},$$

where $\mathbf{e}_j = (0,\ldots,1,\ldots,0) \in \mathbb{Z}^D$, the 1 being in the jth position, and $\Psi(\mathbf{e}_j) := \big(\psi_1(\mathbf{e}_j),\ldots,\psi_t(\mathbf{e}_j)\big) \in \mathbb{R}^t$.

Comparing (3.8) and (3.9) and using the fact that η is a homomorphism on G^Ψ, we obtain

$$\alpha_{\mathbf{i}} = \frac{1}{i_1!\ldots i_D!}\eta\big(g_i^{\Psi(\mathbf{e}_1)^{i_1}\ldots\Psi(\mathbf{e}_D)^{i_D}}\big).$$

Thus, for each $\mathbf{i}$ with $|\mathbf{i}| = |i_1| + \cdots + |i_D| = i$, we have

$$(3.10)\qquad \Big\|\eta\big(g_i^{\Psi(\mathbf{e}_1)^{i_1}\ldots\Psi(\mathbf{e}_D)^{i_D}}\big)\Big\|_{\mathbb{R}/\mathbb{Z}} \ll_\delta N^{-i}$$

To obtain the desired contradiction with the (A, N)-irrationality hypothesis and thus complete the proof, it suffices (after taking A sufficiently large depending on δ) to establish that for at least one choice of $\mathbf{i}$, the map $\xi_{\mathbf{i}} : G_{(i)} \to \mathbb{R}$ defined by

$$\xi_{\mathbf{i}}(g) := \eta\big(g^{\Psi(\mathbf{e}_1)^{i_1}\ldots\Psi(\mathbf{e}_D)^{i_D}}\big)$$

is a nontrivial horizontal i-character of complexity $O_\delta(1)$.

The complexity bound follows from the fact that the coefficients of the forms ψ_i are integers of size $O(1)$ and the Baker–Campbell–Hausdorff formula (Appendix C). That at least one of these maps is nontrivial follows from that fact that η is nontrivial on $G^\Psi_{(i)}$ and the fact that the vectors $\Psi(\mathbf{e}_1)^{i_1}\cdots\Psi(\mathbf{e}_D)^{i_D}$, $i_1+\cdots+i_D = i$, span $\Psi^{[i]}$ (a consequence of Lemma 3.3).

Furthermore $\xi_{\mathbf{i}}$ always annihilates $\Gamma^\Psi_{(i)}$ and $G^\Psi_{(i+1)}$ (by the asserted maximality of i). To qualify as an i-horizontal character we must also show that it vanishes on $\big[G^\Psi_{(j)}, G^\Psi_{(i-j)}\big]$ for each $0 \leqslant j \leqslant i$. To this end, note that we may factor

$$\Psi(\mathbf{e}_1)^{i_1}\cdots\Psi(\mathbf{e}_D)^{i_D} = ww',$$

where $w \in \Psi^{[j]}$ and $w' \in \Psi^{[i-j]}$. Indeed, we may take

$$w = \Psi(\mathbf{e}_1)^{j_1} \cdots \Psi(\mathbf{e}_D)^{j_D}, \qquad w' = \Psi(\mathbf{e}_1)^{i_1 - j_1} \cdots \Psi(\mathbf{e}_D)^{i_D - j_D}$$

for any indices $j_1, \dots, j_D$ with $j_l \leqslant i_l$ and $j_1 + \dots + j_D = j$, whereupon the relevant containments follow from Lemma 3.3. Now if $g \in G^{\Psi}_{(j)}$ and $g' \in G^{\Psi}_{(i-j)}$ are arbitrary then we have

$$[g^w, g'^{w'}] \equiv [g, g']^{ww'} \pmod{G^{\Psi}_{(i+1)}}$$

by the Baker-Campbell–Hausdorff formula (C.2). Applying η, which is trivial on $G^{\Psi}_{(i+1)}$ by assumption, we obtain

$$\xi_{\mathbf{i}}([g, g']) = \eta([g, g']^{ww'}) = \eta([g^w, g'^{w'}]) = 0,$$

the last step being a consequence of the fact that η has abelian image and hence vanishes on $[G^{\Psi}, G^{\Psi}]$. This concludes the proof of the counting lemma, Theorem 1.11.

4. Generalised von Neumann Type Theorems

In this section we recall a number of results asserting the connection between Gowers norms and various types of linear configuration. These results are collectively known in the literature as "generalised von Neumann theorems". The connection between Gowers norms (not called by that name, of course) and linear configurations was first made in [17]. A fairly general result of this type, which appears in [31], is the following.

Theorem 4.1 (Generalised von Neumann Theorem). *Let $\Psi = (\psi_1, \dots, \psi_t)$ be a collection of linear forms $\psi_1, \dots, \psi_t : \mathbb{Z}^D \to \mathbb{Z}$ for some $t, D \geqslant 1$, any two of which are linearly independent. Then there exists an integer $s = s(\Psi)$ with the property that one has the inequality*

$$\left| \mathbb{E}_{\mathbf{n} \in [N]^d} \prod_{i=1}^{t} f_i(\psi_i(\mathbf{n})) \right| \ll_{t,D,\Psi} \inf_{1 \leqslant i \leqslant m} \|f_i\|_{U^{s+1}[N]} \tag{4.1}$$

for all $N \geqslant 1$ and all $f_1, \dots, f_m : [N] \to \mathbb{C}$ bounded in magnitude by 1.

Remarks. A natural value of $s(\Psi)$ comes from the proof in [31], which proceeds via s applications of the Cauchy–Schwarz inequality. For this reason Gowers and Wolf [22] call $s(\Psi)$ the *Cauchy–Schwarz complexity* of the system Ψ. There is a linear-algebra recipe for computing $s(\Psi)$ which is not especially enlightening but sufficiently simple that we can give it here (see the introduction to [31] for more details). If $1 \leqslant i \leqslant t$ and $s \geqslant 0$ then we say that Ψ has i-complexity at most s if one can cover the $t-1$ forms $\{\psi_j : j \in [t] \setminus \{i\}\}$ by $s+1$ classes, such that ψ_i does not lie in the linear span of the forms in any one of these classes. Then $s(\Psi)$ is the smallest s for which the system has i-complexity at most s for all $1 \leqslant i \leqslant t$. Note, then, that the Cauchy–Schwarz complexity of the system $\Psi = \{n_1, n_1+n_2, \dots, n_1+(k-1)n_2\}$ corresponding to a k-term arithmetic progression is $k-2$. As a final remark, let us note that Theorem 4.1, as proved in [31, Appendix C], is regrettably somewhat difficult to understand as we had to establish a more general result in which the functions f_i were bounded by an arbitrary pseudorandom measure, and this is notationally heavy. For a gentle explanation of the special case $\Psi = \{n_1, n_1+n_2, n_1+2n_2, n_1+3n_2\}$ (where $s=2$) the reader may consult [26, Proposition 1.11]. A sketch of the proof of Theorem 4.1 is also given in [22, §2]. See also [5] for a variant of these notions of complexity in the ergodic setting, and for polynomial forms instead of linear ones.

We will need a *twisted* version of the Generalised von Neumann inequality, in which an additional nilsequence of lower degree is inserted. We shall not need it for general linear forms, so we formulate just the special case we need.

Lemma 4.2 (Twisted generalised von Neumann theorem). *Let $k \geqslant 3$, let $f_0, \dots, f_{k-1} : [N] \to \mathbb{C}$ be bounded in magnitude by 1, let $c_0, \dots, c_{k-1}$ be distinct integers, and let $F(g(n)\Gamma)$ be a degree $\leqslant (k-2)$ nilsequence of complexity at most M. Then*

$$\Big|\mathbb{E}_{n\in[N], d\in[-N,N]} F\big(g(d)\Gamma\big) \prod_{i=0}^{k-1} f_i(n+c_i d)\Big| \ll_{k,M,c_0,\dots,c_{k-1}} \inf_{0\leqslant i\leqslant k-1} \|f_i\|_{U^{k-1}[N]}.$$

Proof. We induct on k, starting with the case $k=3$. The underlying nilmanifold G/Γ is then a torus $(\mathbb{R}/\mathbb{Z})^m$ with $m = O_M(1)$, and $g(n) = \theta n + \theta_0$ may be taken to be linear. By a standard Fourier decomposition we may assume that $F(x) = e(\xi \cdot x)$ for some $\xi \in \mathbb{Z}^m$ with $|\xi| = O_M(1)$, in which case we may rewrite the estimate to be proven as

$$\big|\mathbb{E}_{n\in[N]}\mathbb{E}_{d\in[-N,N]} f_0(n+c_0 d) f_1'(n+c_1 d) f_2'(n+c_2 d)\big| \ll_{k,M} \inf_{i=0,1,2} \|f_i\|_{U^2[N]},$$

where

$$f_1'(n) = f_1(n) e\big(-(c_2-c_1)^{-1}\xi\cdot\theta n\big) \text{ and } f_2'(n) = f_2(n) e\big((c_2-c_1)^{-1}\xi\cdot\theta n\big).$$

However it is easy to establish the invariance properties $\|f_1\|_{U^2} = \|f_1'\|_{U^2}$ and $\|f_2\|_{U^2} = \|f_2'\|_{U^2}$, and so the result follows immediately from Theorem 4.1.

Now suppose that $k \geqslant 4$ and the claim has already been proven for smaller k. By permuting indices and then translating n, it suffices to show that

(4.2)
$$\Big|\mathbb{E}_{n\in[N];d\in[-N,N]} F\big(g(d)\Gamma\big) \prod_{i=0}^{k-1} f_i(n+c_i d)\Big| \ll_{k,M,c_0,\dots,c_{k-1}} \|f_{k-1}\|_{U^{k-1}[N]}$$

under the assumption that $c_0 = 0$.

Recall from [30] that we define a *vertical character* to be a continuous homomorphism $\xi : G_{(k-2)}/\big(G_{(k-2)}\cap\Gamma\big) \to \mathbb{R}/\mathbb{Z}$. We say that F has *vertical frequency* ξ if one has $F(g_{k-2}x) = e\big(\xi(g_{k-2})\big) F(x)$ for all $x \in G/\Gamma$ and $g_{k-2} \in G_{(k-2)}$. By a standard Fourier decomposition in the vertical direction (e.g. by arguing exactly as in [30, Lemma 3.7]) we may assume without loss of generality that F has a vertical frequency ξ.

Applying the Cauchy–Schwarz inequality, we can bound the left-hand side of (4.2) by

$$\ll \Big|\mathbb{E}_{n\in[N];h,d\in[-N,N]} F\big(g(d+h)\Gamma\big)\overline{F\big(g(d)\Gamma\big)} \times \prod_{i=0}^{k-1} f_i(n+c_i d+c_i h)\overline{f_i(n+c_i d)}\Big|^{1/2}.$$

Because F has a vertical frequency, $F\big(g(d+h)\big)\overline{F\big(g(d)\Gamma\big)}$ is a degree $\leqslant (k-3)$ nilsequence of complexity $O_{M,k}(1)$ (see [30, Proposition 7.2]).

Applying the induction hypothesis, we may thus bound the above expression by

$$\ll_{M,k,c_0,\dots,c_{k-1}} \left(\mathbb{E}_{h\in[-N,N]}\|\Delta_{c_i h} f_i\|^2_{U^{k-2}[N]}\right)^{1/2}$$

which by Hölder's inequality can be bounded by

$$\ll_{M,k,c_0,\dots,c_{k-1}} \left(\mathbb{E}_{h\in[-|c_i|N,|c_i|N]}\|\Delta_{h} f_i\|^{2^{k-2}}_{U^{k-2}[N]}\right)^{1/2^{k-2}}$$

and the claim follows from the recursive definition of the Gowers norms. ■

Remark. The above argument is very similar to the short proof presented in [33, Appendix G] that s-step nilsequences obstruct uniformity in the U^{s+1}-norm (that is, the inverse conjecture GI(s) is an if-and-only if statement).

5. On a Conjecture of Bergelson, Host, and Kra

We now apply the arithmetic regularity and counting lemmas to establish Theorem 1.12, the proof of the conjecture of Bergelson, Host and Kra. Our strategy here can be viewed as a finitary analogue of the ergodic theory arguments in [4], however there are some slight differences in our approach which we comment on at the end of this section.

It will suffice to prove the following claim.

Theorem 5.1. *Let $k = 1, 2, 3$ or 4, and suppose that $0 < \alpha < 1$ and $\varepsilon > 0$. Then for any $N \geqslant 1$ and any subset $A \subseteq [N]$ of density $|A| \geqslant \alpha N$, one can find a function $\mu : \mathbb{Z} \to \mathbb{R}^+$ such that*

$$\mathbb{E}_{d\in[-N,N]}\mu(d) = 1 + O(\varepsilon) \tag{5.1}$$

and

$$\sup_{d\in[-N,N]} \mu(d) \ll_{\alpha,\varepsilon} 1 \tag{5.2}$$

such that

$$\mathbb{E}_{n\in[N];d\in[-N,N]} 1_A(n)1_A(n+d)\dots 1_A(n+(k-1)d)\,\mu(d) \geqslant \alpha^k - O(\varepsilon). \tag{5.3}$$

Indeed, from (5.1), (5.3), we see that we have

$$\mathbb{E}_{n\in[N]}1_A(n)1_A(n+d)\dots 1_A\big(n+(k-1)d\big) \geqslant \alpha^k - O(\varepsilon)$$

for all d in a subset E of $[-N,N]$ with $\mathbb{E}_{d\in[-N,N]}1_E(d)\mu(d) \gg_{\alpha,\varepsilon} 1$. (Here we crucially use the trivial but fundamental fact that 1_A is nonnegative.) From (5.2) we conclude that $|E| \gg_{\alpha,\varepsilon} N$, and Theorem 1.12 follows (after shrinking ε by an absolute constant). Conversely, it is not difficult to deduce Theorem 1.12 from Theorem 5.1.

It remains to establish Theorem 5.1. We may assume that N is large depending on α, ε as the claim is trivial otherwise (just take μ to be the Kronecker delta function at 0).

For $k=1$ one can simply take $\mu \equiv 1$. For $k=2$, we first observe that

$$\mathbb{E}_{n\in[N]}\mathbb{E}_{h\in[-\varepsilon N,\varepsilon N]}1_A(n+h) = \alpha + O(\varepsilon);$$

applying Cauchy–Schwarz we conclude that

$$\mathbb{E}_{h,h'\in[-\varepsilon N,\varepsilon N]}\mathbb{E}_{n\in[N]}1_A(n+h)1_A(n+h') \geqslant \alpha^2 - O(\varepsilon).$$

The claim then follows, with μ being the probability density function of $h-h'$ as h, h' range uniformly in $[-\varepsilon N,\varepsilon N]$.

Now we turn to the cases $k=3,4$. Here, one has to be more sophisticated about how one chooses μ (for instance, by using a Behrend set construction it is not hard to see that the previous choices of μ do not always work). Let $\mathcal{F}:\mathbb{R}^+\to\mathbb{R}^+$ be a sufficiently rapidly growing function depending on α,ε in a manner to be specified later. We apply Theorem 1.2 with $s:=k-2$ to obtain a quantity $M=O_{\varepsilon,\mathcal{F}}(1)$ and a decomposition

$$1_A(n) = f_{\mathrm{nil}}(n) + f_{\mathrm{sml}}(n) + f_{\mathrm{unf}}(n) \tag{5.4}$$

such that

(i) $f_{\mathrm{nil}}(n)$ is a $\big(\mathcal{F}(M),N\big)$-irrational degree $\leqslant k-2$ virtual nilsequence of complexity at most M and scale N;

(ii) f_{sml} has an $L^2[N]$ norm of at most $\varepsilon/100$;

(iii) f_{unf} has an $U^{k-1}[N]$ norm of at most $1/\mathcal{F}(M)$;

(iv) $f_{\mathrm{nil}}, f_{\mathrm{sml}}, f_{\mathrm{unf}}$ are all bounded in magnitude by 1; and

(v) f_{nil} and $f_{\mathrm{nil}}+f_{\mathrm{sml}}$ are non-negative.

It is clear that $\left|\mathbb{E}_{n\in[N]}f_{\mathrm{sml}}(n)\right| = O(\varepsilon)$, and furthermore, by Theorem 4.1 (setting all but one of the functions equal to 1) we also have $\left|\mathbb{E}_{n\in[N]}f_{\mathrm{unf}}(n)\right| = O(\varepsilon)$ if $\mathcal{F}$ grows rapidly enough. Therefore

$$\mathbb{E}_{n\in[N]}f_{\mathrm{nil}}(n) \geqslant \alpha - O(\varepsilon). \tag{5.5}$$

The heart of the matter is the following proposition.

Proposition 5.2 (Bergelson–Host–Kra for f_{nil}). *Let $k = 3, 4$. Then there exists a non-negative $(k-2)$-step nilsequence $\mu : \mathbb{Z} \to \mathbb{R}^+$ of complexity $O_{\alpha,\varepsilon,M}(1)$ obeying the normalisation*

$$\mathbb{E}_{d\in[N]}\mu(d) = 1 + O(\varepsilon) \tag{5.6}$$

and such that

$$\mathbb{E}_{n,d\in[N]}f_{\mathrm{nil}}(n)f_{\mathrm{nil}}(n+d)\dots f_{\mathrm{nil}}\big(n+(k-1)d\big)\,\mu(d) \geqslant \alpha^k - O(\varepsilon). \tag{5.7}$$

Deduction of Theorem 5.1 from Proposition 5.2. Using (5.4), one can expand the left-hand side of (5.3) into 3^k terms, one of which is (5.7). As for the other terms, any term involving at least one copy of f_{unf} is of size $O_{\alpha,\varepsilon,M}\big(1/\mathcal{F}(M)\big)$ by Lemma 4.2 and the U^{k-1} norm bound on f_{unf}. Finally, consider a term that involves at least one copy of f_{sml}. Suppose first that we have a term that involves $f_{\mathrm{sml}}(n)$. Then after performing the average in d using (5.6), we see that this term is $O\big(\mathbb{E}_{n\in[N]}\left|f_{\mathrm{sml}}(n)\right|\big)$, which is $O(\varepsilon)$ by the $L^2[N]$ bound on f_{sml} and the Cauchy–Schwarz inequality. Similarly for any term that involves $f_{\mathrm{sml}}(n+id)$, after making a change of variables $(n',d) := (n+id,d)$. Putting all this together we obtain the result. ■

It remains, of course, to establish Proposition 5.2. We may assume that N is sufficiently large depending on α, ε, M, as the claim is trivial otherwise by taking μ to be a delta function.

We first establish the proposition in the easier of the two cases, namely the case $k = 3$. This was previously considered in [25]. In this case it is actually easier to work with the (easier) weak regularity lemma, Proposition 2.7, in which the degree 1 polynomial sequence $g(n)$ is not required to be irrational. Note that we have not made any use of irrationality so far, though we shall do so later when discussing the case $k = 4$. We may identify G/Γ with $(\mathbb{R}/\mathbb{Z})^m$ for some $m = O_M(1)$ and, by modulating F if necessary, we may suppose that $g(n) = \theta n$ is linear with no constant term, where $\theta \in \mathbb{R}^m$. Then

$$f_{\mathrm{nil}}(n) = F(n\theta),$$

where $F : (\mathbb{R}/\mathbb{Z})^m \to \mathbb{C}$ has Lipschitz norm $O_M(1)$.

Let $\varepsilon' > 0$ be a small number depending on ε and M to be chosen later, and let $B_1, B_2 \subseteq [-N, N]$ denote be the two Bohr sets

$$B_1 := \big\{d \in [-\varepsilon' N, \varepsilon' N] \,:\, \mathrm{dist}_{(\mathbb{R}/\mathbb{Z})^m}(\theta d, 0) \leqslant \varepsilon'\big\}$$

and

$$B_2 := \big\{d \in [-\varepsilon' N, \varepsilon' N] \,:\, \mathrm{dist}_{(\mathbb{R}/\mathbb{Z})^m}(\theta d, 0) \leqslant \varepsilon'/2\big\}.$$

By the usual Dirichlet pigeonhole argument we see that $|B_2| \gg_{\varepsilon', M} N$. Also, from the Lipschitz nature of F, we see that

$$f_{\mathrm{nil}}(n + d) = f_{\mathrm{nil}}(n) + O_M(\varepsilon')$$

whenever $d \in B_1$ and $n \in \big[-(1-\varepsilon')N, (1-\varepsilon')N\big]$. As a consequence, it follows that

$$\mathbb{E}_{n \in [N]} f_{\mathrm{nil}}(n) f_{\mathrm{nil}}(n+d) f_{\mathrm{nil}}(n+2d) = \mathbb{E}_{n \in \mathbb{N}} f_{\mathrm{nil}}(n)^3 + O_M(\varepsilon')$$

for such d. However from (5.5) and Hölder's inequality one has

$$\mathbb{E}_{n \in \mathbb{N}} f_{\mathrm{nil}}(n)^3 \geqslant \alpha^3 - O(\varepsilon). \tag{5.8}$$

Proposition 5.2 (in the case $k = 3$) now follows by taking $\mu(d) = c\psi(\theta d)$, where $\psi : (\mathbb{R}/\mathbb{Z})^m \to [0, 1]$ is an $O_{M,\varepsilon'}(1)$-Lipschitz function which is 1 on B_2 and 0 outside B_1, $c = O_{M,\varepsilon'}(1)$ is a suitable normalisation constant, and by taking ε' to be suitably small.

It is important to note here that the error term $O(\varepsilon)$ in (5.8) is uniform in M, as otherwise the argument would not work (recall that M will depend on ε). The dependence on M is instead manifested where it does not do significant damage to the argument, namely in the complexity of the weight μ.

We now turn to the $k = 4$ case of Proposition 5.2. For simplicity, let us first consider the model case when f_{nil} is a genuine nilsequence and not just a virtual nilsequence, that is to say

$$f_{\mathrm{nil}}(n) = F\big(g(n)\Gamma\big) \tag{5.9}$$

where $(G/\Gamma, G_\bullet)$ is a degree $\leqslant 2$ filtered nilmanifold of complexity $O_M(1)$, and $g \in \mathrm{poly}(\mathbb{Z}, G_\bullet)$ is $(\mathcal{F}(M), N)$-irrational. By Taylor expansion (see Appendix A), we have

$$g(n) = g_0 g_1^n g_2^{\binom{n}{2}}$$

for some $g_0, g_1 \in G$ and $g_2 \in G_{(2)}$. The $(\mathcal{F}(M), N)$-irrationality of g ensures certain irrationality properties on g_1 and g_2, though we will not need these properties explicitly here, as we will only be using them through the counting lemma (Theorem 1.11), which we shall be using as a black box.

Let $\pi : G \to T_1$ be the projection homomorphism to the torus[12] $T := G/(G_{(2)}\Gamma)$. Then

$$\pi\big(g(n)\big) = \pi(g_0)\pi(g_1)^n.$$

Let $\varepsilon' > 0$ be a small quantity depending on ε, M to be chosen later. We set

$$\mu(d) := c 1_{[-\varepsilon' N, \varepsilon' N]}(d) \phi\big(\pi(g_1)^d\big),$$

where, much as in the analysis of the case $k = 3$, $\phi : T_1 \to \mathbb{R}^+$ is a smooth non-negative cutoff to the ball of radius ε' centered at the origin that is not identically zero, and c is a normalisation constant to be chosen shortly. From Theorem 1.11 one has

$$\mathbb{E}_{d \in [-\varepsilon' N, \varepsilon' N]} \phi\big(\pi(g_1)^d\big) = \int_{T_1} \phi + o_{\mathcal{F}(M) \to \infty; \varepsilon', M}(1) + o_{N \to \infty; \varepsilon', M}(1).$$

Thus if we set

$$c := \frac{1}{\int_{T_1} \phi} = O_{\varepsilon', M}(1) \tag{5.10}$$

then we have the normalisation (5.6), if $\mathcal{F}$ is sufficiently rapid, depending on the way in which ε' depends on ε, M, and N is sufficiently large depending on ε, ε', M. From the bound on c we see that μ is a degree $\leqslant 1$ (and hence also degree $\leqslant 2$) nilsequence of complexity $O_{\varepsilon', M}(1)$.

We now apply the counting lemma, Theorem 1.11, to conclude that

$$\begin{aligned} &\mathbb{E}_{n,d \in [N]} f_{\mathrm{nil}}(n) f_{\mathrm{nil}}(n+d) f_{\mathrm{nil}}(n+2d) f_{\mathrm{nil}}(n+3d) \mu(d) \\ &\qquad = \int_{G^\Psi/\Gamma^\Psi} \tilde{F} + o_{\mathcal{F}(M) \to \infty; \varepsilon', M}(1) + o_{N \to \infty; \varepsilon', M}(1) \end{aligned} \tag{5.11}$$

[12]Note this is not quite the same thing as the *horizontal torus,* which is so important in [30], which is $(G/\Gamma)_{\mathrm{ab}} := G/[G,G]\Gamma$.

where $G^\Psi \subseteq G^4$ is the Leibman group associated to the collection $\Psi = (\psi_0, \psi_1, \psi_2, \psi_3) : \mathbb{Z}^2 \to \mathbb{Z}^4$ of linear forms $\psi_i(\mathbf{n}) := n_1 + in_2$, $i = 0,1,2,3$, that is to say the Hall–Petresco group $\mathrm{HP}^4(G)$, and $\tilde{F} : G^\Psi \to \mathbb{C}$ is the function

$$\tilde{F}(x_0, x_1, x_2, x_3) := c\phi(\pi(x_1)\pi(x_0)^{-1})F(x_0)F(x_1)F(x_2)F(x_3)$$

(here we use the identity $\pi\big(g(n+d)\big)^{-1}\pi\big(g(n)\big) = \pi(g_1)^d$, immediately verified from the Taylor expansion).

We now do some calculations in the Hall–Petrseco group very similar to those in [4]. We saw in §3 that

$$G^\Psi = \Big\{(g_0, g_0g_1, g_0g_1^2g_2, g_0g_1^3g_2^3) : g_0, g_1 \in G,\ g_2 \in G_{(2)}\Big\}$$

(note, of course, that $G_{(3)} = \mathrm{id}$ in the case we are considering). For our calculations it is convenient to use the following obviously equivalent representation:

$$G^\Psi = \Big\{\big(g_0g_{2,0}, g_0g_1g_{2,1}, g_0g_1^2g_{2,2}, g_0g_1^3g_{2,3}\big) : g_0, g_1 \in G;$$
$$g_{2,0}, \dots, g_{2,3} \in G_{(2)}; g_{2,0}g_{2,1}^{-3}g_{2,2}^{3}g_{2,3}^{-1} = \mathrm{id}\Big\}.$$

Here we have taken note of the fact that

$$\Psi^{[2]} = \big\{(x_0, x_1, x_2, x_3) \in \mathbb{R}^4 : x_0 - 3x_1 + 3x_3 - x_3 = 0\big\}.$$

This last equation is quite special in that it exhibits a certain "positivity", as we shall see later; this is key to our argument. The lattice Γ^Ψ can be similarly described by requiring $g_0, g_1, g_{2,0}, \dots, g_{2,3}$ to also lie in Γ. As a consequence of this, an arbitrary point of the nilmanifold G^Ψ/Γ^Ψ can be parameterised uniquely as

$$\big(g_0g_{2,0}, g_0g_1g_{2,1}, g_0g_1^2g_{2,2}, g_0g_1^3g_{2,3}\big)\Gamma^\Psi \tag{5.12}$$

where g_0, g_1 lie in a fundamental domain $\Sigma_1 \subset G$ of the horizontal torus T_1 (i.e. a smooth manifold with boundary on which π is a bijection from Σ_1 to T_1), and $g_{2,0}, \dots, g_{2,3}$ lie in a fundamental domain $\Sigma_2 \subset G_{(2)}$ of the *vertical torus* $T_2 := G_{(2)}/\Gamma_{(2)}$ subject to the constraint $g_{2,0}g_{2,1}^{-3}g_{2,2}^{3}g_{2,3}^{-1} \in \Gamma_{(2)}$. For such a point (5.12), the function $\tilde{F}$ takes the value

$$c\phi\big(\pi(g_1)\big)\prod_{j=0}^{3} F\big(g_0g_i^jg_{2,j}\Gamma\big).$$

On the support of ϕ, g_1 is a distance $O_M(\varepsilon')$ from the identity (if the fundamental domain Σ_1 was chosen in a suitably smooth fashion), and so by the Lipschitz nature of F and the boundedness of g_0 we have

$$F(g_0 g_i^j g_{2,j}) = F(g_0 g_{2,j}\Gamma) + O_M(\varepsilon').$$

As a consequence, the integral $\int_{G^\Psi/\Gamma^\Psi} \tilde{F}$ can be expressed as

(5.13)

$$c\int_{g_0\in\Sigma_1}\int_{g_1\in\Sigma_1}\phi(\pi(g_1))\left(\int_{\substack{g_{2,0},\dots,g_{2,3}\in T_2\\ g_{2,0}g_{2,1}^{-3}g_{2,2}^{3}g_{2,3}^{-1}=\mathrm{id}}}\prod_{j=0}^{3}F(g_0g_{2,j}\Gamma)+O_M(\varepsilon')\right)$$

where all integrals are with respect to Haar measure.

Let $\xi \in \hat{T}_2$ be a vertical character, i.e. a continuous homomorphism from T_2 to $\mathbb{R}/\mathbb{Z}$. For any $x \in G/\Gamma$, we can define the *vertical Fourier transform* $\hat{F}(x,\xi)$ to be the quantity

$$\hat{F}(x,\xi) := \int_{g_2\in T_2} e(-\xi(g_2))\,F(g_2x).$$

From the Fourier inversion formula we have

$$\int_{\substack{g_{2,0},\dots,g_{2,3}\in T_2\\ g_{2,0}g_{2,1}^{-3}g_{2,2}^{3}g_{2,3}^{-1}=\mathrm{id}}}\prod_{j=0}^{3}F(g_0g_{2,j}\Gamma) = \sum_{\xi\in\hat{T}_2}|\hat{F}(g_0,\xi)|^2|\hat{F}(g_0,3\xi)|^2.$$

In particular, we have[13]

$$\int_{\substack{g_{2,0},\dots,g_{2,3}\in T_2\\ g_{2,0}g_{2,1}^{-3}g_{2,2}^{3}g_{2,3}^{-1}=\mathrm{id}}}\prod_{j=0}^{3}F(g_0g_{2,j}\Gamma) \geqslant |\hat{F}(g_0,0)|^4.$$

Inserting this bound and (5.10) into (5.13), we conclude that

$$\int_{G^\Psi/\Gamma^\Psi}\tilde{F} \geqslant \int_{g_0\in\Sigma_1}|\hat{F}(g_0\Gamma,0)|^4 - O_M(\varepsilon') - o_{\mathcal{F}(M)\to\infty;\varepsilon',M}(1).$$

From Fubini's theorem we have

$$\int_{g_0\in\Sigma_1}\hat{F}(g_0\Gamma,0) = \int_{G/\Gamma}F$$

[13] This is the "positivity" alluded to earlier. The argument is essentially that used in [4] and it is special to the $k = 4$ case, which is of course consistent with the failure of Theorem 5.1 to extend to $k \geqslant 5$.

and from Theorem 1.11, (5.9) and (5.5) we have

$$\int_{G/\Gamma} F = \alpha + O(\varepsilon) + o_{\mathcal{F}(M)\to\infty;\varepsilon',M}(1) + o_{N\to\infty;\varepsilon',M}(1).$$

Applying Hölder's inequality, we conclude that

$$\int_{G^\Psi/\Gamma^\Psi} \tilde{F} \geqslant \alpha^4 - O(\varepsilon) - O_M(\varepsilon') - o_{\mathcal{F}(M)\to\infty;\varepsilon',M}(1) - o_{N\to\infty;\varepsilon',M}(1),$$

and so (5.7) follows from (5.11), if ε' is sufficiently small depending on ε, M, $\mathcal{F}$ is sufficiently rapid depending on ε, and N is sufficiently large depending on ε', M.

This concludes the proof of the $k = 4$ case of Proposition 5.2 in the special case when $f_{\mathrm{nil}}(n) = F\big(g(n)\Gamma\big)$ with g irrational. Unfortunately Theorem 1.2 requires us to deal with the somewhat more general setting of virtual nilsequences, in which there is dependence on n mod q or n/N. The extra details required are fairly routine but notationally irritating. Let us now suppose, then, that

$$f_{\mathrm{nil}}(n) = F\big(g(n)\Gamma, n \bmod q, n/N\big). \tag{5.14}$$

We let ε' be as before, but modify μ to now be given by

$$\mu(d) := q1_{q|d}c1_{[-\varepsilon' N,\varepsilon' N]}(d)\phi\big(\pi(g_1)^d\big),$$

with c still chosen by (5.10). As before, one can use Theorem 1.11 to establish (5.6).

Now consider the left-hand side of the expression (5.7) we are to bound in Proposition 5.2, that is to say

$$\mathbb{E}_{n,d\in[N]} f_{\mathrm{nil}}(n) f_{\mathrm{nil}}(n+d) f_{\mathrm{nil}}(n+2d) f_{\mathrm{nil}}(n+3d)\mu(d). \tag{5.15}$$

Splitting into residue classes modulo q, we can express this as

$$c\mathbb{E}_{r\in[q]}\mathbb{E}_{n\in[N/q]}\mathbb{E}_{d\in[-\varepsilon' N/q,\varepsilon' N/q]}\prod_{i=0}^{3} F\big(g(qn+qid+r)\Gamma, r,$$
$$q(n+ir)/N\big)\,\phi\big(\pi(g_1)^{qd}\big) + O_{N\to\infty;\varepsilon',M}(1).$$

We partition $[N/q]$ into intervals P of length $\lfloor \varepsilon' N\rfloor$ (plus a remainder of cardinality $O(\varepsilon' N)$). We can then rewrite the above expression as

$$c\mathbb{E}_P\mathbb{E}_{r\in[q]}\mathbb{E}_{n\in P}\mathbb{E}_{d\in[-\varepsilon' N/q,\varepsilon' N/q]}\prod_{i=0}^{3}F\big(g(qn+qid+r)\Gamma, r,$$

$$q(n+ir)/N\big)\,\phi\big(\pi(g_1)^{qd}\big)\;+O(\varepsilon')+O_{N\to\infty;\varepsilon',M}(1).$$

For each such expression, we can use the Lipschitz nature of F to replace $q(n+ir)/N$ by qn_P/N, where n_P is an arbitrary element of P, losing only an error of $O_M(\varepsilon')$. The above expression thus becomes

$$c\mathbb{E}_P\mathbb{E}_{r\in[q]}\mathbb{E}_{n\in P}\mathbb{E}_{d\in[-\varepsilon' N/q,\varepsilon' N/q]}\prod_{i=0}^{3}F\big(g(qn+qid+r)\Gamma, r, qn_P/N\big)$$

$$\times\,\phi\big(\pi(g_1)^{qd}\big)\;+O_M(\varepsilon')+O_{N\to\infty;\varepsilon',M}(1).$$

Because the orbit $n\mapsto g(n)\Gamma$ is $\big(\mathcal{F}(M),N\big)$-irrational, we see from Lemma A.8 that shifted translate $n\mapsto g(q(n+n_P)+r)\Gamma$ is $\big(\gg_M \mathcal{F}(M),N\big)$-irrational. We may then argue as in the previous case and bound the above average below by

$$\geqslant \mathbb{E}_P\mathbb{E}_{r\in[q]}\left|\int_{G/\Gamma}F(\cdot,r,qn_P/N)\right|^4-O(\varepsilon)-O_M(\varepsilon')$$

$$-\,o_{\mathcal{F}(M)\to\infty;\varepsilon',M}(1)-o_{N\to\infty;\varepsilon',M}(1).$$

Using Theorem 1.11 again, we have

$$\mathbb{E}_{n\in P}f_{\mathrm{nil}}(qn+r)=\int_{G/\Gamma}F(\cdot,r,qn_P/N)+o_{\mathcal{F}(M)\to\infty;\varepsilon',M}(1)+o_{N\to\infty;\varepsilon',M}(1)$$

and so (5.15) is at least

$$\geqslant \mathbb{E}_P\mathbb{E}_{r\in[q]}\big|\mathbb{E}_{n\in P}f_{\mathrm{nil}}(qn+r)\big|^4-O(\varepsilon)-O_M(\varepsilon')$$

$$-\,o_{\mathcal{F}(M)\to\infty;\varepsilon',M}(1)-o_{N\to\infty;\varepsilon',M}(1).$$

Now from (5.5) and double-counting one has

$$\mathbb{E}_P\mathbb{E}_{r\in[q]}\mathbb{E}_{n\in P}f_{\mathrm{nil}}(qn+r)=\alpha+O(\varepsilon)$$

and so, from Hölder's inequality, we deduce that (5.15) is

$$\geqslant \alpha^4-O(\varepsilon)-O_M(\varepsilon')-o_{\mathcal{F}(M)\to\infty;\varepsilon',M}(1)-o_{N\to\infty;\varepsilon',M}(1).$$

Proposition 5.2 now follows by once again choosing ε' small enough depending on ε, M, and choosing $\mathcal{F}$ rapid enough depending on ε, and N sufficiently large depending on ε, ε', M.

Remark. Our arguments are similar to, but slightly different from, the ergodic theory arguments in [4]. However it is likely that the argument in [4] can be translated to a finitary setting; we sketch how this would proceed as follows, restricting attention to the $k = 4$ case for concreteness. The goal is to obtain a lower bound $\mathbb{E}_{n\in[N]}f(n)f(n+d)f(n+2d)f(n+3d) \geqslant \alpha^4 - O(\varepsilon)$ for some positive density set of values of d. The analogue to the argument in [4] would proceed by performing the regularity lemma decomposition at step $s = 3$ rather than $s = 2$, so that the error f_{unf} is tiny in the U^4 norm and not just the U^3 norm. From this and Theorem 4.1, one can show that

$$\mathbb{E}_{d\in[N]}\left|\mathbb{E}_{n\in[N]}f_1(n)f_2(n+d)f_3(n+2d)f_4(n+3d)\right|^2$$

is tiny whenever at least one of f_1, f_2, f_3, f_4 is equal to f_{unf}. As a consequence, $\mathbb{E}_{n\in[N]}f_1(n)f_2(n+d)f_3(n+2d)f_4(n+3d)$ is negligible for almost all d. We can thus ignore the contribution of f_{unf}. The remainder of the argument proceeds along similar lines as above, but at one higher step (though the 3-step nilsequences involved can quickly be reduced to 2-step nilsequences, cf. [4, Section 8.1] or Section 7 below).

One of the innovations in this paper is to introduce weights such as $\mu(d)$, controlling the double average $\mathbb{E}_{n,d\in[N]}f(n)f(n+d)f(n+2d)f(n+3d)\mu(d)$ rather than controlling the single average $\mathbb{E}_{n\in[N]}f(n)f(n+d)f(n+2d)f(n+3d)$ for many values d. Thanks to the twisted generalised von Neumann theorem (Lemma 4.2), the "complexity" of such double averages is slightly less than that of the single averages, and in particular our proof of Theorem 1.12 requires only the inverse U^3 theorem from [28] rather than the more difficult inverse U^4 theorem from [33].

6. Proof of Szemerédi's Theorem

We turn now to the proof of Szemerédi's theorem. We deemed this result too famous to state in the introduction but, for the sake of fixing notation, we recall it here now. It is most natural to establish what might be called the "functional" form of the theorem which is *a priori* a stronger statement (though quite easily shown to be equivalent to the standard formulation by an argument of Varnavides [57]).

Theorem 6.1 (Szemerédi's theorem). *Let $0 < \alpha \leqslant 1$, let $k \geqslant 3$, and let $N \geqslant 1$. If $f : [N] \to [0,1]$ is a function with $\mathbb{E}_{n\in[N]} f(n) \geqslant \alpha$ then*

$$\Lambda_k(f, f, \dots, f) \gg_{k,\alpha} 1,$$

where

$$\Lambda_k(f_1, \dots, f_k) := \mathbb{E}_{n\in[N]; d\in[-N,N]} f_1(n) f_2(n+d) \dots f_k\big(n + (k-1)d\big)$$

is the multilinear operator counting arithmetic progressions.

We now prove this theorem. We fix k, α, and allow implied constants to depend on these quantities. Our argument has some similarities with the ergodic theory proof of (a polynomial generalisation of) Szemerédi's theorem in [6], in particular in first reducing the problem to a problem concerning nilsystems, which one then solves by the equidistribution theory of such systems. However, one of the key steps in [6], in which one shows that multiple recurrence is preserved under inverse limits, is more difficult to replicate in the finitary setting than in the ergodic one (see [50]). Our argument thus differs somewhat from [6], most notably by inserting a carefully chosen weight $\mu(n, d)$ before proceeding.

As usual, we begin by applying the regularity lemma, Theorem 1.2. In view of the generalised von Neumann theorem, Theorem 4.1, it is natural to apply this theorem with $s = k - 2$ (which, as remarked in §4, is the Cauchy–Schwarz complexity $s = s(\Psi)$ of the system Ψ of linear forms $n_1, n_1 + n_2, \dots, n_1 + (k-1)n_2$). If we do so, with a small parameter $\varepsilon > 0$ depending on α, k to be chosen later, and a growth function $\mathcal{F}$ depending on α, k, ε to be specified later, we obtain a decomposition

$$f(n) = f_{\mathrm{nil}}(n) + f_{\mathrm{sml}}(n) + f_{\mathrm{unf}}(n) \tag{6.1}$$

where

(i) f_{nil} is a $\big(\mathcal{F}(M), N\big)$-irrational degree $\leqslant k-2$ virtual nilsequence of complexity $\leqslant M$ and scale N;

(ii) f_{sml} has an $L^2[N]$ norm of at most ε;

(iii) f_{unf} has an $U^{k-1}[N]$ norm of at most $1/\mathcal{F}(M)$;

(iv) f_{nil}, f_{sml}, f_{unf} are all bounded in magnitude by 1; and

(v) f_{nil} and $f_{\mathrm{nil}} + f_{\mathrm{sml}}$ are non-negative.

As we shall soon see, the contribution of f_{unf} can be quickly discarded using the generalised von Neumann theorem. If one could also easily discard the contribution of the small term f_{sml}, then matters would simply reduce to verifying that the contribution of f_{nil} is bounded away from zero, which would be an easy consequence of the counting lemma. Unfortunately the small term f_{sml} is only moderately small (of size $O(\varepsilon)$) rather than incredibly small (e.g. of size $O\big(1/\mathcal{F}(M)\big)$), and so one has to take a certain amount of care in dealing with this term, which makes the analysis significantly more delicate[14].

We turn to the details. Much as the key to proving Theorem 1.12 was to establish Proposition 5.2, the key to establishing Szemerédi's theorem is the following proposition.

Proposition 6.2 (Szemerédi for f_{nil}). *Let f_{nil} be as above, and let $\varepsilon > 0$. Then there exists a function $\mu : \mathbb{Z} \times \mathbb{Z} \to \mathbb{R}^+$ supported on the set*

$$\big\{ (n,d) \in \mathbb{Z}\times\mathbb{Z} : d \in [-\varepsilon N, \varepsilon N]; n+id \in [N] \text{ for all } i = 0,\dots,k-1 \big\} \tag{6.2}$$

with

$$\mathbb{E}_{n\in[N];d\in[-\varepsilon N,\varepsilon N]}\mu(n,d) = 1 + O(\varepsilon) \tag{6.3}$$

and with μ bounded in magnitude by $O_{M,\varepsilon}(1)$, such that

$$f_{\mathrm{nil}}(n+id) = f_{\mathrm{nil}}(n) + O(\varepsilon) \tag{6.4}$$

whenever $0 \leqslant i \leqslant k-1$ and $\mu(n,d) \neq 0$, and such that one has the equidistribution property

$$\mathbb{E}_{n\in[N]}\big|\mathbb{E}_{d\in[-\varepsilon N,\varepsilon N]}\mu(n-id,d)\big|^2 = 1 + O(\varepsilon) \tag{6.5}$$

for all $0 \leqslant i \leqslant k-1$.

The crucial feature of Proposition 6.2 for us is that, with the exception of the uniform bound on μ, the error terms here decay as $\varepsilon \to 0$, even if the complexity bound M on f_{nil} is extremely large compared to $1/\varepsilon$.

The reader may benefit from a few words about the role of the function μ. Supposing that $f_{\mathrm{nil}}(n) = F\big(g(n)\Gamma\big)$ is a genuine nilsequence, this function

[14]In the language of ergodic theory, the problem here is that the characteristic factor is not necessarily a nilsystem, but may merely be a *pro-nilsystem* - an inverse limit of nilsystems. A short, but not entirely trivial, argument of Furstenberg [11] shows that multiple recurrence is preserved under inverse limits. This argument was adapted with some difficulty to the finitary setting in [50]; our approach here is different and exploits some additional equidistribution properties of nilsystems, as well as using a carefully chosen weight $\mu(n,d)$.

acts like a kind of "weight" on progressions $(n, n+d, \ldots, n+(k-1)d)$ which are "almost diagonal" in the sense that $g(n)\Gamma \approx \cdots \approx g(n+(k-1)d)\Gamma$ in G/Γ. The condition (6.5) reflects the fact that the weighted number of almost diagonal progressions whose ith point is n is roughly independent of n. This "non-concentration" of almost diagonal progressions ultimately means that the error f_{sml} cannot destroy too many of these progressions, a fact that is crucial for our argument.

Let us assume Proposition 6.2 for now and see how it implies Theorem 6.1. We use (6.1) to expand out the form $\Lambda_k(f, \ldots, f)$ into 3^k terms. By Theorem 4.1, any term that involves f_{unf} will be of size $O(1/\mathcal{F}(M))$, thus

$$\Lambda_k(f, \ldots, f) = \Lambda_k(f_{\mathrm{nil}} + f_{\mathrm{sml}}, \ldots, f_{\mathrm{nil}} + f_{\mathrm{sml}}) + O(1/\mathcal{F}(M)). \tag{6.6}$$

Next, we use the weight μ arising from Proposition 6.2 and the non-negativity of $f_{\mathrm{nil}} + f_{\mathrm{sml}}$ guaranteed by Theorem 1.2 to write

$$\begin{aligned}&\Lambda_k(f_{\mathrm{nil}} + f_{\mathrm{sml}}, \ldots, f_{\mathrm{nil}} + f_{\mathrm{sml}})\\ &\gg_{M,\varepsilon} \mathbb{E}_{n\in[N];d\in[-\varepsilon N,\varepsilon N]}(f_{\mathrm{nil}} + f_{\mathrm{sml}})(n)\ldots(f_{\mathrm{nil}} + f_{\mathrm{sml}})(n+(k-1)d)\,\mu(n,d).\end{aligned}$$

We then expand this latter average into the sum of 2^k terms. The main term is

$$\mathbb{E}_{n\in[N];d\in[-\varepsilon N,\varepsilon N]} f_{\mathrm{nil}}(n)\ldots f_{\mathrm{nil}}(n+(k-1)d)\,\mu(n,d), \tag{6.7}$$

and the other terms are error terms, involving at least one factor of f_{sml}.

Consider one of the error terms, involving the factor $f_{\mathrm{sml}}(n+id)$ (say) for some $0 \leqslant i \leqslant k-1$. We can bound the contribution of this term by

$$\mathbb{E}_{n\in[N];\ d\in[-\varepsilon N,\varepsilon N]}\big|f_{\mathrm{sml}}(n+id)\big|\,\mu(n,d),$$

which by a change of variables $n \mapsto n - id$ we can write as

$$\mathbb{E}_{n\in[N]}\big|f_{\mathrm{sml}}(n)\big|\,\mathbb{E}_{d\in[-\varepsilon N,\varepsilon N]}\mu(n-id,d).$$

By Cauchy–Schwarz, (6.5), and the $L^2[N]$ bound on f_{sml}, this is $O(\varepsilon)$.

Finally, we look at the main term (6.7). Using (6.4) we can approximate

$$f_{\mathrm{nil}}(n)\ldots f_{\mathrm{nil}}(n+(k-1)d) = f_{\mathrm{nil}}(n)^k + O(\varepsilon)$$

and so (using (6.3)) we can write (6.7) as

$$\mathbb{E}_{n\in[N]} f_{\mathrm{nil}}(n)^k\,\mathbb{E}_{d\in[-\varepsilon N,\varepsilon N]}\mu(n,d) + O(\varepsilon).$$

Now, from (6.3) one has

$$\mathbb{E}_{n\in[N]}\mathbb{E}_{d\in[-\varepsilon N,\varepsilon N]}\mu(n,d) = 1 + O(\varepsilon)$$

and hence by (6.5)

$$\mathbb{E}_{n\in[N]}\left|\mathbb{E}_{d\in[-\varepsilon N,\varepsilon N]}\mu(n,d) - 1\right|^2 = O(\varepsilon).$$

In particular, by Chebyshev's inequality, we have

$$\mathbb{E}_{d\in[-\varepsilon N,\varepsilon N]}\mu(n,d) = 1 + O\big(\varepsilon^{1/3}\big)$$

for all $n \in E$, where $E \subseteq [N]$ has cardinality $|E| \geqslant \big(1 - O\big(\varepsilon^{1/3}\big)\big)N$. Thus, for ε small enough, we can bound (6.7) from below by

$$\gg \mathbb{E}_{n\in[N]}1_E(n)f_{\mathrm{nil}}(n)^k - O\big(\varepsilon^{1/3}\big).$$

Now from hypothesis we have $\mathbb{E}_{n\in[N]}f(n) \gg 1$. From Cauchy–Schwarz we have

$$\mathbb{E}_{n\in[N]}f_{\mathrm{sml}}(n) = O(\varepsilon),$$

and from Theorem 4.1 we also have

$$\mathbb{E}_{n\in[N]}f_{\mathrm{unf}}(n) = O(\varepsilon)$$

if $\mathcal{F}$ is rapid enough. Thus if ε is small enough we have $\mathbb{E}_{n\in[N]}f_{\mathrm{nil}}(n) \gg 1$, which implies that $\mathbb{E}_{n\in[N]}1_E(n)f_{\mathrm{nil}}(n) \gg 1$, and hence by Hölder's inequality that $\mathbb{E}_{n\in[N]}1_E(n)f_{\mathrm{nil}}^k(n) \gg 1$. Putting all this together, we conclude that (6.7) is $\gg 1$ if ε is small enough, and thus

$$\Lambda_k(f_{\mathrm{nil}} + f_{\mathrm{sml}}, \ldots, f_{\mathrm{nil}} + f_{\mathrm{sml}}) \gg_{M,\varepsilon} 1.$$

Inserting this bound into (6.6) we obtain the claim, completing the proof of Szemerédi's theorem, if $\mathcal{F}$ is chosen sufficiently rapid.

Proof of Proposition 6.2. Let us first establish this in the easy case $k = 3$. In this case, f_{nil} is essentially quasiperiodic, which will allow us to take $\mu(n,d)$ to be of the form

$$\mu(n,d) = 1_{[2\varepsilon N,(1-2\varepsilon)N]}(n)\mu(d)$$

with $\mu(d)$ normalised by requiring

$$\mathbb{E}_{d\in[-\varepsilon N,\varepsilon N]}\mu(d) = 1 + O(\varepsilon).$$

It is then easy to verify that both (6.3) and (6.5) follow from this normalisation. To establish the remaining claims in Proposition 6.2, we use the degree $\leqslant 1$ nature of the orbit $n \mapsto g(n)\Gamma$ as in Section 5 to write f_{nil} as

$$f_{\mathrm{nil}}(n) = F(n\theta)$$

for some $\theta \in (\mathbb{R}/\mathbb{Z})^D$ with $D = O_M(1)$ and some $F : (\mathbb{R}/\mathbb{Z})^D \to \mathbb{C}$ of Lipschitz constant $O_M(1)$. If one then sets μ to equal

$$\mu(d) := \frac{\big|[-\varepsilon N, \varepsilon N]\big|}{|B|} 1_B(d)$$

where B is the Bohr set

$$\Big\{d \in [-\varepsilon N, \varepsilon N] \,:\, d_{(\mathbb{R}/\mathbb{Z})^D}(d\theta, 0) \leqslant \delta\Big\}$$

and $\delta > 0$ is sufficiently small depending on ε, M, one easily verifies all the required claims.

We now turn to the case $k > 3$, which is harder because f_{nil} is no longer quasiperiodic, and so $\mu(n,d)$ will have to depend more heavily on n and not just on d. By arguing as in the previous section we can normalise $g(0)$ to equal id. We may also assume N is sufficiently large depending on ε, M, since otherwise we may simply take $\mu(n,d) = 1_{[N]}(n)\delta_0(d)$ where δ_0 is the Kronecker delta function at 0. We may of course also assume that $\varepsilon < 1$.

We take an $O_M(1)$-rational Mal'cev basis $X_1, \ldots, X_{\dim(G)}$ for the Lie algebra $\mathfrak{g} = \log G$ adapted to the filtration $G_\bullet$ as described in [30, Appendix A]. For any radius $r > 0$, we define the "ball" B_r in G to be the set of all group elements of the form

$$\exp\Bigg(\sum_{j=1}^{\dim(G)} t_j X_j\Bigg) \tag{6.8}$$

where the t_j are real numbers with $t_j \leqslant r^{s+1-i}$ whenever $1 \leqslant i \leqslant s$ and $j \leqslant \dim(G) - \dim\big(G_{(i)}\big)$. Thus, when r is small, B_r is quite "narrow" (of diameter comparable to r^s) when projected down to $G/G_{(2)}$, but is relatively large when restricted to the top order component $G_{(s)}$ (of diameter comparable to r). This type of eccentricity is necessary in order to make B_r approximately "normal" with respect to conjugations. Indeed, we have

Lemma 6.3 (Approximate normality). *Let $A, \delta > 0$, and let $g \in G$ be such that $d_G(g, \mathrm{id}) \leqslant A$. Then we have the containments*

$$B_{(1-\delta)r} \subseteq gB_rg^{-1} \subseteq B_{(1+\delta)r}. \tag{6.9}$$

whenever $r > 0$ is sufficiently small depending on A, δ, M.

Proof. We prove the second inclusion only, as the first is similar (and can also be deduced from the second). The conjugation action $h \mapsto ghg^{-1}$ on G induces a Lie algebra automorphism $\exp\big(\operatorname{ad}(\log g)\big) : \mathfrak{g} \to \mathfrak{g}$. If we conjugate the group element (6.8) by g, we thus obtain

$$\exp\Big(\sum_{j=1}^{\dim(G)} t_j \exp\big(\operatorname{ad}(\log g)\big)(X_j)\Big).$$

But if $1 \leqslant i \leqslant s$ and $j \leqslant \dim(G) - \dim\big(G_{(i)}\big)$, we see from the Baker–Campbell–Hausdorff formula (C.2) that

$$\exp\big(\operatorname{ad}(\log g)\big)(X_j) = X_j + \sum_{j'=\dim(G)-\dim(G_{(i)})+1}^{\dim(G)} c_{j,j'}X_{j'}$$

for some coefficients $c_{j,j'}$ of size $O_{A,M}(r^{s+1-i})$. Collecting all the coefficients together, we obtain the claim for r small enough. ∎

Let $0 < \delta < 1/10$ be a small quantity (depending on ε, M), let R be a large quantity depending on the same parameters, and let $r_0 > 0$ be an even smaller[15] quantity than δ (depending on $\varepsilon, M, \delta, R$) to be chosen later. For each r with $0 < r < r_0$ take a Lipschitz function $\phi_r : G \to \mathbb{R}^+$ of Lipschitz norm $O_{M,r,\delta}(1)$ which is supported on B_r and equals one on $B_{(1-\delta)r}$, and choose these functions so that $\phi_r \leqslant \phi_{r'}$ pointwise whenever $0 < r < r' < r_0$. For each such r, let $\Phi_r : G/\Gamma \times G/\Gamma \to \mathbb{R}^+$ be the induced function

$$\Phi_r(x, x') := \sum_{g \in G\,:\, gx = x'} \phi_r(g).$$

[15]Readers may find it helpful to keep the hierarchy of scales

$$1 \sim 1/k, \alpha \gg \varepsilon \gg 1/M \gg \delta \gg 1/R \gg r_0 \gg r \gg 1/\mathcal{F}(M) \gg 1/N > 0$$

in mind.

This function Φ_r is supported near the diagonal of $G/\Gamma \times G/\Gamma$; indeed, $\Phi_r(x,x')$ is only non-zero when $x' \in B_r x$, and furthermore if $x' \in B_{(1-\delta)r}x$ then $\Phi_r(x,x') = 1$. If r_0 is chosen sufficiently small depending on M, δ, we conclude from Lemma 6.3 that we have the approximate shift-invariance

$$\Phi_{(1-3\delta)r}(x,x') \leqslant \Phi_r(gx,gx') \leqslant \Phi_{(1+3\delta)r}(x,x') \tag{6.10}$$

whenever $x,x' \in G/\Gamma$ and $g \in G$ is such that $d_G(g,\mathrm{id}) \leqslant R$ (say).

We now define our cutoff function $\mu = \mu_r$ by

(6.11)

$$\mu_r(n,d) := c_r 1_{q|d} 1_{[k\varepsilon N,(1-k\varepsilon)N]}(n) 1_{[-\delta N,\delta N]}(d) \prod_{i=1}^{k-1} \Phi_r\big(g(n)\Gamma, g(n+id)\Gamma\big),$$

where $c_r > 0$ is a normalisation constant to be chosen later. This function, as discussed immediately following the statement of Proposition 6.2, is a smooth cutoff to the set of "almost-diagonal" progressions in G/Γ. Specifically, μ_r is supported in (6.2), and also in the region where $g(n+id)\Gamma \in B_r g(n)\Gamma$, $|d| \leqslant \delta N$, and $q|d$ for $i = 0,\dots,k-1$. From the Lipschitz nature of F we thus have

$$F\big(g(n+id)\Gamma, (n+id)(\mathrm{mod}\ q), (n+id)/N\big)$$
$$= F\big(g(n)\Gamma, n(\mathrm{mod}\ q), n/N\big) + O_M(r_0)$$

for (n,d) in the support of μ_r, which gives (6.4) for μ_r if r_0 is sufficiently small depending on ε, M.

Next, we compute the expectation of $\mu_r(n,d)$, in order to work out what the normalisation constant c_r should be. Observe that

$$\begin{aligned} \text{(6.12)}\quad & \mathbb{E}_{n\in[N],d\in[-\varepsilon N,\varepsilon N]}\mu_r(n,d) \\ &= \frac{\delta}{q\varepsilon}\big(1+O(\varepsilon)\big)\, c_r \\ &\times \mathbb{E}_{n\in[k\varepsilon N,(1-k\varepsilon)N];d\in[-\delta N,\delta N];q|d}\tilde{\Phi}_r\big(g(n)\Gamma,\dots,g(n+(k-1)d)\,\Gamma\big), \end{aligned}$$

where $\tilde{\Phi}_r : (G/\Gamma)^k \to \mathbb{R}^+$ is the function

$$\tilde{\Phi}_r(x_0,\dots,x_{d-1}) := \prod_{i=1}^{k-1} \Phi_r(x_0,x_i). \tag{6.13}$$

Observe that $\tilde{\Phi}$ has a Lipschitz norm of $O_{M,r,\delta}(1)$. Applying Theorem 1.11, we can express (6.12) as

$$\frac{\delta}{q\varepsilon}\big(1+O(\varepsilon)\big)\,c_r\left(\int_{G^\Psi/\Gamma^\Psi}\tilde{\Phi}_r + o_{\mathcal{F}(M)\to\infty;M,r,\delta}(1) + o_{N\to\infty;M,r,\delta}(1)\right),$$

where $G^\Psi \subseteq G^k$ is the kth Hall-Petresco group, that is to say the Leibman group associated to the collection $\Psi = (\psi_0,\dots,\psi_{k-1})$ of linear forms $\Psi^{(i)} := (n,d)\mapsto n+id$ for $i=0,\dots,k-1$.

The group G^Ψ is a $O_M(1)$-rational subgroup of G^k, which itself has complexity $O_M(1)$. Meanwhile, the function $\tilde{\Phi}_r$ equals 1 on a ball of radius $r^{O_M(1)}$ centred at the identity, and is bounded by 1 throughout. We conclude that the quantity

$$v_r := \int_{G^\Psi/\Gamma^\Psi}\tilde{\Phi}_r$$

obeys the bounds

$$r^{O_M(1)} \ll_M v_r \leqslant 1.$$

Furthermore, from the properties of the functions ϕ_r, we have the monotonicity property

$$v_{(1-\delta)r} \leqslant v_r$$

for any $0<r<r_0$. Applying the pigeonhole principle (using the fact that polynomial growth is always slower than exponential growth), and choosing $\delta \gg_{\varepsilon,M} 1$ sufficiently small depending on ε, M, one can thus find a radius

$$r_0 > r \gg_{r_0,\varepsilon,\delta,M} 1$$

such that we have the regularity property

$$\big(1-O(\varepsilon)\big)\,v_r \leqslant v_{(1-3\delta)r} \leqslant v_{(1+3\delta)r} \leqslant \big(1+O(\varepsilon)\big)\,v_r. \tag{6.14}$$

Note that this idea of picking a "regular" radius originates, in additive combinatorics, in Bourgain's paper [8]. Fix from now on a value of r with this property. If we then set

$$c_r := \frac{q\varepsilon}{\delta v_r} \tag{6.15}$$

we conclude that

$$c_r \ll_{M,r_0,\varepsilon} 1 \tag{6.16}$$

and

$$\mathbb{E}_{n\in[N],\ d\in[-\varepsilon N,\varepsilon N]}\mu_r(n,d) = 1+O(\varepsilon)+o_{\mathcal{F}(M)\to\infty;M,\varepsilon,r_0}(1)+o_{N\to\infty;M,\varepsilon,r_0}(1).$$

This will give (6.3) provided that r_0 is chosen to depend on M, ε, δ, that $\mathcal{F}$ is sufficiently rapid depending on ε, and N is sufficiently large depending on M, ε.

Our remaining task, and the most difficult one, is to study the expression in (6.5). That is to say, we fix $0 \leqslant i \leqslant k-1$ and consider

$$\mathbb{E}_{n\in[N]}\left|\mathbb{E}_{d\in[-\varepsilon N,\varepsilon N]}\mu_r(n-id,d)\right|^2. \tag{6.17}$$

Using (6.11), we can write this expression as

$$\begin{aligned}(1+O(\varepsilon))\left(\frac{\varepsilon}{q\delta}c_r\right)^2 &\mathbb{E}_{n\in[k\varepsilon N,(1-k\varepsilon)N]}\mathbb{E}_{d,d'\in[-\delta N,\delta N];\ q|d,d'}\\ &\tilde{\Phi}_r^{\otimes 2}\big(g(n-id)\Gamma,\dots,g\big(n+(k-1-i)d\big)\Gamma,\\ &\qquad g(n-id')\Gamma,\dots,g\big(n+(k-1-i)d'\big)\Gamma\big)\end{aligned}$$

where $\tilde{\Phi}_r^{\otimes 2} : (G/\Gamma)^k \times (G/\Gamma)^k \to \mathbb{R}^+$ is the tensor square

$$\tilde{\Phi}_r^{\otimes 2}(x,x') := \tilde{\Phi}_r(x)\tilde{\Phi}_r(x').$$

Applying Theorem 1.11, we can thus express (6.17) as

$$\begin{aligned}&(1+O(\varepsilon))\left(\frac{\varepsilon}{q\delta}c_r\right)^2\\ &\quad\times\left(\int_{G^{\Psi^{(i)}}/\Gamma^{\Psi^{(i)}}}\tilde{\Phi}_r^{\otimes 2}+o_{\mathcal{F}(M)\to\infty;\varepsilon,M,r_0}(1)+o_{N\to\infty;\varepsilon,M,r_0}(1)\right)\end{aligned} \tag{6.18}$$

where $G^{\Psi^{(i)}} \subset G^{2k}$ is the Leibman group associated to the collection

$$\Psi^{(i)} := (\psi_{0,i},\dots,\psi_{k-1,i},\psi'_{0,i},\dots,\psi'_{k-1,i})$$

of linear forms

$$\psi_{j,i} : (n,d,d') \mapsto n+(j-i)d$$

and

$$\psi'_{j,i} : (n,d,d') \mapsto n+(j-i)d'$$

for $j=0,\dots,k-1$.

We will be establishing the following claim.

Claim 6.4 (Approximate factorisation). *We have*

$$\int_{G^{\Psi^{(i)}}/\Gamma^{\Psi^{(i)}}} \tilde{\Phi}_r^{\otimes 2} = \left(1 + O(\varepsilon)\right) v_r^2. \tag{6.19}$$

Proof of Proposition 6.2 assuming Claim 6.4. Substitute back into (6.18) and use (6.15), (6.16) to conclude that

$$1(6.17) = 1 + O(\varepsilon) + o_{\mathcal{F}(M)\to\infty;\varepsilon,M,r_0}(1) + o_{N\to\infty;\varepsilon,M,r_0}(1).$$

This gives the result upon choosing r_0 sufficiently small depending on ε, M, δ, $\mathcal{F}$ sufficiently rapid depending on ε, and N sufficiently large depending on ε, M.

It remains to establish Claim 6.4. For notational simplicity we establish only the claim $i = 0$ (the others being very similar). The intuition behind this claim (and behind the key assertion that the number of almost-diagonal progressions whose i^{th} term is n does not depend on n) is that the linear forms $(\psi_{0,0}, \dots, \psi_{k-1,0})$ and $(\psi'_{0,0}, \dots, \psi'_{k-1,0})$ are almost independent of each other, except for the fact that they are coupled via the obvious identity $\psi_{0,0} = \psi'_{0,0}$.

One way to encode this formally is to note that the Leibman group $G^{\Psi^{(0)}}$ is given by

$$H := \left\{ (x, x') \in G^{\Psi} \times G^{\Psi} \,:\, x_0 = x'_0 \right\},$$

a product of two copies of the Hall–Petresco group $G^{\Psi} = \mathrm{HP}^k(G)$ fibred over the zeroth coordinate. To prove this, one may note that the containment $G^{\Psi^{(0)}} \subseteq H$ is obvious. On the other hand, one may compute directly using the dimension formula (3.1) that

$$\dim\left(G^{\Psi}\right) = \dim\left(G\right) + \sum_{i=1}^{k-2} \dim\left(G^{(i)}\right)$$

and

$$\dim\left(G^{\Psi^{(0)}}\right) = \dim\left(G\right) + 2\sum_{i=1}^{k-2} \dim\left(G^{(i)}\right)$$

and thus

$$\dim\left(G^{\Psi^{(0)}}\right) = 2\dim\left(G^{\Psi}\right) - \dim\left(G\right) = \dim\left(H\right),$$

and so since both sides are connected, simply-connected nilpotent Lie groups (and so both are homeomorphic to their Lie algebras) we have $G^{\Psi^{(0)}} = H$.

Write J_r for the integral appearing in (6.19), that is to say

$$J_r := \int_{(x,x')\in G^\Psi/\Gamma^\Psi\times G^\Psi/\Gamma^\Psi\,:\,x_0=x_0'} \tilde{\Phi}_r^{\otimes 2}(x,x').$$

Let R be some quantity, and suppose that $\mathrm{dist}_G(g,\mathrm{id}) \leqslant R$. Then by the almost-invariance property (6.10) we have

$$\int_{(x,x')\in G^\Psi/\Gamma^\Psi\times G^\Psi/\Gamma^\Psi\,:\,x_0=gx_0'} \tilde{\Phi}_{r(1+3\delta)}^{\otimes 2}(x,x') \geqslant J_r.$$

Integrate this over the ball $B_R := \{g \in G : \mathrm{dist}_G(g,\mathrm{id}) \leqslant R\}$. Then we obtain

$$\int_{(x,x')\in(G^\Psi/\Gamma^\Psi)^2} \lambda(x,x')\tilde{\Phi}_{r(1+3\delta)}^{\otimes 2}(x,x') \geqslant \mathrm{vol}\,(B_R)J_r,$$

where $\lambda(x,x')$ is the number of $g \in B_R$ for which $x_0 = gx_0' (\mathrm{mod}\ \Gamma)$, or equivalently

$$\lambda(x,x') := |\Gamma \cap x_0^{-1}B_R x_0'|.$$

Choose representatives x_0, x_0' in some fundamental domain with $x_0, x_0' = O_M(1)$. By a volume-packing argument and simple geometry we then have

$$\lambda(x,x') = \mathrm{vol}\,(B_R)\big(1 + o_{R\to\infty;M}(1)\big).$$

Comparing with the above we have

$$v_{r(1-3\delta)}^2 = \int_{(x,x')\in(G^\Psi/\Gamma^\Psi)^2} \tilde{\Phi}_{r(1+3\delta)}^{\otimes 2} \geqslant J_r\big(1 + o_{R\to\infty;M}(1)\big),$$

and so by (6.14) we have

$$J_r \leqslant \big(1 + O(\varepsilon) + o_{R\to\infty;M}(1)\big)\, v_r^2.$$

This gives the upper bound for Claim 6.4. The lower bound is proven similarly. This concludes the proof of Proposition 6.2 and thus Theorem 6.1.

7. On a Theorem of Gowers and Wolf

Our aim in this section is to prove Theorem 1.13, whose statement we recall now.

Theorem 7.1 (Theorem 1.13). *Let $\Psi = (\psi_1, \dots, \psi_t)$ be a collection of linear forms $\psi_1, \dots, \psi_t : \mathbb{Z}^D \to \mathbb{Z}$, and let $s \geqslant 1$ be an integer such that the polynomials $\psi_1^{s+1}, \dots, \psi_t^{s+1}$ are linearly independent. Then for any function $f : [N] \to \mathbb{C}$ bounded in magnitude by 1 (and defined to be zero outside of $[N]$) obeying the bound $\|f\|_{U^{s+1}[N]} \leqslant \delta$ for some $\delta > 0$, one has*

$$\mathbb{E}_{\mathbf{n}\in[N]^D} \prod_{i=1}^{t} f(\psi_i(\mathbf{n})) = o_{\delta\to 0;s,D,t,\Psi}(1).$$

Henceforth we allow all implied constants to depend on d, t, s, Ψ without indicating this explicitly. Let $s' = s'(\Psi)$ be the Cauchy–Schwarz complexity of the linear forms Ψ, as defined in Theorem 4.1. We may of course assume that $s' > s$, as Theorem 1.13 is immediate otherwise. We may also assume that N is large depending on δ, since otherwise the claim is trivial from a compactness argument.

Let $\varepsilon > 0$ be a small number depending on δ to be chosen later, and let $\mathcal{F}$ be a growth function depending on ε to be chosen later. Applying Theorem 1.2 at degree s' (after first decomposing f as a linear combination of $O(1)$ functions taking values in $[0, 1]$), we can find a positive quantity $M = O_{\varepsilon,\mathcal{F}}(1)$ and a decomposition

$$f = f_{\mathrm{nil}} + f_{\mathrm{sml}} + f_{\mathrm{unf}} \tag{7.1}$$

where:

(i) f_{nil} is a $(\mathcal{F}(M), N)$-irrational virtual nilsequence of degree $\leqslant s'$, complexity $\leqslant M$, and scale N;
(ii) f_{sml} has $L^2[N]$ norm at most ε;
(iii) f_{unf} has $U^{s'+1}[N]$ at most $1/\mathcal{F}(M)$;
(iv) All functions $f_{\mathrm{nil}}, f_{\mathrm{sml}}, f_{\mathrm{unf}}$ are bounded in magnitude by $O(1)$.

We apply this decomposition to split the expression

$$\mathbb{E}_{\mathbf{n}\in[N]^D} \prod_{i=1}^{t} f(\psi_i(\mathbf{n})) \tag{7.2}$$

as the sum of 3^t terms, in which each copy of f has been replaced with either f_{nil}, f_{sml}, or f_{unf}.

Any term involving at least one factor of f_{sml} can be easily seen to be of size $O(\varepsilon)$ by crudely estimating all other factors by 1. By (4.1), any term involving at least one factor of f_{unf} is of size $O\big(1/\mathcal{F}(M)\big)$, which is also of size $O(\varepsilon)$ if $\mathcal{F}$ is chosen to be sufficiently rapidly growing depending on ε. We can therefore express (7.2) as

$$\mathbb{E}_{\mathbf{n}\in[N]^D} \prod_{i=1}^{t} f_{\mathrm{nil}}\big(\psi_i(\mathbf{n})\big) + O(\varepsilon).$$

By hypothesis, we can write

$$f_{\mathrm{nil}}(n) = F\big(g(n)\Gamma, n(\mathrm{mod}\ q), n/N\big)$$

for some q with $1 \leqslant q \leqslant M$, some degree $\leqslant s$, $\big(\mathcal{F}(M), N\big)$-irrational, orbit $n \mapsto g(n)\Gamma$ of complexity $\leqslant M$ and some Lipschitz function $F : G/\Gamma \times \mathbb{Z}/q\mathbb{Z} \times \mathbb{R}$ of norm at most M. The mod q and Archimedean behaviour in f_{nil} are nothing more than technical annoyances, and we set about eliminating them now. We encourage the reader to work through the heart of the argument, starting at (7.3) below, in the model case $f_{\mathrm{nil}} = F\big(g(n)\Gamma\big)$. Let ε' be a small quantity depending on ε, M to be chosen later[16]. We partition $[N]$ into progressions P of spacing q and length $\varepsilon' N$, plus a remainder set of size at most $O_M(1)$. We can then rewrite the above expression as

$$\mathbb{E}_{P_1,\dots,P_D} \mathbb{E}_{\mathbf{n}\in P_1\times\dots\times P_D} \prod_{i=1}^{t} f_{\mathrm{nil}}\big(\psi_i(\mathbf{n})\big) + O(\varepsilon).$$

We abbreviate $P_1 \times \ldots \times P_D$ as $\mathbf{P}$. For a given $\mathbf{P}$, observe that as $\mathbf{n}$ ranges in $\mathbf{P}$, the residue class of $\psi_i(\mathbf{n})$ modulo q is equal to a fixed class $a_{\mathbf{P},i}$, and the value of $\psi_i(\mathbf{P})/N$ differs by at most $O_M(\varepsilon')$ from a fixed number $x_{\mathbf{P},i}$. We may assume that $x_{\mathbf{P},i} \in [0,1]$ for each i, otherwise the inner expectation is zero (except for a few "boundary" values of $\mathbf{P}$ which give a net contribution of $O_M(\varepsilon')$).

[16] Readers may find it helpful to keep the hierarchy of scales

$$1 \gg \varepsilon \gg 1/M, 1/q \gg \varepsilon' \gg 1/\mathcal{F}(M) \gg \delta \gg 1/N > 0$$

in mind.

If ε' is small enough depending on ε, M, the $O_M(\varepsilon')$ error in the above discussion can be absorbed in the $O(\varepsilon)$ error, and so we have

$$\mathbb{E}_{\mathbf{n}\in[N]^D}\prod_{i=1}^t f(\psi_i(\mathbf{n})) = \mathbb{E}_{\mathbf{P}}\mathbb{E}_{\mathbf{n}\in\mathbf{P}}\prod_{i=1}^t F\big(g(\psi_i(\mathbf{n}))\Gamma, a_{\mathbf{P},i}, x_{\mathbf{P},i}\big) + O(\varepsilon).$$

We now apply Theorem 1.11, which tells us the the right-hand side here is

$$\mathbb{E}_{\mathbf{P}}\int_{G^\Psi/\Gamma^\Psi}\tilde{F}_{\mathbf{P}} + O(\varepsilon) + o_{\mathcal{F}(M)\to\infty;M,\varepsilon,\varepsilon'}(1) + o_{N\to\infty;M,\varepsilon,\varepsilon'}(1), \tag{7.3}$$

where as usual $G^\Psi \leqslant G^t$ is the Leibman group associated to the system of forms $\Psi = \{\psi_1,\dots,\psi_t\}$, and here $\tilde{F}_{\mathbf{P}} : G^\Psi/\Gamma^\Psi \to \mathbb{C}$ is the function

$$\tilde{F}_{\mathbf{P}}\big((g_1,\dots,g_t)\Gamma^\Psi\big) := \prod_{i=1}^t F\big(g_i\Gamma, a_{\mathbf{P},i}, x_{\mathbf{P},i}\big).$$

The heart of the matter is to obtain an upper bound on the quantity $\mathbb{E}_{\mathbf{P}}\int_{G^\Psi/\Gamma^\Psi}\tilde{F}_{\mathbf{P}}$ appearing in (7.3). To do this, of course, we need to make use of the assumption on the forms $\psi_1,\dots,\psi_t$, as well as the fact that $\|f\|_{U^{s+1}} \leqslant \delta$.

The aforementioned assumption, namely that $\psi_1^{s+1},\dots,\psi_t^{s+1}$ are linearly independent, implies that $\Psi^{[s+1]}$ is the whole of $\mathbb{R}^t$ which, in view of the definition of the Leibman group G^Ψ, implies that $G^t_{(s+1)} \leqslant G^\Psi$. By Fubini's theorem, we thus have

$$\int_{G^\Psi/\Gamma^\Psi}\tilde{F}_{\mathbf{P}} = \int_{G^\Psi/\Gamma^\Psi}\tilde{F}_{\mathbf{P},\leqslant s}$$

where

$$\tilde{F}_{\mathbf{P},\leqslant s}\big((g_1,\dots,g_t)\Gamma^\Psi\big) := \prod_{i=1}^t F_{\leqslant s}\big(g_i\Gamma, a_{\mathbf{P},i}, x_{\mathbf{P},i}\big) \tag{7.4}$$

and $F_{\leqslant s}$ is defined by averaging over cosets of the normal subgroup $G_{(s+1)}$, specifically

$$F_{\leqslant s}(g\Gamma, a, x) := \int_{G_{(s+1)}/\Gamma_{(s+1)}} F(gg_{s+1}\Gamma, a, x)\; dg_{s+1}.$$

Since F was Lipschitz with norm $O_M(1)$, we see that $F_{\leqslant s}$ is Lipschitz with norm $O_M(1)$ also. Also, since F is bounded in magnitude by $O(1)$, so is $F_{\leqslant s}$.

As the forms $\psi_1^{s+1}, \ldots, \psi_t^{s+1}$ are independent, we see in particular that ψ_1 is non-zero. This implies that the projection of G^Ψ to the first coordinate G is surjective. Meanwhile, from (7.4) and the boundedness of $F_{\leqslant s}$ we have the crude upper bound

$$\left|\tilde{F}_{\mathbf{P},\leqslant s}\big((g_1,\ldots,g_t)\Gamma\big)\right| \ll \left|F_{\leqslant s}\big(g_1\Gamma, a_{\mathbf{P},1}, x_{\mathbf{P},1}\big)\right|.$$

From Fubini's theorem, we obtain the bound

$$\left|\int_{G^\Psi/\Gamma^\Psi} \tilde{F}_{\mathbf{P}}\right| \ll \int_{G/\Gamma} \left|F_{\leqslant s}\big(\cdot, a_{\mathbf{P},1}, x_{\mathbf{P},1}\big)\right|. \tag{7.5}$$

To proceed further, we need a crucial smallness estimate on $F_{\leqslant s}$:

Proposition 7.2 ($F_{\leqslant s}$ small in L^2). *For any $a \in \mathbb{Z}/q\mathbb{Z}$ and $x \in [0,1]$, one has*

$$\int_{G/\Gamma} \left|F_{\leqslant s}(\cdot, a, x)\right|^2 \ll O(\varepsilon) + O_M(\varepsilon')$$
$$+ o_{\delta\to\infty;M,\varepsilon,\varepsilon'}(1) + o_{\mathcal{F}(M)\to\infty;M,\varepsilon,\varepsilon'}(1) + o_{N\to\infty;M,\varepsilon,\varepsilon'}(1).$$

Proof. By reflection symmetry we may assume that $x \leqslant 1/2$. We may also round x so that $x = qn_0/N$ for some $n_0 \in [N/2q]$, as the error in doing so can be easily absorbed by the Lipschitz properties of $F_{\leqslant s}$.

By construction, $F_{\leqslant s}$ is invariant on $G_{(s+1)}$-cosets, while $F - F_{\leqslant s}$ integrates to zero on any such coset. In particular, $F_{\leqslant s}(\cdot, a, x)$ and $F - F_{\leqslant s}(\cdot, a, x)$ are orthogonal, and thus

$$\int_{G/\Gamma} \left|F_{\leqslant s}(\cdot, a, x)\right|^2 = \int_{G/\Gamma} F\overline{F_{\leqslant s}}(\cdot, a, x).$$

Applying Theorem 1.11 (really just the special case of this result asserting that $\big(g(n)\Gamma\big)$ is equidistributed, cf. Lemma 3.7) and the Lipschitz nature of $F\overline{F_{\leqslant s}}$, the right-hand side can be written as

$$\mathbb{E}_{n\in[\varepsilon' N]} F\overline{F_{\leqslant s}}\big(g(qn+qn_0+a)\Gamma, a, x\big) + o_{\mathcal{F}(M)\to\infty;M,\varepsilon,\varepsilon'}(1) + o_{N\to\infty;M,\varepsilon,\varepsilon'}(1).$$

Let P be the progression $\left\{qn + qn_0 + a \,:\, n \in [\varepsilon' N]\right\}$. Then by a further use of the Lipschitz properties of F, we can rewrite the above expression as

$$\mathbb{E}_{n\in P} F\big(g(n)\Gamma, n \bmod q, n/N\big)\,\psi(n) + O_M(\varepsilon') \tag{7.6}$$
$$+ o_{\mathcal{F}(M)\to\infty;M,\varepsilon,\varepsilon'}(1) + o_{N\to\infty;M,\varepsilon,\varepsilon'}(1)$$

where

$$\psi(n) := \overline{F_{\leqslant s}}\big(g(n)\Gamma, a, x\big).$$

Note that, as a consequence of the $G_{(s+1)}$-invariance of $F_{\leqslant s}$, $\psi(n)$ is a degree $\leqslant s$ nilsequence of complexity $O_M(1)$. Now by (7.1) we have

$$F\big(g(n)\Gamma, n \bmod q, n/N\big) = f(n) - f_{\mathrm{unf}}(n) - f_{\mathrm{sml}}(n).$$

The contribution of $f_{\mathrm{sml}}(n)$ to (7.6) is $O(\varepsilon)$ by the Cauchy–Schwarz inequality. Now consider the contribution of f. Observe that because $F_{\leqslant s}$ is $G_{(s+1)}$-invariant, ψ is a degree $\leqslant s$ nilsequence of complexity $O_M(1)$. Meanwhile, $\|f\|_{U^{s+1}[N]} \leqslant \delta$ by hypothesis. Applying the converse to the inverse conjecture for the Gowers norms (first established in [28], though for a simple proof see [33, Appendix G]), we see that

$$\mathbb{E}_{n \in P} f(n)\psi(n) = o_{\delta \to 0; M, \varepsilon, \varepsilon'}(1).$$

Similarly, since $\|f_{\mathrm{unf}}\|_{U^{s'+1}[N]} \leqslant 1/\mathcal{F}(M)$ and $s' \geqslant s$, we have

$$\mathbb{E}_{n \in P} f(n)\psi(n) = o_{\mathcal{F}(M) \to 0; M, \varepsilon, \varepsilon'}(1).$$

Putting all of these estimates together, we obtain the claim. ∎

Applying this bound and (7.5), we can thus bound (7.3) in magnitude by

$$O(\varepsilon) + O_M(\varepsilon') + o_{\delta \to \infty; M, \varepsilon, \varepsilon'}(1) + o_{\mathcal{F}(M) \to \infty; M, \varepsilon, \varepsilon'}(1) + o_{N \to \infty; M, \varepsilon, \varepsilon'}(1).$$

Choosing ε' sufficiently small depending on M and ε, and choosing $\mathcal{F}$ sufficiently rapidly growing depending on ε, and then using the bound $M = O_{\varepsilon, \mathcal{F}}(1)$ (and recalling that N can be chosen large depending on δ), we conclude that

$$\Big|\mathbb{E}_{\mathbf{n} \in [N]^D} \prod_{i=1}^{t} f\big(\psi_i(\mathbf{n})\big)\Big| \ll \varepsilon$$

whenever δ is sufficiently small depending on ε. Theorem 1.13 follows.

Remark. It seems certain that one can extend this result to the case when one has t distinct functions $f_1, \dots, f_t : [N] \to \mathbb{C}$ rather than a single function $f : [N] \to \mathbb{C}$. The main change in the argument would be to use a version of the regularity lemma (Theorem 1.2) valid for several functions simultaneously, in which one regularises the $f_1, \dots, f_t$ using the same data M, q, $(G/\Gamma, G_\bullet)$, $g()$ (but allows each function f_i to be given a separate Lipschitz function $F_i : G/\Gamma \times \mathbb{Z}/q\mathbb{Z} \times \mathbb{R} \to \mathbb{C}$). Such a result could be obtained by straightforward modifications to the proof of Theorem 1.2, but we do not pursue this matter here.

Appendix A. Properties of Polynomial Sequences

In this appendix we collect a variety of facts and definitions concerning polynomial sequences in nilpotent groups, all of which were required at some point in the paper proper. We take for granted the definition of filtration $G_\bullet$ and of the group $\mathrm{poly}(\mathbb{Z}^d, G_\bullet)$ of polynomial sequences $g : \mathbb{Z}^d \to G$ adapted to $G_\bullet$; these notions were recalled in the introduction.

Taylor expansions. Polynomial sequences may be described in terms of so-called Taylor expansions. In the lemma that follows we make use of the generalised binomial coefficients $\binom{\mathbf{n}}{\mathbf{i}}$ are the generalised binomial coefficients

$$\binom{(n_1, \dots, n_D)}{(i_1, \dots, i_D)} := \binom{n_1}{i_1} \cdots \binom{n_D}{i_D}$$

where

$$\binom{n}{i} := \frac{n(n-1)\dots(n-i+1)}{i!}.$$

If $\mathbf{i} = (i_1, \dots, i_d) \in \mathbb{N}^D$ is a D-tuple of non-negative integers we define the degree $|\mathbf{i}| := i_1 + \dots + i_D$. Choose an arbitrary ordering on $\mathbb{N}^D$ with the property that $|\mathbf{i}| \geqslant |\mathbf{j}|$ whenever $\mathbf{i} \geqslant \mathbf{j}$.

Lemma A.1 (Taylor expansions). *Suppose that* $g \in \mathrm{poly}(\mathbb{Z}^D, G_\bullet)$. *Then there are unique* Taylor coefficients $g_\mathbf{i} \in G_{|\mathbf{i}|}$ *with the property that*

$$g(\mathbf{n}) = \prod_{\mathbf{i} \in \mathbb{N}^d} g_\mathbf{i}^{\binom{\mathbf{n}}{\mathbf{i}}}$$

for all $\mathbf{n} \in \mathbb{Z}^D$. *Conversely, every Taylor expansion of this type gives rise to a polynomial sequence* $g \in \mathrm{poly}(\mathbb{Z}^D, G_\bullet)$.

Remarks. This is proven in [30, Lemma 6.7]. Note that, since G is nilpotent, this is a finite expansion. In the case $D = 1$ (which will feature most prominently in the paper) the it takes the form

$$g(n) = g_0 g_1^{\binom{n}{1}} \dots g_s^{\binom{n}{s}}.$$

Note how, from the presentation of polynomial sequences as Taylor expansions, it is by no means clear (and somewhat remarkable) that they form a group under pointwise multiplication (Theorem 1.6).

Polynomial sequences that vary slowly, in a certain sense, are called *smooth.* We employ the following definition, which is the same as the one given in the introduction to [30].

Definition A.2 (Smooth sequences). Let A be a positive parameter and let $N \geqslant 1$ be an integer. Let $\beta \in \mathrm{poly}(\mathbb{Z}, G_\bullet)$. We say that β is (A,N)-*smooth* if we have $d_G(\beta(n), \mathrm{id}) \leqslant A$ and $d_G(\beta(n), \beta(n+1)) \leqslant A/N$ for all $n \in [N]$.

Here d_G is a metric on the group G constructed using the Mal'cev basis, see [30, Definition 2.2]. The precise definition of this metric is not terribly important for our analysis.

In counterpoint[17] to the notion of a smooth sequence is that of a *rational* sequence.

Definition A.3 (Rational sequences). Let $A \geqslant 1$ be an integer, and let $(G/\Gamma, G_\bullet)$ be a filtered nilmanifold. Then an element $g \in G$ is A-*rational* if there is some q, $1 \leqslant q \leqslant A$, such that $g^q \in \Gamma$. If $\gamma \in \mathrm{poly}(\mathbb{Z}, G_\bullet)$ is a polynomial sequence then we say that it is A-*rational* if $\gamma(n)$ is A-rational for every integer n.

We have the following basic facts about smooth and rational sequences:

Lemma A.4 (Basic facts). *Let $(G/\Gamma, G_\bullet)$ be a filtered nilmanifold of complexity $\leqslant M_0$. By a "sequence", we mean an element of* $\mathrm{poly}(\mathbb{Z}, G_\bullet)$. *Then:*

(i) *The product of two (A,N)-smooth sequences is $O_{M_0,A}(1)$-smooth;*
(ii) *The product of two A-rational sequences is $O_{M_0,A}(1)$-rational;*
(iii) *Any A-rational sequence is periodic with period $O_{M_0,A}(1)$.*

Proof. For (i), see [30, Lemma 10.1]; for (ii), see [30, Lemma A.11 (v)]; and for (iii), see [30, Lemma A.12 (ii)]. In fact these results hold in the multiparameter setting, with polynomially effective bounds, but we will not need these facts here. ■

We turn now to an important new definition for this paper, that of an *irrational* polynomial sequence. In [30], much emphasis was placed on the

[17] One could take an "adelic" perspective here and view smooth sequences as those that are local to the Archimedean place ∞, while rational sequences are those that are local to finite places p.

notion of an *equidistributed* polynomial sequence $g : \mathbb{Z} \to G$: one for which the orbit $\big(g(n)\Gamma\big)_{n\in[N]}$ is close to equidistributed on G/Γ. The notion of an irrational sequence implies equidistribution (see Lemma 3.7, which is also a special case of Theorem 1.11), but also encodes an assertion that the filtration $G_\bullet$ is in some sense "minimal" for the sequence. To illustrate the difference, let us think about a simple abelian case in which G/Γ is just the unit circle $\mathbb{R}/\mathbb{Z}$ (written additively), and $g : \mathbb{Z} \to \mathbb{R}$ is a polynomial

$$g(n) = \alpha_0 + \alpha_1\binom{n}{1} + \ldots + \alpha_s\binom{n}{s}. \tag{A.1}$$

This sequence is adapted to the filtration in which $G_{(i)} = \mathbb{R}$ for $i \leqslant s$ and $G_{(i)} = \{0\}$ for $i > s$. Qualitatively speaking, g is equidistributed if at least one of $\alpha_1, \ldots, \alpha_s$ is irrational; in contrast, g is irrational with respect to this filtration if it is α_s which is irrational. Note that if $s > 1$ and α_s is rational, then (after removing the periodic component $\alpha_s n^s$ from g) g is now adapted to the filtration $G'_\bullet$ in which $G'_{(i)} = \mathbb{R}$ for $i \leqslant s-1$ and $G'_{(i)} = \{0\}$ for $i > s-1$, which has a strictly smaller total dimension. This basic example is the model for the more sophisticated result in Lemma 2.9.

Let us turn now to the precise definition in the more general setting of Lie group-valued polynomial sequences, in which the role of the α_i is played by the Taylor coefficients of g. We need a preliminary definition.

Definition A.5 (i-horizontal characters). Let $(G/\Gamma, G_\bullet)$ be a filtered nilmanifold of degree $\leqslant s$ with filtration $G_\bullet = \big(G_{(i)}\big)_{i=0}^{\infty}$. Then by an *$i$-horizontal character* we mean a continuous homomorphism from $\xi_i : G_{(i)} \to \mathbb{R}$ which vanishes on $G_{(i+1)}$, $\Gamma_{(i)}$ and on $\big[G_{(j)}, G_{(i-j)}\big]$ for any $0 \leqslant j \leqslant i$. We say that such a character is *non-trivial* if it is not constant. We can assign a notion of complexity by taking a Mal'cev basis adapted to $G_\bullet$, whereupon one has a natural isomorphism $G_{(i)}/G_{(i+1)} \cong \mathbb{R}^k$. Writing $\psi(g_i)$ for the coordinates of g_i (mod $G_{(i+1)}\mathcal{B}$), any i-horizontal character has the form $\xi_i(g_i) = \vec{m}.\psi(g_i)$, for some vector $\vec{m} = (m_1, \ldots, m_k)$ of integers. We may then define the *complexity* of ξ_i to be $|m_1| + \cdots + |m_k|$.

The list of subgroups on which ξ_i is required to vanish looks rather restrictive and slightly unnatural at first sight. Roughly speaking, this list is intended to isolate that behaviour which genuinely "belongs" to the degree i portion of the filtered nilmanifold, as opposed to arising from those terms of higher or lower degree, or which disappear after quotienting out by the lattice Γ.

Definition A.6 (Irrationality). Let $(G/\Gamma, G_\bullet)$ be a filtered nilmanifold of degree $\leqslant s$ with filtration $G_\bullet = \left(G_{(i)}\right)_{i=0}^{\infty}$. Let $g_i \in G_{(i)}$. Let $A, N > 0$. Then we say that g_i is (A,N)-*irrational* in $G_{(i)}$ if for every non-trivial i-horizontal character $\xi_i : G_{(i)} \to \mathbb{R}$ of complexity $\leqslant A$ one has $\left\| \xi_i(g_i) \right\|_{\mathbb{R}/\mathbb{Z}} \geqslant A/N^i$. We say that the sequence $g(n)$ is (A,N)-*irrational* if its ith Taylor coefficient g_i is (A,N)-irrational in $G_{(i)}$ for each i, $1 \leqslant i \leqslant s$.

To understand this definition, it is helpful to consider examples. We leave it as an exercise to check that in the abelian case (A.1) this amounts to stipulating that the top coefficient of g is poorly approximated by rationals, thus $\|q\alpha_s\|_{\mathbb{R}/\mathbb{Z}} \geqslant A'/N^s$ whenever $1 \leqslant q \leqslant A'$.

A second interesting case to examine is that in which $g(n) = g^n$ is a linear polynomial sequence adapted to the lower central series filtration $(G_i)_{i=0}^{\infty}$. For the lower central series filtration there are no nontrivial i-horizontal characters when $i \geqslant 2$, and 1-horizontal characters are the same thing as horizontal characters in the sense of [30, Definition 1.5]. It follows from this and [30, Theorem 1.16] that $g(n)$ is irrational if and only if $\left(g(n)\Gamma\right)_{n\in[N]}$ is equidistributed. Now polynomial sequences that are not linear do not arise naturally in ergodic-theoretic settings such as those considered in [4, 41], and thus the equivalence of the notions of "irrational" and "equidistributed" in this setting explains why the former concept has not appeared in the literature before. The need for it is a new feature of the quantitative world, as is the need for polynomial nilsequences themselves, for reasons explained on [30, §1].

The following third example is also edifying. Take $g(n)$ to be any polynomial sequence on the Heisenberg group, for example $g(n) = \begin{pmatrix} 1 & \alpha n & \gamma n^2 \\ 0 & 1 & \beta n \\ 0 & 0 & 1 \end{pmatrix}$. This sequence is a polynomial sequence adapted to the lower central series filtration $G_0 = G_1 = G$, $G_2 = [G,G]$, $G_3 = \{\mathrm{id}\}$, and it will be equidistributed in that setting for generic α, β, γ. However g is also a polynomial sequence with respect to some much flabbier filtrations, for example the one in which $G_{(0)} = G_{(1)} = G_{(2)} = \cdots = G_{(10)} = G$, $G_{(11)} = \cdots = G_{(100)} = [G,G]$ and $G_{(i)} = \{\mathrm{id}\}$ for $i \geqslant 101$. It is easy to check that g is *not* irrational in this setting, and indeed irrationality is somehow detecting the fact that a given filtration $G_\bullet$ is minimal for g. This point is quite clear in the proof of Lemma 2.9 (which itself depends on Lemma A.7 below), where the failure of a sequence to be irrational is used to create a coarser filtration for a polynomial sequence related to g.

Lemma A.7. *Suppose that $(G/\Gamma, G_\bullet)$ is a filtered nilmanifold of degree $\leqslant s$ with filtration $G_\bullet = \big(G_{(i)}\big)_{i=0}^{\infty}$. Suppose that g is not (A,N)-irrational. Then there is an index i, $1 \leqslant i \leqslant s$, such that the ith Taylor coefficient g_i factors as $\beta_i g_i' \gamma_i$, where $\beta_i, g_i', \gamma_i \in G_{(i)}$, g_i' lies in the kernel of some i-horizontal character $\xi_i : G_{(i)} \to \mathbb{R}$ of complexity at most A, $d_G(\beta_i, \mathrm{id}) = O_{A,M}(N^{-i})$ and γ_i is $O_{A,M}(1)$-rational.*

Proof. The proof is (unsurprisingly) extremely similar to that of [30, Lemma 7.9]. Reversing the definition of irrational polynomial sequence, we see that there is an index i together with an i-horizontal character $\xi_i : G_{(i)} \to \mathbb{R}$ such that $\big\| \xi_i(g_i)\big\|_{\mathbb{R}/\mathbb{Z}} \leqslant A/N^i$. It is convenient at this point to work in a Mal'cev coordinate system adapted to $G_\bullet$, whereby $G_{(i)}/G_{(i+1)}$ may be identified with $\mathbb{R}^k$ and $\Gamma_{(i)}/G_{(i+1)}$ with $\mathbb{Z}^k$. If $g_i \in G_{(i)}$ then, as above, we write $\psi(g) \in \mathbb{R}^k$ for the corresponding coordinates. Then ξ_i has the form $\xi_i(g_i) = \vec{m}.\psi(g)$ for some vector $\vec{m} = (m_1, \dots, m_k)$ of integers with $|m_1| + \cdots + |m_k| \leqslant A$. Now by assumption we have $\big\| \vec{m}.\psi(g_i)\big\|_{\mathbb{R}/\mathbb{Z}} \leqslant A/N^i$, and therefore $\vec{m}.\psi(g_i) = r + O(A/N^i)$ for some integer r. It follows from simple linear algebra that we may write $\psi(g_i) = \vec{t} + \vec{u} + \vec{v}$, where $\vec{m}.\vec{u} = 0$, the coordinates of $\vec{v}$ lie in $\frac{1}{Q}\mathbb{Z}$ for some $Q = O_A(1)$ and each coordinate of $\vec{t}$ is $O_A(1/N^i)$. Now choose $\beta_i \in G_{(i)}$ in such a way that $\psi(\beta_i) = \vec{t}$ and $d_G(\beta_i, \mathrm{id}) = O_{A,M}(1/N_i)$, choose an $O_{A,M}(1)$-rational element $\gamma_i \in G_{(i)}$ with $\psi(\gamma_i) = \vec{v}$, and finally choose g_i' so that $g_i = \beta_i g_i' \gamma_i$. Then one automatically has $\psi(g_i') = \vec{u}$, which means that g_i' lies in the kernel of the i-homomorphism ξ_i. ■

Finally, we record a convenient scaling lemma.

Lemma A.8 (Scaling lemma). *Let $(G/\Gamma, G_\bullet)$ be a filtered nilmanifold of complexity $\leqslant M$. If $g \in \mathrm{poly}\,(\mathbb{Z}, G_\bullet)$ is (A,N)-irrational, $r \in [-N, N]$, and $1 \leqslant q \leqslant M$, then the sequence $n \mapsto g(nq + r)$ is $(\gg_{M,\varepsilon} A, \varepsilon N)$-irrational for any $\varepsilon > 0$.*

Proof. We need to show that the ith Taylor coefficient of $n \mapsto g(nq + r)$ is $(\gg_{M,\varepsilon} A, \varepsilon N)$-irrational for each $i \geqslant 0$. Note that we may assume $i \leqslant M$ since the filtered manifold has degree $\leqslant M$.

Fix i. We may quotient out the nilmanifold by the normal subgroups $G_{(i+1)}$ and $\big[G_{(j)}, G_{(i-j)}\big]$ for $0 \leqslant j \leqslant i$, since these do not affect the irrationality of the ith coefficient. We may then expand g as a Taylor series

$$g(n) = \prod_{j=0}^{i} g_j^{\binom{n}{j}},$$

and thus

$$g(qn+r) = \prod_{j=0}^{i} g_j^{\binom{qn+r}{j}}.$$

Expanding out the binomial coefficient and using many applications of the Baker–Campbell–Hausdorff formula, we obtain

$$g(qn+r) = \left(\prod_{j=0}^{i-1} (g_j')^{\binom{n}{j}}\right) g_i^{q^i\binom{n}{i}}$$

for some $g_j' \in G_{(j)}$; the point being that the Baker–Campbell–Hausdorff term cannot generate any terms involving polynomials in n of degree i or higher due to the fact that the groups $G_{(i+1)}$ and $\left[G_{(j)}, G_{(i-j)}\right]$ have been quotiented out. As a consequence, we see that the ith Taylor coefficient of $n \mapsto g(qn+r)$ is $q^i g_i$, and the claim is easily verified. ∎

Appendix B. A Multiparameter Equidistribution Result

The purpose of this appendix is to prove Theorem 3.6, which we recall here again.

Theorem 3.6. *Suppose that $(G/\Gamma, G_\bullet)$ is a filtered nilmanifold of complexity $\leqslant M$ and that $g \in \operatorname{poly}(\mathbb{Z}^D, G_\bullet)$ is a polynomial sequence for some $D \leqslant M$. Suppose that $\Lambda \subseteq \mathbb{Z}^D$ is a lattice of index $\leqslant M$, that $\mathbf{n_0} \in \mathbb{Z}^D$ has magnitude $\leqslant M$, and that $P \subseteq [-N, N]^D$ is a convex body. Suppose that $\delta > 0$, and that*

$$\left| \sum_{\mathbf{n} \in (\mathbf{n_0}+\Lambda)\cap P} F(g(\mathbf{n})\Gamma) - \frac{\operatorname{vol}(P)}{[\mathbb{Z}^D : \Lambda]} \int_{G/\Gamma} F \right| > \delta N^D \|F\|_{\mathrm{Lip}}$$

for some Lipschitz function $F : G/\Gamma \to \mathbb{C}$. *Then there is a nontrivial homomorphism* $\eta : G \to \mathbb{R}$ *which vanishes on* Γ, *has complexity* $O_M(1)$ *and such that*

$$\|\eta \circ g\|_{C^\infty([N]^D)} = O_{\delta,M}(1).$$

Recall from [30, Definition 8.2] that the norm $\|g\|_{C^\infty([N]^D)}$ of a polynomial sequence $g : [N]^D \to \mathbb{R}$ is given by the formula

$$\|g\|_{C^\infty([N]^D)} = \sup_{\mathbf{i}\in\mathbb{N}^D} N^{-|\mathbf{i}|}\|g_{\mathbf{i}}\|_{\mathbb{R}/\mathbb{Z}}$$

where $g_{\mathbf{i}}$ are the Taylor coefficients of g, thus

$$g(\mathbf{n}) = \sum_{\mathbf{i}\in\mathbb{N}^D} \binom{\mathbf{n}}{\mathbf{i}} g_{\mathbf{i}}.$$

We now prove the theorem, allowing all implied constants to depend on δ and M. We may assume that N is sufficiently large depending on δ, M, since the claim is trivial otherwise. A simple volume packing argument (using [31, Corollary A.2], for example, to control the boundary terms) shows that

$$\left|(\mathbf{n_0} + \Lambda) \cap P\right| = \frac{\mathrm{vol}\,(P)}{[\mathbb{Z}^D : \Lambda]} + o_{N\to\infty}(N^D).$$

As a consequence, for N large enough we may subtract off the mean of F and normalise F to have Lipschitz norm 1 and mean zero, thus

$$\left|\sum_{\mathbf{n}\in(\mathbf{n_0}+\Lambda)\cap P} F(g(\mathbf{n})\Gamma)\right| \gg N^D.$$

As Λ has index $\leqslant M$ in $\mathbb{Z}^D$, it contains the sublattice $q\mathbb{Z}^D$ for some positive integer $q = O(1)$. By the pigeonhole principle, we may thus find $\mathbf{n_1} \in \mathbb{Z}^D$ of magnitude $O(1)$ such that

$$\left|\sum_{\mathbf{n}\in(\mathbf{n_1}+q\mathbb{Z}^D)\cap P} F(g(\mathbf{n})\Gamma)\right| \gg N^D,$$

and thus

$$\left|\sum_{\mathbf{n}\in\mathbb{Z}^D\cap P'} F(g(q\mathbf{n} + \mathbf{n_1})\Gamma)\right| \gg N^D.$$

for some convex body P' contains in a ball of radius $O(N)$ centered at the origin.

By subdividing P' into cubes of sidelength εN for some sufficiently small $\varepsilon > 0$ (and again using [31, Corollary A.2] to control the boundary terms), and then applying the pigeonhole principle, we see that

$$\left| \sum_{\mathbf{n} \in \mathbb{Z}^D \cap \mathbf{n}_2 + [\varepsilon N]^D} F\big(g(q\mathbf{n} + \mathbf{n}_1)\Gamma\big) \right| \gg N^D$$

for some $\varepsilon \gg 1$ and $\mathbf{n}_2 = O(N)$. We can rearrange this as

$$\left| \sum_{\mathbf{n} \in \mathbb{Z}^D \cap [\varepsilon N]^D} F\big(g(q\mathbf{n} + \mathbf{n}_3)\Gamma\big) \right| \gg N^D$$

for some $\mathbf{n}_3 = O(N)$.

We may now invoke [30, Theorem 8.6] to conclude that there exists a nontrivial homomorphism $\eta : G \to \mathbb{R}$ which vanishes on Γ, has complexity $O(1)$ and such that

$$\big\| \eta \circ g(q \cdot + \mathbf{n}_3) \big\|_{C^\infty([N]^D)} \ll 1.$$

Applying [30, Lemma 8.4] we conclude that

$$\big\| Q\eta \circ g(\cdot + \mathbf{n}_3) \big\|_{C^\infty([N]^D)} \ll 1$$

for some non-negative integer $Q = O(1)$. Shifting the Taylor expansion by $\mathbf{n}_3$, we conclude that

$$\|Q\eta \circ g\|_{C^\infty([N]^D)} \ll 1.$$

The claim follows (with η replaced by $Q\eta$).

Appendix C. The Baker–Campbell–Hausdorff Formula

Let G be a connected, simply connected nilpotent Lie group, and let $\exp : \mathfrak{g} \to G$ and $\log : G \to \mathfrak{g}$ be the associated exponential and logarithm maps between G and its Lie algebra $\mathfrak{g}$. The *Baker–Campbell–Hausdorff formula* asserts that

$$\exp(X_1)\exp(X_2) = \exp\left(X_1 + X_2 + \frac{1}{2}[X_1, X_2] + \prod_\alpha c_\alpha X_\alpha\right)$$

for any X_1, X_2, where α is a finite set of labels, c_α are real constants, and X_α are an iterated Lie bracket of $k_1 = k_{1,\alpha}$ copies of X_1 and $k_2 = k_{2,\alpha}$ copies of X_2 where $k_1, k_2 \geqslant 1$ and $k_1 + k_2 \geqslant 2$.

Using this formula, it is a routine matter to see that for any $g_1, g_2 \in G$ and $x \in \mathbb{R}$, we have

$$(g_1 g_2)^x = g_1^x g_2^x \prod_\alpha g_\alpha^{Q_\alpha(x)} \tag{C.1}$$

where α is a finite set of labels, each g_α is an iterated of $k_1 = k_{1,\alpha}$ copies of g_1 and $k_2 = k_{2,\alpha}$ copies of g_2 where $k_1, k_2 \geqslant 1$ and $k_1 + k_2 \geqslant 2$, and the $Q_\alpha : \mathbb{R} \to \mathbb{R}$ are polynomials of degree at most $k_1 + k_2$ with no constant term.

In a similar vein, for any $g_1, g_2 \in G$ and $x_1, x_2 \in \mathbb{R}$, we have the formula

$$\left[g_1^{x_1}, g_2^{x_2}\right] = [g_1, g_2]^{x_1 x_2} \prod_\alpha g_\alpha^{P_\alpha(x_1, x_2)} \tag{C.2}$$

where α is a finite set of labels, each g_α is an iterated commutator of $k_1 = k_{1,\alpha}$ copies of g_1 and $k_2 = k_{2,\alpha}$ copies of g_2 where $k_1, k_2 \geqslant 1$ and $k_1 + k_2 \geqslant 3$, and the $P_\alpha : \mathbb{R} \times \mathbb{R} \to \mathbb{R}$ are polynomials of degree at most k_1 in x_1 and at most k_2 in x_2 which vanish when $x_1 = 0$ or $x_2 = 0$.

References

[1] N. Alon, E. Fischer, M. Krivelevich and B. Szegedy, *Efficient testing of large graphs,* Proc. of 40th FOCS, New York, NY, IEEE (1999), 656–666. Also: *Combinatorica,* **20** (2000), 451–476.

[2] T. Austin, *Deducing the multidimensional Szemerédi Theorem from an infinitary removal lemma,* preprint, arXiv:0808.2267.

[3] T. Austin, *Deducing the Density Hales–Jewett Theorem from an infinitary removal lemma,* preprint, arXiv:0903.1633.

[4] V. Bergelson, B. Host and B. Kra, Multiple recurrence and nilsequences, with an appendix by Imre Ruzsa, *Invent. Math.,* **160** (2005), no. 2, 261–303.

[5] V. Bergelson, A. Leibman and E. Lesigne, Weyl complexity of a system of polynomials and constructions in combinatorial number theory, *J. D'Analyse Mathématique,* **103** (2007), 47–92.

[6] V. Bergelson, A. Leibman and E. Lesigne, Intersective polynomials and the polynomial Szemerédi theorem, *Adv. Math.,* **219** (2008), no. 1, 369–388.

[7] J. Bourgain, A Szemerédi type theorem for sets of positive density in $\mathbb{R}^k$, *Israel J. Math.,* **54** (1986), no. 3, 307–316.

[8] J. Bourgain, On triples in arithmetic progression, *GAFA,* **9** (1999), 968–984.

[9] F. Chung, Regularity lemmas for hypergraphs and quasi-randomness, *Random Struct. Alg.,* **2** (1991), 241–252.

[10] H. Furstenberg, *Recurrence in Ergodic theory and Combinatorial Number Theory,* Princeton University Press, Princeton NJ 1981.

[11] H. Furstenberg, Ergodic behavior of diagonal measures and a theorem of Szemerédi on arithmetic progressions, *J. Analyse Math.,* **31** (1977), 204–256.

[12] H. Furstenberg, From the Erdős–Turán conjecture to ergodic theory—the contribution of combinatorial number theory to dynamics, in: *Paul Erdős and his mathematics, I* (Budapest, 1999), 261–277, Bolyai Soc. Math. Stud., 11, János Bolyai Math. Soc., Budapest, 2002.

[13] H. Furstenberg and Y. Katznelson, A density version of the Hales–Jewett theorem, *J. d'Analyse Math.,* **57** (1991), 64–119.

[14] H. Furstenberg, Y. Katznelson and D. Ornstein, The ergodic theoretical proof of Szemerédi's theorem, *Bull. Amer. Math. Soc. (N.S.),* **7** (1982), no. 3, 527–552.

[15] A. Frieze and R. Kannan, Quick approximation to matrices and applications, *Combinatorica,* **19** (1999), no. 2, 175–220.

[16] W. T. Gowers, Lower bounds of tower type for Szemerédi's uniformity lemma, *GAFA,* **7** (1997), 322–337.

[17] W. T. Gowers, A new proof of Szemerédi's theorem for progressions of length four, *GAFA,* **8** (1998), no. 3, 529–551.

[18] W. T. Gowers, A new proof of Szemerédi's theorem, *GAFA,* **11** (2001), 465–588.

[19] W. T. Gowers, Hypergraph regularity and the multidimensional Szemerédi theorem, *Ann. of Math. (2),* **166** (2007), no. 3, 897–946.

[20] W. T. Gowers, Quasirandomness, counting, and regularity for 3-uniform hypergraphs, *Combin. Probab. Comput.,* **15** (2006), no. 1–2, 143–184.

[21] W. T. Gowers, *Decompositions, approximate structure, transference, and the Hahn–Banach theorem,* preprint.

[22] W. T. Gowers and J. Wolf, The true complexity of a system of linear equations, to appear in *Mathematika,* arXiv:1002.2209.

[23] W. T. Gowers and J. Wolf, Linear forms and quadratic uniformity for functions on $\mathbb{Z}_N$, to appear in *J. d'Analyse,* arXiv:1002.2210.

[24] W. T. Gowers and J. Wolf, *Linear forms and higher-degree uniformity for functions on* $\mathbb{F}_p^n$, preprint, arXiv:1002.2208.

[25] B. J. Green, A Szemerédi-type regularity lemma in abelian groups, *GAFA,* **15** (2005), no. 2, 340–376.

[26] B. J. Green, *Montréal lecture notes on quadratic Fourier analysis,* Additive Combinatorics (Montréal 2006, ed. Granville et al.), CRM Proceedings vol. 43, 69–102, AMS 2007.

[27] B. J. Green and T. C. Tao, The primes contain arbitrarily long arithmetic progressions, *Ann. Math.,* **167** (2008), 481–547.

[28] B. J. Green and T. C. Tao, An inverse theorem for the Gowers $U^3(G)$-norm, *Proc. Edin. Math. Soc.,* **51** (2008), 73–153.

[29] B. J. Green and T. C. Tao, New bounds for Szemerédi's Theorem, I: Progressions of length 4 in finite field geometries, *Proc. Lond. Math. Soc.,* **98** (2009), 365–392.

[30] B. J. Green and T. C. Tao, The quantitative behaviour of polynomial orbits on nilmanifolds, to appear in *Ann. Math.,* arXiv:0709.3562.

[31] B. J. Green and T. C. Tao, Linear equations in primes, to appear in *Ann. Math.,* arXiv:math/0606088.

[32] B. J. Green and T. C. Tao, The Möbius function is strongly orthogonal to nilsequences, to appear in *Ann. Math.,* arXiv:0807.1736.

[33] B. J. Green, T. C. Tao and T. Ziegler, An inverse theorem for the Gowers U^4-norm, to appear in *Glasgow J. Math.,* arXiv:0911.5681.

[34] B. J. Green, T. C. Tao and T. Ziegler, *An inverse theorem for the Gowers $U^{s+1}[N]$-norm* (announcement), arXiv:1006.0205.

[35] B. Host and B. Kra, Nonconventional ergodic averages and nilmanifolds, *Ann. of Math. (2),* **161** (2005), no. 1, 397–488.

[36] J. Komlós and M. Simonovits, Szemerédi's regularity lemma and its applications in graph theory, in: *Combinatorics, Paul Erdős is eighty,* Vol. 2 (Keszthely, 1993), 295–352, Bolyai Soc. Math. Stud., 2, János Bolyai Math. Soc., Budapest, 1996.

[37] A. Leibman, Polynomial sequences in groups, *Journal of Algebra,* **201** (1998), 189–206.

[38] A. Leibman, Polynomial mappings of groups, *Israel J. Math.,* **129** (2002), 29–60.

[39] A. Leibman, Pointwise convergence of ergodic averages of polynomial sequences of translations on a nilmanifold, *Ergodic Theory and Dynamical Systems,* **25** (2005), no. 1, 201–213.

[40] A. Leibman, Pointwise convergence of ergodic averages for polynomial actions of $\mathbb{Z}^d$ by translations on a nilmanifold, *Ergodic Theory and Dynamical Systems,* **25** (2005), no. 1, 215–225.

[41] A. Leibman, Orbit of the diagonal of the power of a nilmanifold, *Transactions of AMS,* **362** (2010), 1619–1658.

[42] L. Lovász and B. Szegedy, Szemerédi's Lemma for the analyst, *Geom. Func. Anal.,* **17** (2007), 252–270.

[43] B. Nagle, V. Rödl and M. Schacht, Note on the 3-graph counting lemma, *Discrete Mathematics,* **308** (19), 4501–4517.

[44] B. Nagle, V. Rödl and M. Schacht, The counting lemma for regular k-uniform hypergraphs, *Random Structures and Algorithms,* **28** (2), 113–179.

[45] D. H. J. Polymath, *A new proof of the density Hales–Jewett theorem,* to appear in this volume.

[46] O. Reingold, L. Trevisan, M. Tulsiani and S. Vadhan, Dense Subsets of Pseudorandom Sets, in: *Proc. of 49th IEEE FOCS,* 2008 ECCC TR08-45.

[47] E. Szemerédi, On sets of integers containing no four elements in arithmetic progression, *Acta Math. Acad. Sci. Hungar.,* **20** (1969), 89–104.

[48] E. Szemerédi, On sets of integers containing no k elements in arithmetic progression, *Acta Arith.,* **27** (1975), 299–345.

[49] E. Szemerédi, Regular partitions of graphs, in *"Problèmes Combinatoires et Théorie des Graphes, Proc. Colloque Inter. CNRS,"* (Bermond, Fournier, Las Vergnas, Sotteau, eds.), CNRS Paris, 1978, 399–401.

[50] T. C. Tao, A quantitative ergodic theory proof of Szemerédi's theorem, *Electron. J. Combin.,* **13** (2006). 1 No. 99, 1–49.

[51] T. C. Tao, *The dichotomy between structure and randomness, arithmetic progressions, and the primes,* 2006 ICM proceedings, Vol. I., 581–608.

[52] T. C. Tao, A variant of the hypergraph removal lemma, *J. Combin. Thy. A,* **113** (2006), 1257–1280.

[53] T. C. Tao, Szemerédi's regularity lemma revisited, *Contrib. Discrete Math.,* **1** (2006), 8–28.

[54] T. C. Tao, *Structure and randomness in combinatorics,* Proceedings of the 48th annual symposium on Foundations of Computer Science (FOCS) 2007, 3–18.

[55] T. C. Tao and V. H. Vu, *Additive Combinatorics,* Cambridge University Press, 2006.

[56] H. Towsner, *A Model Theoretic Proof of Szemerédi's Theorem,* preprint, arXiv:1002.4456.

[57] P. Varnavides, On certain sets of positive density, *J. London Math. Soc.,* **39** (1959), 358–360.

[58] T. Ziegler, Universal Characteristic Factors and Furstenberg Averages, *J. Amer. Math. Soc.,* **20** (2007), 53–97.

Ben Green
Centre for Mathematical Sciences
Wilberforce Road
Cambridge CB3 0WA
England

e-mail: b.j.green@dpmms.cam.ac.uk

Terence Tao
Department of Mathematics
UCLA
Los Angeles, CA 90095
USA

e-mail: tao@math.ucla.edu

BOLYAI SOCIETY
MATHEMATICAL STUDIES, 21

An Irregular Mind
(Szemerédi is 70)
pp. 335–342.

Yet Another Proof of Szemerédi's Theorem

BEN GREEN and TERENCE TAO

To Endre Szemerédi on the occasion of his 70th birthday

Using the density-increment strategy of Roth and Gowers, we derive Szemerédi's theorem on arithmetic progressions from the inverse conjectures GI(s) for the Gowers norms, recently established by the authors and Ziegler in [8].

1. Introduction

In this note we show how Szemerédi's famous theorem [10, 11] on arithmetic progressions follows from the inverse conjecture GI(s) for the Gowers norms, recently established in [8]. This paper is designed as a coda to [6], and in particular we refer the reader to that paper (or to many other places in the literature) for the definition of the following terms, which we shall use without further comment: *filtered nilmanifold of complexity* $\leqslant M$, *polynomial sequence, degree* $\leqslant s$ *polynomial nilsequence of complexity at most* M, *rational polynomial sequence, Gowers norm, generalised von Neumann theorem* and *smoothness norm* $C^{\infty}[N]$.

Our main point is to show that Szemerédi's theorem can actually be derived rather easily from GI(s). We gave a different deduction in [6], designed to illustrate that for a large class of theorems (including Szemerédi's theorem) it essentially suffices to "check the result for nilsequences". That argument was somewhat complicated, not least because it relied heavily on the quantitative distribution results for nilsequences obtained in [4].

The argument we give here is based on the density-increment strategy of Roth [9] and Gowers [1, 2]. In fact our argument is, structurally, the same as that of Gowers except that we use the inverse theorem as a black box rather than prove, as Gowers did, a weaker version of it.

The only remotely new technical result in this note is the following. Here, and elsewhere in the paper, write $\mathrm{diam}_S(f) := \sup_{s_1,s_2\in S} d_X\big(f(s_1), f(s_2)\big)$ whenever $f : S \to X$ is some function from a set S into a metric space (X, d_X).

Theorem 1.1. *Let $\varepsilon > 0$ be a real parameter, let $s \geqslant 1$ be an integer, and let M be a complexity parameter. Then there is a number $\kappa_{s,M} > 0$ with the following property. Let $\big(F(g(n)\Gamma)\big)_{n\in\mathbb{Z}}$ be a degree $\leqslant s$ polynomial nilsequence of complexity at most M, and let $P \subseteq \mathbb{Z}$ be an arithmetic progression. Then we may partition P into a disjoint union of progressions P_i, $i = 1, \dots, m$, each of size $\gg_{M,\varepsilon} |P|^{\kappa_{s,M}}$, such that*

$$\mathrm{diam}_{P_i}\big(F(g(n)\Gamma)\big) \leqslant \varepsilon$$

for all $i = 1, \dots, m$.

Remark. The progressions P_i need not have the same common difference.

We prove this theorem in §2, and deduce Szemerédi's theorem from it and the inverse theorem GI(s) in §3.

2. Nilsequences are Almost Constant on Progressions

In this paper the *degree* s of a nilsequence will not be particularly important, so we suppress most mention of it, recalling that it is nonetheless bounded by the complexity parameter M. The reader may care to note that, as a consequence of this, we do not need the full strength of GI(s) but only a weaker version in which correlation with a nilsequence of degree $O_s(1)$ (rather than s) is obtained. However, we know of no proof this result that is easier than the full-strength version and we also know (with Ziegler) a not especially painful argument for deducing the full version from the weak one.

The goal of this section is to prove Theorem 1.1. By induction on the dimension of the underlying nilmanifold, the result follows very quickly from the following.

Proposition 2.1. *Let the notation be as in Theorem 1.1 above. Then we may partition P into a disjoint union of progressions P_i, $i = 1, \dots, m$, each of length $\gg_{M,\varepsilon} |P|^{\kappa_M}$, and such that the following is true. For each $i = 1, \dots, m$ there is a polynomial nilsequence $\big(F_i(h_i(n)\Lambda_i)\big)_{n\in\mathbb{Z}}$ of complexity $O_M(1)$ whose underlying nilmanifold has dimension strictly less than that of $\big(F(g(n)\Gamma)\big)_{n\in\mathbb{Z}}$, and such that*

$$\operatorname{diam}_{P_i}\big(F(g(n)\Gamma) - F_i(h_i(n)\Lambda_i)\big) \leqslant \varepsilon$$

for all $i = 1, \dots, m$.

We derive this result in turn from three lemmas. The first and its proof are essentially [2, Corollary 5.6], albeit formulated somewhat differently. It can be viewed as an analogue of Theorem 1.1 for polynomial phases.

Lemma 2.2 (Polynomials are almost constant on progressions). *Let $s \geqslant 1$ be an integer. Then there is some $\kappa_s > 0$ with the following property. Let $\phi : \mathbb{R} \to \mathbb{R}/\mathbb{Z}$ be a polynomial phase of degree s, and suppose that $P \subseteq \mathbb{Z}$ is a finite progression. Then we may partition P into progressions $P_1, \dots, P_m$, $|P_i| \gg_s |P|^{\kappa_s}$, such that $\operatorname{diam}_{P_i}(\phi) \leqslant \frac{1}{10}$ for $i = 1, \dots, m$.*

Proof. Suppose that $|P|$ is sufficiently large in terms of s; the result is trivial otherwise, since we may partition into progressions of length 1. It suffices to prove the weaker statement that we may partition P into progressions $P_1, \dots, P_m$, $|P_i| \gg_s |P|^{\kappa_s}$, such that for each i there is a polynomial phase $\phi_i : \mathbb{R} \to \mathbb{R}/\mathbb{Z}$ of degree at most $s - 1$ such that

$$\operatorname{diam}_{P_i}(\phi - \phi_i) \leqslant \frac{1}{100s^2}.$$

We may then work by induction on the degree to obtain the lemma (with a smaller value of κ_s of course), using the fact that

$$\operatorname{diam}(\phi) \leqslant \operatorname{diam}(\phi - \phi_s) + \operatorname{diam}(\phi_s - \phi_{s-1}) + \dots$$
$$+ \operatorname{diam}(\phi_1 - \phi_0) + \operatorname{diam}\phi_0 \leqslant \frac{1}{100}\sum_s \frac{1}{s^2} < \frac{1}{10}.$$

To obtain the weaker statement one invokes the following standard diophantine result essentially due to Weyl: there is some $\delta_s > 0$ such that, for any $\alpha \in \mathbb{R}/\mathbb{Z}$ and any $N \geqslant 1$, there is some $n \leqslant \sqrt{N}$ such that $\|\alpha n^s\|_{\mathbb{R}/\mathbb{Z}} \ll_s N^{-\delta_s}$.

Supposing that $\phi(n) = \theta n^s + \ldots$ and that P has common difference d and length N, we apply this result with $\alpha := \theta d^s$. Subdividing P into subprogressions P_i of length between $N^{\delta_s/2}$ and $N^{2\delta_s/2}$ and common difference dn gives the required statement. Note that such a subdivision is indeed possible since $n \leqslant \sqrt{N}$ and N is sufficiently large in terms of s. ■

Lemma 2.3 (Weyl-type equidistribution theorem). *Suppose that $\phi : \mathbb{R} \to \mathbb{R}/\mathbb{Z}$ is a polynomial phase of degree s, and that $\mathrm{diam}_{[N]}(\phi) \leqslant \frac{1}{10}$. Then there is some $q = O_s(1)$ such that $\|q\phi\|_{C^\infty[N]} = O_s(1)$.*

Proof. This follows immediately from [4, Proposition 4.3] (the proof of which can be read independently of the rest of that paper, which we do not rely on heavily in this note). Observe, however, that it is quite classical and essentially goes back to Weyl, being the statement that a polynomial phase that is not equidistributed has almost rational coefficients. ■

Lemma 2.4 (Factorisation of polynomial sequences). *Let $(G/\Gamma, G_\bullet)$ be a filtered nilmanifold of complexity M, and suppose that $g \in \mathrm{poly}(\mathbb{Z}, G_\bullet)$. Let $\eta : G \to \mathbb{R}/\mathbb{Z}$ be a horizontal character with Lipschitz constant $O_M(1)$. Suppose that P is an arithmetic progression and that $\mathrm{diam}_P(\eta \circ g) \leqslant \frac{1}{10}$. Then there is a factorisation $g = \beta g' \gamma$, where $\beta, \gamma \in \mathrm{poly}(\mathbb{Z}, G_\bullet)$ and:*

(i) *β is smooth in the sense that $d_G(\beta(n), \beta(n')) = O_M(\delta)$ whenever $n, n' \in P$ and $|n - n'| \leqslant \delta|P|$;*

(ii) *g' takes values in a connected $O_M(1)$-rational subgroup $G' \leqslant G$ with $\dim(G') < \dim(G)$;*

(iii) *γ is $O_M(1)$-rational.*

Proof. By rescaling linearly (and noting that if $g(n)$ lies in $\mathrm{poly}(\mathbb{Z}, G_\bullet)$ then so does $g(an + b)$, cf. [6, Lemma A.8]) we may assume that $P = [N]$. Applying Lemma 2.3 and replacing η by $\tilde{\eta} = q\eta$, where $q = O_M(1)$, we may assume that $\|\eta \circ g\|_{C^\infty[N]} = O_M(1)$. The result may now be proved in exactly the same way as [4, Proposition 9.2] (although that result was a little more notationally intensive, formulated as it was for multiparameter sequences). ■

Deduction of Proposition 2.1. Select a nontrivial horizontal character $\eta : G \to \mathbb{R}/\mathbb{Z}$ with Lipschitz constant $O_M(1)$. Apply Lemma 2.2 followed by Lemma 2.4 to $\phi := \eta \circ g$, obtaining a decomposition of P into progressions P_i such that on each P_i we have a factorisation $g = \beta g' \gamma$ of the stated type. Note that this factorisation depends on i, but we suppress this for notational convenience. Suppose that $\gamma(n)$ has period $q = O_M(1)$, so that $\gamma(n)\Gamma = \gamma(n')\Gamma$ whenever $n \equiv n' (\text{mod } q)$. Subdivide P_i into progressions P_j (this is a convenient abuse of notation) whose common difference is a multiple of q and whose length is $c_{\varepsilon,M}|P_i|$, for a constant $c_{\varepsilon,M} > 0$ to be specified shortly. For each j, fix some $\gamma_0 = O_M(1)$ such that $\gamma(n)\Gamma = \gamma_0\Gamma$ for all $n \in P_j$. Then if $n \in P_j$ we have

$$g(n)\Gamma = \beta(n)\gamma_0(\gamma_0^{-1}g'(n)\gamma_0)\Gamma.$$

Set $H_j := \gamma_0^{-1}G'\gamma_0$ and $\Lambda_j := H_j \cap \Gamma$. Then H_j/Λ_j is a nilmanifold of complexity $O_M(1)$, and certainly $\dim(H_j) = \dim(G') < \dim(G)$, and the polynomial sequence $h_j(n) := \gamma_0^{-1}g'(n)\gamma_0$ takes values in H_j. Pick some $n_0 \in P_j$, and define $F_j : H_j/\Lambda_j \to \mathbb{C}$ by

$$F_j(x) := F(\beta(n_0)\gamma_0 x).$$

Then F_j is $O_M(1)$-Lipschitz and

$$\begin{aligned}
&\operatorname{diam}_{P_j}\big(F\big(g(n)\Gamma\big) - F_j\big(h_j(n)\Lambda_j\big)\big)\\
&\qquad = \sup_{n\in P_j}\Big|F\big(\beta(n)\gamma_0\big(\gamma_0^{-1}g'(n)\gamma_0\big)\Gamma\big) - F\big(\beta(n_0)\gamma_0\big(\gamma_0^{-1}g'(n)\gamma_0\big)\Gamma\big)\Big|\\
&\qquad \leqslant \varepsilon,
\end{aligned}$$

the last line following if $c_{\varepsilon,M}$ is sufficiently small from the smoothness of β and the fact that F has Lipschitz constant $O_M(1)$.

Remark. An almost identical argument appears in [5, §2]. There, the reader will find a more careful discussion of the various rather rough assertions we have just made concerning Lipschitz constants and the like.

3. Proof of Szemerédi's Theorem

We now turn to the deduction of Szemerédi's theorem from Proposition 2.1. As described in many places (for example [2]) it follows easily by an iterated application of the following proposition.

Proposition 3.1 (Density increment step). *Suppose that k is an integer and that $\alpha \in (0,1)$ is a parameter. Then there is a number $N_0(k,\alpha)$, a function $\omega_{k,\alpha} : \mathbb{R}^+ \to \mathbb{R}^+$ which tends to infinity and a non-decreasing function $\tau : (0,1) \to \mathbb{R}^+$ such that the following is true. Suppose that P is a progression and that $A \subseteq P$ is a set of size $\alpha|P|$ containing no nontrivial k-term arithmetic progression. Then either $|P| \leqslant N_0(k,\alpha)$, or else there exists another arithmetic progression $P' \subseteq P$, $|P'| \geqslant \omega_{k,\alpha}(|P'|)$, together with a set $A' \subseteq P'$ with $|A'| \geqslant (\alpha + \tau(\alpha))|P'|$ which contains no nontrivial k-term arithmetic progressions.*

In applying this iteratively to establish Szemerédi's theorem, the point is that the second alternative can only occur $O_\alpha(1)$ times before the density of A' inside P' rises above 1, a contradiction.

Proof. In this proof all implied constants are allowed to depend on k and α. By rescaling we may assume that $P = [N]$. Suppose then that $A \subseteq [N]$ is a set with cardinality αN, but that A contains no nontrivial k-term progressions. Define $f := 1_A - \alpha 1_{[N]}$ to be the balanced function of A, thus $\mathbb{E}_{n\in[N]} f(n) = 0$. Write

$$\Lambda_k(f_0, \ldots, f_{k-1}) := \mathbb{E}_{n,d} f_0(n) f_1(n+d) \ldots f_{k-1}\big(n + (k-1)d\big)$$

for the multilinear operator counting k-term arithmetic progressions, and recall the generalised von Neumann theorem, which states that

$$\big|\Lambda_k(f_0, \ldots, f_{k-1})\big| \ll \sup_{i=0,\ldots,k-1} \|f_i\|_{U^{k-1}}.$$

The expression $I := \Lambda_k(1_A, \ldots, 1_A)$ is a normalised count of k-term progressions inside A, and we are supposing that the only such progressions are trivial (that is, have common difference 0). Therefore $I \leqslant 1/N$. On the other hand we may expand this as a sum of 2^k terms, the "main" term $\Lambda_k(\alpha, \ldots, \alpha) = \alpha^k$ plus a sum of $2^k - 1$ other terms, each of which involves at least one copy of f. Supposing that $N > N_0(k,\alpha)$, the main term is much larger than the contribution of $1/N$ from the trivial progressions, and so one of these $2^k - 1$ other terms must be $\gg 1$. By the generalised von Neumann theorem this implies the crucial inequality

$$\|f\|_{U^{k-1}} \gg 1.$$

By the inverse theorem for the Gowers U^{k-1}-norm (classical for $k = 3$, proved in [3] for $k = 4$, in [7] for $k = 5$ and in the forthcoming paper [8] in the general case) this means that there is a degree $\leqslant (k-2)$ polynomial 1-bounded nilsequence $\big(F(g(n)\Gamma)\big)_{n\in\mathbb{Z}}$ of complexity $O(1)$ such that

$$\left|\mathbb{E}_n f(n)F\big(g(n)\Gamma\big)\right| \geqslant \delta,$$

where $\delta \gg 1$. Now we apply Theorem 1.1 to partition $[N]$ into progressions $P_1, \dots, P_m$, each of length $\gg N^c$, such that $\operatorname{diam}_{P_i}\big(F(g(n)\Gamma)\big) \leqslant \delta/2$ for each i. Choose, for each i, some point $n_i \in P_i$. Then

$$\begin{aligned}\delta N &= \sum_i \sum_{n\in P_i} f(n)F\big(g(n_i)\Gamma\big) + \sum_i \sum_{n\in P_i} f(n)\big(F\big(g(n)\Gamma\big) - F\big(g(n_i)\Gamma\big)\big)\\ &\leqslant \sum_i \left|\sum_{n\in P_i} f(n)\right| + \delta N/2,\end{aligned}$$

and therefore $\sum_i \left|\sum_{n\in P_i} f(n)\right| \geqslant \delta N/2$. Adding to this the equality

$$\sum_i \sum_{n\in P_i} f(n) = 0$$

and applying the pigeonhole principle, we conclude that there is at least one progression P_i for which

$$\left|\sum_{n\in P_i} f(n)\right| + \sum_{n\in P_i} f(n) \geqslant \delta|P_i|/2,$$

which means that

$$\sum_{n\in P_i} f(n) \geqslant \delta|P_i|/4.$$

This means that the density of $A' := A \cap P_i$ in P_i is at least $\alpha + \delta/4$, which implies Proposition 3.1. ∎

References

[1] W. T. Gowers, A new proof of Szemerédi's theorem for progressions of length four, *GAFA*, **8** (1998), no. 3, 529–551.

[2] W. T. Gowers, A new proof of Szemerédi's theorem, *GAFA*, **11** (2001), 465–588.

[3] B. J. Green and T. C. Tao, An inverse theorem for the Gowers $U^3(G)$-norm, *Proc. Edin. Math. Soc.*, **51** (2008), 73–153.

[4] B. J. Green and T. C. Tao, The quantitative behaviour of polynomial orbits on nilmanifolds, to appear in *Ann. Math.*, arXiv:0709.3562.

[5] B. J. Green and T. C. Tao, The Möbius function is strongly orthogonal to nilsequences, to appear in *Ann. Math.*, arXiv:0807.1736.

[6] B. J. Green and T. C. Tao, *An arithmetic regularity lemma, associated counting lemma, and applications*, this volume.

[7] B. J. Green, T. C. Tao and T. Ziegler, An inverse theorem for the Gowers U^4-norm, to appear in *Glasgow J. Math.*, arXiv:0911.5681.

[8] B. J. Green, T. C. Tao and T. Ziegler, *An inverse theorem for the Gowers $U^{s+1}[N]$-norm* (announcement), arXiv:1006.0205.

[9] K. F. Roth, On certain sets of integers, *J. London Math. Soc.*, **28** (1953), 104–109.

[10] E. Szemerédi, On sets of integers containing no four elements in arithmetic progression, *Acta Math. Acad. Sci. Hungar.*, **20** (1969), 89–104.

[11] E. Szemerédi, On sets of integers containing no k elements in arithmetic progression, *Acta Arith.*, **27** (1975), 299–345.

Ben Green
Centre for Mathematical Sciences
Wilberforce Road
Cambridge CB3 0WA
England
e-mail: b.j.green@dpmms.cam.ac.uk

Terence Tao
Department of Mathematics
UCLA
Los Angeles, CA 90095
USA
e-mail: tao@math.ucla.edu

BOLYAI SOCIETY
MATHEMATICAL STUDIES, 21

An Irregular Mind
(Szemerédi is 70)
pp. 343–357.

Online Linear Discrepancy of Partially Ordered Sets

MITCHEL T. KELLER, NOAH STREIB and WILLIAM T. TROTTER

This article is dedicated to Professor Endre Szemerédi on the occasion of his 70th *birthday. Among his many remarkable contributions to combinatorial mathematics and theoretical computer science is a jewel for online problems for partially ordered sets: the fact that* $h(h+1)/2$ *antichains are required for an online antichain partition of a poset of height* h.

The linear discrepancy of a poset $\mathbf{P}$ is the least k for which there is a linear extension L of $\mathbf{P}$ such that if x and y are incomparable in $\mathbf{P}$, then $|h_L(x) - h_L(y)| \leq k$, where $h_L(x)$ is the height of x in L. In this paper, we consider linear discrepancy in an online setting and devise an online algorithm that constructs a linear extension L of a poset P so that $|h_L(x) - h_L(y)| \leq 3k - 1$, when the linear discrepancy of P is k. This inequality is best possible, even for the class of interval orders. Furthermore, if the poset P is a semiorder, then the inequality is improved to $|h_L(x) - h_L(y)| \leq 2k$. Again, this result is best possible.

1. Introduction

The concept of linear discrepancy for finite partially ordered sets (posets) was introduced by Tanenbaum, Trenk, and Fishburn in [9] and represents an effort to model a notion of "fairness", i.e., incomparable elements should be placed close together in a linear extension to avoid an implicit comparison being made when one has much greater height than the other.

In this paper, we will consider linear discrepancy in an online setting. We will show that there is an online algorithm that will construct a linear extension L of a poset $\mathbf{P}$ so that any pair of incomparable points in $\mathbf{P}$ are at distance at most $3k-1$ in L when the linear discrepancy of $\mathbf{P}$ is k. When the poset is a semiorder, this distance can be lowered to $2k$. Both results are best possible.

The remainder of this paper is organized as follows. In the next section, we develop essential notation and terminology and discuss briefly some related work that motivates our line of research. The proofs of lower bounds for the inequalities in our main theorems are given in Section 4, while the proofs for upper bounds are given in Section 5.

2. Notation, Terminology and Background

A partially ordered set (poset) $\mathbf{P}$ is a pair (X, P) where X is a finite *ground set* and P is a reflexive, antisymmetric and transitive binary relation on X. Rather than write $(x, y) \in P$, it is more common to write $x \leq y$ in P, and we just write $x \leq y$ when the partial order P is clear from the context. Of course, we write $x < y$ in P when $x \leq y$ in P and $x \neq y$. Distinct points x and y are *comparable* in P when either $x < y$ in P or $y < x$ in P; else they are *incomparable*, in which case we write $x \parallel y$ in P.

We let $D(x) = \{y \in X \mid y < x\}$ and call $D(x)$ the down-set of x. The up-set of x, $U(x)$, is defined dually. We let $\operatorname{Inc}(x) = \{y \in X \mid y \parallel x\}$ and define[1] $\Delta(\mathbf{P}) = \max_{x \in X} |\operatorname{Inc}(x)|$. Throughout this paper, we adopt the standard convention of denoting a poset by a single symbol, so we write for example, $x \in \mathbf{P}$, $x < y$ in $\mathbf{P}$ and $z \parallel w$ in $\mathbf{P}$.

For a positive integer n, let $[n] = \{1, 2, \ldots, n\}$, and let $\mathbf{n}$ denote a linear order on n points, typically labeled as $0 < 1 < 2 < \cdots < n-1$. If $\mathbf{P}$ and $\mathbf{Q}$ are posets on disjoint ground sets, $\mathbf{P} + \mathbf{Q}$ denotes the disjoint union of $\mathbf{P}$ and $\mathbf{Q}$. Also, when $\mathbf{P}$ does not contain a subposet which is isomorphic to $\mathbf{Q}$, we say $\mathbf{P}$ *excludes* $\mathbf{Q}$.

The reader may find it helpful to consult Trotter's monograph [10] and survey article [11] for additional background material on combinatorial problems for partially ordered sets.

[1] This notation is nonstandard. In other settings, $\Delta(\mathbf{P})$ denotes the maximum degree in the comparability graph of $\mathbf{P}$.

2.1. Interval Orders and Semiorders

A poset $\mathbf{P}$ is called an *interval order* when there exists a function I assigning to each element $x \in \mathbf{P}$ a closed interval $I(x) = \big[l(x), r(x)\big]$ of the real line $\mathbb{R}$ such that $x < y$ in P if and only if $r(x) < l(y)$ in $\mathbb{R}$. We call the family $\big\{\big[l(x), r(x)\big] \,:\, x \in \mathbf{P}\big\}$ of intervals an *interval representation* of $\mathbf{P}$. It is easy to see that when $\mathbf{P}$ is an interval order, it has an interval representation with distinct endpoints. It is a well-known result of Fishburn [2] that a poset is an interval order if and only if it excludes $\mathbf{2}+\mathbf{2}$.

A *semiorder* is an interval order having an interval representation in which all intervals have length 1. Scott and Suppes [8] showed that an interval order is a semiorder if and only if it excludes $\mathbf{3}+\mathbf{1}$.

2.2. Linear Discrepancy

Let $\mathbf{P} = (X, P)$ be a poset. A linear order L on X is called a *linear extension* of $\mathbf{P}$ if $x < y$ in L whenever $x < y$ in P. When L is a linear extension of $\mathbf{P}$ and $x \in \mathbf{P}$, the quantity $\big|\{y \in \mathbf{P} \mid y \le x \text{ in } L\}\big|$ is called the *height* of x in L and is denoted $h_L(x)$.

The *linear discrepancy* of a linear extension L of $\mathbf{P}$, denoted $\operatorname{ld}(\mathbf{P}, L)$, is the least non-negative integer k so that $\big|h_L(x) - h_L(y)\big| \le k$ whenever $x \parallel y$ in P. Note that $\operatorname{ld}(\mathbf{P}, L) = 0$ if and only if $\mathbf{P}$ is a linear order. Now let $\mathscr{E}(\mathbf{P})$ denote the family of all linear extensions of $\mathbf{P}$. The *linear discrepancy* of a poset $\mathbf{P}$, denoted $\operatorname{ld}(\mathbf{P})$, is then defined by

$$\operatorname{ld}(\mathbf{P}) := \min\big\{\operatorname{ld}(\mathbf{P}, L) \,:\, L \in \mathscr{E}(\mathbf{P})\big\}$$

We note that the parameter $\operatorname{ld}(\mathbf{P}, L)$ is called the uncertainty of L in [9].

Fishburn, Tannenbaum and Trenk [3] showed that the linear discrepancy of a poset is equal to the bandwidth of its cocomparability graph. The same authors noted in [9] that it follows from the work of Kloks, Kratsch, and Müller [6] on bandwidth that determining whether the linear discrepancy of a poset $\mathbf{P}$ is at most k is NP-complete.

In spite of the fact that the linear discrepancy of a poset is difficult to compute, it is very easy to approximate. The following result (with different notation) is given in [6].

Theorem 1. *If* $\mathbf{P}$ *is a poset, then* $\operatorname{ld}(\mathbf{P}, L) \le 3 \operatorname{ld}(\mathbf{P})$ *for every linear extension* L *of* $\mathbf{P}$.

The inequality of Theorem 1 is tight for all $d \geq 1$. This can be seen by considering the following poset with $3d+1$ points: let A, B, and C be chains with d points, $d-1$ points, and d points, respectively, such that $a < b < c$ for all points $a \in A$, $b \in B$, and $c \in C$. Let x and y be the remaining points such that $x \parallel a$, $x \parallel b$, $x < c$, $y > a$, $y \parallel b$, and $y \parallel c$, for all $a \in A$, $b \in B$, and $c \in C$. The linear extension with $A < x < B < y < C$ has discrepancy d, whereas the linear extension with $x < A < B < C < y$ has discrepancy $3d$.

We note here some key properties of linear discrepancy that will prove useful later in this paper.

Lemma 2.

(1) *Linear discrepancy is monotonic, i.e., if* $\mathbf{P}$ *is a subposet of* $\mathbf{Q}$, *then* $\operatorname{ld}(\mathbf{P}) \leq \operatorname{ld}(\mathbf{Q})$.

(2) *If* $\mathbf{P}$ *is an* n*-element antichain, then* $\operatorname{ld}(\mathbf{P}) = n-1$, *so* $\operatorname{ld}(\mathbf{P}) \geq \operatorname{width}(\mathbf{P}) - 1$.

(3) $\Delta(\mathbf{P})/2 \leq \operatorname{ld}(\mathbf{P}) \leq 2\Delta(\mathbf{P}) - 2$.

The nontrivial upper bound in the third statement of Lemma 2 is proved in [7], and we note that it remains open to settle whether the upper bound here can be improved to $\lfloor (3\Delta(\mathbf{P}) - 1)/2 \rfloor$. Special cases have been resolved in [1, 4, 7, 9].

In discussions to follow, we say a linear extension L of $\mathbf{P}$ is *optimal* if $\operatorname{ld}(\mathbf{P}, L) = \operatorname{ld}(\mathbf{P})$. It is shown in [9] that $\operatorname{ld}(\mathbf{P}) = \operatorname{width}(\mathbf{P}) - 1$, when $\mathbf{P}$ is a semiorder. To see this, just take an interval representation $\{[l(x), l(x)+1] : x \in X\}$ for $\mathbf{P}$ in which all endpoints are distinct. Let L be the linear extension of $\mathbf{P}$ defined by setting $x < y$ in L whenever $l_x < l_y$ in $\mathbb{R}$.

In [4], Keller and Young show that for an interval order $\mathbf{P}$, $\operatorname{ld}(\mathbf{P}) \leq \Delta(\mathbf{P})$ with equality if and only if $\mathbf{P}$ contains an antichain of size $\Delta(\mathbf{P}) + 1$. They also show that this bound is tight even for interval orders of width 2.

3. Online Linear Discrepancy

In this paper, we consider linear discrepancy in an online setting. A *Builder* constructs a poset $\mathbf{P}$ from a class $\mathscr{P}$ of posets, one point at a time, and an *Assigner* assembles a linear extension L of $\mathbf{P}$, one point at a time. Even though the poset **P** and the linear extension L change with time, we use a single symbol for each. When Builder expands the poset **P** by adding a new point x, he will list those points presented previously that are (1) less than x in **P**, and (2) greater than x in **P**. Assigner will then insert x into a legal position in the linear extension L she had previously constructed just before x entered.

Given a class $\mathscr{P}$ of posets and an integer $k \geq 1$, we will investigate strategies for Builder that will enable him to construct a poset **P** from $\mathscr{P}$ with $\text{ld}(\mathbf{P}) \leq k$ so that Assigner will be forced to assemble a linear extension L of P with $\text{ld}(\mathbf{P}, L)$ much larger than k. We will also study algorithms for Assigner that will enable her to assemble a linear extension L of **P** with $\text{ld}(\mathbf{P}, L)$ relatively close to $\text{ld}(\mathbf{P})$. Of course, the inequality of Theorem 1 looms large in our discussion.

We will consider two different ways in which Builder can construct interval orders and semiorders. One way is for Builder to construct the poset one point at a time, just by listing the comparabilities for the new point x. It is easy for Assigner to be assured that Builder stays within the appropriate class by appealing to their characterization in terms of forbidden subposets.

However, we will also discuss the situation where Builder constructs an interval order or a semiorder by providing an interval representation one interval at a time. In this setting, Builder provides a closed interval $[l(x), r(x)]$ (with $r(x) = 1 + l(x)$ when **P** is a semiorder) for the new point. As we will see, Assigner will find this additional information quite valuable in constructing a linear extension which has linear discrepancy close to the optimal value.

With this notation and terminology in place, we can now give a formal statement of our principal theorems.

Theorem 3. *Let k be a positive integer. There is an online algorithm $\mathscr{A}$ for Assigner so that:*

(1) *If Builder constructs an arbitrary poset **P** with $\text{ld}(\mathbf{P}) = k$ and Assigner assembles a linear extension L using Algorithm $\mathscr{A}$, then $\text{ld}(\mathbf{P}, L) \leq 3k - 1$. This inequality is best possible, even if Builder is required to construct an interval order.*

(2) *If Builder constructs an semiorder **P** with $\text{ld}(\mathbf{P}) = k$ and Assigner assembles a linear extension L using Algorithm $\mathscr{A}$, then $\text{ld}(\mathbf{P}, L) \leq 2k$. This inequality is best possible.*

We will also prove the following result when Builder provides an interval representation.

Theorem 4. *Let k be a positive integer. There is an online algorithm $\mathscr{L}$ for Assigner so that:*

(1) *If Builder constructs an interval order* $\mathbf{P}$ *with* $\mathrm{ld}\,(\mathbf{P}) = k$ *by providing an interval representation and Assigner assembles a linear extension* L *using Algorithm* $\mathscr{L}$*, then* $\mathrm{ld}\,(\mathbf{P}, L) \leq 2k$*. This inequality is best possible.*

(2) *If Builder constructs a semiorder* $\mathbf{P}$ *with* $\mathrm{ld}\,(\mathbf{P}) = k$ *by providing an interval representation and Assigner assembles a linear extension* L *using Algorithm* $\mathscr{L}$*, then* L *is optimal.*

Before proceeding to the proofs of these two theorems, we pause to note that there is a substantive difference in the outcome when Builder is required to give an interval representation online. On the other hand, it can be seen in [5] that for online chain partitioning (and online graph coloring), there is no distinction between the two versions.

4. Lower Bounds for Online Linear Discrepancy

In this section, we provide strategies for Builder which establish lower bounds for the inequalities in Theorems 3 and 4.

Lemma 5. *For each* $k \geq 1$*, Builder can construct an interval order* $\mathbf{P}$ *with* $\mathrm{ld}\,(\mathbf{P}) = k$ *so that Assigner will be forced to assemble a linear extension* L *with* $\mathrm{ld}\,(\mathbf{P}, L) \geq 3k - 1$.

Proof. Builder constructs a poset $\mathbf{P}$ as follows. First, he presents a $k+1$-element antichain. After Assigner has linearly ordered these $k+1$ elements, Builder labels the L-least point as x and the others as members of an k-element antichain A. He then presents another $k+1$-element antichain, with all elements of the new antichain larger than all elements of $A \cup \{x\}$. After assigner has extended L, Builder labels the L-largest element as z and the other elements as as members of a k-element antichain D.

Builder then presents two new elements u and y with

(1) $a < u < d$ and $a < y < d$ for all $a \in A$ and $d \in D$, and

(2) $x < u < z$, $x \parallel y$ and $z \parallel y$.

By symmetry (up to duality), we may assume Assigner makes $y > u$ in L.

Next, Builder inserts two antichains, B and C, of sizes $k-1$ and $k-2$, respectively, so that:

(1) $x, a < b < c < u$ in P, for all $b \in B$ and $c \in C$.

(2) $y \parallel b$ and $y \parallel c$, for all $b \in B$ and $c \in C$.

Assigner must then insert all elements of B as a block of $k-1$ consecutive elements immediately over the L-largest element of A. Assigner must also insert all elements of C as a block immediately over the highest element of B. It follows that $h_L(y) - h_L(x) = 3k - 1$.

On the other hand, the linear order: $A < x < B < y < C < u < z < D$ shows that $\mathrm{ld}(\mathbf{P}) \leq k$. Also, it is also easy to see that $\mathbf{P}$ is an interval order with representation

$$A = [1,1], \qquad x = [1,2], \qquad D = [7,7], \qquad z = [6,7],$$

$$y = [2,6], \qquad u = [5,5], \qquad B = [3,3], \qquad C = [4,4],$$

where $A = [1,1]$, for instance, denotes that each element of A is represented by a copy of the interval $[1,1]$. ∎

We note that the interval order $\mathbf{P}$ constructed in Lemma 5 is not a semiorder, since it contains a subposet isomorphic to $\mathbf{3}+\mathbf{1}$. In fact, it contains a subposet isomorphic to $\mathbf{4}+\mathbf{1}$. We do not know whether Assigner can be forced to construct a linear extension L with $\mathrm{ld}(\mathbf{P}, L) = 3k-1$ if Builder is restricted to interval orders that exclude $\mathbf{4}+\mathbf{1}$. We have been able to show that within this class, Builder can force Assigner to assemble a linear extension L with $\mathrm{ld}(\mathbf{P}, L) \geq 2k + (k-1)/2$, but do not know if this is tight.

Lemma 6. *For each* $k \geq 1$, *Builder can construct a semiorder* $\mathbf{P}$ *with* $\mathrm{ld}(\mathbf{P}) = k$ *so that Assigner will be forced to assemble a linear extension* L *with* $\mathrm{ld}(\mathbf{P}, L) \geq 2k$.

Proof. Builder constructs a poset $\mathbf{P}$ as follows. First, he presents an antichain of size $k+1$. When Assigner has linearly ordered these $k+1$ elements, Builder labels the L-least element as x and the remaining elements as members of a k-element antichain A. He then presents a k-element antichain B with

(1) $a < b$ in $\mathbf{P}$, for all $a \in A$ and $b \in B$, and

(2) $x \parallel b$, for all b in B.

Let b be the L-highest element of B. It follows that $h_L(b) - h_L(x) = 2k$, but the linear order $A < x < B$ shows that $\mathrm{ld}\,(\mathbf{P}) \leq 2k$. Also, it is clear that $\mathbf{P}$ is a semiorder. ∎

Lemma 7. *For each $k \geq 1$, Builder can construct an interval representation of an interval order $\mathbf{P}$ with $\mathrm{ld}\,(\mathbf{P}) = k$ so that Assigner will be forced to assemble a linear extension L with $\mathrm{ld}\,(\mathbf{P}, L) = 2k$.*

Proof. Builder will construct a poset which is isomorphic to $\mathbf{2k+1}$, which has linear discrepancy k. He begins by presenting the intervals $[1, 4k-1]$ and $[2k, 2k]$. If Assigner sets $[1, 4k-1] < [2k, 2k]$ in L, then Builder presents $[2k+1, 2k+1]$, $[2k+2, 2k+2]$, ..., $[4k-1, 4k-1]$. Assigner has no choice but to put these $2k-1$ intervals above $[2k, 2k]$ and therefore constructs a linear extension L of (a copy of) $\mathbf{2k+1}$ with $\mathrm{ld}\,(\mathbf{P}, L) = 2k$.

Similarly, if Assigner makes $[2k, 2k] < [1, 4k-1]$ in L, Builder proceeds to present $[1, 1]$, $[2, 2]$, ..., $[2k-1, 2k-1]$, and again Assigner is forced to assemble a linear extension L with $\mathrm{ld}\,(\mathbf{P}, L) = 2k$. ∎

5. Upper Bounds for Online Linear Discrepancy

In this section, we provide algorithms for Assigner to use in assembling a linear extension of a poset constructed by Builder. We note that these algorithms do not need to know anything about the class from which $\mathbf{P}$ arises, and they do not need to know anything about the linear discrepancy of $\mathbf{P}$. We focus first on posets constructed one point at a time, and later consider the case where Builder provides an interval representation of an interval order (or semiorder).

However, in order to motivate the specifics of our algorithm, we pause briefly to give examples showing that other reasonable strategies are non-optimal, even for the class of semiorders.

5.1. Naïve online linear discrepancy algorithms for semiorders

Perhaps the simplest online linear discrepancy algorithm that Assigner might use works as follows. When presented with a new point x, there is always a non-empty set of positions where x could be legally inserted. These positions are always consecutive in the linear extension L. A reasonable strategy is to insert x as close to the middle of these allowable positions as possible, say rounding down when there is an odd number of them. We refer to this algorithm as $\mathscr{M}$, since it places x in the middle of its allowable range.

However, we claim that for each $k \geq 1$, Builder can construct a semiorder $\mathbf{P}$ with $\operatorname{ld}(\mathbf{P}) = k$ so that Assigner is forced to assemble a linear extension L with $\operatorname{ld}(\mathbf{P}, L) = 3k - 1$, provided she uses algorithm $\mathscr{M}$. As in the proof of Lemma 5, Builder starts by presenting two $k+1$-element antichains with all points of one less than all points of the other. Builder labels the L-least element as x and the L-greatest element as z with the remaining elements belonging to $A \cup D$ so that $a < d$ in $\mathbf{P}$, for every $a \in A$ and $d \in D$.

Builder then presents an element y with

(1) $x \parallel y$ and $z \parallel y$ in $\mathbf{P}$, and

(2) $a < y < d$ in $\mathbf{P}$, for all $a \in A$ and $d \in D$.

Next, Builder presents the elements of an $k-1$-element antichain C with

(1) $x < c$, $y \parallel c$ and $z > c$ in $\mathbf{P}$, for all $c \in C$, and

(2) $a < c < d$ in $\mathbf{P}$, for all $a \in A$, $c \in C$ and $d \in D$.

Using Algorithm $\mathscr{M}$, Assigner will place all elements of C as a block of consecutive elements immediately under y. Assigner then presents an antichain B of size $k-1$ with

(1) $x \parallel b$, $y \parallel b$ and $b < z$ in $\mathbf{P}$, for all $b \in B$, and

(2) $a < b < c$ in $\mathbf{P}$, for all $a \in A$, $b \in B$ and $c \in C$.

Assigner must then insert all elements of B as a block in between the highest element of A and the lowest element of C. It follows that $h_L(y) - h_L(x) = 3k - 1$. On the other hand, the linear order $A < x < B < y < C < z < D$ shows that $\operatorname{ld}(\mathbf{P}) \leq k$. Furthermore, $\mathbf{P}$ is easily seen to be a semiorder.

We now consider a second algorithm that seems intuitive yet fails to be optimal. When presented with a new point x, this algorithm inserts x into the position that minimizes the linear discrepancy of the resulting linear extension, breaking ties by placing x as low as possible. Since this algorithm is in some sense greedy in its operation, we denote it by $\mathscr{G}$. Although we do not include the proof, it is straightforward to modify the argument for Algorithm $\mathscr{M}$ to verify the following claim: Builder can construct a semiorder $\mathbf{P}$, with $\mathrm{ld}(\mathbf{P}) = k$, that will force an Assigner using the $\mathscr{G}$ algorithm to order points as $x < A < B' < y < B'' < C < D < w$, where $B' \cup B'' = B$ and $|B'| = \lfloor |B|/2 \rfloor$. This linear extension has linear discrepancy $\lceil (k-1)/2 \rceil + 2k$.

We comment that it is not difficult to construct examples showing that $\mathscr{M}$, when applied to general posets, can be forced to construct linear extensions with linear discrepancy $3k$ (rather than $3k-1$) for posets of linear discrepancy k.

5.2. An optimal online linear discrepancy algorithm

Let $\mathbf{P}$ be a poset and let (x, y) be an ordered pair of elements from $\mathbf{P}$. We call (x, y) a *critical pair* if (1) $x \parallel y$ in $\mathbf{P}$; (2) $D(x) \subseteq D(y)$ in $\mathbf{P}$; and (3) $U(y) \subseteq U(x)$ in $\mathbf{P}$. If (x, y) is a critical pair and (y, x) is not a critical pair, then we call (x, y) a *one-way critical pair.* A linear extension L of a poset $\mathbf{P}$ is said to *reverse* a critical pair (x, y) when $y < x$ in L. The concept of critical pairs first surfaced in dimension theory, as it is easy to see that the dimension of poset $\mathbf{P}$ is the least positive integer t for which there exists a family $\{L_1, L_2, \ldots, L_t\}$ of linear extensions so that for every critical pair (x, y) in $\mathbf{P}$, there is some i for which (x, y) is reversed in L_i. The reader can find much more information on the role played by linear extensions in reversing critical pairs in [10].

By contrast, linear discrepancy is all about *preserving* critical pairs, as the following elementary but important proposition (see [4]) prevails:

Proposition 8. *Let $\mathbf{P}$ be a poset which is not a total order.*

(1) *If L is a linear extension of $\mathbf{P}$ and x and y are incomparable points with $h_L(y) - h_L(x) = \mathrm{ld}(\mathbf{P}, L)$, then (x, y) is a critical pair in $\mathbf{P}$.*

(2) *There exists an optimal linear extension L of $\mathbf{P}$ so that if (x, y) is a critical pair reversed by L, then (y, x) is also a critical pair.*

Accordingly, the online algorithm $\mathscr{A}$ we will define here endeavors to construct a linear extension that reverses few critical pairs. However, we note that in discussing critical pairs in an online setting, it may happen that a pair (x, y) is critical at one moment in time but is no longer critical at a later moment in time. The converse statement cannot hold.

Suppose a new point x enters the poset. Assigner considers the one-way critical pairs of the form (x, u), where u has already entered, as well as the one-way critical pairs of the form (v, x), where again v has already entered. If there are no one-way critical pairs of either type, then x is inserted in any legal position.

Suppose there are only one-way critical pairs of the form (x, u) but none of the other type. Let let u_0 be the lowest element of L for which (x, u_0) is a one-way critical pair. Insert x in any legal position which is under u_0. There is such a position, since $D(x) \subseteq D(u_0)$. Dually, if there are only one-way critical pairs of the form (v, x) but none of the other type, let v_0 be the highest element of L for which (v_0, x) is a one-way critical pair. Insert x in any legal position over v_0.

We are left to consider the case where there are one-way critical pairs of both types. Now let u_0 and v_0 be defined as above. Insert x in any legal position between u_0 and v_0. In making this statement, we note that u_0 and v_0 can occur in either order in L. If $v_0 < u_0$ in L, then some positions between u_0 and v_0 may be illegal, but there is at least one position between them which is legal. On the other hand, if $u_0 < v_0$ in L, then all positions between them are legal.

To analyze the behavior of $\mathscr{A}$, we require the following elementary proposition.

Proposition 9. *If (x, y) and (y, z) are critical pairs in a poset* $\mathbf{P}$*, then either $x < z$ in* $\mathbf{P}$ *or (x, z) is a critical pair in* $\mathbf{P}$.

Next we establish a key lemma concerning how $\mathscr{A}$ handles a configuration we denote by C. This configuration consists of four points x, y, z, w. Among these points, we require

(1) $z > w$, $z > x$, $y > w$, and $x \parallel y$ in $\mathbf{P}$, and

(2) (y, z) and (w, x) are critical pairs in $\mathbf{P}$.

Note that conditions (1) and (2) imply that (y, z) and (w, x) are in fact one-way critical pairs.

Lemma 10. *Algorithm $\mathscr{A}$ never constructs a linear extension L such that points forming a copy of C are ordered as $x < w < z < y$ in L, all points less than x in L are less than y in* $\mathbf{P}$*, and all points greater than y in L are greater than x in* $\mathbf{P}$.

Proof. We argue by contradiction. Consider the first time a copy of C is placed according to the conditions of the lemma. Notice that the last point presented must be one of x, y, w, or z, as the relationships amongst only those points induce C. By duality, it suffices to consider only the cases where x or w is the last point to enter the poset.

We first consider the possibility that x was the last point presented. Since $x < w$ in L and (w, x) is a critical pair, there is a point x' such that (x, x') is a critical pair and $x' < x$ in L. Since (x, x') is a critical pair, we must have that $x' \parallel y$. This contradicts the fact that x, y, w, and z are placed according to the conditions of the lemma.

Now suppose w was the last point presented. Since $x < w$ in L and (w, x) is a critical pair, there is a point w' such that (w', w) is a critical pair but $w < w'$ in L. Now (w', x) is also a critical pair. Since (w', w) is a critical pair, we must have $w' < z$ and $w' < y$. Hence, $\{x, y, z, w'\}$ forms a copy of C placed in the forbidden order at an earlier stage, a contradiction. ■

With Lemma 10 in hand, we are in position to analyze the performance of algorithm $\mathscr{A}$. We first consider the case of arbitrary posets in the following lemma.

Lemma 11. *Let $k \geq 1$. If Builder constructs a poset* $\mathbf{P}$ *with* $\operatorname{ld}(\mathbf{P}) = k$*, and Assigner uses Algorithm $\mathscr{A}$, she will assemble a linear extension L with* $\operatorname{ld}(\mathbf{P}, L) \leq 3k - 1$.

Proof. Let L be the linear extension of $\mathbf{P}$ assembled by Assigner. We show that $\operatorname{ld}(\mathbf{P}, L) \leq 3k - 1$. To the contrary, suppose that $\operatorname{ld}(\mathbf{P}, L) = 3k$. Consider the first moment in time where there are points x and y with $x \parallel y$ and $h_L(y) - h_L(x) = 3k$. Let $S = \{s \mid x < s < y \text{ in } L\}$ and note that $|S| = 3k - 1$. Also note that (x, y) is a critical pair in $\mathbf{P}$.

Let M be an optimal linear extension of $\mathbf{P}$. In view of Proposition 8, we may assume $x < y$ in M. Now let $Z = \{z \in S \mid y < z \text{ in } M\}$ and $W = \{w \in S \mid w < x \text{ in } M\}$. It follows that $z \parallel y$ for all $z \in Z$, so $|Z| \leq k$. Similarly, $|W| \leq k$. On the other hand, since $h_M(y) - h_M(x) \leq k$, we know $\bigl|S - (Z \cup W)\bigr| \leq k - 1$. Thus $|Z| = |W| = k$ and $\bigl|S - (Z \cup W)\bigr| = k - 1$.

Now let z be the M-largest element of Z and let w be the M-least element of W. It follows that $h_M(x) - h_M(w) \geq k$, but since M is optimal, we know $h_M(x) - h_M(w) = k$ and thus (w, x) is a critical pair in $\mathbf{P}$. Similarly, (y, z) is a critical pair in $\mathbf{P}$. Also $z > w$ in $\mathbf{P}$. It follows that x, w, z, and y form a configuration C which L orders as $x < w < z < y$. Furthermore, all points less than x in L are less than y in $\mathbf{P}$ and all points greater than y in L are greater than x in $\mathbf{P}$ since $h_L(y) - h_L(x) = \mathrm{ld}(\mathbf{P}, L)$. This contradicts Lemma 10. ■

Lemma 12. *Let $k \geq 1$. If Builder constructs a semiorder $\mathbf{P}$ with $\mathrm{ld}(\mathbf{P}) = k$, and Assigner uses Algorithm $\mathscr{A}$, she will assemble a linear extension L with $\mathrm{ld}(\mathbf{P}, L) \leq 2k$.*

Proof. Let L be the linear extension of $\mathbf{P}$ assembled by Assigner. We show that $\mathrm{ld}(\mathbf{P}, L) \leq 2k$. To the contrary, suppose that $\mathrm{ld}(\mathbf{P}, L) \geq 2k + 1$. Consider the first moment in time where there are points x and y with $x \parallel y$ and $h_L(y) - h_L(x) \geq 2k + 1$. Let $S = \{s \mid x < s < y \text{ in } L\}$ and note that $|S| = 2k$. Also note that (x, y) is a critical pair in $\mathbf{P}$.

Let M be an optimal linear extension of $\mathbf{P}$. In view of Proposition 8, we may assume $x < y$ in M. Now let $Z = \{z \in S \mid x < z \text{ in } \mathbf{P}\}$ and $W = \{w \in S \mid w < y \text{ in } \mathbf{P}\}$. If $Z = \emptyset$, then $x \parallel s$ for every $s \in S$, and since $x \parallel y$, we know that $\Delta(\mathbf{P}) \geq 2k + 1$ which would imply that $\mathrm{ld}(\mathbf{P}) \geq k + 1$. The contradiction forces Z to be nonempty. Similarly, $W \neq \emptyset$.

Choose $z \in Z$ and $w \in W$. Since $x \parallel y$, we must have $z \parallel y$ and $w \parallel x$ in $\mathbf{P}$. Since $\mathbf{P}$ is a semiorder and a semiorder is a special case of an interval order, we must have $z > w$ in $\mathbf{P}$. Observe that in a semiorder, whenever we have two distinct incomparable points u and v, then (at least) one of (u, v) and (v, u) is a critical pair. However, this implies that both of (y, z) and (w, x) are critical pairs in $\mathbf{P}$, while neither of (z, y) nor (x, w) is a critical pair. It follows that x, w, z, and y form the configuration C in the order forbidden by Lemma 10, a contradiction. ■

5.3. Online interval representations

We now turn our attention to the situation where Builder constructs an interval order (or semiorder) by providing an interval representation, one interval at a time. Now Assigner will use the following algorithm, which we denote $\mathscr{L}$. A new point x comes with an interval $\left[l(x), r(x)\right]$ and this interval is fixed in time. Assigner will then insert x into L so that elements are ordered by the left endpoints of their intervals, i.e., $u < v$ in L whenever $l(u) < l(v)$ in $\mathbb{R}$. Ties can be broken arbitrarily.

Lemma 13. *Let $k \geq 1$. If Builder constructs an interval order $\mathbf{P}$ by providing an interval representation, with $\mathrm{ld}(\mathbf{P}) = k$, and Assigner uses Algorithm $\mathscr{L}$, she will assemble a linear extension L with $\mathrm{ld}(\mathbf{P}, L) \leq 2k$.*

Proof. Let L be the linear extension of $\mathbf{P}$ assembled by Assigner. We show that $\mathrm{ld}(\mathbf{P}, L) \leq 2k$. To the contrary, suppose that $\mathrm{ld}(\mathbf{P}, L) \geq 2k + 1$. Clearly, we may stop when L first violates the conclusion. So we may choose points x and y so that $h_L(y) - h_L(x) = 2k + 1$.

Let $S = \{s \mid x < s < y \text{ in } L\}$. If $S \subseteq \mathrm{Inc}(x)$, then it follows that $\mathrm{ld}(\mathbf{P}) \geq \lceil \Delta(\mathbf{P})/2 \rceil \geq k + 1$. So there is a point $z \in S$ with $x < z$ in $\mathbf{P}$. It follows that $l(x) \leq r(x) < l(z) \leq l(y)$ in $\mathbb{R}$, which is a contradiction, since it implies that $x < y$ in $\mathbf{P}$. ∎

We state the analogous result for semiorders, noting that there is nothing to prove, as we are simply restating the well-known characterization of optimal linear extensions of semiorders.

Lemma 14. *Let $k \geq 1$. If Builder constructs a semiorder $\mathbf{P}$ by providing an interval representation, with $\mathrm{ld}(\mathbf{P}) = k$ and Assigner uses Algorithm $\mathscr{L}$, then she will assemble an optimal linear extension L.*

Acknowledgments. The authors would like to thank David M. Howard and Stephen J. Young for listening to our arguments on this subject and helping us improve our explanations.

References

[1] J.-O. Choi and D. B. West, *Linear discrepancy and products of chains,* submitted, 2008.

[2] P. C. Fishburn, Intransitive indifference with unequal indifference intervals, *J. Math. Psych.,* **7** (1970), 144–149.

[3] P. C. Fishburn, P. J. Tanenbaum, and A. N. Trenk, Linear discrepancy and bandwidth, *Order,* **18** (2001), 237–245.

[4] M. T. Keller and S. J. Young, Degree bounds for linear discrepancy of interval orders and disconnected posets, *Discrete Math.,* to appear (2010).

[5] H. Kierstead and W. T. Trotter, An extremal problem in recursive combinatorics, *Congressus Numerantium,* **33** (1981), 143–153.

[6] T. Kloks, D. Kratsch and H. Müller, Approximating the bandwidth for asteroidal triple-free graphs, *J. Algorithms,* **32, 1** (1999), 41–57.

[7] D. Rautenbach, A note on linear discrepancy and bandwidth, *J. Combin. Math. Combin. Comput.,* **55** (2005), 199–208.

[8] D. Scott and P. Suppes, Foundational aspects of theories of measurement, *J. Symb. Logic,* **23** (1958), 113–128.

[9] P. J. Tanenbaum, A. N. Trenk and P. C. Fishburn, Linear discrepancy and weak discrepancy of partially ordered sets, *Order,* **18, 3** (2001), 201–225.

[10] W. T. Trotter, *Combinatorics and partially ordered sets: Dimension theory,* Johns Hopkins Series in the Mathematical Sciences. Johns Hopkins University Press, Baltimore, MD, 1992.

[11] W. T. Trotter, Partially ordered sets, in: *Handbook of combinatorics,* Vol. 1, 2. Elsevier, Amsterdam, 1995, pp. 433–480.

Mitchel T. Keller
Noah Streib
William T. Trotter

School of Mathematics
Georgia Institute of Technology
Atlanta, Georgia 30332-0160
U.S.A.

e-mails: keller@math.gatech.edu
nstreib3@math.gatech.edu
trotter@math.gatech.edu

BOLYAI SOCIETY
MATHEMATICAL STUDIES, 21

An Irregular Mind
(Szemerédi is 70)
pp. 359–404.

On the Triangle Removal Lemma for Subgraphs of Sparse Pseudorandom Graphs

YOSHIHARU KOHAYAKAWA*†, VOJTĚCH RÖDL†, MATHIAS SCHACHT*‡ and JOZEF SKOKAN

Dedicated to Endre Szemerédi on the occasion of his 70th birthday

We study an extension of the *triangle removal lemma* of Ruzsa and Szemerédi [*Triple systems with no six points carrying three triangles*, Combinatorics (Proc. Fifth Hungarian Colloq., Keszthely, 1976), Vol. II, North-Holland, Amsterdam, 1978, pp. 939–945], which gave rise to a purely combinatorial proof of the fact that sets of integers of positive upper density contain three-term arithmetic progressions, a result first proved by Roth [*On certain sets of integers*, J. London Math. Soc. **28** (1953), 104–109].

We obtain a generalization of the triangle removal lemma for subgraphs of sparse pseudorandom graphs and deduce the following version of Roth's theorem, which applies to sparse sets of integers: *If* $A \subseteq [n] = \{1, \ldots, n\}$ *has the property that all non-trivial Fourier coefficients* λ *of the indicator function* $\mathbb{1}_A : [n] \to \{0,1\}$ *satisfy* $|\lambda| = o(|A|^3/n^2)$, *then any subset* $B \subseteq A$ *that contains no three-term arithmetic progression satisfies* $|B| = o(|A|)$.

*The collaboration of the first and the third author was supported by a CAPES–DAAD collaboration grant.

†The first author was partially supported by FAPESP and CNPq through a Temático-ProNEx project (Proc. FAPESP 2003/09925-5) and by CNPq (Proc. 308509/2007-2, 485671/2007-7 and 486124/2007-0). The second author was supported by the NSF grant DMS 0800070.

‡The third author was partially supported through the Heisenberg Programme of the DFG (grant SCHA 1263/5-1).

1. Introduction

In the mid-seventies Ruzsa and Szemerédi [25] solved a problem of Brown, Erdős, and Sós [5] and established the so-called *triangle removal lemma.* Loosely speaking, the triangle removal lemma asserts that every graph which does not contain many triangles can be made triangle-free by removing only a few edges. More precisely, for a positive integer n we denote by K_n the complete graph with vertex set $[n] = \{1, \ldots, n\}$. For two graphs F and H we say H *is F-free*, if H does not contain a (not necessarily induced) copy of F. For a graph $G = (V, E)$ we denote the minimum number of edges that meet any triangle in G by

$$\tau_3(G) = \min\big\{ |E'| \, : \, E' \subseteq E \text{ and } H = (V, E \setminus E') \text{ is } K_3\text{-free}\big\}.$$

Theorem 1 (Ruzsa & Szemerédi). *For every $\delta > 0$ there exist $c > 0$ and n_0 such that every graph $G = (V, E)$ with $|V| = n \geq n_0$ that contains at most $c\binom{n}{3}$ copies of K_3 satisfies $\tau_3(G) \leq \delta\binom{n}{2}$.* ∎

Theorem 1 stimulated a great deal of research and several generalizations for graphs and hypergraphs are now known (see, e.g., [23] and the references therein and [10] for a new proof of Theorem 1). Moreover, it was already shown in [25] that Theorem 1 yields a new proof of Roth's famous theorem on arithmetic progressions of length three [24] (see [3] for the best known quantitative bound). We say a set of integers is *AP_3-free*, if it does not contain a non-trivial three-term arithmetic progression, i.e., three distinct elements x, y, z such that $x + z = 2y$. For a finite set of integers A we denote the maximum size of an AP_3-free subset of A by

$$r_3(A) = \max\big\{ |B| \, : \, B \subseteq A \text{ and } B \text{ is AP}_3\text{-free}\big\}$$

and we simply write $r_3(n)$ for $r_3\big([n]\big)$.

Theorem 2 (Roth). *For every $\delta > 0$ there exists n_0 such that for all $n \geq n_0$ we have $r_3(n) \leq \delta n$.* ∎

Sparse versions of Theorems 1 and 2 were studied in the context of random discrete structures. In particular, in [16] it was shown that, for any $\delta > 0$, with high probability a random subsets $A \subseteq [n]$ with $|A| = C_\delta n^{1/2}$ satisfies $r_3(A) \leq \delta|A|$, where C_δ depends only on δ (see [26] for generalizations of this result to arithmetic progressions of length $k > 3$

and see [21] for an alternative proof for the case $k = 3$). The techniques developed in [16] can be used to obtain a similar extension of Theorem 1 for subgraphs of random graphs. This extension asserts that for every fixed $\delta > 0$ there exist constants C_δ and $c_\delta > 0$ such that if $p \geq C_\delta n^{-1/2}$ then a random graph $\Gamma \in G(n,p)$ satisfies with high probability the following: every subgraph $G \subseteq \Gamma$ that contains at most $c_\delta p^3\binom{n}{3}$ copies of K_3 satisfies $\tau_3(G) \leq \delta p\binom{n}{2}$. Note that Theorem 1 corresponds to the case $p = 1$ and Theorem 1 implies such a result for any constant $p > 0$.

Another example of a relative version of Roth's theorem appears in the work of Green [12] who showed that, for the set of primes $\mathbb{P}$, we have $r_3(\mathbb{P} \cap [n]) = o(|\mathbb{P} \cap [n]|)$, i.e., any relatively dense subset of the primes contains an arithmetic progression of length three (see [14] for longer arithmetic progressions).

We obtain extensions of Theorems 1 and 2 for subgraphs of sparse pseudorandom graphs (see Theorem 3) and for subsets of sparse pseudorandom subsets of $[n]$ (see Theorem 4). Our proof is based on the *sparse regularity lemma* [15, 17] and the main technical result presented here (see part (a) of Lemma 9 below).

2. New Results

Next we define the notions of pseudorandomness considered here. Roughly speaking, pseudorandom discrete structures "imitate" a truly random object of the same density. The systematic study of pseudorandom graphs was initiated by Thomason [28, 29] and continued by Chung, Graham, and Wilson [7] and we will use a related concept here. In fact, Chung and Graham obtained several generalizations of those results for other discrete structures and one of the properties of pseudorandom subsets of $\mathbb{Z}/n\mathbb{Z}$ studied in [6] is related to our concept of pseudorandom subsets of $[n]$.

2.1. The triangle removal lemma for pseudorandom graphs

We say a graph $\Gamma = (V,E)$ is (p,β)-*bijumbled* for p and $\beta > 0$, if all subsets $X, Y \subseteq V$ satisfy

$$\big|e_\Gamma(X,Y) - p|X|\,|Y|\big| \leq \beta\sqrt{|X|\,|Y|},$$

where $e_\Gamma(X,Y) = \big|\{(x,y) \in X \times Y : \{x,y\} \in E(\Gamma)\}\big|$ denotes the number of edges of Γ with one endvertex in X and the other endvertex in Y, where edges contained in $X \cap Y$ are counted twice.

It follows from Chernoff's inequality that the binomial random graph $G(n,p)$ is with high probability (p,β)-bijumbled for $\beta = C\sqrt{pn}$ for some sufficiently large constant $C > 1$, if $pn - 3\log n \to \infty$. On the other hand, it follows from the work of Erdős and Spencer [9] (see also [8]) that for every (p,β)-bijumbled n-vertex graph we have $\beta = \Omega(\sqrt{p(1-p)n})$ as long as $p(1-p) > 1/n$.

A well known explicit class of (p,β)-bijumbled graphs are the so-called (n,D,λ)-graphs with $D = pn$ and $\lambda = \beta$. An (n,D,λ)-graph Γ is a D-regular n-vertex graph which satisfies $\lambda \geq \{\lambda_2, |\lambda_n|\}$, where $D = \lambda_1 \geq \lambda_2 \geq \cdots \geq \lambda_n$ are the eigenvalues of the adjacency matrix of Γ (see [2, Chapter 9] and [18] for more details).

Our first result asserts that Theorem 1 can be extended to subgraphs $G \subseteq \Gamma$ of (p,β)-bijumbled n-vertex graphs Γ, if $\beta \leq \gamma p^3 n$ for a sufficiently small $\gamma > 0$.

Theorem 3. *For every $\delta > 0$ there exist $c > 0$, $\gamma > 0$, and n_0 such that the following holds.*

Suppose $\Gamma = (V, E_\Gamma)$ is a $(p,\gamma p^3 n)$-bijumbled graph with n vertices and $p \geq 1/\sqrt{n}$ and let $G = (V, E_G) \subseteq \Gamma$ be a (not necessarily induced) subgraph of Γ. If G contains at most $cp^3\binom{n}{3}$ copies of K_3, then $\tau_3(G) \leq \delta e(\Gamma)$.

We remark the following:

(I) It is easy to show that for sufficiently small $\gamma_\xi > 0$ every $(p,\gamma p^2 n)$-bijumbled n-vertex graph Γ contains $(1 \pm \xi)p^3\binom{n}{3}$ copies of K_3. Consequently, Theorem 3 asserts that if a subgraph $G \subseteq \Gamma$ contains only a small fraction of the triangles of the sufficiently pseudorandom host graph Γ, then G can be made triangle-free by removing a small fraction of the edges of Γ. In Theorem 1 the complete graph K_n plays the same rôle as Γ.

(II) Note that Theorem 3 applies only to (p,β)-bijumbled n-vertex graphs Γ with $\beta = \gamma p^3 n$ for sufficiently small γ. Since $\beta = \Omega(\sqrt{pn})$ for every graph with $p(1-p) \geq 1/n$, it follows that Theorem 3 only applies to graphs Γ of density $p = p(n) = \Omega(n^{-1/5})$.

We believe Theorem 3 is also true for (p,β)-bijumbled n-vertex graphs Γ with $\beta = \gamma p^2 n$, which would allow us to consider graphs of density

$p \ge Cn^{-1/3}$. On the other hand, Alon [1] constructed a $(p, 19\sqrt{pn})$-bijumbled n-vertex graph Γ with $p = n^{-1/3}/4$ which is K_3-free. This indicates that such a strengthening of Theorem 3 would be "best possible."

2.2. Roth's theorem for pseudorandom sets

For $A \subseteq [n]$ we consider the discrete Fourier coefficients $\lambda_0(A), \dots, \lambda_{n-1}(A) \in \mathbb{C}$ of A defined by

$$\lambda_k(A) = \sum_{a \in A} \exp\left(\frac{2\pi \mathrm{i}}{n} \cdot ka\right). \tag{1}$$

Clearly, $\lambda_0(A) = |A|$ and we set

$$\lambda(A) = \max\left\{\left|\lambda_k(A)\right| \ : \ k = 1, \dots, n-1\right\}.$$

Theorem 4 extends Theorem 2 to sparse sets $A \subseteq [n]$ for which $\lambda(A)$ is "small." A somewhat related result appears in the work of Green and Tao [13, Proposition 5.1].

Theorem 4. *For every $\delta > 0$ there exist $\gamma > 0$ and n_0 such that for every odd $n \ge n_0$ and every $A \subseteq [n]$ with $\lambda(A) \le \gamma |A|^3/n^2$ we have $r_3(A) \le \delta |A|$.*

Very roughly speaking, the condition on A in Theorem 4 asserts that a certain Cayley-type graph $\Gamma(A)$ generated by A on the vertex set $[n]$ is an $\left(n, |A|, \gamma |A|^3/n^2\right)$-graph and, therefore, a $(p, \gamma p^3 n)$-bijumbled graph for $p = |A|/n$. This allows us to apply similar techniques as in the proof of Theorem 3.

In view of Remark (II) after Theorem 3 we believe that the condition on $\lambda(A)$ in Theorem 4 can be relaxed to $\lambda(A) \le \gamma |A|^2/n$.

2.3. Key technical result

The main technical result behind the proofs of Theorems 3 and 4, namely, Lemma 9, concerns the theory of pseudorandom properties, which are asserted by the sparse regularity lemma. For the statement of Lemma 9 we require a few definitions.

Definition 5 (DISC). Let $G = (V, E)$ be a graph and let X, $Y \subseteq V$ be disjoint. We say (X, Y) *satisfies the discrepancy condition* DISC (q, p, ε) *in* G for some q, p, and $\varepsilon > 0$ if the following holds:

$$\big|e_G(X', Y') - q|X'|\,|Y'|\big| \le \varepsilon p|X|\,|Y| \quad \text{for all } X' \subseteq X \text{ and } Y' \subseteq Y.$$

As a shorthand, we shall sometimes say $(X, Y)_G$ *satisfies* DISC (q, p, ε).

Note that Szemerédi's regularity lemma [27] asserts that the vertex set of every n-vertex graph G can be partitioned into a bounded number of classes $V_1 \dot\cup \ldots \dot\cup V_t$ in such a way that most pairs $(V_i, V_j)_G$ satisfy DISC (q_{ij}, p, ε), where

$$q_{ij} = d_G(V_i, V_j) = \frac{e_G(V_i, V_j)}{|V_i|\,|V_j|} \quad \text{and} \quad p = 1.$$

Since DISC (q, p, ε) only gives useful information on the edge distribution of $(X, Y)_G$ if $q > \varepsilon p$, Szemerédi's regularity lemma is mainly suited for applications to "dense" graphs G with $\Omega\big(|V(G)|^2\big)$ edges. The first two authors observed that Szemerédi's regularity lemma can be extended to subgraphs G of sparse graphs Γ, provided the host graph Γ does not contain "dense spots." In this context p can be chosen to be $e(\Gamma)/\binom{n}{2}$, i.e., it can tend to 0 as n tends to infinity. For example, (p, β)-bijumbled graphs Γ do not contain "dense spots" as long as $\beta \le \gamma pn$ for some small constant $\gamma > 0$. Consequently, the sparse regularity lemma is applicable to the graphs $G \subseteq \Gamma$ satisfying the assumptions of Theorem 3. For the proof of Theorem 3 we will need a corresponding "counting lemma" for triangles in tripartite graph with all three induced bipartite graphs satisfying DISC (see Lemma 11 below). For that we will show that pairs $(X, Y)_G$ satisfying DISC will also satisfy the so called *pair condition.* For (not necessarily distinct) vertices x and $x' \in V$ and a subset $Y \subseteq V$ we write $\deg_G(x, Y)$ for the number of neighbors of x in Y and we write $\deg_G(x, x', Y)$ for the number of joint neighbors of x and x' in Y. Moreover, we denote the sets of such neighbors by $N_G(x, Y)$ and $N_G(x, x', Y)$

Definition 6 (PAIR). Let $G = (V, E)$ be a graph and let X, $Y \subseteq V$ be disjoint. We say (X, Y) satisfies the *pair condition* PAIR (q, p, δ) in G for some q, p, and $\delta > 0$ if the following holds:

$$\sum_{x \in X} \big|\deg_G(x, Y) - q|Y|\big| \le \delta p|X|\,|Y|, \tag{2}$$

$$\sum_{x \in X} \sum_{x' \in X} \big|\deg_G(x, x', Y) - q^2|Y|\big| \le \delta p^2|X|^2|Y|. \tag{3}$$

As a shorthand, we shall sometimes say $(X,Y)_G$ *satisfies* PAIR(q,p,δ).

It is well known (see, e.g., [7]) that DISC and PAIR are equivalent in the dense case, i.e., for $p = 1$.

Theorem 7 (DISC $\Leftrightarrow$ PAIR (dense case)). *For every $\alpha > 0$ and $\delta > 0$ there exist $\varepsilon > 0$ and n_0 such that the following holds.*

Suppose $G = (V,E)$ is a graph and X, $Y \subseteq V$ are disjoint sets with $|X|, |Y| \geq n_0$. Then the following statements hold:

(*a*) *if $(X,Y)_G$ satisfies* DISC$(q,1,\varepsilon)$ *for some q with $\alpha \leq q \leq 1$, then it also satisfies* PAIR$(q,1,\delta)$*;*

(*b*) *if $(X,Y)_G$ satisfies* PAIR$(q,1,\varepsilon)$ *for some q with $\alpha \leq q \leq 1$, then it also satisfies* DISC$(q,1,\delta)$. ■

The key technical lemma presented here asserts that an analogue of Theorem 7 is true for subgraphs of (p,β)-bijumbled graphs Γ if β is sufficiently small. In fact, we can slightly relax the bijumbledness condition and consider a partite version instead. For simplicity we call this property *jumbledness*.

Definition 8 (jumbledness (partite version)). Let $\Gamma = (U \dot\cup V, E_\Gamma)$ be a bipartite graph and $q, \beta > 0$ be real numbers. We say Γ is (q,β)*-jumbled* if

$$\left| e_\Gamma(X,Y) - q|X|\,|Y| \right| \leq \beta\sqrt{|X|\,|Y|} \quad \text{for all } X \subseteq U \text{ and } Y \subseteq V.$$

Moreover, we say a k-partite graph $\Gamma = (V_1 \dot\cup \cdots \dot\cup V_k, E_\Gamma)$ with $k \geq 2$ is (q,β)*-jumbled* if all induced bipartite subgraphs $\Gamma[V_i \dot\cup V_j]$ for $1 \leq i < j \leq k$ are (q,β)-jumbled.

Note that both discrepancy and jumbledness are two measures of pseudorandomness for graphs.

In the context of this paper, discrepancy (DISC(q,p,ε) where $\varepsilon > 0$ and the ratio q/p are constants independent of the size of the graph) is a property of subgraphs of G that can be obtained by an application of the (sparse) regularity lemma.

On the other hand, jumbledness will be a property imposed on the host graph $\Gamma \supseteq G$. Typically the jumbledness assumption is stronger and cannot be ensured by an application of the regularity lemma.

Lemma 9 (DISC $\Leftrightarrow$ PAIR for subgraphs of jumbled graphs). *For every $\alpha \in (0,1]$ and $\delta > 0$ there exists $\varepsilon > 0$ such that for every $\eta > 0$ there exist $\gamma > 0$ and n_0 such that for every $n \geq n_0$ the following holds.*

Suppose

(*i*) $\Gamma = (U \dot{\cup} V, E_\Gamma)$ *is a bipartite (p, β)-jumbled graph with $|U|, |V| \geq n$ and $p \geq 1/\sqrt{n}$,*

(*ii*) $G = (U \dot{\cup} V, E_G)$ *is a subgraph of Γ, and*

(*iii*) $X \subseteq U$ *and* $Y \subseteq V$ *with $|X|, |Y| \geq \eta n$.*

Then the following statements hold:

(*a*) *if $\beta \leq \gamma p^2 n$ and $(X, Y)_G$ satisfies* DISC(q, p, ε) *for some q with $\alpha p \leq q \leq p$, then $(X, Y)_G$ satisfies* PAIR(q, p, δ)*;*

(*b*) *if $\beta \leq \gamma p^{3/2} n$ and $(X, Y)_G$ satisfies* PAIR(q, p, ε) *for some q with $\alpha p \leq q \leq p$, then $(X, Y)_G$ satisfies* DISC(q, p, δ).

The main part of this paper is devoted to the proof of implication (a) of Lemma 9. The proofs of Theorems 3 and 4 rely on this implication. On the other hand, the proof of implication (b) of the lemma can be established along the lines of the argument for the dense case (Theorem 7(b)). We include the proof of part (b) of Lemma 9 for completeness (see Section 5.1). We note that the jumbledness assumption on the host graph Γ in part (b) is less restrictive than in part (a). We believe that the assumption for implication (a) can be weakened (see Section 6)

Organization

In Section 3 we deduce Theorem 3 from Lemma 9(a) and the sparse regularity lemma. Section 4 is devoted to the proof of Theorem 4, which is based on a tripartite variant of Theorem 3 (see Theorem 10 below). The proof of the key technical result, Lemma 9, is deferred to Section 5.

3. The Ruzsa–Szemerédi Theorem for Pseudorandom Graphs

The proof presented here follows the standard proof of the Ruzsa–Szemerédi theorem, which is based on Szemerédi's regularity lemma and the so-called triangle counting lemma. In fact, the proof of Theorem 3 is based on the sparse regularity lemma [15, 17] and an appropriate triangle counting lemma for subgraphs of jumbled graphs (see Lemma 11 below), which we deduce from part (a) of Lemma 9. First we state a tripartite version of Theorem 3, which will be convenient for the proof of Theorem 4.

3.1. Tripartite version of Theorem 3

We shall show that Theorem 3 easily follows from the following tripartite version.

Theorem 10 (tripartite version of Theorem 3). *For every $\delta > 0$ there exist $c > 0$, $\gamma > 0$, and n_0 such that for every $n \geq n_0$ the following holds.*

Suppose $\Gamma = (V_1 \dot\cup V_2 \dot\cup V_3, E_\Gamma)$ is a tripartite $(p, \gamma p^3 n)$-jumbled graph with $|V_i| \geq n$ for $i \in [3]$, $p \geq 1/\sqrt{n}$, and $G = (V_1 \dot\cup V_2 \dot\cup V_3, E_G) \subseteq \Gamma$ is a subgraph of Γ. If G contains at most $cp^3|V_1|\,|V_2|\,|V_3|$ copies of K_3, then $\tau_3(G) \leq \delta e(\Gamma)$.

Proof: Theorem 10 $\Rightarrow$ Theorem 3. For any given $\delta > 0$ we set $\delta' = \delta/6$ and let c' and $\gamma' > 0$ be given by Theorem 10. For Theorem 3 we then set $c = c'$ and $\gamma = \gamma'$. Finally let n be sufficiently large and $p \geq 1/\sqrt{n}$.

Let $\Gamma = (V, E_\Gamma)$ be a $(p, \gamma p^3 n)$-bijumbled graph and let $G = (V, E_G)$ be a subgraph of Γ which contains at most $cp^3\binom{n}{3}$ copies of K_3.

We consider tripartite graphs $\Gamma' = (V', E_{\Gamma'}) \supseteq G' = (V', E_{G'})$ defined as follows

$$V' = V \times [3]$$

and

$$E_{\Gamma'} = \Big\{\{(u,i),(v,j)\} \;:\; \{u,v\} \in E_\Gamma \text{ and } 1 \leq i < j \leq 3\Big\},$$

$$E_{G'} = \Big\{\{(u,i),(v,j)\} \;:\; \{u,v\} \in E_G \text{ and } 1 \leq i < j \leq 3\Big\}.$$

It follows from the definition of G' and Γ' that $e(\Gamma') = 6e(\Gamma)$, $|V'| = 3n$, and the assumptions on Γ and G yield

(a) Γ' is a tripartite $(p, \gamma p^3 n)$-jumbled graph and

(b) G' contains at most $6cp^3\binom{n}{3} \leq c'p^3n^3$ triangles.

Consequently, Γ' and G' satisfy the assumptions of Theorem 10 and, hence, there exists a set of edges $X' \subseteq E_{G'}$ such that $H' = (V', E_{G'} \setminus X)$ is K_3-free and

$$|X'| \leq \delta' e(\Gamma') = \delta e(\Gamma).$$

We consider the set of edges X which is the "pullback" of X' in G, i.e.,

$$X = \big\{\{u,v\} : \big\{(u,i),(v,j)\big\} \in X' \text{ for some } 1 \leq i < j \leq 3\big\}.$$

Clearly $|X| \leq |X'| \leq \delta e(\Gamma)$ and it is easy to check that $H = (V, E_G \setminus X)$ is K_3-free. Therefore, $\tau_3(G) \leq \delta e(\Gamma)$. ■

3.2. Triangle counting lemma for subgraphs of tripartite jumbled graphs

The proof of Theorem 10 outlined here follows the lines of the standard proof of Theorem 1, which is based on Szemerédi's regularity lemma and a corresponding triangle counting lemma. In fact the sparse version of the regularity lemma applies to subgraphs of $(p, \gamma p^3 n)$-jumbled tripartite graphs and below we state an appropriate triangle counting lemma for subgraphs of jumbled graphs.

Lemma 11. *For every $\alpha \in (0,1]$ and $\xi > 0$ there exists $\varepsilon > 0$ such that for all $\eta > 0$ there exist $\gamma > 0$ and n_0 such that for every $n \geq n_0$ the following holds.*

Suppose

(i) *$\Gamma = (V_1 \dot\cup V_2 \dot\cup V_3, E_\Gamma)$ is a tripartite $(p, \gamma p^3 n)$-jumbled graph with $|V_i| \geq n$ for $i \in [3]$ and $p \geq 1/\sqrt{n}$,*

(ii) *$G = (V_1 \dot\cup V_2 \dot\cup V_3, E_G) \subseteq \Gamma$ is a subgraph of Γ, and*

(iii) *$U_i \subseteq V_i$ with $|U_i| \geq \eta n$ for $i \in [3]$.*

If $e_G(U_2, U_3) = q_{23}|U_2|\,|U_3|$ for some q_{23} with $\alpha p \leq q_{23} \leq p$ and for $j = 2$ and 3 the pair $(U_1, U_j)_G$ satisfies DISC $(q_{1,j}, p, \varepsilon)$ for some $q_{1,j}$ with $\alpha p \leq q_{1,j} \leq p$, then the induced subgraph $G[U_1, U_2, U_3]$ contains at least $(1-\xi)q_{12}q_{13}q_{23}|U_1|\,|U_2|\,|U_3|$ copies of K_3.

Theorem 10 can be deduced from the sparse version of the regularity lemma and Lemma 11 by standard arguments and we omit the details here. Below we deduce Lemma 11 from part (a) of Lemma 9. In the proof we use the following well known consequence of the Cauchy–Schwarz inequality.

Lemma 12. *For every $\nu > 0$ there exists $\mu > 0$ such that for every $m \in \mathbb{N}$ and $x_1, \dots, x_m, \bar{x} \in \mathbb{R}$ the following holds. If*

$$\sum_{i\in[m]} x_i \geq (1-\mu)\bar{x}m \quad \text{and} \quad \sum_{i\in[m]} x_i^2 \leq (1+\mu)\bar{x}^2 m,$$

then

$$\left|\left\{i \in [m] : |x_i - \bar{x}| \geq \nu\bar{x}\right\}\right| \leq \nu m. \qquad \blacksquare$$

Proof: Lemma 9(a) $\Rightarrow$ Lemma 11. Note that the quantifications of the involved constants in Lemma 9 and Lemma 11 are very similar. In fact, there are only two differences. For an application of Lemma 9 we have to make an appropriate choice for $\delta > 0$ (depending only on α) and for the proof of Lemma 11 we are given $\xi > 0$.

Assume we are given α and ξ by Lemma 11. First we fix an auxiliary constant

$$\nu = \frac{\alpha\xi}{3} \tag{4}$$

and let $\mu = \mu(\nu)$ be given by Lemma 12. Then we set

$$\delta = \left(\frac{\alpha^6\mu}{1200}\right)^2 \tag{5}$$

and let $\varepsilon = \varepsilon(\alpha, \delta)$ be given by Lemma 9. Following the quantification of Lemma 11 let η be given. For this η Lemma 9 yields an appropriate choice for γ. Without loss of generality we may assume that

$$\gamma \leq \frac{\alpha^4\mu\eta^2}{144}. \tag{6}$$

Finally, let n be sufficiently large and suppose $p \geq 1/\sqrt{n}$.

Let Γ, G, and U_i be as in the statement of Lemma 11. We have to show that $G[U_1, U_2, U_3]$ contains at least $(1-\xi)q_{12}q_{13}q_{23}|U_1|\,|U_2|\,|U_3|$ triangles. We will first show that most pairs $\{u_2, u_3\} \in E_\Gamma(U_2, U_3)$ have a "large" joint neighborhood in the graph G in U_1. For that we study the distribution of the triangles in Γ with one edge in $E_\Gamma(U_2,U_3)$ and the other two edges in G. For a pair $(u_2,u_3)\in U_2\times U_3$ we recall that $\deg_G(u_2,u_3,U_1)$ denotes the number of joint neighbors of u_2 and u_3 in U_1 in the graph G. Below we will show the following two estimates.

$$\sum\bigl\{\deg_G(u_2,u_3,U_1)\,:\,\{u_2,u_3\}\in E_\Gamma(U_2,U_3)\bigr\} \\ \geq \left(1-\frac{\mu}{3}\right) q_{12}q_{13}|U_1|\cdot p|U_2|\,|U_3|, \tag{7}$$

$$\sum\bigl\{\deg_G^2(u_2,u_3,U_1)\,:\,\{u_2,u_3\}\in E_\Gamma(U_2,U_3)\bigr\} \\ \leq \left(1+\frac{\mu}{3}\right) q_{12}^2q_{13}^2|U_1|^2\cdot p|U_2|\,|U_3|. \tag{8}$$

Before we verify (7) and (8), we finish the proof of Lemma 11 based on those estimates.

The $(p,\gamma p^3 n)$-jumbledness of Γ, combined with (6), $|U_2|,|U_3|\geq\eta n$, and $p\leq 1$, yields

$$\bigl|e_\Gamma(U_2,U_3)-p|U_2|\,|U_3|\bigr| \leq \gamma p^3 n\sqrt{|U_2|\,|U_3|} \overset{(6)}{\leq} \frac{\mu}{3}p^3|U_2|\,|U_3| \leq \frac{\mu}{3}p|U_2|\,|U_3|. \tag{9}$$

Note that (7) and (8) show that the assumptions of Lemma 12 are met for $m=e_\Gamma(U_2,U_3)$ and $\bar{x}=q_{12}q_{13}|U_1|$. Therefore

$$\deg_G(u_2,u_3,U_1)\geq(1-\nu)q_{12}q_{13}|U_1|$$

for all but at most $\nu e_\Gamma(U_2,U_3)$ edges $\{u_2,u_3\}\in E_\Gamma(U_2,U_3)$. Removing those exceptional edges from G yields the following lower bound on the number of triangles with all three edges in G:

$$\bigl(e_G(U_2,U_3)-\nu e_\Gamma(U_2,U_3)\bigr)\cdot(1-\nu)q_{12}q_{13}|U_1|.$$

Since $e_G(U_2,U_3)=q_{23}|U_2|\,|U_3|\geq\alpha p|U_2|\,|U_3|$, the choice of ν in (4) combined with (9) yields

$$\bigl(e_G(U_2,U_3)-\nu e_\Gamma(U_2,U_3)\bigr)\cdot(1-\nu)q_{12}q_{13}|U_1|\geq(1-\xi)q_{12}q_{13}q_{23}|U_1|\,|U_2|\,|U_3|,$$

which concludes the proof of Lemma 11 based on estimates (7) and (8). ∎

Proof of (7). Note that the left-hand side of (7) is the number of triangles in $\Gamma[U_1, U_2, U_3]$ with both edges incident to a vertex in U_1 being also present in G and we have

(10)
$$\sum_{\{u_2,u_3\}\in E_\Gamma(U_2,U_3)} \deg_G(u_2,u_3,U_1) = \sum_{u_1\in U_1} e_\Gamma\big(N_G(u_1,U_2), N_G(u_1,U_3)\big).$$

The $(p, \gamma p^3 n)$-jumbledness of Γ yields

(11)
$$e_\Gamma\big(N_G(u_1,U_2), N_G(u_1,U_3)\big) \geq p \deg_G(u_1,U_2)\deg_G(u_1,U_3) - \gamma p^3 n\sqrt{|U_2|\,|U_3|}.$$

Moreover, by Lemma 9(a), the choice of ε and the assumption of Lemma 11 imply that $(U_1, U_2)_G$ and $(U_1, U_3)_G$ satisfy the pair conditions PAIR (q_{12}, p, δ) and PAIR (q_{13}, p, δ), respectively. In particular, for $j = 2$ and 3,

$$\sum_{u_1\in U_1} \big|\deg_G(u_1,U_j) - q_{1,j}|U_j|\big| \leq \delta p|U_1|\,|U_j|.$$

As $q_{1,j} \geq \alpha p$ for $j = 2$ and 3, we have

(12)
$$\big|N_G(u_1,U_j)\big| = \deg_G(u_1,U_j) \geq \big(1-\sqrt{\delta}\,\big)\, q_{1,j}|U_j|$$

for both $j = 2$ and $j = 3$ for all but at most $(1 - 2\sqrt{\delta}/\alpha)|U_1|$ vertices $u_1 \in U_1$. Combining this with (10), (11) and (12) yields

(13)
$$\sum\big\{\deg_G(u_2,u_3,U_1) : \{u_2,u_3\}\in E_\Gamma(U_2,U_3)\big\}$$
$$\geq \big(1-2\sqrt{\delta}/\alpha\big)\,|U_1|\cdot p\big(1-\sqrt{\delta}\,\big)^2 q_{12}q_{13}|U_2|\,|U_3| - |U_1|\gamma p^3 n\sqrt{|U_2|\,|U_3|}.$$

The choice of δ in (5) ensures

$$\big(1-2\sqrt{\delta}/\alpha\big)\big(1-\sqrt{\delta}\,\big)^2 \geq 1-\mu/6$$

and the choice of γ in (6) together with $|U_2|, |U_3| \geq \eta n$ and $q_{12}, q_{13} \geq \alpha p$ gives

$$|U_1|\gamma p^3 n\sqrt{|U_2|\,|U_3|} \leq \frac{\mu}{6} p q_{12} q_{13}|U_1|\,|U_2|\,|U_3|.$$

Plugging the last two estimates into (13) yields (7). ∎

In the proof of (8) we will use the following estimate, which is a consequence of the definition of jumbled graphs.

Fact 13. *Let $\Gamma = (U \dot\cup V, E)$ be a bipartite (p,β)-jumbled graph. Let $X \subseteq U$ and $Y \subseteq V$ and $\xi > 0$. If X satisfies one of the following two conditions: either $\deg_\Gamma(x,Y) \geq (1+\xi)p|Y|$ for all $x \in X$ or $\deg_\Gamma(x,Y) \leq (1-\xi)p|Y|$ for all $x \in X$, then*

$$|X| \leq \frac{\beta^2}{\xi^2 p^2 |Y|}. \qquad \blacksquare$$

Proof of (8)**.** We start with rewriting the left-hand side of (8) by summing over all pairs of vertices in U_1. We obtain

$$\sum \{ \deg^2_G(u_2,u_3,U_1) \,:\, \{u_2,u_3\} \in E_\Gamma(U_2,U_3)\}$$
$$= \sum_{u_1 \in U_1} \sum_{u_1' \in U_1} e_\Gamma\big((N_G(u_1,u_1',U_2), N_G(u_1,u_1',U_3)\big)$$

and, as in the proof of (7), after applying the jumbledness assumption we arrive at

$$\sum \{ \deg^2_G(u_2,u_3,U_1) \,:\, \{u_2,u_3\} \in E_\Gamma(U_2,U_3)\} \tag{14}$$
$$\leq \sum_{u_1 \in U_1} \sum_{u_1' \in U_1} pD_G(u_1,u_1',U_2,U_3) + \gamma p^3 n \sqrt{D_G(u_1,u_1',U_2,U_3)},$$

where we set

$$D_G(u_1,u_1',U_2,U_3) = \deg_G(u_1,u_1',U_2)\deg_G(u_1,u_1',U_3).$$

It is left to obtain appropriate estimates on $\deg_G(u_1,u_1',U_2)$ and $\deg_G(u_1,u_1',U_3)$. For that we will appeal to the pair condition, which yields good estimates for those quantities for "most" pairs $(u_1,u_1') \in U_1 \times U_1$. For the exceptional pairs (u_1,u_1') we will analyze the joint neighborhood of u_1 and u_1' in Γ and obtain the required bounds from the jumbledness of Γ. For that we classify the pairs (u_1,u_1') according to their degrees in G and Γ. More precisely, let $\mathcal{G}$ be the "good" pairs having the "right" joint degree in U_2 and U_3:

$$\mathcal{G} = \Big\{(u_1,u_1') \in U_1 \times U_1 \,:\, \deg_G(u_1,u_1',U_j) \leq \big(1+\sqrt{\delta}\big)\, q_{1,j}^2 |U_j| \text{ for } j=2,3\Big\}.$$

Clearly,

$$\sum_{(u_1,u_1')\in\mathcal{G}} pD_G(u_1,u_1',U_2,U_3)+\gamma p^3 n\sqrt{D_G(u_1,u_1',U_2,U_3)}$$

$$\le |U_1|^2\cdot p\big(1+\sqrt{\delta}\,\big)^2 q_{12}^2 q_{13}^2|U_2|\,|U_3|+|U_1|^2\cdot\gamma p^3 n\sqrt{4q_{12}^2q_{13}^2|U_2|\,|U_3|}$$

$$\le\Big(1+\frac{\mu}{6}\Big)\,pq_{12}^2q_{13}^2|U_1|^2|U_2|\,|U_3|,$$

where we used the choice of δ and γ from (5) and (6) and the assumptions on $q_{1,j}$ and $|U_j|$ for $j=2$ and 3.

Hence, it is left to verify that

$$\sum_{(u_1,u_1')\notin\mathcal{G}} pD_G(u_1,u_1',U_2,U_3)+\gamma p^3 n\sqrt{D_G(u_1,u_1',U_2,U_3)} \le \frac{\mu}{6}pq_{12}^2q_{13}^2|U_1|^2|U_2|\,|U_3|. \tag{15}$$

Lemma 9(a), the choice of ε, and the assumption of Lemma 11 imply that $(U_1,U_2)_G$ and $(U_1,U_3)_G$ satisfy PAIR$\,(q_{12},p,\delta)$ and PAIR$\,(q_{13},p,\delta)$ respectively. Consequently, we have

$$|\mathcal{G}|\ge\big(1-2\sqrt{\delta}/\alpha^2\big)\,|U_1|^2. \tag{16}$$

For the contribution of those $2\sqrt{\delta}|U_1|^2/\alpha^2$ "exceptional" pairs we analyze the joint neighborhoods of u_1 and u_1' in $\Gamma\supseteq G$. Indeed, we will show that (15) stays valid when we replace G by Γ in $D_G(u_1,u_1',U_2,U_3)$ and show

$$\sum_{(u_1,u_1')\notin\mathcal{G}} pD_\Gamma(u_1,u_1',U_2,U_3)+\gamma p^3 n\sqrt{D_\Gamma(u_1,u_1',U_2,U_3)} \le \frac{\mu}{6}pq_{12}^2q_{13}^2|U_1|^2|U_2|\,|U_3|. \tag{17}$$

We split the pairs in $(U_1\times U_1)\setminus\mathcal{G}$ into the following three classes $\mathcal{B}_1\,\dot\cup\,\mathcal{B}_2\,\dot\cup\,\mathcal{B}_3$:

$$\mathcal{B}_1=\big\{(u_1,u_1')\in(U_1\times U_1)\setminus\mathcal{G}\,:\,\deg_\Gamma(u_1,u_1',U_j)\le 4p^2|U_j|\text{ for }j=2,3\big\},$$

$$\mathcal{B}_2 = \big\{ (u_1, u_1') \in (U_1 \times U_1) \setminus (\mathcal{G} \cup \mathcal{B}_1) \,:\, \deg_\Gamma(u_1, U_j) \le 2p|U_j| \\ \text{for } j = 2, 3 \big\},$$

and

$$\mathcal{B}_3 = (U_1 \times U_1) \setminus (\mathcal{G} \cup \mathcal{B}_1 \cup \mathcal{B}_2).$$

Below we bound the contribution to (17) for each class separately.

Contribution of pairs from $\mathcal{B}_1$. The definition of $\mathcal{B}_1$ and (16) yield

$$(18) \qquad \sum_{(u_1,u_1')\in\mathcal{B}_1} pD_\Gamma(u_1, u_1', U_2, U_3) + \gamma p^3 n \sqrt{D_\Gamma(u_1, u_1', U_2, U_3)}$$
$$\le 2\frac{\sqrt{\delta}}{\alpha^2}|U_1|^2 \cdot \big(16p^5|U_2|\,|U_3| + \gamma p^3 n\sqrt{16p^4|U_2|\,|U_3|}\,\big)$$
$$\le \frac{\mu}{18} p q_{12}^2 q_{13}^2 |U_1|^2 |U_2|\,|U_3|,$$

because of the choice of δ in (5) and $\gamma \le \eta/36$ in (6) and the assumptions on q_{1j} and $|U_j|$ for $j = 2$ and 3.

Contribution of pairs from $\mathcal{B}_2$. Fix $j \in \{2, 3\}$ and let $u_1 \in U_1$ be a vertex with $\deg_\Gamma(u_1, U_j) \le 2p|U_j|$. Let $X(u_1, j)$ be the set of vertices $u_1' \in U_1$ with

$$\deg_\Gamma(u_1, u_1', U_j) > 4p^2|U_j|.$$

As Γ is $(p, \gamma p^3 n)$-jumbled, Fact 13 applied to $X(u_1, j)$ and a superset $Y(u_1, j) \supseteq N_\Gamma(u_1, U_j)$ in U_j containing precisely $2p|U_j|$ vertices gives

$$\big|X(u_1, j)\big| \le \frac{\gamma^2 p^6 n^2}{p^2 \cdot 2p|U_j|} = \frac{\gamma^2 p^3 n^2}{2|U_j|}.$$

Consequently, since $\gamma \le \eta^2$ and $|U_j| \ge \eta n$ for $j = 1, 2, 3$ we have

$$|\mathcal{B}_2| \le |U_1| \left(\frac{\gamma^2 p^3 n^2}{2|U_2|} + \frac{\gamma^2 p^3 n^2}{2|U_3|} \right) \le \gamma p^3 |U_1|^2.$$

Finally, since $\deg_\Gamma(u_1, u_1', U_j) \le \deg_\Gamma(u_1, U_j) \le 2p|U_j|$ for all $(u_1, u_1') \in \mathcal{B}_2$ and $j = 2, 3$ we have

$$(19)\qquad \sum_{(u_1,u_1')\in\mathcal{B}_2} pD_\Gamma(u_1,u_1',U_2,U_3)+\gamma p^3 n\sqrt{D_\Gamma(u_1,u_1',U_2,U_3)}$$

$$\leq |\mathcal{B}_2|\cdot\left(4p^3|U_2|\,|U_3|+\gamma p^3 n\sqrt{4p^2|U_2|\,|U_3|}\,\right)$$

$$\leq \frac{\mu}{18}pq_{12}^2q_{13}^2|U_1|^2|U_2|\,|U_3|,$$

where we use the choice of γ from (5) and the assumptions on $q_{1,j}$ and $|U_j|$ for $j=2$ and 3.

Contribution of pairs from $\mathcal{B}_3$. The analysis for those pairs is similar to the one in the last case. For $j\in\{2,3\}$ let X_j be the vertices $u_1\in U_1$ with

$$\deg_\Gamma(u_1,U_j)>2p|U_j|.$$

Fact 13 applied to X_j and $Y=U_j$ yields

$$|X_j|\leq\frac{\gamma^2p^6n^2}{p^2|U_j|}\leq\gamma p^4|U_1|,$$

since $\gamma\leq\eta^2$. Therefore, we have $|\mathcal{B}_3|\leq 2\gamma p^4|U_1|^2$ and since $\deg_\Gamma(u_1,u_1',U_j)$ $\leq|U_j|$ for $j=2$ and 3 we have

$$(20)\qquad \sum_{(u_1,u_1')\in\mathcal{B}_2} pD_\Gamma(u_1,u_1',U_2,U_3)+\gamma p^3 n\sqrt{D_\Gamma(u_1,u_1',U_2,U_3)}$$

$$\leq 2\gamma p^4|U_1|^2\cdot\left(p|U_2|\,|U_3|+\gamma p^3 n\sqrt{|U_2|\,|U_3|}\,\right)$$

$$\leq \frac{\mu}{18}pq_{12}^2q_{13}^2|U_1|^2|U_2|\,|U_3|,$$

where we use $\gamma\leq\alpha^4\mu\eta/72$.

Finally, we note that (18), (19) and (20) imply (17), which concludes the proof of (8). ■

4. Roth's Theorem for Pseudorandom Sets of Integers

We prove Theorem 4 in this section. We shall in fact consider pseudorandom subsets of finite abelian groups. Throughout this section $\mathbb{G}$ is a finite abelian group, of order $n = |\mathbb{G}|$, which *is assumed to be odd.* (Roth's theorem was generalized to this setting by Brown and Buhler [4]. For a proof based on the triangle removal lemma, see [11] and for better quantitative bounds see [20, 19].)

Let $\chi_0, \ldots, \chi_{n-1} : \mathbb{G} \to \mathbb{C}$ be the n characters of $\mathbb{G}$. We suppose χ_0 is the principal character, that is, $\chi_0(g) = 1$ for all $g \in \mathbb{G}$.

Suppose now that we are given a set $A \subseteq \mathbb{G}$. Let

$$\lambda_k(A) = \sum_{a \in A} \chi_k(a) \in \mathbb{C} \tag{21}$$

for all $0 \leq k < n$. Clearly, $\lambda_0(A) = |A|$. If $\big|\lambda_k(A)\big| \leq \lambda$ for all $1 \leq k < n$, we say that A is an (n, λ)*-subset of* $\mathbb{G}$ or an (n, λ)*-set* for short. We let

$$r_3(A) = \max\big\{\, |B| \,:\, B \text{ is an AP}_3\text{-free subset of } A\big\}. \tag{22}$$

We shall prove the following result.

Theorem 14. *For every $\delta > 0$ there exist $\gamma > 0$ and n_0 such that for every odd $n \geq n_0$ the following holds.*

Suppose A is an (n, λ)-subset of an abelian group $\mathbb{G}$ of order n and

$$\lambda \leq \gamma \frac{|A|^3}{n^2}. \tag{23}$$

Then $r_3(A) \leq \delta|A|$.

The proof of Theorem 14 will be given in Section 4.2.

4.1. Proof of Theorem 4

Let us deduce Theorem 4 from Theorem 14. Let us first specialize the latter theorem to $\mathbb{G} = \mathbb{Z}/n\mathbb{Z}$. Let $\vartheta = \exp(2\pi\mathrm{i}/n)$. The characters χ_k $(0 \le k < n)$ of $\mathbb{Z}/n\mathbb{Z}$ are given by

$$\chi_k(x) = \vartheta^{kx} \qquad (x \in \mathbb{Z}/n\mathbb{Z}). \tag{24}$$

Therefore, the $\lambda_k(A)$ in (21) are given by (1). We conclude that the hypothesis in Theorem 4 simply says that A is an (n,λ)-subset of $\mathbb{Z}/n\mathbb{Z}$ satisfying (23). Theorem 14 tells us that, for any $\delta > 0$, for suitable γ and n_0 we have $r_3(A) \le \delta|A|$.

To finish the proof of Theorem 4, we have to deal with the issue that an arithmetic progression in $\mathbb{Z}/n\mathbb{Z}$ is not necessarily an arithmetic progression in the integers. For instance, $(5,2,6)$ is a 3-term arithmetic progression with difference 4 modulo 7, as is $(1,5,2)$. This is easy to handle, by making use of the following easy fact: *let (a,b,c) be a triple of distinct integers with $0 \le a,b,c \le n/2$ or with $n/2 < a,b,c \le n$, and suppose that the residue classes of a, b and c modulo n form a 3-term arithmetic progression, i.e., $a+c \equiv 2b \pmod{n}$. Then (a,b,c) is an arithmetic progression in $[n]$, i.e., $a+c = 2b$.*

Since, for any $B \subseteq [n]$, either $\big|B \cap \{1,\dots,\lfloor n/2\rfloor\}\big| \ge |B|/2$ or else we have $\big|B\cap\{\lceil n/2\rceil,\dots,n\}\big| \ge |B|/2$, Theorem 14 applied to subsets of $\mathbb{Z}/n\mathbb{Z}$, implies the desired result for subsets on $[n]$. ■

4.2. Proof of Theorem 14

We shall present the proof of Theorem 14 in this section.

We shall make use of a certain tripartite graph $\Gamma = \Gamma(A)$, defined in terms of the given (n,λ)-set $A \subseteq \mathbb{G}$.

Definition 15 (Graph $\Gamma = \Gamma(A)$). The tripartite graph $\Gamma = \Gamma(A)$ has vertex classes X_1, X_2, and X_3, where each X_i is a disjoint copy of $\mathbb{G}$. For every $x_1 \in X_1$, we join x_1 to the vertices $x_1 + a \in X_2$ and $x_1 + 2a \in X_3$ for all $a \in A$. Moreover, we join every $x_2 \in X_2$ to the vertex $x_2 + a \in X_3$ for all $a \in A$.

It is clear that the bipartite graph $\Gamma[X_1 \cup X_2]$ induced by $X_1 \cup X_2$ in Γ and the bipartite graph $\Gamma[X_2 \cup X_3]$ induced by $X_2 \cup X_3$ in Γ are isomorphic in a natural way. The bipartite graph $\Gamma[X_1 \cup X_3]$ induced by $X_1 \cup X_3$ in Γ is "similar" to those bipartite graphs, but one uses, so to speak, the set $2A$ instead of A to obtain this graph. In what follows, we study these bipartite graphs $\Gamma_1(A) \simeq \Gamma[X_1 \cup X_2] \simeq \Gamma[X_2 \cup X_3]$ and $\Gamma_2(2A) \simeq \Gamma[X_1 \cup X_3]$ that form Γ.

4.2.1. The graph $\Gamma_1 = \Gamma_1(A)$. The two vertex classes of $\Gamma_1 = \Gamma_1(A)$ are two disjoint copies X and Y of $\mathbb{G}$. A vertex $x \in X$ is joined to a vertex $y \in Y$ if $y - x \in A$. Thus, the edges of $\Gamma_1 = \Gamma_1(A) = (X \dot\cup Y, E_1)$ are of the form $(x, x+a) \in X \times Y$, where $x \in X$ and $a \in A$. It is clear that Γ_1 is $|A|$-regular. The graph $\Gamma = \Gamma(A)$ defined above contains two copies of $\Gamma_1(A)$.

A key fact about $\Gamma_1 = \Gamma_1(A)$ is given in the following lemma, which is a bipartite version of the so called *expander mixing lemma* (see, e.g., [2, Corollary 9.2.5]).

Lemma 16. *Suppose A is an (n, λ)-subset of $\mathbb{G}$ and let $\Gamma_1 = \Gamma_1(A) = (X \dot\cup Y, E_1)$ be as defined above. Then, for any $X' \subseteq X$ and any $Y' \subseteq Y$, we have*

$$\left| e(X', Y') - \frac{|A|}{n} |X'|\,|Y'| \right| \le \lambda \sqrt{|X'|\,|Y'|}. \tag{25}$$

The proof of Lemma 16 is given in Section 4.2.4.

4.2.2. The graph $\Gamma_2 = \Gamma_2(2A)$. Consider $2A = \{2a : a \in A\}$ and let $\Gamma_2 = \Gamma_2(2A)$ be defined as follows: the vertex classes of Γ_2 are two disjoint copies X and Y of $\mathbb{G}$, and a vertex $x \in X$ is joined to a vertex $y \in Y$ if $y - x \in 2A$. Thus, the edges of $\Gamma_2 = \Gamma_2(2A) = (X \dot\cup Y, E_2)$ are of the form $(x, x + 2a) \in X \times Y$, where $x \in X$ and $a \in A$. Since n is odd, the map $g \mapsto 2g$ is a bijection on $\mathbb{G}$, and hence $|2A| = |A|$. Therefore, $\Gamma_2 = \Gamma_2(2A)$ is $|A|$-regular. The graph $\Gamma = \Gamma(A)$ defined above contains a copy of $\Gamma_2 = \Gamma_2(2A)$.

Observe now that, for all k, we have

$$\lambda_k(2A) = \sum_{a \in A} \chi_k(2a) = \sum_{a \in A} \chi_k(a)^2. \tag{26}$$

Recall that the characters χ_k $(0 \le k < n)$ form a group $\hat{\mathbb{G}}$ under pointwise multiplication. As $|\hat{\mathbb{G}}| = n$ is odd, the map $\chi_k \mapsto \chi_k^2$ is a bijection on $\hat{\mathbb{G}}$, fixing the principal character χ_0. Therefore,

$$\{\lambda_k(A) : 1 \le k < n\} = \{\lambda_k(2A) : 1 \le k < n\}. \tag{27}$$

Moreover, $\lambda_0(2A) = |2A| = |A|$. We have just proved the following fact.

Lemma 17. *If $A \subseteq \mathbb{G}$ is an (n, λ)-set, then so is $2A$ and, in particular, inequality* (25) *holds for $\Gamma_2 = \Gamma_2(2A) = (X \dot\cup Y, E_2)$ for any $X' \subseteq X$ and any $Y' \subseteq Y$.* ■

An immediate corollary of Lemmas 16 and 17 is the following crucial fact about the graph $\Gamma = \Gamma(A)$ (see Definition 15).

Corollary 18. *Let $A \subseteq \mathbb{G}$ be an (n, λ)-set. Then $\Gamma = \Gamma(A)$ is a $(|A|/n, \lambda)$-jumbled tripartite graph (see Definition 8).* ■

4.2.3. Proof of Theorem 14. To prove Theorem 14, it suffices to run the well known derivation of Roth's theorem from the triangle removal lemma (see, e.g., [25]) in our context. Theorem 10 and Corollary 18 will be crucial here.

For convenience, let us first make a simple remark, whose proof is included for completeness.

Fact 19. *Let $A \subseteq \mathbb{G}$ be an (n, λ)-set, with $|A| \le 3n/4$. Then $\lambda \ge \sqrt{|A|}/2$.*

Proof. We follow a proof in, e.g., [18]. Let $M = (m_{xy}) \in \{0,1\}^{n\times n}$ be a $\{0,1\}$-matrix with rows and columns indexed by the group elements of $\mathbb{G}$ and

$$m_{xy} = \begin{cases} 1, & \text{if } y - x \in A, \\ 0, & \text{otherwise.} \end{cases}$$

We denote by M^* the conjugate transpose of M. Computing the trace $\operatorname{tr}(M^*M)$ of M^*M in two ways, we obtain that

$$n|A| = \operatorname{tr}(M^*M) = \sum_{0\le k<n} \lambda_k^2 = |A|^2 + \sum_{1\le k<n} \lambda_k^2 \le |A|^2 + \lambda^2(n-1), \tag{28}$$

whence $\lambda^2 \ge |A|(1 - |A|/n) \ge |A|/4$, and the result follows. ■

Proof of Theorem 14. Let $\delta > 0$ be given. We invoke Theorem 10 with $\delta/3$ and obtain positive constants c and γ and let n be a sufficiently large odd integer.

Let $A \subseteq \mathbb{G}$ be an (n, λ)-set satisfying (23). Without loss of generality we may assume that $|A| \leq 3n/4$, since, for any $\xi > 0$ and $|A| \geq \xi n$, Theorem 14 follows from the result of Brown and Buhler [4].

Therefore Fact 19 implies that the (n, λ)-sets A must satisfy $\lambda \geq \sqrt{|A|}/2$. Consequently, (23) implies that $\gamma|A|^3/n^2 \geq \sqrt{|A|}/2$, which gives

$$|A| \geq (2\gamma)^{-2/5} n^{4/5} > \sqrt{n}. \tag{29}$$

Let $p = |A|/n$. We have just seen that $p > 1/\sqrt{n}$.

Suppose $B \subseteq A$ is such that $|B| > \delta|A|$. We shall show that B contains a 3-term arithmetic progression.

In order to do so, we consider the graphs $\Gamma = \Gamma(A)$ and $G = \Gamma(B)$. I.e., G is a tripartite graph with the vertex set $X_1 \dot\cup X_2 \dot\cup X_3$ consisting of three disjoint copies of $\mathbb{G}$ and $\{x, y\}$ is an edge in $G[X_1 \dot\cup X_2]$ and $G[X_2 \dot\cup X_3]$ if $y - x \in B$ and it is an edge in $G[X_1 \dot\cup X_3]$ if $y - x$ in $2B$. Clearly, G is a subgraph of $\Gamma = \Gamma(A)$.

Corollary 18 tells us that $\Gamma = \Gamma(A)$ is a (p, λ)-jumbled tripartite graph. Together with assumption (23), this implies that Γ is a $(p, \gamma p^3 n)$-jumbled tripartite graph. In view of Theorem 10, this implies that $G \subseteq \Gamma$ satisfies one of the following properties:

(*) either G contains more than $cp^3n^3 = c|A|^3$ triangles, or else

$$\tau_3(G) \leq \frac{\delta}{3} e(\Gamma) = \delta n|A|$$

(recall that we defined c and γ invoking Theorem 10 with $\delta/3$).

Clearly, G is a spanning subgraph of Γ and the edges of G are of the form $(x_1, x_1 + b) \in X_1 \times X_2$, $(x_1, x_1 + 2b) \in X_1 \times X_3$, and $(x_2, x_2 + b) \in X_2 \times X_3$, where $x_1 \in X_1$, $x_2 \in X_2$, and $b \in B$.

The graph G contains some triangles that are *trivial*: triangles whose vertex sets are of the form $\{x_1, x_1 + b, x_1 + 2b\}$, where $x_1 \in X_1$, $x_1 + b \in X_2$, and $x_1 + 2b \in X_3$. Clearly, these triangles are pairwise edge-disjoint. Since there are $n|B|$ such triangles, we see that $\tau_3(G) \geq n|B| > \delta n|A|$. In view of assertion (*) above, we deduce that G contains more than $cp^3n^3 = c|A|^3$ triangles. Now note that, because of (29), we have $c|A|^3 \gg n|A| \geq n|B|$, that is, the total number of triangles in G is larger than the number of trivial triangles in G. Therefore G contains a non-trivial triangle. Let the vertex set of such a non-trivial triangle be $\{x_1, x_2, x_3\}$ (with $x_i \in X_i$ for all i). Then, for some b_1, b_2, and $b_3 \in B$ with $b_1 \neq b_3$, we have $x_2 = x_1 + b_1$, $x_3 = x_1 + 2b_2 = x_2 + b_3$. We deduce that $b_1 + b_3 = 2b_2$, and hence (b_1, b_2, b_3) is a 3-term arithmetic progression in B, as required. ∎

4.2.4. Proof of Lemma 16. We follow the proof of the expander mixing lemma in, e.g., [18] (see also [2]). We state and prove an auxiliary lemma before turning to Lemma 16.

Let $M = (m_{xy})_{x\in X,\ y\in Y}$ be the $\{0,1\}$-*bipartite adjacency matrix* of $\Gamma_1 = \Gamma_1(A)$, by which we mean that $m_{xy} = 1$ if x and y are adjacent in Γ_1 and $m_{xy} = 0$ otherwise. Hence $m_{xy} = \mathbb{1}_A(y - x)$. Let us consider the characters χ_k as column vectors, with the xth entry equal to $\chi_k(x)$. Let us also set

$$\mathbf{u}_k = \frac{1}{\sqrt{n}}\chi_k \in \mathbb{C}^n \tag{30}$$

for all $0 \leq k < n$. Moreover, let U be the $n \times n$ matrix whose kth column is $\mathbf{u}_k$. In what follows, we use the standard notation U^* to denote the conjugate transpose of U.

Lemma 20. *Let M, $\mathbf{u}_0, \ldots, \mathbf{u}_{n-1}$, and U be as above. Then the following holds:*

(*i*) *The eigenvalues of M are the $\lambda_k(A)$ $(0 \leq k < n)$, with associated eigenvectors $\mathbf{u}_k$ $(0 \leq k < n)$.*

(*ii*) *The $\mathbf{u}_k$ $(0 \leq k < n)$ form an orthonormal basis of $\mathbb{C}^n$. Thus $U^*U = I_n$, where I_n is the $n \times n$ identity matrix.*

(*iii*) *We have*

$$M = \sum_{0\leq k<n} \lambda_k \mathbf{u}_k \mathbf{u}_k^*. \tag{31}$$

Proof. Let us compute the product $M\chi_k$. For any $x \in \mathbb{G}$, we have

$$(M\chi_k)_x = \sum_{y\in\mathbb{G}} m_{xy}\chi_k(y) = \sum_{y\in\mathbb{G}} \mathbb{1}(y-x)\chi_k(y) = \sum_{a\in A} \chi_k(x+a) \tag{32}$$

$$= \chi_k(x)\sum_{a\in A} \chi_k(a) = \chi_k(x)\lambda_k(A),$$

whence (i) follows. For (ii), we recall that, for any k and ℓ, we have

$$\sum_{x\in\mathbb{G}} \chi_k(x)\overline{\chi_\ell(x)} = \begin{cases} n & \text{if } k=\ell \\ 0 & \text{otherwise.} \end{cases} \tag{33}$$

Therefore (ii) follows. Finally, let us consider (iii). Since the $\mathbf{u}_k$ form a basis of $\mathbb{C}^n$, it suffices to check that M and the matrix on the right-hand side of (31) act the same way on the $\mathbf{u}_k$. For any fixed $0 \le \ell < n$, recalling (ii), we see that

$$\left(\sum_{0\le k<n} \lambda_k \mathbf{u}_k \mathbf{u}_k^*\right)\mathbf{u}_\ell = \lambda_\ell \mathbf{u}_\ell \mathbf{u}_\ell^* \mathbf{u}_\ell = \lambda_\ell \mathbf{u}_\ell = M\mathbf{u}_\ell. \tag{34}$$

Identity (31) follows and (iii) is proved. ■

We are now ready to prove Lemma 16.

Proof of Lemma 16. Using part (iii) of Lemma 20, we can write M as $M_0 + E$, where $M_0 = \lambda_0 \mathbf{u}_0 \mathbf{u}_0^*$ and $E = \sum_{0<k<n} \lambda_k \mathbf{u}_k \mathbf{u}_k^*$. Note that, then,

$$M_0 = \lambda_0 \mathbf{u}_0 \mathbf{u}_0^* = |A|\chi_0\chi_0^*/n = \big(|A|/n\big)\, J_n, \tag{35}$$

where J_n is the $n \times n$ matrix all whose entries are 1.

To prove (25), we let $X' \subseteq X$ and $Y' \subseteq Y$ be fixed. It is clear that

$$e(X',Y') = \mathbb{1}_{X'}^* M \mathbb{1}_{Y'} = \mathbb{1}_{X'}^* M_0 \mathbb{1}_{Y'} + \mathbb{1}_{X'}^* E \mathbb{1}_{Y'}. \tag{36}$$

In view of (35), we have

$$\mathbb{1}_{X'}^* M_0 \mathbb{1}_{Y'} = \frac{|A|}{n} \mathbb{1}_{X'}^* J_n \mathbb{1}_{Y'} = \frac{|A|}{n} |X'|\,|Y'|. \tag{37}$$

Now let us write $\mathbb{1}_{X'}$ and $\mathbb{1}_{Y'}$ in the orthonormal basis formed by the $\mathbf{u}_k$ $(0 \le k < n)$:

$$\mathbb{1}_{X'} = \sum_{0\le k<n} \alpha_k \mathbf{u}_k \tag{38}$$

and

$$\mathbb{1}_{Y'} = \sum_{0\le k<n} \beta_k \mathbf{u}_k. \tag{39}$$

Again, by the orthonormality of the $\mathbf{u}_k$, we have

$$(40)\quad \left|\mathbb{1}_{X'}^* E \mathbb{1}_{Y'}\right| = \left|\Big(\sum_{0\le j<n} \alpha_j \mathbf{u}_j\Big)\Big(\sum_{0<k<n} \lambda_k \mathbf{u}_k \mathbf{u}_k^*\Big)\Big(\sum_{0\le \ell<n} \beta_\ell \mathbf{u}_\ell\Big)\right|$$

$$= \left|\sum_{1\le k<n} \lambda_k \alpha_k \beta_k\right| \le \sum_{1\le k<n} |\lambda_k \alpha_k \beta_k| \le \lambda \sum_{1\le k<n} |\alpha_k \beta_k|,$$

which, by the Cauchy–Schwarz inequality, is at most

$$(41)\quad \lambda \sqrt{\sum_{1\le k<n} |\alpha_k|^2} \sqrt{\sum_{1\le k<n} |\beta_k|^2} \le \lambda \|\mathbb{1}_{X'}\|_2 \|\mathbb{1}_{Y'}\|_2 = \lambda \sqrt{|X'|}\sqrt{|Y'|}.$$

Inequality (25) follows from (36), (37), (40), and (41). ■

5. Proof of the Key Technical Lemma

Section 5.1 is devoted to the proof of part (b) of Lemma 9. The main tool in the proof of part (a) of Lemma 9 is Lemma 21. We discuss Lemma 21 in Section 5.2 and deduce Lemma 9(a) from Lemma 21 in Section 5.3. The proof of Lemma 21 is deferred to Section 5.4.

5.1. PAIR implies DISC

In this section we prove implication (b) of Lemma 9. The proof of this implication "imitates" the well known proof of the dense case, i.e., of the implication (b) of Theorem 7.

Proof of Lemma 9 part (b)**.** Let α and $\delta > 0$ be given. Applying Lemma 12 with $\nu = \delta/4$ we obtain $\mu > 0$ and we set

$$(42)\quad \varepsilon = \alpha^2\delta^2\mu/4.$$

For any given $\eta > 0$, we let

$$(43)\quad \gamma = \delta\eta/2.$$

Finally, let n be sufficiently large and $p \geq 1/\sqrt{n}$.

Let $\Gamma = (U \dot{\cup} V, E_\Gamma)$, $G = (U \dot{\cup} V, E_\Gamma)$, and sets $X \subseteq U$ and $Y \subseteq V$ satisfy assumptions (i)–(iii) of Lemma 9. Moreover, we suppose Γ is $(p, \gamma p^{3/2} n)$-jumbled and $(X, Y)_G$ satisfies PAIR (q, p, ε) for some $q \in [\alpha p, p]$, i.e.,

$$\sum_{x \in X} \big| \deg_G(x, Y) - q|Y| \big| \leq \varepsilon p |X|\,|Y| \tag{44}$$

and

$$\sum_{x \in X} \sum_{x' \in X} \big| \deg_G(x, x', Y) - q^2 |Y| \big| \leq \varepsilon p^2 |X|^2 |Y|. \tag{45}$$

Let $X' \subseteq X$ and $Y' \subseteq Y$. We will show that

$$\big| e_G(X', Y') - q|X'|\,|Y'| \big| \leq \delta p |X|\,|Y|. \tag{46}$$

First we consider the case in which at least one of the sets X' or Y' is small. Suppose $|X'| < \delta|X|/2$ or $|Y'| < \delta|Y|/2$. In this case we have

$$q|X'|\,|Y'| \leq \delta q |X|\,|Y| \leq \delta p |X|\,|Y|.$$

Moreover, we infer from the $(p, \gamma p^{3/2} n)$-jumbledness of Γ combined with $|X|, |Y| \geq \eta n$, and $p \leq 1$ that

$$\begin{aligned} e_G(X', Y') &\leq e_\Gamma(X', Y') \leq p|X'|\,|Y'| + \gamma p^{3/2} n \sqrt{|X'|\,|Y'|} \\ &\leq \frac{\delta}{2} p |X|\,|Y| + \frac{\gamma}{\eta} p |X|\,|Y| \overset{(43)}{\leq} \delta p |X|\,|Y| \leq q|X'|\,|Y'| + \delta p |X|\,|Y|, \end{aligned}$$

which yields (46) for this case.

Now we assume $|X'| \geq \delta|X|/2$ and $|Y'| \geq \delta|Y|/2$. In this case (44) yields

$$\begin{aligned} \sum_{x \in X'} \big| \deg_G(x, Y) - q|Y| \big| &\leq \sum_{x \in X} \big| \deg_G(x, Y) - q|Y| \big| \leq \varepsilon p |X|\,|Y| \\ &\overset{(42)}{\leq} \mu q |X'|\,|Y| \end{aligned}$$

and, therefore,

$$\sum_{y \in Y} \deg_G(y, X') = \sum_{x \in X'} \deg_G(x, Y) \geq (1 - \mu) q |X'|\,|Y|. \tag{47}$$

Similarly, since $X' \subseteq X$, (45) yields

$$\sum_{x \in X'} \sum_{x' \in X'} \left| \deg_G(x, x', Y) - q^2|Y| \right| \leq \varepsilon p^2 |X|^2 |Y| \overset{(42)}{\leq} \mu q^2 |X'|^2 |Y|$$

and, therefore,

$$\sum_{y \in Y} \deg_G^2(y, X') = \sum_{x \in X'} \sum_{x' \in X'} \deg_G(x, x', Y) \leq (1 + \mu) q^2 |X'|^2 |Y|. \tag{48}$$

Owing to (47) and (48), we infer from Lemma 12 that all but at most $\nu|Y|$ vertices $y \in Y$ satisfy

$$\left| \deg_G(y, X') - q|X'| \right| \leq \nu q |X'|. \tag{49}$$

Since we chose $\nu < \delta/2$, we have

$$\begin{aligned} e_G(X', Y') &\geq (1 - \nu)|Y'| \cdot (1 - \nu) q |X'| \geq (1 - \delta) q |X'|\,|Y'| \\ &\geq q|X'|\,|Y'| - \delta p |X|\,|Y|. \end{aligned} \tag{50}$$

On the other hand, the $(p, \gamma p^{3/2} n)$-jumbledness of Γ yields by Fact 13 that all but at most

$$\frac{\gamma^2 p^3 n^2}{p^2 |X'|} \leq \frac{\gamma^2 p n^2}{\delta |X|} \leq \frac{\gamma^2 p n^2}{\delta \eta n} \leq \frac{\gamma^2 p |Y|}{\delta \eta^2} \overset{(43)}{\leq} \frac{\delta}{4} p |Y| \tag{51}$$

vertices $y \in Y'$ we have

$$\deg_G(y, X') \leq \deg_\Gamma(y, X') \leq 2p|X'|. \tag{52}$$

Consequently, owing to (49), (51), and (52) we have

$$\begin{aligned} e_G(X', Y') &\leq |Y'| \cdot (1 + \nu) q |X'| + \nu |Y'| \cdot 2p|X'| + \frac{\delta}{4} p |Y| \cdot |X'| \\ &\leq q|X'|\,|Y'| + 3\nu p |X|\,|Y| + \frac{\delta}{4} p |Y|\,|X| \\ &\leq q|X'|\,|Y'| + \delta p |X|\,|Y|, \end{aligned} \tag{53}$$

where we used $\nu \leq \delta/4$. Finally, (46) follows from (50) and (53). ■

5.2. Inheritance of the pair condition in neighborhoods of jumbled graphs

Lemma 21 asserts that the pair condition of a dense graph (i.e., $p = 1$) is inherited on neighborhoods of a (possibly sparse) sufficiently jumbled graph. More precisely, suppose that $B = (X_1 \dot{\cup} X_2, E_B)$ is a dense bipartite graph that has property PAIR $(\varrho, 1, \mu)$ for some constant $\varrho > 0$ and sufficiently small $\mu > 0$. Furthermore, let $\Gamma = (U \dot{\cup} V, E_\Gamma)$ be a bipartite $(p, \gamma p^2 n)$-jumbled graph with $X_1, X_2 \subseteq U$. Lemma 21 states that if γ and μ are sufficiently small, then for most vertices $v \in V$ the pair condition PAIR $(\varrho, 1, \nu)$ is inherited on the subgraph of B induced on the neighborhood of v in Γ.

Lemma 21. *For every $\varrho_0 > 0$ and $\nu > 0$ there exists $\mu > 0$ such that for every $\eta > 0$ there exist $\gamma > 0$ and n_0 such that for every $n \geq n_0$ the following holds.*

Suppose

(*i*) *$\Gamma = (U \dot{\cup} V, E_\Gamma)$ is a bipartite $(p, \gamma p^2 n)$-jumbled graph with $p \geq 1/\sqrt{n}$,*

(*ii*) *X_1, $X_2 \subseteq U$ and $Y \subseteq V$ with $|X_1|, |X_2|, |Y| \geq \eta n$, and*

(*iii*) *$B = (X_1 \dot{\cup} X_2, E_B)$ is an arbitrary bipartite graph.*

Then the following statements hold:

(*a*) *if $(X_1, X_2)_B$ satisfies* PAIR $(\varrho, 1, \mu)$ *for some ϱ with $\varrho_0 \leq \varrho \leq 1$, then for all but at most $\nu|Y|$ vertices $y \in Y$ the pair*

$$\big(N_\Gamma(y, X_1), N_\Gamma(y, X_2)\big)_B$$

satisfies PAIR $(\varrho, 1, \nu)$*;*

(*b*) *if $(X_1, X_2)_B$ satisfies* DISC $(\varrho, 1, \mu)$ *for some ϱ with $\varrho_0 \leq \varrho \leq 1$, then for all but at most $\nu|Y|$ vertices $y \in Y$ the pair*

$$\big(N_\Gamma(y, X_1), N_\Gamma(y, X_2)\big)_B$$

satisfies DISC $(\varrho, 1, \nu)$.

From the equivalence of DISC and PAIR for dense graphs (see Theorem 7), we infer that statements (a) and (b) are equivalent, and for the proof of Lemma 21 it suffices to verify only one of them. We will apply part (b) of Lemma 21 in the proof of Lemma 9(a) (see Section 5.4 below).

5.3. DISC implies PAIR

The proof of Lemma 9(a) is based on part (b) of Lemma 21 and we briefly outline the main ideas below.

We will study an auxiliary graph with vertex set X and edges corresponding to pairs $\{x, x'\}$ for which $\big|\deg_G(x, x', Y) - q^2|Y|\big| > \xi q^2|Y|$ for some constant ξ depending only on δ. We split those edges into two sets depending on whether $\deg_G(x, x', Y) > (1+\xi)q^2|Y|$ or $\deg_G(x, x', Y) < (1-\xi)q^2|Y|$ and call the resulting graphs B^+ and B^-. The assumption that Lemma 9(a) fails implies that at least one of the graphs B^+ or B^- has $\varrho\binom{|X|}{2}$ edges for some ϱ depending on α and δ. Suppose $e(B^+) = \varrho\binom{|X|}{2}$ (the argument for the other case is very similar). Applying Szemerédi's regularity lemma (or a weaker version ensuring just one pair with the discrepancy property (see Fact 22 below) to B^+ shows that there exist subsets X_1 and $X_2 \subseteq X$ such that $(X_1, X_2)_B$ satisfies DISC $(\varrho, 1, \mu)$ for some small constant $0 < \mu \ll \xi$ and, hence, we can apply part (b) of Lemma 21 to $B = B^+[X_1, X_2]$ and Γ. Since $\big|N_G(y, X_i)\big| \geq (\alpha - \varepsilon)N_\Gamma(y, X_i)$ for most $y \in Y$ and $i = 1, 2$, the conclusion of part (b) of Lemma 21 yields good estimates on $e_B\big(N_G(y, X_1), N_G(y, X_2)\big)$ for most $y \in Y$. Based on this and the right choice of constants we will be able to show that the number of triples $(x_1, x_2, y) \in X_1 \times X_2 \times Y$ with $x_1 \in N_G(y, X_1)$, $x_2 \in N_G(y, X_2)$, and $\{x_1, x_2\} \in E(B)$ is bounded from above by $(1 + \xi)q^2 e(B)|Y|$. On the other hand, the definition of B^+ implies that there are more than $(1 + \xi)q^2 e(B)|Y|$ such triples, which gives the desired contradiction. In the proof we will use the following fact, which asserts that every dense graph contains a pair of linear size that satisfies the discrepancy property.

Fact 22. *Suppose $\varrho_0 > 0$, $\mu > 0$, and $B = (X, E_B)$ is a graph with $|E_B| \geq \varrho_0\binom{|X|}{2}$. Then there exist disjoint subsets $X_1, X_2 \subseteq X$ such that*

(a) *$(X_1, X_2)_B$ satisfies DISC $(\varrho, 1, \mu)$ for some $\varrho \geq \varrho_0$ and*

(b) *$|X_1|, |X_2| \geq \zeta n$ for $\zeta = \varrho_0^{100/\mu^2}/4$.*

Proof. Fact 22 follows from [22, Theorem 1.1] applied with $d = \varrho_0$ and $\varepsilon = 3\mu/4$. ∎

Proof of Lemma 9 part (a). First we define all constants required in this proof. Let $\alpha > 0$ and $\delta > 0$ be given. We set

$$\xi = \frac{\delta}{6} \tag{54}$$

and for the application of Lemma 21 we define

$$\varrho_0 = \frac{\delta}{50} \tag{55}$$

and

$$\nu = \frac{\alpha^2 \xi \varrho_0}{64}. \tag{56}$$

For this choice of ϱ_0 and ν Lemma 21 ensures the existence of some constant $\mu > 0$. Without loss of generality we may assume that

$$\mu \leq \frac{\xi \varrho_0}{4} \tag{57}$$

Preparing for an application of Fact 22 we set

$$\zeta = \frac{\varrho_0^{100/\mu^2}}{4} \tag{58}$$

and for Lemma 9 we define $\varepsilon = \varepsilon(\alpha, \delta)$ by

$$\varepsilon = \min\left\{ \frac{\alpha\delta^2}{36}, \left(\frac{\alpha^3 \xi \varrho_0 \zeta}{64}\right)^2 \right\}. \tag{59}$$

Following the quantification of Lemma 9, we are given some $\eta > 0$. We set

$$\eta_{\mathrm{L.21}} = \eta\zeta. \tag{60}$$

For this choice of $\eta_{\mathrm{L.21}}$ Lemma 21 yields $\gamma_{\mathrm{L.21}} > 0$ and for Lemma 9 we set

$$\gamma = \min\left\{ \gamma_{\mathrm{L.21}}, \eta\sqrt{\delta/3}, \alpha\sqrt{\xi\varrho_0\eta\eta_{\mathrm{L.21}}/24} \right\}. \tag{61}$$

Finally, let n be sufficiently large and suppose $p \geq 1/\sqrt{n}$.

After we fixed all constants involved in the proof of Lemma 9 consider bipartite graphs $\Gamma = (U \dot\cup V, E_\Gamma)$ and $G = (U \dot\cup V, E_G)$ and sets $X \subseteq U$ and $Y \subseteq V$ that satisfy the assumptions of Lemma 9. In particular, $(X, Y)_G$ satisfies $\mathrm{DISC}(q, p, \varepsilon)$ for some $q \geq \alpha p$. We want to infer that $(X, Y)_G$ satisfies $\mathrm{PAIR}(q, p, \delta)$, i.e., we have to verify that $(X, Y)_G$ satisfies (2) and (3) in Definition 6.

Verifying (2) **for** $(X,Y)_G$. As we will see, this is a simple consequence of the given discrepancy property of $(X,Y)_G$ and the jumbledness of Γ. First we note that all but at most $\delta|X|/6$ vertices $x \in X$ satisfy

$$\big|\deg_G(x,Y) - q|Y|\big| \le \frac{\delta}{3} q|Y|.$$

Otherwise there exists a set $X' \subseteq X$ with $|X'| \ge \delta|X|/12$ such that for all $x \in X'$ either $\deg_G(x,Y) > (1+\delta/3)q|Y|$ or $\deg_G(x,Y) < (1-\delta/3)q|Y|$. In either case, we would face

$$\big|e(X',Y) - q|X'|\,|Y|\big| > \frac{\delta^2}{36} q|X|\,|Y| \ge \frac{\delta^2\alpha}{36} p|X|\,|Y| \overset{(59)}{\ge} \varepsilon p|X|\,|Y|,$$

which would contradict the assumption that $(X,Y)_G$ satisfies DISC (q,p,ε).

Fact 13 implies that the number of vertices $x \in X$ with $\deg_\Gamma(x,Y) \ge 2p|Y|$ is bounded from above by $\gamma^2p^2n^2/|Y|$. Consequently,

$$\begin{aligned}\sum_{x\in X} \big|\deg_G(x,Y) - q|Y|\big| &\le |X| \cdot \frac{\delta}{3} q|Y| + \frac{\delta}{6}|X| \cdot 2p|Y| + \gamma^2p^2\frac{n^2}{|Y|} \cdot |Y| \\ &\le \delta p|X|\,|Y|,\end{aligned}$$

where the last inequality follows from $q \le p$, the choice of γ in (61) and assumption (*iii*) of Lemma 9.

Verifying (3) **for** $(X,Y)_G$. For the proof of (3) we proceed by contradiction. We consider the "bad" pairs $(x,x') \in X \times X$ for which $\deg_G(x,x',Y)$ deviates substantially from $q^2|Y|$. First we consider the following sets of bad pairs

$$\mathcal{B}_1 = \big\{ (x,x') \in X \times X \,:\, \deg_\Gamma(x,Y) > 2p|Y| \big\},$$

$$\mathcal{B}_2 = \big\{ (x,x') \in (X \times X) \setminus \mathcal{B}_1 \,:\, \deg_\Gamma(x,x',Y) > 4p^2|Y| \big\}.$$

From the $(p,\gamma p^2 n)$-jumbledness of Γ, the choice γ in (61), and assumption (*iii*) of Lemma 9 combined with Fact 13, it follows that

$$|\mathcal{B}_1| \le \gamma^2p^2\frac{n^2}{|Y|} \cdot |X| \le \frac{\delta}{3}p^2|X|^2.$$

For an upper bound on $|\mathcal{B}_2|$ we consider a vertex $x \in X$ with $\deg_\Gamma(x,Y) \le 2p|Y|$. Applying Fact 13 to a superset of $N_\Gamma(x,Y)$ with $2p|Y|$ elements implies that there are at most $\gamma^2 pn^2/(2|Y|)$ vertices x' such that $\deg_\Gamma(x,x',Y) > 4p^2|Y|$. Consequently,

$$|\mathcal{B}_2| \le |X| \cdot \gamma^2 p \frac{n^2}{2|Y|} \le \frac{\delta}{6} p|X|^2.$$

Combining those estimates on $|\mathcal{B}_1|$ and $|\mathcal{B}_2|$ with the assumption that (3) fails, we infer

$$\sum_{(x,x')\notin \mathcal{B}_1\cup\mathcal{B}_2} \left|\deg_G(x,x',Y) - q^2|Y|\right| \tag{62}$$

$$> \delta p^2|X|^2|Y| - |\mathcal{B}_1|\cdot|Y| - |\mathcal{B}_2|\cdot 2p|Y| \ge \frac{\delta}{3} p^2|X|^2|Y|.$$

In other words we have just shown that ignoring the contribution of pairs $(x,x') \in \mathcal{B}_1 \cup \mathcal{B}_2$ does not affect in an essential way the property drawn from the assumption that (3) fails.

Moreover, note that the total contribution of pairs $(x,x') \notin \mathcal{B}_1 \cup \mathcal{B}_2$ with $\left|\deg_G(x,x',Y) - q^2|Y|\right| \le \delta q^2|Y|/6$ to the sum considered in (62) is at most

$$\frac{\delta}{6} q^2|X|^2|Y| \le \frac{\delta}{6} p^2|X|^2|Y|.$$

Since, furthermore, for all $(x,x') \notin \mathcal{B}_1 \cup \mathcal{B}_2$ we have

$$\left|\deg_G(x,x',Y) - q^2|Y|\right| \le \max\left\{q^2|Y|, (4p^2 - q^2)|Y|\right\} \le 4p^2|Y|,$$

there are at least $\delta|X|^2/24$ pairs $(x,x') \in (X \times X) \setminus (\mathcal{B}_1 \cup \mathcal{B}_2)$ such that

$$\left|\deg_G(x,x',Y) - q^2|Y|\right| > \frac{\delta}{6} q^2|Y| \overset{(54)}{=} \xi q^2|Y|. \tag{63}$$

Next we consider the graphs B^+ and B^- with vertex set X and edges corresponding to those pairs defined by

(64)
$$E(B^+) = \left\{\{x,x'\} : (1+\xi)q^2|Y| < \deg_G(x,x',Y) \le 4p^2|Y| \text{ and } x \ne x'\right\}$$

and

$$E(B^-) = \left\{\{x,x'\} : \deg_G(x,x',Y) < (1-\xi)q^2|Y| \text{ and } x \ne x'\right\}.$$

Since there are at least $\delta|X|^2/24$ ordered pairs (x,x') for which (63) holds, for sufficiently large n we have

$$\max\big\{e(B^+),e(B^-)\big\} \geq \frac{\delta}{96}|X|^2 - |X| \overset{(55)}{\geq} \varrho_0\binom{|X|}{2}.$$

Below we assume $e(B^+) \geq \varrho_0\binom{|X|}{2}$ and remark that the argument for the case $e(B^-) \geq \varrho_0\binom{|X|}{2}$ follows the same lines.

Fact 22, combined with the choice of ζ in (58), gives that there exist disjoint subsets $X_1, X_2 \subseteq X$ such that $(X_1,X_2)_{B^+}$ satisfies DISC $(\varrho,1,\mu)$ for some $\varrho \geq \varrho_0$ and $|X_1|,|X_2| \geq \zeta|X| \geq \zeta\eta n = \eta_{\mathrm{L}.21}n$ (see (60)).

Hence, it follows from part (*b*) of Lemma 21 that for all but at most $\nu|Y|$ vertices $y \in Y$ we have that

$$(65) \qquad \big(N_\Gamma(y,X_1),N_\Gamma(y,X_2)\big)_{B^+} \quad \text{satisfies} \quad \mathrm{DISC}\,(\varrho,1,\nu).$$

Based on (65) we obtain estimates on the number of triplets (x_1,x_2,y) from $X_1\times X_2\times Y$ such that $\{x_1,x_2\}$ is an edge in B^+ contained in the neighborhood of y in G. Let

$$\mathcal{T} = \big\{(x_1,x_2,y) \in X_1\times X_2\times Y :$$
$$x_1 \in N_G(y,X_1),\ x_2 \in N_G(y,X_2),\ \text{and } \{x_1,x_2\}\in E(B^+)\big\}.$$

From the definition of B^+ in (64) and the fact that $(X_1,X_2)_{B^+}$ satisfies DISC $(\varrho,1,\mu)$ we infer

$$(66) \quad |\mathcal{T}| > (1+\xi)q^2|Y|\cdot e_{B^+}(X_1,X_2) \;\geq\; (1+\xi)q^2|Y|\cdot(\varrho-\mu)|X_1|\,|X_2|$$
$$\overset{(57)}{\geq} \left(1+\frac{\xi}{2}\right)\varrho q^2|X_1|\,|X_2|\,|Y|.$$

On the other hand,

$$|\mathcal{T}| = \sum_{y\in Y} e_{B^+}\big(N_G(y,X_1),N_G(y,X_2)\big).$$

Let $Y' = \big\{y \in Y : \deg_\Gamma(y,X_i) \leq 2p|X_i| \text{ for } i=1,2\big\}$. It follows from (65) that for all but at most $\nu|Y|$ vertices $y\in Y'$ we have

$$e_{B^+}\big(N_G(y,X_1),N_G(y,X_2)\big)$$

$$\leq \varrho \big| N_G(y, X_1) \big| \, \big| N_G(y, X_2) \big| + \nu \big| N_\Gamma(y, X_1) \big| \, \big| N_\Gamma(y, X_2) \big|$$

$$\leq \varrho \deg_G(y, X_1) \deg_G(y, X_2) + 4\nu p^2 |X_1| \, |X_2|.$$

Fact 13 gives

$$|Y \setminus Y'| \leq \gamma^2 p^2 \frac{n^2}{|X_1|} + \gamma^2 p^2 \frac{n^2}{|X_2|}.$$

Consequently,

$$|T| \leq \sum_{y \in Y'} \big(\varrho \deg_G(y, X_1) \deg_G(y, X_2) + 4\nu p^2 |X_1| \, |X_2| \big)$$

$$+ \nu |Y| \cdot 4p^2 |X_1| \, |X_2| + \left(\gamma^2 p^2 \frac{n^2}{|X_1|} + \gamma^2 p^2 \frac{n^2}{|X_2|} \right) \cdot |X_1| \, |X_2|$$

and the choices of ν in (56) and γ in (61), combined with $|X_1|, |X_2| \geq \eta_{\text{L.21}} n$ and $|Y| \geq \eta n$ and $\varrho \geq \varrho_0$, yield

$$|T| \leq \varrho \sum_{y \in Y'} \deg_G(y, X_1) \deg_G(y, X_2) + \frac{\xi}{4} \varrho q^2 |X_1| \, |X_2| \, |Y|. \tag{67}$$

Finally, we appeal to the assumption of Lemma 9(a) stating that $(X, Y)_G$ satisfies DISC (q, p, ε). As $|X_1|, |X_2| \geq \zeta |X|$ and $q \geq \alpha p$ we have for $i = 1$ and 2

$$\big| \{ y \in Y : \deg_G(y, X_i) > (1 + \sqrt{\varepsilon}\,) q |X_i| \} \big| \leq \frac{\varepsilon p |X| \, |Y|}{\sqrt{\varepsilon} q |X_i|} \leq \frac{\sqrt{\varepsilon}}{\alpha \zeta} |Y|$$

$$\overset{(59)}{\leq} \frac{\xi \alpha^2 \varrho}{64} |Y|.$$

Since $\sqrt{\varepsilon} \leq \xi / 24$ we can further bound the right-hand side of (67) and obtain

$$|T| \leq \varrho |Y| \cdot (1 + \sqrt{\varepsilon}\,)^2 q^2 |X_1| \, |X_2| + \frac{\xi \alpha^2 \varrho}{32} |Y| \cdot 4p^2 |X_1| \, |X_2|$$

$$+ \frac{\xi}{4} \varrho q^2 |X_1| \, |X_2| \, |Y|$$

$$\leq \left(1 + \frac{\xi}{2} \right) \varrho q^2 |X_1| \, |X_2| \, |Y|,$$

which contradicts (66). ■

5.4. Proof of Lemma 21

Recall that it suffices to prove part (*a*) of Lemma 21 and then part (*b*) follows from Theorem 7.

Proof of Lemma 21. Let ϱ_0 and ν be given. We fix an auxiliary constant

$$\nu' = \frac{\nu^2}{100} \tag{68}$$

and let $\mu'' > 0$ be given by Lemma 12 applied with $\nu'' = \nu'/4$. Moreover, fix μ' and ξ with $0 < \mu' \leq \mu''/2$ and $0 < \xi < \min\{\mu''/6, \nu'/4\}$ so that

$$\frac{(1-\mu')^2}{1+\xi} \geq 1 - \frac{\mu''}{2} \quad \text{and} \quad (1+\xi)\sqrt{1+\mu'} \leq 1 + \frac{\mu''}{2}. \tag{69}$$

Next let μ be sufficiently small, so that

$$\mu \leq \frac{\varrho_0^4 \mu'}{5}, \quad \frac{\sqrt{\mu}}{\varrho^2}(1+\mu)^2(1+\xi)^2 \leq \frac{\mu'\varrho^4}{5}, \quad \text{and} \tag{70}$$

$$(1+\mu)^4\left(1+\sqrt{\mu}\right)^2 \leq 1 + \frac{\mu'}{5}.$$

For any given $\eta > 0$, we let

$$\gamma = \min\left\{\xi\eta\sqrt{\nu/12},\, \mu^{3/2}\eta\varrho_0/2\right\}. \tag{71}$$

Let n be sufficiently large and suppose $p \geq 1/\sqrt{n}$.

Suppose the bipartite graph $\Gamma = (U \mathbin{\dot\cup} V, E_\Gamma)$, the sets $X_1, X_2 \subseteq U$, and $Y \subseteq V$, and the bipartite graph $B = (X_1 \mathbin{\dot\cup} X_2, E_B)$ satisfy assumptions (*i*)–(*iii*) of Lemma 21. Moreover, suppose that $(X_1, X_2)_B$ satisfies PAIR$(\varrho, 1, \mu)$ for some $\varrho \geq \varrho_0$. We have to show that for all but $\nu|Y|$ vertices $y \in Y$ the pair condition is inherited in the subgraph of B induced on the neighborhoods of y in Γ, i.e., $\big(N_\Gamma(y, X_1), N_\Gamma(y, X_2)\big)_B$ satisfies PAIR$(\varrho, 1, \nu)$.

In the first step we exclude all vertices $y \in Y$ for which $\deg_\Gamma(y, X_1)$ or $\deg_\Gamma(y, X_2)$ deviates substantially from its "expectation." For $i = 1$ and 2 we set

$$Y_i^* = \big\{y \in Y : \big|\deg_\Gamma(y, X_i) - p|X_i|\big| > \xi p|X_i|\big\} \quad \text{and} \quad Y^* = Y_1^* \cup Y_2^*.$$

Since Γ is $(p, \gamma p^2 n)$-jumbled and $|X_1|, |X_2|, |Y| \geq \eta n$, Fact 13 yields

$$|Y^*| \leq |Y_1^*| + |Y_2^*| \leq \frac{2\gamma^2 p^4 n^2}{\xi^2 p^2 |X_1|} + \frac{2\gamma^2 p^4 n^2}{\xi^2 p^2 |X_2|} \leq \frac{4\gamma^2 p^2 |Y|}{\xi^2 \eta^2}.$$

Since $p \leq 1$, the choice of γ in (71) yields

$$|Y^*| \leq \frac{\nu}{3}|Y|. \tag{72}$$

For the rest of the proof we will ignore the vertices from Y^* and they will be included in the set of $\nu|Y|$ exceptional vertices.

Next we study the triplets $(x_1, x_2, y) \in X_1 \times X_2 \times (Y \setminus Y^*)$ with $\{x_1, x_2\}$ being an edge of B and $\{x_1, y\}$ and $\{x_2, y\}$ being present in Γ. We set

$$Y' = Y \setminus Y^*$$

and define

$$t_{1,1} = \sum_{y \in Y'} \sum_{x_1 \in N_\Gamma(y, X_1)} \deg_B\left(x_1, N_\Gamma(y, X_2)\right)$$

as well as

$$t_{1,2} = \sum_{y \in Y'} \sum_{x_1 \in N_\Gamma(y, X_1)} \deg_B^2\left(x_1, N_\Gamma(y, X_2)\right),$$

$$t_{2,1} = \sum_{y \in Y'} \sum_{x_2 \in N_\Gamma(y, X_2)} \deg_B^2\left(x_2, N_\Gamma(y, X_1)\right),$$

and

$$t_{2,2} = \sum_{y \in Y'} \sum_{x_1 \in N_\Gamma(y, X_1)} \sum_{x_1' \in N_\Gamma(y, X_1)} \deg_B^2\left(x_1, x_1', N_\Gamma(y, X_2)\right).$$

Below we will verify the following bounds on $t_{1,1}$ and $t_{2,2}$

$$t_{1,1} \geq (1 - \mu')\varrho p^2 |X_1|\,|X_2|\,|Y'| \tag{73}$$

and

$$t_{2,2} \leq (1 + \mu')\varrho^4 p^4 |X_1|^2 |X_2|^2 |Y'|. \tag{74}$$

Before we prove (73) and (74) we deduce the conclusion of Lemma 21 from those estimates.

Since $\big|N_\Gamma(y,X_1)-p|X_1|\big| \leq \xi p|X_1|$ for all $y\in Y'$ the number of summands m considered in $t_{1,1}$ is bounded by $|Y'|(1+\xi)p|X_1|$. Hence, it follows from (73) combined with the Cauchy–Schwarz inequality $\sum_{i=1}^m a_i^2 \geq \big(\sum_{i=1}^m a_i\big)^2/m$ that

$$(75)\quad t_{1,2} \geq \frac{(1-\mu')^2}{1+\xi}\varrho^2p^3|X_1|\,|X_2|^2|Y'| \overset{(69)}{\geq} (1-\mu''/2)\varrho^2p^3|X_1|\,|X_2|^2|Y'|.$$

Moreover, owing to the identity

$$t_{1,1} = \sum_{y\in Y'}\ \sum_{x_2\in N_\Gamma(y,X_2)} \deg_B\big(x_2, N_\Gamma(y,X_1)\big)$$

the same argument gives

$$(76)\qquad t_{2,1} \geq (1-\mu''/2)\varrho^2p^3|X_1|^2|X_2|\,|Y'|.$$

Similarly, since we can rewrite $t_{2,1}$ as

$$t_{2,1} = \sum_{y\in Y'}\ \sum_{x_1\in N_\Gamma(y,X_1)}\ \sum_{x_1'\in N_\Gamma(y,X_1)} \deg_B\big(x_1,x_1',N_\Gamma(y,X_2)\big)$$

the bound in (74) combined with the Cauchy–Schwarz inequality yields

(77)

$$t_{2,1} \leq (1+\xi)\sqrt{1+\mu'}\varrho^2p^3|X_1|^2|X_2|\,|Y'| \overset{(69)}{\leq} (1+\mu''/2)\varrho^2p^3|X_1|^2|X_2|\,|Y'|.$$

In a very similar way we obtain

$$(78)\qquad t_{1,2} \leq (1+\mu''/2)\varrho^2p^3|X_1|\,|X_2|^2|Y'|.$$

Summarizing (75)–(78), we showed that (73) and (74) yield

$$(79)\qquad \big|t_{1,2}-\varrho^2p^3|X_1|\,|X_2|^2|Y'|\big| \leq \frac{\mu''}{2}\varrho^2p^3|X_1|\,|X_2|^2|Y'|$$

and

$$(80)\qquad \big|t_{2,1}-\varrho^2p^3|X_1|^2|X_2|\,|Y'|\big| \leq \frac{\mu''}{2}\varrho^2p^3|X_1|^2|X_2|\,|Y'|.$$

In particular, (73) gives

$$t_{1,1} \geq (1-\mu'')(1+\xi)\varrho p^2|X_1|\,|X_2|\,|Y'|$$

and, similarly, (79) gives

$$t_{1,2} \le (1+\mu'')(1-\xi)\varrho^2 p^3 |X_1|\,|X_2|^2 |Y'|.$$

Recalling the definition of Y', by Lemma 12 we see that those estimates yield

$$\begin{aligned}(81)\qquad &\Big|\big|N_B\big(x_1, N_\Gamma(y,X_2)\big)\big| - \varrho|N_\Gamma(y,X_2)|\Big| \\ &\le \Big|\big|N_B\big(x_1, N_\Gamma(y,X_2)\big)\big| - \varrho p|X_2|\Big| + \xi\varrho p|X_2| \\ &\le \left(\frac{\nu'}{4}+\xi\right)\varrho p|X_2| \le \nu'\big|N_\Gamma(y,X_2)\big|\end{aligned}$$

for all but at most $\nu'(1+\xi)p|X_1|\,|Y'|$ pairs $(y,x_1)\in Y\times X_1$ with $x_1\in N_\Gamma(y,X_1)$. Let $Y^{**}\subseteq Y'$ contain those vertices y for which there are at least $\nu\big|N_\Gamma(y,X_1)\big|/2$ vertices $x_1\in N_\Gamma(y,X_1)$ such that (81) fails. Since (81) fails for at most $\nu'(1+\xi)p|X_1|\,|Y'|$ pairs and each $y\in Y^{**}$ is contained in at least $\nu\big|N_\Gamma(y,X_1)\big|/2$ such pairs we have

$$(82)\qquad |Y^{**}| \le \frac{2\nu'(1+\xi)p|X_1|\,|Y'|}{\nu(1-\xi)p|X_1|} \overset{(68)}{\le} \frac{\nu}{3}|Y|.$$

Recalling that for $y\notin Y^*\cup Y^{**}$ a "typical" vertex $x_1\in N_\Gamma(y,X_1)$ satisfies (81) and that there are at most $\nu\big|N_\Gamma(y,X_1)\big|/2$ "atypical" vertices we infer

$$\begin{aligned}(83)\qquad &\sum_{x_1\in N_\Gamma(y,X_1)} \Big|\big|N_B\big(x_1,N_\Gamma(y,X_2)\big)\big| - \varrho\big|N_\Gamma(y,X_2)\big|\Big| \\ &\le \left(\nu'+\frac{\nu}{2}\right)\big|N_\Gamma(y,X_1)\big|\,\big|N_\Gamma(y,X_2)\big| \le \nu\big|N_\Gamma(y,X_1)\big|\,\big|N_\Gamma(y,X_2)\big|,\end{aligned}$$

which verifies the first part of PAIR$(\varrho,1,\nu)$ (i.e., (2)) for $\big(N_\Gamma(y,X_1), N_\Gamma(y,X_2)\big)_B$ for all $y\in Y\setminus(Y^*\cup Y^{**})$.

Next we deduce the second part of PAIR$(\varrho,1,\nu)$ (i.e., (3)) for "most" vertices $y\in Y\setminus(Y^*\cup Y^{**})$ from (80) and (74) in a very similar manner.

In fact, (80), (74), and $\xi\le\mu''/6$ imply

$$t_{2,1} \ge (1-\mu'')(1+\xi)^2\varrho^2 p^3|X_1|^2|X_2|\,|Y'|$$

and

$$t_{2,2} \leq (1+\mu'')(1-\xi)^2\varrho^4 p^4 |X_1|^2 |X_2|^2 |Y'|.$$

By the definition of Y', Lemma 12 yields

$$\begin{aligned}(84)\qquad & \Big| \big| N_B\big(x_1, x_1', N_\Gamma(y, X_2)\big) \big| - \varrho^2 |N_\Gamma(y, X_2)| \Big| \\ &\leq \Big| \big| N_B\big(x_1, x_1', N_\Gamma(y, X_2)\big) \big| - \varrho^2 p |X_2| \Big| + \xi \varrho^2 p |X_2| \\ &\leq \left(\frac{\nu'}{4} + \xi\right) \varrho^2 p |X_2| \leq \nu' \big| N_\Gamma(y, X_2) \big|\end{aligned}$$

for all but at most $\nu'(1+\xi)^2 p^2 |X_1|^2 |Y'|$ triplets $(y, x_1, x_1') \in Y \times X_1 \times X_1$ with $x_1, x_1' \in N_\Gamma(y, X_1)$. Let $Y^{***} \subseteq Y'$ contain those vertices y for which there are at least $\nu \big| N_\Gamma(y, X_1) \big|^2/2$ vertices $x_1, x_1' \in N_\Gamma(y, X_1)$ such that (84) fails. It follows from the definition of Y^{***} that

$$(85)\qquad |Y^{***}| \leq \frac{2\nu'(1+\xi)^2 p^2 |X_1|^2 |Y'|}{\nu(1-\xi)^2 p^2 |X_1|^2} \overset{(68)}{\leq} \frac{\nu}{3}|Y|.$$

Finally, we note that for all vertices $y \in Y \setminus (Y^* \cup Y^{***})$ we have

$$\begin{aligned}(86)\qquad & \sum_{x_1 \in N_\Gamma(y, X_1)} \sum_{x_1' \in N_\Gamma(y, X_1)} \Big| \big| N_B\big(x_1, x_1', N_\Gamma(y, X_2)\big) \big| - \varrho^2 \big| N_\Gamma(y, X_2) \big| \Big| \\ &\leq \left(\nu' + \frac{\nu}{2}\right) \big| N_\Gamma(y, X_1) \big|^2 \big| N_\Gamma(y, X_2) \big| \leq \nu \big| N_\Gamma(y, X_1) \big|^2 \big| N_\Gamma(y, X_2) \big|,\end{aligned}$$

which verifies (3) for $\big(N_\Gamma(y, X_1), N_\Gamma(y, X_2)\big)_B$ for all $y \in Y \setminus (Y^* \cup Y^{***})$.

Inequalities (83) and (86) imply that for every $y \notin Y^* \cup Y^{**} \cup Y^{***}$ the pair $\big(N_\Gamma(y, X_1), N_\Gamma(y, X_2)\big)_B$ satisfies PAIR$\,(\varrho, 1, \nu)$. Moreover, owing to (72), (82), and (85), we have

$$|Y^* \cup Y^{**} \cup Y^{***}| \leq \nu |Y|.$$

This concludes the proof of part (a) of Lemma 21. It remains to verify (73) and (74). ■

Proof of (73). In order to verify (73) we have to estimate the number of triplets $(x_1, x_2, y) \in X_1 \times X_2 \times Y'$ such that $\{x_1, x_2\}$ is an edge of B and x_1 and x_2 are neighbors of y in Γ.

We recall the assumption that $(X_1, X_2)_B$ satisfies PAIR$(\varrho, 1, \mu)$ and, therefore,

$$\sum_{x_1 \in X_1} \big|\deg_B(x_1, X_2) - \varrho|X_2|\big| \leq \mu|X_1|\,|X_2|.$$

Consequently,

$$e_B(X_1, X_2) \geq (\varrho - \mu)|X_1|\,|X_2|. \tag{87}$$

Moreover, owing to Fact 13 all but at most

$$\frac{\gamma^2 p^2 n^2}{\mu^2 |Y'|} \leq \frac{\gamma^2 p^2 n^2}{\mu^2 (1 - \nu/3)|Y|} \overset{(71)}{\leq} \mu|X_1| \tag{88}$$

vertices $x_1 \in X_1$ satisfy

$$\deg_\Gamma(x_1, Y') \geq (1 - \mu)p|Y'|. \tag{89}$$

Finally, for every vertex x_1 for which (89) holds another application of Fact 13 yields that

$$\deg_\Gamma\big(x_2, N_\Gamma(x_1, Y')\big) \geq (1 - \mu)p \deg_\Gamma(x_1, Y') \geq (1 - \mu)^2 p^2 |Y'| \tag{90}$$

for all but at most

(91)

$$\frac{\gamma^2 p^2 n^2}{\mu^2 \deg_\Gamma(x_1, Y')} \leq \frac{\gamma^2 p^2 n^2}{\mu^2 \cdot (1 - \mu)p|Y'|} \leq \frac{\gamma^2 p^2 n^2}{\mu^2 \cdot (1 - \mu)(1 - \nu/3)p|Y|} \overset{(71)}{\leq} \mu|X_2|$$

vertices $x_2 \in X_2$. Summarizing (88)–(91), we infer that for all but at most $2\mu|X_1|\,|X_2|$ pairs of vertices $(x_1, x_2) \in X_1 \times X_2$ we have

$$\deg_\Gamma(x_1, x_2, Y') \geq (1 - \mu)^2 p^2 |Y'|.$$

Combining this with (87) yields

$$\begin{aligned} t_{1,1} &\geq \big(e_B(X_1, X_2) - 2\mu|X_1|\,|X_2|\big) \cdot (1 - \mu)^2 p^2 |Y'| \\ &\geq (\varrho - 3\mu)|X_1|\,|X_2| \cdot (1 - \mu)^2 p^2 |Y'| \overset{(70)}{\geq} (1 - \mu')\varrho p^2 |X_1|\,|X_2|\,|Y'|, \end{aligned}$$

which concludes the proof of (73). ∎

Proof of (74). In order to verify (74) we will find an upper bound on the number of 5-tuples $(x_1, x_1', y, x_2, x_2')$ satisfying $y \in Y'$, x_1, $x_1' \in N_\Gamma(y, X_1)$, x_2, $x_2' \in N_\Gamma(y, X_2)$, with x_1, x_1', x_2, and x_2' spanning a cycle of length 4 in B.

First we appeal to the assumption that $(X_1, X_2)_B$ satisfies PAIR$(\varrho, 1, \mu)$ and, hence,

$$\sum_{x_1 \in X_1} \sum_{x_1' \in X_1} \left|\deg_B(x_1, x_1', X_2) - \varrho^2|X_2|\right| \le \mu |X_1|^2 |X_2|.$$

Consequently, all but at most

$$\frac{\sqrt{\mu}}{\varrho^2}|X_1|^2 \tag{92}$$

pairs $(x_1, x_1') \in X_1 \times X_1$ satisfy

$$\left|\deg_B(x_1, x_1', X_2) - \varrho^2|X_2|\right| \le \sqrt{\mu}\varrho^2 |X_2|. \tag{93}$$

Three applications of Fact 13 yield the following estimates:

(A) All but at most

$$\frac{2\gamma^2 p^4 n^2}{\mu^2 p^2 |Y'|} \overset{(71)}{\le} \mu p^2 |X_1| \tag{94}$$

vertices $x_1 \in X_1$ satisfy

$$\left|\deg_\Gamma(x_1, Y') - p|Y'|\right| \le \mu p |Y'|. \tag{95}$$

(B) For every $x_1 \in X_1$ satisfying (95), all but at most

$$\frac{\gamma^2 p^4 n^2}{\mu^2 p^2 \left|N_\Gamma(x_1, Y')\right|} \overset{(95)}{\le} \frac{\gamma^2 p^2 n^2}{\mu^2 (1-\mu) p |Y'|} \overset{(71)}{\le} \mu p |X_1| \tag{96}$$

vertices $x_1' \in X_1$ satisfy

$$\deg_\Gamma(x_1, x_1', Y') \le (1+\mu) p \left|N_\Gamma(x_1, Y')\right| \le (1+\mu)^2 p^2 |Y'|. \tag{97}$$

(C) For every $(x_1, x_1') \in X_1 \times X_1$ satisfying (93), all but at most

$$\frac{\gamma^2 p^4 n^2}{\mu^2 p^2 \left|N_B(x_1, x_1', X_2)\right|} \overset{(93)}{\le} \frac{\gamma^2 p^2 n^2}{\mu^2 (1-\sqrt{\mu}) \varrho^2 |X_2|} \overset{(71)}{\le} \mu p^2 |Y| \tag{98}$$

vertices $y \in Y'$ satisfy

$$\deg_\Gamma\big(y, N_B(x_1, x_1', X_2)\big) \le (1+\mu)p \deg_B(x_1, x_1', X_2). \tag{99}$$

For the proof of the upper bound on $t_{2,2}$ we consider the following "exceptional" 5-tuples $(x_1, x_1', y, x_2, x_2') \in X_1^2 \times Y' \times X_2^2$:

$$\begin{aligned}
\mathcal{T}_1 &= \big\{ (x_1, x_1', y, x_2, x_2') : \text{(95) fails for } x_1 \text{ and } x_2, x_2' \in N_\Gamma(y, X_2) \big\},\\
\mathcal{T}_2 &= \big\{ (x_1, x_1', y, x_2, x_2') : \text{(95) holds for } x_1 \text{ but (97) fails for } x_1',\\
&\qquad y \in N_\Gamma(x_1, Y'), \text{ and } x_2, x_2' \in N_\Gamma(y, X_2) \big\},\\
\mathcal{T}_3 &= \big\{ (x_1, x_1', y, x_2, x_2') : \text{(95) holds for } x_1, \text{ (97) holds for } x_1',\\
&\qquad \text{but (93) fails for } (x_1, x_1'),\ y \in N_\Gamma(x_1, x_1', Y'),\\
&\qquad \text{and } x_2, x_2' \in N_\Gamma(y, X_2) \big\},
\end{aligned}$$

and

$$\begin{aligned}
\mathcal{T}_4 &= \big\{ (x_1, x_1', y, x_2, x_2') : \text{(95) holds for } x_1, \text{ (97) holds for } x_1',\\
&\qquad \text{(92) holds for } (x_1, x_1'), \text{ but (99) fails for } y \text{ and } x_2, x_2' \in N_\Gamma(y, X_2) \big\}.
\end{aligned}$$

We note that

$$\begin{aligned}
t_{2,2} &\le |\mathcal{T}_1| + |\mathcal{T}_2| + |\mathcal{T}_3| + |\mathcal{T}_4|\\
&\qquad + |X_1|^2 \cdot (1+\mu)^2 p^2 |Y'| \cdot (1+\mu)^2 p^2 \big(1+\sqrt{\mu}\big)^2 \varrho^4 |X_2|^2,
\end{aligned} \tag{100}$$

where the last term stands for the 5-tuples $(x_1, x_1', y, x_2, x_2')$ for which (93), (95), (97), and (99) hold. Since $\big|N_\Gamma(y, X_2)\big| \le (1+\xi)p|X_2|$ for every $y \in Y'$ the estimates from (92)–(99) imply

$$\begin{aligned}
|\mathcal{T}_1| &\overset{(94)}{\le} \mu p^2 |X_1| \cdot |X_1| \cdot |Y'| \cdot (1+\xi)^2 p^2 |X_2|^2,\\
|\mathcal{T}_2| &\overset{(96)}{\le} |X_1| \cdot \mu p |X_1| \cdot (1+\mu) p |Y'| \cdot (1+\xi)^2 p^2 |X_2|^2,\\
|\mathcal{T}_3| &\overset{(92)}{\le} \frac{\sqrt{\mu}}{\varrho^2} |X_1|^2 \cdot (1+\mu)^2 p^2 |Y'| \cdot (1+\xi)^2 p^2 |X_2|^2,
\end{aligned}$$

and

$$|\mathcal{T}_4| \overset{(98)}{\leq} |X_1|^2 \cdot \mu p^2 |Y'| \cdot (1+\xi)^2 p^2 |X_2|^2.$$

Applying these estimates to (100) gives

$$\begin{aligned} t_{2,2} &\leq \mu p^2 |X_1| \cdot |X_1| \cdot |Y'| \cdot (1+\xi)^2 p^2 |X_2|^2 \\ &\quad + |X_1| \cdot \mu p |X_1| \cdot (1+\mu) p |Y'| \cdot (1+\xi)^2 p^2 |X_2|^2 \\ &\quad + \frac{\sqrt{\mu}}{\varrho^2} |X_1|^2 \cdot (1+\mu)^2 p^2 |Y'| \cdot (1+\xi)^2 p^2 |X_2|^2 \\ &\quad + |X_1|^2 \cdot \mu p^2 |Y'| \cdot (1+\xi)^2 p^2 |X_2|^2 \\ &\quad + |X_1|^2 \cdot (1+\mu)^2 p^2 |Y'| \cdot (1+\mu)^2 p^2 \left(1+\sqrt{\mu}\right)^2 \varrho^4 |X_2|^2 \\ &\leq (1+\mu')\varrho^4 p^4 |X_1|^2 |X_2|^2 |Y'|, \end{aligned}$$

where the last inequality follows from the choice of μ from (70), which ensures $\sqrt{\mu}(1+\mu)^2(1+\xi)^2/\varrho^2 \leq \mu'\varrho^4/5$ and $(1+\mu)^4\left(1+\sqrt{\mu}\right)^2 \leq 1+\mu'/5$. This concludes the proof of (74). ■

6. Concluding Remarks

In this paper, we prove an extension of the well known theorem of Ruzsa and Szemerédi for subgraphs of $(p, \gamma p^3 n)$-bijumbled graphs Γ with sufficiently small $\gamma > 0$ (see Theorem 3). It would be interesting to find weaker assumptions for the host graph Γ which allow one to prove the same result. In particular, one can ask if the jumbledness condition in Theorem 3 can be replaced by (p, β)-bijumbledness for some $\beta > \gamma p^3 n$. Our approach taken here is based on an appropriate triangle counting lemma (Lemma 11), for the proof of which we needed $\beta = \gamma p^3 n$. We note that the triangle counting lemma fails to be true if $\beta = 145 p^2 n$. Indeed, Alon [1] has constructed (p, β)-bijumbled graphs with n vertices for $p = n^{-1/3}/4$ and $\beta = 145 p^2 n$ which contain no triangle at all. Using a blowup-type construction one obtains $(p, 145p^2 n)$-bijumbled graphs Γ_p for any p with $n^{-1/3} \ll p = o(1)$ which are triangle-free. Since all subgraphs of Γ_p are triangle-free, the conclusion of

Lemma 11 can hold only for (p, β)-bijumbled graphs with $\beta < 145p^2n$. We believe that under the assumption $\beta = o(p^2n)$ Lemma 11 is true. Such a result would imply that Theorem 3 holds even if $\gamma p^3 n$ is replaced by $\gamma p^2 n$ and the condition on $\lambda(A)$ in Theorem 4 is replaced by $\lambda(A) \leq \gamma |A|^2/n$.

In the proof of the triangle counting lemma, we make use of the implication "DISC implies PAIR" (see part (a) of Lemma 9), which was established under the condition that $\beta \leq \gamma p^2 n$. Note that the reverse implication (see Lemma 9(b)) holds under a less restrictive assumption, namely, $\beta \leq \gamma p^{3/2} n$. Finding the thresholds for β under which these implications are valid remains an open problem.

References

[1] N. Alon, Explicit Ramsey graphs and orthonormal labelings, *Electronic Journal of Combinatorics,* **1** (1994), Research Paper 12, approx. 8 pp. (electronic).

[2] N. Alon and J. H. Spencer, *The probabilistic method,* third ed., Wiley–Interscience Series in Discrete Mathematics and Optimization, John Wiley & Sons Inc., Hoboken, NJ, 2008, With an appendix on the life and work of Paul Erdős.

[3] J. Bourgain, Roth's theorem on progressions revisited, *J. Anal. Math.,* **104** (2008), 155–192.

[4] T. C. Brown and J. P. Buhler, A density version of a geometric Ramsey theorem, *J. Combin. Theory Ser. A,* **32** (1982), no. 1, 20–34.

[5] W. G. Brown, P. Erdős and V. T. Sós, *Some extremal problems on r-graphs,* New directions in the theory of graphs (Proc. Third Ann Arbor Conf., Univ. Michigan, Ann Arbor, Mich, 1971), Academic Press, New York, 1973, pp. 53–63.

[6] F. R. K. Chung and R. L. Graham, Quasi-random subsets of Z_n, *Journal of Combinatorial Theory, Series A,* **61** (1992), no. 1, 64–86.

[7] F. R. K. Chung, R. L. Graham and R. M. Wilson, Quasi-random graphs, *Combinatorica,* **9** (1989), no. 4, 345–362.

[8] P. Erdős, M. Goldberg, J. Pach and J. Spencer, Cutting a graph into two dissimilar halves, *J. Graph Theory,* **12** (1988), no. 1, 121–131.

[9] P. Erdős and J. Spencer, Imbalances in k-colorations, *Networks,* **1** (1971/72), 379–385.

[10] J. Fox, *A new proof of the triangle removal lemma,* preprint.

[11] P. Frankl, R. L. Graham and V. Rödl, On subsets of abelian groups with no 3-term arithmetic progression, *J. Combin. Theory Ser. A,* **45** (1987), no. 1, 157–161.

[12] B. Green, Roth's theorem in the primes, *Annals of Mathematics. Second Series,* **161** (2005), no. 3, 1609–1636.

[13] B. Green and T. Tao, *Restriction theory of the Selberg sieve, with applications,* Journal de Théorie des Nombres de Bordeaux **18** (2006), no. 1, 147–182.

[14] B. Green and T. Tao, The primes contain arbitrarily long arithmetic progressions, *Annals of Mathematics. Second Series,* **167** (2008), no. 2, 481–547.

[15] Y. Kohayakawa, *Szemerédi's regularity lemma for sparse graphs,* Foundations of computational mathematics (Rio de Janeiro, 1997), Springer, Berlin, 1997, pp. 216–230.

[16] Y. Kohayakawa, T. Łuczak and V. Rödl, Arithmetic progressions of length three in subsets of a random set, *Acta Arithmetica,* **75** (1996), no. 2, 133–163.

[17] Y. Kohayakawa and V. Rödl, *Szemerédi's regularity lemma and quasi-randomness,* Recent advances in algorithms and combinatorics, CMS Books Math./Ouvrages Math. SMC, vol. 11, Springer, New York, 2003, pp. 289–351.

[18] M. Krivelevich and B. Sudakov, Pseudo-random graphs, in: *More sets, graphs and numbers,* Bolyai Soc. Math. Stud., vol. 15, Springer, Berlin, 2006, pp. 199–262.

[19] V. F. Lev, Progression-free sets in finite abelian groups, *J. Number Theory,* **104** (2004), no. 1, 162–169.

[20] R. Meshulam, On subsets of finite abelian groups with no 3-term arithmetic progressions, *J. Combin. Theory Ser. A,* **71** (1995), no. 1, 168–172.

[21] H. H. Nguyen, *On distribution of three-term arithmetic progressions in sparse subsets of* $\mathbf{F}_p^n$, `arXiv:0905.3890v2`, 16 pp., 2009.

[22] Y. Peng, V. Rödl and A. Ruciński, Holes in graphs, *Electronic Journal of Combinatorics,* **9** (2002), no. 1, Research Paper 1, 18 pp. (electronic).

[23] V. Rödl and M. Schacht, Generalizations of the removal lemma, *Combinatorica,* **29** (2009), no. 4, 467–501.

[24] K. F. Roth, On certain sets of integers, *J. London Math. Soc.,* **28** (1953), 104–109.

[25] I. Z. Ruzsa and E. Szemerédi, Triple systems with no six points carrying three triangles, in: *Combinatorics* (Proc. Fifth Hungarian Colloq., Keszthely, 1976), Vol. II, Colloq. Math. Soc. János Bolyai, vol. 18, North-Holland, Amsterdam, 1978, pp. 939–945.

[26] M. Schacht, *Extremal results for random discrete structures,* submitted, 2009, 27 pp.

[27] E. Szemerédi, Regular partitions of graphs, in: *Problèmes combinatoires et théorie des graphes* (Colloq. Internat. CNRS, Univ. Orsay, Orsay, 1976), Colloq. Internat. CNRS, vol. 260, CNRS, Paris, 1978, pp. 399–401.

[28] A. Thomason, Pseudorandom graphs, in: *Random graphs '85* (Poznań, 1985), North-Holland Math. Stud., vol. 144, North-Holland, Amsterdam, 1987, pp. 307–331.

[29] A. Thomason, Random graphs, strongly regular graphs and pseudorandom graphs, in: *Surveys in combinatorics 1987* (New Cross, 1987), London Math. Soc. Lecture Note Ser., vol. 123, Cambridge Univ. Press, Cambridge, 1987, pp. 173–195.

Yoshiharu Kohayakawa
Instituto de Matemática e Estatística
Universidade de São Paulo
Rua do Matão 1010
05508-090 São Paulo
Brazil
e-mail: yoshi@ime.usp.br

Vojtěch Rödl
Department of Mathematics and
Computer Science
Emory University
Atlanta, GA 30322
USA
e-mail: rodl@mathcs.emory.edu

Mathias Schacht
Fachbereich Mathematik
Universität Hamburg
Bundesstraße 55
D-20146 Hamburg
Germany
e-mail:
schacht@math.uni-hamburg.de

Jozef Skokan
Department of Mathematics
London School of Economics
Houghton Street
London WC2A 2AE
United Kingdom
e-mail: J.Skokan@lse.ac.uk

BOLYAI SOCIETY
MATHEMATICAL STUDIES, 21

An Irregular Mind
(Szemerédi is 70)
pp. 405–414.

Almost All H-Free Graphs Have the Erdős–Hajnal Property

MARTIN LOEBL, BRUCE REED, ALEX SCOTT,
ANDREW THOMASON and STÉPHAN THOMASSÉ

Erdős and Hajnal conjectured that, for every graph H, there exists a constant $\varepsilon(H) > 0$ such that every H-free graph G (that is, not containing H as an induced subgraph) must contain a clique or an independent set of size at least $|G|^{\varepsilon(H)}$.

We prove that there exists $\varepsilon(H)$ such that almost every H-free graph G has this property, meaning that, amongst the H-free graphs with n vertices, the proportion having the property tends to one as $n \to \infty$.

1. Introduction

Szemerédi's Regularity Lemma is a powerful tool with applications in many fields. This paper discusses one of its applications in extremal graph theory.

A class of graphs $\mathcal{P}$ is said to have the Erdős–Hajnal property if there is a positive constant $\varepsilon = \varepsilon(\mathcal{P})$ such that every graph $G \in \mathcal{P}$ contains a *homogeneous* set of size at least $|G|^{\varepsilon}$, where a homogeneous set is either a clique or an independent set. Let $\mathrm{Forb}\,(H)$ be the class of graphs not containing the graph H as an induced subgraph. Erdős and Hajnal [7] conjectured that $\mathrm{Forb}\,(H)$ has the Erdős–Hajnal property.

Given a class of graphs $\mathcal{P}$, we write $\mathcal{P}^n$ for the members of $\mathcal{P}$ having vertex set $\{1, \dots, n\}$. In particular, we will focus on $\mathcal{P}_H = \mathrm{Forb}\,(H)$, which we will sometimes simply write as $\mathcal{P}$. Our intention in this note is to prove the following theorem.

Theorem 1. *For any graph H, there is a sub-class $\mathcal{Q}_H \subset \mathcal{P}_H$ which has the Erdős–Hajnal property, with $|\mathcal{Q}_H^n|/|\mathcal{P}_H^n| \to 1$ as $n \to \infty$.*

The Erdős–Hajnal conjecture itself remains open except in a few cases. Erdős and Hajnal themselves proved that the conjecture holds for those graphs H obtainable recursively from K_1 by disjoint union and complementation. They also proved it for the path of length three. Alon, Pach and Solymosi [3] showed that the class of graphs for which the conjecture holds is closed under replacement; this means if $H, F_1, \ldots, F_k$ satisfy the conjecture and $V(H) = \{v_1, \ldots, v_k\}$, then so does the graph $H(F_1, \ldots, F_k)$, obtained from disjoint copies of $F_1, \ldots, F_k$ by joining every vertex in F_i to every vertex in F_j precisely if $v_iv_j \in E(H)$ (for instance, if H satisfies the conjecture then so does any graph obtained by blowing up the vertices of H into cliques or independent sets). By a very different method, Chudnovsky and Safra [6] proved the conjecture for the *bull*, the self-complementary graph of order 5 comprising a triangle and two pendant edges.

The size of $\mathcal{P}_H^n$ for arbitrary H has received a lot of attention. Letting $\chi(H)$ denote the chromatic number of H, we see that any graph G which can be partitioned into $\chi(H) - 1$ stable sets obviously contains no induced copies of H. Considering a fixed partition of n vertices into $\chi(H) - 1$ parts each of which has size within 1 of $\frac{n}{\chi(H)-1}$, and counting all the graphs where the only edges go between partition elements we see that there are at least $2^{\left(1-\frac{1}{\chi(H)-1}+o(1)\right)\binom{n}{2}}$ graphs in $\mathcal{P}_H^n$. In the same vein, if G cannot be partitioned into a cliques and b stable sets, then a similar argument shows that $|\mathcal{P}_H^n| \geq 2^{\left(1-\frac{1}{a+b}+o(1)\right)\binom{n}{2}}$

Prömel and Steger [9] showed that this lower bound is not too far from the truth.

Definition 2. The *colouring number* $\tau(H)$ of a graph H is the smallest integer $k \geq 1$ such that, for every integer $0 \leq \ell \leq k$, the vertices of H can be partitioned into ℓ cliques and $k - \ell$ independent sets.

Prömel and Steger [9] proved:

$$\mathcal{P}_H^n = 2^{\left(1-\frac{1}{\tau(H)-1}+o(1)\right)\binom{n}{2}}.$$

Let $P(a, b)$ be the class of graphs whose vertices can be partitioned into a cliques and b independent sets. Another way of writing the Prömel–Steger theorem is that, if $\mathcal{P} = \operatorname{Forb}(H)$ then $|\mathcal{P}^n| = 2^{(1-1/t+o(1))\binom{n}{2}}$, where

$$t = \max\left\{a + b : P(a, b) \subseteq \mathcal{P} \text{ for some } a,\, b\right\}.$$

The theorem in this form was extended to all hereditary properties $\mathcal{P}$ by Alexeev [1] and by Bollobás and Thomason [5][1].

For $H = C_4$ and $H = C_5$, Prömel and Steger [8, 10] proved a much sharper result by finding a very well-structured class $\mathcal{Q}$ with $\mathcal{Q} \subset \mathcal{P} = \text{Forb}\,(H)$ and $|\mathcal{Q}^n|/|\mathcal{P}^n| \to 1$ as $n \to \infty$. Note that our aim in Theorem 1 is to do just this where the class $\mathcal{Q}$ has the Erdős–Hajnal property. In the case $H = C_4$ they showed that $\mathcal{Q} = P(1,1)$ works. Clearly the class $P(1,1)$ has the Erdős–Hajnal property, and indeed if $G \in P(1,1)$ then G can be partitioned into two homogeneous sets, one of which must have size linear in $|G|$. For $H = C_5$ they showed that the class $\mathcal{Q}$ of *generalized split graphs* works. A graph G is a generalized split graph either if its vertices can be partitioned into two classes U and W with $G[U]$ being a disjoint union of cliques and $G[W]$ being a single clique, or else the complement $\overline{G}$ of G has this property. Here again the class $\mathcal{Q}$ has the Erdős–Hajnal property, although since $\mathcal{Q}$ is dominated by graphs in which the cliques in $G[U]$ have size around $\log|G|$, we do not get a partition of G into two pieces each of which contains a homogeneous set of size linear in their order. Nevertheless, almost every generalized split graph does have a linear size homogenous set since in a typical generalized split graph, both W and U will have about half the vertices (see [10]).

It would be of interest to determine for which graphs H the following property holds:

(∗) almost every graph in $\mathcal{P}_H$ has a homogeneous set of linear size.

We will not attack this problem here. We do however make a few remarks. First we note that a graph has no induced path on three vertices precisely if it is a disjoint union of cliques. So, as discussed above, for $H = P_3$, we have that for almost every graph G without H as an induced subgraph, the largest homogeneous set in G has size $\Theta\big(|V(G)|/\log|V(G)|\big)$. We do not know of any graph other than P_3 (and possibly P_4) which does not satisfy (∗). These two graphs are exceptional, in that they satisfy $\tau = 2$. We wonder if (∗) holds for all other graphs. To this end we note that McDiarmid and Reed claim that for almost every graph H, almost every graph in $\mathcal{P}_H$ contains a homogeneous set of linear size. Finally we note that our approach to proving that a graph H satisfies (∗) is to show that almost every

[1]A class $\mathcal{P}$ of graphs that is closed under induced subgraphs is called *hereditary*. Clearly Forb (H) is hereditary, as is Forb $(\mathcal{H})$, the class of graphs none of whose induced subgraphs is in the class $\mathcal{H}$. Every hereditary property $\mathcal{P}$ is of the form $\mathcal{P} = \text{Forb}\,(\mathcal{H})$ for some $\mathcal{H}$ (just take $\mathcal{H}$ to be those graphs not in $\mathcal{P}$).

graph in $\mathcal{P}_H$ has a partition into $\tau(H) - 1$ linear sized pieces, one of which contains a homogeneous set which has size linear in its order.

Clearly, if H is a graph for which we can take $\mathcal{Q}$ to be the union of the $P(a,b) \subset \mathcal{P}_H$ with $a + b = t$, then Theorem 1 and (*) hold for H. Very recently, Balogh and Butterfield [4] have characterized the graphs H for which this is possible: they call such H "critical". The remarks above show that C_4 is critical but C_5 is not. Curiously it turns out that, for $\ell \geq 6$, C_ℓ is not critical if ℓ is even, but C_ℓ is critical if ℓ is odd.

Given that Theorem 1 is weaker than the Erdős–Hajnal conjecture and that it is known in special cases, we aim to give a proof that is short. In particular, we make no effort to optimize $\varepsilon(H)$. The results mentioned above all begin with applications of Szemerédi's Regularity Lemma, together with the Erdős–Stone theorem and Ramsey's theorem and perhaps the Erdős–Simonovits stability theorem. (The exception to this is Alexeev [1], who uses only an extension of the Sauer–Shelah lemma.) We shall not use this machinery apart from one of the basic consequences of Szemerédi's Lemma common to all the cited papers. Our proof also relies on a new observation about partitioning H into $\tau(H) - 1$ sets (Lemma 3). Surprisingly, in order to prove this observation, we need the fact from [6] that the Erdős–Hajnal conjecture is true for the bull.

Because it rests on Lemma 3, our proof of Theorem 1 does not immediately extend to all hereditary properties $\mathcal{P}$. Alon, Balogh, Bollobás and Morris [2] have recently described, for any hereditary property $\mathcal{P}$, a property $\mathcal{Q}$ with $\mathcal{Q} \subset \mathcal{P}$ and $|\mathcal{Q}^n|/|\mathcal{P}^n| \to 1$ as $n \to \infty$. The graphs in $\mathcal{Q}$ have a partition into t sets each of which is "somewhat homogenous", in a well-defined way. However, it is not evident whether the Erdős–Hajnal property can be derived from this description.

2. Proof of Theorem 1

The proof combines the following simple lemma with some, by now standard, regularity lemma machinery.

Lemma 3. *Let $t \geq 1$. Then there is a finite set $\mathcal{F}_t$ of graphs, such that Forb(F) has the Erdős–Hajnal property for all $F \in \mathcal{F}_t$, and the vertices of any graph H with $\tau(H) = t+1$ can be partitioned into t sets each inducing a graph in $\mathcal{F}_t$.*

Proof. Let H be any graph with $\tau(H) = t+1$. By the definition of $\tau(H)$, the vertices of H can be partitioned into $t+1$ independent sets $I_1, \ldots, I_{t+1}$ and also into $t+1$ sets $C_1, \ldots, C_{t+1}$ inducing cliques. For any independent set I of H, $|C_i \cap I| \leq 1$ holds for all $1 \leq i \leq t+1$, and so $|I| \leq t+1$. In particular this is true for $|I_j|$ for all j, and so $|H| \leq (t+1)^2$. Symmetrically no clique of G has more than $t+1$ elements.

If $|H| < (t+1)^2$ we may assume that $I_{t+1} = \{v_1, \ldots, v_s\}$ where $s \leq t$. Then put $V_j = I_j \cup \{v_j\}$ for $1 \leq j \leq s$ and put $V_j = I_j$ for $s < j \leq t$.

If $|H| = (t+1)^2$, since $\tau(H) = t+1$ there is a partition of the vertices of H into t independent sets $J_1, \ldots, J_t$ and a set D inducing a clique. Since $|J_j| \leq t+1$ for all j and $|D| \leq t+1$, it follows that all these sets have size exactly $t+1$, so write $D = \{w_1, \ldots, w_{t+1}\}$. Then put $V_j = J_j \cup \{w_j\}$ for $1 \leq j < t$ and let $V_t = J_t \cup \{w_t, w_{t+1}\}$.

In each case, we obtain a partition $V(H) = V_1 \cup \cdots \cup V_t$; let F_j be the subgraph induced by V_j. For $j < t$ the graph F_j consists of a star together with isolated vertices. Note that the star $K_{1,s}$ equals $K_2(K_1, \overline{K_s})$ (here we are using the replacement notation from the introduction) and $K_{1,s}$ together with i isolated vertices equals $\overline{K}_2(K_{1,s}, \overline{K_i})$, so each graph F_j, $j < t$ satisfies the Erdős–Hajnal property. Let $F_t = F$ together with i isolated vertices, where F is connected. Then F is obtained from the bull by replacing vertices with (possibly empty) independent sets, and then $F_t = \overline{K}_2(F, \overline{K_i})$. Thus F_t also satisfies the Erdős–Hajnal property.

Finally, let $\mathcal{F}_t$ be the set of all graphs that can arise in the procedure above. ∎

In the case when $|H| = (t+1)^2$ we could instead have obtained F_t by distributing I_{t+1} amongst $I_1, \ldots, I_t$, but this would require the path of length 4 to satisfy the Erdős–Hajnal property, which is currently unknown. Note that graphs with $\tau(H) = t+1$ and $|H| = (t+1)^2$ do exist, at least if $t+1$ is a prime power. Take $V(H) = \mathbb{F}_{t+1}^2$. Each pair of vertices lies in exactly one line, whose gradient is one of $0, 1, \ldots, t, \infty$. The $t+1$ parallel lines of each gradient form a partition of the vertex set. For finite gradients $m \leq t$ make m of these lines cliques and the other $t+1-m$ lines independent sets; for $m = \infty$ make all the lines cliques.

We remarked earlier that Szemerédi's Lemma is fundamental in the study of the number of H-free graphs. In fact by making use of earlier work we can avoid many of the technicalities involved. Lemma 3.5 in the pioneering work of Prömel and Steger [9] is entirely adequate for our purposes but we borrow instead [5, Theorem 3.1], which is very similar. It is phrased in terms of the density of bipartite graphs, which has the natural meaning, and η-regularity, which has its usual meaning in relation to Szemerédi's Lemma: however, it is not necessary to know this meaning in order to follow the proof of Theorem 1. A *coloured partition* π is a colouring of the edges of the complete graph K_m with colours red, blue, green and grey, where m is the *order* of π, denoted $|\pi|$. Given a graph G and constants $0 < \lambda, \eta < 1$ we say that a partition of the vertices of G into $|\pi|$ classes $V_1, \dots, V_{|\pi|}$ *satisfies* π with respect to λ and η if $|V_1| \le |V_2| \le \cdots \le \left|V_{|\pi|}\right| \le |V_1| + 1$ and the pair (V_i, V_j) is not η-regular only if ij is grey, and otherwise $0 \le d(V_i, V_j) \le \lambda$, $\lambda < d(V_i, V_j) < 1 - \lambda$ or $1 - \lambda \le d(V_i, V_j) \le 1$ according as ij is red, green or blue. Here $d(V_i, V_j)$ stands for the density of the bipartite graph between V_i and V_j. We say that G *satisfies* π if there is a partition of G satisfying π.

Szemerédi's Regularity Lemma [11] asserts that, given λ, η and some integer ℓ, there exists an integer $L = L(\ell, \eta)$ such that any graph G satisfies some coloured partition π with respect to λ and η, where $\ell \le |\pi| < L$ and where π has at most $\eta\binom{|\pi|}{2}$ grey edges. The following proposition, roughly speaking, states that if the proportion of green edges in π exceeds $1-1/t$ then G contains every small member of $P(a,b)$ for some (a,b) with $a+b=t+1$.

Proposition 4 ([5, Theorem 3.1])**.** *Let $t, h \in \mathbb{N}$ and $0 < \lambda, \nu < 1$ be given. Then there exist positive constants ℓ_0, η_0, and n_0 with the following property. Let π be a coloured partition with $|\pi| \ge \ell_0$, having at most $\eta_0\binom{|\pi|}{2}$ grey edges and at least $(1-1/t+\nu)\binom{|\pi|}{2}$ green edges. Then there are integers a and b with $a + b = t + 1$, such that every graph of order at least n_0 that satisfies π with respect to λ and η_0 contains every member of $P(a,b)$ with at most h vertices as an induced subgraph.*

With our lemma and the regularity lemma in hand, we are in a position to finish the proof of Theorem 1.

Proof of Theorem 1. Let $\mathcal{P} = \text{Forb}\,(H)$ and let $t = \tau(H) - 1$. Note that, as in the proof of Lemma 3, $|H| \le (t+1)^2$, so we may assume $t \ge 1$. Furthermore, if $\tau(H) = 2$ then (as the Erdős–Hajnal property holds for K_2, $K_2 \cup K_1$, P_4 and C_4) we are done, so we may assume $t \ge 2$. By the definition of $\tau(H)$ there exist a, b with $a+b = t$ and $P(a,b) \subset \mathcal{P}$. The graphs of order n formed by adding edges between a cliques and b independent sets are all in $\mathcal{P}$, and hence $\left|\mathcal{P}^n\right| \ge 2^{(1-1/t+o(1))\binom{n}{2}}$.

Let $\mathcal{F}_t$ be the class given by Lemma 3. For each F in $\mathcal{F}_t$ let ε_F be such that for every graph G, either G contains F as an induced subgraph or G contains a homogenous set of size at least $|G|^{\varepsilon_F}$. Let ε be the minimum of the ε_F.

Let $\mathcal{Q}$ be the class of graphs $G \in \mathcal{P}$ containing a homogeneous set of size $|G|^{\varepsilon/2}$, so $\mathcal{Q}$ has the Erdős–Hajnal property. Let $\mathcal{R} = \mathcal{P} - \mathcal{Q}$. We shall show that $|\mathcal{R}^n|/|\mathcal{P}^n| \to 0$ as $n \to \infty$; that is, $|\mathcal{Q}^n|/|\mathcal{P}^n| \to 1$ as required.

Our approach is straightforward. Szemerédi's Regularity Lemma tells us that for large enough n, every graph in $\mathcal{R}^n$ satisfies one of a certain class of coloured partitions. We count the number of elements of $\mathcal{R}^n$ by summing the number of elements satisfying each partition. Forthwith the details.

We set $h = |H|$ and pick λ, ν to certify certain inequalities given below. Choose η_0, ℓ_0 and n_0 satisfying Proposition 4 with respect to h, t, λ, and ν. We choose $\ell \geq \ell_0$ and $\eta \leq \eta_0$ which satisfy some inequalities given below. We choose L satisfying the Szemerédi Regularity Lemma for this choice of λ, η and ℓ. Let $G \in \mathcal{P}^n$ for some $n \geq n_0$. Then G satisfies some π with respect to λ, η where $\ell \leq |\pi| < L$ and π has at most $\eta\binom{|\pi|}{2}$ grey edges.

We want to bound the number of graphs of $\mathcal{R}^n$ satisfying a particular partition π with respect to a given λ and η. We actually bound the number of partitions of graphs in $\mathcal{R}^n$ which satisfy π, which is larger. We do so by summing over each partition of V, the number of graphs in $\mathcal{R}^n$ for which this partition satisfies π with respect to λ, η. If V_i, V_j corresponds to a green edge or a grey edge then there are at most $2^{|V_i||V_j|}$ ways to join V_i to V_j. But for red and blue edges there are at most $2^{c|V_i||V_j|}$ where $c \to 0$ as $\lambda \to 0$. Furthermore, there are at most $\binom{n}{2}/|\pi|$ edges within the partition classes. So letting n_g be the proportion of green edges of the partition we see that the total number of graphs G such that this partition satisfies π is at most

$$2^{\left(n_g+\eta+c+\frac{1}{|\pi|}\right)\binom{n}{2}}. \tag{1}$$

Since we know that $\mathcal{P}^n$ has at least $2^{\left(1-\frac{1}{t}+o(1)\right)\binom{n}{2}}$ elements, an easy computation yields that for large n the number of graphs satisfying partitions with $n_g < 1 - \frac{1}{t} - \eta - 2c - \frac{1}{\ell}$ is $o\left(|\mathcal{P}^n|\right)$.

In counting these graph partitions satisfying π for which n_g is larger, we need to exploit the fact that we are only counting graphs in $\mathcal{R}^n$. Since $H \notin P(a,b)$, Proposition 4 therefore implies that we need only consider π with at most $(1 - 1/t + \nu)\binom{|\pi|}{2}$ green edges.

The proof of the following two lemmas is postponed to the end of the proof.

Lemma 5. *If π contains s edge disjoint cliques of size t all of whose edges are green then the total number of graphs in $\mathcal{R}^n$ for which a given partition satisfies π is at most $2^{\left(n_g-\frac{s}{16t^2|\pi|^2}+\eta+c+\frac{1}{|\pi|}\right)\binom{n}{2}}$.*

Lemma 6. *If n_g exceeds $1-\frac{1}{t-1}+\frac{1}{2t^2}$ and $|\pi|$ is sufficiently large then there are at least $|\pi|^2/4t^4$ edge disjoint cliques of size t in π all of whose edges are green.*

With these two lemmas in hand, it is straightforward to prove the theorem. We choose λ so that c is less than $\frac{1}{1000t^6}$. We choose $\nu=\frac{1}{1000t^6}$. We choose ℓ such that $\ell \geq \max\left(l_0, 1000t^6\right)$ and so that Lemma 6 applies for $|\pi| \geq \ell$, and we choose $\eta \leq \min\left(\eta_0, \frac{1}{1000t^6}\right)$.

Now, we have that $\eta+c+\frac{1}{|\pi|} \leq \frac{3}{1000t^6}$. So, if π is a partition for which $n_g<1-\frac{1}{t-1}+\frac{1}{2t^2}$ then there are at most $2^{\left(1-\frac{1}{t-1}+\frac{2}{3t^2}\right)\binom{n}{2}}$ graphs G in $\mathcal{R}^n$ for which a given partition satisfies π. On the other hand for any partition π with $n_g \geq 1-\frac{1}{t-1}+\frac{1}{2t^2}$ we know by Proposition 4 that $n_g \leq 1-\frac{1}{t}+\nu$. Hence, combining Lemmas 6 and 5, we see that at most $2^{\left(1-\frac{1}{t}-\frac{1}{64t^6}+\frac{4}{1000t^6}\right)\binom{n}{2}}$ graphs in $\mathcal{R}^n$ satisfy π. So, in either case, there are at most $2^{\left(1-\frac{1}{t}-\frac{1}{100t^6}\right)\binom{n}{2}}$ graphs in $\mathcal{R}^n$ for which a particular partition satisfies π. But the number of choices for π is independent of n, and the number of partitions of the vertex set is L^n which is $o\left(2^{\binom{n}{2}}\right)$. So for large n, the number of elements of $\mathcal{R}^n$ is $o\left(2^{\left(1-\frac{1}{t}\right)\binom{n}{2}}\right)$ and hence $|\mathcal{R}^n|=o\left(|\mathcal{P}^n|\right)$. This completes the proof of the lemma and the theorem. ■

It remains only to prove the two lemmas.

The second one is straightforward. We simply greedily rip out the edges of a green clique of size t in π until no such cliques remain. Turan's theorem tells us that when we stop, only $\left(1-\frac{1}{t-1}\right)\binom{|\pi|}{2}$ green edges can remain. But by assumption π has at least $\left(1-\frac{1}{t-1}+\frac{1}{2t^2}\right)\binom{|\pi|}{2}$ green edges. So we must have ripped out at least $\left(1+o(1)\right)\binom{|\pi|}{2}/2t^2$ edges and hence at least $\left(1+o(1)\right)\binom{|\pi|}{2}/t^4$ cliques which is at least $\binom{|\pi|}{2}/4t^4$ for large enough $|\pi|$.

With the proof of Lemma 6 completed, we turn to the proof of Lemma 5.

Proof of Lemma 5. For each green clique C, we let m_C be the sum of $|V_i|\,|V_j|$ over every two partition classes V_i and V_j in C. We claim that for each of the s cliques, having fixed the edges within the partition classes, there are at most $2^{m_C - \frac{n^2}{16t^2|\pi|^2}}$ ways to pick the edges within the green edges of the clique. Combining this with our earlier analysis leading to (1) then yields the desired result.

It remains to prove the claim. By Lemma 3, there is a partition $V(H) = V_1 \cup \cdots \cup V_t$ such that each V_i induces a subgraph F_i that belongs to $\mathcal{F}_t$. Let G be any graph in $\mathcal{R}^n$ and let C be a green clique, say with vertices corresponding to $V_1, \ldots, V_t$. Let p be a prime between $n/4t|\pi|$ and $n/3t|\pi|$ (which can be found provided n is large enough). Because G contains no large homogeneous set, and each F_i satisfies the Erdős–Hajnal property with constant $\varepsilon_{F_i} \geq \varepsilon$, we can pick out p vertex-disjoint copies $F_i^{(1)}, \ldots, F_i^{(p)}$ of F_i in $G[V_i]$ for each i. For $1 \leq r, s \leq p$, consider the t-tuple $\left(F_1^{(r)}, F_2^{(r+s)}, \ldots, F_t^{(r+(t-1)s)}\right)$, where indices are taken modulo p. For each t-tuple, there is at least one way to join the classes to obtain a copy of H; on the other hand, as p is prime, no pair of t-tuples coincide in more than one coordinate, and so no edge between classes is spanned by more than one t-tuple. As there are p^2 t-tuples, it follows that the number of ways of filling in the edges between $V_1, \ldots, V_t$ is at most $2^{m_C - p^2} \leq 2^{m_c - n^2/16t^2|\pi|^2}$. The desired result follows. ■

This completes the proof of the lemma and our theorem.

References

[1] V. E. Alekseev, On the entropy values of hereditary classes of graphs, *Discrete Math. Appl.,* **3** (1993), 191–199.

[2] N. Alon, J. Balogh, B. Bollobás and R. Morris, The structure of almost all graphs in a hereditary property (preprint).

[3] N. Alon, J. Pach and J. Solymosi, Ramsey-type theorems with forbidden subgraphs, *Combinatorica,* **21** (2001), 155–170.

[4] J. Balogh and J. Butterfield, Excluding induced subgraphs: critical graphs (preprint).

[5] B. Bollobás and A. Thomason, Hereditary and monotone properties of graphs, in: "The Mathematics of Paul Erdős II" (R. L. Graham and J. Nešetřil, eds.) *Algorithms and Combinatorics,* **14** Springer-Verlag (1997), 70–78.

[6] M. Chudnovsky and S. Safra, The Erdős–Hajnal conjecture for bull-free graphs, *J. Combin. Theory Ser. B,* **98** (2008), 1301–1310.

[7] P. Erdős and A. Hajnal, Ramsey-type theorems, *Discrete Applied Mathematics,* **25** (1989), 37–52.

[8] H.-J. Prömel and A. Steger, Excluding induced subgraphs: quadrilaterals, *Random Structures and Algorithms,* **2** (1991), 55–71.

[9] H.-J. Prömel and A. Steger, Excluding induced subgraphs III: extremal graphs, *Random Structures and Algorithms,* **3** (1992), 19–31.

[10] H.-J. Prömel and A. Steger, Almost all Berge graphs are perfect, *Combinatorics, Probability and Computing,* **1** (1992), 53–79.

[11] E. Szemerédi, Regular partitions of graphs, in: *Problèmes combinatoires et théorie des graphes,* Colloq. Internat. CNRS, Univ. Orsay, Orsay, 1976, CNRS, Paris, 1978, pp. 399–401.

Martin Loebl
Department of Applied Mathematics
Malostranske namesti 25
118 00 Praha 1
Czech Republic
e-mail: `loebl@kam.mff.cuni.cz`

Bruce Reed
School of Computer Science
McGill University
3480 University
Montreal, Quebec
Canada H3A 2A7
e-mail: `breed@cs.mcgill.ca`

Alex Scott
Mathematical Institute
University of Oxford
24-29 St Giles'
Oxford OX1 3LB
United Kingdom
e-mail: `scott@maths.ox.ac.uk`

Andrew Thomason
DPMMS
University of Cambridge
Wilberforce Road
Cambridge CB3 0WB
United Kingdom
e-mail:
`a.g.thomason@dpmms.cam.ac.uk`

Stéphan Thomassé
LIRMM
161 rue Ada
34392 Montpellier Cedex 5
France
e-mail: `monnom@lirmm.fr`

BOLYAI SOCIETY
MATHEMATICAL STUDIES, 21

An Irregular Mind
(Szemerédi is 70)
pp. 415–446.

Regularity Partitions and the Topology of Graphons

LÁSZLÓ LOVÁSZ and BALÁZS SZEGEDY

Dedicated to Endre Szemerédi on the occasion of his 70th birthday

In this paper we highlight a topological aspect of the graph limit theory. We introduce the representation of a graphon on a unique metric space and we relate the dimension of this metric space to the size of regularity partitions. We prove that if a graphon has an excluded induced sub-bigraph then the underlying metric space is compact and has finite packing dimension. It implies in particular that such graphons have regularity partitions of polynomial size.

1. Introduction

One can define convergence of a growing graph sequence [4, 3, 5], and construct a limit object to such a sequence [11] in the form of a symmetric measurable function $W : J \times J \to [0,1]$, where J is any probability space (one may assume here that $J = [0,1]$ with the Lebesgue measure, but this is not always convenient). We call the pair (J, W) a *graphon.*

The goal of this paper is to show that one can introduce also a topology on J (in fact, a metric), and that topological properties of this space are related to combinatorial properties of the graphon (or of the graphs whose limit it represents). A related metric was introduced in [12], and the topology on J was used in [13].

The theory of graph limits is tied to the Regularity Lemma of Szemerédi [14, 15] in several ways. In [12] it was shown that the Regularity Lemma is equivalent to the compactness of the space of graphons in an appropriate metric, and also to a "dimensionality" of particular graphons. This paper relates to the latter result.

The metric in question is simply the L_1 metric on functions $W(x,.)$, $x \in J$. This metric itself can be weird (it may not even be defined on all points of J). We show in Section 3 that every graphon is "equivalent" (technically: weakly isomorphic, see the end of Section 2) to a graphon (J, W) with special properties: J is a complete separable metric space, and the probability measure on J has full support. We call such graphons *pure.* We also prove that the pure version of a graphon is uniquely determined up to changing the function W on a 0-set in each row. We define another metric in which J is compact, and characterize the cases when the two define the same topology. We prove that several important functions defined on J are continuous in this topology, which shows that it is indeed the "right" topology to define on J.

In Section 4 we show that topological properties of pure graphons are related to their graph-theoretic properties. Our main result states that *if we exclude any bipartite graph from the graphon, then J must be compact and finite dimensional.*

In [12] it was shown that weak regularity partitions of a graphon (J, W) (which generalize weak regularity partitions of graphs in a natural way) correspond to covering J with sets of small diameter. In Section 5 we give a stronger and cleaner version of this result. Combined with the results in Section 4, we obtain the following fact: If a graph does not contain a fixed bipartite graph F as an induced sub-bigraph, then it has polynomial size strong regularity partitions (in the error bound ε).

A motivation for our paper comes from extremal combinatorics. In [13] we study the structure of graphons that arise as unique solutions of extremal problems involving the densities of finitely many subgraphs (we call such graphons *finitely forcible*). Such graphons come up naturally in extremal graph theory. Quite interestingly, all the examples of finitely forcible graphons produced in [13] have a compact and finite dimensional underlying metric space. The question arises whether every extremal problem (involving a finite number of subgraph densities) has a solution of this type.

Finally we mention that graph limit theory has a close connection to the theory of dynamical systems. Probability spaces with measure preserving actions can often be endowed by a natural topology in which the action is continuous. The corresponding theory is called topological dynamics. Informally speaking, we can say that the relationship between graphons and topological graphons is similar to the relationship between dynamics and topological dynamics.

2. Preliminaries

We make a technical but useful distinction between bipartite graphs and bigraphs. A *bipartite graph* is a graph (V, E) whose node set has a partition into two classes such that all edges connect nodes in different classes. A *bigraph* is a triple (U_1, U_2, E) where U_1 and U_2 are finite sets and $E \subseteq U_1 \times U_2$. So a bipartite graph becomes a bigraph if we fix a bipartition and specify which bipartition class is first and second. On the other hand, if $F = (V, E)$ is a graph, then (V, V, E') is an associated bigraph, where $E' = \{(x, y) : xy \in E\}$. This bigraph is obtained from F by a standard construction of doubling the nodes.

If $G = (V, E)$ is a graph, then an *induced sub-bigraph* of G is determined by two subsets $S, T \subseteq V$, and its edge set consists of those pairs $(x, y) \in S \times T$ for which $xy \in E$ (so this is an induced subgraph of the bigraph associated with G).

Let $J_i = (\Omega_i, \mathcal{A}_i, \pi_i)$ $(i = 1, 2)$ be (standard) probability spaces. A measurable function $W : J_1 \times J_2 \to [0, 1]$ is called a *bigraphon*. A *graphon* is a special bigraphon where $J_1 = J_2 = J$ and W is symmetric: $W(x, y) = W(y, x)$ for all $x, y \in J$.

For a fixed probability space J, graphons can be considered as elements of the space $L_\infty(J \times J)$. The norm that it most important in their is study is, however, not the L_∞ norm, but the *cut-norm*, defined by

$$\|W\|_\square = \sup_{S,T \subseteq J} \left| \int_{S\times T} W(x, y)\, dx\, dy \right|.$$

We will also use the L_1 norm

$$\|W\|_1 = \int_{J\times J} |W(x, y)|\, dx\, dy.$$

A graphon (J, W) is called a *stepfunction*, if there is a partition of J into a finite number of measurable sets $S_1, \ldots, S_n$ so that W is constant on every $S_i \times S_j$. The partition classes will be called the *steps* of the stepfunction.

Every graph $F=(V,E)$ can be considered as a graphon, if we consider V as a finite probability space with the uniform measure, and E, as the indicator function of adjacency. We can resolve the atoms into intervals of length $1/|V|$, to get a graphon $\big([0,1],W_F\big)$ (which is a stepfunction). More explicitly, we split $[0,1]$ in $|V|$ equal intervals L_i, and define $W_F(x,y)=E(i,j)$ for $ix\in L_i$ and $y\in L_j$. This graphon is weakly isomorphic to (V,E) (see below).

In a similar way, every bigraph can be considered as a finite bigraphon, and defines a bigraphon $\big([0,1],[0,1],W_F\big)$.

Remark 2.1. We could consider a further version of this notion, when $J_1=J_2$ but W is not necessarily symmetric. Such a structure arises as the limit object of a convergent sequence of directed graphs with no parallel edges, and therefore can be called a *digraphon.* We do not need them in this paper.

Every bigraphon (J_1,J_2,W) can be considered as a linear kernel operator $L_1(J_1)\to L_\infty(J_2)$, defined by

$$f\mapsto\int\limits_J W(.,y)f(y)\,dy.$$

Of course, this operator remains well-defined if we increase the subscript in L_1 in the domain and lower the subscript in L_∞ in the range. In the case of a graphon (J,W), it is useful to consider it as an operator $L_2(J)\to L_2(J)$, since it is then a Hilbert-Schmidt operator, and a rich theory is applicable. In particular, we know that it has a discrete spectrum.

If (J_1,J_2,U) and (J_2,J_3,W) are two bigraphons, we can define their *operator product* $(J_1,J_3,U\circ W)$ by

$$(U\circ W)(x,y)=\int\limits_{J_2} U(x,z)W(z,y)\,dz.$$

(We will write dz instead of $d\pi_2(z)$, where π_2 is the measure on J_2: integrating over J_2 means that we integrate with respect to the probability measure of J_2.)

The notion of the density of a graph in a graphon has been introduced in [7]. Here we need several versions, which unfortunately leads to some messy notation. For a graphon (J, W) and graph $F = (V, E)$, we associate a variable $x_v \in J$ with every node $v \in V$, and define

$$t(F, W; x) = \prod_{uv \in E(F)} W(x_u, x_v), \qquad t(F, W) = \int_{J^V} t(F, W; x)\, dx.$$

We can think of $t(F, W)$ as "counting subgraphs isomorphic to F". We also need the induced version:

$$t_{\text{ind}}(F, W; x) = \prod_{uv \in E(F)} W(x_u, x_v) \prod_{\substack{u,v \in V \\ uv \notin E(F)}} \bigl(1 - W(x_u, x_v)\bigr)$$

$$t_{\text{ind}}(F, W) = \int_{J^V} t_{\text{ind}}(F, W; x)\, dx.$$

For any subset $S \subseteq V$, we define $t_S(F, W; .) : J^S \to \mathbb{R}$ by integrating only over variables corresponding to $V \setminus S$: If x' and x'' denote the restrictions of $x \in J^V$ to S and $V \setminus S$, respectively, then

$$t_S(F, W; x') = \int_{J^{V \setminus S}} t(F, W; x)\, dx''.$$

Note that $t_\emptyset(F, W) = t(F, W)$ and $t_V(F, W; .) = t(F, W; .)$.

These quantities have obvious analogues for bigraphs and bigraphons. For a bigraphon (J_1, J_2, W) and bipartite graph (U_1, U_2, E), we introduce variables $x_u \in J_1$ $(u \in U_1)$ and $y_v \in J_2$ $(v \in U_2)$, and define

$$t^{\mathsf{b}}(F, W; x, y) = \prod_{uv \in E(F)} W(x_u, y_v), \quad t^{\mathsf{b}}(F, W) = \int_{J_1^{U_1}} \int_{J_2^{U_2}} t^{\mathsf{b}}(F, W; x, y)\, dy\, dx.$$

Again, we define an induced version:

$$t^{\mathsf{b}}_{\text{ind}}(F, W; x, y) = \prod_{ij \in E(F)} W(x_i, y_j) \prod_{\substack{i \in U_1, j \in U_2 \\ ij \notin E(F)}} \bigl(1 - W(x_i, y_j)\bigr)$$

$$t^{\mathsf{b}}_{\text{ind}}(F, W) = \int_{J_1^{U_1}} \int_{J_2^{U_2}} t^{\mathsf{b}}_{\text{ind}}(F, W; x, y)\, dy\, dx.$$

Assume that subsets $S_i \subseteq U_i$ are specified. We define the function

$$t^{\mathsf{b}}_{S_1,S_2}(F,W;.) : J_1^{S_1} \times J_2^{S_2} \to \mathbb{R}$$

by

$$t^{\mathsf{b}}_{S_1,S_2}(F,W;x',y') = \int\limits_{J_1^{U_1\setminus S_1}} \int\limits_{J_2^{U_2\setminus S_2}} t^{\mathsf{b}}(F,W;x,y)\, dy''\, dx'',$$

where, similarly as above, x' and x'' denote the restrictions of $x \in J_1^{U_1}$ to S_1 and $U_1 \setminus S_1$, respectively, and similarly for y. We can define $t^{\mathsf{b}}_{\mathrm{ind};S_1,S_2}(F,W)(x',y')$ analogously.

Two graphons (J,W) and (J',W') are *weakly isomorphic* if for every graph F, $t(F,W) = t(F,W')$. Various characterizations of weak isomorphism were given in [2]. Every graphon is weakly isomorphic to a graphon on $[0,1]$ (with the Lebesgue measure), and also to a (possibly different) graphon which is twin-free in the sense that $W(x,.)$ and $W(x',.)$ differ on a set of positive measure for all $x \neq x'$.

3. The Topology of Graphons

3.1. The neighborhood distance

Let (J,W) be a graphon. We can endow the space J with a distance function by

$$r_W(x,y) = \big\| W(x,.) - W(y,.) \big\|_1.$$

This function is defined for almost all pairs x,y; we can delete those points from J where $W(x,.) \notin L_1(W)$ (a set of measure 0), to have r_W defined on all pairs. It is clear that r_W is a pseudometric (it is symmetric and satisfies the triangle inequality). We call r_W the *neighborhood distance* on W.

We also define metrics on bigraphons, endowing the spaces J_1 and J_2 with distance functions by

$$r_1(x,y) = \big\| W(x,.) - W(y,.) \big\|_1 \qquad (x,y \in J_1),$$

$$r_2(x,y) = \big\| W(.,x) - W(.,y) \big\|_1 \qquad (x,y \in J_2).$$

These functions are defined for almost all pairs x, y.

Example 1. Let S^k denote the unit sphere in $\mathbb{R}^{k+1}$, consider the uniform probability measure on it, and let $W(x,y) = 1$ if $x \cdot y \geq 0$ and $W(x,y) = 0$ otherwise. Then (S^k, W) is a graphon, in which the neighborhood distance of two points $a, b \in S^k$ is just their spherical distance (normalized by dividing by π).

Example 2. Let (M, d) be a metric space, and let π be a Borel probability measure on M. Assume that the diameter of M is at most 1. Then d can be viewed as a graphon on (M, d). For $x, y \in M$, we have

$$r_d(x,y) = \int\limits_M \left| d(x,z) - d(y,z) \right| d\pi(z) \leq \int\limits_M d(x,y)\, d\pi(z) = d(x,y),$$

so the identity map $(M, d) \to (M, r_d)$ is contractive. This implies that if (M, d) is compact, and/or finite dimensional (in many senses of dimension), then so is (M, r_d). For most "everyday" metric spaces like (like segments, spheres, or balls) $r_d(x,y)$ can be bounded from below by $\Omega\big(d(x,y)\big)$, in which case (M, d) and (M, r_d) are homeomorphic.

More generally, if $F : [0,1] \to [0,1]$ is a continuous function, then $W(x,y) = F\big(d(x,y)\big)$ defines a graphon, and the identity map $(M, d) \to (M, r_W)$ is continuous.

Example 3. Finitely forcible graphons, mentioned in the introduction, give interesting examples, for whose details we refer to [13]. One class is stepfuctions (equivalent to finite weighted graphs), which were proved to be finitely forcible by Lovász and Sós [10]; for these, the underlying metric space is finite. Other examples introduced in [13] provide as underlying topologies an interval, the Cantor set, and the one-point compactification of $\mathbb{N}$.

3.2. Pure [bi]graphons

A bigraphon (J_1, J_2, W) is *pure* if (J_i, r_i) is a complete separable metric space and the probability measure has full support (i.e., every open set has positive measure). This definition includes that $r_i(x,y)$ is defined for all $x, y \in J_i$ and $r_i(x,y) > 0$ if $x \neq y$, i.e., the bigraphon has no "twin points". We say that a graphon is *pure*, if the underlying metric probability space is complete, separable and the probability measure has full support.

Theorem 3.1. *Every [bi]graphon is weakly isomorphic to a pure [bi]-graphon.*

Remark 3.2. It was shown in [2] that every graphon is weakly isomorphic to a graphon on a standard probability space with no parallel points, which means that for any two points $x, x' \in J$, $W(x,.)$ and $W(x',.)$ differ on a set of positive measure. Theorem 3.1 can be considered as a strengthening of this result.

Proof. We give the proof for bigraphons; the case of graphons is similar. We assume that J_1 and J_2 are standard probability spaces; this can be achieved similarly as for graphons. Let T_1 be the set of functions $f \in L_1[J_2]$ such that for every L_1-neighborhood U of f, the set $\{x \in J_1 : W(x,.) \in U\}$ has positive measure.

Claim 3.3. *For almost every point $x \in J_1$, $W(x,.) \in T_1$.*

Indeed, it is clear that for almost all $x \in J_1$, $W(x,.) \in L_1[J_2]$. Every function $g \in L_1[J_2] \setminus T_1$ has an open neighborhood U_g in $L_1[J_2]$ such that $\pi_1\{x \in J_1 : W(x,.) \in U_g\} = 0$. Let $U = \bigcup_{g \notin T_1} U_g$. Since $L_1[J_2]$ is separable, U equals the union of some countable subfamily $\{U_{g_i} : i \in \mathbb{N}\}$ and thus $\pi_1\{x \in J_1 : W(x,.) \in U\} = 0$. Since if $W(x,.) \notin T_1$ then $W(x,.) \in U$, this proves the Claim.

Clearly T_1 inherits a metric from $L_1[J_2]$, and it is complete and separable in this metric. The functions $W(x,.)$ are everywhere dense in $T_1(W)$ and have measure 1. It also inherits a probability measure π_1' from J_1 through

$$\pi_1'(X) = \pi_1\{x \in J_1 : W(x,.) \in X\}.$$

So T_1 is a complete separable metric space with a probability measure on its Borel sets. It also follows from the definition of T_1 that every open set has positive measure.

Define $\widetilde{W} : T_1 \times J_2 \to [0,1]$ by $\widetilde{W}(f,y) = f(y)$ for $f \in T_1$ and $y \in J_2$. Then we can replace J_1 by T_1 and W by $\widetilde{W}$, to get a weakly isomorphic graphon. Similarly, we can replace J_2 by T_2. ∎

We say that two graphons (J, W) and (J', W') are *isometric* if there is an isometric bijection $\phi : J \to J'$ that is measure preserving, and $W'(\phi(x), \phi(y)) = W(x,y)$ for almost all $x, y \in J$. The definition for bigraphons is slightly more complicated: two bigraphons (J_1, J_2, W) and (J_1', J_2', W') are *isometric* if there are isometric, measure preserving bijections $\phi_1 : J_1 \to J_1'$ and $\phi_2 : J_2 \to J_2'$ such that $W'(\phi_1(x), \phi_2(y)) = W(x,y)$ for almost all $(x,y) \in J_1 \times J_2$.

Theorem 3.4. *If two pure [bi]graphons are weakly isomorphic, then they are isometric.*

Proof. We describe the proof for graphons. Theorem 2.1 (a) in [2] says that if two graphons (J, W) and (J', W') are weakly isomorphic, and they have no twins, then one can delete 0-sets $S \subseteq J$ and $S' \subseteq J'$ such that there is a bijective measure preserving map $\phi : J \setminus S \to J' \setminus S'$ such that $W'(\phi(x), \phi(y)) = W(x, y)$ for almost all $(x, y) \in J \times J$. We may even assume that for every $x \in J \setminus S$, $W'(\phi(x), \phi(y)) = W(x, y)$ holds for almost all y (and vice versa), since this can be achieved by deleting further 0-sets. Clearly ϕ preserves the metric.

We also know that $J \setminus S$ is dense in J (since (J, W) is pure and so its probability measure has full support), and so J is the completion of $J \setminus S$ (and similarly for J'). Hence ϕ extends to an isometry between J and J', which shows that (J, W) and (J', W') are isometric graphons. ■

Remark 3.5. Is purity the ultimate normalization of a graphon? There is still some freedom left: we can change the value of W on a symmetric subset of $J \times J$ that intersects every fiber $J \times \{v\}$ in a set of measure 0. We can take the integral of W (which is a measure ω on J), and then the derivative of ω wherever this exists. This way we get back W almost everywhere, and a well defined value for some further points. What is left undefined is the set of "essential discontinuity" of W (of measure 0). It would be interesting to relate this set to combinatorial properties of W.

3.3. Density functions on pure [bi]graphons

The following technical Lemma will be very useful in the study of r_W and related distance functions.

Lemma 3.6. (a) *Let (J, W) be a graphon, F, a graph, and $S \subseteq V$, an independent set of nodes. Then the function $t = t_S(F, W; .) : J^S \to \mathbb{R}$ satisfies*

$$\big|t(x) - t(x')\big| \le |E| \max_{i \in S} r_W(x_i, x_i').$$

(b) *Let (J_1, J_2, W) be a bigraphon, let $F = (U_1, U_2, E)$ be a bigraph, and let $S_i \subseteq U_i$ be such that no edge connects S_1 to S_2. Then the function $t = t^{\mathsf{b}}_{S_1,S_2}(F, W, .) : J_1^{S_1} \times J_2^{S_2} \to \mathbb{R}$ satisfies*

$$\big|t(x, y) - t(x', y')\big| \le |E| \max \Big\{ \max_{i \in S_1} r_1(x_i, x_i'), \max_{j \in S_2} r_2(y_j, y_j') \Big\}.$$

Remark 3.7. (i) It follows that the functions t in (a) and (b) are Lipschitz (and hence continuous).

(ii) In both parts (a) and (b) of the Lemma, the graph F could have multiple edges.

Proof. We describe the proof of (a); the proof of (b) is similar. For each $i \in U \setminus S$, let $x_i = x'_i$ be a variable and $y = (x_i : i \in U \setminus S)$. Let $E = \{u_1v_1, \dots u_mv_m\}$, where we may assume that $v_i \in U \setminus S$. Then

$$t(x) - t(x') = \int\limits_{J^{U\setminus S}} \prod_{i=1}^{m} W(x_{u_i}, x_{v_i})\, dy - \int\limits_{J^{U\setminus S}} \prod_{i=1}^{m} W(x'_{u_i}, x'_{v_i})\, dy$$

$$= \sum_{j=1}^{m} \int\limits_{J^{U\setminus S}} \prod_{i<j} W(x_{u_i}, x_{v_i})\big(W(x_{u_j}, x_{v_j}) - W(x'_{u_j}, x'_{v_j})\big) \prod_{j>i} W(x'_{u_i}, x'_{v_i}), dy$$

and hence

$$\big|t(x) - t(x')\big| \le \sum_{j=1}^{m} \int\limits_{J^{U\setminus S}} \big|W\big(x_{u_j}, x_{v_j}\big) - W\big(x'_{u_j}, x'_{v_j}\big)\big|\, dy.$$

By the assumption that $v_i \in U \setminus S$, we have $x_{v_j} = x'_{v_j}$ for every j, and so

$$\big|t(x) - t(x')\big| \le \sum_{j=1}^{m} r_W(x_{u_j}, x'_{u_j}) \le |E| \max_{i\in S} r_W(x_i, x'_i),$$

which proves the assertion. ∎

Lemma 3.6 has an important corollaries for pure graphons, which are closely related to Lemma 2.8 in [13]. We do not formulate all versions, just a few that we need.

Corollary 3.8. *Let (J, W) be a pure graphon, and let F be a graph and let $S \subseteq V$, where S is independent. Then $t_S(F, W; x)$ is a continuous function of $x \in J^S$.*

Applying this when F is a path of length 2, we get:

Corollary 3.9. *For every pure graphon (J, W), $W \circ W$ is a continuous function on J.*

Another application of Corollary 3.8 gives:

Corollary 3.10. *Let (J,W) be a pure graphon, and let $F_1,\dots,F_m$ be graphs whose node set contains a common set S, which is independent in each. Let $T \subseteq S$, and let $a_1,\dots,a_m$ be real numbers. Let $x \in J^T$, and assume that the equation*

$$\sum_{i=1}^{m} a_i t_S(F_i,W;x,y) = 0 \tag{1}$$

holds for almost all $y \in J^{S\setminus T}$. Then it holds for all $y \in J^{S\setminus T}$.

Proof. By Corollary 3.8, the left hand side of (1) is a continuous function of (x,y), and so it remains a continuous function of y if we fix x. Hence the set where it is not 0 is an open subset of $J^{S\setminus T}$. Since the graphon is pure, it follows that this set is either empty of has positive measure. ■

We formulate one similar corollary for bigraphons.

Corollary 3.11. *Let (J_1,J_2,W) be a pure bigraphon, and let $F_1,\dots,F_m$ be bigraphs with the same bipartition classes U_1 and U_2. Let $a_1,\dots,a_m$ be real numbers. Assume that the equation*

$$\sum_{i=1}^{m} a_i t^{\mathsf{b}}_{U_1}(F_i,W;x) = 0 \tag{2}$$

holds for almost all $x \in J_1^{U_1}$. Then it holds for all $x \in J_1^{U_1}$.

3.4. The similarity distance

It turns out (it was already noted in [12]) that the distance function $r_{W\circ W}$ defined by the operator square of W is also closely related to combinatorial properties of a graphon. We call this the *similarity distance* (for reasons that will become clear later). In explicit terms, we have

$$r_{W\circ W}(a,b) = \int_J \left| \int_J W(a,y)W(y,x)\,dy - \int_J W(b,y)W(y,x)\,dy \right| dx \tag{3}$$

$$= \int_J \left| \int_J W(x,y)\big(W(y,a) - W(y,b)\big)\, dy \right| dx.$$

Example 4. In Example 1, we have seen that the neighborhood distance is just the spherical distance. It is not hard to see that $1 - 2(W \circ W)(x,y)$ is just the spherical distance of x and y, and from here, the similarity distance is within constant factors of the neighborhood distance.

Remark 3.12. Let $\mathbf{X}, \mathbf{Y}, \mathbf{Z}$ be independent uniform random points from J, then we can rewrite the definitions of these distances as

(4) $$r_W(a,b) = \mathsf{E}_{\mathbf{X}} \big| W(\mathbf{X},a) - W(\mathbf{X},b) \big|,$$

(5) $$r_{W \circ W}(a,b) = \mathsf{E}_{\mathbf{X}} \Big| \mathsf{E}_{\mathbf{Y}} \big(W(\mathbf{X},\mathbf{Y}) \big(W(\mathbf{Y},a) - W(\mathbf{Y},b) \big) \big) \Big|.$$

This formulation shows that this distance can be computed with arbitrary precision from a bounded size sample. We do not go into the details of this.

Lemma 3.13. *If (J,W) is a pure graphon, then the similarity distance $r_{W \circ W}$ is a metric.*

So $(J, r_{W \circ W})$ is a metric space, and hence Hausdorff. We will show later that it is always compact.

Proof. The only nontrivial part of this lemma is that $r_{W \circ W}(x,y) = 0$ implies that $x = y$. The condition $r_{W \circ W}(x,y) = 0$ implies that for almost all $u \in J$ we have $(W \circ W)(x,u) = (W \circ W)(y,u)$, or more explicitly

$$\int_J \big(W(x,z) - W(y,z)\big)\, W(z,u)\, dz = 0.$$

Using that (J,W) is pure, Corollary 3.11 implies that this holds for every $u \in J$. in particular, it holds for $u = x$ and $u = y$. Taking the difference, we get that

$$\int_J \big(W(x,z) - W(y,z)\big) \big(W(z,x) - W(z,y)\big)\, dz = 0,$$

and hence $W(x,z) = W(y,z)$ almost everywhere. Using again that (J,W) is pure, we get that $x = y$. ■

For every $x \in J$, the function $W(x,.)$ is in $L_\infty(J)$, and hence the weak topology of $L_1(J)$ gives a topology on J. It is well known that when restricted to $L_\infty(J)$, this topology is the weak-$*$ topology on $L_\infty(J)$, and hence it is metrizable, and the unit ball of $L_\infty(J)$ is compact in it (Alaoglu's Theorem). A sequence of points (x_n) is convergent in this topology if and only if

$$\int_A W(x_n, y)\, dy \to \int_A W(x,y)\, dy$$

for every measurable set $A \subseteq J$. We call this the *weak topology* on J. We need this name only temporarily, since we are going to show that $r_{W\circ W}$ gives a metrization of the weak topology.

Theorem 3.14. *For any pure graphon, the metric $r_{W\circ W}$ defines exactly the weak topology.*

Proof. First we show that the weak topology is finer than the topology of $(J, r_{W\circ W})$. Suppose that $x_n \to x$ in the weak topology, and consider

$$r_{W\circ W}(x_n, x) = \int\limits_J \left| \int\limits_J \big(W(x_n, y) - W(x,y)\big)\, W(y,z)\, dy \right| dz.$$

Here the inner integral tends to 0 for every z, by the weak convergence $x_n \to x$. Since it also remains bounded, it follows that the outer integral tends to 0. This implies that $x_n \to x$ in $(J, r_{W\circ W})$.

From here, the equality of the two topologies follows by general arguments: the weak topology is compact, and the coarser topology of $r_{W\circ W}$ is Hausdorff, which implies that they are the same. ■

Corollary 3.15. *For every pure graphon (J, W), the space $(J, r_{W\circ W})$ is compact.*

To compare the topology of (J, r_W) with these, note that for any two points $x, y \in J$, we have

$$r_{W\circ W}(x,y) \le r_W(x,y), \tag{6}$$

which implies that the topology of (J, r_W) is finer than the topology of $(J, r_{W\circ W})$.

3.5. Compact Graphons

Graphons for which the finer space (J, r_W) is also compact seem to have a special importance in combinatorics. Let us call such a graphon a **compact graphon.**

Proposition 3.16. *A pure graphon (J, W) is compact if and only if (J, r_W) and $(J, r_{W\circ W})$ define the same topologies.*

Proof. If the topologies (J, r_W) and $(J, r_{W\circ W})$ are the same, then (J, r_W) is compact by Corollary 3.15. Conversely, if (J, r_W) is compact then, by the argument used before in the proof of Theorem 3.14, the coarser Hausdorff topology of $(J, r_{W\circ W})$ must be the same. ■

Example 5. Let $J = [0,1]$, $f(y) = \lfloor \log(1/y) \rfloor$, and define

$$W(x,y) = \begin{cases} x_{f(y)}, & \text{if } x > 1/2 \text{ and } y \le 1/2, \\ y_{f(x)}, & \text{if } x \le 1/2 \text{ and } y > 1/2, \\ 0, & \text{otherwise,} \end{cases}$$

where $x = 0.x_1x_2\ldots$ and $y = 0.y_1y_2\ldots$ are the binary expansions of x and y, respectively. Then selecting one point from each interval $\left[2^{-k+1}, 2^{-(k)}\right]$, we get an infinite number of points in $\big([0,1], r_2\big)$ mutually at distance $1/4$, so (J, r_W) is not compact, but by Corollary 3.15, $(J, r_{W\circ W})$ is compact. So the two topologies are different.

We conclude this section with an observation relating the topology of J to spectral theory.

Lemma 3.17. *Let (J, W) be a pure graphon. Then every eigenfunction $f \in L_2(J)$ of W as a kernel operator belonging to a nonzero eigenvalue is continuous in the metric $r_{W\circ W}$ (and therefore also in r_W).*

Proof. It suffices to prove that f is continuous in (J, r_W), since we can apply the argument to the graphon $(J, W \circ W)$, which also has f as an eigenvector.

First, we have

$$|f(x)| = \frac{1}{|\lambda|}\left|\int_J W(x,y)f(y)\,dy\right| \le \frac{1}{|\lambda|}\|f\|_1 \le \frac{1}{|\lambda|}\|f\|_2,$$

and so f is bounded. We know by Corollary 3.9 that $W \circ W$ is continuous in (J, r_W), and hence so is

$$f = \frac{1}{\lambda^2} \int_J (W \circ W)(x,y) f(y)\, dy. \qquad \blacksquare$$

4. Thin Graphons

4.1. The main theorem

We say that a bigraphon W is *thin* if there is a bigraph F such that $t^{\mathsf{b}}_{\text{ind}}(F, W) = 0$. Trivially, if W is thin, then so is its complementary bigraphon $1 - W$.

We call a graphon *thin* if it is thin as a bigraphon. (Note: for this, it is not enough to require $t_{\text{ind}}(F, W) = 0$ for some bipartite graph F. For example, consider the graphon $U : [0,1]^2 \to [0,1]$ defined by $U(x,y) = U(y,x) = 1/2$ if $x \in [0,1/2]$ and $y \in (1/2,1]$, and $U(x,y) = 1$ otherwise. As a bigraphon, this is not thin, but satisfies $t_{\text{ind}}(F, W) = 0$ for every bigraph with at least 3 nodes in one of the classes.

The *(upper) packing dimension* of a metric space (M, d) is defined as

$$\limsup_{\varepsilon \to 0} \frac{\log N(\varepsilon)}{\log(1/\varepsilon)},$$

where $N(\varepsilon)$ is the maximum number of points in M mutually at distance at least ε. So this dimension is finite if and only if there is a $d \geq 0$ such that every set of points mutually at distance at least ε has at most ε^{-d} elements. It is easy to see that we could use instead of $N(\varepsilon)$ the minimum number of sets of diameter at most ε covering the space.

Our main goal is to prove:

Theorem 4.1. *If a pure bigraphon (J_1, J_2, W) is thin, then* (a) *$W(x,y) \in \{0,1\}$ almost everywhere,* (b) *J_1, J_2 are compact, and* (c) *J_1, J_2 have finite packing dimension.*

Remark 4.2. The proof will show that if $t_{\text{ind}}(F, W) = 0$ for a bigraph F with k nodes, then the packing dimension of J_i is bounded by $2k$.

Before giving the proof, we describe a class of examples, and then recall some facts about the Vapnik–Červonenkis dimension.

Example 6. Let V be a finite or countable set, π, a probability measure on V, and define $J_1 = [0,1]^V$, $J_2 = [0,1] \times V$, with the power measure μ_1 on J_1 and the product measure μ_2 on J_2. We define a bigraphon on $J_1 \times J_2$ by

$$W(x,y) = \mathbf{1}_{t \le x_i}$$

for $x = (x_i : i \in S)$ and $y = (t,i)$. We can metrize this bigraphon by

$$r_1(x,x') = \sum_{i \in V} \pi(i)|x_i - x_i'|$$

for $x = (x_i : i \in S)$, $x' = (x_i' : i \in S) \in J_1$, and

$$r_2(y,y') = \begin{cases} |t-t'| & \text{if } i = 1', \\ t + t' - 2tt' & \text{otherwise.} \end{cases}$$

for $y = (t,i)$, $y' = (t',i') \in J_2$.

If V is finite, then (J_1, r_1) has dimension $|V|$, while (J_2, r_2) has dimension 1, and both are compact. These facts also follow if we observe that W is thin. Indeed, if F denotes the matching with $|V|+1$ edges, then $t^{\mathsf{b}}_{\mathrm{ind}}(F,W) = 0$, since among any $|V|+1$ points in J_2, there are two points of the form $y = (t,i)$ and $y' = (t',i)$ with $t < t'$, and then $W\big(.,(t,i)\big) \ge W\big(.,(t',i)\big)$.

If V is infinite, then (J_1, r_1) is infinite dimensional but compact, while (J_2, r_2) is not compact.

Example 7. Let $J_1 = J_2 = [0,1]$, and let $W(x,y) = x_{f(y)}$, where $x = 0.x_1x_2\dots$ is the binary expansion of x, and $f(y) = \lceil \log(1/y) \rceil$. Then for $x = 0.x_1x_2\dots$ and $x' = 0.x_1'x_2'\dots$ we have $r_1(x,x') = \sum_{k=1}^\infty 2^{-k}|x_k - x_k'|$, and from here is easy to see that $\big([0,1], r_1\big)$ is compact. Furthermore, if $S \subseteq [0,1]$ is a set of points mutually more than 2^{-n} apart, then any two elements of S must differ in one of their first n digits, and so their number is at most 2^n. Hence the packing dimension of $\big([0,1], r_1\big)$ is 1.

On the other hand, selecting a point $y_k \in \big[2^{-k}, 2^{-(k-1)}\big]$, we get an infinite number of points in $\big([0,1], r_2\big)$ mutually at distance $1/2$, so this space is not compact and infinite dimensional.

4.2. Vapnik–Červonenkis dimension

For any set V and family of subsets $\mathcal{H} \subseteq 2^V$, a set $S \subseteq V$ is called *shattered*, if for every $X \subseteq S$ there is a $Y \in \mathcal{H}$ such that $X = Y \cap S$. The *Vapnik–Červonenkis dimension* or *VC-dimension* $\dim_{VC}(\mathcal{H})$ of a family of sets is the supremum of cardinalities of shattered sets [16]. For us, S will be always finite.

Let V be a probability space and $\mathcal{H}$, a family of measurable subsets of V. A finite subfamily $\mathcal{H}'$ is *qualitatively independent* if all the $2^{|\mathcal{H}'|}$ atoms of the set algebra they generate have positive measure. The *dual essential Vapnik–Červonenkis dimension,* or briefly *DE-dimension,* of $\mathcal{H}$ is a supremum of all cardinalities of qualitatively independent subfamilies of $\mathcal{H}$.

We recall two basic facts about VC-dimension:

Lemma 4.3 (Sauer–Shelah Lemma). *If a family $\mathcal{H}$ of subsets of an m-element set has VC-dimension k, then*

$$|\mathcal{H}| \leq 1 + m + \cdots + \binom{m}{k}.$$

For a family $\mathcal{H}$ of sets, we denote by $\tau(\mathcal{H})$ the minimum cardinality of a set meeting every member of $\mathcal{H}$. The following basic fact about VC-dimension was proved by Komlós, Pach and Woeginger [9], based on the results of Vapnik and Červonenkis [16] (we do not state it in its sharpest form):

Theorem 4.4. *Let J be a probability space and, $\mathcal{H}$ a family of measurable subsets of J such that every $A \in \mathcal{H}$ has measure at least ε. Suppose that $\mathcal{H}$ has finite VC-dimension k. Then*

$$\tau(\mathcal{H}) \leq 8k\frac{1}{\varepsilon}\log\frac{1}{\varepsilon}.$$

We need a couple of further facts. For a family $\mathcal{H}$ of sets, let $\mathcal{H}(\triangle)\mathcal{H} = \{A\triangle B : A, B \in \mathcal{H}\}$.

Lemma 4.5. *For every family of sets,* $\dim_{VC}\big(\mathcal{H}(\triangle)\mathcal{H}\big) \leq 10\dim_{VC}(\mathcal{H})$.

Proof. Set $k = \dim_{VC}(\mathcal{H})$. Let S be a subset of $V = \cup\mathcal{H}$ with m elements that is shattered by $\mathcal{H}(\triangle)\mathcal{H})$. Then every $X \subseteq S$ arises as $X = (A\triangle B)\cap S$, where $A, B \in \mathcal{H}$. Since $(A\triangle B) \cap S = (A \cap S)\triangle(B \cap S)$, the number of different sets of the form $A \cap S$ is at least $2^{m/2}$. By the Sauer–Shelah Lemma, this implies that

$$2^{m/2} \leq 1 + m + \cdots + \binom{m}{k},$$

whence $m \leq 10k$ follows by standard calculation. ■

Lemma 4.6. *Let $\mathcal{H}$ be a family of measurable sets in a probability space with VC-dimension k such that $\pi(A \triangle B) \geq \varepsilon$ for all $A, B \in \mathcal{H}$. Then $|\mathcal{H}| \leq (80k)^k \varepsilon^{-2k}$.*

Proof. Consider the family $\mathcal{H}' = \mathcal{H}(\triangle)\mathcal{H}$. Every $A \in \mathcal{H}'$ has $\pi(A) \geq 1/\varepsilon$, and $\dim_{VC}(\mathcal{H}') \leq 10k$ by Lemma 4.5. Hence by Theorem 4.4, we have

$$\tau(\mathcal{H}') \leq 80k\frac{1}{\varepsilon}\ln\frac{1}{\varepsilon}.$$

Let $S \subseteq \cup\mathcal{H}$ be a set of size $\tau(\mathcal{H}')$ meeting every symmetric difference $A \triangle B$ $(A, B \in \mathcal{H})$. Then the sets $S \cap A$, $A \in \mathcal{H}$ are all different. By the Sauer–Shelah Lemma, this implies that

$$|\mathcal{H}| \leq 1 + |S| + \cdots + \binom{|S|}{k} < |S|^k \leq \left(80k\frac{1}{\varepsilon}\ln\frac{1}{\varepsilon}\right)^k < (80k)^k \varepsilon^{-2k} \quad ■$$

4.3. VC-dimension and graphons

Lemma 4.7. *Let (J_1, J_2, W) be a pure 0-1 valued bigraphon. Then W is thin if and only if the DE-dimension of the family $\mathcal{R}_W = \{\text{supp}\,(W(x,.)) : x \in T_1\}$ is finite.*

Proof. Suppose that this dimension is infinite. We claim that $t^{\mathsf{b}}_{\text{ind}}(F, W) > 0$ for every bipartite graph $F = (U, U', E)$. Let $S \subseteq J_1$ be a set such that the subfamily $\{\text{supp}\,(W(x,.)) : x \in T_1\}$ is qualitatively independent. To each $i \in U$, assign a value $x_i \in S$ bijectively. By Corollary 3.11, the set of points $y \in J_2$ such that $\text{supp}\,(W(.,y)) \cap S = \{x_i : i \in N(j)\}$ has positive measure for each $j \in U'$. Hence $t^{\mathsf{b}}_{\text{ind}}(F, W) > 0$.

Conversely, suppose that $k = \dim(\mathcal{R}_W)$ is finite. Let F denote the bipartite graph with $k+1$ nodes in one class U and 2^{k+1} nodes in the other class U', in which the nodes in U' have all different neighborhoods. Then $t^{\mathsf{b}}_{\text{ind}}(F, W) = 0$. ■

Remark 4.8. The proof above in fact gives the following quantitative result: $t^{\mathsf{b}}_{\mathrm{ind}}(F,W)=0$ for some bigraph F with k nodes in its smaller bipartition class if and only if $\dim_{\mathrm{DE}}(\mathcal{R}_W)<k$.

Proof of Theorem 4.1.

(a) Suppose that the bigraph $F=(U_1,U_2,E)$ satisfies $t^{\mathsf{b}}_{\mathrm{ind}}(F,W)=0$. Then for almost all $x\in J_1^{U_1}$, we have $t^{\mathsf{b}}_{U_1,\mathrm{ind}}(F,W;x)=0$. By Corollary 3.11, it follows that $t^{\mathsf{b}}_{U_1,\mathrm{ind}}(F,W;x)=0$ for every x. In particular, $t^{\mathsf{b}}_{U_1,\mathrm{ind}}(F,W;x_0,\dots,x_0)=0$ for all $x_0\in J_1$. But for this substitution,

$$t^{\mathsf{b}}_{U_1,\mathrm{ind}}(F,W;x_0,\dots,x_0)=\int\limits_{J_2^{V_2}}\prod_{j\in J_2}W(x_0,y_j)^{d_F(j)}\big(1-W(x_0,y_j)\big)^{|U_1|-d_F(j)},$$

and so for every x_0 we must have $W(x_0,y_0)\in\{0,1\}$ for almost all y_0.

(b) By Theorem 3.16 it suffices to prove that if $W(x_n,.)$, $n=1,2,\dots$ weakly converges to f, i.e.,

$$\lim_{n\to\infty}\int\limits_S W(x_n,y)\,dy\to\int\limits_S f(y)\,dy$$

for every measurable set $S\subseteq J_2$, then it is also convergent in L_1.

Claim 4.9. *The weak limit function f is almost everywhere 0-1 valued.*

Suppose not, then there is an $\varepsilon>0$ and a set $Y\subseteq J_2$ with positive measure such that $\varepsilon\le f(x)\le 1-\varepsilon$ for $x\in Y$. Let $S_n=\operatorname{supp}\big(W(x_n,.)\big)\cap Y$. We select, for every $k\ge 1$, k indices $n_1,\dots n_k$ so that the Boolean algebra generated by $S_{n_1},\dots S_{n_k}$ has 2^k atoms of positive measure. If we have this for some k, then for every atom A of the Boolean algebra

$$\lambda(A\cap S_n)=\int\limits_A W(x,y_n)\,dx\longrightarrow\int\limits_A f(x)\,dx\qquad(n\to\infty),$$

and so if n is large enough then

$$\frac{\varepsilon}{2}\lambda(A)\le\lambda(A\cap S_n)\le\Big(1-\frac{\varepsilon}{2}\Big)\lambda(A).$$

If n is large enough, then this holds for all atoms A, and so S_n cuts every previous atom into two sets with positive measure, and we can choose $n_{k+1}=n$.

But this means that the DE-dimension of the supports of the $W(x,.)$ is infinite, contradicting Lemma 4.7. This proves Claim 4.9.

So we know that $f(x) \in \{0,1\}$ for almost all x, and hence

$$\|f - W(.,y_n)\|_1 = \int\limits_{\{f=1\}} (1 - W(x,y_n))\,dx + \int\limits_{\{f=0\}} W(x,y_n)\,dx \longrightarrow 0.$$

Thus $W(.,y_n) \to f$ in L_1, which we wanted to prove.

(c) Let $F = (U_1, U_2, E)$ be a bigraph such that $t^{\mathsf{b}}_{\mathrm{ind}}(F,W) = 0$, and let $U_i = [k_i]$. We show that the packing dimension of J_1 is at most $2k_2$. To this end, we show that if any two elements of a finite set $Z \subseteq J_1$ are at a distance at least ε, then $|Z| \le c(k_2)\varepsilon^{-2k_2}$. Let $\mathcal{H} = \{\mathrm{supp}\,(W(x,.)) \,:\, x \in Z\}$, then

$$\pi_2(X \triangle Y) \ge \varepsilon \tag{7}$$

for any two distinct sets $X, Y \in \mathcal{H}$.

Let A be the union of all atoms of the set algebra generated by $\mathcal{H}$ that have measure 0. Clearly A itself has measure 0, and hence the family $\mathcal{H}' = \{X \setminus A \,:\, X \in \mathcal{H}\}$ still has property (7).

We claim that $\mathcal{H}'$ has VC-dimension less than k_2. Indeed, suppose that $J_2 \setminus A$ contains a shattered set S with $|S| = k_2$. To each $j \in U_2$, assign a point $q_j \in S$ bijectively. To each $i \in U_1$, assign a point $p_i \in Z$ such that $q_j \in \mathrm{supp}\,(W(p_i,.))$ if and only if $ij \in E$. This is possible since S is shattered. Now fixing the p_i, for each j there is a subset of J_2 of positive measure whose points are contained in exactly the same members of $\mathcal{H}'$ as q_j, since $q_j \notin A$. This means that the function $t = t^{\mathsf{b}}_{J_1,\mathrm{ind}}(F,W;.) \,:\, V_1^{J_1} \to \mathbb{R}$ satisfies $t(p) > 0$. Corollary 3.11 implies that $t(x) > 0$ for a positive fraction of the choices of $x \in J_1^{V_1}$, and hence $t^{\mathsf{b}}_{\mathrm{ind}}(F,W) > 0$, a contradiction.

Applying Lemma 4.6 we conclude that $|Z| = |\mathcal{H}| \le (80k_2)^{k_2}\varepsilon^{-2k_2}$.

4.4. Hereditary properties and thin bigraphons

A graph property $\mathcal{P}$ is a class of finite graphs closed under isomorphism. The property is called *hereditary,* if whenever $G \in \mathcal{P}$, then every induced subgraph is also in $\mathcal{P}$.

Let $\mathcal{P}$ be any graph property. We denote by $\overline{\mathcal{P}}$ its *closure,* i.e., the class of graphons (J,W) that arise as limits of graph sequences in $\mathcal{P}$. For every graphon W, let $\mathcal{I}(W)$ denote the set of those graphs F for which $t_{\text{ind}}(F, W) > 0$. Clearly, $\mathcal{I}(W)$ is a hereditary graph property.

Let $\mathcal{P}$ be a hereditary property of graphs. Then

$$\cup_{W \in \overline{\mathcal{P}}} \mathcal{I}(W) \subseteq \mathcal{P}. \tag{8}$$

Indeed, if $F \notin \mathcal{P}$, then $t_{\text{ind}}(F, G) = 0$ for every $G \in \mathcal{P}$, since $\mathcal{P}$ is hereditary. This implies that $t_{\text{ind}}(F, W) = 0$ for all $W \in \overline{\mathcal{P}}$, and so $F \notin \mathcal{I}(W)$.

Equality does not always hold in (8). For example, we can always add a bigraph G and all its induced subgraphs to $\mathcal{P}$ without changing $\overline{\mathcal{P}}$. As a less trivial example, consider all bigraphs with degrees bounded by 10. This property is hereditary, and $\overline{\mathcal{P}}$ consists of a single bigraphon (the identically 0 function).

Proposition 4.10. *For a hereditary property $\mathcal{P}$ of graphs equality holds in* (8) *if and only if for every graph $G \in \mathcal{P}$ and $v \in V(G)$, if we add a new node v' and connect it to all neighbors of v, then at least one of the two graphs obtained by joining or not joining v and v' has property $\mathcal{P}$.*

Proof. Suppose that this condition holds. Let $F \in \mathcal{P}$ have n nodes, and let $F(k)$ denote a graph in $\mathcal{P}$ obtained from F by a repetition of this operation so that each original node has k copies. Then $t_{\text{ind}}\big(F, F(k)\big) \ge 1/n^n$. Let W be the limit graphon of some subsequence of the $F(k)$ $(k \to \infty)$, then $W \in \overline{\mathcal{P}}$. Furthermore, clearly $t_{\text{ind}}(F, W) > 0$, and so $F \in \mathcal{I}(W)$.

Conversely, assume that $F = (V, E) \in \mathcal{I}(W)$ for some $W \in \overline{\mathcal{P}}$, so that $t_{\text{ind}}(F, W) > 0$. Let F' and F'' be the two graphs obtained from F by doubling a node v ($vv' \notin E(F')$, but $vv' \in E(F'')$), then

$$\int\limits_{J^V} t_{\text{ind}}(F, W; x)\, dx > 0$$

implies that there is a positive measure of choices for the values of x_u $(u \in V(F) \backslash v)$, for which the set X of the choices of x_v with $t_{\text{ind}}(F, W; x) > 0$ has positive measure. Clearly either $W(x, y) < 1$ for a positive measure of choices of $(x, y) \in Y$ or this holds for $W(x, y) > 0$. One or the alternative, say the first one, holds for a positive measure of choices for the values of x_u $(u \in V(F) \setminus v)$. But then $t(F', W) > 0$. ■

All of the above notions and simple facts extend to bigraphs and bigraphons trivially.

Let us turn to thin graphons and bigraphons. The significance of thin bigraphons is supported by the following observation:

Proposition 4.11. *Let $\mathcal{P}$ be a hereditary bigraph property that does not contain all bigraphs. Then every bigraphon in its closure is thin.*

Proposition 4.11 and Theorem 4.1 imply:

Corollary 4.12. *Let $\mathcal{P}$ be a hereditary bigraph property that does not contain all bigraphs. Then for every pure bigraphon $(J_1, J_2, , W)$ in its closure, W is 0-1 valued almost everywhere, and J_1 and J_2 are compact and their dimension is bounded by a finite number depending on $\mathcal{P}$ only.*

By this corollary, we can define, for every nontrivial hereditary property of bigraphs, a finite dimension. It would be interesting to find further combinatorial properties of this dimension.

The natural analogue of this corollary for graph properties fails to hold.

Example 8. Let $\mathcal{P}$ be the property of a graph that it is triangle-free. Then every bipartite graphon is in its closure, but such graphons need not be 0-1 valued, and their topology need not be finite dimensional or compact.

However, if we include the (seemingly) simplest of the conclusions of Corollary 4.12 as a hypothesis, then we can extend it to all graphs. A graph property $\mathcal{P}$ is *random-free,* if every $W \in \overline{\mathcal{P}}$ is 0-1 valued almost everywhere.

Theorem 4.13. *Let $\mathcal{P}$ be a hereditary random-free graph property. Then for every pure graphon (J, W) in its closure, J is compact and finite dimensional.*

Before proving this theorem, we need some preparation.

Lemma 4.14. *For a hereditary graph property $\mathcal{P}$, the following are equivalent:*

(i) *$\mathcal{P}$ is random-free;*

(ii) *there is a bigraph F such that $t_{\text{ind}}(F, W) = 0$ for all $W \in \overline{\mathcal{P}}$;*

(iii) *there is a bipartite graph F with bipartition (U_1, U_2) such that no graph obtained from F by adding edges within U_1 and U_2 has property $\mathcal{P}$.*

Proof. (i) $\Rightarrow$ (iii): Assume that (iii) does not hold, then for every bigraph F there is a graph $\hat{F} \in \mathcal{P}$ and a partition $V(\hat{F}) = \{U_1(\hat{F}), U_2(\hat{F})\}$ such that the bigraph between $U_1(\hat{F})$ and $U_2(\hat{F})$ is isomorphic to F. We want to show that $\mathcal{P}$ is not random-free.

Let $(F_1, F_2, \dots)$ be a quasirandom sequence of bigraphs with edge density $1/2$, with the same number of nodes in each bipartition class. Consider the graphs $\hat{F}_n$, and let F'_n and F''_n denote the subgraphs of $\hat{F}_n$ induced by $U_1(\hat{F}_n)$ and $U_2(\hat{F}_n)$, respectively. By selecting a subsequence we may assume that the graph sequences $(F'_1, F'_2, \dots)$ $(F''_1, F''_2, \dots)$ are convergent. By Lemma 4.16 in [5], we can order the nodes of F'_n so that $W_{F'_n}$ converges to a graphon $\big([0,1], W'\big)$ in the cut norm $\|.\|_\square$, and similarly, $W_{F''_n}$ converges to a graphon $\big([0,1], W''\big)$ in the cut norm. We order the nodes of $\hat{F}_n$ so that the nodes in F'_n precede the nodes of F''_n, and keep the above ordering otherwise. Then trivially $W_{\hat{F}_n}$ converges to the graphon

$$U(x,y) = \begin{cases} W'(2x, 2y) & \text{if } x, y < 1/2, \\ W''(2x-1, 2y-1) & \text{if } x, y > 1/2, \\ 1/2 & \text{otherwise.} \end{cases}$$

So $U \in \overline{\mathcal{P}}$ is not 0-1 valued, and so $\mathcal{P}$ is not random-free.

(ii) $\Rightarrow$ (i): Suppose that $\mathcal{P}$ is not random-free, and let $(J, W) \in \overline{\mathcal{P}}$ be a graphon that is not 0-1 valued almost everywhere. Then by Theorem 4.1, it is not thin as a bigraphon, which means that for every bigraph $F = (U_1, U_2, E)$, $t^{\mathsf{b}}_{\text{ind}}(F, W) > 0$, so (ii) is not satisfied.

(iii) $\Rightarrow$ (ii): Consider a bigraph $F = (U_1, U_2, E)$ as in (iii), and consider it as a bipartite graph on $V = U_1 \cup U_2$ (we assume that $U_1 \cap U_2 = \emptyset$). Suppose that it does not satisfy (ii), then there is a graphon $W \in \overline{\mathcal{P}}$ such that $t_{\text{ind}}(F, W; x) > 0$ for a positive measure of choices of the $x \in J^V$. For every such choice, we define a graph F' by connecting those pairs $\{i, j\}$ of nodes of F for which $W(x_i, x_j) > 0$ and either $i, j \in U_1$ or $i, j \in U_2$. The same supergraph F' will occur for a positive measure of choices of the x_i, and for this F' we have $t_{\text{ind}}(F', W) > 0$, so using (8), we get $F' \in \mathcal{I}(W) \subseteq \mathcal{P}$, a contradiction. ∎

Proof of Theorem 4.13. By Lemma 4.14, there is a bigraph F such that $t^{\mathsf{b}}(F, W) = 0$ for all $W \in \overline{\mathcal{P}}$. Thus Theorem 4.1 implies the assertion.

5. Regularity Partitions

5.1. Weak and strong regularity partitions

The Regularity Lemma of Szemerédi [14, 15], and various weaker and stronger versions of it are basic tools in the study of large graphs and graphons [12]. Our goal is to show that it is also closely related to the topology of graphons.

Let (J, W) be a graphon and $\mathcal{P}$, a partition of J into measurable sets with positive measure. For $x \in J$, let $S(x)$ denote the partition class containing x. Define

$$f_{\mathcal{P}}(x) = \frac{1}{\pi\big(S(x)\big)} \int\limits_{S(x)} f(x)\,dx$$

for a function $f \in L_1(J)$, and

$$W_{\mathcal{P}}(x,y) = \frac{1}{\pi\big(S(x)\big)\,\pi\big(S(y)\big)} \int\limits_{S(x)\times S(y)} W(x,y)\,dx.$$

We say that $\mathcal{P}$ is a *weak regularity partition* with error ε, if $\|W - W_{\mathcal{P}}\|_\square \le \varepsilon$.

We define a *Szemerédi partition* of a graphon with error ε as a partition $\mathcal{P} = \{S_1 \cup \dots \cup S_k\}$ of J into measurable sets such that

$$\big|\langle W - W_{\mathcal{P}}, H\rangle\big| \le \varepsilon \tag{9}$$

for every function $H : J \times J \to [0,1]$ that is 0-1 valued and whose support is the union of product sets $R_{ij} = R'_{ij} \times R''_{ij} \subseteq S_i \times S_j$ $(i, j \in [k])$. To relate this to the weak partitions, we note that $\|W - W_{\mathcal{P}}\|_\square \le \varepsilon$ can be expressed as (9) for all functions h of the form $\mathbf{1}_{S\times T}$. (The formulation above is not a direct generalization of Szemerédi's definition, but it is closest in our setting; cf. [12].)

A *strong regularity partition* of a graph was introduced by Alon, Fischer, Krivelevich and M. Szegedy [1]. Here the error is specified by an infinite sequence $\mathcal{E} = (\varepsilon_0, \varepsilon_1, \dots)$ of positive numbers. Again recasting it in our setting, $\mathcal{P}$ is a strong regularity partition with error $\mathcal{E}$ of a graphon (J, W) if there is a graphon (J, U) such that

$$\|W - U\|_1 \le \varepsilon_0 \qquad \text{and} \qquad \|U - W_{\mathcal{P}}\|_\square \le \varepsilon_{|\mathcal{P}|}.$$

Even stronger would be, of course, to require that $\|W - W_{\mathcal{P}}\|_1 \le \varepsilon$ (equivalently, (9) holds for all measurable functions $H : J \times J \to [-1, 1]$). In this case we call $\mathcal{P}$ an *ultra-strong regularity partition* with error ε.

The following result is a graphon version of the original Szemerédi's Regularity Lemma [14, 15], its "weak" form due to Frieze and Kannan [8], and its strong form due to Alon, Fischer, Krivelevich and M. Szegedy [1]. It was proved for graphons in [12].

Theorem 5.1. *Let (J, W) be a graphon on an atomless probability space. Then*

(a) *for every $\varepsilon > 0$ (J, W) has a Szemerédi partition with error ε into no more than $T(\varepsilon)$ classes, where $T(\varepsilon)$ depends only on ε;*

(b) *for every $\varepsilon > 0$ (J, W) has a weak regularity partition with error ε into no more than $2^{2/\varepsilon^2}$ classes.*

(c) *for every sequence $\mathcal{E} = (\varepsilon_0, \varepsilon_1, \dots)$ of positive numbers, (J, W) has a strong regularity partition of (J, W) with error $\mathcal{E}$ into no more than $T(\mathcal{E})$ classes, where $T(\mathcal{E})$ depends only on $\mathcal{E}$.*

Remark 5.2. (i) We note that every graphon has an ultra-strong partition with error ε by standard results in analysis, but the number of classes cannot be bounded uniformly by any function of ε.

(ii) In the usual formulation, partitions in the Regularity Lemma are equitable, i.e., the partition classes are as equal as possible. For graphons on atomless probability spaces, the classes can be required to have the same measure. In fact, it is easy to see that the partitions constructed e.g. in Corollary 5.4 and Theorem 5.8 below can be repartitioned so that the classes will be as equal as possible, the error is at most doubled, and the number of classes is increased by a factor of at most $\lceil 1/\varepsilon \rceil$.

Several other analytic aspects and versions of the Regularity Lemma were proved in [12]. One of these results made a connection between regularity partitions and partitions of J into sets with small diameter in the $r_{W \circ W}$ metric. Here we prove a stronger, cleaner version of that result, and then show how to combine it with our results on thin graphons to get better bounds on the number of partition classes in weak regularity partitions of this graphons.

5.2. Voronoi cells and regularity partitions

We show that Voronoi cells in the metric spaces (J, R_W) and $(J, R_{W\circ W})$ are intimately related to different versions of the Regularity Lemma.

Let (J, d) be a metric space and let π be a probability measure on its Borel sets. We say that a set $S \subseteq J$ is an *average ε-net*, if $\int_J d(x, S)\, d\pi(x) \leq \varepsilon$.

Let $S \subseteq J$ be a finite set and $s \in S$. The *Voronoi cell* of S with center s is the set of all points $x \in J$ for which $d(x, s) \leq d(x, y)$ for all $y \in S$. Clearly, the Voronoi cells of S cover J. (We can break ties arbitrarily to get a partition.)

Theorem 5.3. *Let (J, W) be a graphon, and let $\varepsilon > 0$.*

(a) *Let S be an average ε-net in the metric space $(S, r_{W\circ W})$. Then the Voronoi cells of S form a weak regularity partition with error at most $8\sqrt{\varepsilon}$.*

(b) *Let $\mathcal{P} = \{J_1, \dots, J_k\}$ be a weak regularity partition with error ε. Then there are points $v_i \in J_i$ such that the set $S = \{v_1, \dots, v_k\}$ is an average (4ε)-net in the metric space $(S, r_{W\circ W})$.*

Proof. (a) Let $\mathcal{P}$ be the partition into the Voronoi cells of S. Let us write $R = W - W_{\mathcal{P}}$. We want to show that $\|R\|_\square \leq 8\sqrt{\varepsilon}$. It suffices to show that for any 0-1 valued function f,

$$\langle f, Rf\rangle \leq 2\sqrt{\varepsilon}. \tag{10}$$

Let us write $g = f - f_{\mathcal{P}}$, where $f_{\mathcal{P}}(x)$ is obtained by replacing $f(x)$ by the average of f over the class of $\mathcal{P}$ containing x. Clearly $\langle f_{\mathcal{P}}, Rf_{\mathcal{P}}\rangle = 0$, and so

$$\langle f, Rf\rangle = \langle g, Rf\rangle + \langle f_{\mathcal{P}}, Rf\rangle = \langle f, Rg\rangle + \langle f_{\mathcal{P}}, Rg\rangle \leq 2\|Rg\|_1 \leq 2\|Rg\|_2. \tag{11}$$

For each $x \in J$, let $\varphi(x) \in S$ be the center of the Voronoi cell containing x, and define $W'(x, y) = W\big(x, \phi(y)\big)$ and similarly $R'(x, y) = R\big(x, \phi(y)\big)$. Then using that $(W - R)g = W_{\mathcal{P}}g = 0$, $W - W' = R - R'$ and $R'g = 0$, we get

$$\begin{aligned}\|Rg\|_2^2 &= \langle Rg, Rg\rangle = \big\langle Wg, (R - R')g\big\rangle \\ &= \big\langle Wg, (W - W')g\big\rangle = \big\langle g, W(W - W')g\big\rangle\end{aligned}$$

$$\leq \|W(W-W')\|_1 = \int_{J^2} \left| \int_J W(x,y)(W(y,z) - W\big(y,\varphi(z)\big)\, dy \right| dx\, dz$$

$$= \int_J r_W\big(z,\varphi(z)\big) = \mathsf{E}_{\mathbf{X}}\big(r_W(\mathbf{X},S)\big) \leq \varepsilon.$$

This proves (10).

(b) Suppose that $\mathcal{P}$ is a weak Szemerédi partition with error ε. Let $R = W - W_{\mathcal{P}}$, then we know that $\|R\|_\square \leq \varepsilon$.

For every $x \in J$, define

$$F(x) = \int_J \left| \int_J R(x,y)W(y,z)\, dy \right| dz.$$

Then we have

$$\int_J F(x)\, dx = \int_{J^3} s(x,z)R(x,y)W(y,z)\, dx\, dy\, dz,$$

where $s(x,z)$ is the sign of $\int R(x,y)W(y,z)\, dy$. For every $z \in J$,

$$\int_{J^2} s(x,z)R(x,y)W(y,z)\, dx\, dy \leq 2\|R\|_\square \leq 2\varepsilon,$$

and hence

$$\int_J F(x)\, dx \leq 2\varepsilon. \tag{12}$$

Let $x, y \in J$ be two points in the same partition class of $\mathcal{P}$. Then $W_{\mathcal{P}}(x,s) = W_{\mathcal{P}}(y,s)$ for every $s \in J$, and hence

$$r_{W\circ W}(x,y) = \int_J \left| \int_J \big(W(x,s) - W(y,s)\big)\, W(s,z)\, ds \right| dz$$

$$= \int_J \left| \int_J \big(R(x,s) - R(y,s)\big)\, W(s,z)\, ds \right| dz$$

$$\leq \int_J \left| \int_J R(x,s)W(s,z)\,ds \right| dz + \int_J \left| \int_J R(y,s)W(s,z)\,ds \right| dz$$

$$= F(x) + F(y).$$

For every set $T \in \mathcal{P}$, let $v_T \in T$ be a point "below average" in the sense that

$$F(v_T) \leq \frac{1}{\pi(T)} \int_T F(x)\,dx,$$

and let $S = \{v_T : T \in \mathcal{P}\}$. Then using (12),

$$\mathsf{E}_{\mathbf{X}} d(\mathbf{X}, S) \leq \sum_{T \in \mathcal{P}} \int_T d(x, v_T)\,dx \leq \sum_{T \in \mathcal{P}} \int_T \big(F(x) + F(v_T)\big)\,dx$$

$$\leq \int_J F(x)\,dx + \sum_{T \in \mathcal{P}} \lambda(T) F(v_T) \leq 2 \int_J F(x)\,dx \leq 4\varepsilon.$$

This proves the Theorem. ∎

Theorems 5.3 and 4.1 imply the following Corollary (we prove a stronger result in the next section).

Corollary 5.4. *For every bigraph $F = (V, E)$ there is a constant $c_F > 0$ such that if G is a graph not containing F as an induced sub-bigraph, then for every $\varepsilon > 0$, G has a weak regularity partition with error ε with at most $c_F \varepsilon^{-2|V|}$ classes.*

Remark 5.5. The conclusion does not remain true if the subgraph we exclude is nonbipartite. Any bipartite graph will then satisfy the condition, and some bipartite graphs are known to need an exponential (in $1/\varepsilon$) number of classes in their weak regularity partitions.

5.3. Edit distance

We conclude with deriving bounds on the size of the Szemerédi partitions and approximations in L_1, using the packing dimension of (J, r_W). In the graph theoretic case, this corresponds to approximation in edit distance.

Lemma 5.6. *Let W be a graphon such that (J, r_W) can be covered by m balls of radius ε. Then there is a stepfunction U with $m(1/\varepsilon)^m$ steps such that $\|W - U\|_1 \le 2\varepsilon$.*

Remark 5.7. If W is 0-1 valued, then the bound on the number of classes can be improved to $m2^m$.

Proof. Let $\mathcal{P} = \{J_1, J_2, \dots, J_m\}$ be a partition of J into measurable sets such that for every i there is $x_i \in J$ with $\big\| W(x_i, .) - W(x, .)\big\|_1 \le \varepsilon$ for every $x \in J_i$. Let $W'(x, y) = W(x_i, y)$ for $x \in J_i$, then trivially $\|W - W'\|_1 \le \varepsilon$. Let $\mathcal{Q}_i$ be a partition of J into $1/\varepsilon$ measurable classes so that $W(x_i, .)$ varies at most ε on each class of $\mathcal{Q}_i$. For $x \in J_i$ and $y \in S \in \mathcal{Q}_i$, define

$$U(x, y) = \frac{1}{\pi(S)} \int_S W'(x, z)\, dz.$$

Then clearly $\big|U(x, y) - W'(x, y)\big| \le \varepsilon$ for all $x, y \in J$, and hence $\|U - W\|_1 \le \|U - W'\|_1 + \|W - W'\|_1 \le 2\varepsilon$. It is obvious that U is a stepfunction in the partition generated by $\mathcal{P}$ and $\mathcal{Q}_1, \dots, \mathcal{Q}_m$, which has at most $m(1/\varepsilon)^m$ classes. ∎

We obtain from this lemma:

Theorem 5.8. *Let W be a graphon such that (J, r_W) has packing dimension d, then for every $0 < \varepsilon < 1$ it has an ultra-strong partition with error ε and with at most $\varepsilon^{-O(\varepsilon^{-d})}$ classes.*

Proof. Consider a maximal packing in (J, r_W) of balls with radius $\varepsilon/8$; this consists of $m = O(\varepsilon^{-d})$ balls. The balls with the same centers and with radius $\varepsilon/4$ cover J, so Lemma 5.6 there is a stepfunction U with $m(4/\varepsilon)^m \le \varepsilon^{-c\varepsilon^{-d}}$ steps such that $\|W - U\|_1 \le \varepsilon/2$. For the partition $\mathcal{P}$ into the steps of U, we have

$$\|W - W_{\mathcal{P}}\|_1 \le 2\|W - U\|_1 \le \varepsilon$$

(the first inequality follows by easy computation). ∎

For thin graphons, we get a stronger bound.

Theorem 5.9. *Let W be a thin graphon in which a bigraph $F = (V, E)$ is excluded as an induced sub-bigraph. Then for every $0 < \varepsilon < 1$, it has an ultra-strong partition with error ε and with $O\big(\varepsilon^{-2|V|^2}\big)$ classes.*

Proof. Theorem 4.1 implies that W is 0-1 valued and it has finite packing dimension at most $2|V|$. Similarly to the proof of lemma 5.6, let $\mathcal{P} = \{J_1, J_2, \dots, J_m\}$ be a partition of J with $m = O(\varepsilon^{-|V|})$ into measurable sets such that for every i there is an $x_i \in J$ with $\big\| W(x_i, .) - W(x, .)\big\|_1 \le \varepsilon$ for every $x \in J_i$. Let $W'(x, y) = W(x_i, y)$ for $x \in J_i$, then $\|W' - W\|_1 \le \varepsilon$. Let S_i be the support of the function $W(x_i, .)$, and let A be the set of atoms of the Boolean algebra generated by $\{S_1, S_2, \dots, S_m\}$ with positive measure. For every atom $a \in A$, let $F_a \subseteq [m]$ denote the index set $\{i \mid a \subseteq S_i\}$ and let $\mathcal{F}$ denote the set system $\{F_a \mid a \in A\}$. Since F is not an induced sub-bigraph, $\mathcal{F}$ has VC-dimension less than $|V|$, and so by lemma 4.3 we obtain that $|A| \le O\big(m^{|V|-1}\big)$. The joint refinement $\mathcal{P}_2$ of A and $\mathcal{P}$ is of size at most $O\big(\varepsilon^{-2|F|^2}\big)$. This completes the proof since W' is a stepfunction with steps in $\mathcal{P}_2$. ■

It is easy to see that in the definition of ultra-strong regularity partitions of 0-1 valued graphons, we can replace $W_{\mathcal{P}}$ by a 0-1 valued stepfunction with the same steps, at the cost of doubling the error. Together with Remark 5.2, we can apply this to a (large) finite graph G. To state the result, we need a definition. Let H be a simple graph, and let us replace each node v of H by a set S_v of "twin" nodes, where two nodes $x \in S_u$ and $y \in S_v$ are connected if and only if $uv \in E(H)$. For each $u \in V(H)$, either connect all pairs of nodes in S_u, or none of them. We call every graph obtained this way a *blow-up* of H.

Corollary 5.10. *For every bigraph F there is a constant $c_F > 0$ such that if G is a graph not containing F as an induced sub-bigraph, then for every $\varepsilon > 0$, we can change $\varepsilon|G|^2$ edges of G so that the resulting graph is a blow-up of a graph with at most $c_F\varepsilon^{-2|F|^2}$ nodes.*

Let us say that a graphon W has *polynomial L_1-complexity* if there is a $d > 0$ such that for every $\varepsilon > 0$ there is a stepfunction W' with $O(\varepsilon^{-d})$ steps satisfying $\|W - W'\|_1 \le \varepsilon$. We can define *polynomial $\square$-complexity* analogously. As we have pointed out, polynomial $\square$-complexity corresponds to the finite dimensionality of the metric space of $W \circ W$. Theorem 5.9 implies that every thin graphon has polynomial L_1-complexity.

If W has polynomial complexity, then the structure of W can be described by a polynomial number (in $1/\varepsilon$) of real parameters with an error ε in the appropriate norm. The set of graphons with polynomial complexity is closed under many natural operations such as operator product, tensor product, etc.

It could be interesting to study other aspects of this complexity notion. We offer a conjecture relating our complexity notion to extremal combinatorics. It is supported by examples in [13].

Conjecture 5.11. Let $F_1, F_2, \dots, F_n$ be a set of finite graphs, $t_1, t_2, \dots, t_m$ be real numbers in $[0,1]$ and $\mathcal{S}$ be the set of graphons W with $t(F_i, W) = t_i$ for $1 \le i \le n$. Then $\mathcal{S}$ is either empty or it contains a graphon of polynomial L_1-complexity.

References

[1] N. Alon, E. Fischer, M. Krivelevich and M. Szegedy, Efficient testing of large graphs, *Combinatorica,* **20** (2000), 451–476.

[2] C. Borgs, J. Chayes and L. Lovász, Moments of Two-Variable Functions and the Uniqueness of Graph Limits, *Geometric and Functional Analysis* (to appear). `http://www.cs.elte.hu/~lovasz/limitunique.pdf`

[3] C. Borgs, J. Chayes, L. Lovász, V. T. Sós and K. Vesztergombi, Counting graph homomorphisms, in: *Topics in Discrete Mathematics* (ed. M. Klazar, J. Kratochvil, M. Loebl, J. Matoušek, R. Thomas, P. Valtr), Springer (2006), 315–371.

[4] C. Borgs, J. T. Chayes, L. Lovász, V. T. Sós, B. Szegedy and K. Vesztergombi, Graph Limits and Parameter Testing, Proc. 38th Annual ACM Symp. on Theory of Computing 2006, 261–270.

[5] C. Borgs, J. T. Chayes, L. Lovász, V. T. Sós and K. Vesztergombi, Convergent Graph Sequences I: Subgraph frequencies, metric properties, and testing, *Advances in Math.* (2008), 10.1016/j.aim.2008.07.008.

[6] P. Erdős, L. Lovász and J. Spencer, Strong independence of graphcopy functions, in: *Graph Theory and Related Topics,* Academic Press (1979), 165–172.

[7] M. Freedman, L. Lovász and A. Schrijver, Reflection positivity, rank connectivity, and homomorphisms of graphs, *J. Amer. Math. Soc.,* **20** (2007), 37–51.

[8] A. Frieze and R. Kannan, Quick approximation to matrices and applications, *Combinatorica,* **19**, 175–220.

[9] J. Komlós, J. Pach and G. Woeginger, Almost Tight Bounds for epsilon-Nets, *Discr. Comput. Geometry,* **7** (1992), 163–173.

[10] L. Lovász, V.T. Sós, Generalized quasirandom graphs, *J. Comb. Th. B,* **98** (2008), 146–163.

[11] L. Lovász, B. Szegedy, Limits of dense graph sequences, *J. Comb. Theory B,* **96** (2006), 933–957.

[12] L. Lovász and B. Szegedy, Szemerédi's Lemma for the analyst, *Geom. Func. Anal.,* **17** (2007), 252–270.

[13] L. Lovász and B. Szegedy, Finitely forcible graphons (submitted). `http://arxiv.org/abs/0901.0929`

[14] E. Szemerédi, On sets of integers containing no k elements in arithmetic progression, *Acta Arithmetica,* **27** (1975) 199-245.

[15] E. Szemerédi, Regular partitions of graphs, *Colloque Inter. CNRS* (J.-C. Bermond, J.-C. Fournier, M. Las Vergnas and D. Sotteau, eds.) (1978) 399–401.

[16] V. Vapnik and A. Chervonenkis, On the uniform convergence of relative frequencies of events to their probabilities, *Theor. Prob. Appl.,* **16** (1971), 264–280.

László Lovász
Eötvös University
Department of Computer Science
Pázmány Péter sétány 1/C
Budapest 1117, Hungary
e-mail: lovasz@cs.elte.hu

Balázs Szegedy
University of Toronto
Department of Mathematics
40 St George St.
Toronto, ON, M5S 2E4, Canada
e-mail: szegedy@utsc.utoronto.ca

BOLYAI SOCIETY
MATHEMATICAL STUDIES, 21

An Irregular Mind
(Szemerédi is 70)
pp. 447–490.

Extremal Problems for Sparse Graphs

JAROSLAV NEŠETŘIL* and PATRICE OSSONA DE MENDEZ

Dedicated to Endre Szemerédi on the (unbelievable) occasion of his seventieth birthday

We survey some of the recent results related to the study of sparse graphs using the nowhere dense – somewhere dense dichotomy. Particularly we extend known results related to property testing, sublinear expanders, Ramsey numbers and FO model checking. All this is done under the same umbrella of nowhere dense and bounded expansion classes in many of their incarnations.

We concentrate on extremal (mostly graph theory) results leaving algorithmic and structural aspects to other occasions.

1. Introduction

Consider the following (admittedly too general and imprecise) problems:

1. Does a graph have a small separator?

2. Do graphs present a good model for the neighborhood statistics?

3. Are the ("generalized") Ramsey numbers of graphs linear?

4. Is there a simple procedure to check the validity of a First Order formula on graphs?

Although these are questions from various areas and none of them uses the word "extremal", in fact all of these questions may be viewed as various

*Supported by grant 1M0545 of the Czech Ministry of Education.

instances of extremal problems: they ask for special partitions, homomorphism and asymptotic. It is also easy to see that in all of these problems we have to put some restriction on the graphs to be considered: in the full generality for finite graphs (and in case **4** for structures) the answers to all of our questions are known to be negative. But not only that the answer tends to be negative even for graphs which have many (or "moderately many") edges, what is usually described by the term dense graphs. In the context of this paper dense graphs have not only $O(n^2)$ edges but even $O(n^{1+\varepsilon})$ edges (n is the number of vertices). Even such edge sizes do not guarantee a positive answer to the above questions.

Dense graphs have been extensively studied in the context of extremal graph theory. The prime example here is the outstanding Szemerédi Regularity Lemma (shortly SRL) [88] which states that any dense network has properties which are close to the ones of a random graph. In particular, a large dense network cannot be too irregular. This structural result is one of the cornerstones of contemporary combinatorics (and one would like to say of mathematics in general). It also led to manifold applications and generalizations, see e.g. [51, 50, 62, 90, 35]. We believe that this result and various aspects of it are covered on many places in this volume. Thus we can be brief. But we want to mention explicitly that SRL was a motivation for a large part of our research covered in this paper.

For positive answers to the above problems we have to look at instances with very few edges. For example Ramsey problem (**3**) has a positive answer for bounded degree graphs (see [24]) and it is an old problem whether the degeneracy (or the Maximal Average Degree, MAD) suffices here. We present results on this in Section 5. In contrast, the Separator Problem (i.e. **1**) has the negative answer even for cubic graphs. In this later case the answer is positive for planar graphs [57], for graphs with a fixed genus [41] and, more generally, for graphs which exclude a fixed minor [6, 5]. This aspect will be treated in a more general setting in Section 6. Also for other three problems the positive answer requires special instances which could be described as sparse (for example, in Problem **3** one considers bounded degree planar graphs, see [15, 36]).

It seems that the very notion of sparsity of graphs depends on the particular problem considered. Yet in this paper we present a classification of graph classes which clarifies the boundary between sparse and dense instances and which proved to be useful in many concrete applications and, particularly, in all of the above problems. For us, a class of graphs will always mean a class of non-isomorphic graphs.

As such sparsity seems to be an elusive and typically "fuzzy" notion. Perhaps the best way to define it is by means of stability with respect to some operations. This approach is taken in this paper. We employ the mixture of geometric and combinatorial approach and define (time) resolution of a graph and of a class of graphs. This in turn leads to the surprisingly general dichotomy of classes of graphs – there are nowhere dense classes and classes which are somewhere dense. This is stated in Section 3 after we introduce all the relevant notions in Section 2. In Section 4 we show how this dichotomy, which may appear on the first glance arbitrary, can be described in several very different ways. In fact all the basic extremal combinatorial parameters seem to be suitable for the description of this dichotomy. Thus in Section 3.1 we deal with edge densities, in Section 4.1 with the chromatic number χ, in Section 4.2 with the independence number α, in Section 4.3 with orderings and coloring numbers. All this shows that the nowhere dense – somewhere dense dichotomy is not just accident or a combinatorial curiosity but rather a natural dichotomy which is stable and robust.

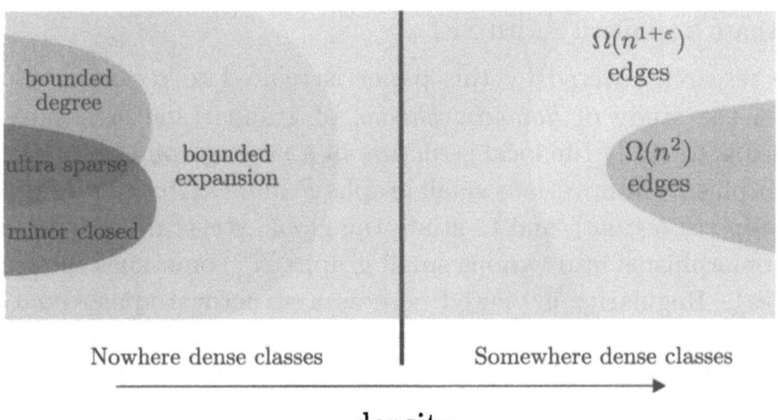

There is further evidence which goes beyond the α, χ, ω. Very recently this list was complemented by the counting densities (which can be seen as a possible answer to the above Problem **2**; see Section 10) and also by the results in mathematical logic: the nowhere dense – somewhere dense dichotomy relates to the dividing line between (hereditary) classes of graphs for which the model checking problem for First Order formulas is Fixed Parameter Tractable and those classes for which the problem is hard (and in fact with undecidable theory), see [27, 30]. This provides a nearly satisfactory answer to Problem **4**.

Some of these applications will be mentioned in Section 5, 6, 7 in greater detail. The core of these applications is a possibility to approximate (with arbitrary precision) any graph in a nowhere dense class by a graph defined by finitely many data. Technically this takes the form of Low Tree Depth Decomposition which, as an illustration, we formulate here for the case of a bounded expansion class $\mathcal{C}$:

Theorem 1 (Low Tree Depth Decomposition – LTDD). *For every bounded expansion class $\mathcal{C}$ and for every positive integer p there exists an integer $N = N(p, \mathcal{C})$ such that every graph $G = (V, E) \in \mathcal{C}$ has a decomposition $V = V_1 \cup \ldots \cup V_N$ with the following property:*

G restricted to any set $\bigcup(V_i : i \in I)$ where $|I| \leq p$, has tree depth at most $|I|$.

(See Section 4.1 for more details.)

As there are only finitely many core graphs with tree depth $\leq p$, a Low Tree Depth Decomposition can be approximated by a finite set of data and this decomposition is much in the spirit of SRL. Also many applications of LTDD share a similarity with SRL.

The research covered by this paper is related to recent developments based on the study of *homomorphisms* of graphs (and structures). The main idea is to study the local structure of a large graph G by counting the homomorphisms from various small graphs F into G (this relates to the area called *property testing*), and to study the global structure of G by *counting* its homomorphisms into various small graphs H (sometimes interpreted as *templates*). Regularity is viewed here as a structural approximation in a proper metric and also as a convergence. For a survey of this development see [20]. This approach proved to be very fruitful and relates (among others) to the notion of *quasi-random* graphs, see e.g. [23], and to the full characterizations of testable graph properties, see e.g. [7, 20].

In this paper we take a different approach via the homomorphism order. We shall see that in this setting, at a proper level of generality, some of the results for dense graphs can be extended to the world of sufficiently sparse classes of graphs. Moreover, in Section 10 (and also in Section 4.4) we shall see that these two approaches are actually not so different and that the nowhere dense – somewhere dense dichotomy can be characterized by counting functions.

Although in most of this paper we deal with graphs, the results can be extended to oriented graphs, hypergraphs and to general relational structures by means of appropriate constructions (such as Gaifman graph, 2-section, or bigraph of incidence).

2. Geometric Derivation – Graph Resolution

2.1. Resolution in time

Let us review some basic notions which will be used.

We work with finite simple undirected graphs, except when explicitly stated otherwise and we denote by $\mathcal{G}$ the class of all such graphs.

We use standard graph theory terminology. However we find it useful to introduce the following: for a graph $G = (V, E)$, we denote by $|G|$ the *order* of G (that is: $|V|$) and by $\|G\|$ the *size* of G (that is: $|E|$).

The *distance* in a graph G between two vertices x and y is the minimum length of a path linking x and y (or ∞ if x and y do not belong to the same connected component of G) and is denoted by $\mathrm{dist}_G(x, y)$.

A class $\mathcal{C}$ of graphs is *hereditary* if every induced subgraph of a graph in $\mathcal{C}$ belongs to $\mathcal{C}$, and it is *monotone* if every subgraph of a graph in $\mathcal{C}$ belongs to $\mathcal{C}$.

As mentioned above, the notion of sparsity of graphs is a fuzzy notion. First it does not relate to any particular graph but rather to a set, or sequence, or a class of graphs. Secondly the notion should be (certainly from naive point of view) invariant to some small changes in a graph. Third, to be a sparse graph is clearly a global property and the property should be hereditary. Combining these observations and motivated by numerous particular cases we are led to the following definitions:

For any graphs H and G and any integer d, the graph H is said to be a *shallow minor* of G at *depth* d ([78] attribute this notion, called then *low depth minor* to Ch. Leiserson and S. Toledo) if there exists a subset $\{x_1, \ldots, x_p\}$ of G and a collection of disjoint subsets $V_1, \ldots, V_p$ of vertices of G, each inducing a connected subgraph of G, such that $x_i \in V_i$, every vertex in V_i is at distance at most d from x_i in the subgraph of G induced by V_i, and so that H is a subgraph of the graph obtained from G by contracting each V_i into x_i and removing loops and multiple edges (see Fig. 1).

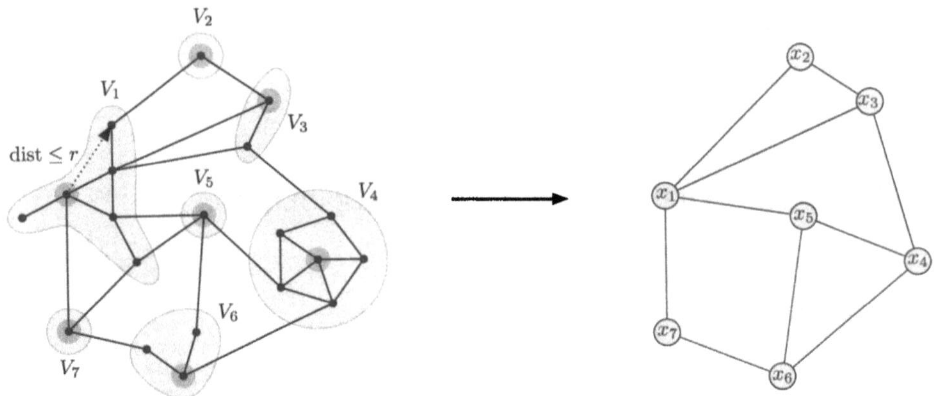

Fig. 1. A shallow minor of depth r of a graph G is a simple subgraph of a minor of G obtained by contracting vertex disjoint subgraphs with radius at most r

The set of all shallow minors of G at depth d is denoted by $G \triangledown d$. In particular, $G \triangledown 0$ is the set of all subgraphs of G. Hence we have the following non decreasing sequence of classes:

$$G \in G \triangledown 0 \subseteq G \triangledown 1 \subseteq \cdots \subseteq G \triangledown d \subseteq \cdots \subseteq G \triangledown \infty.$$

We extend this definition to arbitrary class of graphs $\mathcal{C}$ by:

$$\mathcal{C} \triangledown d = \bigcup_{G \in \mathcal{C}} G \triangledown d.$$

Consequently, we have the following hierarchy of classes

$$\mathcal{C} \subseteq \mathcal{C} \triangledown 0 \subseteq \mathcal{C} \triangledown 1 \subseteq \cdots \subseteq \mathcal{C} \triangledown d \subseteq \cdots \subseteq \mathcal{C} \triangledown \infty.$$

We call this sequence *minor resolution* of the class $\mathcal{C}$ and denote it by $\mathcal{C}^{\triangledown}$. Note that $\mathcal{C} \triangledown 0$ is the monotone closure of $\mathcal{C}$ and that $\mathcal{C} \triangledown \infty$ is the minor closed class generated by $\mathcal{C}$.

2.2. The Nowhere Dense – Somewhere Dense Dichotomy

The minor resolution of a class leads to a classification of classes and to their interesting properties. The following are the key definitions of this paper:

Definition 1. An infinite class of graphs $\mathcal{C}$ is *somewhere dense* if there exists an integer d such that $\mathcal{C} \triangledown d = \mathcal{G}$. In other words: $\mathcal{C}$ is somewhere dense if every graph is a bounded depth shallow minor of a graph in $\mathcal{C}$.

If an infinite class is not somewhere dense, it is *nowhere dense.*

It follows directly from the definition of the minor resolution that a class $\mathcal{C}$ is nowhere dense if and only if for every i the supremum of $\omega(G)$ for $G \in \mathcal{C} \triangledown i$ is finite (here $\omega(G)$ is the the *clique number* of graph G, i.e. the maximal order of a complete graph in G). (It is perhaps surprising, as we shall see in Section 3.3, that nowhere dense classes may be defined by their independence number as well.)

For relational structures we can define analogous notions using incidence graphs (or Gaifman graphs). Of course, for oriented graphs we can consider the underlying undirected graphs. For the sake of simplicity, in this paper we illustrate our results mostly on classes of undirected graphs.

We think of the graph resolution as evolving in time. Comparing the classes in the resolution $\mathcal{C}^{\triangledown}$ of a class $\mathcal{C}$ with the whole class $\mathcal{G}raph$ of all simple graphs, three different situations may occur.

The first possibility is that there exists a time $\tau = \tau(\mathcal{C})$ such that $\mathcal{C} \triangledown \tau = \mathcal{C} \triangledown \infty = \mathcal{G}raph$:

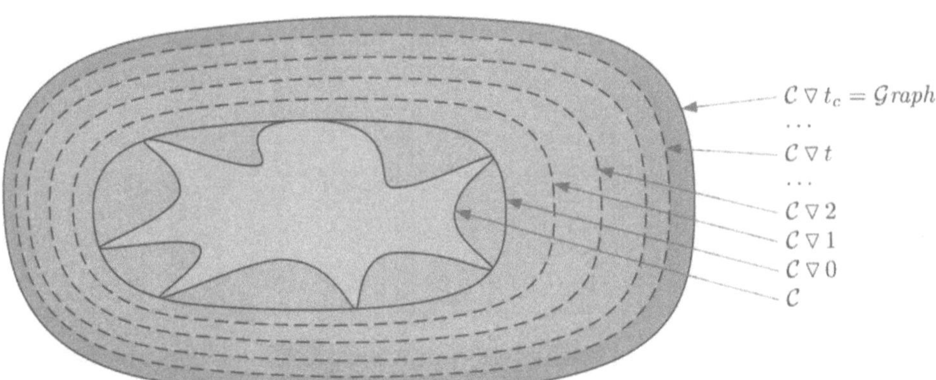

This is the case when $\mathcal{C}$ is somewhere dense. Such a situation suggests that properties of the class $\mathcal{C}$ could be obtained by transporting general properties of dense graphs backwards.

The second possibility is that the class $\mathcal{C} \triangledown \infty$ is equal to $\mathcal{G}raph$, although $\mathcal{C} \triangledown t$ is strictly included in $\mathcal{G}raph$ for each t:

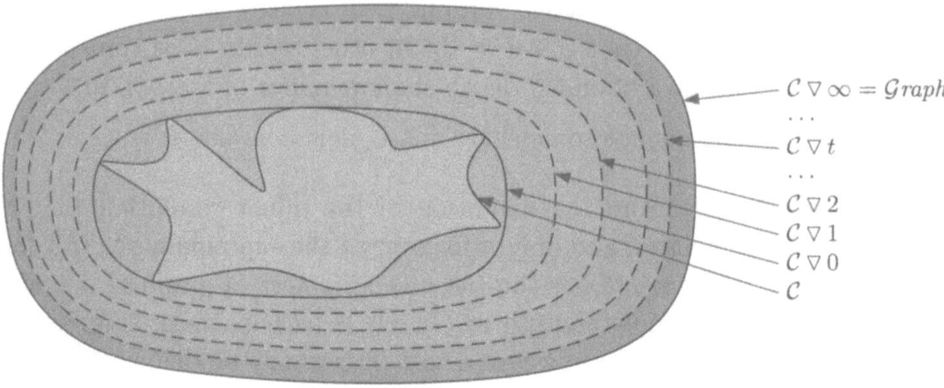

This is the case when $\mathcal{C}$ is nowhere dense but not included in a proper minor closed class. Such a situation allows us to "measure" every graph in the light of the minor resolution of $\mathcal{C}$, by associating to each graph G the minimum integer t such that $G \in \mathcal{C} \triangledown t$. The interest of this scaling stands in its universality (it applies to every graph) and its extension (its set of values is not finite).

The third possibility is that the class $\mathcal{C} \triangledown \infty$ is strictly included in $\mathcal{G}raph$:

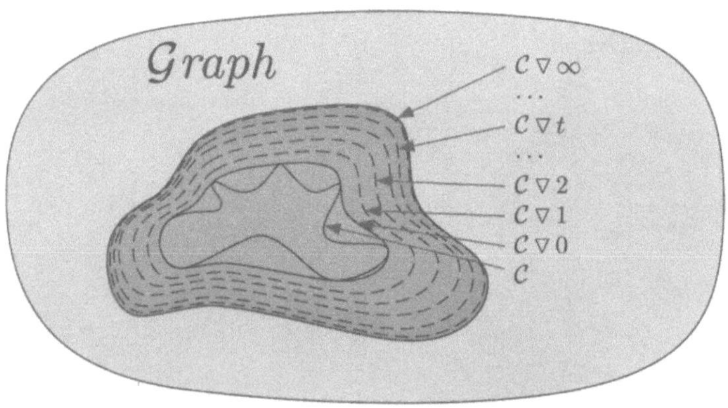

This is the case when $\mathcal{C}$ is included in a proper minor closed class. The minor resolution may then be used to get finer informations about a subclass with smaller density.

3. Trichotomy

3.1. Classification by Edge Densities

Let $\mathcal{C}$ be an infinite class of graphs and let $f : \mathcal{C} \to \mathbb{R}$ be a graph invariant. Let $\operatorname{Inj}(\mathbb{N}, \mathcal{C})$ be the set of all injective mappings from $\mathbb{N}$ to $\mathcal{C}$. Then we define:

$$\limsup_{G \in \mathcal{C}} f(G) = \sup_{\phi \in \operatorname{Inj}(\mathbb{N},\mathcal{C})} \limsup_{i \to \infty} f\big(\phi(i)\big)$$

Notice that $\limsup_{G \in \mathcal{C}} f(G)$ always exists and is either a real number or $\pm\infty$.

Theorem 2 (Trichotomy theorem). *Let $\mathcal{C}$ be an infinite class of graphs (asymptotically not all edgeless). Then the limit*

$$\overline{\ell\mathrm{dens}}\,(\mathcal{C}^{\triangledown}) = \lim_{i \to \infty} \limsup_{G \in \mathcal{C} \triangledown i} \frac{\log \|G\|}{\log |G|}$$

may take only three values, namely 0, 1 *and* 2. *Moreover, we have:*

$$\overline{\ell\mathrm{dens}}\,(\mathcal{C}^{\triangledown}) = \begin{cases} 0, & \text{iff } \sup_{G \in \mathcal{C}} \|G\| < \infty, \\ 0 \text{ or } 1, & \text{iff } \mathcal{C} \text{ is nowhere dense,} \\ 2, & \text{iff } \mathcal{C} \text{ is somewhere dense.} \end{cases}$$

For a proof see [75]. It can be seen easily that $\overline{\ell\mathrm{dens}}\,(\mathcal{C}^{\triangledown}) = 0$ if and only if the class $\mathcal{C}$ contains only graphs with at most k_0 edges. These *essentially finite* classes are interesting. A prime example is the class of all core graphs with tree depth bounded (see Section 3.1 for the definition of the tree depth).

3.2. Classifications by topological resolution

A graph G' is a *subdivision* of a graph G if G' arises from G by adding vertices (of degree 2) on the edges of G. Thus in the "topological sense" we have the same graph: all edges of G are replaced by simple openly disjoint paths. If all these paths have length $\leq 2d+1$ we say that G' is a p-shallow subdivision of G. Conversely, we say that H is *topological shallow minor at depth* d of a graph G if there exists a subgraph H' of G such that H' is a shallow subdivision of H at depth d.

Having defined this we can proceed similarly as for the shallow minors and define the notion of topological minor resolution:

The set of all topological shallow minors of G at depth d is denoted by $G \mathbin{\widetilde{\nabla}} d$. In particular, $G \mathbin{\widetilde{\nabla}} 1$ is the set of all subgraphs of G. Hence we have the following non decreasing sequence of classes:

$$G \in G \mathbin{\widetilde{\nabla}} 0 \subseteq G \mathbin{\widetilde{\nabla}} 1 \subseteq G \mathbin{\widetilde{\nabla}} 2 \subseteq \cdots \subseteq G \mathbin{\widetilde{\nabla}} d \subseteq \cdots \subseteq G \mathbin{\widetilde{\nabla}} \infty.$$

We extend this definition to arbitrary graph class $\mathcal{C}$ by:

$$\mathcal{C} \mathbin{\widetilde{\nabla}} d = \bigcup_{G \in \mathcal{C}} G \mathbin{\widetilde{\nabla}} d.$$

Consequently, we have the following hierarchy of graph classes

$$\mathcal{C} \subseteq \mathcal{C} \mathbin{\widetilde{\nabla}} 0 \subseteq \mathcal{C} \mathbin{\widetilde{\nabla}} 1 \subseteq \mathcal{C} \mathbin{\widetilde{\nabla}} 2 \subseteq \cdots \subseteq \mathcal{C} \mathbin{\widetilde{\nabla}} d \subseteq \cdots \subseteq \mathcal{C} \mathbin{\widetilde{\nabla}} \infty.$$

We call this sequence *topological minor resolution* of class $\mathcal{C}$ and denote it by $\mathcal{C}^{\widetilde{\nabla}}$. Note that $\mathcal{C} \mathbin{\widetilde{\nabla}} 1$ is the monotone closure of $\mathcal{C}$ and that $C \nabla \infty$ is the topological minor closed class generated by $\mathcal{C}$.

Theorem 3 (Trichotomy theorem (topological version)). *Let $\mathcal{C}$ be an infinite class of graphs (asymptotically not all edgeless). Then the limit*

$$\overline{\ell\mathrm{dens}}\,(\mathcal{C}^{\widetilde{\nabla}}) = \lim_{i \to \infty} \limsup_{G \in \mathcal{C} \widetilde{\nabla} i} \frac{\log \|G\|}{\log |G|}$$

may only take three values, namely $0, 1$ *and* 2. *Moreover, we have:*

$$\overline{\ell\mathrm{dens}}\,(\mathcal{C}^{\widetilde{\nabla}}) = \begin{cases} 0, & \textit{iff } \sup_{G \in \mathcal{C}} \|G\| < \infty, \\ 0 \text{ or } 1, & \textit{iff } \mathcal{C} \textit{ is nowhere dense}, \\ 2, & \textit{iff } \mathcal{C} \textit{ is somewhere dense}. \end{cases}$$

Surprisingly, not only the limit

$$\lim_{i \to \infty} \limsup_{G \in \mathcal{C} \widetilde{\nabla} i} \frac{\log \|G\|}{\log |G|}$$

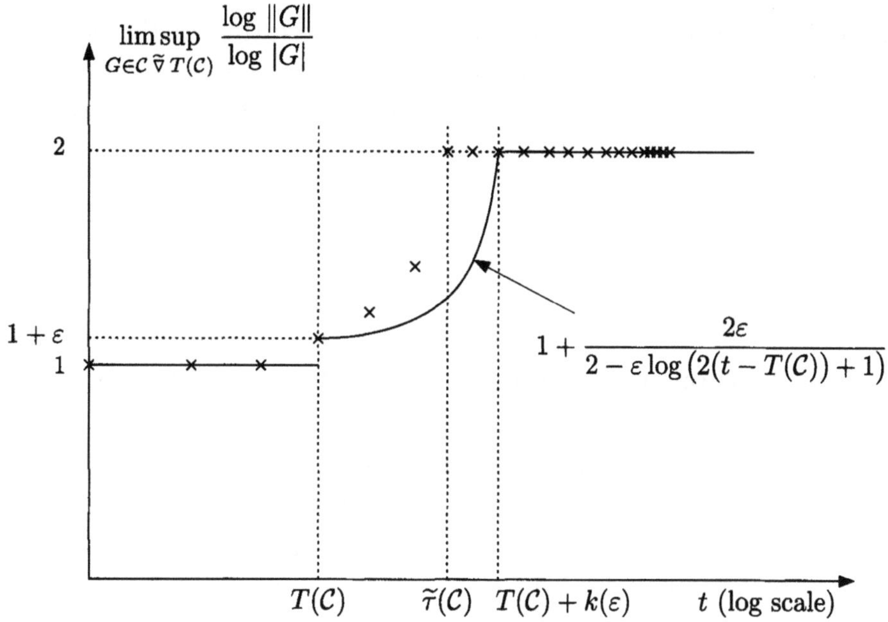

Fig. 2. Evolution of the upper logarithmic density $\overline{\ell\mathrm{dens}}\,(\mathcal{C}^{\widetilde{\nabla}})$ of the topological resolution of a typical somewhere dense class $\mathcal{C}$

can take only three values (namely $0, 1$ and 2) but this limit is the same as the one defined by the minor resolution. For a proof of Theorem 3 see [75]. (This extends work of Zdeněk Dvořák [28, 29].)

Also, the property that there exists a critical value $\widetilde{\tau}(\mathcal{C})$ at which the topological resolution stabilizes to $\mathcal{G}raph$ is equivalent to the existence of a critical value $\tau(\mathcal{C})$ at which the minor resolution stabilizes to $\mathcal{G}raph$. Notice that, according to Theorem 3, the existence of a critical value $\widetilde{\tau}(\mathcal{C})$ is equivalent to the existence of a value $T(\mathcal{C})$ such that there exists $\varepsilon > 0$ with

$$\limsup_{G\in\mathcal{C}\widetilde{\nabla}T(\mathcal{C})} \frac{\log\|G\|}{\log|G|} = 1+\varepsilon.$$

Moreover, the difference between $\widetilde{\tau}(\mathcal{C})$ and $T(\mathcal{C})$ is actually bounded by a function of ε (see Fig .2).

Note that for nowhere dense classes, the asymptotic behavior of the resolution varies. For instance, the class $\mathcal{D}_3$ of graphs with maximum degree at most 3 is such that $\mathcal{D}_3 \nabla \infty = \mathcal{G}raph$ but $\mathcal{D}_3 \,\widetilde{\nabla}\, \infty = \mathcal{D}_3$.

Why do we state this topological variant of shallow minors, when we then claim just analogous results? The main reason is that this connection is surprising and non-trivial. The fact that minors and topological minors lead to the same classification of classes is interesting in the context of graph-minor theory where minors and topological minors lead often to very different results (as demonstrated for example by Hajós and Hadwiger's conjectures), see [75, 71] for more details.

4. Equivalences – Everything Goes

4.1. Classification by Decomposition – Chromatic Numbers

The building blocks of our decompositions will be induced by trees.

A *rooted forest* is a disjoint union of rooted trees. The *height* of a vertex x in a rooted forest F is the number of vertices of the path from the root (of the tree to which x belongs to) to x and is noted height (x, F). The *height* of F is the maximum height of the vertices of F. Let x, y be vertices of F. The vertex x is an *ancestor* of y in F if x belongs to the path linking y and the root of the tree of F to which y belongs to. The *closure* clos (F) of a rooted forest F is the graph with vertex set $V(F)$ and edge set $\big\{\{x, y\} : x$ is an ancestor of y in F, $x \neq y\big\}$. A rooted forest F defines a partial order on its set of vertices: $x \leq_F y$ if x is an ancestor of y in F. The comparability graph of this partial order is obviously clos (F).

The *tree-depth* td (G) of a graph G is the minimum height of a rooted forest F such that $G \subseteq$ clos (F) [67] (see Fig. 3).

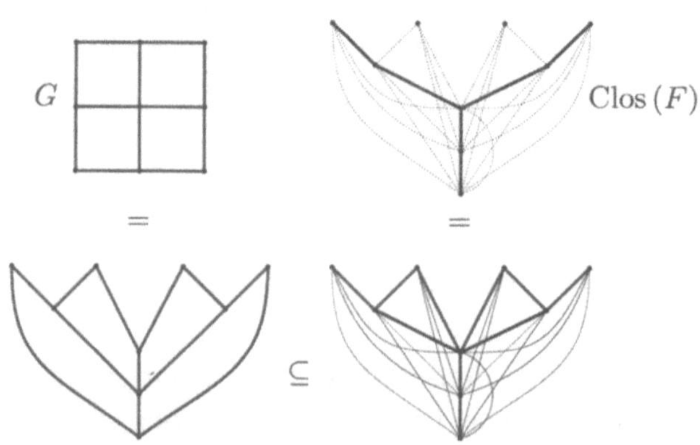

Fig. 3. The tree-depth of the 3 × 3 grid is 4

A principal property of the class of all graphs with td $(G) \leq k$ is that this class is finite when restricted to *core* graphs (or core structures, see e.g. [47]). This holds more generally for colored graphs and for relational structures in general. This has also a number of consequences. For example the class of all graphs with $\mathrm{td}(G) \leq k$ is well quasi ordered with respect to induced subgraph ordering. Nevertheless one should remark that the number of core graphs with $\mathrm{td}(G) \leq k$ has an Ackermann growth.

In [67] we introduced the following parametrized generalization of the chromatic number: for any integer p, $\chi_p(G)$ denotes the minimum number of colors one shall use to color the vertices of G in such a way that for every subset I of at most p colors, the subgraph G_I of G induced by the vertices with color in I has tree-depth at most $|I|$. Thus χ_1 is the usual chromatic number of a graph (i.e. no edge is monochromatic) and χ_2 is a minimal coloring with the property that no path with 4 vertices gets less than 3 colors.

These generalized chromatic numbers characterize nowhere dense classes ([69, 75]):

Theorem 4. *Let $\mathcal{C}$ be an infinite class of graphs. Then the following conditions are equivalent:*

- *$\mathcal{C}$ is nowhere dense,*
- *for every integer p,* $\limsup_{G\in\mathcal{C}} \dfrac{\log \chi_p(G)}{\log |G|} = 0$

Thus any graph G in a (fixed) nowhere dense class $\mathcal{C}$ can be decomposed into a small number of classes such that the subgraphs induced by any $\leq p$ classes of the partition have components of only finitely many (homomorphism) types. Such a decomposition is called *Low Tree Depth Decomposition – LTDD* and it has been mentioned in the Introduction. The parameter p expresses the precision of the decomposition. Moreover such a low tree-depth decomposition can be found in almost linear number of steps. This has a number of algorithmic consequences which are not covered here, see [66, 69].

4.2. Classification by Independence

In the context of relativizations of first-order homomorphism preservation theorems to specific classes of structures, Anush Dawar [26] introduced the following notion of quasi-wideness:

Let $r \geq 1$ be an integer. A subset A of vertices of a graph G is r-*independent* if the distance between any two distinct elements of A is strictly greater than r. Note that if we denote by $\alpha_r(G)$ the maximum size of an r-independent set of G, then $\alpha_1(G)$ is the usual independence number $\alpha(G)$ of graph G.

A class of graphs $\mathcal{C}$ is *quasi-wide* if there is a function $s : \mathbb{N} \to \mathbb{N}$ such that for every integers d and m, every sufficiently big graph $G \in \mathcal{C}$ (i.e. of order at least $F(d, m)$) contains a subset S of size at most $s = s(d)$ so that the graph $G - S$ contains a d-independent set of vertices of size at least m (see Fig. 4).

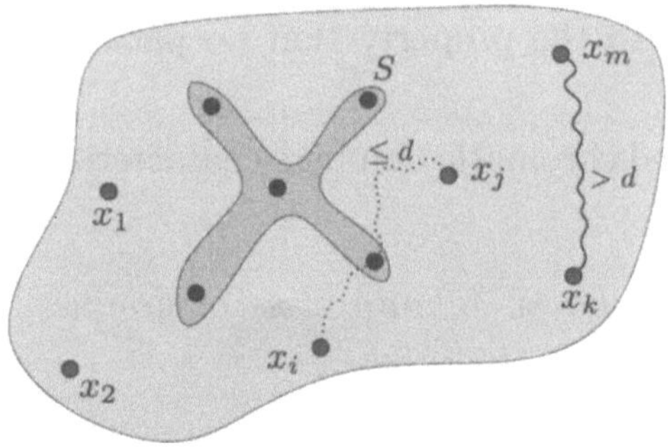

Fig. 4. For every m, every sufficiently large $G \in \mathcal{C}$ contains a subset S of size at most $s(d)$ so that $G - S$ has a d-independent set of size m

The quasi-wide property is not hereditary. Thus we introduce the following, stronger version:

A class of graphs $\mathcal{C}$ is *uniformly quasi-wide* if there is a function $s : \mathbb{N} \to \mathbb{N}$ such that for every integers d and m, every sufficiently big subset A of vertices of a graph $G \in \mathcal{C}$ (i.e. such that $|A| \geq F(d, m)$) is such that G contains a subset S of size at most $s = s(d)$ so that $G - S$ contains a d-independent set of vertices of size at least m included in A (see Fig. 5)

It appears that uniform quasi-wideness is strongly related to our classification:

Theorem 5. *Let $\mathcal{C}$ be an infinite class of graphs. Then the following conditions are equivalent:*

- *$\mathcal{C}$ is nowhere dense,*
- *the hereditary closure of $\mathcal{C}$ is quasi-wide,*
- *$\mathcal{C}$ is uniformly quasi-wide.*

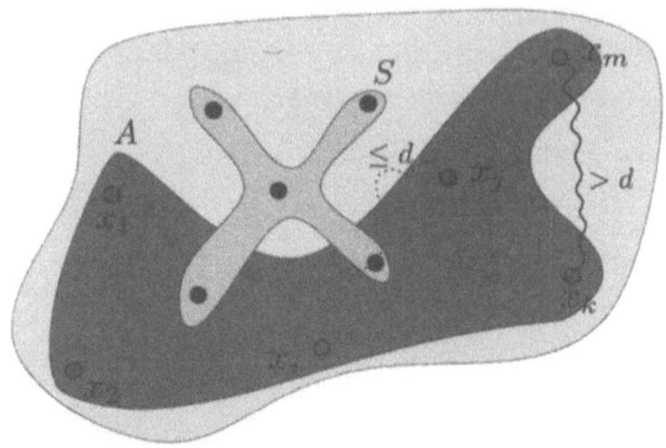

Fig. 5. For every m, every sufficiently large subset A of a graph $G \in \mathcal{C}$ includes a d-independent set of size m after the deletion of at most $s(d)$ vertices of G

This is a non-trivial result with several consequences, see [72].

4.3. Classification by Vertex Ordering

As a generalization of notions of arrangeability and coloring number Kierstead and Yang introduced in [49] two new series of invariants col_k and wcol_k, that is: the *coloring number* of rank k and the *weak coloring number* of rank k.

Let L be a linear order on the vertex set of a graph G, and let x, y be vertices of G. We say y is *weakly k-accessible from* x if $y <_L x$ and there exists an x-y-path P of length at most k (i.e. with at most k edges) with minimum vertex y with respect to $<_L$ (see Fig. 6). The vertex y is *k-accessible* from x if $y <_L x$ and there exists an x-y-path P of length at most k with minimum vertex y and second minimum vertex x with respect to $<_L$.

Let $Q_k(G, L, x)$ and $R_k(G, L, x)$ be the sets of vertices that are respectively weakly k-accessible and k-accessible from x:

$$Q_k(G, L, x) = \{y : \exists x\text{–}y \text{ path } P \text{ such that } \min P = y\}$$

$$R_k(G, L, x) = \big\{y : \exists x\text{–}y \text{ path } P \text{ such that } \min P = y \text{ and } \min(P - y) = x\big\}$$

The *weak k-coloring number* $\mathrm{wcol}_k(G)$ and the *k-coloring number* $\mathrm{col}_k(G)$ of G are defined by:

$$\mathrm{wcol}_k(G) = 1 + \min_L \max_{v \in V(G)} \big|Q_k(G, L, v)\big|,$$

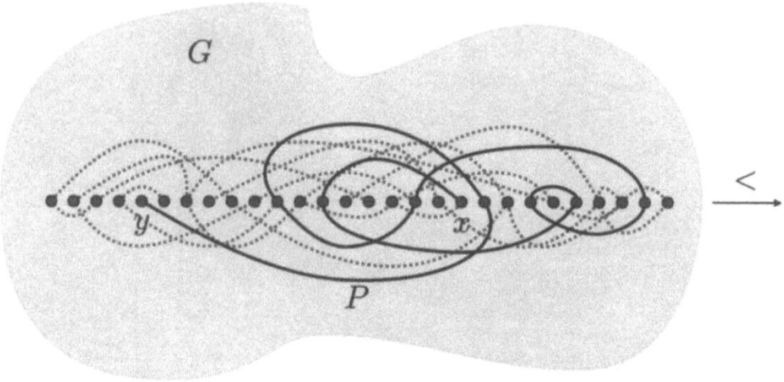

Fig. 6. The vertex y is weakly 8-accessible from x

$$\mathrm{col}_k(G) = 1 + \min_L \max_{v \in V(G)} \left| R_k(G, L, v) \right|.$$

Of course for $k = 1$ we have $\mathrm{wcol}_k(G) = \mathrm{col}_k(G)$ in which case these two notions are equal 1+ the degeneracy of graph G.

It is easy to see ([49]) that these two graph invariants are polynomially dependent:

$$\mathrm{col}_k(G) \leq \mathrm{wcol}_k(G) \leq \left(\mathrm{col}_k(G)\right)^k$$

These parameters form two non-decreasing sequences. The sequence of weak-coloring numbers has the tree-depth as its maximum:

$$\begin{aligned} \mathrm{col}\,(G) &= \mathrm{wcol}_1(G) \leq \mathrm{wcol}_2(G) \leq \cdots \leq \mathrm{wcol}_k(G) \leq \cdots \leq \mathrm{wcol}_\infty(G) \\ &= \mathrm{td}\,(G). \end{aligned}$$

Generalized coloring numbers are strongly related to the maximum density of shallow minors: it has been proved by X. Zhu that there exists polynomials F_k such that the following holds:

Theorem 6 ([91]). *For every half integer*[1] *k and every graph G:*

$$1 + \nabla_k(G) \leq \mathrm{wcol}_{2k+1}(G) \leq F_{2k+1}\big(\nabla_k(G)\big)$$

This polynomial dependence leads to the following characterization:

Theorem 7. *Let $\mathcal{C}$ be an infinite class of graphs. Then the following conditions are equivalent:*

[1] When k is not an integer, an appropriate generalization of $G \triangledown k$ has to be used [76, 91].

- *$\mathcal{C}$ is nowhere dense,*
- *for every integer p,* $\limsup_{G \in \mathcal{C}} \dfrac{\log \mathrm{col}_p(G)}{\log |G|} = 0,$
- *for every integer p,* $\limsup_{G \in \mathcal{C}} \dfrac{\log \mathrm{wcol}_p(G)}{\log |G|} = 0.$

4.4. Classification by Counting

The trichotomy theorem (Theorem 2) is related to counting the numbers of copies of K_2 in a graph. This may be extended (using the decomposition theorem) if we consider homomorphism or induced copies of any non-trivial graph F. (Recall that $\hom(F, G)$ denotes the number of homomorphisms from F to G and that $(\#F \subseteq G)$ denotes the number of induced subgraphs of G which are isomorphic to F.)

Theorem 8. *Let F be a (connected) non trivial graph (i.e. with at least one edge). Then the following limits*

$$\lim_{i\to\infty} \limsup_{G \in \mathcal{C} \triangledown i} \frac{\log \hom(F, G)}{\log |G|}, \qquad \lim_{i\to\infty} \limsup_{G \in \mathcal{C} \widetilde{\triangledown} i} \frac{\log \hom(F, G)}{\log |G|},$$

$$\lim_{i\to\infty} \limsup_{G \in \mathcal{C} \triangledown i} \frac{\log (\#F \subseteq G)}{\log |G|}, \qquad \text{and} \qquad \lim_{i\to\infty} \limsup_{G \in \mathcal{C} \widetilde{\triangledown} i} \frac{\log (\#F \subseteq G)}{\log |G|}$$

can only take the values $-\infty, 0, 1, \ldots, \alpha(F)$ and $|F|$, where $\alpha(F)$ stands for the independence number of F. Moreover, $\mathcal{C}$ is somewhere dense if and only if the limit is $|F|$.

For a proof, see [74]. There is more to this than meets the eye. The recent theory of graph limits developed by Lovász with his coauthors, see e.g. [59, 60, 62, 61] deals with counting of homomorphisms from small graphs, or alternatively, with probabilities that a random map is a homomorphism. In this context the last alternative description is very pleasing.

There are other alternative descriptions of nowhere dense – somewhere dense dichotomy. We refer to the forthcoming book [76].

5. Application I: Small Ramsey Numbers

Ramsey theory is a domain of (very) large numbers, see e.g. [46], [65]. However there are exceptions: one of them are game versions of Ramsey problems, see [14], and the other one is the detailed analysis of (generalized) Ramsey numbers defined for an arbitrary graph G as the least integer $r(G)$, the *Ramsey number of* G, so that for every graph H of order at least $r(G)$ either H or its complement contains G as a subgraph. When the graph G is sparse then we can expect small Ramsey numbers (and in many cases exact results). Such results often belong more to graph theory than to Ramsey theory. But this is not the case with the linear Ramsey numbers where the analysis involves techniques from the very heart of Ramsey theory.

In this paper we will only consider simple loopless graphs. Let G be a graph. It follows from Ramsey's theorem that there exists a least integer $r(G)$, the *Ramsey number of* G, so that for every graph H of order at least $r(G)$ either H or its complement contains G as a subgraph. A family of graphs $\mathcal{F}$ is a *Ramsey linear family* if there exists a constant $c = c(\mathcal{F})$ such that $r(G) \leq cn$ for every $G \in \mathcal{F}$ of order n. In 1973, Burr and Erdős [21] formulated the following conjecture.

Conjecture 1. *For each positive integer p, there exists a constant c_p so that, if G is a p-degenerate graph on n vertices, then $r(G) < c_p n$.*

In 1983 Chvátal, Rödl, Szemerédi, and Trotter [24] proved (in one of the first applications of SRL) that the conjecture holds for graphs with bounded maximum degree (improved in [44], tight bounds for bipartite case in [45]). This result has been extended to p-arrangeable graphs by Chen and Schelp [22].

Recall that a graph G is *p-arrangeable* (concept introduced in [22]) if its vertices can be ordered as $v_1, v_2, \ldots, v_n$ in such a way that $\left|N_{L_i}\left(N_{R_i}(v_i)\right)\right| \leq q$ for each $1 \leq i \leq b_1$, where $L_i = \{v_1, v_2, \ldots, v_i\}$, $R_i = \{v_{i+1}, v_{i+2}, \ldots, v_n\}$, and $N_A(B)$ denotes the neighbors of B which lie in A.

In [22], the authors proved that planar graphs are p-arrangeable for some p. In [83], Rödl and Thomas prove that graphs containing no subdivision of K_q are p-arrangeable for some p depending on q. The bound on the Ramsey number of p-arrangeable graphs was improved by Eaton [31] and then by Graham, Rödl and Ruciński [44].

The Burr–Erdős conjecture is known to hold for subdivided graphs [3] (improved in [56]). Moreover, some further progress toward the conjecture may also be found in [53], [54], [55]. A general survey of what is known on Ramsey numbers may be found in [79].

In [73] we give a new sufficient condition for a graph to be p-arrangeable using the extremal edge densities in the beginning of the resolution, more precisely using

$$\nabla_0(G) = \sup_{H \in G \triangledown 0} \frac{\|H\|}{|H|} \quad \text{and} \quad \nabla_1(G) = \sup_{H \in G \triangledown 1} \frac{\|H\|}{|H|}.$$

In [68] we defined these invariants ∇_0, ∇_1, and generally

$$\nabla_d(G) = \sup_{H \in G \triangledown 1} \frac{\|H\|}{|H|}.$$

These invariants are called *grads* (from "Greatest Reduced Average Density"). Using this notation we can formulate the Burr–Erdős conjecture and our results in a concise form.

Conjecture 2 (Alternate form of Burr–Erdős conjecture). *There exists a function $f : \mathbb{R} \to \mathbb{R}$ such that for any graph G of order n:*

$$\frac{r(G)}{n} < f\big(\nabla_0(G)\big)$$

In [73] we show that for any graph G its arrangeability can be actually bounded as a function of $\nabla_1(G)$ only. Combining this with [44] we obtain:

Theorem 9. *There exists a function $f : \mathbb{R} \to \mathbb{R}$ such that for any graph G of order n:*

$$\frac{r(G)}{n} < f\big(\nabla_1(G)\big)$$

More precisely:

$$\log_2\left(\frac{r(G)}{n}\right) = O\Big(\big(\nabla_0(G)\nabla_1(G)\log\nabla_1(G)\big)^2\Big)$$

Also, combining our results with those of [86, 87] we get a slightly stronger off-diagonal version:

Corollary 10. *There exists a function* $g : \mathbb{R}^2 \to \mathbb{R}$ *such that for any graphs* G_1, G_2*:*

$$\frac{r(G_1, G_2)}{\max\left(\left|V(G_1)\right|, \left|V(G_2)\right|\right)} \leq g\left(\nabla_0(G_1), \nabla_1(G_2)\right)$$

Figure 7 indicates inclusion of some of the graph classes related to the Erdős–Burr conjecture.

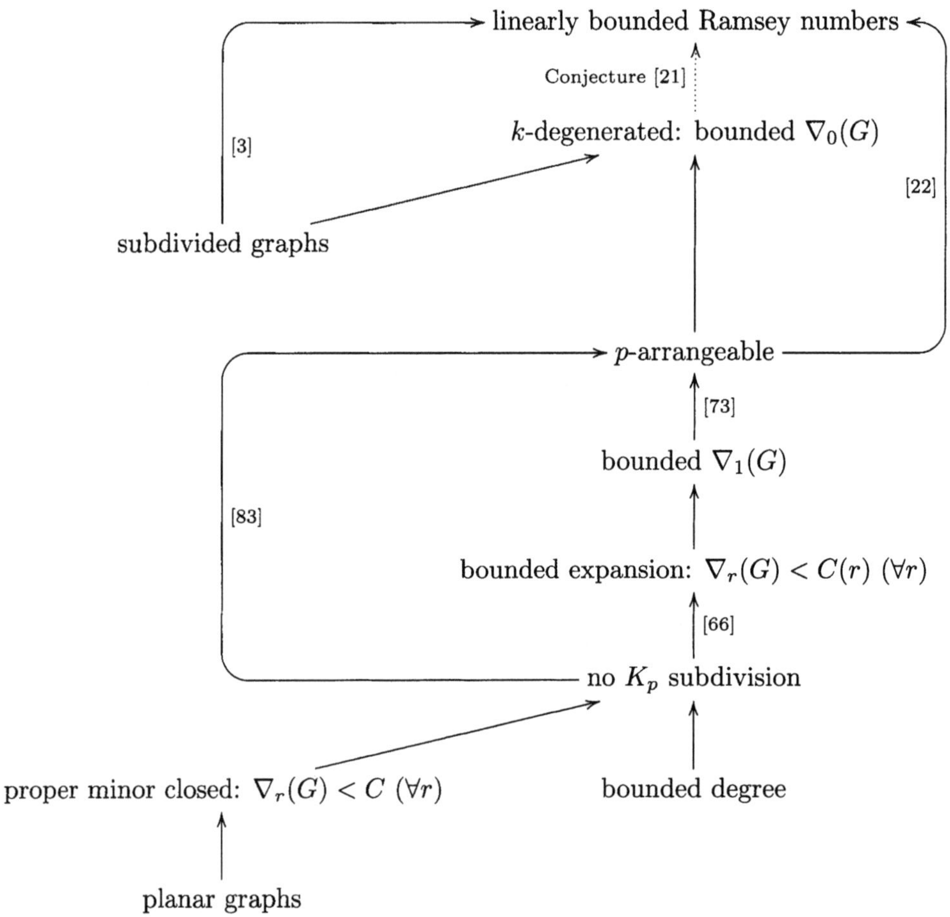

Fig. 7. Inclusion of linear Ramsey graph classes

6. Application II: Vertex Separators

Let G be a graph of order n. Recall that an *α-vertex separator* of G is a subset S of vertices such that every connected component of $G-S$ contains at most αn vertices.

6.1. Sub-exponential ω-expansion

A celebrated theorem of Lipton and Tarjan [57] states that any planar graph has a separator of size $O(\sqrt{n})$. Alon, Seymour and Thomas [5] showed that excluding K_h as a minor ensures the existence of a separator of size at most $O(h^{3/2}\sqrt{n})$. Gilbert, Hutchinson, and Tarjan [41] further proved that graphs with genus g have a separator of size $O(\sqrt{gn})$ (this result is optimal). Plotkin et al. [78] introduced the concept of *limited-depth minor* exclusion and have shown that exclusion of small limited-depth minors implies the existence of a small separator.

Precisely, Plotkin et al. prove in [78] that any graph excluding K_h as a depth l minor (i.e. any graph G such that $K_h \notin G \triangledown l$) has a separator of size $O(lh^2 \log n + n/l)$ hence proving that excluding a K_h minor ensures the existence of a separator of size $O(h\sqrt{n}\log n)$. They proved the following:

Theorem 11 ([78]). *Given a graph with m edges and n nodes, and integers l and h, there is an $O(mn/l)$ time algorithm that will either produce a K_h-minor of depth at most $l \log n$ or will find a separator of size at most $O(n/l + 4lh^2 \log n)$.* ■

The *ω-expansion* of a class $\mathcal{C}$ is the mapping (from $\mathbb{N}$ to $\mathbb{N} \cup \{\infty\}$)

$$i \mapsto \sup_{G \in \mathcal{C} \triangledown i} \omega(G),$$

where $\omega(G)$ stands for the *clique number* of G, i.e. the order of the largest complete subgraph of G. Notice that a class has bounded ω-expansion if and only if it is nowhere dense.

A class $\mathcal{C}$ has sub-exponential ω-expansion if

$$\limsup_{i \to \infty} \sup_{G \in \mathcal{C} \triangledown i} \frac{\log \omega(G)}{i} = 0.$$

Theorem 12. *Let $\mathcal{C}$ be a class of graphs with sub-exponential ω-expansion.*

Then the graphs of order n in $\mathcal{C}$ have separators of size $s(n) = o(n)$ which may be computed in time $O\big(ns(n)\big) = o(n^2)$.

As random cubic graphs almost surely have bisection width at least $0.101n$ [52], they have almost surely no separator of size smaller than $n/20$ It follows that if $\log f(x) = (\log 2)x$, the graphs have no sublinear separators any more. This shows the optimality of Theorem 12.

7. Application III: Property Testing, Weak Hyperfiniteness

7.1. Property testing

Property testing has been introduced by Blum, Luby and Rubinfeld [17] and Rubinfeld and Sudan [84] (in the context of program testing), and by Arora, Lund, Motwani, Sudan, and M. Szegedy [10] and Arora and Safra [11] (in the context of probabilistically checkable proofs). Testing graph properties was first investigated by Goldreich, Goldwasser, and Ron [42]. From a "mathematical" point of view, the main ingredients of property testing are:

- a random sampling of a large structure,
- a suitable notion of distance between objects.

Let $\mathcal{P}$ be a class of graphs (called *graph property* in this context). A graph G is said to have property $\mathcal{P}$ if $G \in \mathcal{P}$; it is said to be ε*-far* for satisfying $\mathcal{P}$ if no graph at distance at most ε from G satisfies $\mathcal{P}$. A testing algorithm (or *tester*) for graph property $\mathcal{P}$ and accuracy ε is an algorithm that distinguishes with probability at least $2/3$ between graphs satisfying $\mathcal{P}$ from graphs that are ε-far from satisfying it. More precisely, the property testing algorithm

- should `accept` with probability at least $2/3$ every input graph that belongs to $\mathcal{P}$,
- should reject with probability at least $2/3$ every input graph that has distance more than ε from any graph in $\mathcal{P}$, i.e. if its ε-far from satisfying $\mathcal{P}$.

Here, the probabilities are taken over the coin tosses of the tester.

However, from the original "computational" point of view, the notion of property testing is a bit trickier: instead of defining a distance between objects, we first have to define an encoding Λ of the objects, and to consider the distance of two objects given by the ratio of the edit distance between the encodings by the "length" of the encoding. Also, on has to precise what "local" queries are allowed to the tester.

A graph property $\mathcal{P}$ is *testable* if for any $\varepsilon > 0$, there is a constant time randomized algorithm that can distinguish with high probability between graphs satisfying $\mathcal{P}$ from those that are ε-far from satisfying it.

One should notice that the introduction of the parameter ε will make some properties impossible to distinguish. Precisely, two properties $\mathcal{P}$ and $\mathcal{Q}$ are *indistinguishable* if for every $\varepsilon > 0$ there exists $N = N(\varepsilon)$ such that:

- for every graph $G \in \mathcal{P}$ with order at least N there exists $H \in \mathcal{Q}$ with the same order such that $\operatorname{dist}(G, H) < \varepsilon$,
- for every graph $H \in \mathcal{Q}$ with order at least N there exists $G \in \mathcal{P}$ with the same order such that $\operatorname{dist}(G, H) < \varepsilon$.

As proved in [4] (in the context of dense graphs, but easily extended to the general case), if two properties are indistinguishable then either they are both testable or none of them is testable.

7.1.1. Dense graphs. In the context of dense graphs, the standard encoding is given by the adjacency matrix of the graph, and the local queries correspond to checking the adjacency of two sample vertices. The encoding of a graph of order n is then $\binom{n}{2}$, and two graphs G and G' of order n will be ε-far if one has to change at least $\varepsilon\binom{n}{2}$ adjacencies in G to get a graph isomorphic to G'.

Despite the apparent symmetry of the definition, it should be noticed that the fact that a property P is testable is not related to the fact that $\neg P$ is testable. For instance, Alon et al. [4] proved that any first-order property of the form "$\exists\forall$" is testable while there exist some first-order properties of the form "$\forall\exists$" which are not testable.

The study of property testing in the context of dense graphs is usually based on SRL [89]. Using this structural lemma, Alon and Shapira proved that every hereditary property is testable [9, 7] and then extended this result to monotone properties [8]. The generalization of Szemerédi's regularity lemma and of the removal lemma to hypergraphs [82, 81] allowed Rödl and Schacht to prove that every hereditary hypergraph property is testable [80] and also, with Avart, that every monotone 3-graph property is testable [13]. The testability of hereditary properties has been further extended to partite hypergraph properties by Ishigami [48] and to multiple directed polychromatic graphs and hypergraphs by Austin and Tao [12].

7.1.2. Ultra sparse graphs. In the context of sparse graphs, instead of coding graphs by an adjacency matrix one can use adjacency lists. This is particularly suitable when studying sparse graphs [43]. In this latter case, two graphs of order n are at distance ε if one can be obtained from the other by means of at most $\varepsilon(n+m)$ edge deletions or edge additions.

In the case where the input graphs are bound to have maximum degree at most Δ the ratio of the Hamming distance between the adjacency lists of two graphs of order n and size Δn actually defines the distance between the two graphs. In this setting, properties that were testable in the dense case may not be testable in the sparse case, and conversely. For instance, some properties which are trivially testable in the dense case (such as acyclicity or connectivity) are still testable in the bounded degree model but in a non obvious way, and other properties which are testable in the dense model, such as bipartiteness, are not testable in the bounded degree model [43]. In particular, the fact that bipartiteness is not testable in the bounded degree model shows that not every monotone property is testable in this model. If one seeks for an analog of the result of Rödl and Schacht in the bounded degree model, one will have to add further restrictions to the tested property.

7.2. Hyperfinite Classes

Asymptotic properties of classes of finite graphs with bounded degree can be studied using a notion of convergence which is the local weak convergence, as studied in [15] and [2]. In [85], Schramm extended the study of the local weak convergence of graphs by noting its properties in relation to hyperfinite classes of graphs. A class $\mathcal{C}$ of (finite) graphs is *hyperfinite* if for every positive real $\varepsilon > 0$ there exists a positive integer $K(\varepsilon)$ such that every graph $G \in \mathcal{C}$ has a subset of at most $\varepsilon\,|G|$ edges whose deletion leaves no connected component of order greater than $K(\varepsilon)$. Although this notion appeared implicitly in the literature (i.e., [58]), as far as we know Elek [32, 33, 34] was the first to give it a name and propose its systematic study.

In [16], Benjamini, Schramm and Shapira showed that every minor-closed graph property can be tested with a constant number of queries in the bounded degree model (see also [25]). For instance, planarity is testable in the bounded degree model. Actually, they prove a much stronger theorem:

Theorem 13 ([16]). *Every hyperfinite monotone graph property is testable.*

Using a detailed analysis of bounded expansion classes with an sub-exponential growth we will now extend the range of applications of this result.

7.3. Weakly hyperfinite classes

A class $\mathcal{C}$ of graphs is *weakly hyperfinite* if for any $\varepsilon > 0$ there exists $K(\varepsilon)$ such that every $G \in \mathcal{C}$ has a subset of at most $\varepsilon|G|$ vertices whose deletion leaves no connected component of order greater than K.

Although it is obvious that in order to be hyperfinite a monotone class of graphs needs to bound the maximum degrees of its elements. However, weakly hyperfinite classes may contain graphs with unbounded maximum degrees. Moreover, it is clear that any hyperfinite class is also weakly hyperfinite and that these two notions coincide for classes of graphs with bounded maximum degrees

The relation between the two notions will be made precise by the following theorem:

Theorem 14. *For a positive integer D, denote by Δ_D the class of the graphs having maximum degree at most D. Let $\mathcal{C}$ be a monotone class of graphs with bounded average degree.*

The class $\mathcal{C}$ is weakly hyperfinite if and only if for every integer D the class $\mathcal{C} \cap \Delta_D$ is hyperfinite.

A key advantage of the notion of weak hyperfinite class is its connection with the existence of sublinear vertex separators.

7.3.1. Tree-depth, Tree-width and Vertex Separators. The following inequalities between tree-width, pathwidth and tree-depth are easy to prove [18]: for any graph G it holds

$$\operatorname{tw}(G) \le \operatorname{pw}(G) \le \operatorname{td}(G).$$

Lemma 15. *Let G be a graph of order n and let $s_G : \{1,\dots,n\} \to \mathbb{N}$ be defined by*

$$s_G(i) = \max_{\substack{|A|\le i,\\ A\subseteq V(G)}} \min\left\{|S| \,:\, S \text{ is a } \frac{1}{2}\text{-vertex separator of } G[A]\right\}$$

Then:

$$s_G(n) \le \operatorname{td}(G) \le \sum_{i=0}^{\log_2 n} s_G\left(\frac{n}{2^i}\right)$$

Relative sizes of vertex separators for graphs in a class $\mathcal{C}$ may be studied through the function $\varsigma : \mathbb{N} \to \mathbb{N}$ defined by:

$$\varsigma(n) = \sup_{G\in\mathcal{C},\ |G|\le n} \min\left\{|S| \,:\, S \text{ is a } \frac{1}{2}\text{-vertex separator of } G\right\}.$$

Hence, if $\mathcal{C}$ is hereditary we have

$$\varsigma(n) = \sup_{G\in\mathcal{C}} s_G(n).$$

Instead of ς, it will be convenient to consider a real valued concave sublinear approximation of ς. This will be achieved by the following standard construction: The *convex conjugate* of a lower semi-continuous function $\phi : X \to \mathbb{R}\cup\{\infty\}$ is the function $\phi^\star : X^\star \to \mathbb{R}\cup\{\infty\}$ (where X is a real normed vector space and X^* is its dual space) defined by

$$\phi^\star(x^\star) = \sup\left\{\langle x^\star, x\rangle - \phi(x) \,:\, x \in X\right\}.$$

The *convex biconjugate* $\phi^{\star\star}$ of ϕ (i.e. the convex conjugate of the convex conjugate of ϕ) is also the *closed convex hull* of ϕ, i.e. the largest lower semi-continuous convex function smaller than ϕ. Hence if ϕ is is convex and lower semi-continuous then $\phi^{\star\star} = \phi$.

For a non-decreasing function $f : \mathbb{N} \to \mathbb{R}^+$ we define $\hat{f}(x) = -g^{\star\star}(-x)$, where

$$g(x) = f(\lfloor x\rfloor) + (x - \lfloor x\rfloor)\big(f(\lceil x\rceil) - f(\lfloor x\rfloor)\big).$$

The function $\hat{f}$ is then the smallest upper continuous concave function greater or equal to f. It is non-decreasing if f is non-decreasing.

Hence, according to Lemma 15 we have for a hereditary class $\mathcal{C}$:

$$\begin{aligned}\varsigma(n) = \sup_{G\in\mathcal{C}} s_G(n) &\leq \sup_{G\in\mathcal{C},\ |G|\leq n} \operatorname{td}(G)\\ &\leq \sup_{G\in\mathcal{C}} \sum_{i=0}^{\log_2 n} s_G\left(\frac{n}{2^i}\right) \leq \sum_{i=0}^{\log_2 n} \varsigma\left(\frac{n}{2^i}\right) \leq \sum_{i=0}^{\log_2 n} \hat{\varsigma}\left(\frac{n}{2^i}\right)\\ &\leq \hat{\varsigma}(n).\end{aligned}$$

Moreover, as ς and $\hat{\varsigma}$ are bounded by the same linear functions, $\hat{f}$ is sublinear if and only if f is sublinear. This is the basis of the following:

Proposition 16. *Let $\mathcal{C}$ be a monotone class of graphs. The following properties are equivalent:*

1. *the graphs in $\mathcal{C}$ have sublinear vertex separators:*

$$\limsup_{G\in\mathcal{C}} \frac{\min\left\{|S|\ :\ S \text{ is a } \tfrac{1}{2}\text{-vertex separator of } G\right\}}{|G|} = 0;$$

2. *the graphs in $\mathcal{C}$ have sublinear s_G:*

$$\lim_{n\to\infty} \sup_{G\in\mathcal{C}} \frac{s_G(n)}{n} = 0;$$

3. *the function ς defined by $\mathcal{C}$ is sublinear:*

$$\limsup_{n\to\infty} \frac{\varsigma(n)}{n} = 0;$$

4. *the graphs in $\mathcal{C}$ have sublinear tree-width:*

$$\limsup_{G\in\mathcal{C}} \frac{\operatorname{tw}(G)}{|G|} = 0;$$

5. *the graphs in $\mathcal{C}$ have sublinear tree-depth:*

$$\limsup_{G\in\mathcal{C}} \frac{\operatorname{td}(G)}{|G|} = 0.$$

Let $\mathcal{C}$ be a monotone class of graphs, such that each $G \in \mathcal{C}$ has a $\frac{1}{2}$-vertex separator of size at most $\varsigma(|G|) = o(|G|)$. We shall state a consequence of the existence of sublinear vertex-separators. Our result will follow from an extension of an optimization result of Lipton and Tarjan [57], which is stated in Lemma 18 and which is based on the property of $\hat{\varsigma}$ expressed by the following lemma.

Lemma 17. *Let $G \in \mathcal{C}$ have order n and let $\mu : V(G) \to [0,1]$ be a probability measure. Then $V(G)$ can be split into parts A, B, C such that $|C| \leq \hat{\varsigma}(2n)$, $\max(|A|,|B|) \leq 2n/3$ and $\max(\mu(A),\mu(B)) \leq 2/3\,\mu(G)$ and no vertex in A has a neighbor in B.*

Lemma 18. *Let $G \in \mathcal{C}$ have order n, let $\mu : V(G) \to [0,1]$ be weights with total sum at most 1 (i.e. $\sum_{v\in V(G)} \mu(v) \leq 1$) and let $0 < \varepsilon < 1$ be a positive real.*

Then there exists a set C of cardinality at most $3\hat{\varsigma}(2\varepsilon n/3)/\varepsilon$ such that no connected component of $G - C$ has a weight greater than ε.

As a consequence we deduce the following results about hyperfiniteness.

Theorem 19. *Let $\mathcal{C}$ be a monotone class of graphs with sublinear vertex separators. Then, for every positive real $\varepsilon > 0$ there exists an integer K such that every graph $G \in \mathcal{C}$ has a subset of vertices S of cardinality at most $\varepsilon|G|$ whose deletion leaves no connected component of order greater than K.*

Corollary 20. *Let $\mathcal{C}$ be a monotone class of graphs with sublinear vertex-separators and bounded average degree and let D be a positive integer.*

Then the subclass of $\mathcal{C}$ including those graphs in $\mathcal{C}$ which have maximum degree at most D is hyperfinite.

Combining with our results about vertex separators we arrive to the following:

Theorem 21. *Let $\mathcal{P}$ be a monotone class of graphs with sub-exponential ω-expansion.*

Then the property $G \in \mathcal{P}$ is testable in the bounded degree model.

Proof. This is a direct consequence of Theorem 13, Theorem 12 and Corollary 20. ■

8. Bounded Expansion Classes

A specific example of classes which are nowhere dense are classes with bounded expansion. These classes have been introduced in [66]. A class $\mathcal{C}$ has *bounded expansion* if there exists a function $f : \mathbb{N} \to \mathbb{R}$ (called *expansion function*) such that

$$\forall d \in \mathbb{N} \qquad \sup_{G \in \mathcal{C} \triangledown d} \frac{\|G\|}{|G|} \leq f(d).$$

For an extensive study of bounded expansion classes we refer the reader to [68], [69], [70], [28], [29], [77].

(See [77] for the definition of stack and queue numbers. This paper contains further examples of bounded expansion classes.)

As for nowhere dense classes, several equivalent characterizations exist for classes with bounded expansion. In fact most of these characterizations are miniature of characterizations of nowhere dense classes in the sense that nearly linear bounds (resp. $n^{o(1)}$ bounds) are replaced by linear ones (resp. constant bounds). We give a sample of characterizations (extending Theorem 1).

Theorem 22. *Let $\mathcal{C}$ be a class of graphs. The following properties are equivalent:*

- *$\mathcal{C}$ has bounded expansion;*
- *for every integer p, $\sup_{G \in \mathcal{C}} \chi_p(G) < \infty$;*
- *for every integer p, $\sup_{G \in \mathcal{C}} \mathrm{col}_p(G) < \infty$;*
- *for every integer p, $\sup_{G \in \mathcal{C} \triangledown p} \chi(G) < \infty$;*
- *for every integer p there exists a graph U_p with odd-girth at least $2p+1$ such that every subdivision with odd-girth at least $2p+1$ of a graph in $\mathcal{C}$ has a homomorphism to U_p.*

Thus any graph G in a (fixed) bounded expansion class $\mathcal{C}$ can be decomposed into a fixed number $N_p(G)$ of classes such that the subgraphs induced by any $\leq p$ classes of the partition have components of only finitely many

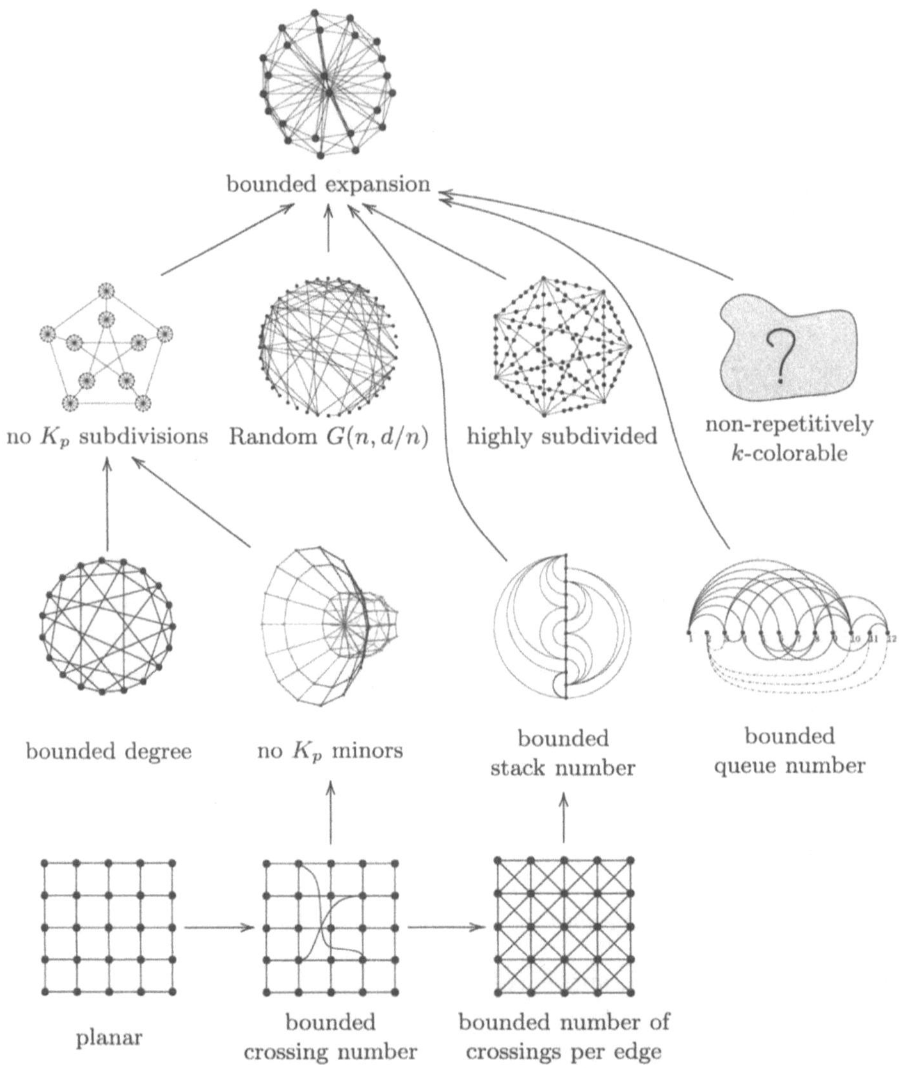

Fig. 8. Classes with Bounded Expansion

(homomorphism) types. Thus p is then parameter expressing the precision of such decomposition. Moreover such decomposition can be found in a linear number of steps. Not surprisingly, this has a number of algorithmic consequences ([66, 69].

8.1. Random Graphs via Bounded Expansion

Bounded expansion classes not only gather most of the usual sparse classes of graphs, but they also capture sparse random graphs:

The $G(n,p)$ model of random graphs is the most common random graph model, see e.g. [19]. In this model, a graph with n vertices is built, where each edge appears independently with probability p. It is frequently considered that p may be a function of n, hence the notation $G\big(n,p(n)\big)$.

On the one hand, the order of the largest complete (topological) minor in $G(n,p/n)$ is well-studied. Recall that a property of random graphs holds *asymptotically almost surely* (*a.a.s.*) if, over a sequence of draws, its probability converges to 1. It is known since the work of Łuczak, Pittel and Wierman [63] that random graphs $G\big(n,p(n)\big)$ with $p(n)-1/n \ll n^{-4/3}$ are asymptotically almost surely planar, whereas those with $p(n)-1/n \gg n^{-4/3}$ asymptotically almost surely contain unbounded clique minors. Fountoulakis et al. [39] proved that for every $c>1$ there exists a constant $\delta(c)$ such that asymptotically almost surely the maximum order $h\big(G(n,c/n)\big)$ of a complete minor of a graph in $G(n,c/n)$ satisfies the inequality $\delta(c)\sqrt{n} \leq h\big(G(n,c/n)\big) \leq 2\sqrt{cn}$. Also, Ajtai, Komlós and Szemerédi [1] proved that as long as the expected degree $(n-1)p$ is at least $1+\varepsilon$ and is $o\big(\sqrt{n}\big)$, then asymptotically almost surely the order of the largest complete topological minor of $G(n,p)$ is almost as large as the maximum degree, which is $\Theta(\log n/\log\log n)$.

It is known that the number of short cycles of $G(n,d/n)$ is bounded. More precisely, the expected number of cycles of length t in $G(n,d/n)$ is at most $(e^2d/2)^t$. It follows that the expected value $\mathrm{E}\big(\omega(G \,\widetilde{\nabla}\, r)\big)$ of the clique size of a shallow topological minor of G at depth r is bounded by approximately $(Cd)^{2r}$.

Fox and Sudakov [40] proved that $G(n,d/n)$ is asymptotically almost surely $(16d,16d)$-degenerate, where a graph H is said to be (d,Δ)-*degenerate* if there exists an ordering $v_1,\dots,v_n$ of its vertices such that for each v_i, there are at most d vertices v_j adjacent to v_i with $j<i$, and there are at most Δ subsets of the form $N(v_j)\cap\{v_1,\dots,v_i\}$ for some neighbor v_j of v_i with $j>i$ (recall that the neighborhood $N(v_j)$ is the set of vertices that are adjacent to v_j). A modified version of this proof has been used by the authors and D. R. Wood to prove:

Theorem 23 ([77]). *For each integer d there exists a bounded expansion class $\mathcal{R}_d$ such that $G(n,d/n)$ almost surely belongs to $\mathcal{R}_d$.*

9. Selected Examples

In this section we give a few examples of nowhere dense classes.

9.1. Classical Sparse Classes

Fig. 9 shows the inclusion map of some important hereditary nowhere dense classes which were studied in combinatorial as well as algorithmic context.

9.2. Simplicial Graphs

A *k-dimensional simplex*, or *k-simplex*, is the convex hull of $k+1$ affinely independent points in $\mathbb{R}^d$ space. A *d-dimensional simplicial complex* is a collection of k-simplexes, $k \leq d$, closed under sub-simplex and intersection. For example, a 3-dimensional simplicial complex is a collection of cells (3-simplexes), faces (2-simplexes), edges (1-simplexes) and vertices (0-simplexes). A *d-dimensional simplicial graph* is the collection of edges and vertices of a d-dimensional simplicial complex. The *aspect ratio* of a body is its diameter divided dth root of its volume [64]. The volume of a regular d-simplex, d-cube, and d-ball of unit diameter are respectively $2^{-d/2}\sqrt{d+1}/d!, d^{-d/2}$ and $2^{-d}\pi^{d/2}/(d/2)!$. Hence the aspect ratios of a d-simplex, d-cube, and d-ball are respectively $\alpha_s = 2^{1/2}(d!)^{1/d}(d+1)^{-1/(2d)} \sim \sqrt{2}d/e, \alpha_c = \sqrt{d}$, and $\alpha_b = 2\pi^{-1/2}(d/2)!^{1/d} \sim \sqrt{2d/(e\pi)}$. A simplicial graph of aspect ratio α means a simplicial graph coming from a complex in which every d-simplex has aspect ratio at most α.

Classes of simplicial graphs with bounded aspect ratio exclude big shallow complete minors as proved by Plotkin, Rao and Smith [78]. It follows that such classes are nowhere dense. However we don't know if they have bounded expansion.

A natural question arises: which classes are nowhere dense but do not have bounded expansion. This problem relates to Erdős–Hajnal problem about high girth high chromatic subgraphs in high chromatic graphs. This is going to appear elsewhere.

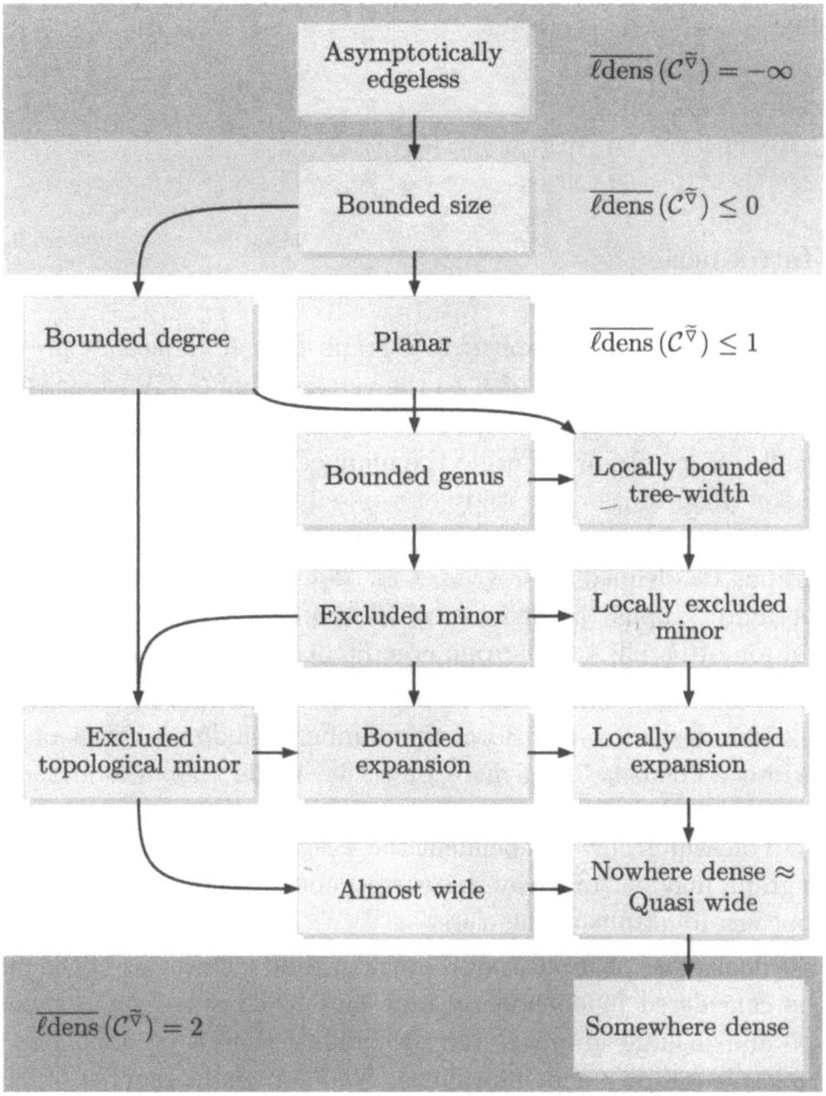

Fig. 9. Inclusion map of some important hereditary nowhere dense classes

9.3. High Girth Graphs

A standard example of a monotone nowhere dense class of graphs is the class of the graphs whose maximum degree does not exceed some function of the girth, i.e. $\mathcal{B}_\phi = \big\{G : \Delta(G) \le \phi\big(\mathrm{girth}\,(G)\big)\big\}$. When ϕ is unbounded, such a class contains graphs with order n, girth at least $1/f(n)$, and degrees

$k \approx n^{f(n)}$ for some unbounded function f hence fails to have bounded expansion.

10. Counting – Degrees of Freedoms

10.1. Introduction

A homomorphism from a graph F to a graph G is an adjacency preserving mapping from the vertex set of F to the vertex set of G. When considering a series of larger and larger graphs, counting the number of homomorphisms from small test graphs or counting the number of induced copies of a small pattern is a main tool in the study of a possible limit of the sequence. For instance, this is the case when one considers the convergence criteria for dense graphs (as defined by Lovász et al. [20], linked to SRL) and the one for hyperfinite graphs (as defined by Elek and Lippner [37], linked to a finitarization of Farell–Varadarajan ergodic decomposition theorem).

One of the main differences between these two approaches lies in the normalization needed to transform the number of induced copies of a fixed pattern into a "density" and more precisely in the exponent of the order of the graph which is used to divide the number of copies. This exponent intuitively measures how independent the assignments of the vertices of the pattern graph may be. In a way this corresponds to the "degree of freedom" of the pattern in graphs of the class.

In the dense case, it is natural to consider that each vertex of the pattern could be considered independently and thus to consider, for a small test graph F and a large graph G the probability that a random map from $V(F)$ to $V(G)$ will be a homomorphism. Notice that the number of induced copies of F in G may be easily derived from the number of homomorphisms from F to G but that the notion of homomorphism can benefits from the independence assumption to ease the estimations and the computations. The considered density is thus

$$t(F,G) = \frac{\hom(F,G)}{|G|^{|F|}}, \tag{1}$$

(Recall that $\hom(F,G)$ stands for the number of homomorphisms from F to G).

Now consider the ultra-sparse case – for instance in the case of bounded degree graphs excluding a minor. A random map from a test graph F to a large ultra-sparse graph G is unlikely to be a homomorphism (except if F is edgeless). However, there are obviously only finitely many isomorphism types of the balls $B_r(v)$ for $v \in V(G)$, and it easily follows that the number of copies of F in G will be at most linear in the order of G and that the considered density should be

$$\text{dens}\,(F,G) = \frac{(\#F \subseteq G)}{|G|}, \tag{2}$$

where $(\#F \subseteq G)$ stands for the number of induced copies of F in G.

When studying large graphs belonging to an infinite class of graphs $\mathcal{C}$ and fixing a small test graph F, the natural question is to determine whether there could exist a "natural" exponent $\mathrm{f}(F,\mathcal{C})$ and a corresponding natural density function

$$\text{dens}_{\mathcal{C}}\,(F,G) = \frac{(\#F \subseteq G)}{|G|^{\mathrm{f}(F,\mathcal{C})}}. \tag{3}$$

This motivates a study of the *asymptotic upper logarithmic density* of F in $\mathcal{C}$, that is of the limit

$$\limsup_{G \in \mathcal{C}} \frac{\log\,(\#F \subseteq G)}{\log\,|G|}.$$

In general, this value is not an integer. For instance, consider the class $\mathcal{C}_0$ of girth at least 5 (i.e. the class of C_3- and C_4-free graphs) and the class $\mathcal{C}_1$ which are subgraphs of the Cartesian product of a graph in $\mathcal{C}_0$ and the complete graph K_2 (see Fig. 10). Recall that the *Cartesian product* $G \,\square\, H$ of two graphs G and H is the graph with vertex set $V(G) \times V(H)$ in which two vertices (u,x) and (v,y) are adjacent if $u = v$ and $\{x,y\} \in E(H)$ or $x = y$ and $\{u,v\} \in E(G)$. Obviously, both $\mathcal{C}_0$ and $\mathcal{C}_1$ are addable (i.e. closed by disjoint union) and monotone (i.e. closed under subgraphs) and we have

$$\limsup_{G \in \mathcal{C}_0} \frac{\log\,(\#K_2 \subseteq G)}{\log\,|G|} = \frac{3}{2}, \tag{4}$$

from which follows

$$\limsup_{G \in \mathcal{C}_1} \frac{\log\,(\#C_4 \subseteq G)}{\log\,|G|} = \frac{3}{2}. \tag{5}$$

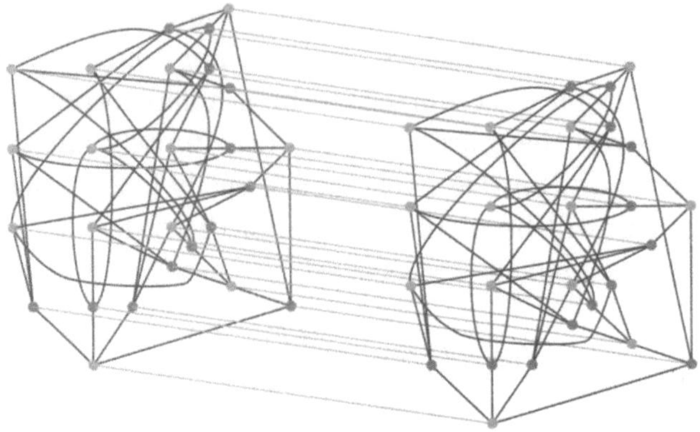

Fig. 10. Construction of a graph, all subgraphs of which have $O(n^{3/2})$ edges and $O(n^{3/2})$ C_4's

Of course, we suspect that the asymptotic upper logarithmic density of graphs in $\mathcal{C}_1$ will play an important role but the relation might be not obvious as if one considers the class $\mathcal{P}$ of planar graphs we have

$$(6) \qquad \limsup_{G\in\mathcal{P}} \frac{\log(\#C_4 \subseteq G)}{\log|G|} = 2 > \frac{3}{2} = \limsup_{G\in\mathcal{C}_1} \frac{\log(\#C_4 \subseteq G)}{\log|G|},$$

although

$$(7) \qquad \limsup_{G\in\mathcal{P}} \frac{\log\|G\|}{\log|G|} = 1 < \frac{3}{2} = \limsup_{G\in\mathcal{C}_1} \frac{\log\|G\|}{\log|G|}.$$

However (to complement Theorem 8) there is nevertheless a difference in the behavior of the limit

$$\limsup_{G\in\mathcal{C}} \frac{\log(\#F \subseteq G)}{\log|G|}$$

for nowhere dense and somewhere dense hereditary classes $\mathcal{C}$:

10.2. Nowhere dense classes

Theorem 8 characterizes nowhere dense classes by means of the integrality and bounding of the upper logarithmic densities in the resolution. The essential part of the proof of Theorem 8 is of an independent interest. It relates to the class itself (not the resolution limit) and expresses a necessary integrality condition for a class to be nowhere dense (showing, for instance, that the class $\mathcal{C}_1$ above is not nowhere dense).

Theorem 24. *Let $\mathcal{C}$ be an infinite nowhere dense hereditary class of graphs and let F be a fixed graph with at least one edge and stability number $\alpha(F)$. Then the limit*

$$\limsup_{G\in\mathcal{C}} \frac{\log(\#F\subseteq G)}{\log|G|} \tag{8}$$

belongs to $\{-\infty,0,1,\ldots,\alpha(F)\}$.

Although we know that the values of the limit (8) need not to be an integer (for a somewhere dense class), we are naturally led to the question of determining its possible values for somewhere dense hereditary classes $\mathcal{C}$. For instance, it is easy to see that in the case where $F=K_2$, the above limit can take any value in $\{-\infty,0\}\cup[1,2]$ (for a suitable choice of a monotone somewhere dense class $\mathcal{C}$).

We can prove a rigidity result for Theorem 24: every sufficiently large graph G in a nowhere dense class $\mathcal{C}$ with many copies of F contains a large "regular" substructure with at least the same logarithmic density of copies of F. This substructure is a simple generalization of the notion of sunflower (or delta-system) introduced by Erdős and Rado [38].

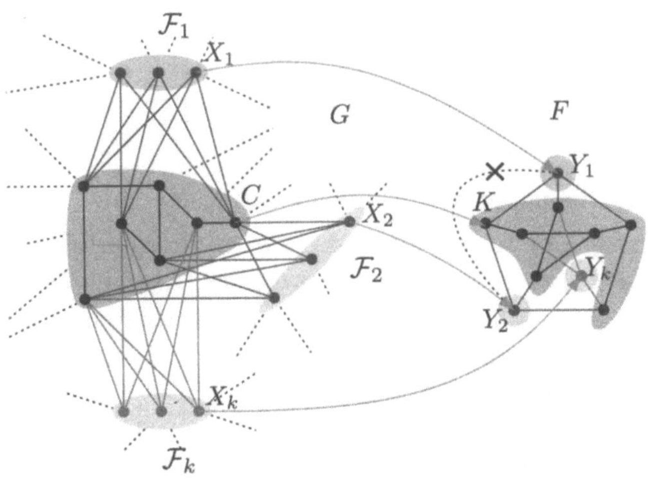

Fig. 11. A generalized $(3,F)$-sunflower

We formulate this more precisely using the following notion:

Let F, G be graphs. A (k, F)-*sunflower* in G is a $(k+1)$-tuple $(C, \mathcal{F}_1, \ldots, \mathcal{F}_k)$, such that $C \subseteq V(G)$, $\mathcal{F}_i$ is a family of subsets of $V(G)$, such that C and all the sets in the $\mathcal{F}_i$'s are pairwise disjoint, and such that there exists a partition $(K, Y_1, \ldots, Y_k)$ of $V(F)$ so that (see Fig. 11):

- there is no edge between vertices in Y_i and vertices in Y_j for $i \neq j$,
- there exists an isomorphism $\iota_0 : G[C] \to F[K]$ from the subgraph of G induced by C to the subgraph of F induced by K,
- for each $1 \leq i \leq k$ and each $X_i \in \mathcal{F}_i$ there exists an isomorphism $\iota_{X_i} : G[X_i] \to F[Y_i]$,
- for any choice of $X_1 \in \mathcal{F}_1, \ldots, X_k \in \mathcal{F}_k$ the mapping $\iota_{X_1,\ldots,X_k}$ from $C \cup \bigcup_i X_i$ to $V(F)$ whose restriction to C is ι_0 and whose restriction to X_i is ι_{X_i} (for $1 \leq i \leq k$) is an isomorphism from $G\big[C \cup \bigcup_i X_i\big]$ to F.

We prove that every sufficiently large sparse graph which contains many copies of a fixed graph F includes a large (F, k)-sunflower (where sparsity is measured in terms of generalized chromatic numbers χ_p). This is expressed by the following theorem:

Theorem 25 (Clearing and Stepping up [74]). *Let F be a graph of order p and let $0 < \varepsilon < 1$. Then there exist positive reals c and τ (depending on both F and ε) such that every graph G which contains more than $|G|^{k+\varepsilon}$ copies of F and which is such that $\chi_p(G) < c|G|^{\varepsilon/p}$ actually contains a $(k+1, F)$-sunflower $(C, \mathcal{F}_1, \ldots, \mathcal{F}_{k+1})$ where*

$$\min |\mathcal{F}_i| \geq \left(\frac{|G|}{\chi_p(G)^{p/\varepsilon}}\right)^{\tau}.$$

Let us end this paper with the following result.

The property not to include a shallow subdivision of a large complete graph is characteristic to nowhere dense classes. Also, nowhere dense classes are uniformly quasi-wide, what implies that for these classes the existence of a long path in a graph of the class is sufficient to guarantee the existence of a long induced path. Also, the non-existence of a long path in a graph is characteristic to classes with bounded tree-depth. Finally, classes with bounded tree-depth have the property that every sufficiently large graph admits a non-trivial symmetry.

Putting together these properties leads us to the following (admittedly inprecise) theorem (see Fig. 12 and [76] for details):

Theorem 26. *Every large graph either has a non-trivial symmetry or it contains a long induced path or it contains a shallow subdivision of a large complete graph.*

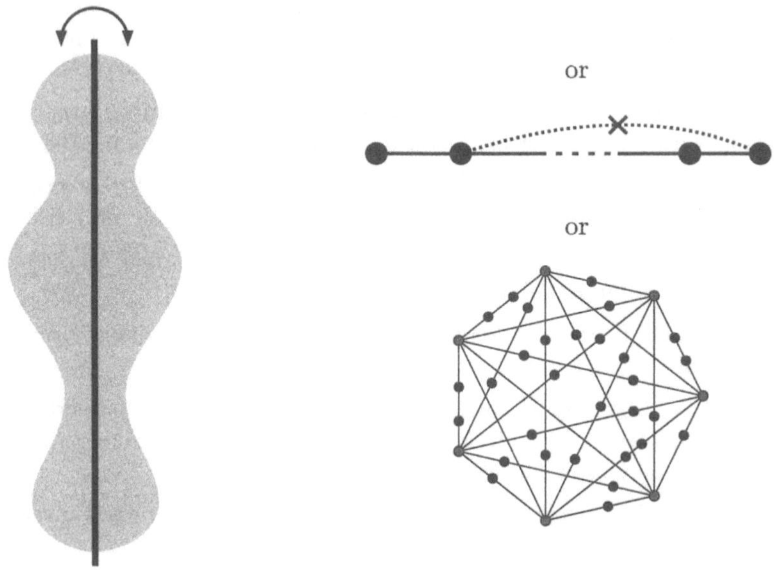

Fig. 12. Every large graph either has a non trivial automorphism, or includes a long induced path, or includes a shallow subdivision of a large complete graph

Perhaps in sparse extremal graph theory we have to deal with symmetries too.

References

[1] M. Ajtai, J. Komlós and E. Szemerédi, Topological complete subgraphs in random graphs, *Studia Sci. Math. Hungar.*, **14** (1979), 293–297.

[2] D. Aldous and R. Lyons, *Processes on unimodular random networks,* arXiv:math/0603062, 2006.

[3] N. Alon, Subdivided graphs have linear ramsey numbers, *J. Graph Theory,* **18** (1994), no. 4, 343–347.

[4] N. Alon, E. Fischer, M. Krivelevich and M. Szegedy, Efficient testing of large graphs, *Combinatorica,* **20** (2000), 451–476.

[5] N. Alon, P. D. Seymour and R. Thomas, *A separator theorem for graphs with excluded minor and its applications,* Proceedings of the 22nd Annual ACM Symposium on Theory of Computing, 1990, pp. 293–299.

[6] N. Alon, P. D. Seymour and R. Thomas, A separator theorem for nonplanar graphs, *J. Amer. Math. Soc.*, **3** (1990), 801–808.

[7] N. Alon and A. Shapira, *A characterization of the (natural) graph properties testable with one-sided error,* Proc. 46th IEEE FOCS, 2005, pp. 429–438.

[8] N. Alon and A. Shapira, *Every monotone graph property is testable,* Proceedings of the thirty-seventh annual ACM symposium on Theory of computing (STOC), ACM Press, 2005, pp. 128–137.

[9] N. Alon and A. Shapira, A characterization of the (natural) graph properties testable with one-sided error, *SIAM J. Comp.*, **37** (2008), no. 6, 1703–1727.

[10] S. Arora, C. Lund, R. Motwani, M. Sudan and M. Szegedy, Proof verification and the hardness of approximation problems, *Journal of the ACM,* **45** (1998), 501–555.

[11] S. Arora and S. Safra, Probabilistic checking of proofs: a new characterization of NP, *Journal of the ACM,* **45** (1998), 70–122.

[12] T. Austin and T. Tao, *On the testability and repair of hereditary hypergraph properties,* 2009.

[13] C. Avart, V. Rödl and M. Schacht, Every monotone 3-graph property is testable, *Electronic Notes in Discrete Mathematics,* **22** (2005), 539–542, 7th International Colloquium on Graph Theory.

[14] J. Beck, *Combinatorial games: Tic-Tac-Toe theory,* Cambridge University Press, 2008.

[15] I. Benjamini and O. Schramm, Recurrence of distibutional limits of finite planar graphs, *Electron. J. Probab.,* **6** (2001), no. 23, 13pp.

[16] I. Benjamini, O. Schramm and A. Shapira, *Every minor-closed property of sparse graphs is testable,* Proceedings of the 40th annual ACM symposium on Theory of computing, 2008, pp. 393–402.

[17] M. Blum, M. Luby and R. Rubinfeld, Self-testing/correcting with applications to numerical problems, *JCSS,* **47** (1993), 549–595.

[18] H. L. Bodlaender, J. R. Gilbert, H. Hafsteinsson and T. Kloks, Approximating tree-width, pathwidth, frontsize, and shortest elimination tree, *J. Algorithms,* **18** (1995), 238–255.

[19] B. Bollobás, *Random graphs,* Cambridge University Press, 2001.

[20] C. Borgs, J. Chayes, L. Lovász, V. T. Sós and K. Vesztergombi, Counting graph homomorphisms, *Topics in Discrete Mathematics* (M. Klazar, J. Kratochvíl, M. Loebl, J. Matoušek, R. Thomas and P. Valtr, eds.), Algorithms and Combinatorics, vol. 26, Springer Verlag, 2006, pp. 315–371.

[21] S. A. Burr and P. Erdős, *On the magnitude of generalized Ramsey numbers for graphs,* Colloquia Mathematica Societatis János Bolyai 10, Infinite and Finite Sets, vol. 1, 1975, pp. 214–240.

[22] G. T. Chen and R. H. Schelp, Graphs with linearly bounded ramsey numbers, *J. Combin. Theory, Ser. B,* **57** (1993), no. 1, 138–149.

[23] F. R. K. Chung, R. L. Graham and R. M. Wilson, Quasi-random graphs, *Combinatorica,* **9** (1989), no. 4, 345–362.

[24] V. Chvátal, V. Rödl, E. Szemerédi and W. T. Trotter Jr., The Ramsey number of a graph with bounded maximum degree, *J. Combin. Theory, Ser. B,* **34** (1983), no. 3, 239–243.

[25] A. Czumaj, A. Shapira and C. Sohler, Testing hereditary properties of nonexpanding bounded-degree graphs, *SIAM J. Comput.,* **38** (2009), no. 6, 2499–2510.

[26] A. Dawar, *Finite model theory on tame classes of structures,* Mathematical Foundations of Computer Science 2007 (L. Kučera and A. Kučera, eds.), Lecture Notes in Computer Science, vol. 4708, Springer, 2007, pp. 2–12.

[27] A. Dawar and S. Kreutzer, *Parametrized complexity of first-order logic,* Tech. Report 131, Electronic Colloquium on Computational Complexity, 2009.

[28] Z. Dvořák, *Asymptotical structure of combinatorial objects,* Ph.D. thesis, Charles University, Faculty of Mathematics and Physics, 2007.

[29] Z. Dvořák, On forbidden subdivision characterizations of graph classes, European *J. Combin.,* **29** (2008), no. 5, 1321–1332.

[30] Z. Dvořák, D. Kráľ and R. Thomas, *A linear time for deciding first-order properties on classes with bounded expansion,* personal communication, 2009.

[31] N. Eaton, Ramsey numbers for sparse graphs, *Discrete Math.,* **185** (1998), 63–75.

[32] G. Elek, *The combinatorial cost,* arXiv:math/0608474, 2006.

[33] G. Elek, *L^2-spectral invariants and convergent sequences of finite graphs,* arXiv:0709.1261, 2007.

[34] G. Elek, *A regularity lemma for bounded degree graphs and its applications: Parameter testing and infinite volume limits,* arXiv:0711.2800, 2007.

[35] G. Elek, Weak convergence of finite graphs, integrated density of states and a Cheeger type inequality, *J. Comb. Theory, Ser. B,* **98** (2008), no. 1, 62–68 (English).

[36] G. Elek, *Parameter testing in bounded degree graphs of subexponential growth,* arXiv:0711.2800v3[math.CO], July 2009.

[37] G. Elek and G. Lippner, *An analogue of the Szemerédi Regularity Lemma for bounded degree graphs,* ArXiv e-prints (2008).

[38] P. Erdős and R. Rado, Intersection theorems for systems of sets, *J. London Math. Soc.,* **35** (1960), 85–90.

[39] N. Fountoulakis, D. Kühn and D. Osthus, The order of the largest complete minor in a random graph, *Random Structures and Algorithms,* **33** (2008), no. 2, 127–141.

[40] J. Fox and B. Sudakov, Two remarks on the Burr–Erdős conjecture, *European. J. Combin.,* **30** (2009), no. 7, 1630–1645.

[41] J. R. Gilbert, J. P. Hutchinson and R. E. Tarjan, A separator theorem for graphs of bounded genus, *J. Algorithms* (1984), no. 5, 375–390.

[42] O. Goldreich, S. Goldwasser and D. Ron, Property testing and its connection to learning and approximation, *JACM,* **45** (1998), no. 4, 653–750.

[43] O. Goldreich and D. Ron, Property testing in bounded-degree graphs, *Algorithmica,* **32** (2002), 302–343.

[44] R. L. Graham, V. Rödl and A. Ruciński, On graphs with linear Ramsey numbers, *J. Graph Theory,* **35** (2000), 176–192.

[45] R. L. Graham, V. Rödl and A. Ruciński, On bipartite graphs with linear Ramsey numbers, Paul Erdős and his mathematics, *Combinatorica,* **21** (2001), 199–209.

[46] R. L. Graham, B. L. Rothschild and J. Spencer, *Ramsey theory,* Wiley, New-York, 1980.

[47] P. Hell and J. Nešetřil, *Graphs and homomorphisms,* Oxford Lecture Series in Mathematics and its Applications, vol. 28, Oxford University Press, 2004.

[48] Y. Ishigami, *Removal lemma for infinitely-many forbidden hypergraphs and property testing,* 2008.

[49] H.A. Kierstead and D. Yang, *Orderings on graphs and game coloring number,* Order **20** (2003), 255–264.

[50] Y. Kohayakawa and V. Rödl, *Szemerédi's regularity lemma and quasi-randomness,* Recent Advances in Algorithmic Combinatorics (B. Reed and C. Linhares-Sales, eds.), CMS Books Math./Ouvrages Math. SMC, vol. 11, Springer, New-York, 2003, pp. 289–347.

[51] J. Komlós and M. Simonovits, Szemerédi's regularity lemma and its applications in graph theory, *Combinatorics, Paul Erdős is Eighty,* vol. 2, János Bolyai Math. Soc., 1993, pp. 295–352.

[52] A. V. Kostochka and L. S. Melnikov, On bounds of the bisection width of cubic graphs, *Fourth Czechoslovakian Symposium on Combinatorics, Graphs and Complexity* (J. Nesetril and M. Fiedler, eds.), Elsevier, 1992, pp. 151–154.

[53] A. V. Kostochka and V. Rödl, On graphs with small Ramsey numbers, *J. Graph Theory,* **37** (2001), 198–204.

[54] A. V. Kostochka and V. Rödl, On graphs with small Ramsey numbers II, *Combinatorica,* **24** (2004), no. 3, 389–401.

[55] A. V. Kostochka and B. Sudakov, On Ramsey numbers of sparse graphs, *Combin. Probab. Comput.,* **12** (2003), 627–641.

[56] Yusheng Li, C. C. Rousseau and L. Soltés, Ramsey linear families and generalized subdivided graphs, *Discrete Math.* (1997), 269–275.

[57] R. Lipton and R. E. Tarjan, A separator theorem for planar graphs, *SIAM Journal on Applied Mathematics,* **36** (1979), no. 2, 177–189.

[58] R.J. Lipton and R. E. Tarjan, Applications of a planar separator theorem, *SIAM J. Comp.,* **9** (1980), no. 3, 615–627.

[59] L. Lovász and M. L. Marx, A forbidden subgraph characterization of Gauss codes, *Bull. Am. Math. Soc.,* **82** (1976), 121–122.

[60] L. Lovász and V. T. Sós, Generalized quasirandom graphs, *J. Combin. Theory, Ser. B,* **98** (2008), 146–163.

[61] L. Lovász and B. Szegedy, Limits of dense graph sequences, *J. Combin. Theory, Ser. B,* **96** (2006), 933–957.

[62] L. Lovász and B. Szegedy, Szemerédi lemma for the analyst, *Geom. Func. Anal.*, **17** (2007), 252–270.

[63] T. Łuczak, B. Pittel and J. C. Wierman, The structure of a random graph at the point of the phase transition, *Trans. AMS*, **341** (1994), 721–748.

[64] G. L. Miller, S.-H. Teng, W. Thurston and S. A. Vavasis, Geometric separators for finite-element meshes, *SIAM J. Sci. Comput.*, **19** (1998), no. 2, 364–386 (English).

[65] J. Nešetřil, Ramsey theory, *Handbook of Combinatorics* (R. L. Graham, M. Grötschel and L. Lovász, eds.), Elsevier, 1995.

[66] J. Nešetřil and P. Ossona de Mendez, *Linear time low tree-width partitions and algorithmic consequences*, STOC'06. Proceedings of the 38th Annual ACM Symposium on Theory of Computing, ACM Press, 2006, pp. 391–400.

[67] J. Nešetřil and P. Ossona de Mendez, Tree depth, subgraph coloring and homomorphism bounds, *European Journal of Combinatorics*, **27** (2006), no. 6, 1022–1041.

[68] J. Nešetřil and P. Ossona de Mendez, Grad and classes with bounded expansion I. decompositions, *European Journal of Combinatorics*, **29** (2008), no. 3, 760–776.

[69] J. Nešetřil and P. Ossona de Mendez, Grad and classes with bounded expansion II. algorithmic aspects, *European Journal of Combinatorics*, **29** (2008), no. 3, 777–791.

[70] J. Nešetřil and P. Ossona de Mendez, Grad and classes with bounded expansion III. restricted graph homomorphism dualities, *European Journal of Combinatorics*, **29** (2008), no. 4, 1012–1024.

[71] J. Nešetřil and P. Ossona de Mendez, Structural properties of sparse graphs, *Building Bridges Between Mathematics and Computer Science* (Martin Grötschel and Gyula O. H. Katona, eds.), Bolyai Society Mathematical Studies, vol. 19, Springer, 2008, pp. 369–426.

[72] J. Nešetřil and P. Ossona de Mendez, First order properties on nowhere dense structures, *The Journal of Symbolic Logic* (2009), accepted.

[73] J. Nešetřil and P. Ossona de Mendez, Fraternal augmentations, arrangeability and linearly bounded Ramsey numbers, *European Journal of Combinatorics*, **30** (2009), no. 7, 1696–1703.

[74] J. Nešetřil and P. Ossona de Mendez, How many F's are there in G?, *European Journal of Combinatorics* (2010), submitted.

[75] J. Nešetřil and P. Ossona de Mendez, On nowhere dense graphs, *European Journal of Combinatorics* (2010), accepted.

[76] J. Nešetřil and P. Ossona de Mendez, *Sparsity (graphs, structures, and algorithms)*, Springer, 2010, In preparation.

[77] J. Nešetřil, P. Ossona de Mendez and D. R. Wood, Characterizations and examples of graph classes with bounded expansion, *European Journal of Combinatorics* (2009), submitted.

[78] S. Plotkin, S. Rao and W. D. Smith, *Shallow excluded minors and improved graph decomposition*, 5th Symp. Discrete Algorithms, SIAM, 1994, pp. 462–470.

[79] S. Radziszowski, Small Ramsey numbers, *Electron. J. Combin.*, **DS1** (2004), dynamic survey; `http://www.combinatorics.org/Surveys/ds1.pdf`.

[80] V. Rödl and M. Schacht, Property testing in hypergraphs and the removable lemma, *Proceedings of the thirty-ninth annual ACM symposium on Theory of computing,* 2007, pp. 488–495.

[81] V. Rödl and M. Schacht, Regular partitions of hypergraphs: Regularity lemmas, *Combin. Probab. Comput.,* **16** (2007), 833–885.

[82] V. Rödl and M. Schacht, Generalizations of the removal lemma, *Combinatorica,* **29** (2009), no. 4, 467–501.

[83] V. Rödl and R. Thomas, Arrangeability and clique subdivisions, *The mathematics of Paul Erdős* (R. L. Graham and J. Nešetřil, eds.), Springer, 1991, pp. 236–239.

[84] R. Rubinfeld and M. Sudan, Robust characterization of polynomials with applications to program testing, *SIAM J. Comp.,* **25** (1996), 252–271.

[85] O. Schramm, Hyperfinite graph limits, *Electron. Res. Announc. Math. Sci.,* **15** (2008), 17–23.

[86] L. Shi, *Numbers and topologies: Two aspects of Ramsey theory,* Ph.D. thesis, Mathematisch-Naturwissenschaftlichen Fakultät II, Humboldt-Universität zu Berlin, 2003.

[87] L. Shi, Linear ramsey numbers of sparse graphs, *J. Graph Theory,* **50** (2005), no. 3, 175–185.

[88] E. Szemerédi, *Regular partitions of graphs,* Problémes combinatoires et théorie des graphes, CNRS, 1976, pp. 399–401.

[89] E. Szemerédi, Regular partitions of graphs, *Colloq. Int. CNRS,* **260** (1978), 399–401.

[90] T. Tao, The dichotomy between structure and randomness, arithmetic progression, and the primes, *Proceedings of the International Congress of Mathematicians* (Madrid, 2006) (European Math. Society, ed.), vol. 1, 2007, pp. 581–608.

[91] X. Zhu, Colouring graphs with bounded generalized colouring number, *Discrete Math.,* **309** (2009), no. 18, 5562–5568.

Jaroslav Nešetřil
Department of Applied Mathematics and Institute of Theoretical Computer Science (ITI)
Charles University
Malostranské nám. 25
11800 Praha 1
Czech Republic

e-mail: nesetril@kam.ms.mff.cuni.cz

Patrice Ossona de Mendez
Centre d'Analyse et de Mathématiques Sociales
CNRS, UMR 8557
54 Bd Raspail
75006 Paris
France

e-mail: pom@ehess.fr

BOLYAI SOCIETY
MATHEMATICAL STUDIES, 21

An Irregular Mind
(Szemerédi is 70)
pp. 491–524.

Squares in Sumsets

HOI H. NGUYEN and VAN H. VU*

To Anna and Endre

A finite set A of integers is square-sum-free if there is no subset of A sums up to a square. In 1986, Erdős posed the problem of determining the largest cardinality of a square-sum-free subset of $\{1,\dots,n\}$. Answering this question, we show that this maximum cardinality is of order $n^{1/3+o(1)}$.

1. Introduction

Let A be a set of numbers. We denote by S_A the collection of finite partial sums of A,

$$S_A := \left\{ \sum_{x\in B} x;\ B\subset A,\ 0<|B|<\infty \right\}.$$

For a positive integer $l\le |A|$ we denote by l^*A the collection of partial sums of l elements of A,

$$l^*A := \left\{ \sum_{x\in B} x;\ B\subset A,\ |B|=l \right\}.$$

Let $[x]$ denote the set of positive integers at most x. In 1986, Erdős [5] raised the following question:

Question 1.1. *What is the maximal cardinality of a subset A of $[n]$ such that S_A contains no square?*

*The authors are partially supported by NSF and AFORS grants.

We denote by $SF(n)$ the maximal cardinality in question. Erdős observed that

$$(1) \qquad SF(n) = \Omega\big(n^{1/3}\big).$$

To see this, consider the following example

Example 1.2. Let p be a prime and k be the largest integer such that $kp \leq n$. We choose p of order $n^{2/3}$ such that $k = \Omega\big(n^{1/3}\big)$ and $1+\cdots+k < p$. Then the set $A := \{p, 2p, \dots, kp\}$ is square-sum-free.

Remark 1.3. The fact that p is a prime is not essential. The construction still works if we choose p to be a square-free number, namely, a number of the form $p = p_1 \dots p_l$ where p_i are different primes.

Erdős [5] conjectured that $SF(n)$ is close to the lower bound in (1). Shortly after Erdős' paper, Alon [1] proved the first non-trivial upper bound

$$(2) \qquad SF(n) = O\left(\frac{n}{\log n}\right).$$

Next, Lipkin [9] improved to

$$(3) \qquad SF(n) \leq n^{3/4+o(1)}.$$

In [2], Alon and Freiman improved the bound further to

$$(4) \qquad SF(n) \leq n^{2/3+o(1)}.$$

The latest development was due to Sárközy [11], who showed

$$(5) \qquad SF(n) = O\big(\sqrt{n \log n}\,\big).$$

In this paper, we obtain the asymptotically tight bound

$$(6) \qquad SF(n) \leq n^{1/3+o(1)}.$$

Theorem 1.4. *There is a constant C such that for all $n \geq 2$*

$$(7) \qquad SF(n) \leq n^{1/3}(\log n)^C$$

In fact, we are going to prove the following (seemingly) more general theorem

Theorem 1.5. *There is a constant C such that the following holds for all sufficiently large n. Let d be positive integer less than $n^{2/3}(\log n)^{-C}$ and A be a subset of cardinality $n^{1/3}(\log n)^{C}$ of $[n/d]$. Then there exists an integer z such that $dz^2 \in S_A$.*

Theorem 1.4 is the special case when $d = 1$. Furthermore, Theorem 1.4 implies many special cases of Theorem 1.5. To see this, choose A to have the form $A := \{db \mid b \in B\}$ where B is a subset of $[n/d]$ and d is a square-free-number. Then finding a square in S_A is the same as finding a number of the form dz^2 in S_B.

If one replaces squares by higher powers, then the problem becomes easier and asymptotic bounds have been obtained earlier (see next section).

Notations. We use Landau asymptotic notation such as O, Ω, Θ, o throughout the paper, under the assumption that $n \to \infty$. Notation such as $\Theta_c(.)$ means that the hidden constant in Θ depends on a (previously defined) quantity c. We will also omit all unnecessary floors and ceilings. All logarithms have natural base. As usual, $e(x)$ means $\exp(2\pi i x) = \cos 2\pi x + i \sin 2\pi x$.

2. The Main Ideas

The general strategy for attacking Question 1.1 is as follows. One first tries to show that if $|A|$ is sufficiently large, then S_A should contain a large additive structure. Next, one would argue that a large additive structure should contain a square.

In previous works [1, 2, 9, 11], the additive structure was a (homogeneous) arithmetic progression. (An arithmetic progression is homogeneous if it is of the form $\{ld, (l+1)d, \dots, (l+k)d\}$.) It is easy to show that if P is a homogeneous AP of length $C_0 m^{2/3}$ in $[m]$, for some large constant C_0, then P contains a square. Notice that the set S_A is a subset of $[m]$ where $m := |A|n$. Thus, if one can show that S_A contains a homogeneous AP of length $C_0 m^{2/3}$, then we are done. Sárközy could prove that this is indeed the case, given $|A| \geq C_1\sqrt{n \log n}$ for a properly chosen constant C_1. This also solves (asymptotically) the problem when squares are replaced by higher powers, since in these cases, the lower bound (which can be obtained by modifying Example 1.2) is $\Omega(\sqrt{n})$.

Unfortunately, $\sqrt{n}$ is the limit of this argument, since there are examples of a subset A of $[n]$ of size $\Omega(\sqrt{n})$ where the longest AP in S_A is much shorter than $(|A|n)^{2/3}$. In order to present such an example, we will need the following definition (which will play a crucial role in the rest of the paper).

Definition 2.1 (Generalized arithmetic progression-GAP). A generalized arithmetic progression of rank r is a set of the form

$$Q = \{a_0 + x_1a_1 + \cdots + x_ra_r \mid 0 \le x_i \le L_i\}.$$

If all the sums $x_1a_1 + \cdots + x_da_d$ are distinct, we say that Q is *proper*. If $a_0 = 0$, we say that Q is *homogeneous*. (Homogeneous arithmetic progression thus corresponds to the case $r = 1$.) We call $L_1, \ldots, L_r$ the sizes of Q and $a_1, \ldots, a_r$ its steps.

Example 2.2. Consider

$$A := \{q_1x_1 + q_2x_2 \mid 1 \le x_i \le N\}$$

where $q_1 \approx q_2 \approx n^{3/4}$ are different primes and $N = \frac{1}{100}n^{1/4}$. It is easy to show that A is a proper GAP of rank 2 and S_A is contained in the proper GAP

$$\{q_1x_1 + q_2x_2 \mid 1 \le x_i \le 1 + \cdots + N\}.$$

Thus, the longest AP in S_A has length at most $1 + \cdots + N = \Theta(n^{1/2})$, while A has cardinality $\Theta(n^{1/2})$.

The key fact that enables us to go below $\sqrt{n}$ and reach the optimal bound $n^{1/3}$ is a recent theorem of Szemerédi and Vu [12] that showed that if $|A| \ge Cn^{1/3}$ for some sufficiently large constant C, then S_A does contain a large proper GAP of rank at most 2.

Lemma 2.3 ([12]). *There are positive constants C and c such that the following holds. If A is a subset of $[n]$ of cardinality at least $Cn^{1/3}$, then S_A contains either an AP Q of length $c|A|^2$ or a proper GAP Q of rank 2 and cardinality at least $c|A|^3$.*

Ideally, the next step would be showing that a large proper GAP Q (which is a subset of $[|A|n]$) contains a square. Thanks to strong tools from number theory, this is not too hard (though not entirely trivial) if Q is homogeneous. However, we do not know how to force this assumption.

The assumption of homogeneity is essential, as without this, one can easily run into local obstructions. For example, if Q is a GAP of the form

$$\{a_0 + a_1x_1 + a_2x_2 \mid 0 \le x_i \le L\}$$

where both a_1 and a_2 are divisible by 6, but $a_0 \equiv 2 \pmod 6$, then clearly Q cannot contain a square, as 2 is not a square modulo 6.

In order to overcome this obstacle, we need to add several twists to the plan. First, we are going to use only a small subset A' of A to create a large GAP Q. Assume that Q has the form

$$\{a_0 + a_1x_1 + a_2x_2 \mid 0 \le x_i \le L\}.$$

(Q can also have rank one but that is the simpler case.) Let q be the g.c.d of a_1 and a_2. If a_0 is a square modulo q, then there is no local obstruction and in principle we can treat Q as if it was homogeneous.

In the next move, we try to add the remaining elements of A (from $A'' := A \backslash A'$) to a_0 to make it a square modulo q. This, however, faces another local obstruction. For instance, if in the above example, all elements of A'' are divisible by 6, then a_0 will always be 2 (mod 6) no matter how we add elements from A'' to it.

Now comes a key point. A careful analysis reveals that having all elements of A'' divisible by the same integer (larger than one, of course) is the *only* obstruction. Thus, we obtain a useful dichotomy: either S_A contains a square or there is an integer $d > 1$ which divides all elements of a large subset A'' of A.

Now we keep working with A''. We can write this set as $\{db \mid b \in B\}$ where B is a subset of $[n/d]$. In order to show that $S_{A''}$ contains a square, it suffices to show that S_B contains a number of the form dz^2. This explains the necessity of Theorem 1.5.

A nice feature of the above plan is that it also works for the more general problem considered in Theorem 1.5. We are going to iterate, setting new $A := A''$ of the previous step. Since the number of iterations (i.e., the number of p's) is only $O(\log n)$, if we have $|A''| \ge \big(1 - \frac{1}{(\log n)^c}\big)|A|$ in each step, for a sufficiently large constant c, then the set A'' will never be empty and this guarantees that the process should terminate at some point, yielding the desired result.

In the next lemma, which is the main lemma of the paper, we put these arguments into a quantitative form.

Lemma 2.4. *The followings holds for any sufficiently large constant C. Let p be positive integer less than $n^{2/3}(\log n)^{-C}$ and A be a subset of $[n/p]$ of cardinality $n^{1/3}(\log n)^C$. Then there exists $A' \subset A$ of cardinality $|A'| \leq n^{1/3}(\log n)^{C/3}$ such that one of the followings holds (with $A'' := A \backslash A'$)*

- *$S_{A'}$ contains a GAP*
$$Q = \{r + qx \mid 0 \leq x \leq L\}$$
where $L \geq n^{2/3}(\log n)^{C/4}$ and $q \leq \frac{n^{2/3}(\log n)^{C/12}}{p}$ and $r \equiv pz^2 \pmod q$ for some integer z.
- *$S_{A'}$ contains a proper GAP*
$$Q = \big\{r + q(q_1x_1 + q_2x_2) \mid 0 \leq x_1 \leq L_1,\ 0 \leq x_2 \leq L_2,\ (q_1, q_2) = 1\big\}$$
such that $\min(L_1, L_2) \geq n^{1/3}(\log n)^{C/4}$, $L_1L_2 \geq n(\log n)^{C/2}$, $q \leq \frac{n^{1/3}}{(\log n)^{C/6}p}$ and $r \equiv pz^2 \pmod q$ for some integer z.
- *There exists an integer $d > 1$ such that $d|a$ for all $a \in A''$.*

Given this lemma, we can argue as before and show that after some iterations, one of the first two cases must occur. We show that in these cases the GAP Q should contain a number of the form pz^2, using classical tools from number theory (see Section 9 and Section 10).

The proof of Lemma 2.4 is technical and requires a preparation involving tools from both combinatorics and number theory. These tools will be the focus of the next two sections.

3. Tools from Additive Combinatorics

This section contains tools from additive combinatorics, which will be useful in the proof of Lemmas 3.7 and 2.4. Let X, Y be two sets of numbers. We define
$$X + Y := \{x + y \mid x \in X,\ y \in Y\}; \quad X - Y := \{x - y \mid x \in X,\ y \in Y\}.$$
A translate of a set X is a set X' of the form $X' := \{a + x \mid x \in X\}$. For instance, every GAP is a translate of a homogeneous GAP.

The first tool is the so-called Covering lemma, due to Ruzsa (see [10] or [14, Lemma 2.14]).

Lemma 3.1 (Covering Lemma). *Assume that X, Y are finite sets of integers. Then X is covered by at most $|X+Y|/|Y|$ translates of $Y-Y$.*

The second tool is the powerful inverse theorem of Freiman [6], [14, Chapter 5]

Lemma 3.2 (Freiman's inverse theorem). *Let γ be a given positive number. Let X be a set in $\mathbf{Z}$ such that $|X+X| \leq \gamma|X|$. Then there exists a proper GAP P of rank at most $d = d(\gamma)$ and cardinality $O_\gamma\big(|X|\big)$ that contains X.*

Assume that Q is a GAP, $Q = \{a_0 + x_1a_1 + \ldots x_ra_r \,:\, 0 \leq x_i \leq L_i,\ 1 \leq i \leq r\}$. For any $t > 0$, denote by tQ the set

$$\{ta_0 + x_1a_1 + \cdots + x_ra_r \,:\, 0 \leq x_i \leq tL_i,\ 1 \leq i \leq r\}.$$

We say that Q is t-proper if tQ is proper. In general, a GAP is not necessarily t-proper. However, one can embed it into a t-proper one with some small loss (see [3, 4], [14, Theorem 3.40]).

Lemma 3.3 (Embedding into proper GAP). *Let Q be a GAP of rank r in a torsion-free group G, and let $t \geq 1$. Then there exists a t-proper GAP Q' with rank at most r and $|Q'| \leq (2t)^r r^{6r^2}|Q|$. Furthermore, if Q is not proper, we may choose Q' to have rank at most $r-1$.*

Next, assume that A is a dense subset of a GAP Q, then the iterated sumsets kA contains a structure similar to Q (see [12, Lemma 4.4], [13, Lemma B3]).

Lemma 3.4 (Sárközy-type theorem in progressions). *Let Q be a proper GAP in a torsion-free group of rank r. Let $X \subset Q$ be a subset such that $|X| \geq \delta|Q|$ for some $0 < \delta < 1$. Then there exists a positive integer $1 \leq m \ll_{\delta,r} 1$ such that mX contains a GAP Q' of rank r and size $\Theta_{\delta,r}\big(|Q|\big)$. Furthermore, the generators of Q' are bounded multiples of the generators of Q. If Q and X are symmetric, then Q' can be chosen to be symmetric.*

A more general result holds when we replace one set by many sets from the same GAP (see [12, Lemma 5.5]).

Lemma 3.5 (Sárközy-type theorem in progressions, generalized form). *Let Q be a proper GAP in a torsion-free group of rank r. Let $0 < \delta \leq 1$ be a given constant. Then there exists a positive integer $1 \leq m \ll_{\delta,r} 1$ such that the following holds. If $X_1, \ldots, X_m \subset Q$ and $|X_i| \geq \delta|Q|$, then $X_1 + \cdots + X_m$ contains a GAP Q' of rank r and size $\Theta_{\delta,r}\big(|Q|\big)$. Furthermore, the generators of Q' are bounded multiples of the generators of Q.*

We say that a GAP $Q = \{a_0 + x_1 a_1 + \dots x_d a_d \mid 0 \le x_i \le L_i\}$ is *positive* if its steps a_i's are positive. A useful observation is that if the elements of Q are positive, then Q itself can be brought into a positive form.

Lemma 3.6. *A GAP with positive elements can be brought into a positive form.*

Proof. (Proof of Lemma 3.6.) Assume that

$$Q = \{a_0 + x_1 a_1 + \cdots + x_d a_d \mid 0 \le x_i \le L_i\}.$$

By setting $x_i = 0$, we can conclude that $a_0 > 0$. Without loss of generality, assume that $a_1, \dots, a_j < 0$ and $a_{j+1}, \dots, a_d > 0$. By setting $x_i = 0$ for all $i > j$ and $x_i = L_i, i \le j$, we have

$$a_0' := a_0 + a_1 L_1 + \dots a_j L_j > 0.$$

Now we can rewrite Q as

$$Q := \left\{a_0' + x_1(-a_1) + \cdots + x_j(-a_j) + x_{j+1}a_{j+1} + \cdots + x_d a_d \mid 0 \le x_i \le L_i\right\},$$

completing the proof. ■

Since we only deal with positive integers, this lemma allows us to assume that all GAPs arising in the proof are in positive form.

Using the above tools and ideas from [12], we will prove Lemma 3.7 below, which asserts that if a set A of $[n/d]$ is sufficiently dense, then there exists a small set $A' \subset A$ whose subset sums contain a large GAP Q of small rank. Furthermore, the set $A'' = A \backslash A'$ is contained in only a few translates of Q. This lemma will serve as a base from which we will attack Lemma 2.4, using number theoretical tools discussed in the next section.

Lemma 3.7. *The following holds for all sufficiently large constants C. Let p be positive integer less than $n^{2/3}(\log n)^{-C}$ and A be a subset of $[n/p]$ of cardinality $n^{1/3}(\log n)^C$. Then there exists a subset A' of A of cardinality $|A'| \le n^{1/3}(\log n)^{C/3}$ such that one of the followings holds (with $A'' := A \backslash A'$):*

- *$S_{A'}$ contains an AP*

$$Q = \{r + qx \mid 0 \le x \le L\}$$

where $L \ge n^{2/3}(\log n)^{C/2}$ and there exist $m = O(1)$ different numbers $s_1, \dots, s_m$ such that $A'' \subset \{s_1, \dots, s_m\} + Q$.

- $S_{A'}$ *contains a proper GAP*

$$Q = \{r + a_1x_1 + a_2x_2 \mid 0 \le x_1 \le L_1,\ 0 \le x_2 \le L_2\}$$

such that $L_1L_2 \ge n(\log n)^{C/2}$ *and there exist* $m = O(1)$ *numbers* $s_1, \dots, s_m$ *such that* $A'' \subset \{s_1, \dots, s_m\} + Q$.

Remark. The proof actually gives a better lower bounds for L_1L_2 in the second case ($2C/3$ instead of $C/2$), but this is not important in applications.

4. Tools from Number Theory

Fourier Transform and Poisson summation. Let f be a function with support on $\mathbf{Z}$. The Fourier transform $\widehat{f}$ is defined as

$$\widehat{f}(w) := \int_{\mathbf{R}} f(t)e(-wt)\,dt.$$

The classical Poisson summation formula asserts that

$$\sum_{n=-\infty}^{\infty} f(t+nT) = \frac{1}{T}\sum_{m=-\infty}^{\infty} \widehat{f}\left(\frac{2\pi m}{T}\right) e(mt/T). \tag{8}$$

For more details, we refer to [8, Section 4.3].

Smooth indicator functions. We will use the following well-known construction (see for instance [7, Theorem 18] for details).

Lemma 4.1. *Let* $\delta < 1/16$ *be a positive constant and let* $[M, M+N]$ *be an interval. Then there exists a real function* f *satisfying the following*

- $0 \le f(x) \le 1$ *for any* $x \in \mathbf{R}$.
- $f(x) = 0$ *if* $x \le M$ or $x \ge M+N$.
- $f(x) = 1$ *if* $M + \delta N \le x \le M + N(1-\delta)$.
- $\big|\widehat{f}(\lambda)\big| \le 16\widehat{f}(0)\exp\big(-\delta|\lambda N|^{1/2}\big)$ *for every* λ.

A Weyl type estimate. Next, we need a Weyl type estimate for exponential sums.

Lemma 4.2. *For any positive constant ε there exist positive constants $\alpha = \alpha(\varepsilon)$ and $c(\varepsilon)$ such that the following holds. Let a, q be co-prime integers, θ be a real number, and I be an interval of length N. Let M be a positive number such that $MN \geq q^{1+\varepsilon}$. Then,*

$$\sum_{\substack{|m|\leq M\\ m\neq 0}} \left| \sum_{z\in I} e\left(\frac{amz^2}{q} + \theta mz\right)\right| \leq c\left(M\sqrt{N} + \frac{MN}{\sqrt{q}}\right)(\log MN)^{\alpha}.$$

Quadratic residues. Finally, and most relevant to our problem, we need the following lemma, which shows the existence of integer solutions with given constrains for a quadratic equation.

Lemma 4.3. *There is an absolute constants D such that the following holds. Let $a_1, \dots, a_d, r, p, q$ be integers such that $p, q > 0$ and $(a_1, \dots, a_d, q) = 1$. Then the equation*

$$a_1x_1 + \cdots + a_dx_d + r \equiv pz^2 \ (\mathrm{mod} q) \tag{9}$$

has an integer solution $(z, x_1, \dots x_d)$ satisfying $0 \leq x_i \leq (pq)^{1/2}(\log q)^D$.

The rest of the paper is organized as follows. The proof of the combinatorial statement, Lemma 3.7, comes first in Section 5. We then start the number theoretical part by giving a proof for Lemma 4.2. The verification of Lemma 4.3 comes in Section 7. After all these preparations, we will be able to establish Lemma 2.4 in Section 8. The proof of the main result, Theorem 1.5, is presented in Sections 9 and 10.

5. Proof of Lemma 3.7

We repeat some arguments from [12] with certain modifications. The extra information we want to get here (compared with what have already been done [12]) is the fact that the set A'' is covered by only few translates of Q.

5.1. An algorithm

Let A' be a subset of cardinality $|A'| = n^{1/3}(\log n)^{C/3}$ and let $A'' := A \backslash A'$. By a simple combinatorial argument (see [12, Lemma 7.9]), we can find in A' disjoint subsets $A'_1, \dots, A'_{m_1}$ such that $|A'_i| \le 20 \log_2 |A'|$ and $|l_1^* A'_i| \ge |A'|/2$ where

$$l_1 \le 10 \log_2 |A'| \quad \text{and} \quad m_1 = |A'|/(40 \log_2 |A'|). \tag{10}$$

(For the definition of $l^* A$ see the beginning of the introduction.)

Without loss of generality, we can assume that m_1 is a power of 4. Let $B_1, \dots, B_{m_1}$ be subsets of cardinality $b_1 = |A'|/2$ of the sets $l_1^* A'_1, \dots, l_1^* A'_{m_1}$ respectively. Following [12, Lemma 7.6]), we will run an algorithm with the B_i's as input. The goal of this algorithm is to produce a GAP which has nice relations with A'' (while still not as good as the GAP we wanted in the lemma). In the next few paragraphs, we are going to describe this algorithm.

At the first step, set $B_1^1 := B_1, \dots, B_{m_1}^1 := B_{m_1}$ and let $\mathfrak{B}^1 = \{B_1^1, \dots, B_{m_1}^1\}$. Let h be a large constant to be determined later.

At the $(t+1)$-th step, we choose indices i, j and elements $a_1, \dots, a_h \in A''$ that maximizes the cardinality of $\bigcup_{d=1}^h (B_i^t + B_j^t + a_d)$ (if there are many choices, choose one arbitrarily). Define $B_1^{t+1'}$ to be the union. Delete from A'' the used elements $a_1, \dots, a_h$, and remove from $\mathfrak{B}^t$ the used sets B_i^t, B_j^t. Find the next maximum union $\bigcup_{k=1}^h B_i^t + B_j^t + a_k$ with respect to the updated sets $\mathfrak{B}^t$ and A''.

Assume that we have created $m_{t+1} := m_t/4$ sets $B_1^{t+1'}, \dots, B_{m_{t+1}}^{t+1\ '}$. By the algorithm, we have

$$|B_1^{t+1'}| \ge \cdots \ge |B_{m_{t+1}}^{t+1\ '}| := b_{t+1}.$$

Now for each $1 \le i \le m_{t+1}$ we choose a subset B_i^{t+1} of cardinality exactly b_{t+1} in $B_i^{t+1'}$. These m_{t+1} sets (of the same cardinality) from a collection $\mathfrak{B}^{t+1}$, which is the output of the $(t+1)$-th step.

Since $m_{t+1} = m_t/4$, there are still $m_t/2$ unused sets B_i^t left in $\mathfrak{B}^t$. Without loss of generality, assume that those are $B_1^t, \dots, B_{m_t/2}^t$. With a slight abuse of notation, we use A'' at every step, although this set loses a few elements each time. (The number of deleted elements is very small compared to the size of A''.)

Let $l_{t+1} := 2l_t + 1$. Observe that

- $l_t \le 2^t l_1$ (by definition);
- $b_t \le l_t n/p$ (since $\bigcup_{d=1}^h \left(B_i^{t-1} + B_j^{t-1} + a_d\right) \subset [l_t n/p]$);
- (11) $$\left|\bigcup_{d=1}^h B_i^t + B_j^t + a_d\right| \le b_{t+1}$$

  for all $1 \le i < j \le m_t/2$ and $a_1, \dots, a_h \in A''$ (by the algorithm, as it always chooses a union with maximum size).

Now let c be a large constant and k be the largest index such that $b_i \ge cb_{i-1}$ for all $i \le k$. Then we have

$$c^k b_1 \le b_k \le l_k n/p.$$

Since $b_1 = |A'|/2$ and $l_k \le 2^k l_1$, we deduce an upper bound for k,

$$k \le \log_{c/2} \frac{l_1 n}{b_1 p}.$$

Next, by the definition of k, we have $b_{k+1} \le cb_k$. By (11), the following holds for all unused sets B_i^k, B_j^k (with $1 \le i \le j \le m_k/2$) and for all $a_1, \dots, a_h \in A''$:

$$\left|\bigcup_{d=1}^h \left(B_i^k + B_j^k + a_d\right)\right| \le b_{k+1} \le cb_k = c\left|B_i^k\right|.$$

In particular

$$\left|B_1^k + B_i^k\right| \le c\left|B_1^k\right|$$

holds for all $2 \le i \le m_k/2$.

By Plunnecke–Ruzsa estimate (see [14, Corollary 6.28]), we have

$$\left|B_1^k + B_1^k\right| \le c^2\left|B_1^k\right|.$$

It then follows from Freiman's theorem, Theorem 3.2, that there exists a proper GAP R of rank $O_c(1)$, of size $O_c(1)\left|B_1^k\right|$ such that R contains B_1^k. Furthermore, by Lemma 3.1, B_i^k is contained in c translates of $B_1^k - B_1^k$, thus B_i^k is also contained in $O_c(1)$ translates of R.

Before continuing, we would like to point out that the parameter h has not yet played any role in the arguments. The freedom of choosing h will be important in what follows. We are going to obtain the desired GAP Q (claimed in the lemma) from R by a few additional operations.

5.2. Creation of many similar GAPs

One problem with R is that its cardinality can be significantly smaller than the bounds on Q in Lemma 3.7. We want to obtain larger GAPs by adding many translates of R. While we cannot do exactly this, we can do nearly as good by the following argument, which creates many GAPs which are translates of each other and have cardinalities comparable to that of R.

By the pigeon hole principle, for $i \leq m_k/2$, we can find a set $B_i' \subset B_i^k$ with cardinality $\Theta_c(1)b_k$ which is contained in one translate of R.

By Lemma 3.5, there exists $g = O_c(1)$ such that $B_1' + \cdots + B_g'$ contains a proper GAP Q_1 of cardinality $\Theta_c(1)|R|$. Create Q_2 by summing $B_{g+1}', \dots, B_{2g}'$, and so on. At the end we obtain $\frac{m_k}{2g} = \Theta_c(1)m_k$ such GAPs. We can require the Q_i's to have the properties below

- $\operatorname{rank}(Q_i) = \operatorname{rank}(R) = O_c(1)$;
- $|Q_i| = \Theta_c(1)|R| = \Theta_c(1)b_k$;
- each Q_i is a subset of a translate of gR. Thus by Lemma 3.1, R is contained in $O_c(1)$ translates of $Q_i - Q_i$;
- the j-th size of Q_i is different from j-th size of R by a (multiplicative) factor of order $\Theta_c(1)$, for all j;
- the j-th step of Q_i is a multiple of the j-th step of R for all j;

Thus, by the pigeon hole principle and truncation (if necessary) we can obtain $m' = \Theta_c(m_k)$ GAPs, say, $Q_1, \dots, Q_{m'}$, which are translates of each other. An important remark here is that since the Q_i are obtained from summing different B's, the sum $Q_1 + \cdots + Q_{m'}$ is a subset of $S_{A'}$. The desired GAP Q will be a subset of this sum.

5.3. Embedding A''

In this step, we embed A'' in a union of few translates of a GAP Q_1 of constant rank.

We set the (so far untouched) parameter h to be sufficiently large so that

$$\Theta_c(1) = h > c|B_1^k|/|B_1'|.$$

Let d be the largest number such that there are d elements $a_1, \ldots, a_d$ of A'' for which the sets $B_1' + B_2' + a_i$ are disjoint. Assume for the moment that $d \geq h$, then we would have

$$\left| \bigcup_{i=1}^{h} \left(B_1' + B_2' + a_i\right) \right| = h\left| B_1' + B_2' \right| \geq h|B_1'| > c\left| B_1^k \right|$$

However, this is impossible because

$$\bigcup_{i=1}^{h} \left(B_1' + B_2' + a_i\right) \subset \bigcup_{i=1}^{h} \left(B_1^k + B_2^k + a_i\right)$$

and the latter has cardinality less than $c\left| B_1^k \right|$ by definition. Thus we have $d < h$. So $d = O_c(1)$.

Let us fix d elements $a_1, \ldots, a_d$ from A'' which attained the disjointness in the definition of d. By the maximality of d, for any $a \in A''$ there exists a_i so that $\left(B_1' + B_2' + a\right) \cap \left(B_1' + B_2' + a_i\right) \neq \emptyset$. Hence

$$a - a_i \in B_1^k + B_2^k - \left(B_1^k + B_2^k\right) = \left(B_1^k - B_1^k\right) + \left(B_2^k - B_2^k\right) \subset 2R - 2R.$$

Thus A'' is covered by at most $d = O_c(1)$ translates of $2R - 2R$. On the other hand, since R is contained in $O_c(1)$ translates of $Q_1 - Q_1$, $2R - 2R$ is contained in $O_c(1)$ translates of $4Q_1 - 4Q_1$. It follows that that A'' is covered by $O_c(1)$ translates of Q_1.

The remaining problem here is that Q_1 does not yet have the required rank and cardinality. We will obtain these by adding the Q_i together (recall that these GAPs are translates of each other) and using a rank reduction argument, following [12] (see also [14, Chapter12]).

5.4. Rank reduction

Let P be the homogenous translate of Q_1(and also of $Q_2, \ldots, Q_{m'}$). Recall that

$$|P| = |Q_1| = \Theta_c(b_k) = \Omega_c\left(c^k b_1\right).$$

and also

$$m' = \Theta_c(m_k) = \Theta_c\left(\frac{b_1}{4^k}\right), \quad \text{and} \quad l_{k+1} \leq 2^{k+1} l_1.$$

Set $l := \min\{m', |A'|/2l_{k+1}\}$. Recall that $|A'| = n^{1/3}(\log n)^{C/2}$, $l_1 \leq 10\log_2|A'|$ and $b_1 = |A'|/2$. By choosing c and C sufficiently large, we can guarantee that

$$(12) \qquad l|P| \geq n^{2/3}(\log n)^{C/2}; \quad l^2|P| \geq n(\log n)^{2C/3}.$$

and also

$$(13) \qquad l^3|P| \geq n^{4/3}(\log n)^C$$

Now we invoke Lemma 3.3 to find a large GAP in lP. Assume, without loss of generality, that $l = 2^s$ for some integer s. We start with $P_0 := P$ and $\ell_0 := l$. If $2^s P_0$ is proper, then we stop. If not, then there exists a smallest index i_1 such that $2^{i_1}P_0$ is proper but $2^{i_1+1}P_0$ is not.

By Lemma 3.3 (applying to $2^{i_1}P_0$) we can find a GAP S which contains $2^{i_1}P_0$ such that $\text{rank}\,(S) < r := \text{rank}\,(2^{i_1}P_0)$.

By Lemma 3.4, there is a constant $g = \Theta_c(1)$ such that the set $2^g(2^{i_1}P_0)$ contains a proper GAP P_1 of rank equals $\text{rank}\,(S)$ and cardinality $\Theta_c(1)\big|2^{i_1}P_0\big|$. Set $\ell_1 := \ell_0/2^{i_1+g}$ if $\ell_0/2^{i_1+g} \geq 1$ and proceed with P_1, ℓ_1 and so on. Otherwise we stop.

Observe that if $2^{i_j}P_j$ is proper, then $\big|2^{i_j}P_j\big| = \big(1+o(1)\big)\,2^{i_j r_j}|P_j|$, where r_j is the rank of P_j.

As the rank of P_0 is $O_c(1)$, and $r_{j+1} \leq r_j - 1$, we must stop after $\Theta_c(1)$ steps. Let Q' be the proper GAP Q' obtained when we stop. It has rank d', for some integer $d' < r$ and cardinality at least $\Theta_c(1)\ell_0^{d'}|P_0| = \Theta_c(1)l^{d'}|P|$. On the other hand, since a translate of lP is contained in $S_{A'}$, $|Q'| \leq |A'|n/p \leq |A'|n$, that is $\Theta_c(1)l^{d'}|P| \leq |A'|n$. Because of (13), this holds only if $d' \leq 2$.

5.5. Properties of Q

By the Covering Lemma 3.1 and by the definition of P_j's, P_i is contained in $O_c(1)$ translates of P_{i+1} for all $i \geq 0$. At the starting point, we know that A'' is contained in $O_c(1)$ translates of P_0. Since there are only $O_c(1)$ different P_j's, at the last step we conclude that A'' is covered by $O_c(1)$ translates of Q'.

Furthermore, Q' is a subset of lP. Thus a translate Q of Q' lies in $Q_1 + \cdots + Q_{m'} \subset S_{A'}$. This Q has rank $1 \leq d' \leq 2$ and cardinality $|Q| =$

$|Q'| \geq \Theta(1)l^{d'}|A'|$. (The right hand side satisfies the lower bounds claimed in Lemma 3.7, thanks to (12).) This is the GAP claimed in Lemma 3.7 and our proof is complete.

6. Proof of Lemma 4.2

If q is a prime, the lemma is a corollary of the well known Weyl's estimate (see [8]. We need to add a few arguments to handle the general case. The following lemma will be useful.

Lemma 6.1. *Let $\tau(n)$ be the number of positive divisors of n. For any given $k \geq 3$ there exists a positive constant $\beta(k)$ such that the following holds for every n.*

$$\tau(n) = O_k\Bigg(\sum_{\substack{d|n \\ d\leq n^{1/k}}} \tau(d)^{\beta(k)} \Bigg).$$

Proof of Lemma 6.1. We can set $\beta(k) = k\log{(k+1)}$. We factorize n in the following specific way

$$n = \prod_{i=1}^{u} p_i^{a_i} \prod_{j=1}^{v} q_j^{b_j}$$

where $p_1 \leq \cdots \leq p_u$, $q_1 \leq \cdots \leq q_v$ are primes and $a_i \geq k > b_j \geq 1$. Set

$$d := \prod_{i=1}^{u} p_i^{\lfloor \frac{a_i}{k} \rfloor} \prod_{j\leq \lfloor \frac{v}{k} \rfloor} q_j.$$

Then $d \leq n^{1/k}$ by definition and

$$(k+1)^k\tau(d)^{\beta(k)} = (k+1)^k 2^{\lfloor \frac{v}{k} \rfloor k\log{(k+1)}} \prod_{i=1}^{u} \left(\left\lfloor \frac{a_i}{k} \right\rfloor + 1\right)^{k\log{(k+1)}}$$

$$\geq (k+1)^v \prod_{i=1}^{u} (1+a_i) \geq \tau(n),$$

completing the proof. ■

Now we start the proof of Lemma 4.2. Let

$$S := \sum_{\substack{|m| \le M \\ m \ne 0}} \left| \sum_{z \in I} e\left(\frac{amz^2}{q} + \theta mz\right)\right|.$$

Following Weyl's argument, we use Cauchy–Schwarz and the triangle inequality to obtain

$$S^2 \le 2M \sum_{\substack{|m| \le M \\ m \ne 0}} \sum_{z_1, z_2 \in I} e\left(\frac{am(z_1 - z_2)(z_1 + z_2)}{q} + \theta m(z_1 - z_2)\right).$$

For convenience, we change the variables, setting $u := z_1 - z_2$, $v := z_2$, then

$$S^2 \le 2M \sum_{\substack{|m| \le M \\ m \ne 0}} \sum_{|u| \le N} e\left(\frac{amu^2}{q} + \theta mu\right) \sum_{v \in I,\ v \in I - u} e\left(\frac{2amuv}{q}\right)$$

$$\le 2M \sum_{\substack{|m| \le M \\ m \ne 0}} \sum_{|u| \le N} \left| \sum_{v \in I,\ v \in I - u} e\left(\frac{2amuv}{q}\right)\right|.$$

Next, using the basic estimate (see [8, Section 8.2], for instance)

$$\left| \sum_{K_0 < k \le K_0 + K} e(\omega k)\right| \le \min\left(K, \frac{1}{\|2\omega\|}\right)$$

we obtain that

$$S^2 \le 2M \sum_{\substack{|m| \le M \\ m \ne 0}} \sum_{|u| \le N} \min\left(N, \frac{1}{\|2amu/q\|}\right).$$

To estimate the right hand side, let N_r be the number of pairs (m, u) such that $2amu \equiv r \pmod q$. (In what follows, it is useful to keep in mind that a and q are co-primes.) We have

$$S(M, N, q)^2 \le 2M\left(N_0 N + \sum_{1 \le r \le q/2} (N_r + N_{q-r})\frac{q}{r}\right). \tag{14}$$

To finish the proof, we are going to derive a (uniform) bound for the N_r's. For $0 \le r \le q-1$ let $0 \le r_a \le q-1$ be the only number such that $ar_a \equiv r \pmod q$. Thus $2amu \equiv r \pmod q$ is equivalent with $2mu \equiv r_a \pmod q$.

First we consider the case $r \neq 0$, thus $r_a \neq 0$. Write $2mu = r_a + sq$. It is clear that $r_a + sq \neq 0$ for all s. Since $2mu \le 2MN$, we have $|s| \le 2MN/q$. For each given s the number of such pairs (m,u) is bounded by $\tau(r_a+sq)$.

Choose $k = \max\left(\frac{1}{\varepsilon}+2, 3\right)$, then $MN/q \ge (MN)^{2/k}$ by the assumption $MN \ge q^{1+\varepsilon}$. It follows from Lemma 6.1 that, for $r \neq 0$,

$$N_r \le \sum_{|s|\le 2MN/q} \tau(r_a+sq) = O_\varepsilon\Bigg(\sum_{d\le (MN)^{1/k}} \tau(d)^{\beta(k)}\Bigg(\sum_{\substack{|s|\le 4MN/q\\ d\mid r_a+sq}} 1\Bigg)\Bigg)$$

$$= O_\varepsilon\Bigg(\sum_{d\le (MN)^{1/k}} \tau(d)^{\beta(k)}\left(\frac{4MN}{qd}+O(1)\right)\Bigg)$$

$$= O_\varepsilon\Bigg(\frac{MN}{q}\sum_{d\le (MN)^{1/k}} \frac{\tau(d)^{\beta(k)}}{d} + O\big((MN)^{1/k+o(1)}\big)\Bigg)$$

$$= O_\varepsilon\Bigg(\frac{MN}{q}\sum_{d\le (MN)^{1/k}} \frac{\tau(d)^{\beta(k)}}{d}\Bigg).$$

Notice that $\sum_{d\le x} \tau(d)^{\beta(k)} \ll x\log^{\beta'(k)} x$ for some positive constant $\beta'(k)$ depending on $\beta(k)$ (see [8, Section 1.6], for instance). By summation by parts we deduce that

$$N_r = O_\varepsilon\left(\frac{MN}{q}\log^{\beta''(k)}(MN)\right)$$

for some positive constant $\beta''(k)$ depending on $\beta'(k)$.

Now we consider the case $r = 0$. The equation $2mu = sq$ has at most $\tau(sq)$ solution pairs (m,u), except when $s = 0$, the case that has $2M$ solutions $\big\{(m,0);\, |m| \le 2M,\, m \neq 0\big\}$. Thus we have

$$N_0 \le 2M + \sum_{|s|\le 2MN/q,\ s\neq 0} \tau(sq),$$

and hence,

$$N_0 = O_\varepsilon\left(2M + \frac{MN}{q}\log^{\beta''(k)}(MN)\right).$$

Combining these estimates with (14), we can conclude that

$$S(M,N,q) \ll_\varepsilon \left(M\sqrt{N} + MN/\sqrt{q}\right)\log^{\alpha}(MN)$$

for some sufficiently large constant $\alpha = \alpha(\varepsilon)$.

7. Proof of Lemma 4.3

We are going to need the following simple fact.

Fact 7.1. Let $a_1, \ldots, a_m, q$ be integers such that $(a_1, \ldots, a_m, q) = 1$. Then we can select a decomposition $q = q_1 \ldots q_l$ of q and l different numbers $a_{i_1}, \ldots, a_{i_l}$ of $\{a_1, \ldots, a_m\}$ (for some $l \geq 1$) such that

$$(q_i, q_j) = 1 \text{ for every } i \neq j \text{ and } (a_{i_j}, q_j) = 1 \text{ for every } j.$$

Proof of Fact 7.1. Let $q = q'_1 \ldots q'_k$ be the decomposition of q into prime powers. For each q'_i we assign a number a'_i from $\{a_1, \ldots, a_m\}$ such that $(q'_i, a'_i) = 1$ (the same a_i may be assigned to many q'_j). Let a_{i_j}'s be the collection of the a'_i's without multiplicity. Set q_j to be the product of all q'_i assigned to a_{i_j}. ■

The core of the proof of Lemma 4.3 will be the following proposition, which is basically the case of one variable in a slightly more general setting.

Proposition 7.2. *There is a constants D such that the following holds. For given integers g, h, p, t, z_1; $g, h, p > 0$ there exist integers $x \in [0, (ph)^{1/2}(\log h)^D]$ and z_2 such that $gx + pz_1^2 + tk \equiv pz_2^2 \pmod{h}$, where $k = (g, h)$.*

Lemma 4.3 follows from Fact 7.1 and Proposition 7.2 by an inductive argument. Indeed, by the above fact we may assume that $q = q_1 \ldots q_l$ where $(a_i, q_i) = 1$, and so

$$(a_l, q) \mid q_1 \cdots q_{l-1}.$$

Now if Lemma 4.3 is true for $l-1$ variables, i.e. there are appropriate $x_1, \ldots, x_{l-1}$ such that $a_1x_1 + \cdots + a_{l-1}x_{l-1} + r = pz_1^2 + tq_1 \ldots q_{l-1}$. Then we apply Proposition 7.2 for $q = h$, $g = a_l$ to find x_l. It thus remains to justify Proposition 7.2.

Proof of Proposition 7.2. Without loss of generality we assume that $h \geq 3$. As $k = (g, h)$, we can write $g = ka, h = kq$ where $(a, q) = 1$. We shall find a solution in the form $z_2 = z_1 + zk$. Plugging in z_2 in this form and simplifying by k, we end up with the equation

$$ax + t \equiv pkz^2 + 2pz_1z \ (\mathrm{mod} q).$$

or equivalently,

$$x \equiv \bar{a}pkz^2 + 2\bar{a}pz_1z - \bar{a}t \ (\mathrm{mod} q) \tag{15}$$

where $\bar{a}$ is the reciprocal of a modulo q, $a\bar{a} \equiv 1 \pmod q$.

Our task is to find $x \in \left[0, (ph)^{1/2}(\log h)^D\right]$ such that (15) holds for some integer z. Notice that if q is small and D is large then $(ph)^{1/2}(\log h)^D \geq (\log 3)^D$, therefore the interval $\left[0, (ph)^{1/2}\right]$ contains every residue class modulo q; as a result, (15) holds trivially. From now on we can assume that q is large,

$$q \geq \exp\left(16\left(6(\alpha+1)/e\right)^{\alpha+1}\right) \tag{16}$$

where c, α are constants arising from Lemma 4.2 with $\varepsilon = 1/3$.

Let $s = (pk, q)$; so we can write $pk = sp', q = sq'$ with $(p', q') = 1$.

Let D be a large constant (to be determined later) and set

$$L := (sq)^{1/2}(\log q)^D/2 \quad \text{and} \quad I := [L, 2L].$$

Note that

$$ph = pkq = sp'q \geq sq.$$

Thus we have

$$I \subset \left[0, (ph)^{1/2}(\log h)^D\right].$$

Let f be a smooth function defined with respect to the interval I (as in Lemma 4.1). For fixed $z \in [1, q]$ the numbers of x in $\left[0, (sq)^{1/2}\log^D q\right]$ satisfying (15) is at least

$$N_z := \sum_{m \in \mathbf{Z}} f\left(\bar{a}pkz^2 + 2\bar{a}pz_1z - \bar{a}t + mq\right).$$

By Poisson summation formula (8)

$$N_z = \sum_{m \in \mathbf{Z}} \frac{1}{q}\widehat{f}\left(\frac{m}{q}\right) e\left(\frac{(\bar{a}pkz^2 + 2\bar{a}pz_1z - \bar{a}t)m}{q}\right).$$

By summing over $z \in [1, q]$ we obtain

$$N := \sum_{z=1}^{q} N_z = \frac{1}{q} \sum_{m \in \mathbf{Z}} \widehat{f}\left(\frac{m}{q}\right) \sum_{z=1}^{q} e\left(\frac{(\bar{a}pkz^2 + 2\bar{a}pz_1 z - \bar{a}t)m}{q}\right).$$

To conclude the proof, it suffices to show that $N > 0$. We are going to show (as fairly standard in this area) that the sum is dominated by the contribution of the zero term.

By the triangle inequality, we have

$$\left|N - \widehat{f}(0)\right| \leq \frac{1}{q} \sum_{m \in \mathbf{Z},\ m \neq 0} \left|\widehat{f}\left(\frac{m}{q}\right)\right| \left|\sum_{z=1}^{q} e\left(\frac{(\bar{a}pkz^2 + 2\bar{a}pz_1 z)m}{q}\right)\right|.$$

Let γ_1, γ_2 be a sufficiently large constant and let

$$L' := \frac{\gamma_1 q (\log q)^{\gamma_2}}{L}.$$

Set

$$S_1 := \frac{1}{q} \sum_{|m| \geq L'} \left|\widehat{f}\left(\frac{m}{q}\right)\right| \left|\sum_{z=1}^{q} e\left(\frac{(\bar{a}pkz^2 + 2\bar{a}pz_1 z)m}{q}\right)\right|$$

and

$$S_2 := \frac{1}{q} \sum_{\substack{|m| \leq L' \\ m \neq 0}} \left|\widehat{f}\left(\frac{m}{q}\right)\right| \left|\sum_{z=1}^{q} e\left(\frac{(\bar{a}pkz^2 + 2\bar{a}pz_1 z)m}{q}\right)\right|.$$

We then have

$$\left|N - \widehat{f}(0)\right| \leq S_1 + S_2.$$

In what follows, we show that both S_1 and S_2 are less than $\widehat{f}(0)/4$.

Estimate for S_1. It is not hard to show that

$$\sum_{k \in \mathbf{Z}} \exp\left(-\sqrt{x|k|}\right) < \frac{5}{x} \quad \text{for} \quad 0 < x < 1.$$

To see this, observe that

$$\sum_{k \geq 1} \exp\left(-\sqrt{xk}\right) \leq \int_0^\infty \exp\left(-\sqrt{xt}\right) dt = \frac{2}{x},$$

where the integral is evaluated by changing variable and integration by parts.

Thus

$$\sum_{|k|\ge k_0} \exp\big(-\sqrt{x|k|}\big) < \sum_{k\in\mathbf{Z}} \exp\left(-\sqrt{x}\left(\frac{\sqrt{|k|}+\sqrt{k_0}}{2}\right)\right) \le \frac{20}{x}\exp\left(-\frac{\sqrt{xk_0}}{2}\right). \tag{17}$$

From the property of f (Lemma 4.1) we can deduce that

$$S_1 \le 16\widehat{f}(0) \sum_{|m|\ge \frac{\gamma_1 q(\log q)^{\gamma_2}}{L}} \exp\big(-\delta\sqrt{|Lm/q|}\big),$$

which, via (17) and since $q \ge 3$, implies

$$S_1 \le 16\widehat{f}(0)\frac{20}{Lq^{-1}}\exp\left(-\frac{\delta\big(\gamma_1(\log q)^{\gamma_2}\big)^{1/2}}{2}\right) \le \widehat{f}(0)/4,$$

given that we choose γ_1, γ_2 sufficiently large.

Estimate for S_2. We have

$$S_2 = \frac{\widehat{f}(0)}{q}\sum_{\substack{|m|\le L'\\ m\ne 0}}\left|\sum_{z=1}^{q} e\left(\frac{\bar{a}p'z^2}{q'} + \frac{2\bar{a}pz_1zm}{q}\right)\right|.$$

We shall choose $D > \gamma_2$.

Set

$$\gamma_1 := \left(\frac{6(D-\gamma_2)}{e}\right)^{D-\gamma_2}.$$

First, we observe that

$$L'q = \frac{2\gamma_1 q^2(\log q)^{\gamma_2}}{(sq)^{1/2}(\log q)^D} = \frac{2\gamma_1 q^{3/2}}{s^{1/2}(\log q)^{D-\gamma_2}} = \frac{2\gamma_1 q'^{1/2}q}{(\log q)^{D-\gamma_2}} \ge q'^{4/3}\frac{\gamma_1 q^{1/6}}{(\log q)^{D-\gamma_2}}.$$

It is not hard to show that the function $q^{1/6}/(\log q)^{D-\gamma_2}$, where $q \ge 3$, attains its minimum at $q = \exp\big(6(D-\gamma_2)\big)$. Therefore, by the choice of γ_1, we have

$$L'q \ge q'^{4/3}.$$

Next, Lemma 4.2 applied for $\varepsilon = 1/3$ (and with the mentioned c and α) yields

$$S_2 = \frac{\widehat{f}(0)}{q} \sum_{\substack{|m| \leq L' \\ m \neq 0}} \left| \sum_{z=1}^{q} e\left(\frac{\bar{a}p'z^2}{q'} + \frac{2\bar{a}pz_1zm}{q}\right) \right|$$

$$\leq c\frac{\widehat{f}(0)}{q}\left(\frac{L'q}{\sqrt{q'}} + L'\sqrt{q}\right)(\log q)^{\alpha}$$

$$\leq 2c\frac{\widehat{f}(0)}{q}\frac{L'q}{\sqrt{q'}}(\log q)^{\alpha} = 2c\frac{\widehat{f}(0)L'}{\sqrt{q'}}(\log q)^{\alpha}.$$

It follows that

$$S_2 \leq \frac{4c\gamma_1 q(\log q)^{\alpha+\gamma_2}}{\left(\sqrt{sq}\log^D q\right)\sqrt{q'}}\widehat{f}(0) = \frac{4c\gamma_1(\log q)^{\alpha+\gamma_2}}{(\log q)^D}\widehat{f}(0).$$

Now we choose D, γ_2 so that $D - \gamma_2 - \alpha = 1$. Thus $\gamma_1 = \big(6(\alpha+1)/e\big)^{\alpha+1}$, and

$$S_2 \leq \frac{4c\gamma_1(\log q)^{\alpha+\gamma_2}}{(\log q)^D}\widehat{f}(0) = \frac{4c\big(6(\alpha+1)/e\big)^{\alpha+1}}{\log q}\widehat{f}(0) \leq \widehat{f}(0)/4$$

where the last inequality comes from (16). ■

Remark 7.3. We can also use Burgess estimate to have an alternative proof with a slightly better bound. However, an improvement in this section does not improve the main theorem.

8. Proof of Lemma 2.4

We first apply Lemma 3.7 to obtain a large proper GAP Q of rank 1 or 2. By this lemma, we have $A'' \subset \{s_1, \ldots, s_m\} + Q$, where m is a constant.

Let $S_i = A'' \cap (s_i + Q)$ for $1 \leq i \leq m$. We would like to guarantee that all S_i are large by the following argument.

If S_i is smaller than $n^{1/3}(\log n)^{3C/10}$, then we delete it from A'' and add to A'. The new sets A', A'' and Q still satisfy the claim of Lemma 3.7.

On the other hand, that the total number of elements added to A' is only $O(n^{1/3}(\log n)^{3C/10} = o(|A'|)$, thus the sizes of A' and A'' hardly changes.

From now on, we assume that $|S_i| \geq n^{1/3}(\log n)^{3C/10}$ for all i.

For convenience, we let

$$s_i' := s_i + r.$$

Thus every element of S_i is congruent with s_i' modulo q.

8.1. Q has rank one

In this subsection, we deal with the (easy) case when Q has rank one. We write $Q = \{r + qx \mid 0 \leq x \leq L\}$ where $L \geq n^{2/3}(\log n)^{C/2}$.

Since $Q \subset S_{A'} \subset \left[\frac{n}{p}|A'|\right]$, we have

$$q \leq \frac{|A'|n}{pL} \leq \frac{n^{2/3}}{(\log n)^{C/6}p}.$$

By setting C (of Lemma 3.7) sufficiently large compared to D (of Lemma 4.3), we can guarantee that

$$(pq)^{1/2}(\log q)^D \leq n^{1/3}. \tag{18}$$

Let $d := (s_1 + r, \ldots, s_m + r, q) = (s_1', \ldots, s_m', q)$. If $d > 1$ then all elements of A'' are divisible by d, since A'' are covered by $\{s_1, \ldots, s_m\} + Q$. Thus we reach the third case of the lemma and are done.

Assume now that $d = 1$. By Lemma 4.3, we can find $0 \leq x_i \leq (pq)^{1/2}(\log q)^D$ such that

$$s_1'x_1 + \cdots + s_m'x_m + r \equiv pz^2 \ (\mathrm{mod} q). \tag{19}$$

Pick from S_i's exactly x_i elements and add them together to obtain a number s. The set $s + Q$ is a translate of Q which satisfies the first case of Lemma 2.4 and we are done.

8.2. Q has rank two

In this section, we consider the (harder) case when Q has rank two. The main idea is similar to the rank one case, but the technical details are somewhat more tedious. We write

$$Q = r + q(q_1 x + q_2 y) \mid 0 \le x \le L_1, \quad 0 \le y \le L_2$$

where $L_1 L_2 = |Q| \ge n \log^{2C/3} n$.

As Q is proper, either $q_1 \ge L_2$ or $q_2 \ge L_1$ holds. Thus $qL_1L_2 \le |A'|n/p$, which yields (with room to spare)

$$q \le \frac{n^{1/3}}{(\log n)^{C/6} p}. \tag{20}$$

We consider two cases. In the first (simple) case, both L_1 and L_2 are large. In the second, one of them can be small.

Case 1. $\min(L_1, L_2) \ge n^{1/3}(\log n)^{C/4}$. Define $d := (s'_1, \ldots, s'_m, q)$ and argue as in the previous section. If $d > 1$, then we end up with the third case of Lemma 2.4. If $d = 1$ then apply Lemma 4.3. The fact that q is sufficiently small (see (20)) and that $|S_i|$ is sufficiently large guarantee that we can choose x_i elements from S_i. At the end, we will obtain a GAP of rank 2 which is a translate of Q and satisfies the second case of Lemma 2.4.

Case 2. $\min(L_1, L_2) \le n^{1/3}(\log n)^{C/4}$. In this case the sides of GAP Q are unbalanced and one of them is much larger than the other. We are going to exploit this fact to create a GAP of rank one (i.e., an arithmetic progression) which satisfies the first case of Lemma 3.7, rather than trying to create a GAP of rank two as in the previous case.

Without loss of generality, we assume that $L_1 \le n^{1/3}(\log n)^{C/4}$. By the lower bound on L_1L_2, we have that $L_2 \ge n^{2/3}(\log n)^{C/4}$. This implies

$$qq_2 \le \frac{|A'|n}{pL_2} \le \frac{n^{2/3}}{(\log n)^{C/12} p}.$$

Again by setting C sufficiently large compared to D, we have

$$(pqq_2)^{1/2}(\log qq_2)^D \le n^{1/3}(\log n)^{C/5}. \tag{21}$$

Creating a long arithmetic progression. In the rest of the proof we make use of A'' and Q to create an AP of type $\{r' + qq_2x_2 \mid 0 \le x_2 \le L_2, r' \equiv pz^2$

(mod qq_2)$\}$. This gives the first case in Lemma 3.7 and thus completes the proof of this lemma.

Let S be an element of $\{S_1, \dots, S_m\}$. Since S is contained in a translate of Q, there is a number s such that any $a \in S$ satisfies $a \equiv s + tqq_1 \pmod{qq_2}$ for some $0 \le t \le L_1$ (for instance, if $a \in S_i$ then $a \equiv s_i' + tqq_1 \pmod{qq_2}$). Let T denote the *multiset* of t's obtained this way. Notice that T could contain one element of multiplicity $|S|$. Also recall that $|S| \ge n^{1/3}(\log n)^{3C/10}$.

For $0 \le l \le |S|/2$, let m_l and M_l (respectively) be the minimal and maximal values of the sum of l elements of T. Since $0 \le t \le L_1$ for every $t \in T$, by swapping summands of m_l with those of M_l, we can obtain a sequence $m_l = n_0 \le \dots \le n_l = M_l$ where each $n_i \in l^*T$ and $n_{i+1} - n_i \le L_1$ for all relevant i.

By construction, we have

$$[m_l, M_l] \subset \{n_0, \dots, n_l\} + [0, L_1] \subset l^*T + [0, L_1]. \tag{22}$$

Next we observe that if l is large and $M_l - m_l$ is small, then T looks like a sequence of only one element with high multiplicity. We will call this element the *essential* element of T.

Proposition 8.1. *Assume that* $\frac{1}{4}(n^{1/3}(\log n)^{3C/10} \le l \le \frac{1}{2}n^{1/3}(\log n)^{3C/10}$ *and* $M_l - m_l < \frac{1}{4}n^{1/3}(\log n)^{3C/10}$. *Then all but at most* $\frac{1}{2}n^{1/3}(\log n)^{3C/10}$ *elements of* T *are the same.*

Proof of Proposition 8.1. Let $t_1 \le t_2 \le \dots \le t_l$ be the l smallest elements of T and $t_1' \le \dots \le t_l'$ be the l largest. By the upper bound on l and lower bound on $|S| = |T|$, $t_1' \ge t_l$. On the other hand, $M_l - m_l = (t_1' - t_1) + \dots + (t_l' - t_l)$. Thus if $M_l - m_l < \frac{1}{4}n^{1/3}(\log n)^{3C/10} \le l - 1$ then $t_i' = t_i$ for some i. The claim follows. ■

The above arguments work for any S among $S_1, \dots, S_m$. We now associate to each S_i a multiset T_i, for all $1 \le i \le m$.

Subcase 2.1. *The hypothesis in Proposition* 8.1 *holds for all* T_i. In this case we move to A' those elements of S_i whose corresponding parts in T_i is not the essential element. The number of elements moved is only $O\big(n^{1/3}(\log n)^{3C/10}\big)$, which is negligible compared to both $|A'|$ and $|A''|$. Furthermore, the properties claimed in Lemma 3.7 remain unchanged and the size of new S_i are now at least $\frac{1}{2}n^{1/3}(\log n)^{3C/10}$.

Now consider the elements of A'' with respect to modulo qq_2. Since each T_i has only the essential element, the elements of A'' produces at most m residues $u_i = s'_i + t_i qq_1$, each of multiplicity at least

$$|S_i| \geq \frac{1}{2} n^{1/3} (\log n)^{3C/10} \geq (pqq_2)^{1/2} (\log qq_2)^D$$

where the last inequality comes from (21). Define $d = (u_1, \ldots, u_m, qq_2)$ and proceed as usual, applying Lemma 4.3.

Subcase 2.2. *The hypothesis in Proposition* 8.1 *does not hold for all* T_i. We can assume that, with respect to T_1, $M_l - m_l \geq \frac{1}{4} n^{1/3} (\log n)^{3C/10}$ for all $\frac{1}{4} n^{1/3} (\log n)^{3C/10} \leq l \leq \frac{1}{2} n^{1/3} (\log n)^{3C/10}$. From now on, fix an l in this interval.

Next, for a technical reason, we extract from S_1 a very small part S'_1 of cardinality $n^{1/3} (\log n)^{C/5}$ and set $S''_1 = S_1 \backslash S'_1$. Let T be the multiset associated with S''_1. We can assume that T satisfies the hypothesis of this subcase.

Define $d := (s'_1, \ldots, s'_m, q)$. As usual, the case $d > 1$ leads to the third case of Lemma 2.4, so we can assume $d = 1$. By Lemma 4.3, there exist integers

$$0 \leq x_i \leq (pq)^{1/2} (\log n)^D \leq n^{1/3} (\log n)^{C/5} \leq |S_i|$$

and k, z_1 such that

$$s'_1 x_1 + \cdots + s'_m x_m + (l s'_1 + r) = p z_1^2 + kq. \tag{23}$$

For $i \geq 2$ we pick from S_i exactly x_i elements $a^i_1, \ldots, a^i_{x_i}$, and for $i = 1$ we pick x_1 elements $a^1_1, \ldots, a^1_{x_1}$ from S'_1 and add them together. By (23) the following holds for some integer k',

$$\sum_{i=1}^{m} \sum_{j=1}^{x_i} a^i_j + (l s'_1 + r) = p z_1^2 + k' q. \tag{24}$$

Furthermore, by Proposition 7.2, as $q = (qq_1, qq_2)$, there exist $0 \leq x \leq (pqq_2)^{1/2} \log^D (qq_2)$ and k'', z_2 such that

$$qq_1 x + p z_1^2 + (k' + m_l q_1) q = p z_2^2 + k'' qq_2, \tag{25}$$

$$p z_1^2 + k' q + (x + m_l) qq_1 = p z_2^2 + k'' qq_2.$$

As $(pqq_2)^{1/2}\log^D(qq_2) \le n^{1/3}\log^{C/5} n$ and $n^{1/3}\log^{C/5} n \le M_l - m_l$, we have

$$m_l \le x + m_l \le M_l.$$

On the other hand, recall that $[m_l, M_l] \subset l^*T + [0, L_1]$ (see (22)), we have

$$\{ls_1' + r + [m_l, M_l]qq_1\} \subset l^*S_1'' + r + [0, L_1]qq_1 \pmod{qq_2}.$$

Thus

$$ls_1' + r + (x + m_l)qq_1 \in l^*S_1'' + r + [0, L_1]qq_1 \pmod{qq_2}. \tag{26}$$

Combining (24),(25) and (26) we infer that there exist l elements $a_1, \dots, a_l$ of S_1'', and there exist $0 \le u \le L_1$ and v such that

$$\sum_{i=1}^{m}\sum_{j=1}^{x_i} a_j^i + a_1 + \cdots + a_l + r + uqq_1 = pz_2^2 + vqq_2.$$

Hence, $\sum_{i=1}^{m}\sum_{j=1}^{x_i} a_j^i + a_1 + \cdots + a_l + Q$ contains the AP $\{(pz_2^2 + vqq_2) + qq_2x_2 \mid 0 \le x_2 \le L_2\}$, completing Subcase 2.2.

Finally, one checks easily that the number of elements of A'' involved in the creation of pz_2^2 in all cases is bounded by $O(n^{1/3}\log^{C/5} n) = o(|A'|)$, thus we may put all of them to A' without loss of generality.

9. Proof of Theorem 1.5: The Rank One Case

Here we consider the (easy) case when Q (in Lemma 2.4) has rank one. In this case, $S_{A'}$ contains an AP $Q = \{r + qx \mid 0 \le x \le L\}$, where $L \ge n^{2/3}(\log n)^{C/4}$ as in the first statement of Lemma 2.4. We are going to show that Q contains a number of the form pz^2.

Write $r = pz_0^2 + tq$ for some $0 \le z_0 \le q$. Since r is a sum of some elements of A', we have

$$0 \le r \le |A'|(n/p) \le \frac{n^{4/3}(\log n)^{C/3}}{p}.$$

Thus

$$-pq \le t \le \frac{n^{4/3}(\log n)^{C/3}}{pq}. \tag{27}$$

The interval $[t/pq, (t+L)/pq]$ contains at least two squares because

$$\left(\frac{L}{pq}\right)^2 \geq \frac{n^{4/3}(\log n)^{C/2}}{(pq)^2} \geq 10\frac{t}{pq} + 20.$$

Thus, we can find an integer $x_0 \geq 0$ such that $\frac{t}{pq} < x_0^2 < (x_0+1)^2 \leq \frac{t+L}{pq}$. It is implied that (since $0 \leq z_0 \leq q$)

$$t \leq pqx_0^2 + 2pz_0x_0 \leq t + L. \tag{28}$$

Set $z := z_0 + qx_0$. We have

$$pz^2 = pz_0^2 + q(pqx_0^2 + 2pz_0x_0).$$

On the other hand, by (28), the right hand side belongs to

$$pz_0^2 + q[t, t+L] = pz_0^2 + tq + q[0, L] = r + q[0, L] = Q.$$

Thus, Q contains pz^2, completing the proof for this case.

10. Proof of Theorem 1.5: The Rank Two Case

In this case, we assume that $S_{A'}$ contains a proper GAP as in the second statement of Lemma 2.4. We can write

$$Q = \left\{r + q(q_1x_1 + q_2x_2) \mid 0 \leq x_1 \leq L_1,\ 0 \leq x_2 \leq L_2,\ (q_1, q_2) = 1\right\}$$

where

- $\min(L_1, L_2) \geq n^{1/3}(\log n)^{C/4}$,
- $L_1L_2 \geq n(\log n)^{C/2}$,
- $q \leq \frac{n^{1/3}(\log n)^{-C/6}}{p}$,
- and $r = pz_0^2 + tq$ for some integers t and $0 \leq z_0 \leq q$.

Since r is a sum of some elements of A', we have $0 \leq r \leq \frac{n^{4/3}(\log n)^{C/3}}{p}$, and so

$$-pq \leq t \leq \frac{n^{4/3}(\log n)^{C/3}}{pq}.$$

Without loss of generality, we assume that $q_2L_2 \geq q_1L_1$. Because Q is proper, either $q_2 \geq L_1$ or $q_1 \geq L_2$. On the other hand, if $q_2 < L_1$ then $L_2 \leq q_1$, which is impossible by the assumption. Hence,

$$q_2 \geq L_1.$$

Now we write $Q = \{pz_0^2 + q(q_1x_1 + q_2x_2 + t) \mid 0 \leq x_1 \leq L_1, 0 \leq x_2 \leq L_2, (q_1, q_2) = 1\}$ and notice that if we set $w := z_0 + zq$ then

$$pw^2 - pz_0^2 = p(z_0 + qz)^2 - pz_0^2 = q(pqz^2 + 2pz_0z).$$

Thus if there is an integer z satisfies

$$pqz^2 + 2pz_0z \in \{q_1x + q_2y + t \mid 0 \leq x \leq L_1,\ 0 \leq y \leq L_2\} \tag{29}$$

then $pw^2 \in Q$, and we are done with this case. The rest of the proof is the verification of the following proposition, which shows the existence of a desired z.

Proposition 10.1. *There exists an integer z which satisfies* (29).

Proof of Proposition 10.1. The method is similar to that of Lemma 4.3, relying on Poisson summation.

Set $a := pq$ and $b := 2pz_0$. Notice that since $0 \leq z_0 \leq q$, $0 \leq b \leq 2pq = 2a$. Our task is to find a z such that

$$az^2 + bz - q_1x - t = q_2y \quad \text{for some} \quad 0 \leq x \leq L_1, \quad 0 \leq y \leq L_2.$$

Define (with foresight; see (31)) $I_x := [L_1/8, L_1/4]$ and

$$I_z := \left[\left(\frac{q_1L_1/4 + t}{a}\right)^{1/2} + 1, \left(\frac{q_2L_2 + q_1L_1/8 + t}{a}\right)^{1/2} - 1\right].$$

(Notice the that the lower bounds on L_1, L_2 and the upper bound on pq guarantee that the expressions under the square roots are positive.)

Since $r + qq_1L_1 + qq_2L_2 = pz_0^2 + tq + q(q_1L_1 + q_2L_2) \in Q$, it follows that (with max (Q) denoting the value of the largest element of Q)

$$q_2L_2 + q_1L_1/8 + t \leq \max(Q)/q \leq \frac{p^{-1}n^{4/3}(\log n)^{C/3}}{q} = \frac{n^{4/3}(\log n)^{C/3}}{a}.$$

Thus

$$|I_z| \geq \frac{1}{4}\frac{(q_2L_2 - q_1L_1/4)a^{-1}}{\sqrt{\frac{q_2L_2+q_1L_1/8+t}{a}}} \tag{30}$$

$$|I_z| = \Omega\left(\frac{q_2L_2}{n^{2/3}(\log n)^{C/6}}\right).$$

By the definitions of I_x and I_z, we have, for any $x \in I_x$ and $z \in I_z$

$$0 \leq az^2 + bz - q_1x - t \leq a(z+1)^2 - q_1x - t \leq q_2L_2. \tag{31}$$

Thus, for any such pair of x and z, if $az^2 + bz - q_1x - t$ is divisible by q_2, then $y := (az^2 + bz - q_1x - t)/q_2$ is an integer in $[1, L_2]$. We are now using the ideas from Section 7, with respect to modulo q_2 and the intervals I_x, I_z.

Let $\bar{q_1}$ be the reciprocal of q_1 modulo q_2 (recall that $(q_1, q_2) = 1$). Let f be a function given by Lemma 4.1 with respect to the interval I_x. For a given $z \in I_z$, the number of $x \in I_x$ satisfying (29) is at least N_z, where

$$N_z := \sum_{m\in\mathbf{Z}} f(\bar{q_1}az^2 + \bar{q_1}bz - \bar{q_1}t + mq_2).$$

By applying Poisson summation formula (8) and summing over z in I_z we obtain

$$N := \sum_{z\in I_z} N_z = \sum_{m\in\mathbf{Z}} \frac{1}{q_2}\widehat{f}\left(\frac{m}{q_2}\right)\sum_{z\in I_z} e\left(\frac{(\bar{q_1}az^2 + \bar{q_1}bz - \bar{q_1}t)m}{q_2}\right).$$

It suffices to show that $N > 0$. Similar to the proof of Lemma 4.3, we will again show that the right hand side is dominated by the contribution at $m = 0$. By triangle inequality, we have

$$\left|N - \frac{1}{q_2}\widehat{f}(0)|I_z|\right| \leq \sum_{\substack{m\in\mathbf{Z}\\ m\neq 0}} \frac{1}{q_2}\left|\widehat{f}\left(\frac{m}{q_2}\right)\right|\left|\sum_{z\in I_z} e\left(\frac{(\bar{q_1}az^2 + \bar{q_1}bz - \bar{q_1}t)m}{q_2}\right)\right|.$$

Let γ be a sufficiently large constant and let

$$L' := \frac{8q_2(\log q_2)^\gamma}{L_1}.$$

We have

$$\left|N - \frac{1}{q_2}\widehat{f}(0)|I_z|\right| \leq S_1 + S_2$$

where

$$S_1 := \sum_{|m|\geq L'} \frac{1}{q_2}\left|\widehat{f}\left(\frac{m}{q_2}\right)\right|\left|\sum_{z\in I_z} e\left(\frac{(\bar{q_1}az^2 + \bar{q_1}bz - \bar{q_1}t)m}{q_2}\right)\right|$$

and

$$S_2 := \sum_{\substack{|m|\leq L' \\ m\neq 0}} \frac{1}{q_2}\left|\widehat{f}\left(\frac{m}{q_2}\right)\right|\left|\sum_{z\in I_z} e\left(\frac{(\bar{q_1}az^2 + \bar{q_1}bz - \bar{q_1}t)m}{q_2}\right)\right|.$$

To conclude the proof, we will show that both S_1 and S_2 are $o\big(\frac{\widehat{f}(0)|I_z|}{q_2}\big)$.

Estimate for S_1. By the property of f,

$$S_1 \leq \frac{\widehat{f}(0)|I_z|}{q_2} \sum_{|m|\geq \frac{8q_2(\log q_2)^\gamma}{L_1}} \exp\Big(-\delta\sqrt{|mL_1/(8q_2)|}\Big).$$

By (17), and as q_2 is large ($q_2 \geq L_1 > n^{1/3}$), the inner sum is $o(1)$, so

$$S_1 = o\left(\frac{\widehat{f}(0)|I_z|}{q_2}\right) \tag{32}$$

as desired.

Estimate for S_2. Let $q' = (\bar{q_1}a, q_2)$. We can write

$$\bar{q_1}a = q'q_1', \quad q_2 = q'q_2' \quad \text{with} \quad (q_1', q_2') = 1. \tag{33}$$

Then

$$S_2 \leq \frac{\widehat{f}(0)}{q_2} \sum_{\substack{|m|\leq L' \\ m\neq 0}} \left|\sum_{z\in I_z} e\left(\frac{q_1'mz^2}{q_2'} + \frac{(\bar{q_1}bz - \bar{q_1}t)m}{q_2}\right)\right|.$$

By Lemma 4.2 there are absolute constants c, α such that

$$S_2 \leq c\frac{\widehat{f}(0)}{q_2}\left(L'\sqrt{|I_z|}(\log n)^\alpha + \frac{L'|I_z|(\log n)^\alpha}{\sqrt{q_2'}}\right).$$

To show that $S_2 = o\big(\frac{\widehat{f}(0)|I_z|}{q_2}\big)$, it suffices to show that

(34) $$L'(\log n)^\alpha = o\big(\sqrt{|I_z|}\big)$$

and

(35) $$L'(\log n)^\alpha = o(q_2')$$

To verify (34), notice that by (30), we have

$$|I_z|L_1^2 = \Omega\left(\frac{L_1^2 q_2 L_2}{n^{2/3}(\log n)^{C/6}}\right).$$

Thus

$$\frac{|I_z|}{L'^2(\log n)^{2\alpha}} = \Omega\left(\frac{|I_z|L_1^2}{q_2^2(\log n)^{2\alpha+2\gamma}}\right) = \Omega\left(\frac{L_1^2 L_2^2}{L_2 q_2 n^{2/3}(\log n)^{C/6+2\alpha+2\gamma}}\right).$$

Since $(L_1L_2)^2 \geq \big(n(\log n)^{C/2}\big)^2 = n^2\log^C n$ and $L_2 q_2 = O\big(\max(Q)\big) = O\big(p^{-1}n^{4/3}(\log n)^{C/3}\big)$, the last formula is $\omega(1)$ if we set C sufficiently large compared to α and γ. This proves (34).

As a result,

$$\frac{\widehat{f}(0)}{q_2}L'\sqrt{|I_z|}(\log n)^\alpha = o\big(\widehat{f}(0)|I_z|/q_2\big).$$

Now we turn to (35). Recall that $q_2 = q'q_2'$ and $q' = (\bar{q}_1 a, q_2) = (a, q_2)$ (as q_1 and q_2 are co-primes). Thus

$$q_2' \geq \frac{q_2}{a} = \frac{q_2}{pq}.$$

To show (35), it suffices to show that

$$\frac{q_2}{pq} = \omega\big(L'^2(\log n)^{2\alpha}\big)$$

which (taking into account the definition of L') is equivalent to

$$q_2 L_1^2 = \omega\big(pq q_2^2(\log n)^{2\alpha+2\gamma}\big).$$

Multiplying both sides with $L_2 q_2^{-1}$, it reduces to

$$L_1^2 L_2 = \omega\big(pq q_2 L_2(\log n)^{2\alpha+2\gamma}\big).$$

Now we use the fact that $qq_2L_2 = O\big(\max(Q)\big) = O\big(p^{-1}n^{4/3}(\log n)^{C/3}\big)$ and the lower bounds $L_1L_2 \geq n(\log n)^{C/2}$ and $L_1 \geq n^{1/3}(\log n)^{C/4}$. The claim follows by setting C sufficiently large compared to α and γ, as usual. Our proof is completed.

Acknowledgements. The authors would like to thank Henryk Iwaniec for helpful discussions.

References

[1] N. Alon, Subset sums, *Journal of Number Theory,* **27** (1987), 196–205.

[2] N. Alon and G. Freiman, On sums of subsets of a set of integers, *Combinatorica,* **8** (1988), 297–306.

[3] Y. Bilu, *Structure of sets with small sumset,* Structure theory of set addition. Asterisque, **258** (1999), xi, 77–108.

[4] M. C. Chang, Generalized arithmetical progressions and sumsets, *Acta Math. Hungar.,* **65** (1994), no. 4, 379–388.

[5] P. Erdős, *Some problems and results on combinatorial number theory,* Proc. 1st. China Conference in Combinatorics (1986).

[6] G. Freiman, *Foundations of a structural theory of set addition,* translated from the Russian, Translations of Mathematical Monographs, Vol. 37. American Mathematical Society, Providence, R. I., 1973.

[7] B. Green, *An exposition on triples in Arithmetic progression,* `http://www.dpmms.cam.ac.uk/~ bjg23/papers/bourgain-roth.pdf`.

[8] H. Iwaniec and E. Kowalski, *Analytic number theory,* American Mathematical Society, Colloquium publications, Volume 53.

[9] E. Lipkin, On representation of r-powers by subset sums, *Acta Arithmetica,* **52** (1989), 114–130.

[10] I. Ruzsa, *An analogue of Freiman's theorem in group,* Structure theory of set addition. Asterisque, **258** (1999), 323–326.

[11] A. Sárközy, Finite addition theorems, II, *Journal of Number Theory,* **48** (1994), 197–218.

[12] E. Szemerédi and V. H. Vu, Long arithmetic progressions in sumsets: Thresholds and Bounds, *Journal of the A.M.S,* **19** (2006), no. 1, 119–169.

[13] T. Tao, *Freiman's theorem in solvable groups,* `http://arxiv.org/abs/0906.3535`.

[14] T. Tao and V. H. Vu, *Additive combinatorics,* Cambridge University Press, 2006.

Hoi H. Nguyen
Department of Mathematics
Rutgers University
Piscataway
NJ 08854
USA

e-mail: hoi@@math.rutgers.edu

Van H. Vu
Department of Mathematics
Rutgers University
Piscataway
NJ 08854
USA

e-mail: vanvu@@math.rutgers.edu

BOLYAI SOCIETY
MATHEMATICAL STUDIES, 21

An Irregular Mind
(Szemerédi is 70)
pp. 525–559.

Are There Arbitrarily Long Arithmetic Progressions in the Sequence of Twin Primes?

JÁNOS PINTZ*

Dedicated to Endre Szemerédi on the occasion of his 70th birthday

1. Introduction

The problem in the title seemed to be out of reach of any methods before 2004. We have still no answer for it, and it is no surprise that we will not answer it in the present work either. However, in the last few years the following developments arose in connection with the above problem.

Theorem A (Green and Tao [10]). *The primes contain arbitrarily long arithmetic progressions.*

Theorem B (Goldston, Pintz, Yıldırım [8]). *If the level ϑ of distribution of primes exceeds $1/2$, then there exists a positive $d \le C_1(\vartheta)$, such that there are infinitely many generalized twin prime pairs n, $n+d$. If $\vartheta > 0.971$ we have $C_1(\vartheta) = 16$.*

Let us call a k-tuple $\mathcal{H} = \{h_i\}_{i=1}^k$ consisting of non-negative integers *admissible* if it does not cover all residue classes modulo any prime p. Theorem B was the consequence of the sharper result that if $\vartheta > 1/2$ and $k \ge C_2(\vartheta)$, then any admissible k-tuple $\mathcal{H}$, that is the set $n+\mathcal{H}$ contains at least two primes for infinitely many values of n.

*Supported by OTKA Grants K72731, K67676 and ERC-AdG.228005.

We say that ϑ is a level of distribution of primes if for every $A > 0$ and $\varepsilon > 0$ we have

$$\sum_{q \le N^{\vartheta-\varepsilon}} \max_{\substack{a \\ (a,q)=1}} \left| \sum_{\substack{p \le N \\ p \equiv a(q)}} \log p - \frac{N}{\varphi(q)} \right| \ll_{\varepsilon,A} \frac{N}{(\log N)^A}. \tag{1.1}$$

The fact that $\vartheta = 1/2$ is an admissible level of distribution of primes is the celebrated Bombieri–Vinogradov theorem. This just missed an unconditional proof of the existence of infinitely many generalized twin prime pairs. However, their theorem was crucial in the proof of

$$\Delta = \liminf_{n\to\infty} \frac{p_{n+1} - p_n}{\log p_n} = 0 \qquad [8], \tag{1.2}$$

and in its improvement ($\log_2 x = \log\log x$)

$$\liminf_{n\to\infty} \frac{p_{n+1} - p_n}{\sqrt{\log p_n}(\log_2 p_n)^2} < \infty \quad [9]. \tag{1.3}$$

Assuming the Elliott–Halberstam conjecture [1] to be abbreviated later by EH, which states that $\vartheta = 1$ is an admissible level, or, just assuming that $\vartheta > 0.971$ we obtained in [8] infinitely many gaps of size at most 16 in the sequence of primes, that is $C_1(0.971) = 16$. In the following let p' denote the prime following p.

The aim of the present work is to combine the methods of [10] and [8] in order to show, even in a stronger form with consecutive primes p, $p' = p+d$, the following result.

Theorem 1. *If the level ϑ of distribution of primes exceeds $1/2$, then there exists a positive $d \le C_1(\vartheta)$ so that there are arbitrarily long arithmetic progressions of primes p such that $p' = p + d$ is the prime following p for each element of the progression. If $\vartheta > 0.971$ then the above holds for some d with $d \le 16$.*

In such a way we can show a positive answer to a weaker form of the question mentioned in the title (where twin primes are substituted by generalized twin primes) under the unproved condition that the exponent $1/2$ in the Bombieri–Vinogradov theorem can be improved to a $\vartheta = 1/2 + \delta$, where δ is an arbitrarily small, but fixed positive number. We have to note, however, that such a quantitatively tiny improvement is probably very difficult. For example, the Generalized Riemann Hypothesis (GRH) trivially implies that $\vartheta = 1/2$ is admissible but it does not provide any admissible level beyond that.

Our analysis, more exactly, Theorem 5 will show that the answer for the question in the title would be positive if one could show $\pi_2(x) \geq cx/\log^2 x$ for the number of twin primes up to x.

2. The Methods of Green–Tao and Goldston–Pintz–Yildirim

At the first sight it seems that there is no serious problem in combining the methods of [10] and [8] in order to show Theorem 1, since [10] uses weights from Selberg's sieve

$$\Lambda_R(n) := \sum_{d \leq R,\ d|n} \mu(d) \log \frac{R}{d}, \tag{2.1}$$

and applies the method of Goldston and Yıldırım [11] to construct a pseudo-random measure where the set of primes has positive density. Similar weights are used in the proof of Theorem B in [8]. There are, however, very important differences as well.

1) The weights (2.1) and the method used by Goldston and Yıldırım [11] led only to the weaker result $\Delta = 1/4$. The value of the parameter R was a very small power of the size of the primes $p \asymp N$. In this approach of Goldston and Yıldırım primes were searched in admissible tuples (a k-tuple $\mathcal{H}$ is called admissible if it does not cover all residue classes mod p for any prime p)

$$\mathcal{H} = \{h_1, \ldots, h_k\}, \quad 0 \leq h_1 < h_2 < \cdots < h_k, \quad h_i \in \mathbb{Z} \tag{2.2}$$

and they used the weights

$$\Lambda_R(n; \mathcal{H}) := \prod_{i=1}^{k} \Lambda_R(n + h_i). \tag{2.3}$$

2) In the work [8] yielding Theorem B and (1.2), it was crucial to use the maximal possible value of R satisfying $R^2 \leq N^{\vartheta - \varepsilon}$ (that is, $R = N^{\frac{\vartheta - \varepsilon}{2}}$)

allowed by the known information (1.1) about primes. Additionally, the weights (2.3) had to be replaced by (the factor $1/(k+l)!$ is insignificant here, it serves just for normalization)

(2.4)

$$\Lambda_R(n;\mathcal{H},k+l) := \frac{1}{(k+l)!}\sum_{\substack{d\mid P_{\mathcal{H}}(n)\\ d\le R}} \mu(d)\log^{k+l}\frac{R}{d}, \quad P_{\mathcal{H}}(n) := \prod_{i=1}^{k}(n+h_i)$$

with l and k being bounded but arbitrarily large, $l = o(k)$.

If one imposed any upper bound on at least one of k or l, our method would not lead to $\Delta = 0$ in (1.2). Similarly if we had at our disposal just any level $\vartheta < 1/2$ (say $\vartheta = 0.49999$) we would not be able to deduce $\Delta = 0$ (it would lead however to $\Delta \le c_0(\vartheta)$, with $c_0(\vartheta) \to 0$ as $\vartheta \to 1/2$). In case of Theorem B, if one uses k and l with $\min(l,k) \le C_0$, then beyond $\vartheta > 1/2$ we would need $\vartheta > 1/2 + c(C_0)$.

3) The main difficulty is that the number of generalized twin primes up to X, produced by Theorem B did not yield a lower bound for them beyond the very weak implicit bound

$$N^{1-C(k)/\log\log N}. \tag{2.5}$$

This is far from the expected order of magnitude (the singular series $\mathfrak{S}(d) = \mathfrak{S}(\{0,d\})$ is defined in (5.5))

$$\frac{\mathfrak{S}(d)N}{\log^2 N} \sim \frac{\mathfrak{S}(d)\pi(N)}{\log N}, \tag{2.6}$$

which is still just a density of size $c/\log N$ among the primes. We also do not know whether the primes p_n satisfying $p_{n+1} - p_n < \eta \log p_n$ have a positive density if $\eta < 1/4$. It will turn out, however, that the first part of the proof (Section 6) can help to answer this question positively too. We shall return to this problem in a later work.

3. The Principal Idea of the Proof

The seemingly simplest solution, to work with the same weight function during the whole proof, seems to be impossible, due to the reasons mentioned in Section 2. So we will use two different weight functions:

(i) the first one of type (2.4) to produce the generalized twin prime pairs (with an $R = N^{\vartheta/2-\varepsilon}$), afterwards

(ii) the second one of type (2.3) with a different $R = N^{\delta}$, with a small value of δ.

However, this still does not solve our problem that we are not able to produce a good lower bound for the number of generalized twin primes. How do we find the set (or, more precisely the pseudorandom measure ν) where the generalized twin primes produced in the first step are contained with positive density? In case of the primes in Green–Tao's theorem this set was the set of "almost primes", more precisely the measure (2.1).

It is of crucial importance that the generalized twin prime pairs $n, n+d$ ($n \in \mathcal{N}_1$) were produced in [8] as two primes within a set of type

$$\{n + h_i\}_{i=1}^{k} := n + \mathcal{H}, \quad \mathcal{H} \text{ admissible.} \tag{3.1}$$

The main idea is to embed these twin primes into the set of almost prime k-tuples with pattern $\mathcal{H}$; more precisely to consider just that part $\mathcal{N}_2 \subset \mathcal{N}_1$ for which the other components $n + h_j$ are almost primes (we could simply say that all components are almost primes since primes are considered themselves as almost primes). In the usual literature almost primes P_r are considered as numbers with a bounded number of prime factors. (A P_r number is by definition a number with at most r prime factors.) The sieve methods producing almost primes usually automatically produce almost primes satisfying

$$P^{-}(n) > n^{c}, \qquad c > 0, \text{ fixed,} \tag{3.2}$$

where $P^{-}(n)$ denotes the least prime factor of n. On the other hand, numbers with property (3.2) are automatically P_r almost primes for any $r \geq \lfloor 1/c \rfloor$. It will be important in the sequel that talking about almost primes we need in the proofs always the stronger sense (3.2). Concerning almost primes we have to mention an analogue of the Dickson–Hardy–Littlewood prime k-tuple conjecture for almost primes, proved first by Halberstam and Richert [12] (cf. Theorem 7.4) and later in a sharper form by Heath-Brown [13]. Let $\Omega(n)$ denote the number of prime divisors of n.

Theorem C. *For any admissible k-tuple $\mathcal{H}$ we have infinitely many values of n such that $P_{\mathcal{H}}(n) = \prod_{i=1}^{k}(n + h_i)$ is a P_K almost prime with $K \leq C_3(k)$, that is, $\Omega\big(P_{\mathcal{H}}(n)\big) \leq C_3(k)$.*

Halberstam and Richert proved additionally

(i) $P^-(n+h_i) \geq n^{c/k}$ $(i=1,2,\dots,k)$, $C_3(k) = \bigl(1+o(1)\bigr) k \log k$,

where $c = 2/5$ can be chosen if k is large enough, while Heath-Brown showed that there are infinitely many n values with

(ii) $\max\limits_{i \leq k} \Omega(n+h_i) \leq \bigl(1+o(1)\bigr) \dfrac{2}{\log 2} \log k$.

Theorem C raises the problem, whether we can combine it with Theorem B in order to show the existence of infinitely many n values such that all elements $n+h_i$ are almost primes and at least two of them are primes under the hypothesis $\vartheta > 1/2$. It turns out that it is "easier" to prove a common generalization of Theorems A, B, and C than just of Theorems A and B. In fact our strategy in proving Theorem 1 will be as follows.

Step 1. We show that the procedure of Theorem B yields in fact for most of the "good" values of n not just two primes in the k-tuple considered, but also almost primes for each component $n+h_i$, $i \leq k$.

Step 2. We show that although the produced generalized twin primes probably do not form a set of positive density within the set of all generalized twin primes, however, the produced n values with at least two primes $n+h_i$ form a set of positive density among all n's satisfying $P^-(n+h_i) > n^{c_1(k)}$ for each $h_i \in \mathcal{H}$.

Step 3. To show that the measure $\overline{\nu}(n)$ derived from the weight function $\Lambda_R(n;\mathcal{H})$ with the new parameter $R = N^\delta$ (with a small positive constant δ) is a pseudorandom measure and the produced tuples form a set of positive measure.

In this way we obtain a common generalization of Theorems A, B and C, thereby a sharper form of Theorem 1 as follows.

Theorem 2. *Let us suppose that $\vartheta = 1/2+\delta > 1/2$ is a level of distribution of primes. Let $\mathcal{H} = \{h_i\}_{i=1}^k$ be an admissible k-tuple with $k \geq C_0(\vartheta) = \bigl(2\lceil 1/2\delta \rceil + 1\bigr)^2$. Then we have with some values $b_i \leq C_3(k)$, with at least two of the b_i's being equal to 1, arbitrarily long arithmetic progressions of n values such that $\Omega(n+h_i) = b_i$, $P^-(n+h_i) > n^{c_1(k)}$. If $\vartheta \geq 0.971$ then this is true for $k \geq 6$. Further, the same is true with at least two consecutive primes in the set $n+\mathcal{H}$.*

Remark. In other words, for sufficiently large $k \geq C_0(\vartheta)$ the same multiplicative structure of $n+\mathcal{H}$ (i.e. $\Omega(n+h_i) = b_i$) containing at least two primes and almost primes elsewhere, appears arbitrarily many times with equal distances among the neighboring constellations (i.e. the n values, thereby the primes $n+h_i$ and $n+h_j$ forming an arithmetic progression). We can even require that $n+h_i$ should have the same exponent pattern, where the multiset $(\alpha_1, \dots, \alpha_j)$ is defined as the exponent pattern of n (see [6]) if

$$n = \Pi p_{\nu_i}^{\alpha_i}, \quad p_{\nu_i} \in \mathcal{P}. \tag{3.3}$$

Remark. We can not determine in advance the places of the two primes in the constellation, neither the values of b_ν apart from $b_\nu \le C_3(k)$. However, they will be the same for each translated copy, that is, for each element n of the arithmetic progressions.

Remark. If we had $b_i = 1 \quad (i = 1, \dots, k)$, that is, $C_3(k) = 1$, this would contain Dickson's conjecture about prime k-tuples, also called the Hardy–Littlewood prime k-tuple conjecture.

Remark. The lower bound $k \ge \left(2\lceil 1/2\delta\rceil + 1\right)^2$ can be improved if δ is not too small.

4. Further Results about Generalized Twin Primes

The execution of Steps 1 and 2 reveals already interesting properties of the weights (2.4) and (coupled with other arguments in some cases) yields or helps to yield important consequences, such as the positive proportion of small gaps of size at most $\eta \log p$ between consecutive primes p and p' for any $\eta > 0$. Therefore it is worth formulating the result of these steps as the following separate theorem.

Theorem 3. *Suppose that the level of distribution of primes is $\vartheta = \frac{1}{2} + \delta > \frac{1}{2}$. If $\mathcal{H} = \{h_i\}_{i=1}^k$ is any admissible k-tuple with $k \ge C_0(\vartheta) = \left(2\lceil 1/2\delta\rceil + 1\right)^2$, then the number of $n \le N$ for which $n + \mathcal{H}$ contains at least two consecutive primes and almost primes in each component satisfying $P^-(n + h_\nu) > n^{c_1(k)}$ is at least*

$$c_1'(\mathcal{H}) \frac{N}{\log^k N} \tag{4.1}$$

for a suitable sequence $N = N_\nu \to \infty$ *with some* $c_1'(\mathcal{H})$, *depending on* $\mathcal{H}$. *Choosing a simple k-tuple with a small diameter we obtain at least*

$$c_2(k)\frac{N}{\log^k N}, \qquad k = \left(2\left\lceil\frac{1}{2\delta}\right\rceil + 1\right)^2 \tag{4.2}$$

generalized twin prime pairs n, $n+d$ *with a difference*

$$d \le C^*(\vartheta) := \big(1+o(1)\big)\,k\log k = \big(2+o(1)\big)\,\delta^{-2}\log\frac{1}{\delta}. \tag{4.3}$$

If the Elliott–Halberstam conjecture or at least $\vartheta > 0.971$ *holds then we obtain at least*

$$c_1\frac{N}{\log^6 N} \tag{4.4}$$

generalized twin prime pairs n, $n+d$ *with a positive* $d \le 16$.

Remark. It is actually the number (4.1) of the obtained k-tuples which enables the use of the procedure of Green and Tao (with a parameter $R < N^{c(k,m)}$, where m is the number of terms in the arithmetic progression) to assure the existence of arbitrarily long arithmetic progressions among the k-tuples $n + \mathcal{H}$ with the above properties.

The earliest known written formulation of the twin prime conjecture seems to be due to de Polignac [14] from the year 1849. This conjecture was already about prime pairs with a general even difference d. This also indicates that the original twin prime conjecture arose probably already earlier. If $d = 2$ or 4 then a pair of primes n, $n + d$ must be clearly a consecutive prime-pair if $n > 3$. On the other hand, this is probably not true for all prime pairs with $d \ge 6$. This would follow for example from the special case $k = 3$ of the Dickson–Hardy–Littlewood prime k-tuple conjecture, stating that any admissible k-tuple contains infinitely many prime k-tuples. The original de Polignac conjecture [14] stated that for any even d

(4.5) $\quad n,\ n+d$ are consecutive primes for infinitely many values n.

A weaker form of this conjecture would be that for any even d

(4.6) $\quad n,\ n+d$ are both primes for infinitely many values n.

As long as in general nearly nothing was known about either (4.5) or (4.6) there was not much reason to discuss the difference between (4.5) and (4.6). Now, in view of our Theorem B, under the plausible hypothesis $\vartheta > 1/2$ it seems to be worth to discuss these aspects. To formulate the results more easily we introduce the following definitions.

Definition 1. We will call an even number d a de Polignac number in the strong sense (briefly strong de Polignac number) if (4.5) is true for it.

Definition 2. We will call an even number d a de Polignac number in the weak sense (briefly weak de Polignac number) if (4.6) is true for it.

Let us denote the set of all strong and weak de Polignac numbers, respectively, by $\mathcal{D}_s$ and $\mathcal{D}_w$. The fact that for $k > C_0(\vartheta)$ every admissible k-tuple $\mathcal{H}$ contains infinitely many times at least two primes implies that the set $\mathcal{D}_w$ of weak de Polignac numbers has a positive lower density depending on ϑ. However, the earlier results could not show that apart from the smallest element d_0 of $\mathcal{D}_w$ any of the others would satisfy (4.5) due to the possible existence of another prime between n and $n + d$, if $d \in \mathcal{D}_w$, $d > d_0$. In such a way we did not have any more information about $\mathcal{D}_s$ beyond $\mathcal{D}_s \neq \emptyset$. This is still highly non-trivial (and still not known unconditionally) since without any hypothesis the following three assertions are clearly equivalent:

(i) there exists at least one strong de Polignac number ($\mathcal{D}_s \neq \emptyset$),

(ii) there exists at least one weak de Polignac number ($\mathcal{D}_w \neq \emptyset$),

(iii) there are infinitely many bounded gaps between consecutive primes, that is, $\liminf_{n\to\infty}(p_{n+1} - p_n) < \infty$.

Now, Theorem 3 changes this great difference about our present knowledge of the cardinality of the weak and strong de Polignac numbers, namely, that under the assumption $\vartheta > 1/2$

$$\underline{d}(\mathcal{D}_w) \geq c(\vartheta), \quad \text{whereas} \quad |\mathcal{D}_s| \geq 1, \tag{4.7}$$

where $\underline{d}(X)$ denotes the lower density, $|X|$ the cardinality of a set X.

In fact, Theorem 3 implies with some relatively easy elementary arguments the following

Theorem 4. *Let us suppose that the primes have a level of distribution* $\vartheta = \frac{1}{2} + \delta > \frac{1}{2}$. *Let* $k = \left(2\lceil 1/2\delta\rceil + 1\right)^2$, $P := P(k) := \prod_{p\leq k} p$. *Then Theorem 3 is true in the stronger form that* $n + \mathcal{H}$ *contains at least two consecutive primes.*

Corollary 1. *We have (φ is Euler's totient function)*

$$d(\mathcal{D}_w) \geq d(\mathcal{D}_s) \geq \frac{\varphi(P)}{Pk(k-1)} \sim \frac{e^{-\gamma}}{k^2 \log k} \sim \frac{\delta^4}{2e^{\gamma} \log(1/\delta)} \tag{4.8}$$

as $\delta \to 0$. If EH or at least $\vartheta > 0.971$ is true then

$$\underline{d}(\mathcal{D}_w) \geq \underline{d}(\mathcal{D}_s) \geq \frac{2}{225}. \tag{4.9}$$

Remark. The above theorem shows that under the Elliott–Halberstam conjecture at least approximately 1.78 percent of all even numbers appear infinitely often as the difference of two consecutive primes, thereby satisfying the original de Polignac conjecture (4.5). Furthermore, the set of such even numbers has a positive lower density for any level $\vartheta > 1/2$ of the distribution of primes.

5. Preparation for Step 1. Notation

In this section we will make preparations for the proof of Theorem 3, which is a refinement of Theorem B. Since we will follow the version of [7] we will reintroduce here its notation and describe the necessary changes to obtain the same result with the extra requirement that (in case of $\vartheta > 1/2$) apart from the two primes we obtain almost primes with $P^-(n + h_i) > n^{c_1(k)}$ in all components. We call the reader's attention here that a different notation will be used in Section 10, when we describe the changes in the procedure of [10], since unfortunately the same variables $k, h_1, \ldots, h_k$ refer to different quantities in the works [7] and [10].

Let $n \sim N$ mean $N < n \leq 2N$, let $\mathcal{P}$ denote the set of primes, and let k, l be arbitrary bounded integers with $k \geq 1$, $l \geq 0$, $\log_2 N = \log\log N$,

$$H \ll \log N \ll \log R \ll \log N, \qquad H \to \infty, \tag{5.1}$$

$$\mathcal{H} = \{h_i\}_{i=1}^k \subseteq [0, H], \quad h_i < h_{i+1}, \quad h_i \in \mathbb{Z}, \tag{5.2}$$

where $\mathcal{H}$ is an admissible k-tuple. For the aim of later use in other works we allow here $H \to \infty$, whereas for the present work it would be enough to suppose $H \leq C(k)$. Constants c, C, ε may be different at different occurrences and they might depend on k, l and ϑ, as well as the constants implied by $\ll$, and O symbols (without indicating the dependence); $\log_\nu x$ denotes the ν-fold iterated logarithmic function. We use the notation $\Omega_{\mathcal{H}}(p)$ and its multiplicative extension $\Omega_{\mathcal{H}}(d)$ for squarefree d to denote the set of residue classes n mod d for which

$$d \mid P_{\mathcal{H}}(n) := \prod_{i=1}^{k}(n+h_i) \Longleftrightarrow n \in \Omega_{\mathcal{H}}(d), \tag{5.3}$$

where we write simply $n \in \Omega_{\mathcal{H}}(d)$ instead of $n(\bmod d) \in \Omega_{\mathcal{H}}(d)$. Further, in accordance with (2.4) we define

(5.4)

$$\Lambda_R(n;\mathcal{H},a) := \sum_{\substack{d \\ n\in\Omega_{\mathcal{H}}(d)}} \lambda_R(d;a), \qquad \lambda_R(d;a) := \frac{\mu(d)}{a!}\left(\left(\log\frac{R}{d}\right)_+\right)^a$$

where $y_+ = y$ for $y \geq 0$ and 0 otherwise. Let $\theta(n) = \log n$ if $n \in \mathcal{P}$ and 0 otherwise. The singular series is defined, as usual, by

$$\mathfrak{S}(\mathcal{H}) := \prod_p \left(1 - \frac{|\Omega_{\mathcal{H}}(p)|}{p}\right)\left(1-\frac{1}{p}\right)^{-k}, \tag{5.5}$$

where $|X|$ denotes the cardinality of the set X. Theorem B was the immediate consequence of Lemma 1 and the case $h \in \mathcal{H}$ of Lemma 2 of [7]. The following two lemmas were proved in [7] for arbitrary but bounded $k \geq 1$, $l \geq 0$.

Lemma 1. *For a sufficiently large $C > C(k,l)$ and $R \leq \sqrt{N}/(\log N)^C$ we have*

(5.6)

$$\begin{aligned} S_0(\mathcal{H}) :&= \sum_{n\sim N} \Lambda_R(n;\mathcal{H},k+l)^2 \\ &= \frac{\mathfrak{S}(\mathcal{H})}{(k+2l)!}\binom{2l}{l} N(\log R)^{k+2l} + O\big(N(\log N)^{k+2l-1}(\log_2 N)^c\big). \end{aligned}$$

Lemma 2. *Suppose the level of distribution of primes is ϑ and $\varepsilon > 0$. Then for $R \leq N^{(\vartheta-\varepsilon)/2}$ we have for $h \leq H$, $\kappa = \kappa(\mathcal{H},h) = \mathbb{1}_{\mathcal{H}}(h) = 1$ if $h \in \mathcal{H}$, $\kappa = 0$ if $h \notin \mathcal{H}$*

$$S_1(\mathcal{H},h) := \sum_{n\sim N} \theta(n+h)\Lambda_R(n;\mathcal{H},k+l)^2 \tag{5.7}$$

$$= \frac{\mathfrak{S}(\mathcal{H} \cup \{h\})}{(k+2l+\kappa)!} \binom{2(l+\kappa)}{l+\kappa} N(\log R)^{k+2l+\kappa}$$

$$+ O\big(N(\log N)^{k+2l+\kappa-1}(\log_2 N)^c\big).$$

Remark. Although [7] contains the condition $l > 0$, but the proof works for $l = 0$, too.

The proof of Theorem B follows from these lemmas by

$$\sum_{n\sim N}\Big(\sum_{h\in\mathcal{H}} \theta(n+h) - \log 3N\Big)\Lambda_R(n;\mathcal{H},k+l)^2 \tag{5.8}$$

$$= \frac{\mathfrak{S}(\mathcal{H})}{(k+2l)!}\binom{2l}{l} N \log N (\log R)^{k+2l}$$

$$\times \left(\frac{k\cdot 2(2l+1)}{(k+2l+1)(l+1)} \cdot \frac{\vartheta - \varepsilon}{2} - 1 + o(1)\right) > 0$$

if the constants ϑ, k, l satisfy the crucial inequality

$$\frac{k}{k+2l+1}\,\frac{2l+1}{l+1}\vartheta > 1, \tag{5.9}$$

since ε can be chosen arbitrarily small after k and l are chosen.

The crucial property of the weights $\Lambda_R(n;\mathcal{H},k+l)^2$ to be proved is that it is concentrated so strongly for almost prime k-tuples satisfying $P^-\big(P_{\mathcal{H}}(n)\big) > R^\eta$ for any $\eta < c(k,l)$ that the sum of those weights $\Lambda_R(n;\mathcal{H},k+l)^2$ $(n \sim N)$ for which $P^-\big(P_{\mathcal{H}}(n)\big) < R^\eta$, is negligible compared with the total sum for all $n \sim N$ if $N \to \infty$, $\eta \to 0$ (k, l, $\mathcal{H}$ being fixed). The same is true for the weighted sum of primes (cf. Lemmas 2 and 5), although this is not needed to prove Theorems 1–4.

6. The Execution of Step 1

The mentioned property of the sieve weights $\Lambda_R(n;\mathcal{H},k+l)$ can be expressed by the following lemmas. The signs $\ll$ and o, O will always be allowed to depend on k and l.

Lemma 3. *Let $N^{c_0} < R \le \sqrt{N/q}(\log N)^{-C}$, $q \in \mathcal{P}$, $\beta := \log q/\log R$, $\beta < c_0$, where c_0 and C are suitably chosen constants depending on k and l. Then we have*

$$\sum_{\substack{n\sim N\\ q|P_{\mathcal{H}}(n)}} \Lambda_R(n;\mathcal{H},k+l)^2 \ll \frac{\beta}{q}\sum_{n\sim N}\Lambda_R(n;\mathcal{H},k+l)^2. \tag{6.1}$$

This immediately implies

Lemma 4. *Let $N^{c_0} < R \le N^{1/(2+\eta)}(\log N)^{-C}$, $\eta > 0$. Then we have*

$$\sum_{\substack{n\sim N\\ (\mathcal{P}_{\mathcal{H}}(n),P(R^\eta))>1}} \Lambda_R(n;\mathcal{H},k+l)^2 \ll \eta\sum_{n\sim N}\Lambda_R(n;\mathcal{H},k+l)^2. \tag{6.2}$$

Remark 1. In some cases we need to use an analogue of (6.2) with the product of the weights $\Lambda_R(n;\mathcal{H},k+l_1)$ and $\Lambda_R(n;\mathcal{H},k+l_2)$ in place of $\Lambda_R(n;\mathcal{H},k+l)^2$ with different values of l_1 and l_2. However, a simple use of Cauchy's inequality reduces the estimates of these quantities to (6.2). The same applies to Lemma 6, which is a simple consequence of Lemma 4.

If the above sum is twisted by primes we can only prove an analogue of Lemma 4.

Lemma 5. *Let $N^{c_0} \le R \le N^{(\vartheta-\varepsilon)/(2+\eta)}(\log N)^{-C}$, $0 < \eta < c_0$, $\varepsilon > 0$. Let $h \le H$, and $\kappa = 1$ if $h \in \mathcal{H}$, $\kappa = 0$ if $h \notin \mathcal{H}$. Then*

$$\begin{aligned}\sum_{\substack{n\sim N\\ (\mathcal{P}_{\mathcal{H}}(n),P(R^\eta))>1}} &\theta(n+h)\Lambda_R(n;\mathcal{H},k+l)^2 \\ &\ll \eta\sum_{n\sim N}\theta(n+h)\Lambda_R(n;\mathcal{H},k+l)^2 \\ &\quad + O\Big(N\big((\log_2 N)^c\big((\log N)^{k+2l+\kappa-1} + (\log N)^{k+l-\frac{1}{2}}\big)\big)\Big).\end{aligned} \tag{6.3}$$

Remark 2. We will not investigate in the present work the dependence of the sign $\ll$ on k and l, although it has some significance for some applications.

Remark 3. In the present applications Lemma 5 will not be used. However, it has a significance in other applications and also here if the dependence on k and l is also considered. At any rate the proof of Lemmas 3–5 will be very similar.

Proof of Lemma 3. We will follow the proof of [7] and just point out the differences. In evaluating the expression (1.4) in [7] we have to take into consideration the extra condition $q \mid \mathcal{P}_{\mathcal{H}}(n) \Leftrightarrow n \in \Omega_{\mathcal{H}}(q)$ which has to be added to the conditions $n \in \Omega_{\mathcal{H}}(d_1)$, $n \in \Omega_{\mathcal{H}}(d_2)$. Therefore the critical quantity $\mathcal{T}$ in the main term $N\mathcal{T}$ will take now the form $N\mathcal{T}'_q$

$$\mathcal{T}'_q = \sum_{d_1,d_2} \frac{\bigl|\Omega_{\mathcal{H}}\bigl([d_1,d_2,q]\bigr)\bigr|}{[d_1,d_2,q]} \lambda_R(d_1;k+l)\lambda_R(d_2;k+l). \tag{6.4}$$

Due to the multiplicative property of $\Omega_{\mathcal{H}}$ this will mean that the main term will be now

$$\frac{N\bigl|\Omega_{\mathcal{H}}(q)\bigr|}{q}\widetilde{\mathcal{T}}_q \quad \text{with} \quad \mathcal{T}'_q = \frac{\bigl|\Omega_{\mathcal{H}}(q)\bigr|}{q}\widetilde{\mathcal{T}}_q, \tag{6.5}$$

where $\widetilde{\mathcal{T}}_q$ can be expressed similarly to [7] by the new generating function $\widetilde{F}_q = F_q^{\#} \cdot F_q$, where $\mathbf{s} = (s_1, s_2)$ and $F_q(\mathbf{s})$ is up to the missing term for $p = q$ the same as F in [7]:

$$\begin{aligned} F_q^{\#}(\mathbf{s}) &:= \left(1 - \frac{1}{q^{s_1}}\right)\left(1 - \frac{1}{q^{s_2}}\right), \\ F_q(\mathbf{s}) &:= \prod_{p \neq q} \left(1 - \frac{\bigl|\Omega_{\mathcal{H}}(p)\bigr|}{p}\left(\frac{1}{p^{s_1}} + \frac{1}{p^{s_2}} - \frac{1}{p^{s_1+s_2}}\right)\right). \end{aligned} \tag{6.6}$$

Analogously to (1.5) of [7] we can define now

$$\widetilde{G}_q(\mathbf{s}) := \widetilde{F}_q(\mathbf{s}) \left(\frac{\zeta(s_1+1)\zeta(s_2+1)}{\zeta(s_1+s_2+1)}\right)^k := F_q^{\#}(\mathbf{s})G_q(\mathbf{s}). \tag{6.7}$$

The appearance of the term $F_q^{\#}(\mathbf{s})$ in $\widetilde{F}_q(\mathbf{s})$ causes additional difficulties in evaluating the expression $\widetilde{\mathcal{T}}_q$ compared to that of the analogous quantity $\mathcal{T}$ of [7]. We have, namely

$$\begin{aligned} \widetilde{\mathcal{T}}_q := \frac{1}{(2\pi i)^2} \int\limits_{(1)} \int\limits_{(1)} G_q(s_1,s_2) \left(\frac{\zeta(s_1+s_2+1)}{\zeta(s_1+1)\zeta(s_2+1)}\right)^k \\ \times \frac{R^{s_1+s_2} - (R/q)^{s_1}R^{s_2} - R^{s_1}(R/q)^{s_2} + (R/q)^{s_1+s_2}}{(s_1 s_2)^{k+l+1}} \, ds_1\, ds_2, \end{aligned} \tag{6.8}$$

where $\int_{(\sigma)}$ means integration over the vertical line $\operatorname{Re} z = \sigma$.

This means that due to the numerator of the last term above we need to evaluate (or at least estimate) integrals of the form

$$\mathcal{T}_q(R_1, R_2) := \frac{1}{(2\pi i)^2} \int\limits_{(1)} \int\limits_{(1)} Z_q(s_1, s_2) \frac{R_1^{s_1} R_2^{s_2}}{(s_1+s_2)^k (s_1 s_2)^{l+1}}\, ds_1\, ds_2, \tag{6.9}$$

where analogously to [7] we define

$$Z_q(\mathbf{s}) = G_q(\mathbf{s}) \left(\frac{(s_1+s_2)\zeta(s_1+s_2+1)}{(s_1\zeta(s_1+1)s_2\zeta(s_2+1))} \right)^k; \tag{6.10}$$

however, in contrast to [7], where $R_1 = R_2 = R$ we have here now, $R_1 = R_2 q^j$ with $j = \pm 1$, so

$$R_1 := R_2^a, \quad a := 1 + \alpha, \quad -2c_0 < \alpha < 2c_0. \tag{6.11}$$

First we will consider the changes in the contribution of the main term corresponding to the residue $s_1 = s_2 = 0$. This term is now, by an argument similar to [7],

$$\begin{aligned} \mathcal{T}_{q,0}(R_1, R_2) = \operatorname*{Res}_{s_2=0} \operatorname*{Res}_{s_1=0} &= \frac{1}{(2\pi i)^2} \int\limits_{C_2} \int\limits_{C_1} \frac{Z_q(s_1, s_2) R_2^{a s_1 + s_2}}{(s_1+s_2)^k (s_1 s_2)^{l+1}}\, ds_1\, ds_2 \\ &= \frac{1}{(2\pi i)^2} \int\limits_{C_3} \int\limits_{C_1} \frac{Z_q(s, s\xi) R_2^{s(a+\xi)}}{(\xi+1)^k \xi^{l+1} s^{k+2l+1}}\, ds\, d\xi \end{aligned} \tag{6.12}$$

where we wrote $s_1 = s$, $s_2 = s\xi$, and C_1, C_2, C_3 are circles with $|s_1| = |s| = \varrho$, $|s_2| = \varrho/2$, $|\xi| = 1/2$, respectively, with a small $\varrho > 0$. Using the analogue of (1.6) of [7] for $G_q(\mathbf{s})$ we can write $\mathcal{T}_{q,0}$ as

(6.13)
$$\mathcal{T}_{q,0}(R_1, R_2) = \frac{Z_q(0,0)}{(k+2l)!} (\log R_2)^{k+2l} \mathcal{T}_{q,1}(a) + O\big((\log N)^{k+2l-1} (\log_2 N)^c \big),$$

since $\frac{\partial^j}{\partial s^j} Z_q(0,0) \ll (\log_2 N)^c$ if $j \le C(k,l)$. The main term is

$$\mathcal{T}_{q,1}(a) := \frac{1}{2\pi i} \int\limits_{C_3} \frac{(a+\xi)^{k+2l}}{(\xi+1)^k \xi^{l+1}}\, d\xi \tag{6.14}$$

$$= \frac{1}{l!} \left[\left(\frac{d}{d\xi} \right)^l \left\{ \left(1 + \frac{\alpha}{1+\xi} \right)^k (1+\alpha+\xi)^{2l} \right\} \right]_{\xi=0}$$

$$= \binom{2l}{l} (1+\alpha)^{k+l} + \sum_{j=1}^{l} \mathcal{T}_j(\alpha),$$

where

$$\mathcal{T}_j(\alpha) := \binom{l}{j} \frac{(2l)!}{l!(l+j)!} (1+\alpha)^{l+j} \left[\left(\frac{d}{d\xi} \right)^j \left(1 + \frac{\alpha}{1+\xi} \right)^k \right]_{\xi=0} . \tag{6.15}$$

We remark here that the simpler case $R_1 = R_2 \Leftrightarrow a = 1$ yielded immediately $\mathcal{T}_{q,1}(1) = \binom{2l}{l}$ in [7]. Due to $j \geq 1$ the jth derivative has the form

$$\left(\frac{d}{d\xi} \right)^{j-1} \left\{ k\alpha \left(1 + \frac{\alpha}{1+\xi} \right)^{k-1} \frac{(-1)}{(1+\xi)^2} \right\}, \tag{6.16}$$

so, after taking again derivatives $j-1$ (≥ 0) times we obtain

$$\mathcal{T}_j(\alpha) \ll \alpha \quad \text{for} \quad j = 1, 2, \ldots, l \tag{6.17}$$

which, finally, by (6.14)–(6.15), yields

$$\mathcal{T}_{q,1}(a) = \mathcal{T}_{q,1}(1+\alpha) = \binom{2l}{l} + O(\alpha). \tag{6.18}$$

Remark. As mentioned after (6.15), (6.14) implies immediately

$$\mathcal{T}_{q,1}(1) = \frac{1}{2\pi i} \int_{C_3} \frac{(1+\xi)^{2\ell}}{\xi^{l+1}} d\xi = \binom{2l}{l}. \tag{6.19}$$

We will show below that with a little extra effort we can show the explicit dependence on k and l:

$$T := \mathcal{T}_{q,1}(1+\alpha) - \mathcal{T}_{q,1}(1) \ll (k + l^{3/2})\alpha \quad \text{if} \quad \alpha \leq \frac{1}{4k}. \tag{6.20}$$

We have, namely for $1 \leq l \leq k$, $C_4 = \left\{ \xi;\ |\xi| = 1 - \frac{1}{2l} \right\}$

$$T := \frac{1}{2\pi i} \int_{C_4} \left(\frac{(1+\alpha+\xi)^k \{ (1+\alpha+\xi)^{2l} - (1+\xi)^{2l} \}}{\xi^{l+1}(1+\xi)^k} \right. \tag{6.21}$$

$$\left. + \frac{(1+\xi)^{2l} \{ (1+\alpha+\xi)^k - (1+\xi)^k \}}{\xi^{l+1}(1+\xi)^k} \right) d\xi$$

$$= \frac{1}{2\pi i}\int\limits_{C_4}\left(1+\frac{\alpha}{1+\xi}\right)^k \frac{(1+\xi+\alpha)^{2l}-(1+\xi)^{2l}}{\xi^{l+1}}\,d\xi$$

$$+\frac{1}{2\pi i}\int\limits_{C_4}\frac{(1+\xi)^{2l}}{\xi^{l+1}}\left\{\left(1+\frac{\alpha}{1+\xi}\right)^k-1\right\}d\xi$$

$$:= T_{1,1}+T_{2,1}+T_{1,2}+T_{2,2},$$

where the first index refers to the first and 2nd integral, resp., the 2nd index refers to $|1+\xi| \ge 1/2$ and $|1+\xi| \le 1/2$, resp. Now we have by $\left(1-\frac{1}{2l}\right)^{-(l+1)} \ll 1$, $\alpha \le 1/4k$, $\binom{2l}{l} \asymp \frac{2^l}{\sqrt{l}}$

(6.22)
$$T_{1,1}+T_{2,1} \ll l\alpha\cdot 2^{2l} + k\alpha\cdot 2^{2l}\sum_{i=1}^{l}\left(1-\frac{i^2}{l^2}\right)^{2l}\cdot\frac{1}{l}$$
$$\ll \binom{2l}{l}\alpha\left(l^{3/2}+k\right),$$

(6.23)
$$T_{1,2}+T_{2,2} \ll \frac{1}{l}\sum_{i=1}^{l|2}\left\{e^{\frac{k\alpha l}{i}} l\alpha\left(\frac{i}{l}\right)^{2l-1}+\left(\frac{i}{l}\right)^{2l}\left(e^{\frac{k\alpha l}{i}}-1\right)\right\}$$
$$\ll \alpha + k\alpha \ll k\alpha.$$

(6.22)–(6.23) imply an explicit form of (6.18) as

(6.24)
$$T_{q,1}(1+\alpha) = \binom{2l}{l}\left(1+O\left(\alpha(k+l^{3/2})\right)\right).$$

Hence an easy calculation shows that the contribution of the residue at $s_1 = s_2 = 0$ in the value of $\widetilde{T}_q$ in (6.8) is

(6.25)
$$\ll \alpha\frac{G_q(0,0)}{(k+2l)!}(\log R)^{k+2l}\binom{2l}{l} + O\left((\log R)^{k+2l-1}\log_2^c N\right),$$

where $\alpha = -\log q/\log R = -\beta$ or $\log q/\log(R/q) = \beta/(1-\beta)$, and

(6.26)
$$G_q(0,0) = \mathfrak{S}(\mathcal{H})\left(1-\frac{|\Omega_{\mathcal{H}}(q)|}{q}\right)^{-1} \ll \mathfrak{S}(\mathcal{H}),$$

since $\left|\Omega_{\mathcal{H}}(q)\right| \le \min(q-1,k)$ because $\mathcal{H}$ is admissible.

We need also a more careful treatment in the error term estimation, since the earlier relation $R^{s_1+s_2} \ll 1$ for $|s_1+s_2| = (\log N)^{-1}$ (cf. (1.7)–(1.9) of [7]) is no longer true if $R^{s_1+s_2}$ is replaced by $R_1^{s_1}R_2^{s_2}$ with $R_1 \ne R_2$ ($R_1 = R$, $R_2 = R/q$ or reversed).

However, due to the symmetry of s_1 and s_2 in (6.9) we may suppose during the estimation of (6.9) that $R_2 \le R_1$. In this case (1.7) of [7] remains valid since in the shifted integral for $s_1 \in L_3$, $s_2 \in L_2$ (L_2 and L_3 are defined as in [7] by $L_2 : c_2/(2\log U) + it$, $|t| \le U/2$, $L_3 : -c_2/\log U + it$, $|t| \le U$ with $U = \exp\left(\sqrt{\log N}\right)$) we have, similarly to [7], but now by $\operatorname{Re} s_2 > 0$

$$\left|{R_1}^{s_1}{R_2}^{s_2}\right| \le \left|{R_1}^{s_1+s_2}\right| \le {R_1}^{-c_2/(2\log U)} \ll \exp\left(-c\sqrt{\log N}\right). \tag{6.27}$$

Further, in the secondary term, writing $C(s_2) : |s_1+s_2| = (\log N)^{-1}$ we obtain again, with the same notation as in [7],

(6.28)

$$I := \frac{1}{2\pi i}\int_{L_2}\left\{\operatorname*{Res}_{s_1=-s_2}\right\} ds_2$$

$$= \frac{1}{(2\pi i)^2}\int_{L_2}\int_{C(s_2)} G_q(s_1,s_2)\left(\frac{\zeta(s_1+s_2+1)}{\zeta(s_1+1)\zeta(s_2+1)}\right)^k \frac{R_1^{s_1}R_2^{s_2}}{(s_1s_2)^{k+l+1}}\, ds_1\, ds_2,$$

and the crucial quantity $R_1^{s_1}R_2^{s_2}$ is for $s_1 \in C(s_2)$, $s_2 \in L_2$

$$\left|R_1^{s_1}R_2^{s_2}\right| \le \left|R_1^{s_1+s_2}\right| \le e^{\log R_1/\log N} < e, \tag{6.29}$$

since $R_2 \le R_1$, $\operatorname{Re} s_2 = c_2/(2\log U) > 0$, and everything else remains valid; yielding the same estimate as in (1.9) of [7]:

$$I \ll (\log N)^{k+l-1/2}(\log_2 N)^c. \tag{6.30}$$

During the proof we used that the only difference between the present $G_q(\mathbf{s})$ and $G(\mathbf{s})$ of [7] (and similarly with F and Z) is that (due to $\left|\Omega_{\mathcal{H}}(q)\right| \le \min(k,q-1)$)

(6.31)

$$\frac{Z_q(\mathbf{s})}{Z(\mathbf{s})} = \frac{G_q(\mathbf{s})}{G(\mathbf{s})} = \frac{F_q(\mathbf{s})}{F(\mathbf{s})} = \left(1 - \frac{\left|\Omega_{\mathcal{H}}(q)\right|}{q}\left(\frac{1}{q^{s_1}} + \frac{1}{q^{s_2}} - \frac{1}{q^{s_1+s_2}}\right)\right)^{-1} \ll 1$$

in both regions $\operatorname{Re} s_1, \operatorname{Re} s_2 \in (-c_3, c_3)$ and $\operatorname{Re} s_1, \operatorname{Re} s_2 \geq 0$, say, if $c_3 < c_0(k)$ is chosen sufficiently small.

Finally the last integral, $s_1 \in L_3$, $s_2 \in L_4 : -c_2/\log U + it$, $|t| \leq U/2$ remains to be $O\big(\exp\big(-c\sqrt{\log N}\big)\big)$ as in [7], which finishes the proof of Lemma 3.

Summation over all primes $q \leq R^\eta$ gives an upper estimate for the LHS of (6.2) and yields Lemma 4, in view of (6.4), (6.5), (6.25)–(6.27), (6.30), (6.31), since

$$\sum_{q \leq R^\eta} \frac{|\Omega_{\mathcal{H}}(q)|}{q} \frac{\log q}{\log R} \ll \eta. \qquad \blacksquare \tag{6.32}$$

The proof of Lemma 5 runs again similarly to that of the analogous Lemma 2 of [7]. The needed changes are essentially the same as described above, so we will be brief. First we remark that during the applications of the Bombieri–Vinogradov theorem or its hypothetical improvement (1.1) the existence of a prime $q \leq R^\eta$ with the extra condition $q \mid \mathcal{P}_{\mathcal{H}}(n)$ means that we need now the stronger condition

$$[q, d_1, d_2] \leq R^{2+\eta} \leq N^{\vartheta - \varepsilon}, \tag{6.33}$$

which, however, appears in the statement of our Lemma 5.

With this change in our assumption the substitution of the contribution of the primes by its expected contribution, the analogues of formulae (2.4)–(2.6) and the displayed inequality following (2.6) in [7] remain valid with the change that $[d_1, d_2]$ has to be replaced always by $[q, d_1, d_2]$. After this substitution, we arrive again at the analogous quantity

(6.34)

$$\prod_{p \mid [q,d_1,d_2]} \Bigg(\sum_{\substack{b (\bmod p) \\ b \in \Omega_{\mathcal{H}}(p)}} \delta\big((b+h,p)\big) \Bigg) = \big(|\Omega^+(q)| - 1\big) \prod_{\substack{p \mid [d_1,d_2] \\ p \neq q}} \big(|\Omega^+(p)| - 1\big),$$

where Ω^+ corresponds to the set $\mathcal{H}^+ = \mathcal{H} \cup \{h\}$ (and as remarked in [7], $|\Omega^+(p)| = p$ can occur now already) and $\delta(m) = 1$ if $m = 1$, $\delta(m) = 0$ if $m \neq 1$.

This yields now, similarly to (6.4)–(6.5) to the slightly modified analogue of $\mathcal{T}^*$ in (2.7) of [7], to the expressions

$$\frac{N\big(|\Omega^+(q)| - 1\big)}{q - 1} \mathcal{T}_q^* \tag{6.35}$$

with

$$(6.36)\qquad \mathcal{T}_q^* = \frac{1}{(2\pi i)^2} \int\limits_{(1)} \int\limits_{(1)} \widetilde{F}_q^*(s_1, s_2) \frac{R^{s_1+s_2}}{(s_1 s_2)^{k+l+1}}\, ds_1\, ds_2,$$

where, with the same $F_q^{\#}(\mathbf{s}) = (1 - q^{-s_1})(1 - q^{-s_2})$ as in (6.6) we have now

(6.37)

$$\widetilde{\mathcal{F}}_q^* = F_q^{\#} \cdot F_q^*, \quad F_q^*(\mathbf{s}) = \prod_{p \neq q} \left(1 - \frac{|\Omega^+(p)| - 1}{p-1} \left(\frac{1}{p^{s_1}} + \frac{1}{p^{s_2}} - \frac{1}{p^{s_1+s_2}}\right)\right).$$

The whole treatment of the error terms is the same as in case of Lemma 3, the only change being in the main term and in the singular series. We have to distinguish two cases (although as mentioned earlier Case 2 is not needed for the present work).

Case 1. $h \in \mathcal{H}$. In this case $\mathcal{H}^+ = \mathcal{H} \cup \{h\} = \mathcal{H}$, $m = 1$, $\Omega^+(d) = \Omega_{\mathcal{H}}(d)$ for every d, the singular series is according to [7], (6.31) and (6.37) now

$$(6.38)\qquad G_q^*(0,0) = \mathfrak{S}_q(\mathcal{H}^+) = \prod_{p \neq q} \left(1 - \frac{|\Omega^+(p)|}{p}\right) \left(1 - \frac{1}{p}\right)^{-(k+1)}$$

$$\ll \mathfrak{S}(\mathcal{H}^+) = \mathfrak{S}(\mathcal{H})$$

and the same reasoning as in [7], the translation $k \to k-1$, $l \to l+1$ gives the result, since, if $n + h \in \mathcal{P}$, then

$$(6.39)\qquad d \mid P_{\mathcal{H}}(n) \Longleftrightarrow d \mid P_{\mathcal{H} \setminus \{h\}}(n).$$

Case 2. $h \notin \mathcal{H}$. In this case $\mathcal{H}^+ = \mathcal{H} \cup \{h\}$, $m = 0$, $\Omega^+(p) = k+1$ for $p > k$ and $\mathcal{H}^+$ is not necessarily admissible. However, $\Omega^+(p) = p$ may occur only for $p \leq k+1$ since $\Omega^+(p) \leq k+1$. If $\Omega^+(p) = p$ is the case for some $p \neq q$, then as remarked in [7] the corresponding Euler product vanishes at $s_1 = 0$ or $s_2 = 0$, the main term lacks $\big(G_q^*(0,0) = \mathfrak{S}_q(\mathcal{H}^+) = 0\big)$ and the error term is the same or actually smaller. Finally if the only prime for which $\Omega^+(p) = p$ holds is $p = q \leq k+1$, then by $\mathfrak{S}\ \big(\mathcal{H} \cup \{h\}\big) = 0$ we use the trivial consequence of Lemma 2:

(6.40)

$$\sum_{\substack{n \sim N \\ (P_{\mathcal{H}}(n), P(R^\eta)) > 1}} \theta(n+h) \Lambda_R(n; \mathcal{H}, k+l)^2 \leq \sum_{n \sim N} \theta(n+h) \Lambda_R(n; \mathcal{H}, k+l)^2$$

$$\ll N (\log R)^{k+2l-1} (\log_2 N)^c.$$

Remark. In most applications we can replace Lemma 5 with the following slightly weaker assertion, which is a trivial consequence of Lemma 4.

Lemma 6. *Let $N^{c_0} < R \le N^{1/(2+\eta)}(\log N)^{-C}$. Then we have for any $h \le H$*

$$\sum_{\substack{n\sim N\\(P_{\mathcal{H}}(n),P(R^\eta))>1}} \theta(n+h)\Lambda_R(n;\mathcal{H},k+l)^2 \ll \eta\log N\sum_{n\sim N}\Lambda_R(n;\mathcal{H},k+l)^2. \tag{6.41}$$

Lemma 6 can relatively well substitute for Lemma 5 if $h \in \mathcal{H}$ and the dependence of the constants on k and l in the $\ll$ symbol is not investigated, since by Lemmas 1 and 2 we have in fact for $h \in \mathcal{H}$

$$\log R\sum_{n\sim N}\Lambda_R(n;\mathcal{H};k+l)^2 \sim C(k,l)\sum_{n\sim N}\theta(n+h)\Lambda_R(n;\mathcal{H},k+l)^2 \tag{6.42}$$

with a constant $C(k,l)$ depending only on k and l, so the right-hand sides of (6.41) and (6.42) are really the same order of magnitude as a function of R and N, since $\log R \asymp \log N$ by (5.1).

7. The Execution of Step 2. Partial Proof of Theorem 3

As mentioned already in the previous section, Lemma 4 and its trivial consequence Lemma 6, together with Lemmas 1 and 2 provide already sufficient information about primes in almost prime k-tuples to prove Theorems 1 and 2.

We have, namely, similarly to (5.8), by Lemmas 1, 2, 4 and 6, for $R = N^{(\vartheta-\varepsilon)/(2+\eta)} > (3N)^{1/4}$

$$\sum_{\substack{n\sim N\\(P_{\mathcal{H}}(n),P(R^\eta))=1}} \left(\sum_{h\in\mathcal{H}}\theta(n+h) - \log 3N\right)\Lambda_R(n;\mathcal{H},k+l)^2 \tag{7.1}$$

$$= \frac{\mathfrak{S}(\mathcal{H})}{(k+2l)!}\binom{2l}{l}N\log N(\log R)^{k+2l}$$

$$\times\left(\frac{k}{k+2l+1}\cdot\frac{2(2l+1)}{l+1}\cdot\frac{\vartheta-\varepsilon}{2+\eta}+O(\eta)-1+o(1)\right).$$

It is easy to see that for any given $\vartheta = \frac{1}{2}+\delta > \frac{1}{2}$, if l and k/l are chosen sufficiently large, then

$$\frac{k}{k+2l+1}\cdot\frac{2l+1}{l+1}\left(\frac{1}{2}+\delta\right) > 1. \tag{7.2}$$

Now we can choose ε and $\eta = c(k,l,\vartheta)$ sufficiently small as to have

$$\sum_{\substack{n\sim N\\ (P_{\mathcal{H}}(n),P(R^{\eta}))=1}}\left(\sum_{h\in H}\theta(n+h)-\log 3N\right)\Lambda_R(n;\mathcal{H},k+l)^2 \tag{7.3}$$
$$\gg_{k,l,\mathcal{H},\vartheta} N\log N(\log R)^{k+2l}.$$

However, if $\left(P_{\mathcal{H}}(n),P(R^{\eta})\right)=1$, $R>(3N)^{1/4}$, then any $P_{\mathcal{H}}(n)$ has at most $k\cdot\frac{4}{\eta}$ prime divisors, so we have

$$\left|\Lambda_R(n;\mathcal{H},k+l)\right|^2 \le \left(\frac{2^{4k/\eta}}{(k+l)!}(\log R)^{k+l}\right)^2. \tag{7.4}$$

Now if we have at most one prime among $n+h_i$ $(i=1,2,\dots,k)$, then $\sum_{h\in\mathcal{H}}\theta(n+h)-\log 3N<0$, so we obtain for the number of n's in $[N,2N]$ with at least two primes among $(n+h_i)$ and almost primes in all coordinates $n+h_j$ with $P^-(n+h_j)>n^{\eta/4}$ the lower estimate

$$c(k,l,\mathcal{H},\vartheta)\frac{N}{(\log R)^k} > c'(k,l,\mathcal{H},\vartheta)\frac{N}{(\log N)^k} \tag{7.5}$$

as required by (4.1) of Theorem 3. We remark that the dependence on l and $\mathcal{H}$ can be omitted, since for $k\to\infty$ we will choose $l=\left(\sqrt{k}-1\right)/2$ (cf. (8.1)), further we have for any admissible k-tuple $\mathcal{H}$

$$\mathfrak{S}(\mathcal{H}) := \prod_p\left(1-\frac{|\Omega_{\mathcal{H}}(p)|}{p}\right)\left(1-\frac{1}{p}\right)^{-k} \tag{7.6}$$
$$\ge \prod_{p\le 2k}\frac{1}{p}\cdot\prod_{p>2k}\left(1-\frac{k}{p}\right)\left(1-\frac{1}{p}\right)^{-k} \ge c_3(k).$$

The extra assertion that we have at least two primes $n+h_i$ and $n+h_j$ in some position (i,j) and we have the same number b_s of prime divisors of $n+h_s$ ($s\neq i,j$, $1\le s\le k$) for all elements n of the progression, is a trivial consequence of the fact that the number of the possible vectors $\mathbf{b}=(b_1,\dots,b_k)$ is bounded (by $\left(1/c_1(k)\right)^k$) if all $n+h_s$ components are free of prime factors below $n^{1/c_1(k)}$. This means that at least one configuration, that is, one vector $\mathbf{b}$ (with at least two entries equal to 1) occurs at least $c_4(k)N/\log^k N$ times, fully describing the multiplicative pattern of $n+\mathcal{H}$ by $\Omega(n+h_s)=b_s$, where $\Omega(n)$ denotes here the number of prime divisors of n. We may mention that we could require beyond $\Omega(n+h_s)=b_s$ for all n also the stronger property that the exponent pattern $A_s=\left\{\alpha_{s1},\alpha_{s2},\dots,\alpha_{sj_s}\right\}$ of $n+h_s$ should be the same for all elements n of the progression. Namely, due to the trivial relation $b_s=\alpha_{s1}+\cdots+\alpha_{sj_s}$, any vector $\mathbf{b}=(b_1,\dots,b_k)$ gives rise only to a bounded number of possibilities for the values α_{st}. Hence at least one of them has to appear at least $c_5(k)N/(\log N)^k$ times for n's up to N.

8. How to Choose the Parameters k, l and a Small $\mathcal{H}_k$ for a Given Distribution Level ϑ? Continuation of the Proof of Theorem 3

In order to prove Theorem B, further our present Theorems 1 and 2 the values of the parameters k,l could be optimized to yield a minimal k for a given $\vartheta=1/2+\delta>\frac{1}{2}$ by the aid of computers as long as δ is not too small ($\delta\ge 1/10$, for example, see the table after (3.4) on p. 832 in [8]). The crucial inequality to be satisfied is our (7.2).

In view of the above, we will focus our attention to small values of δ (which means large values of k and l), although our argument holds for any $\delta\in(0,1/2]$. An easy calculation gives that if we did not require l to be an integer, then for a given k the expression on the left-hand side of (7.2) would be maximal for $l=\left(\sqrt{k}-1\right)/2$, i.e. $k=(2l+1)^2$ and then its value is for $l\ge(2\delta)^{-1}$

$$\text{(8.1)}\qquad \ge\frac{(2l+1)^2}{2(l+1)(2l+1)}\cdot\frac{2l+1}{l+1}\cdot\frac{l+1}{2l}=\frac{k}{k-1}.$$

We remark that if $\delta=1/2$ ($\vartheta=1$) for example, then this argument would give $l=1$, $k=9$, whereas $l=1$, $k=7$, $\vartheta>20/21$ already satisfies (7.2). A further improvement is possible in this case (at least for $\vartheta > 4\left(8-\sqrt{19}\right)/15 = 0.97096\ldots$) by choosing instead of the single optimal $l=1$ a linear combination of the weight functions $\Lambda_R(n;\mathcal{H},k+l)(\log R)^{-l}$ for $l=0$ and $l=1$. Then the argument works for $k=6$ already as shown in Section 3 (cf. (3.11)–(3.16)) of [8]. Since all our earlier arguments remain essentially valid if instead of a single weight function $\Lambda_R(n;\mathcal{H},k+l)$ we choose a linear combination of them (using Remark 1 after Lemma 4), the arguments (6.11)–(6.16) of [8] together with our present ones in Sections 5–7 prove Theorems 1–4 for $\vartheta \geq 0.971$.

In order to construct an admissible k-tuple $\mathcal{H}_k = \{h_i\}_{i=1}^k$ with a small diameter $d(\mathcal{H}_k) := h_k - h_1$, we can again obtain help from computers for relatively small values of k (cca. $k < 100$) as shown by the table after (3.4) in [8]. However, for any value of k we can choose $\mathcal{H}$ as the first k primes exceeding k, $\{p_{\nu+1},\ldots,p_{\nu+k}\}$, where $p_\nu \leq k < p_{\nu+1}$. This set clearly does not cover the residue class 0 for $p \leq k$, while for $p > k = |\mathcal{H}|$ it clearly can not cover all residue classes mod p. On the other hand the diameter of $\mathcal{H}_k$ is by the prime number theorem, that is, by $p_n \sim n\log n$

$$(8.2)\quad d(\mathcal{H}_k) = p_{\nu+k} - p_{\nu+1} = \left(1+o(1)\right)\left\{\left(k+\frac{k}{\log k}\right)\log k - k\right\} \sim k\log k$$

if $k\to\infty$ (which occurs for $\delta\to 0$). This is asymptotically probably close to optimal, since in general a set $\mathcal{H}$ of numbers up to X avoiding at least one residue class mod p for any $p \leq k$ is heuristically of size at most

$$(8.3)\qquad X\prod_{p\leq k}\left(1-\frac{1}{p}\right) \sim X\frac{1}{e^\gamma \log k},$$

whereas our set above has a somewhat larger density $\sim 1/\log p_{\nu+k} \sim 1/\log k$.

9. How do we Get Strong de Polignac Numbers? Completion of the Proof of Theorem 3

In this section we will show that we obtain at least $c_2(k)N/\log^k N$ numbers n up to N, where $n+h_i$ and $n+h_j$ are *consecutive* primes. Let with a fixed sufficiently small $c_1(k)$

(9.1)

$$\mathcal{B}(i,j,N) = \{n \le N;\ n+h_i \in \mathcal{P},\ n+h_j \in \mathcal{P},\ P^-\big(P_{\mathcal{H}}(n)\big) > n^{c_1(k)}\},$$

$$\mathcal{T} = \left\{(i,j);\ j>i,\ \limsup_{N\to\infty} \frac{\big|\mathcal{B}(i,j,N)\big| \log^k N}{N} > 0\right\}, \tag{9.2}$$

and let us choose any given pair $\{s,t\} \in \mathcal{T}$ with minimal value of $t-s$. Then for any $h_\mu \in (h_s, h_t)$ that is for $s < \mu < t$ we must have clearly

$$\limsup_{N\to\infty} \frac{\big|\mathcal{B}(\mu,t,N)\big| \log^k N}{N} = 0, \tag{9.3}$$

so all components $n + h_\mu$ between $n + h_s$ and $n + h_t$ are almost always composite if $n \in \mathcal{B}(s,t,N)$ as $N \to \infty$.

On the other hand, if we have an arbitrary $h \in (h_s, h_t)$, $h \notin \mathcal{H}$, then the assumption $n + h \in \mathcal{P}$ implies for $\mathcal{H}^+ = \mathcal{H} \cup h$

$$P^-\big(P_{\mathcal{H}^+}(n)\big) > n^{c_1(k)}. \tag{9.4}$$

However, by the Selberg sieve (see Theorem 5.1 of [12], or our present Lemma 1, with the choice $R = n^{c_1(k)}$) the number of such $n \le N$ is for all N

$$\ll_{k,c_1} \frac{\mathfrak{S}\big(\mathcal{H}\cup\{h\}\big)N}{\log^{k+1} N} \ll_{k,c_1} \frac{\mathfrak{S}(\mathcal{H})N\log h_k}{\log^{k+1} N} \ll_{k,c_1,\mathcal{H}} \frac{N}{\log^{k+1} N}, \tag{9.5}$$

which means that for a given fixed $\mathcal{H}$, this case might happen also rarely. This, together with (9.3) shows that the number of $n \le N$ where Theorem 3 is true with two consecutive primes is, similarly to (4.1), at least

$$\big(c_1(k,\mathcal{H}) + o(1)\big)\frac{N}{\log^k N}, \tag{9.6}$$

for some sequence $N = N_\nu \to \infty$.

10. Application of the Method of Green and Tao. Proofs of Theorems 1 and 2

Since Theorem 2 is a more general form of Theorem 1, it is clearly sufficient to prove just Theorem 2. This is now relatively straightforward as we have already proved Theorem 3. So we have for any admissible $\mathcal{A}$ with $r \geq \left(2\lceil 1/2\delta\rceil + 1\right)^2$ elements a set $\mathcal{N}^*(\mathcal{A}) = \mathcal{N}^* \subset \mathbb{N}$ at our disposal with the properties that with some $i, j \in \{1, \dots, r\}$ and some $b_s \leq C(r)$ we have for $n \in \mathcal{N}^*$

$$(10.1) \qquad n + a_i \ \text{ and } \ n + a_j \ \text{ are consecutive primes,}$$

$$(10.2) \qquad \Omega(n + a_s) = b_s, \ P^-(n + a_s) \geq n^{c_1(r)} \text{ for } s \in \{1, \dots, r\},$$

$$(10.3) \qquad \left|\{n \sim N;\ n \in \mathcal{N}^*\}\right| \geq c_3'(\mathcal{A}) \frac{N}{\log^r N}$$

for some sequence $N = N_\nu \to \infty$. As remarked at the end of Section 7 the condition $\Omega(n + a_s) = b_s$ can be replaced by the even stronger condition that the exponent pattern of $n + a_s$ should be $\boldsymbol{\alpha}_s = \left\{\alpha_{s1}, \dots, \alpha_{sj_s}\right\}$.

This set $\mathcal{N}^*$ has a positive upper density in the set $\widetilde{\mathcal{N}}$ of all integers satisfying

$$(10.4) \qquad P^-\left(P_{\mathcal{A}}(n)\right) \geq n^{c_1(r)},$$

due to the already mentioned Theorem 5.1 of [12], or our Lemma 1 (cf. (9.4)–(9.5)).

Remark. The above formulation shows that the generalization of the somewhat heuristic description, appearing in many works of Green and Tao that during their proof the primes are embedded into the set of almost primes with positive (upper) density can be proved in an exact form ((10.1)–(10.4)) in our case as well.

The proof now follows closely that of Green and Tao (cf. Sections 9–10 and the Appendix of [10]). Our task is made even easier by the recent work of Binbin Zhou [16], where he proved the existence of arbitrarily long arithmetic progressions of Chen primes, where for the sake of convenience he defined p to be a Chen prime if

$$p \in \mathcal{P}, \quad \Omega(p + 2) \leq 2, \quad P^-(p + 2) \geq p^{1/10}$$

and used the lower bound $CN/\log^2 N$ for the number of Chen primes below N.

In fact we can formulate our result in the following general form.

Theorem 5. *Let $\mathcal{A} = \{a_1, \dots, a_r\} \subseteq [0, A] \cap \mathbb{Z}$, $\mathcal{P}_{\mathcal{A}}(n) = \prod_{i=1}^{r}(n + a_i)$. Let $P^-(n)$ denote the least prime divisor of n. Let the set $\mathcal{N}_0$ satisfy*

$$(10.5)\quad \mathcal{N}_0 \subseteq \left\{n; P^-\big(P_{\mathcal{A}}(n)\big) \geq n^{c_1(r)}\right\}, \quad \left|\{n \leq X;\ n \in \mathcal{N}_0\}\right| \geq \frac{c_6(\mathcal{A})X}{\log^r X},$$

with $c_1(r), c_6(\mathcal{A}) > 0$ for a sequence $X = X_\nu \to \infty$. Then $\mathcal{N}_0$ contains k-term arithmetic progressions for any $k > 0$.

Remark. This is clearly a generalization of the results of Green–Tao ($r = c_1 = 1$) and Zhou ($r = 2$, $c_1 = 1/10$).

Remark. The terms of the arithmetic progression of length k are below N if $N > N_0(c_1, c_6, r, A, k)$ and their total number is at least $c_7(c_1, c_6, r, A, k)N^2/\log^{rk} N$.

Remark. Theorem 5 trivially shows that the twin primes really contain arbitrarily long arithmetic progressions if their number up to x, $\pi_2(x) \gg x/\log^2 x$. This result is implicitly contained in [16] as well.

Remark. $P^-\big(P_{\mathcal{A}}(n)\big) \geq n^{c_1}$ implies that $\mathcal{A}$ is admissible, since otherwise $P_{\mathcal{A}}(n)$ would have a fixed prime divisor $p \leq r$.

Remark. Since the proof is analogous to that in [16], which in fact is analogous to that in [10] we will point out only the essential differences.

Remark. The above said strong analogy is only true if $\mathcal{A}$ is considered to be fixed, more precisely if

$$(10.6)\qquad \mathcal{A} \subseteq [1, A] \quad \text{with a fixed } A.$$

Otherwise, when A is allowed to increase with N, serious difficulties may occur with the linear form property unless A is relatively small compared to N.

As the reader observed we changed our set $\mathcal{H}$ to $\mathcal{A}$, the elements h_i to a_i and the size k to r, compared with Sections 1–9. This is necessary since $\mathcal{H}$ and k are used in [10] (and [16]) to denote other quantities, namely our aim is to show the existence of k-term arithmetic progressions in $\mathcal{N}^*$.

The definition of $W = W(N)$ and $w = w(N)$ remain the same,

$$W = \prod_{p \le w} p, \tag{10.7}$$

where $w = w(N)$ and thereby $W = W(N)$ is a function of N, sufficiently slowly growing to infinity with N.

In the following we will suppose that $n \in \mathcal{N}^*$. In this case $(n + a_i, W) = 1$. Following Zhou (Section 2) we will choose a b mod W with $(Wm + b + a_i, W) = 1$, through first choosing a b_p mod p for every prime $p \mid W$ with $b_p \not\equiv -a_i(\text{ mod } p)$ and then applying the chinese remainder theorem to obtain $b \equiv b_p(\text{ mod } p)$ for each $p \mid W$. Since we have for any p exactly $p - |\Omega_{\mathcal{A}}(p)|$ possibilities for b_p (where $\Omega_{\mathcal{A}}(p) = \{-a_i \text{ mod } p\}_{i=1}^r$, as in Section 5), we obtain for the cardinality of the set X_W of possible choices of b mod W the quantity (we note that we can suppose $r < A < w \to \infty$)

$$\begin{aligned} |X_W| &= W \prod_{p \le A} \left(1 - \frac{|\Omega_{\mathcal{A}}(p)|}{p}\right) \prod_{A < p \le w(n)} \left(1 - \frac{r}{p}\right) \\ &\le WC(A) \prod_{p \mid W} \left(1 - \frac{1}{p}\right)^r = WC(A) \left(\frac{\varphi(W)}{W}\right)^r \end{aligned} \tag{10.8}$$

for every admissible set $\mathcal{A}$ with a uniform constant $C(A)$. If we do not indicate in the following dependence on either A, r or k any more, then we obtain by (10.3)

$$\sum_{b \in X_W} \left| n \in [\varepsilon_k N, 2\varepsilon_k N];\ Wn + b \in \mathcal{N}^* \right| \gg \frac{\varepsilon_k N}{\log^r N}, \tag{10.9}$$

where $\varepsilon_k = 1/(2^k(k+4)!)$, since $P^-\big(P_{\mathcal{A}}(Wn + b)\big) > n^c > W$ implies $(Wn + b + a_i, W) = 1 \Leftrightarrow b \in X_W$. Thus by (10.8) we can choose a fixed residue class b mod W (depending on $\mathcal{A}$), $0 \le b < W$, such that the set

$$|X| := \left|\{ n \in [\varepsilon_k N, 2\varepsilon_k N];\ Wn + b \in \mathcal{N}^* \}\right| \gg \frac{\varepsilon_k N}{\log^r N} \cdot \left(\frac{W}{\varphi(W)}\right)^r. \tag{10.10}$$

Our measure ν is similarly to (2.5) of [16] defined now on $\mathbb{Z}_N$ by

$$(10.11)\quad \nu(n) := \begin{cases} \left(\frac{\varphi(W)}{W}\right)^r \prod_{i=1}^{r} \frac{\Lambda_R(Wn+b+a_i)^2}{\log R} & \text{if } n \in [\varepsilon_k N, 2\varepsilon_k N] \\ 1 & \text{otherwise} \end{cases},$$

where Λ_R is given in (2.1), $R < N^{c_1(r)}$ will be chosen as a sufficiently small power of N, thereby satisfying

$$(10.12)\qquad \left(Wn+b+a_i, \prod_{p \le R} p\right) = 1 \quad \text{for } n \in X;$$

due to the crucial condition (10.2) and the definition of X in (10.10), taking into account that $W\varepsilon_k > 1$ by $W \to \infty$. Now (10.12) implies trivially for $n \in [\varepsilon_k N, 2\varepsilon_k N]$ and $Wn + b \in \mathcal{N}^*$ that is for $n \in X$

$$(10.13)\qquad \Lambda_R(Wn+b+a_i) = \log R, \quad \text{so} \quad \nu(n) = \left(\frac{\varphi(W)}{W}\right)^r (\log R)^r.$$

This means that defining (in analogy with (2.1) of [16])

$$(10.14)\qquad \widetilde{\Lambda}_{\mathcal{A}}(n) := \begin{cases} \frac{\varphi(W)}{W} \log(Wn+b) & \text{if } Wn+b \in \mathcal{N}^* \\ 0 & \text{otherwise} \end{cases},$$

we have

$$(10.15)\qquad \nu(n) \ge f(n) := \left[k^{-1}2^{-k-5}\widetilde{\Lambda}_{\mathcal{A}}(n)\right]^r \quad \text{for } n \in [\varepsilon_k N, 2\varepsilon_k N].$$

The proof that $\nu(n)$ satisfies the k-pseudorandomness property follows that of [16], which again follows the proof of [10] which are essentially the special cases $r = 2$ and $r = 1$ of our case. The fact that ν is a measure, that is, $E(\nu) = 1 + o(1)$ is the special case of the linear form property ($m = 1$, $b = 1$, $\psi_1(x) = x_1$, $B = [\varepsilon_\kappa N, 2\varepsilon_\kappa N]$). The proof of the linear form condition runs completely analogously to that of [16] (which is nearly the same as that of [10]), a crucial point being here that if a prime $p > W$ would satisfy

$$(10.16)\quad p \mid W\left(\sum_{l=1}^{t} L_{il}x_l + b_l\right) + b + a_u \quad \text{and} \quad p \mid W\left(\sum_{l=1}^{t} L_{il}x_l + b_l\right) + b + a_v,$$

then obviously $p \mid a_u - a_v$, which is a contradiction since

(10.17)
$$\mathcal{A} = \{a_1, \dots, a_r\} \subseteq [1, A] \quad A \text{ is bounded, } w = w(N) \to \infty \text{ as } N \to \infty.$$

Hence, for $u \neq v$ $\theta_i(\mathbf{x}) + a_u \equiv \theta_i(\mathbf{x}) + a_v \equiv 0 \pmod p$ is impossible. (This step is not valid if we allow $A \to \infty$ with $A \geq w(N)$.) Concerning the analogue of Proposition 2.10 of [16], or (9.10) of [10] according to which the measure ν satisfies the 2^{k-1}-correlation condition, the main difference is that the role of Δ is played here (cf. Proposition 2.6 of [16] or Proposition 9.6 of [10]) by the quantity

$$(10.18)\quad \Delta = \prod_{1\leq i<j\leq m} (h_i - h_j) \prod_{1\leq u<v\leq r} (W(h_i - h_j) + a_u - a_v) \ll N^{3\binom{m}{2}\binom{r}{2}}$$

which, however, still obeys the estimate (10.15) of [10], namely

$$(10.19)\qquad \Delta \ll R^{O_{m,r}(1)}.$$

Afterwards, the analogue of Lemma 2.9 of [16] or Lemma 9.9 of [10] is here again slightly more difficult, in the sense that we need the generalized Hölder's inequality instead of the standard case, applied in [16], to infer that with the notation $I = [1, N]$, $S(n) = \prod_{1\leq u<v\leq r} (Wn + a_u - a_v)$ we have

$$(10.20)\qquad E\Bigg(\prod_{p\mid nS(n),\ p>w} \left(1 + p^{-1/2}\right)^{O_{m,r,q}(1)} \mid I \Bigg) = O_{m,r,q}(1).$$

The quantity on the left-hand side is, namely, with the notation $B = \binom{r}{2} + 1$

(10.21)
$$\leq E\Bigg(\prod_{\substack{p\mid n\\ p>w}} \left(1 + p^{-\frac{1}{8}}\right) \mid I \Bigg)^{1/B} \prod_{1\leq u<v\leq r} E\Bigg(\prod_{\substack{p\mid Wn+a_u-a_v\\ p>w}} \left(1 + p^{-\frac{1}{8}}\right) \mid I \Bigg)^{1/B}$$

since $\left(1 + p^{-1/2}\right)^{BO_{m,r,q}(1)} \leq 1 + p^{-1/8}$ if $p > w \to \infty$ as $N \to \infty$. Then the rest of Lemma 2.9 of [16] can be followed using again the crucial property (10.17).

So we obtain, similarly to [16] and [10] for every value of k actually $cN^2/(\log N)^{rk}$ k-term arithmetic progressions in the set $\mathcal{N}^*(\mathcal{A})$, where c depends on r, A, k, and $\mathcal{A} = \{a_i\}_{i=1}^r$. This proves our Theorem 5, and consequently Theorems 1 and 2, too.

11. The Density of the de Polignac Numbers

Let us suppose that any admissible k-tuple $\mathcal{H}$ produces at least two consecutive primes infinitely often. Then this phenomenon clearly occurs infinitely often for prime pairs $n+h_i$, $n+h_j$ in the same position, i.e. we have at least one strong de Polignac number d among $h_i - h_j$ $(i > j)$. The question is: at least how many different d's do we get by choosing all possible admissible sets $\mathcal{H}$ with elements at most H, if H is large $(H \to \infty)$. Let

$$P := P(k) := \prod_{p \le k} p, \tag{11.1}$$

where we can clearly suppose $P|H$, as the size of k and thereby P remains fixed and $H \to \infty$.

In order to show admissibility we may suppose that we choose all elements of all k-tuples $\mathcal{H}$ from the set

(11.2)

$$\mathcal{M} := \{m \le H;\ (m, P) = 1\}, \text{ having } M := |\mathcal{M}| = \frac{\varphi(P)}{P} H \text{ elements;}$$

thereby excluding the residue class 0 for each $p \le k$. Since every admissible k-tuple $\mathcal{H}$ gives rise to at least one strong de Polignac number d, we obtain at least $\binom{M}{k}$ strong de Polignac numbers below H, counted with multiplicity according to the k-tuples $\mathcal{H}$. A fixed value d might appear as the difference of two elements of $\mathcal{H}$ with at most $M - 1$ choices for the larger element and afterwards the smaller element is determined uniquely. Furthermore, we have $\binom{M-2}{k-2}$ choices for the remaining $k - 2$ elements of $\mathcal{H}$. This implies that we obtain at least

$$\frac{\binom{M}{k}}{(M-1)\binom{M-2}{k-2}} = \frac{M}{k(k-1)} = \frac{H\varphi(P)}{Pk(k-1)} \tag{11.3}$$

different strong de Polignac numbers d until H, which proves (4.8).

In case of $\vartheta \ge 0.971$ we can work with $k = 6$ tuples, so $P = 30$, $\varphi(30) = 8$, which proves (4.9), hence Theorem 4 is proved completely.

12. Further Problems

It is clear from our work that Steps 1 and 2 (Lemmas 3–6) led to some new information about primes in tuples, in particular about the frequency of the occurrence of two primes in any admissible k-tuple for $\vartheta > 1/2$ (Theorem 3) and that under the same condition we have more than one even d, in fact a positive proportion of all numbers, which appear infinitely many times as the difference of two consecutive primes. We mentioned that Lemma 4 might help to deduce unconditionally that a positive proportion of gaps between consecutive primes are less than $\eta \log p$ for any fixed $\eta > 0$. The question still arises: does the combination of the two methods of [10] and [8]–[9] yield also some unconditional results? The answer is yes. We mention a few of them.

Theorem 6. *Let $\mathcal{H} = \{h_i\}_{i=1}^k$ be an admissible k-tuple for any $k \geq 1$. Then there exist arbitrarily long arithmetic progressions of primes n such that all $n + h_i$'s are almost primes and with some vectors $\boldsymbol{\alpha}_i = (\alpha_{i1}, \dots, \alpha_{ij_i})$*

$$(12.1) \qquad P^-(n + h_i) > n^{c_1(k)}, \quad n + h_i \ \text{has exponent pattern} \ \boldsymbol{\alpha}_i,$$

for all values of n in the progression. (Here $b_i = \sum_{1 \leq s \leq j_i} \alpha_{is} \leq 1/c_1(k)$.)

It is not a consequence of the stated results but using the method of [4]–[6] in combination with that of [10] one can show the following unconditional results.

Theorem 7. *Let $\mathcal{H} = \{h_i\}_{i=1}^k$ be any admissible k-tuple for any $k \geq 3$. Then there exist arbitrarily long arithmetic progressions of n values such that for some $i, j \in \{1, \dots, k\}$, $i \neq j$, $n + h_i$ and $n + h_j$ are semiprimes (that is product of two different primes, i.e. having exponent pattern $(1,1)$) and there exist some vectors $\boldsymbol{\alpha}_s$ for $s \in \{1, \dots, k\}$, $s \neq i, j$ such that*

$$(12.2) \qquad P^-(n + h_s) > n^{c_1(k)}, \quad n + h_s \ \text{has exponent pattern} \ \boldsymbol{\alpha}_s.$$

Corollary. *There exist arbitrarily long arithmetic progressions of generalized twin semiprime pairs $(q, q + d)$ where $d = 2, 4$ or 6. The same is true for $d = 6D$ or $12D$ for any integer D.*

Theorem 8. *Let $\boldsymbol{\alpha} = (\alpha_1, \dots, \alpha_j)$ be any exponent pattern which includes at least one $\alpha_i = 2$ and at least three different entries equal to 1. Then there exist arbitrarily long arithmetic progressions of numbers n such that* *all n and $n+1$ in the progression have exponent pattern $\boldsymbol{\alpha}$. In particular we have arbitrarily long arithmetic progressions of integers n satisfying simultaneously*

(12.3) $\omega(n) = \omega(n+1) = 4, \quad \Omega(n) = \Omega(n+1) = 5, \quad d(n) = d(n+1) = 24.$

The above assertion is true if the triplet $(4, 5, 24)$ is substituted by $(4+B, 5+B, 24 \cdot 2^B)$ or $\big(5, 5+B, 24 \cdot (B+1)\big)$ for any $B \geq 0$. The same is true (not necessarily simultaneously) for any of the equations

(12.4)
$$\omega(n) = \omega(n+1) = A, \quad \Omega(n) = \Omega(n+1) = B, \quad d(n) = d(n+1) = C,$$

where A, B, C are any integers with $A \geq 3$, $B \geq 4$, $24 \mid C$.

The above is a far-reaching generalization of three conjectures of Erdős [2] and Erdős–Mirsky [3], respectively, asking whether there exists an infinite set of numbers n satisfying (not necessarily simultaneously)

(i) $\omega(n) = \omega(n+1)$ ($\omega(n)$ is the number of distinct prime divisors of n),
(ii) $\Omega(n) = \Omega(n+1)$ ($\Omega(n)$ is the total number of prime divisors of n),
(iii) $d(n) = d(n+1)$ ($d(n)$ is the number of divisors of n).

We mention that Theorems 6–9 need apart from a variant of the Green–Tao method sketched in our present Section 10 also a variant of the method of [8], due to S. W. Graham, D. Goldston, C. Yıldırım and the present author. In contrast to this, in Theorem 6 – which is a generalization of Rényi's result [15] about the existence of infinitely many primes with $p+2 = P_K$ for some large K – the method of [8]–[9], our Lemma 4, can be replaced by a result contained implicitly in Theorem 10.7 of Halberstam and Richert [12].

Although we do not know whether any given number d appears infinitely many times as the difference of two semiprimes, we are able to prove the following rather general

Theorem 9. *There exists an admissible k-tuple $\mathcal{H} = \{0, h_1, \dots, h_{k-1}\}$ for any $k \geq 1$ such that there are arbitrarily long arithmetic progressions of semiprimes q_j with the property that all numbers $q_j + h_i$ $(i = 1, 2, \dots, k-1)$ are semiprimes, too.*

Some further results, which are not connected with the Green–Tao method, but which form a part of the proof of Theorems 7–9, are the following

Theorem 10. *Let $\mathcal{H}$ be an admissible k-tuple with $k \geq 3$. The number of those n's up to N for which $n + \mathcal{H}$ contains at least two semiprimes and almost primes in all other components $n + h_s$ is*

$$\geq c_2'(\mathcal{H}) \frac{N}{\log^k N}. \tag{12.5}$$

Theorem 11. *The number of integers n satisfying any of the equations*

$$\omega(n) = \omega(n+1) = A, \qquad A \text{ fixed, } A \geq 3, \tag{12.6}$$

$$\Omega(n) = \Omega(n+1) = B, \qquad B \text{ fixed, } B \geq 4, \tag{12.7}$$

$$d(n) = d(n+1) = C, \qquad C \text{ fixed, } 24 \mid C \tag{12.8}$$

as well as the number of integers having an exponent pattern including $\{2, 1, 1, 1\}$ is

$$\gg \frac{N}{\log^3 N}, \tag{12.9}$$

where the constant implied by the $\gg$ symbol depends on A, B or C, respectively.

In particular, (12.3) has at least $cN/\log^3 N$ solutions below N. We remark that the expected number of solutions would be $c_6 N (\log_2 N)^{c_7}/\log^2 N$.

To the proof of Theorems 7–11 we shall return in a later work. Theorem 6 actually follows from the results of Sections 6 and 10 of our present work, more precisely from Lemmas 1, 2, 4, 6 and Theorem 5.

References

[1] P. D. T. A. Elliott and H. Halberstam, *A conjecture in prime number theory,* Symposia Mathematica, **4** INDAM, Rome, 59–72, Academic Press, London, 1968/69.

[2] P. Erdős, Some problems on number theory, in: Analytic and elementary number theory (Marseille, 1983), *Publ. Math. Orsay,* **86**-1 (1983), 53–57.

[3] P. Erdős and L. Mirsky, The distribution of values of the divisor function $d(n)$, *Proc. London Math. Soc. (3),* **2** (1952), 257–271.

[4] D. A. Goldston, S. W. Graham, J. Pintz and C. Y. Yıldırım, Small gaps between primes or almost primes, *Trans. Amer. Math. Soc.,* **36** (2009), 5285–5330.

[5] D. A. Goldston, S. W. Graham, J. Pintz and C. Y. Yıldırım, Small gaps between products of two primes, *Proc. London Math. Soc.,* **98** (2009), 741–774.

[6] D. A. Goldston, S. W. Graham, J. Pintz and C. Y. Yıldırım, Small gaps between almost primes primes, the parity problem, and some conjectures of Erdős on consecutive integers, *Int. Math. Res. Notes,* to appear.

[7] D. A. Goldston, Y. Motohashi, J. Pintz and C. Y. Yıldırım, Small gaps between primes exist, *Proc. Japan Acad.,* **82A** (2006), 61–65.

[8] D. A. Goldston, J. Pintz and C. Yıldırım, Primes in Tuples, *Annals of Math.* (2), **170** (2009), 819–862.

[9] D. A. Goldston, J. Pintz and C. Yıldırım, Primes in Tuples II, *Acta Math.,* **204** (2010), 1–47.

[10] B. Green and T. Tao, The primes contain arbitrarily long arithmetic progressions, *Ann. of Math.* (2), **167** (2008), 481–547.

[11] D. A. Goldston and C. Y. Yıldırım, Higher correlations of the divisor sums related to primes III. Small gaps between primes, *Proc. London Math. Soc. (3),* **95** (2007), no. 3, 653–686.

[12] H. Halberstam and H.-E. Richert, *Sieve Methods,* Academic Press, London, 1974.

[13] D. R. Heath-Brown, Almost-prime k-tuples, *Mathematika,* **44** (1997), 245–266.

[14] A. de Polignac, Six propositions arithmologiques déduites du crible d'Ératosthène, *Nouv. Ann. Math.,* **8** (1849), 423–429.

[15] A. Rényi, On the representation of an even number as the sum of a single prime and a single almost-prime number, *Izv. Akad. Nauk SSSR.,* **12** (1948), 57–78 (Russian).

[16] Binbin Zhou, The Chen primes contain arbitrarily long arithmetic progressions, *Acta Arith.,* **138** (2009), no. 4, 301–315.

János Pintz
Rényi Mathematical Institute of the
Hungarian Academy of Sciences
Budapest
Reáltanoda u. 13–15
H-1053 Hungary

e-mail: pintz@renyi.hu

BOLYAI SOCIETY
MATHEMATICAL STUDIES, 21

An Irregular Mind
(Szemerédi is 70)
pp. 561–590.

Dirac-type Questions for Hypergraphs – a Survey (or More Problems for Endre to Solve)

VOJTECH RÖDL* and ANDRZEJ RUCIŃSKI†

Dedicated to Endre Szemerédi on the occasion of his 70th birthday

In 1952 Dirac [8] proved a celebrated theorem stating that if the minimum degree $\delta(G)$ in an n-vertex graph G is at least $n/2$ then G contains a Hamiltonian cycle. In 1999, Katona and Kierstead initiated a new stream of research devoted to studying similar questions for hypergraphs, and subsequently, for perfect matchings. A pivotal role in achieving some of the most important results in both these areas was played by Endre Szemerédi. In this survey we present the current state-of-art and pose some open problems.

1. Introduction

A k-uniform hypergraph, or *k-graph* for short, is a pair $H=(V,E)$, where $V:=V(H)$ is a finite set of vertices and $E:=E(H)\subseteq\binom{V}{k}$ is a family of k-element subsets of V. Whenever convenient we will identify H with $E(H)$. *A matching* in H is a set of disjoint edges of H, and a matching containing all vertices of H is called *perfect.*

There are several notions of a hypercycle. Berge [2] defined a hypercycle of length m in a hypergraph H as an alternating sequence of m vertices and m edges $x_1, e_1, x_2, e_2, \ldots, x_m, e_m, x_1$ such that $\{x_i, x_{i+1}\}\subseteq e_i$ for all

*Research supported by NSF grant DMS-080070.

†Research supported by Polish grant N201036 32/2546. Part of research performed at Emory University, Atlanta.

$i = 1, 2, \ldots, m$, where $x_{m+1} = x_1$. Observe that there may be other vertices than $x_1, \ldots, x_m$ in the edges of a Berge cycle and that there are several nonisomorphic Berge hypercycles of length m. Bermond et al. [3] studied the existence of Hamiltonian Berge cycles under some degree conditions. Also, more recently, there has been some interest in Ramsey-type questions for Hamiltonian Berge cycles (see, e.g., [10].)

However, following the paper by Katona and Kierstead [13], another notion of a hypergraph cycle has become gradually more and more popular.

Definition 1.1. For $0 \le l \le k-1$ *a* (k,l)*-cycle* is a k-graph whose vertices can be ordered cyclically in such a way that the edges are segments of that cyclic order and every two consecutive edges share exactly l vertices (see Figure 1 [1]). *A Hamiltonian* l*-cycle* in a k-graph H is then defined as a (k,l)-cycle in H containing all vertices of H.

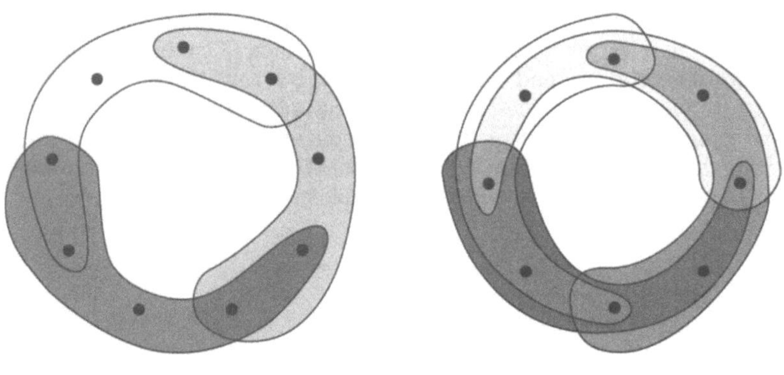

Fig. 1. A $(5,2)$-cycle and a $(5,3)$-cycle

The notion of a (k,l)-cycle, unlike the Berge hypercycle, is unique up to isomorphism. Let us denote by $C_s^{k,l}$ the (k,l)-cycle on s vertices. Observe that s must be divisible by $k-l$ and the cycle has $s/(k-l)$ edges. Furthermore, if we write $k = t(k-l) + r$, where $1 \le t \le k$ and $0 \le r \le k-l-1$ are uniquely determined by k and l, then $s/(k-l) \ge t+1$. In particular, $s \ge k+1$ for $l = k-1$ while $s \ge 2(k-l)$ for $l < k/2$.

If, in addition, $k-l$ divides k then a (k,l)-cycle is regular of degree $k/(k-l)$. Otherwise, its minimum degree is $\lfloor k/(k-l) \rfloor$ and maximum degree is $\lceil k/(k-l) \rceil$. Note also that for $l = 0$ an l-cycle reduces to a matching.

[1] All figures prepared electronically by Emory students, Domingos Dellamonica Jr. and Sangjune Lee

Given a k-graph H and d vertices $v_1, \dots, v_d \in V(H)$, $1 \le d \le k-1$, we denote by $\deg_H(v_1, \dots, v_d)$ the degree of the d-tuple $\{v_1, \dots, v_d\}$ in H, that is, the number of edges of H which contain $v_1, \dots, v_d$. For a vertex $v \in V(H)$, let $H(v)$ denote *the link of v in H* that is,

$$H(v) = \left\{ e \in \binom{V \setminus \{v\}}{k-1} : e \cup \{v\} \in H \right\}.$$

In particular, $|H(v)| = \deg_H(v)$.

Further, let

$$\delta_d(H) := \delta_d = \min\left\{ \deg_H(v_1, \dots, v_d) : \{v_1, \dots, v_d\} \subset V(H) \right\}.$$

For $d = 1$, $\delta_d(H)$ is the ordinary minimum vertex degree in H. Observe that $\delta_d(H) \le \binom{n-d}{k-d}$.

Definition 1.2. Let d, k, l, and n satisfy $1 \le d \le k-1$ and $k-l$ divide n. We define $h_d^l(k, n)$ to be the smallest integer h such that every n-vertex k-graph H satisfying $\delta_d(H) \ge h$ contains a Hamiltonian l-cycle.

As mentioned before, for $l = 0$, a Hamiltonian l-cycle in a k-graph H becomes a perfect matching in H. Moreover, any Hamiltonian $(k-1)$-cycle contains a matching of size $\lfloor n/k \rfloor$. Hence, not surprisingly, the results for Hamiltonian cycles and perfect (or almost perfect) matchings are related.

To our knowledge, the first result relating the minimum degree and the existence of a large (though, far from perfect) matching in a k-graph was obtained by Bollobás, Daykin, and Erdős in [4]. It was further extended to perfect matchings by Daykin and Häggkvist in [7].

Definition 1.3. Let d, k, r, and n satisfy $1 \le d \le k-1$ and k divide $n - r$. We define $m_d^r(k, n)$ to be the smallest integer m such that every n-vertex k-graph H satisfying $\delta_d(H) \ge m$ contains a matching M with $|V(M)| = n - r$.

In Sections 2 and 3, respectively, we summarize what we know about the parameters $h_d^l(k, n)$ and $m_r^r(k, n)$. We present both, asymptotic and exact results, some with sketches of proofs, as well as pose several open questions. We also discuss the k-partite case and some other related topics.

Throughout the paper we will be giving a particular interest to the cases when $d = k-1$, $l = k-1$, and/or $r = 0$. We will be then suppressing the subscript or the superscript, or both, respectively. For instance, $m_d(k,n) = h_d^0(k,n)$ will stand for the smallest integer m such that every k-graph on n vertices with n divisible by k and $\delta_d \geq m$ contains a perfect matching. For future references we summarize our notation here.

Summary of notation: For n divisible by $k-l$

- $h_d^l(k,n) = \min\{h : \delta_d(H) \geq h \Rightarrow H \text{ contains a Hamiltonian } l\text{-cycle}\}$
- $h^l(k,n) = h_{k-1}^l(k,n)$
- $h_d(k,n) = h_d^{k-1}(k,n)$
- $h(k,n) = h_{k-1}^{k-1}(k,n)$,

and for $n-r$ divisible by k

- $m_d^r(k,n) = \min\{m : \delta_d(H) \geq m \Rightarrow H \text{ contains a matching } M, |V(M)| = n-r\}$
- $m^r(k,n) = m_{k-1}^r(k,n)$
- $m_d(k,n) = m_d^0(k,n)$
- $m(k,n) = m_{k-1}^0(k,n)$.

The parameters $h_d^l(k,n)$ and $m_d^r(k,n)$ are often referred to as *Dirac-type* thresholds. So far, all known results and conjectures indicate that the Dirac thresholds are asymptotic to $c\binom{n-d}{k-d}$, for some $0 < c < 1$. Therefore, the following observation can be useful.

Remark 1.4. Since, by simple averaging,

$$\delta_{d-1}(H) \geq \frac{n-d+1}{k-d+1} \times \delta_d(H),$$

we have for every $c > 0$ that

$$\delta_d(H) \geq c\binom{n-d}{k-d} \quad \text{implies} \quad \delta_{d-1}(H) \geq c\binom{n-(d-1)}{k-(d-1)}.$$

Consequently,

$$h_d^l(k,n) \geq c\binom{n-d}{k-d} \quad \text{implies} \quad h_{d-1}^l(k,n) \geq c\binom{n-(d-1)}{k-(d-1)}$$

and

$$h^l_{d-1}(k,n) \le c\binom{n-(d-1)}{k-(d-1)} \qquad \text{implies} \qquad h^l_d(k,n) \le c\binom{n-d}{k-d},$$

and similar implications hold for the parameter $m^r_d(k,n)$ as well.

2. Hamilton Cycles

For most of this section we will deal with the case $d = k-1$ and $l = k-1$ and set $h(k,n) = h^{k-1}_{k-1}(k,n)$ for convenience (see Summary of notation in Section 1). Also for convenience, we will call Hamiltonian $(k-1)$-cycles just Hamiltonian cycles, and k-graphs containing such cycles – Hamiltonian.

In 1952 Dirac [8] proved that $h(2,n) = \lceil n/2 \rceil$. The two following graphs show that this result is tight: the union of two complete graphs $2K_{\lceil n/2\rceil}$ (with one vertex in common when n is odd) and the complete bipartite graph $K_{\lceil n/2\rceil-1,\lfloor n/2\rfloor+1}$. The first Dirac-type result for hypergraphs was obtained by Katona and Kierstead who proved in [13] that

$$\left\lfloor \frac{n-k+3}{2} \right\rfloor \le h(k,n) \le \left(1 - \frac{1}{2k}\right) n + O_k(1).$$

As a proof of the lower bound they provided the following construction of an extremal k-graph H_0.

Construction 2.1 ([13]). Let $V = V' \cup \{v\}$, $|V| = n \ge k^2+1$. Split $V' = X \cup Y$, where, $|X| = \left\lfloor \frac{n-1}{2} \right\rfloor$ and $|Y| = \left\lceil \frac{n-1}{2} \right\rceil$. The edges of H_0 are all k-element subsets S of V such that $|X \cap S| \ne \left\lfloor \frac{k}{2} \right\rfloor$ or $v \in S$. It is shown in [13] that H_0 is not Hamiltonian, while $\delta_{k-1}(H_0) \ge \left\lfloor \frac{n-k+1}{2} \right\rfloor$. Thus,

$$h(k,n) \ge \delta_{k-1}(H_0) + 1 = \left\lfloor \frac{n-k+1}{2} \right\rfloor + 1 = \left\lfloor \frac{n-k+3}{2} \right\rfloor.$$

Katona and Kierstead (implicitly) conjectured that their lower bound is the correct value of $h(k,n)$. Recently, this has been confirmed for $k=3$, first asymptotically [22], then exactly [27], solving also the corresponding Hamiltonian path problem.

Theorem 2.2 ([27]). *Let H be a 3-graph on n vertices, where n is sufficiently large.*

1. *If $\delta_2(H) \geq \lfloor n/2 \rfloor$ then H has a Hamiltonian cycle. Moreover, for every n there exists a 3-graph H_n such that $\delta(H_n) = \lfloor n/2 \rfloor - 1$ and H_n does not have a Hamiltonian cycle. In other words, $h(3,n) = \left\lfloor \frac{n}{2} \right\rfloor$.*

2. *If $\delta_2(H) \geq \lceil n/2 \rceil - 1$ then H has a Hamiltonian path. Moreover, for every n there exists a 3-graph H_n such that $\delta(H_n) = \lceil n/2 \rceil - 2$ and H_n does not have a Hamiltonian path.*

An analogous question regarding the Dirac threshold for Hamiltonian cycles in k-graphs remains open.

Problem 2.3. *Prove that $h(k,n) = \left\lfloor \frac{n-k+3}{2} \right\rfloor$ for all $k \geq 4$.*

As a step toward solving this problem, it was proved in [24] that $h(k,n) \sim \frac{1}{2}n$, that is, $h(k,n) = \left(1 + o(1)\right)\frac{1}{2}n$, for all $k \geq 3$.

Theorem 2.4 ([24]). *Let $k \geq 3$, $\gamma > 0$, and let H be a k-graph on n-vertices, where n is sufficiently large. If $\delta_{k-1}(H) \geq (1/2 + \gamma)n$ edges, then H is Hamiltonian. In other words, $h(k,n) \sim \frac{1}{2}n$.*

A sketch of the proof of Theorem 2.4 from [24] is presented in Section 2.2.

2.1. Dirac thresholds for loose(r) Hamiltonian cycles

For two integers, a and b, let us write $a|b$ if a divides b. As an (almost) immediate consequence of Theorem 2.4 we can asymptotically determine the value $h^l(k,n)$ of the Dirac threshold for Hamiltonian l-cycles for all $1 \leq l \leq k-1$ satisfying the congruence $(k-l) \mid k$.

Corollary 2.5 ([19]). *If $(k-l) \mid k$ and $(k-l) \mid n$, then $h^l(k,n) \sim \frac{1}{2}n$.*

Proof. We will show first that $h^l(k,n) \leq \left(\frac{1}{2} + o(1)\right)n$. Since $(k-l) \mid k$ and $(k-l) \mid n$, every Hamiltonian $(k-1)$-cycle $C_n^{k,k-1}$ contains a Hamiltonian l-cycle $C_n^{k,l}$ (indeed, take every $(k-l)$th edge of $C_n^{k,k-1}$). Thus, we have

$$h^l(k,n) \leq h(k,n) = \left(\frac{1}{2} + o(1)\right)n,$$

where the equation follows from Theorem 2.4.

For the lower bound, assume first that, in addition to $(k-l) \mid k$ and $(k-l) \mid n$, we also have $k|n$. Then, by taking every $\frac{k}{k-l}$th edge of $C_n^{k,l}$, we can find a perfect matching $C_n^{k,0}$ inside $C_n^{k,l}$. Thus, in this case,

$$m(k,n) = h^0(k,n) \le h^l(k,n).$$

By the lower bound (2) given in Section 3 we know that

$$m(k,n) \ge \frac{1}{2}n - k,$$

which completes the proof if $k|n$.

If k does not divide n then still $h^l(k,n) \sim \frac{1}{2}n$ because $h^l(k,n) \ge \frac{1}{2}n - k$ by a simple argument from [19] which uses the following constructions.

Construction 2.6. Let $H_1 = (V,E)$ where $V = A \cup B$, $\frac{1}{2}n - 1 \le |A| \le \frac{1}{2}n + \frac{1}{2}$, $|A|$ is odd, and E consists of all $e \in \binom{V}{k}$ such that $|e \cap V|$ is even. Let $H_2 = (V,E)$ where $V = A \cup B$, $|A| = \lceil \frac{1}{2}n \rceil$, and E consists of all $e \in \binom{V}{k}$ such that $|e \cap V|$ is odd. It is easy to check that $\delta_{k-1}(H_i) \ge n/2 - k$, $i = 1,2$. Moreover, it follows by a parity argument that H_1 contains no Hamiltonian l-cycle if $\frac{k}{k-l}$ is odd, while H_2 contains no Hamiltonian l-cycle if $\frac{k}{k-l}$ is even and $\frac{n}{k-l}$ is odd. The remaining case, when $\frac{k}{k-l}$ and $\frac{n}{k-l}$ are even, can be reduced to one of the two previous cases. ■

In the meantime, the value of $h^l(k,n)$ has been determined asymptotically for all $0 \le l \le k-1$, that is, also when $k-l$ does not divide k. First, Kühn and Osthus proved in [17] that $h^1(3,n) \sim \frac{1}{4}n$ and conjectured that $h^1(k,n) \sim \frac{1}{2(k-1)}n$. This conjecture was proved in [14], and independently in [12], where Hán and Schacht generalized it further, obtaining the asymptotic formula $h^l(k,n) \sim \frac{1}{2(k-l)}n$ for all $1 \le l < \frac{1}{2}k$. In turn, Hán and Schacht conjectured the right result for all values of l which was finally proved by Kühn, Mycroft, and Osthus in [15].

Theorem 2.7 ([15]). *If $k-l$ does not divide k and $(k-l) \mid n$, then*

$$h^l(k,n) \sim \frac{n}{\lceil \frac{k}{k-l} \rceil (k-l)}.$$

(Note that $\lceil \frac{k}{k-l} \rceil = 2$ for $l < k/2$.)

So, the situation is quite peculiar as our next example shows. Let $k = 10$. Then the asymptotic values of $h^l(10,n)$ for $l = 0,1,2,\dots,9$ are $\frac{1}{2}$, $\frac{1}{18}$, $\frac{1}{16}$, $\frac{1}{14}$, $\frac{1}{12}$, $\frac{1}{2}$, $\frac{1}{12}$, $\frac{1}{12}$, $\frac{1}{2}$, $\frac{1}{2}$.

The lower bound in the above theorem comes from the following construction which sheds some light on the origin of the cumbersome formula.

Construction 2.8. Let $H_3 = (V,E)$ where $V = A \cup B$,

$$|A| = \left\lceil \frac{n}{\lceil \frac{k}{k-l} \rceil (k-l)} \right\rceil - 1 \quad \text{and} \quad E = \left\{ e \in \binom{V}{k} : |e \cap A| \neq \emptyset \right\}.$$

It follows that $\delta_{k-1}(H_3) = |A|$. Recall that every Hamiltonian l-cycle has $m = n/(k-l)$ edges and maximum degree $\Delta = \lceil \frac{k}{k-l} \rceil$. If there was a Hamiltonian l-cycle in H_3, then A would be its vertex cover. However,

$$|A| \times \Delta = \left(\left\lceil \frac{n}{\lceil \frac{k}{k-l} \rceil (k-l)} \right\rceil - 1 \right) \times \left\lceil \frac{k}{k-l} \right\rceil < n/(k-l),$$

a contradiction.

It seems that it will be very hard to pinpoint the value of $h^l(k,n)$ precisely.

Problem 2.9. *Determine the exact value of $h^l(k,n)$ for all $k \geq 3$, $0 \leq l \leq k-1$ and all (sufficiently large) n.*

So far this has been solved for $k = 3$, $l = 2$ in [27] (see Theorem 2.2 above) and for $k \geq 3$, $l = 0$ in [23] (see Theorem 3.4 in Section 3).

2.2. An outline of the proof of Theorem 2.4

In this section we assume that $\delta_{k-1}(H) \geq (1/2 + \gamma)n$ for $\gamma > 0$ and sufficiently small with respect to k. The proof in [24] is built around the notion of an absorbing path. A *k-uniform (tight) path P of length s* is a k-graph with s vertices and $s - k + 1$ edges whose vertices can be ordered $v_1, \dots, v_s$ in such a way that every k consecutive vertices form an edge (each path has exactly two such orderings). The sequences $(v_1, \dots, v_{k-1})$ and $(v_s, \dots, v_{s-k+1})$ are called *the ends* of P, and we say that P *connects* them.

Lemma 2.10 (Absorbing Lemma, [24]). *There exists a path A in H (called* absorbing*) with $|V(A)| \leq 16k\gamma^{k-1}n$ such that for every subset $U \subset V \setminus V(A)$ of size $|U| \leq 2^{k-4}\gamma^{2k}n$ there is a path A_U in H with $V(A_U) = V(A) \cup U$ and such that A_U has the same ends as A.*

In other words, the above lemma asserts that there is *one*, not too long path such that *every* not too large subset of vertices can be "absorbed" into the "interior" of this path.

The idea of the proof of Theorem 2.4 can be described in three steps (see Figure 2).

Outline of proof of Theorem 2.4.

1. Fix an absorbing path A guaranteed by Lemma 2.10.
2. Build a cycle C of length at least $n - 2^{k-4}\gamma^{2k}n$ containing A.
3. Applying the absorbing property of A to the set $U = V(H) \setminus V(C)$, insert U into A, obtaining a Hamiltonian cycle $C_{\mathbf{HAM}}$ in H.

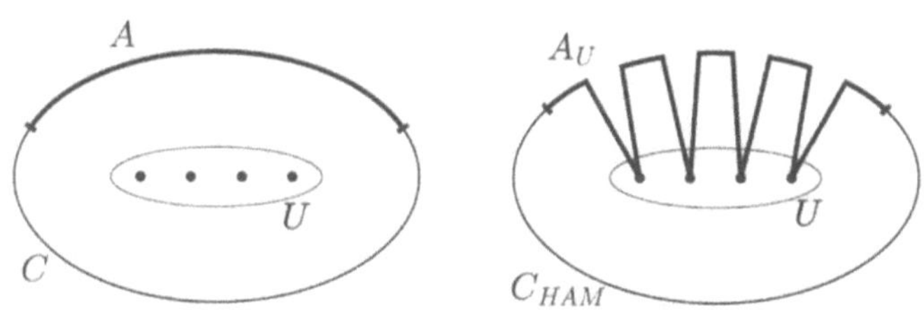

Fig. 2. A bird's view of the proof of Theorem 2.4

Below we explain how these three steps are implemented.

Step 1. The absorbing path will be constructed from absorbing sequences.

Definition 2.11. Given a vertex v, we say that a $(2k-2)$-element sequence of vertices $\mathbf{x} = (x_1, \ldots, x_{2k-2})$ *absorbs* v in H if

- for every $i = 1, \ldots, k-1$ we have $\{x_i, x_{i+1}, \ldots, x_{i+k-1}\} \in H$ (that is, $\mathbf{x}$ spans a path in H) and
- for every $i = 1, \ldots, k$ we also have $\{x_i, x_{i+1}, \ldots, x_{i+k-2}, v\} \in H$ (that is, $\mathbf{x}$ spans a $(k-1)$-uniform path in the link $H(v)$ of v in H).

If $\mathbf{x}$ is actually a segment of a path P and v is not a vertex of P, then P can "absorb" v by replacing the edges $\{x_i, x_{i+1}, \dots, x_{i+k-1}\}$, $i = 1, \dots, k-1$, by $\{x_i, x_{i+1}, \dots, x_{i+k-2}, v\}$, $i = 1, \dots, k$. This way, the segment $\mathbf{x}$ of P is replaced by the new segment $\mathbf{x}' = (x_1, \dots, x_{k-1}, v, x_k, \dots, x_{2k-2})$.

A key feature of absorbing sequences is that there are plenty of them.

Claim 2.12. *For every $v \in V(H)$, there are at least*

$$2^{k-2}\gamma^{k-1}n^{2k-2}$$

sequences absorbing v in H.

Proof. While constructing a v-absorbing sequence $\mathbf{x} = (x_1, \dots, x_{2k-2})$, there is no restriction on the vertices $x_1, \dots, x_{k-2}$ other than they should be different from v. Thus, $x_1, \dots, x_{k-2}$ can be chosen in precisely $(n-1)_{k-2}$ ways. By the degree assumption applied to the set $\{x_1, \dots, x_{k-2}, v\}$, there are at least $(1/2 + \gamma)n$ vertices x_{k-1} such that $\{x_1, \dots, x_{k-1}, v\} \in H$.

By the degree assumption applied to the sets $\{x_1, \dots, x_{k-1}\}$ and $\{x_2, \dots, x_{k-1}, v\}$, there are at least $2\gamma n + k - 2 > 2\gamma n$ vertices x_k such that

$$\{x_1, \dots, x_k\} \in H \quad \text{and} \quad \{x_2, \dots, x_k, v\} \in H.$$

(See Fact 3.1 in [24] for details.) Similarly, for each $i = k+1, \dots, 2k-2$, there are at least $2\gamma n + k - 2$ vertices x_i such that

$$\{x_{i-k+1}, \dots, x_i\} \in H \quad \text{and} \quad \{x_{i-k+2}, \dots, x_i, v\} \in H.$$

Among them, at least $2\gamma n + k - 2 - (i-k) \geq 2\gamma n$ satisfy $x_i \neq x_1, \dots, x_{i-k}$. Altogether, this implies that there are at least

$$(n-1)_{(k-2)}(1/2+\gamma)n(2\gamma n)^{k-1} > 2^{k-2}\gamma^{k-1}n^{2k-2}$$

sequences $\mathbf{x} = (x_1, \dots, x_{2k-2})$ absorbing v. ■

The construction of an absorbing path consists of two phases:

1(a) Selecting a small number of *disjoint*, absorbing sequences such that each vertex is absorbed by many of them;

1(b) Connecting these sequences into one path.

Phase 1(a). We select randomly, with probability $p = \gamma^{k+1}/n^{2k-3}$, a family $\mathcal{R}$ of $(2k-2)$-element sequences $\mathbf{x}$ of vertices. By standard probabilistic argument and by Claim 2.12 it follows that with positive probability $\mathcal{R}$ contains a subfamily $\mathcal{F}$ of at most $2\gamma^{k+1}n$ *disjoint* sequences such that for every vertex v at least $2^{k-4}\gamma^{2k}n$ of these sequences are v-absorbing (see [24] for details).

Phase 1(b). This phase is executed with the help of the connecting lemma from [24], the proof of which is omitted here.

Lemma 2.13 (Connecting Lemma, [24]). *If $\delta_{k-1}(H) \geq (1/2+\gamma)n$ then, for every two disjoint $(k-1)$-element sequences of vertices of H, there is a path in H of length at most $2k/\gamma^2$ which connects them.*

We use Lemma 2.13, but with $\gamma/2$ instead of γ, to connect, one by one, all sequences of $\mathcal{F}$ obtaining an absorbing path. This is possible, because the whole path will have at most

$$|\mathcal{F}| \times (8k/\gamma^2) \leq 2\gamma^{k+1}n \times (8k/\gamma^2) = 16k\gamma^{k-1}n$$

vertices, and thus, at any given time of the connecting procedure, the subhypergraph H^* spanned by the remaining vertices will have

$$\delta_{k-1}(H^*) \geq (1/2+\gamma)n - 16k\gamma^{k-1}n > (1/2+\gamma/2)n > (1/2+\gamma/2)\big|V(H^*)\big|,$$

for sufficiently small $\gamma > 0$.

Step 2. The process of finding a long cycle containing A, can be broken up into three phases:

2(a) Selecting a small "reservoir set" R such that $|R| = 2^{k-5}\gamma^{2k}n$, $R \cap V(A) = \emptyset$, and $H[R]$ inherits the degree property of the entire k-graph H, scaled down to its size.

2(b) Constructing, via The Weak Regularity Lemma, a constant size collection of long, disjoint paths in $H' = H\big[V \setminus \big(V(A) \cup R\big)\big]$, covering all but at most $2^{k-5}\gamma^{2k}\big|V(H')\big|$ vertices of H'.

2(c) Connecting these paths and the absorbing path A into one cycle, utilizing a small chunk of R.

Phase 2(a) is necessary, since toward the end of the connecting phase 2(c), there will be only few vertices left outside the path under construction, and thus available for connecting. We make sure, however, that this residual part of H' will contain a small "copy" of H, namely $H[R]$ or its large portion $H[R']$, $R' \subset R$, and so, we will be in position to apply an analog of Lemma 2.13 to it.

Phase 2(a).

Lemma 2.14 (Reservoir Lemma). *There exists a subset $R \subset V \setminus V(A)$ of size $|R| = \lfloor 2^{k-5}\gamma^{2k}n \rfloor$ such that for every $(k-1)$-element set $S \subset V$ we have*

$$|N_H(S) \cap R| \geq (1/2 + \gamma/2)|R|. \tag{1}$$

Proof. Select R randomly. By Chernoff's bound, with high probability, the set R will satisfy (1). ∎

Phase 2(b).

Lemma 2.15 (Path Cover Lemma). *All but at most $2^{k-5}\gamma^{2k}|V(H')|$ vertices of $H' = H\big[V \setminus \big(V(A) \cup R\big)\big]$ can be covered by at most $m = m(\gamma)$ vertex-disjoint paths $P_1, \dots, P_m$.*

Proof. See [24]. ∎

Phase 2(c). In this final phase of Step 2, we use a lemma which was implicitly proved in [24].

Lemma 2.16 (Restricted connecting Lemma). *Let R be as in Lemma 2.14. Then for every two disjoint, $(k-1)$-element sequences $(x_1, \dots, x_{k-1})$ and $(y_1, \dots, y_{k-1})$ of vertices of H, there is a path P in H of length at most $8k/\gamma^2 + 2(k-1)$, which connects them and such that*

$$V(P) \setminus \{x_1, \dots, x_{k-1}, y_1, \dots, y_{k-1}\} \subset R.$$

Proof. By property (1) there exist distinct vertices $u_1, \dots, u_{k-1} \in R$ and $v_1, \dots, v_{k-1} \in R$ such that $Q_x = (x_1, \dots, x_{k-1}, u_1, \dots, u_{k-1})$ and $Q_y = (y_1, \dots, y_{k-1}, v_1, \dots, v_{k-1})$ form paths in H. Now, we can apply Lemma 2.13 with $\gamma/2$ to the k-graph $H[R]$ and the sequences $(u_1, \dots, u_{k-1})$ and $(v_1, \dots, v_{k-1})$, obtaining a path Q of length $8k/\gamma^2$ connecting them. Then, the path $P = Q_xQQ_y$ connects $(x_1, \dots, x_{k-1})$ with $(y_1, \dots, y_{k-1})$ and has length $|V(Q)| + 2(k-1)$. ∎

Now, we are ready to prove the main lemma of this phase.

Lemma 2.17 (Long Cycle Lemma). *There is in H a cycle C of length at least $n - 2^{k-5}\gamma^{2k}n$ containing A.*

Proof. We perform $m+1$ applications of Lemma 2.16, with $\gamma/3$ instead of $\gamma/2$, to large subsets $R' \subseteq R$, and connect all paths $P_1, \ldots, P_m$, as well as the absorbing path A, into one long cycle C in H. Suppose that at some point we are to connect an end $(x_1, \ldots, x_{k-1})$ of one path with an end $(y_1, \ldots, y_{k-1})$ of another path. Let us denote the yet unused portion of R by R'. Since we connect only $O(1)$ times, each time using only $O(1)$ vertices of R, throughout the procedure we maintain that $|R'| = |R| - O(1) > (1-\gamma/6)|R|$, and thus, by property (1) of R, for every $(k-1)$-element set $S \subset V$ we still have

$$\big|N_H(S) \cap R'\big| \geq (1/2+\gamma/2)|R| - \big(|R| - |R'|\big) > (1/2+\gamma/3)|R| > (1/2+\gamma/3)|R'|.$$

Hence, we apply Lemma 2.16 with $\gamma/3$ instead of $\gamma/2$, and so, the obtained connecting paths are of lengths at most $18k/\gamma^2 + 2(k-1)$.

Let T be the set of vertices of H' not covered by the paths $P_1, \ldots, P_m$. Only a subset R' of R and the set T are uncovered by the cycle C. The union of these two sets has size at most $|R| + |T| \leq 2^{k-4}\gamma^{2k}n$ (see Figure 3). ■

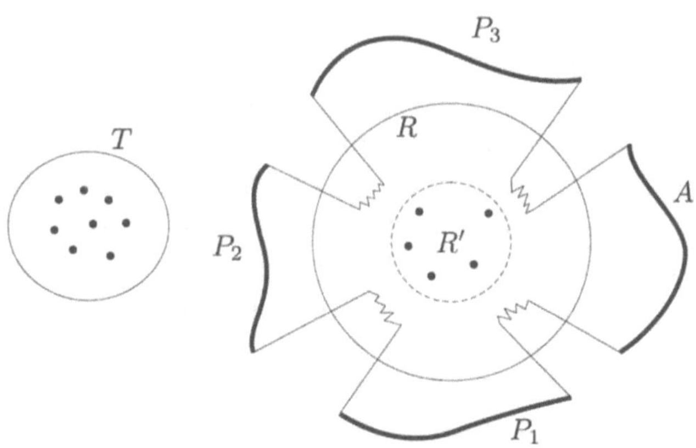

Fig. 3. Phase 2(c) of the proof of Theorem 2.4

Step 3. Let $U = R' \cup T$. Note that $|U| \leq 2^{k-4}\gamma^{2k}n$. Let A_U be the path as defined in Lemma 2.10. Then, replacing A with A_U in C yields a Hamiltonian cycle in H.

This completes the outline of the proof of Theorem 2.4.

2.3. Hamilton cycles in hypergraphs with large vertex minimum degree

There are virtually no results on $h_d^l(k,n)$ for $d \le k-2$. Here we consider the smallest unsolved case: $k=3$ and $d=1$.

Two constructions set the bound $h_1(3,n) \ge \left(\frac{5}{9}+o(1)\right)\binom{n-1}{2}$. One is obtained by modifying the hypergraph H_0 from Construction 2.1. We now take $V = X \cup Y$, where $|Y| \sim 2|X|$ (instead of $|Y| \sim |X|$) and all triples S of vertices with $|S \cap X| \ne 1$ as the edges. Let H_0' be the obtained 3-graph. Then

$$\begin{aligned}\delta_1(H_0') &= \max\left(\binom{|Y|-1}{2}+\binom{|X|}{2}, \binom{|X|-1}{2}+(|X|-1)\,|Y|\right)\\ &\sim \frac{5}{9}\binom{n-1}{2}.\end{aligned}$$

and, likewise in H_0, there is no Hamiltonian cycle in H_0'.

The other construction is very similar to the hypergraph H_3 described in Construction 2.8. We define H_4 as a hypergraph on the vertex set $V = X \cup Y$, where $|X| = n/3 - 1$, and with the edge set consisting of all triples intersecting X. Then, again, $\delta_1(H_4) \sim \frac{5}{9}\binom{n-1}{2}$ and H_4 has no Hamiltonian cycle.

Note that (for n divisible by 3) the hypergraph H_4 does not even have a perfect matching. As we will see in Section 3 (see Theorem 3.4 below, proved in [11]), the threshold $m_1(3,n)$ for the existence of a perfect matching is, in fact, $\left(\frac{5}{9}+o(1)\right)\binom{n-1}{2}$. Judging by the similarities between Dirac thresholds for perfect matchings and Hamiltonian cycles in various situations, it was tempting to conjecture that $h_1(3,n) \sim m_1(3,n)$. However, even showing that $h_1(3,n) \le c\binom{n-1}{2}$ for some $c<1$ does not seem to be completely trivial. In our preliminary reconnaissance of this problem, by adapting the original proof from [24] and using Theorem 3.4 along the way, we were able to obtain only the upper bound $h_1(3,n) \le \left(\frac{11}{12}+\gamma\right)\binom{n-1}{2}$. Very recently we learned from Endre that he knows how to prove that, indeed, $h_1(3,n) \sim m_1(3,n)$.

Endre's insight and the existing results showing that $h(k,n) \sim m(k,n)$ for all k suggest that the same is true in general.

Conjecture 2.18. For all $1 \le d \le k-1$,

$$h_d(k,n) \sim m_d(k,n).$$

Note that formula (4) and Conjecture 3.6 in Section 3.2 specify the value of $m_d(k,n)$.

2.4. The k-partite case

Unlike matchings (see the next section) there are very few results on the Hamiltonicity of partite hypergraphs. For graphs, Moon and Moser [20] extended Dirac's theorem to bipartite graphs. Later, the authors of [5] provided a generalization to balanced k-partite graphs. Here we treat briefly the case of k-partite k-graphs, $k \geq 3$.

A k-graph H is k-partite if its vertices can be partitioned into k classes, $V(H) = V_1 \cup \cdots \cup V_k$, in such a way that for every edge $e \in H$ and each $i = 1, \dots, k$, we have $|e \cap V_i| = 1$. Given such a partition, we call a set of vertices S *legal* if for each $i = 1, \dots, k$, $|S \cap V_i| \leq 1$. We denote by $\delta'(H) := \delta'_{k-1}(H)$ the minimum of $\deg_H(S)$ taken over all legal $(k-1)$-tuples S in H.

An adaptation of the proof of Theorem 2.4 leads to the following result, which, in turn, implies Theorem 2.4 by taking a random k-partition.

Proposition 2.19. *Let $k \geq 3$, $\gamma > 0$, and let H be a k-partite k-graph on kn vertices with a given equitable partition $V_1, \dots, V_k$, $|V_i| = n$, where n is sufficiently large. If $\delta'_{k-1}(H) \geq (1/2 + \gamma)n$ edges, then H is Hamiltonian. Moreover, there is a k-partite k-graph H_0 on kn vertices and with an equitable partition such that $\delta'_{k-1}(H_0) \geq \left\lfloor \frac{1}{2}n \right\rfloor$ and H_0 does not have a Hamiltonian cycle.*

To obtain H_0, we modify Construction 2.1.

Construction 2.20. Given k and n, let $X = X_1 \cup \cdots \cup X_k$, $Y = Y_1 \cup \cdots \cup Y_k$, and $V_i = X_i \cup Y_i$, $i = 1, \dots, k$, where all sets X_i and Y_i are pairwise disjoint, $\lfloor kn/2 \rfloor \leq |X|, |Y| \leq \lceil kn/2 \rceil$, $|X| + |Y| = kn$, and, for $i = 1, \dots, k$, $\lfloor n/2 \rfloor \leq |X_i|, |Y_i| \leq \lceil n/2 \rceil$, and $|V_i| = n$.

Let H_0 be a k-graph with $V = V_1 \cup \cdots \cup V_k = X \cup Y$ whose edge set consists of all k-element subsets S of V such that $|X \cap S| \neq \left\lfloor \frac{k}{2} \right\rfloor$ and $\left|S \cap (V_i)\right| \leq 1$, $i = 1, 2, \dots, k$. Being a subhypergraph of the k-graph from Construction 2.1, this new H_0 is not Hamiltonian either. Moreover, for every $(k-1)$-element subset S of V, if $|X \cap S| \in \left\{ \left\lfloor \frac{k}{2} \right\rfloor - 1, \left\lfloor \frac{k}{2} \right\rfloor \right\}$, then $\deg_{H_0}(S) \in \{|X_i|, |Y_i|\} = \left\{ \lfloor n/2 \rfloor, \lceil n/2 \rceil \right\}$, while if $|X \cap S| \notin \left\{ \left\lfloor \frac{k}{2} \right\rfloor - 1, \left\lfloor \frac{k}{2} \right\rfloor \right\}$ then $\deg_{H_0}(S) = |V_i| = n$, where i is the unique index such that $S \cap V_i = \emptyset$.

The proof of the main part of Proposition 2.19 follows the lines of the proof of Theorem 2.4 from [24], outlined in Section 2.2. It only needs to be substantially altered in the construction of the absorbing path. Below we provide details of this modified absorbing scheme.

Note that on every path or cycle the cyclical order in which the first edge meets the sets $V_1, \dots, V_k$ is maintained by all subsequent edges. Without loss of generality, we choose $V_1, V_2, \dots, V_k$ as the canonical order, and will be assuming that the absorbing path we build as well as the final Hamiltonian cycle will follow that order.

We will use two different absorbing strategies depending on whether a given set of k vertices which is to be absorbed forms an edge in H or not.

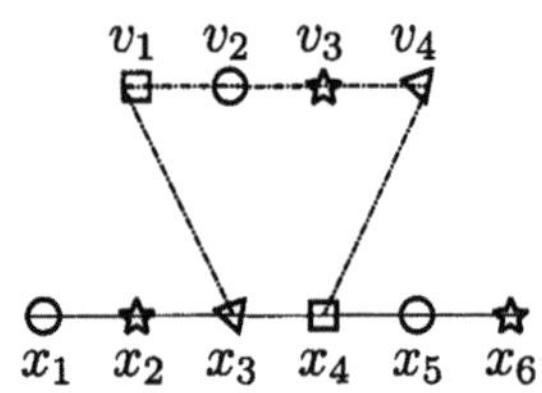

Fig. 4. Absorbing sequence, $k = 4$, the partition sets marked by different symbols

Definition 2.21. For an edge $e = \{v_1, \dots, v_k\} \in H$, where $v_i \in V_i$, $i = 1, \dots, k$, we say that a $(2k-2)$-element sequence of vertices $\mathbf{x} = (x_1, \dots, x_{2k-2})$, *absorbs* e in H if

(a) $x_1 \in V_2, x_2 \in V_3, \dots, x_{k-1} \in V_k, x_k \in V_1, \dots, x_{2k-2} \in V_{k-1}$,

(b) for every $i = 1, \dots, k-1$, we have $\{x_i, x_{i+1}, \dots, x_{i+k-1}\} \in H$ (that is, $\mathbf{x}$ spans a path in H),

(c) for every $i = 1, \dots, k-1$, we have $\{x_i, \dots, x_{k-1}, v_1, \dots, v_i\} \in H$, and

(d) for every $i = 2, \dots, k$, we have $\{v_i, \dots, v_k, x_k, \dots, x_{k-2+i}\} \in H$. (Properties (c) and (d) together imply that the sequence $\mathbf{x}' = (x_1, \dots, x_{k-1}, v_1, \dots, v_k, x_k, \dots, x_{2k-2})$ spans a path in H.)

If $\mathbf{x}$ is actually a segment of a path P and $v_1, v_2, \dots, v_k$ are not on P, then P can "absorb" all these vertices by replacing the segment $\mathbf{x}$ with the new segment $\mathbf{x}'$ (see Fig. 4).

In the final stage of the proof of Proposition 2.19 the above absorbing technique can be used for as long as there are edges induced by the vertices remaining outside the long cycle. When the set of such vertices becomes independent, we use a swapping device which will exchange some k vertices outside the cycle with a set of k vertices which form an edge of H, allowing us to use again the absorbing device and absorb the released vertices back into the cycle.

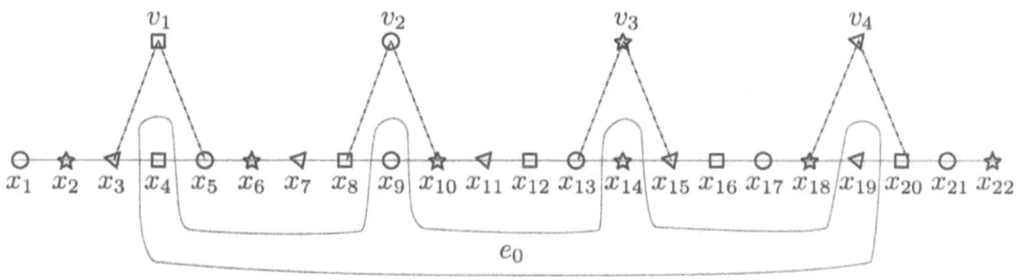

Fig. 5. Swapping sequence, $k = 4$, the partition sets marked by different symbols

Definition 2.22. For a set $S = \{v_1, \ldots, v_k\} \subset V(H)$, where $v_i \in V_i$, $i = 1, \ldots, k$, we say that a $(k^2 + 2k - 2)$-element sequence of vertices $\mathbf{x} = (x_1, \ldots, x_{k^2+2k-2})$, *is edge-swapping for* S if

(a) $x_1 \in V_2, x_2 \in V_3, \ldots, x_{k-1} \in V_k, x_k \in V_1, \ldots, x_{k^2+2k-2} \in V_{k-1}$,

(b) the sequence $\mathbf{x}$ spans a path P_1 in H,

(c) the sequence $\mathbf{x}$ with each x_{ik+i-1} replaced by v_i, $i = 1, \ldots, k$, spans a path P_2 in H, and

(d) $e_0 := \{x_k, x_{2k+1}, \ldots, x_{k^2+k-1}\} \in H$.

If $\mathbf{x}$ is actually a segment of a path P and $v_1, v_2, \ldots, v_k$ are not on P, then P can "swap" the vertices $x_k, x_{2k+1}, x_{3k+2} \ldots, x_{k^2+k-1}$ for $v_1, v_2, \ldots, v_k$ by replacing P_1 with P_2, and thus, releasing the vertices of e_0 from P (see Fig. 5).

So, our absorbing strategy is as follows: create two, disjoint, not too long paths: an absorbing path A containing many absorbing sequences for each *edge* of H, and a swapping path B containing many edge-swapping sequences for each k-element set of vertices of H.

To successfully complete this task all we need are two statements analogous to Claim 2.12. Let us begin with counting, for a given edge $\{v_1, \ldots, v_k\} \in H$, the number of absorbing sequences.

Claim 2.23. *For every edge* $\{v_1, \ldots, v_k\} \in H$, *there are at least* $\gamma^{k-1}n^{2k-2}$ *absorbing sequences in* H.

Proof. As for each $i = k-1, k-2, \ldots, 1$, $\deg_H(x_{i+1}, \ldots, x_{k-1}, v_1, \ldots, v_i) \geq (1/2+\gamma)n$, there are at least $(n/2)^{k-1}$ choices of $x_{k-1}, \ldots, x_1$, selected in that order. Then, each of $x_k, \ldots, x_{2k-2}$ must be a common neighbor of two $(k-1)$-tuples of already existing vertices, and so there are at least, roughly, $(2\gamma n)^{k-1}$ choices of these vertices. Altogether, we have at least $\gamma^{k-1}n^{2k-2}$ such sequences. ■

Claim 2.24. *For every set $S = \{v_1, \ldots, v_k\} \subset V(H)$, there are at least $2^{k^2-k}\gamma^{k^2}n^{k^2+2k-2}$ edge-swapping sequences in H.*

Proof. For a given set $S = \{v_1, \ldots, v_k\} \subset V(H)$, we will proceed systematically and count, for each $i = 1, \ldots, k^2+2k-2$, the number of choices of x_i, given that $x_1, \ldots, x_{i-1}$ have been already selected. There are, roughly, n choices for each of $x_1, \ldots, x_{k-2}$ as there are no constraints on them. The vertex x_{k-1} must be a neighbor of $\{x_1, \ldots, x_{k-2}, v_1\}$ and then, x_k must be a neighbor of $\{x_1, \ldots, x_{k-1}\}$, yielding at least $n/2$ choices of each. The vertices $x_{k+1}, \ldots, x_{2k-1}$ are each a common neighbor of two $(k-1)$-tuples of already existing vertices, one on the path P_1, the other on P_2. This is also true for x_{2k}, although for a different reason. Indeed, the paths P_1 and P_2 run together between x_{k+1} and x_{2k}, however x_{2k} must be a common neighbor of $\{x_{k+1}, \ldots, x_{2k-1}\}$ and $\{x_{k+2}, \ldots, x_{2k-1}, v_2\}$. Then, x_{2k+1} has to be a neighbor of only one $(k-1)$-tuple (the one on P_1, namely $\{x_{k+1}, \ldots, x_{2k}\}$). This pattern continues for the next $k-2$ intervals of length $k+1$, until we reach x_{k^2+k-1} which, in addition, has to be a neighbor of $x_k, x_{2k+1}, \ldots, x_{k^2-2}$. It is crucial for the success of our construction that no vertex needs to be a common neighbor of three or more already existing $(k-1)$-tuples.

Hence, altogether, there are at least

$$n^{k-2} \times (n/2)^k \times (2\gamma n)^{k^2} = 2^{k^2-k}\gamma^{k^2}n^{k^2+2k-2}$$

choices of the entire edge-swapping sequence. ■

The rest of the proof of a k-partite version of the absorbing lemma follows mutatis mutandis the proof from [24] described in Section 2.2, except that we need to be careful to maintain the canonical order on all paths we build.

3. Perfect Matchings

There are several results on Dirac-type degree thresholds for perfect matchings in k-graphs. To some extent, they resemble the results for Hamiltonian cycles and are often obtained by methods based on similar ideas, most notably, the idea of absorption. In this section we give an overview of such results.

Recall that the Dirac-type threshold $m_d^r(k,n)$ has been introduced in Defintion 1.3 and that we suppress the subscript d when $d = k-1$ as well as we suppress the superscript r when $r = 0$, that is, when we consider perfect matchings (see the Summary of notation in Section 1).

For graphs, an easy argument shows that $m(2,n) = n/2$. Since, for n divisible by k, every Hamiltonian cycle contains a perfect matching, it follows from [24] that $m(k,n) \le n/2 + o(n)$. In [16], Kühn and Osthus sharpened this bound to $m(k,n) \le n/2 + 3k^2\sqrt{n \log n}$, using a result for the k-partite case which they had shown first (see Subsection 3.4). This was further improved in [23] to $m(k,n) \le n/2 + C \log n$, using the idea of absorption. The authors of [25] found a fairly simple proof of the inequality $m(k,n) \le n/2 + k/4$, based on a beautiful idea of Aharoni, Georgakopoulos, and Sprüssel [1] (see Subsection 3.4).

This last bound is very close to the true value of $m(k,n)$. Indeed, constructions presented in [26] yield the lower bound

(2)
$$m(k,n) \ge t(n,k) := \begin{cases} n/2 + 3 - k & \text{if } k/2 \text{ is even and } n/k \text{ is odd,} \\ n/2 + 5/2 - k & \text{if } k \text{ is odd and } (n-1)/2 \text{ is odd,} \\ n/2 + 3/2 - k & \text{if } k \text{ is odd and } (n-1)/2 \text{ is even,} \\ n/2 + 2 - k & \text{otherwise.} \end{cases}$$

Moreover, the main result of [26] shows that, in fact, there is equality in (2).

Theorem 3.1 ([26]). *For all $k \ge 2$, $m(k,n) = t(k,n)$, where $t(k,n)$ is given by (2).*

When comparing with Problem 2.3, we see that the conjectured Dirac threshold for a Hamiltonian cycle and the above threshold for a perfect matching differ only by an additive term of about $k/2$. In fact, we know that they coincide for $k = 2$ and differ by at most one for $k = 3$ (see Theorem 2.2).

To prove Theorem 3.1, two cases are separately considered in [26]. When H is "close" to one of the critical k-graphs yielding the lower bound (2), one can find a perfect matching in H by "brute force" If, on the other hand, H is far from the critical k-graphs, we apply a version of the absorbing technique.

The absorbing configurations used in [23] and [26] (as well as in [11]), although different from each other, follow the same pattern: given a set $S \subset V(H)$, $|S| = k$, a matching M_1 is *S-absorbing* if the vertex set $V(M_1) \cup S$ spans in H a matching M_2 of size $|M_1| + 1$. Consider a matching M and a set S, $S \cap V(M) = \emptyset$. If M contains an S-absorbing matching M_1, then one can absorb S into M by swapping M_1 for M_2.

The idea of the proofs in [23] and in the "far-from-critical" case in [26] is now transparent and similar to the idea described in the Outline of the proof of Theorem 2.4:

- Find a relatively small matching M_A such that for every set $S \subset V(H)$, $|S| = k$, there is an S-absorbing matching in M_A.
- Build a matching M' in $H' = H - V(M_A)$ which leaves only a set S of k vertices unmatched.
- Apply the absorbing procedure to S.

Building the almost perfect matching M' requires itself a version of the absorbing technique which works for as long as there are more than k vertices uncovered. Adding the last edge represent a more significant difficulty. In the next subsection we will see that if we allow even one vertex to be uncovered the threshold drops significantly. A matching of size $n/k - k + 2$ can be, however, constructed by a standard greedy approach.

3.1. Almost perfect matchings

Here we present results about $m^r(k, n)$ for $r > 0$. The following construction yields the lower bound

$$m^r(k, n) \geq \frac{n - r}{k}$$

for all $r > 0$.

Construction 3.2. With $n = r \pmod{k}$, let A and B be disjoint sets of sizes $|A| = \frac{n-r}{k} - 1$ and $|B| = n - |A|$. Let H_r be a k-graph on $V = A \cup B$ consisting of all k-element subsets S of vertices which intersect A. Then the largest matching of H_r has size at most $|A|$, and thus, it has at most $k|A| < n - r$ vertices.

In [26] we established that

$$m^r(k,n) = \frac{n-r}{k} \tag{3}$$

holds for all $r \geq k(k-2)$. This was shown by a fairly simple argument involving a greedy algorithm.

Also, using a version of the absorption method, with the sets S of size $k+1$, it was proved in [26] that for all $r > 0$

$$\frac{n-r}{k} \leq m^r(k,n) \leq \frac{n}{k} + O(\log n).$$

This result stands in a striking contrast with Theorem 3.1, where the threshold is around $n/2$. Hence, from the Dirac threshold perspective, an almost perfect matching appears much sooner than a prefect one.

Note that for $0 < r < k$ we have $\frac{n-r}{k} = \left\lfloor \frac{n}{k} \right\rfloor$ which is the size of the largest matching one can possibly have if n is not divisible by k. We feel that the $O(\log n)$ term, brought in by the technicalities of the absorption method, should not be there.

Problem 3.3. *Prove (or disprove) that $m^r(k,n) = \left\lfloor \frac{n}{k} \right\rfloor$ for all $0 < r < k$. In particular, is it true that if $n \neq 0 \pmod 3$ and $\delta_2(H) \geq \left\lfloor \frac{n}{3} \right\rfloor$ then there is a matching in H of size $\left\lfloor \frac{n}{3} \right\rfloor$?*

3.2. The parameter $m_d^r(k,n)$ for $1 \leq d \leq k-2$

Pikhurko [21] proved that for all $d \geq k/2$

$$m_d(k,n) \sim \frac{1}{2}\binom{n-d}{k-d}. \tag{4}$$

His proof is in part based on the ideas from [16]. Similarly as in [16] he proved first a related result for k-partite k-graphs (see Subsection 3.4).

Moreover, in view of Remark 1.4 it was sufficient to prove the lower bound in (4) only for $d = k-1$ and the upper bound in (4) only for $d = \lceil k/2 \rceil$.

The case $d < k/2$ seems to be much harder. The constructions yielding (2) together with Construction 3.2 applied with $r = 0$ give the following, general lower bound:

$$(5) \qquad m_d(k,n) \geq \left(\max \left\{ \frac{1}{2}, 1 - \left(\frac{k-1}{k} \right)^{k-d} \right\} + o(1) \right) \binom{n-d}{k-d}.$$

As for the upper bound, Hàn, Person, and Schacht [11], by a similar method as in [21], proved first that for all $0 \leq d \leq k-1$,

$$(6) \qquad m_d^{k(d-1)}(k,n) \leq \left(\frac{k-d}{k} + o(1) \right) \binom{n-d}{k-d}.$$

That is, if for a k-graph H, $\delta_d(H)$ is at least as large as the R-H-S of (6) then H contains a matching covering all but $k(d-1)$ vertices. Then, combining (6) with the absorption method, they improved (6) in the lower range of d by showing that for $1 \leq d < k/2$

$$(7) \qquad m_d(k,n) \leq \left(\frac{k-d}{k} + o(1) \right) \binom{n-d}{k-d}.$$

Note that for $d = k-1$, (6) is asymptotically the same result as (3). For $d = 1$, on the other hand, (6) is asymptotically equivalent to an old result of Daykin and Häggvist [7].

In the same paper [11] the authors improved (7) in the smallest case of $k = 3$, $d = 1$, achieving asymptotically the lower bound (5):

Theorem 3.4 ([11]).

$$m_1(3,n) \sim \frac{5}{9} \binom{n-1}{2}$$

A crucial ingredient of the proof in [11] was a strong version of the Absorbing Lemma for matchings, an analog of Lemma 2.10 from Section 2.2.

Lemma 3.5 ([11], Lemma 10). *For all $\gamma > 0$ and integers $k > d > 0$ there is an n_0 such that for all $n > n_0$ the following holds: Suppose that H is a k-graph on n vertices with $\delta_d(H) \geq (1/2 + 2\gamma)\binom{n-d}{k-d}$, then there exists a matching $M := M_{abs}$ in H such that*

(i) $|M| < \gamma^k n/k$, and

(ii) *for every set* $W \subset V \setminus V(M)$ *of size at most* $|W| \leq \gamma^{2k} n$ *and divisible by* k *there exists a matching in* H *covering exactly the vertices of* $V(M) \cup W$.

This success prompted Hàn, Person, and Schacht to conjecture that (5) is the correct asymptotics of $m_d(k,n)$.

Conjecture 3.6 ([11]). For all $1 \leq d < k/2$,

$$m_d(k,n) \sim \max\left\{\frac{1}{2}, 1 - \left(\frac{k-1}{k}\right)^{k-d}\right\}\binom{n-d}{k-d}.$$

Observe that with $d = 1$ the above coefficient equals $\frac{5}{9}$ for $k = 3$, $\frac{37}{64}$ for $k = 4$, and $\frac{369}{625}$ for $k = 5$. However, for $d = 2$ and $k = 5$ it is $\frac{1}{2}$.

Very recently, Markström and the second author [19] lowered slightly the general bound (7) by using some ideas behind Theorem 3.4. They proved that for all $1 \leq d < k/2$

$$m_d(k,n) \leq \left(\frac{k-d}{k} - \frac{1}{k^{k-d}} + o(1)\right)\binom{n-d}{k-d}. \tag{8}$$

In the smallest unknown case, $k = 4$, inequality (7) yields a bound $m_1(4,n) \leq \left(\frac{48}{64} + o(1)\right)\binom{n-1}{3}$. It follows from (8) that $m_1(4,n) \leq \left(\frac{47}{64} + o(1)\right)\binom{n-1}{3}$. By some tedious case by case analysis the coefficient can be lowered further to $\frac{42}{64}$ (see [19]), still far from the conjectured $\frac{37}{64}$.

3.3. Fractional perfect matching

A relaxation of the notion of a perfect matching can be obtained by allowing the inclusion of fractional edges into a matching. *A fractional perfect matching* in a k-graph $H = (V,E)$ is a function $w : E \to [0,1]$ such that for each $v \in V$ we have $\sum_{e \ni v} w(e) = 1$. It follows that if an n-vertex k-graph has a fractional perfect matching then $\sum_{e \in H} w(e) = \frac{n}{k}$, which justifies the name.

For every $1 \leq d \leq k-1$, let

$$m_d^*(k,n) = \min\{m : \delta_d(H) \geq m \implies H \text{ contains a fractional perfect matching}\}.$$

It was proved in [23] that $m^*_{k-1}(k,n) \le \lceil n/k \rceil$, so, again, the threshold is much lower than that for perfect matchings. Moreover, Construction 3.2 with $|A| = \lceil n/k \rceil - 1$ provides an n-vertex k-graph with $\delta_{k-1} = \lceil n/k \rceil - 1$ which has no fractional perfect matching. Hence, we have the following result.

Theorem 3.7 ([23]). $m^*_{k-1}(k,n) = \lceil n/k \rceil$.

The proof of Theorem 3.7 utilizes the Farkas Lemma (see, e.g., [6] or [18]) which asserts that a system of equations $\mathbf{yA} = \mathbf{b}, \mathbf{y} \ge \mathbf{0}$, is solvable if and only if the system $\mathbf{Ax} \ge \mathbf{0}, \mathbf{bx} < 0$, is unsolvable.

Let $\mathbf{A} := \mathbf{A_H}$ be the incidency matrix of a hypergraph H with rows representing the edges and columns representing the vertices of H. We applied Farkas' Lemma with this $\mathbf{A}$ and with $\mathbf{b} = \mathbf{1}$ – the vector of length n whose all entries are equal to 1, and showed that, under the assumption $\delta_{k-1}(H) \ge \lceil n/k \rceil$ the system of inequalities $\mathbf{Ax} \ge \mathbf{0}, \mathbf{1x} < 0$, has no solutions. Hence, there is a solution to $\mathbf{yA} = \mathbf{1}, \mathbf{y} \ge \mathbf{0}$, which determines a fractional perfect matching $w(e) = y_e$ for all $e \in H$.

It turns out that fractional matchings can be used to give an alternative proof of Theorem 3.4, and possibly even to settle Conjecture 3.6 in full generality. Indeed, the following relation holds.

Theorem 3.8. *For every* $1 \le d \le k-1$ *and every* $\alpha > 0$

$$\frac{m_d(k,n)}{\binom{n-d}{k-d}} \le \max\left(\frac{1}{2}, \frac{m^*_d(k,n)}{\binom{n-d}{k-d}}\right) + \alpha$$

for sufficiently large n.

Observe that, trivially, $m^*_d(k,n) \le m_d(k,n)$. Therefore, if $m^*_d(k,n) \ge \frac{1}{2}\binom{n-d}{k-d}$ then

$$m^*_d(k,n) \sim m_d(k,n) \tag{9}$$

The proof of Theorem 3.8 is based on Theorem 1.1 in [9]. An immediate corollary of that result asserts the existence of an almost perfect matching in a k-graph with all degrees almost equal and all pair degrees much smaller than the vertex degrees (see the Remark after Theorem 1.1 in [9]). Here we formulate this corollary in the following lemma.

Lemma 3.9 ([9]). *For all k, $\varepsilon > 0$ and $a > 3$ there exists $\tau = \tau(\varepsilon)$ and $n_0 = n_0(\tau)$ such that if $n > n_0$ and H is an n-vertex k-graph satisfying*

1. $(1-\tau)D < \deg_H(v) < (1+\tau)D$ *for some D and all $v \in V$, and*
2. $\delta_2(H) < D/(\log n)^a$

then H contains a matching M_{alm} covering all but at most εn vertices.

The second tool is the Strong Absorbing Lemma 3.5 (see previous section).

Sketch of Proof of Theorem 3.8. Assume that there exists a constant $0 < c < 1$ such that $m_d^*(k,n) \sim c\binom{n-d}{k-d}$. This is not a restriction at all, as we know by (5) that $m_d^*(k,n) = \Theta\left(\binom{n-d}{k-d}\right)$. For any $\alpha > 0$ consider an n-vertex k-graph H, n large, with

$$\delta_d(H) > (c+\alpha)\binom{n-d}{k-d}.$$

Set $\gamma = \alpha/2$ and $\varepsilon = (\alpha/2)^{2k}$. The proof consists of four steps.

1. Find an absorbing matching M_{abs} satisfying properties (i) and (ii) of Lemma 3.5. Set $H' = H \backslash V(M_{abs})$. Note that $\delta_d(H') \geq (c+\alpha/2)\binom{n-d}{k-d}$.
2. Select a spanning subhypergraph H'' of H' satisfying the assumptions of Lemma 3.9 with $D = n^{0.2}$, $\tau = o(1)$ any $a > 0$, and $n \geq n_0(a)$.
3. Find an almost perfect matching M_{alm} in H'' by applying Lemma 3.9. Note that $|V(M_{alm})| \geq (1-\varepsilon)|V(H')|$ and thus, $|V(M_{alm} \cup M_{abs})| \geq (1-\varepsilon)n$.
4. Extend $M_{alm} \cup M_{abs}$ to a perfect matching of H by using the absorbing property (ii) of M_{abs} with respect to $W = V(H') \setminus V(M_{alm})$. ∎

In view of relation (9), in order to prove Conjecture 3.6 it is sufficient to show that

$$m_d^*(k,n) \sim \left(1 - \left(\frac{k-1}{k}\right)^{k-d}\right)\binom{n-d}{k-d}.$$

This is work in progress. We have heard from Endre that he knows how to determine $m_1(3,n)$ exactly.

3.4. The k-partite case

Recall from Section 2.4 that $\delta'(H) := \delta'_{k-1}(H)$ is the minimum of $\deg_H(S)$ taken over all *legal* $(k-1)$-tuples of vertices S in a k-partite k-graph H. Throughout this subsection, we asume that the k-partition $V(H) = V_1 \cup \cdots \cup V_k$ satisfies $|V_1| = \cdots = |V_k| = n$.

In [16], Kühn and Osthus showed that if

$$\delta'_{k-1}(H) \geq n/2 + \sqrt{2n \log n}$$

then H has a perfect matching. Improving this result, Aharoni, Georgakopoulos, and Sprüssel obtained in [1] a surprisingly strong result.

Theorem 3.10 ([1]). *If for every $(k-1)$-tuple of vertices $(v_1, \ldots, v_{k-1}) \in V_1 \times \cdots \times V_{k-1}$ we have $\deg_H(v_1, \ldots, v_{k-1}) > n/2$ and for every $(v_2, \ldots, v_k) \in V_2 \times \cdots \times V_k$ we have $\deg_H(v_2, \ldots, v_k) \geq n/2$, then H has a perfect matching. Consequently, if $\delta'(H) > n/2$ then H contains a perfect matching.*

There is an example in [16] (see also Example 1 in [1]) of a k-partite k-graph H_0 with k even and $n = 2 \pmod 4$, such that $\delta'_{k-1}(H_0) = n/2$ and H_0 does not have a perfect matching. For all other values of k and n one can provide similar constructions with $\delta'_{k-1}(H_0) \geq n/2 - 1$, leaving open the possibility that the result from [1] can be strengthen even further.

Problem 3.11. *Assume that k is even or $n \neq 2 \pmod 4$. Is it true that if $\delta'_{k-1}(H) \geq n/2$ then H has a perfect matching? If so, is it sufficient to impose this degree bound only on two types of legal $(k-1)$-tuples, similar to Theorem 3.10?*

In [1] several other open problems and conjectures are posed. We just quote two of them here. The first one is related to $m_1(k, n)$ in the non-partite case. Note that $1 - (1 - 1/k)^{k-1} < 1 - 1/e$ and compare with Problem 3.6 above.

Problem 3.12 ([1]). *Is it true that if $\delta'_1(H) \geq (1 - 1/e)n^{k-1}$ then there is a perfect matching in H?*

Another problem from [1] is to prove the following conjecture. For a subset $I \subseteq [k]$ of indices, let us call a subset S of vertices of H an I-tuple if $|S| = |I|$ and $S \cap V_i \neq \emptyset$ if and only if $i \in I$. (Observe that if S is an I-tuple then, in fact, for all $i \in I$, we have $|S \cap V_i| = 1$.)

Conjecture 3.13 ([1]). Let I be a subset of $[k]$. If $\deg'(S) > \frac{1}{2}n^{k-|I|}$ for every I-tuple S, and $\deg'(S) > \frac{1}{2}n^{|I|}$ for every $\big([k]\setminus I\big)$-tuple S, then H has a perfect matching.

This conjecture was asymptotically verified by Pikhurko in [21], while its fractional version was proved in [1] (cf. Section 3.3 for the definition).

For $d < k-1$, there are also Dirac-type results relating δ'_d with perfect and almost perfect matchings. Already in 1981, Daykin and Häggvist proved that

$$\delta'_1(H) \geq \frac{k-1}{k}\big(n^{k-1}-1\big)$$

guarantees a perfect matching. This was extended in [11]: if

$$\delta'_d(H) > \frac{k-d}{k}n^{k-d} + kn^{k-d-1}$$

then H contains a matching covering all but $k(d-1)$ vertices, and so, a perfect matching for $d=1$.

The other extreme case, $d = k-1$, has been also studied in [16]. It was proved there that if

$$\delta'_{k-1}(H) \geq \lceil n/k \rceil$$

then there is a matching in H covering at least $n-(k-2)$ vertices from each partition class V_i, $i = 1, \dots, k$. It is, perhaps, interesting to compare this result with the results of Subsection 3.1 and consider the following analogue of Problem 3.3.

Problem 3.14. *Is it true for every k-partite k-graph H that if $\delta'_{k-1}(H) \geq \lceil n/k \rceil$ then H has a matching covering at least $n-1$ vertices from each partition class?*

3.5. Other packings

In this section we briefly discuss F-packings, that is, tilings of a hypergraph with vertex disjoint copies of F. Given two hypergraphs, F and H, an *F-packing in H* is a set of vertex disjoint copies of F in H. An F-packing is *perfect* if it covers all vertices of H. For n divisible by $\big|V(F)\big|$, let $p_d(k,n;F)$ be the smallest integer p such that whenever a k-graph H on n vertices, with n divisible by $\big|V(F)\big|$, satisfies $\delta_d(H) \geq p$ then H contains a perfect F-packing. In particular, when $F = K_k^{(k)}$ is a single edge, then $p_d(k,n;K_k^{(k)}) = m_d(k,n)$ is the Dirac threshold for perfect matchings.

Unlike for graphs, there are very few results about degree conditions guaranteeing perfect F-packings in hypergraphs. Below we present two problems, both assuming that $k = 3$ and $d = 2$.

In [17] the authors study, among other things, packings of copies of a $(3,1)$-cycle $C_s^{(3,1)}$ on s vertices, s even (see Definition 1.1). In the smallest case of $s = 4$, that is, the unique 3-graph with 4 vertices and 2 edges, they show that

$$p_2(3,n;C_4^{(3,1)}) \sim n/4. \tag{10}$$

This seems surprising, since the obtained threshold is about twice smaller than the threshold for perfect matchings. For $s \geq 6$, the value of $p_2\big(3,n;C_s^{(3,1)}\big)$ remains unknown, except for large s when $p_2\big(3,n;C_s^{(3,1)}\big) \sim n/4$, but unlike in (10), here the asymptotics is also as $s \to \infty$ (see Theorem 1.2 in [17]).

As for the lower bound, a construction provided in [17] yields that

$$p_2\big(3,n;C_s^{(3,1)}\big) \geq \frac{\lceil s/4 \rceil}{s} n.$$

This is quite interesting, since it shows that for a fixed s not divisible by 4, the threshold constant is strictly larger than $\frac{1}{4}$ (e.g., it is at least $\frac{1}{3}$ for $s = 6$).

Problem 3.15. *Determine* $p_2\big(3,n;C_s^{(3,1)}\big)$, $s \geq 6$, s *even.*

Similar lower bounds are claimed in [17] for $k > 3$ with $\frac{1}{4}$ replaced by $\frac{1}{2(k-1)}$.

In [21], Pikhurko investigated a challenging problem of determinig $p_2\big(3,n;K_4^{(3)}\big)$, where $K_4^{(3)}$ is the complete 3-graph on 4 vertices, and obtained bounds

$$\frac{3}{4}n - 2 \leq p_2\big(3,n;K_4^{(3)}\big) \leq \frac{2+\sqrt{10}}{6}n + O\big(\sqrt{n \log n}\,\big),$$

where the upper bound was also proved, independently, by Keevash and Sudakov (unpublished). There is some indication that the truth may lie at the lower end. Indeed, another result from [21] states that for $n \geq 15$, if

$$\delta_2(H) \geq \frac{3}{4}n - \frac{5}{4},$$

then there is a $K_4^{(3)}$-packing in H covering all but at most 14 vertices. However, one should remember that divisibility has a big impact on the Dirac thresholds for (almost) perfect matchings; compare, for instance, the values of $m(3, n)$ and $m^1(3, n)$.

Problem 3.16. *Determine* $p_2\big(3, n; K_4^{(3)}\big)$.

Acknowledgements. Many thanks to Endre for his longstanding friendship, generosity, and the good time we have had together while working on the Dirac Project. We also thank Domingos Dellamonica for his valuable remarks on the manuscript.

References

[1] R. Aharoni, A. Georgakopoulos and P. Sprüssel, Perfect matchings in r-partite r-graphs, *European J. Combin.*, **30(1)** (2009), 39–42.

[2] C. Berge, Nombres de coloration de l'hypergraphe h-parti complet, *Springer Lecture Notes in Math.*, Vol. **411** (1975), 13–20.

[3] J. C. Bermond et al., Hypergraphes hamiltoniens, *Prob. Comb. Theorie Graph Orsay,* **260** (1976), 39–43.

[4] B. Bollobás, D. E. Daykin and P. Erdős, Sets of independent edges of a hypergraph, *Quart. J. Math. Oxford Ser. (2),* **27(105)** (1976), 105 25–32.

[5] G. Chen, R. J. Faudree, R. J. Gould, M. S. Jacobson and L. Lesniak, Hamiltonicity in Balanced k-Partite Graphs, *Graphs and Combinatorics,* **11** (1995), 221–231.

[6] V. Chvátal, *Linear Programming,* W. H. Freeman, New York (1983)

[7] D. E. Daykin and R. Häggkvist, Degrees giving independent edges in a hypergraph, *Bull. Austral. Math. Soc.,* **23(1)** (1981), 103–109.

[8] G. A. Dirac, Some theorems of abstract graphs, *Proc. London Math. Soc.,* **3** (1952), 69–81.

[9] P. Frankl and V. Rödl, Near perfect coverings in graphs and hypergraphs, *European J. Combin.,* **6(4)** (1985), 317-326.

[10] A. Gyárfás, G. Sárközy and E. Szemerédi, Monochromatic Hamiltonian 3-tight Berge cycles in 2-colored 4-uniform hypergraphs." *Journal of Graph Theory,* **63** (2010), 288–299.

[11] H. Hàn, Y. Person and M. Schacht, On perfect matchings in uniform hypergraphs with large minimum vertex degree, *SIAM J. Discrete Math.,* **23(2)** (2009), 732–748.

[12] H. Hàn and M. Schacht, Dirac-type results for loose Hamilton cycles in uniform hypergraphs, *J. of Combin. Theory, Ser. B,* **100(3)** (2010), 332–346.

[13] G. Y. Katona and H. A. Kierstead, Hamiltonian chains in hypergraphs, *J. Graph Theory,* **30** (1999), 205–212.

[14] D. Kühn, P. Keevash, R. Mycroft and D. Osthus, *Loose Hamilton cycles in hypergraphs,* submitted.

[15] D. Kühn, R. Mycroft and D. Osthus, Hamilton l-cycles in k-graphs, *J. of Combin. Theory, Ser. A,* to appear.

[16] D. Kühn and D. Osthus, Matchings in hypergraphs of large minimum degree, *J. Graph Theory,* **51** (2006), 269–280.

[17] D. Kühn and D. Osthus, Loose Hamilton cycles in 3-uniform hypergraphs of high minimum degree, *J. Combin. Theory, Ser. B,* **96(6)** (2006), 767–821.

[18] L. Lovász & M. D. Plummer, *Matching theory,* North-Holland Mathematics Studies 121, Annals of Discrete Mathematics 29, North-Holland Publishing Co., Amsterdam; Akadémiai Kiadó, Budapest, 1986.

[19] K. Markstrom and A. Ruciński, *Perfect matchings and Hamilton cycles in hypergraphs with large degrees,* submitted.

[20] J. Moon and L. Moser, On Hamiltonian bipartite graphs, *Israel J. Math.,* **1** (1963), 163–165.

[21] O. Pikhurko, Perfect matchings and K_4^3-tilings in hypergraphs of large codegree, *Graphs Combin.,* **24(4)** (2008), 391–404.

[22] V. Rödl, A. Ruciński and E. Szemerédi, A Dirac-type theorem for 3-uniform hypergraphs *Combin. Probab. Comput.,* **15(1–2)** (2006), 229–251.

[23] V. Rödl, A. Ruciński and E. Szemerédi, Perfect matchings in uniform hypergraphs with large minimum degree, *Europ. J. Combin.,* **27** (2006), 1333–1349.

[24] V. Rödl, A. Ruciński and E. Szemerédi, An approximate Dirac-type theorem for k-uniform hypergraphs, *Combinatorica,* **28(2)** (2008), 229–260.

[25] V. Rödl, A. Ruciński, M. Schacht and E. Szemerédi, A note on perfect matchings in uniform hypergraphs with large minimum collective degree, *Commentationes Mathematicae Universitatis Carolinae,* **49(4)** (2008), 633–636.

[26] V. Rödl, A. Ruciński and E. Szemerédi, Perfect matchings in large uniform hypergraphs with large minimum collective degree, *J. Combin. Theory, Ser. A,* **116** (2009), 613–636.

[27] V. Rödl, A. Ruciński and E. Szemerédi, *Dirac-type conditions for hamiltonian paths and cycles in 3-uniform hypergraphs,* submitted.

Vojtech Rödl
Emory University
Atlanta, GA
e-mail: rodl@mathcs.emory.edu

Andrzej Ruciński
A. Mickiewicz University
Poznań, Poland
e-mail: rucinski@amu.edu.pl

BOLYAI SOCIETY
MATHEMATICAL STUDIES, 21

An Irregular Mind
(Szemerédi is 70)
pp. 591–605.

Towards a Noncommutative Plünnecke-Type Inequality

IMRE Z. RUZSA*

To Endre Szemerédi

We relax the assumption of commutativity in certain Plünnecke-type inequalities.

1. Introduction

Plünnecke [3] developed a graph-theoretic method to estimate the density of sumsets $A+B$, where A has a positive density and B is a basis. I published a simplified version of his proof [5, 6]. Accounts of this method can be found in Malouf [1], Nathanson [2], Tao and Vu [11] and the author's lecture notes [8].

The simplest instance of Plünnecke's inequality for finite sets goes as follows.

Theorem 1.1. *Let $l < h$ be integers, A, B sets in a commutative group and write $|A| = m$, $|A + lB| = \alpha m$. There exists an $X \subset A$, $X \neq \emptyset$ such that*

$$|X + hB| \leq \alpha^{h/l}|X|. \tag{1.1}$$

Plünnecke deduced his results from a property of the directed graph built on the sets A, $A + B$, $\ldots$, $A + hB$ as vertices (in $h + 1$ different

*Author was supported by ERC–AdG Grant No. 228005 and Hungarian National Foundation for Scientific Research (OTKA), Grant No. 61908 and No. 72731.

copies of the group), where from an $x \in A + iB$ edges go to each $x + b \in A + (i+1)B$. This property (which he called "commutativity") is the following. If $x, y, z_1, \dots, z_n$ are distinct vetices such that there is an edge from x to y, and from y to each z_i, then there are distinct vertices $y_1, \dots, y_n$ such that there are vertices from x to each y_i, and from y_i to z_i; also, the same property is required for the graph obtained by reversing the direction of all vertices. The fact that the addition graph has this property follows from the possibility of replacing a path from x to $x + b + b'$ through $x + b$ by a path through $x + b'$, so commutativity of addition and the assumption that we add the same set B repeatedly seemed to be central ingredients of this method. Still, it is possible to relax these assumptions. In this paper we discuss the possibility to relax commutativity.

To motivate our approach first we mention some of the most frequently used consequences that miserably fail in a noncommutative setting. Our examples use a free group, which is "very noncommutative"; it is possible that for groups "nearer" to commutative ones in some sense some results can be extended.

The case $l = 1$ of (1.1) tells us

$$|X + hB| \le \alpha^h |X| \le \alpha^h m,$$

consequently

$$|hB| \le \alpha^h m.$$

For $h = 2$ and $B = -A$ this means that if $|A| = m$, $|A - A| \le \alpha m$, then $|2A| \le \alpha^2 m$. This fails in noncommutative groups. To see this, take a free group with generators a, b and put

$$A = \{ia + b : 1 \le i \le m\}. \tag{1.2}$$

Then both difference sets $A - A$ and $-A + A$ have $2m - 1$ elements, while $|2A| = m^2$.

For $h = 3$ and $B = A$ we get that if $|A| = m$, $|2A| \le \alpha m$, then $|3A| \le \alpha^3 m$. This also fails. To see this, in the same group put

$$A = \{a, 2a, \dots, ma, b\}. \tag{1.3}$$

We have $|A| = m + 1$, $|2A| = 4m$ and $|3A| > m^2$ since all the elements $ia + b + ja$, $1 \le i, j \le m$ are distinct.

I asked whether an estimate of $|3A|$ can be given under the (clearly necessary) additional assumption that

$$\max_{a \in A} |A + a + A| \leq \alpha m.$$

An affirmative answer is given by Tao [10] in the form of $\alpha^c m$. Tao works in a more general setting and he is not interested in estimating the exponent c; it would be of some interest to specify a value of c here.

Next we mention some results that do not require commutativity. An inequality of mine [4] asserts that

$$|X|\,|Y - Z| \leq |X - Y|\,|X - Z|. \tag{1.4}$$

This has the following consequences: if $|A| = m$, $|2A| \leq \alpha m$, then $|-A + A| \leq \alpha^2 m$ and $|A - A| \leq \alpha^2 m$.

We remark that the separation of the two cases in the above corollary is not superfluous, in a noncommutative group $|-A + A|$ and $|A - A|$ can be very different (of course, without the assumption on $2A$).

Indeed, in the same free group put

$$A = \{ia + b \,:\, 1 \leq i \leq m\} \cup \{ia \,:\, 1 \leq i \leq m\}.$$

Then $|A| = 2m$ and

$$-A = \{-b - ja \,:\, 1 \leq j \leq m\} \cup \{-ja \,:\, 1 \leq j \leq m\}.$$

Here $A - A$ contains the $2m^2$ different elements $ia \pm b - ja$, while

$$-A+A = \{\,(i-j)a\} \cup \{\,(i-j)a+b\} \cup \{\,-b+(i-j)a\} \cup \{\,-b+(i-j)a+b\},$$

altogether $4m$ elements.

So if the sumset is small, both difference sets are small without commutativity, but there is no converse.

Between double and triple sums we have the following inequality without commutativity [7]:

$$|X + Y + Z|^2 \leq |X + Y|\,|Y + Z| \max_{y \in Y} |X + y + Z|. \tag{1.5}$$

In a commutative group, the middle y at the end does not matter and we do not need to take the maximum. In general, however, the maximum cannot be omitted or be replaced by an average, even in the case of identical sets. An example is given by $X = Y = Z = A$ with the set A of (1.3).

We have $|A| = m + 1$, $|2A| = 4m$ and $|3A| > m^2$ as mentioned. Furthermore, from the $m+1$ sets $A + y + A$, $y \in A$ only one is of size m^2, namely the one with $y = b$, all the others have $O(m)$ elements.

For reasons that will be apparent soon, our noncommutative inequalities will use different summands. An extension of the case $l = 1$ of Theorem 1.1 to different summands (in the commutative case) can already be found in my paper [5] and sounds as follows.

Theorem 1.2. *Let A, $B_1, \dots, B_h$ be finite sets in a commutative group and write $|A| = m$, $|A + B_i| = \alpha_i m$, for $1 \le i \le h$. There exists an $X \subset A$, $X \ne \emptyset$ such that*

$$|X + B_1 + \dots + B_h| \le \alpha_1 \alpha_2 \dots \alpha_h |X|. \tag{1.6}$$

Our noncommutative attempts start from this result. The simplest, and possibly most interesting case was already included in the lecture note [8] and sounds as follows.

Theorem 1.3. *Let A, L, R be sets in a (typically noncommutative) group G and write $|A| = m$, $|L + A| = \alpha m$, $|A + R| = \beta m$. There is an $X \subset A$, $X \ne \emptyset$ such that*

$$|L + X + R| \le \alpha\beta|X|. \tag{1.7}$$

The reason that this result is true is that there is a kind of commutativity hidden here. An addition from the left commutes with an addition from the right: $(x + y) + z = x + (y + z)$; this sort of commutativity is called "associativity". The obstacle to further extension seems that there are no more directions, we cannot add from above and below. To handle more summands we need an extra condition.

Definition 1.4. A collection of sets $B_1, \dots, B_k$ in a (noncommutative) group is *exocommutative,* if for all $x \in B_i$, $y \in B_j$ with $i \ne j$ we have $x + y = y + x$.

Theorem 1.5. *Let $A, L_1, L_2, \dots, L_k, R_1, R_2, \dots, R_l$ be sets in a group G and write $|A| = m$, $|L_i + A| = \alpha_i m$, $i = 1, \dots, k$, $|A + R_j| = \beta_j m$ $j = 1, \dots, l$. Assume that both collections $L_1, \dots, L_k$ and $R_1, \dots, R_l$ are exocommutative. There is a set $X \subset A$, $X \ne \emptyset$ such that*

$$|L_1 + \ldots + L_k + X + R_i + \ldots + R_l| \le \alpha_1 \dots \alpha_k \beta_1 \dots \beta_l |X|. \tag{1.8}$$

The seemingly different role of the sets L_i and R_i can be unified as follows. The following $k+l$ sets, as sets of operations on G, act in an exocommutative way:

– multiplication of elements of L_i from the left,

– multiplication of elements of R_i from the right.

It is the set A that has a really different position. We express this by putting the previous results in a different context. We do not need any structure on the set A; the important thing is that the other sets *act* on A. To express this, we take an (unstructured) fundamental set G, a set $A \subset G$; a group P, which is the group of permutations of G, and sets $B_i \subset P$. We write

$$B(A) = \{ b(a) \, : \, b \in B, \; a \in A \};$$

the composition of permutations will be denoted by simple juxtaposition, so that

$$b_1 b_2(a) = b_1 \big(b_2(a) \big), \quad B_1 B_2 = \{ b_1 b_2 \, : \, b_1 \in B_1, \; b_2 \in B_2 \}.$$

Theorem 1.6. *Let G be a set, and let P be the group of permutations of G. Let $A \subset G$, and let $B_1, \dots, B_h$ be an exocommutative collection of subsets of P. Write $|A| = m$, $\big| B_i(A) \big| = \alpha_i m$. Then there is an $X \subset A$, $X \neq \emptyset$ such that*

$$\big| B_1 \dots B_h(X) \big| \leq \alpha_1 \dots \alpha_h |X|. \tag{1.9}$$

The previous theorem can be obtained as follows. Any $g \in G$ induces two permutations corresponding to addition from left and right,

$$l_g(x) = g + x, \quad r_g(x) = x + g.$$

From the sets L_i and R_i we make subsets of P in a natural way,

$$L_i' = \{ l_g \, : \, g \in L_i \}, \quad R_i' = \{ r_g \, : \, g \in R_i \},$$

and Theorem 1.6 yields Theorem 1.5.

In the last sections we will show how to find a large subset X with a similar property, and contrast the problems of passing from two summands to three, and from three to more.

2. Plünnecke's Commutative Graphs

In this section we list those results on graphs that we shall apply later. Proofs can be found in [5, 6], [1], [2], [8]. (Plünnecke's original work [3] proceeds slightly differently.)

We consider directed graphs $\mathcal{G} = (V, E)$, where V is the set of vertices and E is that of the edges. If there is an edge from x to y, then we also write $x \to y$. A graph is *semicommutative*, if for every collection $(x; y; z_1, z_2, \ldots, z_k)$ of distinct vertices such that $x \to y$ and $y \to z_i$ there are distinct vertices $y_1, \ldots, y_k$ such that $x \to y_i$ and $y_i \to z_i$. $\mathcal{G}$ is *commutative*, if both $\mathcal{G}$ and the graph $\hat{\mathcal{G}}$ obtained by reversing the direction of every edge of $\mathcal{G}$ are semicommutative.

Our graphs will be of a special kind we call *layered*. By an *h-layered* graph we mean a graph with a fixed partition of the set of vertices

$$V = V_0 \cup V_1 \cup \cdots \cup V_h$$

into $h+1$ disjoint sets (layers) such that every edge goes from some V_{i-1} into V_i.

For $X, Y \subset V$, we define the *image* of X in Y as

$$\operatorname{im}(X, Y) = \{y \in Y : \text{ there is a directed path from some } x \in X \text{ to } y\}.$$

The *magnification ratio* is defined by

$$\mu(X, Y) = \min\left\{ \frac{|\operatorname{im}(Z, Y)|}{|Z|} : Z \subset X, \ Z \neq \emptyset \right\}.$$

For a layered graph we write

$$\mu_j(\mathcal{G}) = \mu(V_0, V_j).$$

Now Plünnecke's main result can be stated as follows.

Lemma 2.1 (Plünnecke [3]). *In a commutative layered graph $\mu_j^{1/j}$ is decreasing.*

That is, for $j < h$ we have $\mu_h \le \mu_j^{h/j}$. An obvious (and typically the only available) upper estimate for μ_j is $|V_j|/|V_0|$. This yields the following corollary (in fact, an equivalent assertion).

Lemma 2.2. *Let $j < h$ be integers, $\mathcal{G}$ a commutative layered graph on the layers $V_0, \dots, V_h$. Write $|V_0| = m$, $|V_j| = s$. There is an $X \subset V_0$, $X \neq \emptyset$ such that*

$$\big|\operatorname{im}(X, V_h)\big| \leq (s/m)^{h/j}|X|.$$

Definition 2.3. Let $\mathcal{G}' = (V', E')$ and $\mathcal{G}'' = (V'', E'')$ be h-layered graphs with layers V_i' and V_i'', resp. Their *layered product* is the h-layered graph on the layers $V_i = V_i' \times V_i''$, where two vertices $(x', x'') \in V_i$ and $(y', y'') \in V_{i+1}$ are connected if both $x' \to y'$ and $x'' \to y''$. This graph will be denoted by $\mathcal{G} = \mathcal{G}'\mathcal{G}''$. For repeated products with identical factors the usual power notation $\mathcal{G}^n$ will be used.

Observe that this is a proper subgraph of the usual product of these graphs.

Lemma 2.4. *The layered product of commutative graphs is commutative as well.*

Lemma 2.5. *Magnification ratios are multiplicative: if $\mathcal{G}$, $\mathcal{G}'$, $\mathcal{G}''$ are h-layered graphs with magnification ratios μ_i, μ_i', μ_i'', resp., and $\mathcal{G} = \mathcal{G}'\mathcal{G}''$, then $\mu_i = \mu_i'\mu_i''$ for all i.*

3. Proof of Theorem 1.6

We build an h-layered graph on 2^h copies of the fundamental set G. A typical vertex will be a pair (x, I), where $x \in G$ and $I \subset H = \{1, 2, \dots, h\}$. Such a vertex will reside at level $|I|$. Thus, level j is composed of $\binom{h}{j}$ parts, corresponding to the j-element subsets of H. From a vertex (x, I) there is an edge into (y, J) if J is of the form $J = I \cup \{j\}$ with $j \notin I$ and $y = b(x)$ with some $b \in B_j$. Our graph will not contain all such vertices and edges; at level 0 we keep exactly the vertices corresponding to elements of our set A, and at higher levels those that can be reached through a directed path. At level j this means elements of the form

$$\left(b_{i_1} b_{i_2} \dots b_{i_j}(a), \{i_1, \dots, i_j\}\right),$$

where $a \in A$, $1 \leq i_1 < \dots < i_j$, $b_{i_u} \in B_{i_u}$. Note that such an element arises from $j!$ paths, corresponding to the permutations of the subscripts i_u. Our aim is to prove that the magnification ratio of this graph satisfies

$$\mu_h \leq \alpha_1 \dots \alpha_h.$$

This will proceed in three stages. First, we deduce *some* estimate in terms of the α_i. Next, we improve this estimate to one that will be a constant worse than the desired bound. Finally, we remove the constant.

The first estimate will follow from Plünnecke's theorem, Lemma 2.2 of the previous section. In order to apply this result we need to show that our graph is commutative, and to estimate $|V_1|$.

To see commutativity, take a typical edge, which goes from an (x, I) to a (y, J), $J = I \cup j$, $y = b(x)$, $b \in B_j$. From y edges bo to elements of the form $\big(b'(y), J \cup k\big)$, where $k \notin J$ and $b' \in B_k$. We replace such a sequence

$$(x, I) \to \big(b(x), I \cup \{j\}\big) \to \big(b'b(x), I \cup \{j,k\}\big)$$

of edges by

$$(x, I) \to \big(b'(x), I \cup \{k\}\big) \to \big(bb'(x), I \cup \{k,j\}\big).$$

We have $bb' = b'b$ by exocommuativity, so the last elements are the same as before. The middle elements are clearly all different. Observe that they may be in different "components" of the level, and the component of the original middle element (y, J) is avoided.

We also have to check this possibility of replacement for the inverse graph. This is performed in a completely analogous manner.

By definition, V_1 is the disjoint union of sets of size $B_i(A)$, hence

$$|V_1| = \sum |B_i(A)| = \sum \alpha_i m.$$

Lemma 2.2 gives our first estimate

$$\mu_h \le \Big(\sum \alpha_i\Big)^h. \tag{3.1}$$

If all the α_i are equal, this is h^h times the desired bound; if they are different, it is worse. In the next stage we reduce the general case to the case of equal α_i.

To this end take h positive integers $n_1, \dots, n_h$ and embed our set G into

$$G' = G \times \mathbb{Z}_{n_1} \times \cdots \times \mathbb{Z}_{n_h}.$$

We replace A by its natural image

$$A' = \big\{(a, 0, \dots, 0) : a \in A\big\}.$$

We replace each $b \in B_i$ by n_i new elements. For $t \in \mathbb{Z}_{n_i}$ this will be

$$b'_t(a, x_1, \dots, x_h) = \big(b(a), x_1, \dots, x_{i-1}, x_i + t, x_{i+1}, \dots, x_h\big)$$

and we put

$$B'_i = \{b'_t : b \in B_i,\ t \in \mathbb{Z}_{n_i}\}.$$

Clearly

$$B'_i(A') = B(A) \times \{0\} \times \cdots \times \mathbb{Z}_{n_i} \times \cdots \times \{0\},$$

so

$$\big|B'_i(A')\big| = \alpha'_i m, \quad \alpha'_i = \alpha_i n_i.$$

For each $X' \subset A'$, which is the image of an $X \subset A$, the set $B'_1 \dots B'_h(X')$ will have elements with the first coordinate in $B_1 \dots B_h(X)$ and the others arbitrary, so

$$\big|B'_1 \dots B'_h(X')\big| = n_1 \dots n_h \big|B_1 \dots B_h(X)\big|.$$

Consequently the new h'th magnification ratio will be

$$\mu'_h = n_1 \dots n_h \mu_h.$$

Applying (3.1) for the sets $A', B'_1, \dots, B'_h$ we obtain

$$n_1 \dots n_h \mu_h \le \Big(\sum n_i \alpha_i\Big)^h. \tag{3.2}$$

If we take integers n_i so that all products $n_i \alpha_i$ are equal, say $n_i = n/\alpha_i$ with a suitable n, then (3.2) becomes

$$\mu_h \le h^h \alpha_1 \dots \alpha_h. \tag{3.3}$$

This completes the second stage.

In the third stage we remove the constant h^h. This will be done by the method of taking direct powers. We will apply Lemma 2.5 to a certain graph. This will be an "extract" of the graph previously used.

Write $B = B_1 \dots B_h$. Consider the following 1-layered (bipartite) directed graph $\mathcal{G}^1$. The first collection of vertices V_0 are the elements of set A, and the second collection of vertices V_1 are the elements of set $B(A)$ (taken in two different copies of the fundamental set G to make them disjoint). The edges go from any $a \in A$ to each $b_1 \dots b_h(a)$, $b_i \in B_i$. The image of an $X \subset A$ in V_1 is exactly $B(X)$, consequently the magnification ratio $\mu_1(\mathcal{G}^1)$ of this graph is exactly our μ_h above.

Consider now the layered power $\mathcal{G}^r = \mathcal{G}^1\mathcal{G}^1 \dots \mathcal{G}^1$. By Lemma 2.4, the magnification ratio of this graph is

$$\mu_1(\mathcal{G}^r) = \mu_1(\mathcal{G}^1)^r = \mu_h^r. \tag{3.4}$$

On the other hand, this graph is isomorphic to a graph constucted from powers of our set. Let P_r be the group of permutations of G^r. Strictly speaking, the direct power P^r of the set of permutations is a rather different object than a permutation of G^k, but there is a natural homomorphism $\varphi : P^r \to P_r$ given by

$$\varphi(f_1, \dots, f_k)(x_1, \dots, x_r) = \big(f_1(x_1), \dots, f_r(x_r)\big), \quad f_i \in P,\ x_i \in G.$$

Clearly for any $X \subset G$ and $Y \subset P$ we have

$$\varphi(Y^r)(X^r) = \big(Y(X)^r\big).$$

In particular, we have

$$\big|\varphi(B_i^r)(A^r)\big| = \alpha_i^r m^r.$$

We can apply the second stage result to the sets A^r, $\varphi(B_i^r)$ to conclude that there is an $X \subset A^r$ such that

$$\big|\varphi(B^r)(X)\big| \le h^h(\alpha_1 \dots \alpha_h)^r |X|.$$

This means that the magnification ratio of the 1-layered graph $\mathcal{G}_r$ built using A^r and $\varphi(B^r)$ in the place of A and B satisfies

$$\mu_1(\mathcal{G}_r) \le h^h(\alpha_1 \dots \alpha_h)^r. \tag{3.5}$$

On the other hand it is easy to see that this graph $\mathcal{G}_r$ is isomorphic to $\mathcal{G}^r$. (This would not hold if we used the complete h-layered graphs instead.)

Combining (3.4) and (3.5) and taking r'th roots we obtain

$$\mu_h \le h^{h/r}\alpha_1 \dots \alpha_r.$$

Since this is true for arbitrary r, we conclude

$$\mu_h \le \alpha_1 \dots \alpha_r$$

as wanted.

4. Finding a Large Subset

Typically the set X whose existence is asserted in our theorems is a proper subset of the starting set A. However, once we can find some subset, by repeating the selection we can find a subset that contains 99% of the elements of A.

Theorem 4.1. *Let G be a set, and let P be the group of permutations of G. Let $A \subset G$, and let $B_1, \dots, B_h$ be an exocommutative collection of subsets of P. Write $|A| = m$, $|B_i(A)| = \alpha_i m$. Write $B = B_1 \dots B_h$, $\vartheta = \alpha_1 \dots \alpha_h$.*

(a) *Let an integer a be given, $1 \le a \le m$. There exists an $X \subset A$, $|X| \ge a$ such that*

$$|B(X)| \le \vartheta m^h \bigl(m^{-h} + (m-1)^{-h} + \dots + (m-a+1)^{-h} + (|X| - a)(m-a+1)^{-h}\bigr). \tag{4.1}$$

(b) *Let a real number t be given, $0 \le t < m$. There exists an $X \subset A$, $|X| > t$ such that*

$$|B(X)| \le \vartheta m^h \left(\frac{1}{h-1}\bigl((m-t)^{1-h} - m^{1-h}\bigr) + (|X| - t)(m-t)^{-h}\right). \tag{4.2}$$

(c) *Let a real number ε be given, $0 \le \varepsilon < m$. There exists an $X \subset A$, $|X| > (1-\varepsilon)m$ such that*

$$|B(X)| \le \vartheta |X| \bigl(h\varepsilon^{1-h} - h + 1\bigr). \tag{4.3}$$

Proof. To prove (a), we use induction on a. The case $a = 1$ is Theorem 1.6. Now suppose we know it for a; we prove it for $a+1$. The assumption gives us a set X, $|X| \ge a$ with a bound on $|B(X)|$ as given by (4.1). We want to find a set X' with $|X'| \ge a+1$ and

$$|B(X')| \le \vartheta m^h \bigl(m^{-h} + (m-1)^{-h} + \dots + (m-a)^{-h} + (|X'| - a - 1)(m-a)^{-h}\bigr). \tag{4.4}$$

If $|X| \geq a+1$, we can put $X' = X$. If $|X| = a$, we apply Theorem 1.6 to the sets $A' = A \setminus X$, $B_1, \ldots, B_k$. In doing this the numbers α_i should be replaced by

$$\alpha_i' = \frac{|B_i(A')|}{|A'|} \leq \frac{|B_i(A)|}{|A'|} = \alpha_i \frac{m}{m-a}.$$

This gives us a set $Y \subset A \setminus X$ such that

$$|B(Y)| \leq \vartheta' |Y|$$

with

$$\vartheta' = \prod \alpha_i' \leq \vartheta \left(\frac{m}{m-a}\right)^h$$

and we put $X' = X \cup Y$.

To prove part (b) we apply (4.1) with $a = [t]+1$. The right side of (4.2) can be written as $\vartheta m^h \int_0^{|X|} f(x)\,dx$, where $f(x) = (m-x)^{-h}$ for $0 \leq x \leq t$, and $f(x) = (m-t)^{-h}$ for $t < x \leq |X|$. Since f is increasing, the integral is $\geq f(0) + f(1) + \cdots + f(|X|-1)$. This exceeds the right side of (4.1) by a termwise comparison.

To prove part (c) we divide the right hand side of (4.2) by $|X|$, and by easy calculus find that the quotient is an increasing function of $|X|$. Hence it assumes its maximum at $|X| = m$; we make this substitution, and then substitute $t = (1-\varepsilon)m$ to reduce it to form (4.3). ∎

We formulate separately the case corresponding to Theorem 1.3.

Theorem 4.2. *Let A, L, R be sets in a group G and write $|A| = m$, $|L+A| = \alpha m$, $|A+R| = \beta m$.*

(b) *Let a real number t be given, $0 \leq t < m$. There exists an $X \subset A$, $|X| > t$ such that*

$$|L+X+R| \leq \alpha\beta m \frac{m|X| - t^2}{(m-t)^2}. \tag{4.5}$$

(c) *Let a real number ε be given, $0 \leq \varepsilon < m$. There exists an $X \subset A$, $|X| > (1-\varepsilon)m$ such that*

$$|L+X+R| \leq \alpha\beta |X| \left(\frac{2}{\varepsilon} - 1\right). \tag{4.6}$$

Part (c) of the above statement for the case of identical sets sounds as follows.

Corollary 4.3. *Let A be a finite set in a group G and write $|A| = m$, $|A + A| = \alpha m$.*

Let a real number ε be given, $0 \le \varepsilon < m$. There exists an $X \subset A$, $|X| > (1 - \varepsilon)m$ such that

$$|3X| \le |A + X + A| \le \alpha\beta|X| \left(\frac{2}{\varepsilon} - 1\right). \tag{4.7}$$

We shall see in the next section that from three summands we can pass to any larger number, so in particular $|kX|$ will be small. With a minimal extra effort we can estimate some mixed sums involving A and X, but only those involving at most two copies of A.

5. Two, Three, Many

To deduce a cardinality estimate for a triple sum using double sums is difficult in the commutative case and impossible in the noncommutative one. However, from three to larger numbers it is easy and can be done just by iterated application of my inequality (1.4) which we repeat here:

$$|X|\,|Y - Z| \le |X - Y|\,|X - Z|. \tag{5.1}$$

This observation was first done in Ruzsa-Turjányi [9]. It was formulated only for sets of integers; the proof works for any commutative group and gives the following. If $|A| = m$, $|3A| \le \alpha m$, then

$$|kA - lA| \le \alpha^{k+l} m.$$

The authors were unaware of Plünnecke's result, by which this conclusion can be drawn from the weaker assumption $|2A| \le \alpha m$. They were not interested in noncommutative generalizations either. The argument given there automatically gives some noncommutative estimates; however, in the noncommutative setting there are many sum-difference expressions to estimate, and this requires some modifications, which we describe below.

A less explicit form can be found in Tao [10].

To express our results in a concise way, we introduce the following notation. For an integer vector

$$\mathbf{k} = (k_1, \dots, k_h)$$

we write

$$\mathbf{k}X = k_1X + \dots + k_hX;$$

negative coordinates are interpreted as

$$(-k)X = k(-X) = -(kX).$$

Theorem 5.1. *Let A, B be finite sets in a group and write*

$$m = \min\left\{ |A|, |B| \right\}.$$

Let $\mathbf{k} = (k_1, \dots, k_h)$ be a vector with $k_i = \pm 1$.

(a) *If $|A + B - A| \le \alpha m$, then*

$$|A + \mathbf{k}B - A| \le \alpha^h m. \tag{5.2}$$

(b) *If $|A + 2B| \le \alpha m$, then*

$$|A + \mathbf{k}B - A| \le \alpha^{2h} m. \tag{5.3}$$

In particular, by putting $B = A$ or $B = -A$ we obtain the following. Let $|A| = m$, and assume that the size of one of the triple sum-differences $\pm A \pm A \pm A$ is at most αm. Then, for six of the possible eight combinations of signs, we conclude that any h-fold sum-difference combination has cardinality at most $\alpha^{2h} m$. The two cases not covered are $A - A + A$ and $-A + A - A$, and for these such a conclusion cannot be drawn. Indeed, consider example (1.2). For this set these sums are small, as well as all alternating sums $A - A + A - A + \dots$, but $A + A$ is large.

Proof. First we prove part (a). We use induction on h. The case $h = 1$ is the assumption (observe that $A - B - A = -(A + B - A)$, so it is small).

Suppose we know it for $h - 1$. Put $\mathbf{k}' = (k_1, \dots, k_{h-1})$ and apply (5.1) for $X = A$, $Y = \mathbf{k}'B$, $Z = -k_hB$.

To prove part (b) we first apply (5.1) with $X = -B$, $Y = A+B$, $Z = A$. Then

$$Y - Z = A + B - A,$$

$$X - Y = -B - B - A = -(A + 2B)$$

and

$$X - Z = -B - A - (A + B).$$

Since clearly

$$|A + B| \le |A + 2B| \le \alpha m,$$

we have the assumption of part (a) with α^2 in the place of α, and part (b) follows from part (a). ∎

References

[1] J. L. Malouf, On a theorem of Plünnecke concerning the sum of a basis and a set of positive density, *J. Number Theory,* **54**.

[2] M. B. Nathanson, *Additive number theory: Inverse problems and the geometry of sumsets,* Springer, 1996.

[3] H. Plünnecke, Eine zahlentheoretische Anwendung der Graphtheorie, *J. Reine Angew. Math.,* **243** (1970), 171–183.

[4] I. Z. Ruzsa, *On the cardinality of $A+A$ and $A-A$,* Combinatorics (Keszthely 1976), Coll. Math. Soc. J. Bolyai, vol. 18, North-Holland – Bolyai Társulat, Budapest, 1978, pp. 933–938.

[5] I. Z. Ruzsa, An application of graph theory to additive number theory, *Scientia, Ser. A,* **3** (1989), 97–109.

[6] I. Z. Ruzsa, Addendum to: An application of graph theory to additive number theory, *Scientia, Ser. A,* **4** (1990/91), 93–94.

[7] I. Z. Ruzsa, *Cardinality questions about sumsets,* Additive Combinatorics (Providence, RI, USA) (A. Granville, M. B. Nathanson, and J. Solymosi, eds.), CRM Proceedings and Lecture Notes, vol. 43, American Math. Soc., 2007, pp. 195–205.

[8] I. Z. Ruzsa, *Sumsets and structure,* Combinatorial number theory and additive group theory, Advanced courses in mathematics, CRM Barcelona, Birkhäuser, Basel – Boston – Berlin, 2009, pp. 87–210.

[9] I. Z. Ruzsa and S. Turjányi, A note on additive bases of integers, *Publ. Math. Debrecen,* **32** (1985), 101–104.

[10] T. Tao, *Product set estimates for non-commutative groups,* arXiv:math/0601431v2.

[11] T. Tao and V. H. Vu, *Additive combinatorics,* Cambridge University Press, Cambridge, 2006.

Imre Z. Ruzsa

Alfréd Rényi Institute of Mathematics
Budapest, Pf. 127
H-1364 Hungary

e-mail: ruzsa@renyi.hu

BOLYAI SOCIETY
MATHEMATICAL STUDIES, 21

An Irregular Mind
(Szemerédi is 70)
pp. 607–617.

Quasirandom Multitype Graphs

JOEL SPENCER

1. Quasirandomness

The notion of *quasirandom* graphs has received much attention in the last decades. The idea is that a graph need only satisfy certain properties. With these properties a graph will share many other properties with suitably defined random graphs. This very general idea applies to many other structures, but here we restrict ourselves to graphs, without direction, loops, or multiple edges.

Definition 1. Let $H = \big(V(H), E(H)\big)$ and $G = \big(V(G), E(G)\big)$ be graphs. A map $\phi : V(H) \to V(G)$ is a *homomorphism* if

$$\{x, y\} \in E(H) \Rightarrow \big\{\phi(x), \phi(y)\big\} \in E(G)$$

We set $T(H, G)$ equal the number of such homomorphisms. Letting $n = \big|V(G)\big|$, we set $t(H, G) = T(H, G)n^{-|V(H)|}$.

Let $p \in [0, 1]$. Let G_n be a sequence of graphs, G_n having n vertices. Limits below are always for $G = G_n$ as $n \to \infty$.

Theorem 1.1. *The following are equivalent:*

1. $t(H, G) \to p^{|E(G)|}$ *for every* H *with at most four vertices.*
2. $t(H, G) \to p^{|E(G)|}$ *for every* H.

A sequence G_n is called *quasirandom* if it has the above properties. Such a sequence has many of the properties of randomly generated $G(n, p)$. The above result was known to many in the 1980s, and perhaps before. A paper of Chung, Graham and Wilson [1] played a key role in the popularization of this idea. There these two properties and several more were given, and it was shown that they were all equivalent, thus making the notion of quasirandomness a robust one. For this author the "surprise" was that if the G_n emulated $G(n,p)$ with respect to $t(H,G)$ for the finite number of H with at most four vertices then it would emulate $G(n,p)$ with respect to any $t(H,G)$.

1.1. Multitype graphs

There is a natural generalization of $G(n,p)$ to multitype graphs. Let $\mathtt{DATA}$ consist of

1. A positive integer K
2. $\alpha_i \in (0,1]$, $1 \le i \le K$, with $\sum_{i=1}^{K} \alpha_i = 1$
3. $p_{ij} \in [0,1]$ for $1 \le i,j \le K$, with $p_{ij} = p_{ji}$

Then the random graph $G(n;\mathtt{DATA})$ consists of n vertices split into sets V_i of size $n\alpha_i$ (technically, of course, there is a roundoff here) such that when $x \in V_i$ and $y \in V_j$ the pair x,y is adjacent with probability p_{ij}.

For any fixed H, we let $t(H,\mathtt{DATA})$ denote the limiting value of the expected value $t\big(H,G(n;\mathtt{DATA})\big)$. Letting $V(H) = \{1,\ldots,s\}$, for each choice of $1 \le \psi(1),\ldots,\psi(s) \le K$ the expected number of ϕ with $\phi(i) \in V_{\psi(i)}$ would be the product of the sizes $n\alpha_{\psi(i)}$ times the product of the $p_{\psi(i),\psi(j)}$ over all $\{i,j\} \in E(H)$. Normalizing, $t(H,\mathtt{DATA})$ is the sum over all such choices of the product of the $\alpha_{\psi(i)}$ times the product of the $p_{\psi(i),\psi(j)}$ over all $\{i,j\} \in E(H)$.

The celebrated Szemerédi Regularity Lemma states that, speaking roughly, any graph can be approximately estimated by a $G(n;\mathtt{DATA})$ where K (the number of types) is bounded.

1.2. The Lovász–Sós Result

Lovász and Sós [2] showed the following deep generalization of Theorem 1.1.

Theorem 1.2. *For each choice of* DATA *there is a constant L such that the following are equivalent:*

1. $t(H,G) \to t(H,\mathtt{DATA})$ *for every H with at most L vertices.*

2. $t(H,G) \to t(H,\mathtt{DATA})$ *for every H.*

The object in this note is to give an alternate proof to Theorem 1.2 in a quite general, but not totally general, case.

Definition 2. We say that DATA is *nice* if the values

$$d_i = \sum_{j=1}^{K} \alpha_j p_{ij}, \quad 1 \le i \le K$$

are distinct.

Note that $d_i n$ will be the expected degree of a vertex in V_i in the random graph $G(n;\mathtt{DATA})$. The assumption that DATA is *nice* will allow us to separate the vertices. We shall show Theorem 1.2 under this additional assumption. We will restrict our counts $t(H,G)$ to rather specific H.

1. For $a \ge 0$, STAR (a) consists of a vertex v and a neighbors.

2. For $a_1, a_2 \ge 0$, EDGESTAR (a_1, a_2) consists of vertices v_1, v_2, a_1 neighbors of v_1, a_2 neighbors of v_2, with v_1, v_2 adjacent.

3. For $a_1, a_2, a_3, a_4 \ge 0$, CYCLESTAR (a_1, a_2, a_3, a_4) consists of vertices v_1, v_2, v_3, v_4, a_1 neighbors of v_1, a_2 neighbors of v_2, a_3 neighbors of v_3, a_4 neighbors of v_4, with v_1, v_2, v_3, v_4 forming a 4-cycle in that order.

With these definitions, we can state our result:

Theorem 1.3. *Let* DATA *be* nice. *Assume that $t(H,G) \to t(H,\mathtt{DATA})$ for every H of the forms*

1. STAR (a_1) *for* $0 \le a_1 \le 2K$

2. EDGESTAR (a_1, a_2) *for* $0 \le a_1, a_2 \le 2K-2$

3. CYCLESTAR (a_1, a_2, a_3, a_4) *for* $0 \le a_1, a_2, a_3, a_4 \le 2K-2$

Then $t(H,G) \to t(H,\mathtt{DATA})$ for every H.

Remark. This result shows that we may take $L = 8K - 4$ in Theorem 1.2, given the additional condition that DATA is *nice*. It remains an open question as to the best value of L as a function of K, both with and without the assumption that DATA is *nice*. When $K = 1$, standard quasirandomness, it is known that $L = 4$ is the correct result.

2. The Sieve

Let X be a real valued random variable. Let $a_1, \ldots, a_T$ be distinct real numbers.

Theorem 2.1. *If*

$$E\left[\prod_{i=1}^{T}(X - a_i)^2\right] = 0 \tag{1}$$

then, almost surely, X takes on one of the values $a_1, \ldots a_T$.

This is immediate as the variable is nonnegative. For $1 \le i \le T$ set

$$g_i(x) = \prod_{j \ne i} \frac{(x - a_j)^2}{(a_i - a_j)^2} \tag{2}$$

The function g_i acts as a sieve, taking the value one when $x = a_i$ and the value zero when $x = a_j$, $j \ne i$.

Theorem 2.2. *If* (1) *holds then*

$$E\big[g_i(X)\big] = \Pr[X = a_i]$$

In application we will deal asymptotically with a sequence X_n.

Theorem 2.3. *Let X_n be uniformly bounded. Let $a_1, \ldots, a_T$ be distinct real numbers. If*

$$E\left[\prod_{i=1}^{T}(X_n - a_i)^2\right] = o(1) \tag{3}$$

then, asymptotically almost surely, $X_n = a_i + o(1)$ for some $1 \le i \le T$.

Theorem 2.4. *Under the assumptions of Theorem 2.3, with g_i given by (2), if $E\big[g_i(X_n)\big]$ approaches a limit p_i then $\Pr\big[X_n = a_i + o(1)\big] = p_i + o(1)$.*

3. Degrees

Let $\texttt{NDEG}(v)$, the normalized degree of v, be given by $n^{-1}\deg(v)$. Then, for any integer $a_1 \geq 0$,

$$n^{-1} \sum_{v \in V(G)} \texttt{NDEG}(v)^{a_1} = t(\texttt{STAR}(a_1), G) \tag{4}$$

Thus for any polynomial $f(x) = \sum_{i=0}^{2K} \beta_i x^i$,

$$n^{-1} \sum_{v \in V(G)} f(\texttt{NDEG}(v)) = \sum_{i=0}^{2K} \beta_i t(\texttt{STAR}(i), G) \tag{5}$$

Let G_n satisfy assumption 1.3 of Theorem 1.3. Then

$$\lim_{n\to\infty} n^{-1} \sum_{v \in V(G_n)} f(\texttt{NDEG}(v)) = \lim_{n\to\infty} E\left[n^{-1} \sum_v f(\texttt{NDEG}(v))\right] \tag{6}$$

where v on the right ranges over the vertices of $G(n; \texttt{DATA})$. Now set $f(x) = \prod_{i=1}^{K} (x - d_i)^2$, the d_i given by Definition 2. In $G(n; \texttt{DATA})$ asymptotically almost surely all $v \in V_i$ have $\texttt{NDEG}(v) = d_i + o(1)$, so that $f(\texttt{NDEG}(v)) = o(1)$, and hence the right hand limit of (6) is zero. Thus the left hand limit of (6) is zero. From Theorem 2.3 we deduce:

Lemma 3.1. *Almost all $v \in G_n$ have $\texttt{NDEG}(v) = d_i + o(1)$ for some $1 \leq i \leq K$.*

For $1 \leq u \leq K$ we now set V_u equal those vertices v with $\texttt{NDEG}(v) = d_u + o(1)$. We let V_{ext} denote all other vertices, so that $|V_{\text{ext}}| = o(n)$.

Now, for $1 \leq i \leq K$, set $f = g_i$ in (6) where we set a_j equal the normalized degree d_j. The right hand limit of (6) is α_i so the left hand limit of (6) is α_i. The value $g_i(\texttt{NDEG}(v))$ is $1 + o(1)$ when $v \in V_i$, $o(1)$ when $v \in V_j$ with $j \neq i$, and $O(1)$ for the $o(n)$ other vertices v. Thus the left hand side is $n^{-1}|V_i| + o(1)$. We deduce:

Lemma 3.2. *For $1 \leq i \leq K$, $|V_i| \sim n\alpha_i$.*

4. Counting Edges

Here we shall count the edges $\{v, w\}$ with $v \in V_i$, $w \in V_j$, $i \neq j$.

Let $\chi(v, w)$ be one if v, w are adjacent, zero otherwise. For any $0 \leq a_1, a_2 \leq 2K - 2$,

$$t\big(\texttt{EDGESTAR}\,(a_1, a_2), G\big) = n^{-2} \sum \texttt{NDEG}\,(v)^{a_1}\,\texttt{NDEG}\,(w)^{a_2} \chi(v, w)$$

where the sum is over $v, w \in V(G)$. For any polynomial $f(x, y) = \sum_{r,s=0}^{2K-2} \beta_{rs} x^r y^s$,

$$(7)\quad n^{-2} \sum f\big(\texttt{NDEG}\,(v), \texttt{NDEG}\,(w)\big)\, \chi(v, w) = \sum_{r,s=0}^{2K-2} \beta_{rs} t\big(\texttt{EDGESTAR}\,(r, s), G\big)$$

We apply this with a sieve polynomial $f(x, y) = g_i(x) g_j(y)$ with g_i, g_j given by Definition (2) where, again, the $a_1, \ldots, a_K$ are the normalized degrees $d_1, \ldots, d_K$. Then $f\big(\texttt{NDEG}\,(v), \texttt{NDEG}\,(w)\big) = 1 + o(1)$ when $v \in V_i$ and $w \in V_j$, $f\big(\texttt{NDEG}\,(v), \texttt{NDEG}\,(w)\big) = o(1)$ when $v \in V_{i'}$ and $w \in V_{j'}$ for any $(i', j') \neq (i, j)$, and $f\big(\texttt{NDEG}\,(v), \texttt{NDEG}\,(w)\big) = O(1)$ whenever $v \in V_{\text{ext}}$ or $w \in V_{\text{ext}}$. Therefore

$$(8)\qquad n^{-2} \sum f\big(\texttt{NDEG}\,(v), \texttt{NDEG}\,(w)\big)\, \chi(v, w) \sim n^{-2} \sum_{v \in V_i} \sum_{w \in V_j} \chi(v, w)$$

The right hand side is n^{-2} times the number of edges of $V_i \times V_j$. In $G(n; \texttt{DATA})$ this has expected value $\alpha_i \alpha_j p_{ij}$. From Assumption 1.3 of Theorem 1.3 we deduce:

Lemma 4.1. *For $1 \leq i < j \leq K$, $V_i \times V_j$ has $|V_i| \cdot |V_j| p_{ij}\big(1 + o(1)\big)$ edges.*

The same argument works for $i = j$, taking care to not doublecount edges. The result:

Lemma 4.2. *For $1 \leq i \leq K$, V_i has $\big(|V_i|^2/2\big)\, p_{ii}\big(1 + o(1)\big)$ edges.*

5. Counting Cycles

Here we shall count the cycles $\{v_1, w_1, v_2, w_2\}$ with $v_1, v_2 \in V_i$, $w_1, w_2 \in V_j$, $i \neq j$. Let $\chi(v_1, w_1, v_2, w_2)$ be one if v_1, w_1, v_2, w_2 forms a cycle (in that order), zero otherwise. For any $0 \leq a_1, a_2, a_3, a_4 \leq 2K-2$,

$$(9) \quad t\big(\texttt{CYCLESTAR}\,(a_1, a_2, a_3, a_4), G\big)$$

$$= n^{-4} \sum \texttt{NDEG}\,(v_1)^{a_1}\,\texttt{NDEG}\,(w_1)^{a_2}\,\texttt{NDEG}\,(v_2)^{a_3}\,\texttt{NDEG}\,(w_2)^{a_4} \chi(v_1, w_1, v_2, w_2)$$

where the sum is over $v_1, w_1, v_2, w_2 \in V(G)$. For any polynomial $f(x, y, z, w) = \sum_{r,s,r',s'=0}^{2K-2} \beta_{rsr's'} x^r y^s z^{r'} w^{s'}$,

$$(10) \quad n^{-4} \sum f\big(\texttt{NDEG}\,(v_1), \texttt{NDEG}\,(w_1), \texttt{NDEG}\,(v_2), \texttt{NDEG}\,(w_2)\big)\, \chi(v_1, w_1, v_2, w_2)$$

$$= \sum_{r,s,r',s'=0}^{2K-2} \beta_{rsr's'} t\big(\texttt{EDGESTAR}\,(r, s, r', s'), G\big)$$

We apply this with a sieve polynomial $f(x, y, z, w) = g_i(x) g_j(y) g_i(z) g_j(w)$ with g_i, g_j given by Definition (2). Set

(11)
$$X = X(v_1, w_1, v_2, w_2) = f\big(\texttt{NDEG}\,(v_1), \texttt{NDEG}\,(w_1), \texttt{NDEG}\,(v_2), \texttt{NDEG}\,(w_2)\big)$$

Then $X = 1 + o(1)$ when $v_1, v_2 \in V_i$ and $w_1, w_2 \in V_j$; $X = o(1)$ when none of v_1, w_1, v_2, w_2 are in V_{ext} but the above does not occur; $X = O(1)$ when one (or more) of v_1, w_1, v_2, w_2 are in V_{ext}. Therefore, with probabilities and expectations over the uniform indepedent choice of vertices $v_1, w_1, v_2, w_2 \in V(G)$,

$$(12) \quad n^{-4} \sum f\big(\texttt{NDEG}\,(v_1), \texttt{NDEG}\,(w_1), \texttt{NDEG}\,(v_2), \texttt{NDEG}\,(w_2)\big)\, \chi(v_1, w_1, v_2, w_2)$$

$$= E[X]$$

$$\sim \Pr\big[v_1, v_2 \in V_i, w_1, w_2 \in V_j, \chi(v_1, w_1, v_2, w_2) = 1\big]$$

In $G(n; \texttt{DATA})$ (12) has expected value $\alpha_i^2 \alpha_j^2 p_{ij}^4$. From Assumption 1.3 of Theorem 1.3 we deduce:

Lemma 5.1. *For $1 \leq i < j \leq K$, $V_i \times V_j$ has $|V_i|^2 |V_j|^2 p_{ij}^4 \big(1 + o(1)\big)$ cycles. Here a cycle is an ordered (v_1, w_1, v_2, w_2) with $v_1, v_2 \in V_i$, $w_1, w_2 \in V_j$ and adjacencies $\{v_1, w_1\}$, $\{w_1, v_2\}$, $\{v_2, w_2\}$ and $\{w_2, v_1\}$.*

The same argument works for $i = j$, taking care not to doublecount edges. The result:

Lemma 5.2. *For $1 \le i \le K$, V_i has $|V_i|^4 p_{ii}^4(1+o(1)$ cycles. Here a cycle is an ordered (v_1, w_1, v_2, w_2) with $v_1, v_2, w_1, w_2 \in V_i$ and adjacencies $\{v_1, w_1\}$, $\{w_1, v_2\}$, $\{v_2, w_2\}$ and $\{w_2, v_1\}$.*

6. The Final Steps

Here we show that Theorem 1.3 follows from Lemmas 3.1, 3.2, 4.1, 4.2, 5.1, 5.2. The methods are standard, we include them for completeness. First consider the implications of Lemmas 4.1, 5.1 on the restriction of G to $V_i \times V_j$. Let us call the sets top and bottom, T, B respectively with $|T| \sim n\alpha$, $|B| \sim n\beta$, so that $T \times B$ has $\sim |T| \cdot |B| \cdot p$ edges $\{t_1, b_1\}$ and $\sim |T|^2 \cdot |B|^2 \cdot p^4$ ordered cycles (t_1, b_1, t_2, b_2). Then $\sum_{v \in T} \deg(v) \sim |T| \cdot |B| p$ so

$$\sum_{t \in T} \deg^2(t) \ge |T| \cdot |B|^2 p^2 \big(1 + o(1)\big) \tag{13}$$

For $b_1, b_2 \in B$ let $N(b_1, b_2)$ denote the number of $t \in T$ adjacent to them both. Then

$$\sum N(b_1, b_2) = \sum_t \deg^2(t) \ge |T| \cdot |B|^2 p^2 \big(1 + o(1)\big)$$

so that

$$\sum N^2(b_1, b_2) \ge |B|^2 \big[|T| p^2\big]^2 = |T|^2 |B|^2 p^4 \big(1 + o(1)\big) \tag{14}$$

But $\sum N^2(b_1, b_2)$ counts pairs t_1, t_2, both adjacent to b_1, b_2, which is a count on ordered 4-cycles t_1, b_1, t_2, b_2 and thus, from Lemma 5.1 is $\sim |T|^2 |B|^2 p^4$. Thus inequalities (13, 14) must be asymptotic equalities from which we deduce that asymptotically almost all $t \in T$ have $\deg(t) \sim |B|$ and that asymptotically almost all $b_1, b_2 \in T$ have $N(b_1, b_2) \sim |T| p^2$. Reversing the roles of T, B give two analogous results. Similar arguments work in the $i = j$ (internal edges) case. Summarizing:

Lemma 6.1. *Under the assumptions of Lemmas* 3.1, 3.2, 4.1, 4.2, 5.1, 5.2, *letting V_i, $1 \le i \le K$, be given by Lemma* 3.2 *we have*

1. *($i \neq j$) All but $o(n)$ $v \in V_i$ are adjacent to $\sim p_{ij}|V_j|$ $w \in V_j$.*

2. *($i \neq j$) All but $o(n^2)$ distinct $v_1, v_2 \in V_i$ are adjacent to $\sim p_{ij}^2|V_j|$ $w \in V_j$.*

3. *All but $o(n)$ $v \in V_i$ are adjacent to $\sim p_{ii}|V_i|$ $w \in V_i$.*

4. *All but $o(n^2)$ distinct $v_1, v_2 \in V_i$ are adjacent to $\sim p_{ii}^2|V_i|$ $w \in V_i$.*

Now fix a graph H on vertices $1, \ldots, u$. We need show $t(H, G) \rightarrow t(H, \mathtt{DATA})$. For each $1 \leq \psi(1), \ldots, \psi(u) \leq K$ let $t(H, G, \psi)$ denote n^{-u} times the number of homomorphisms $\phi : H \rightarrow G$ with $\phi(i) \in V_{\psi(i)}$. As $|V_{\text{ext}}| = o(n)$, the contribution to $t(H, G)$ from other ϕ is $o(1)$. Similarly let $t(H, \mathtt{DATA}, \psi)$ refer to n^{-u} times the expected number of such restricted homomorphisms. Thus it suffices to

$$t(H, G, \psi) \rightarrow t(H, \mathtt{DATA}, \psi)$$

for an arbitrary fixed ψ.

We count the homomorphism one vertex at a time. Suppose $v_j = \phi(j)$, $1 \leq i < s$ have been chosen. We need select $w = v_s$. It must be in the particular class V_b, with $b = \psi(s)$. Further, it must have adjacencies to certain of the previously chosen vertices. Renumbering for convenience suppose $w = v_s$ need be adjacent to $v_1, \ldots, v_R$.

Call $(v_1, \ldots, v_R)$ *nice* if $v_i \in V_{\psi(i)}$ for $1 \leq i \leq R$. For nice $(v_1, \ldots, v_R)$ set $X(v_1, \ldots, v_R)$ equal the proportion of $w \in V_b$ adjacent to all $v_1, \ldots, v_R$. In G satisfying the Lemmata we consider X a random variable, with v_i ranging uniformly over $V_{\psi(i)}$. It now suffices to show:

Lemma 6.2. *Under the above assumptions*

$$X = X(v_1, \ldots, v_R) = \prod_{i=1}^{R} p_{b,\psi(i)} + o(1) \tag{15}$$

for all but $o(n^R)$ choices of nice $v_1, \ldots, v_R$.

To show Lemma 6.2 it suffices to show

$$E[X] = \prod_{i=1}^{R} p_{b,\psi(i)} + o(1) \tag{16}$$

and

$$E[X^2] = \prod_{i=1}^{R} p_{b,\psi(i)}^2 + o(1) \tag{17}$$

as then Var $[X] = o(1)$ and the desired concentration follows.

For $w \in V_b$ set $Y = Y(w)$ equal the proportion of nice $v_1, \ldots, v_R$ with w adjacent to each v_i. Consider Y as a random variable with w chosen uniformly from V_b. Then

$$E[X] = E[Y] \tag{18}$$

as both represent the probability that w is adjacent to each v_i when $w \in V_b$ and $v_i \in V_{\psi(i)}$ are selected uniformly. (This essential step can be thought of as counting $K_{1,R}$ in two ways, over the choice of nice $v_1, \ldots, v_R$ or over the choice of $w \in V_b$.) From Lemma 6.1, all but $o(n)$ choices of $w \in V_b$ have a proportion $p_{b,r} + o(1)$ of neighbors in V_r for each $1 \leq r \leq K$. For these v, $Y(w) \sim \prod p_{b,\psi(i)} + o(1)$. As $0 \leq Y \leq 1$ tautologically, this implies $E[Y] \sim \prod p_{b,\psi(i)} + o(1)$ so that (18) gives (16).

For $w_1, w_2 \in V_b$ set $Z = Z(w_1, w_2)$ equal the proportion of nice $v_1, \ldots, v_R$ with both w_1, w_2 adjacent to each v_i. Consider Z as a random variable with w_1, w_2 chosen uniformly from V_b. Then

$$E[X^2] = E[Z] \tag{19}$$

as both represent the probability that w_1, w_2 are both adjacent to each v_i when $w_1, w_2 \in V_b$ and $v_i \in V_{\psi(i)}$ are selected uniformly. (This essential step can be thought of as counting $K_{2,R}$ in two ways, over the choice of nice $v_1, \ldots, v_R$ or over the choice of $w_1, w_2 \in V_b$.) From Lemma 6.1, all but $o(n^2)$ choices of $w_1, w_2 \in V_b$ have a proportion $p_{b,r}^2 + o(1)$ of common neighbors in V_r for each $1 \leq r \leq K$. For these w_1, w_2, $Z(w_1, w_2) = \left[\prod p_{b,\psi(i)}\right]^2 + o(1)$. As $0 \leq Z \leq 1$ tautologically, this implies $E[Z] \sim \left[\prod p_{b,\psi(i)}\right]^2$ so that (19) gives (17).

References

[1] F. Chung, R. Graham and R. Wilson, Quasi-random graphs, *Combinatorica,* vol. **9** (1989), 345–362.

[2] L. Lovász and V. Sós, Generalized Quasirandom Graphs, *J. Combinatorial Theory (B),* vol. **98** (2008), 146–163.

Joel Spencer
Courant Institute
New York

e-mail: `spencer@cims.nyu.edu`

BOLYAI SOCIETY
MATHEMATICAL STUDIES, 21

An Irregular Mind
(Szemerédi is 70)
pp. 619–650.

Pseudorandomness in Computer Science and in Additive Combinatorics

LUCA TREVISAN*

Dedicated to Endre Szemerédi on the occasion of his 70th birthday

Notions of pseudorandomness and (implicitly) of indistinguishability arise in several key results in additive combinatorics. In this expository paper, we show how several results can be translated from the analytic language of norms, decompositions, and transference to the computer science language of indistinguishability, simulability and pseudoentropy. Some of these results, once so reformulated, can be given "computer science proofs" which are quantitatively better in some respects, and which have some applications.

We discuss variants of the Szemerédi regularity lemma for graphs; the inverse theorems for the Gowers uniformity norms; a key step in the "transference" results of Green, Tao and Ziegler; and various "decomposition" results.

1. Introduction

We outline how the computer science notions of "pseudorandomness", "indistinguishability" and "simulability", which arise in computational complexity theory and in the foundations of cryptography, can be used to reformulate several results in additive combinatorics, which are usually stated in the analytic language of norms and decompositions.

We discuss the following points.

- The notion of "quasirandomness" of a graph, introduced by Chung, Graham and Wilson [6] and Thomason [41], can be defined in terms

*This material is based upon work supported by the National Science Foundation under grant No. CCF-0729137 and by the BSF under grant 2006060.

of the computer science notion of pseudorandomness of distributions: a graph is quasirandom if and only if the uniform distribution over its edge set is indistinguishable from the uniform distribution over all vertex pairs by the family of adversaries defined by characteristic functions of partial cuts.

Similarly, the notion in which a regular partition of a graph (as given by the Szemerédi regularity lemma [35, 36]) gives an "approximation" of the graph can be formulated in terms of indistinguishability of the given graph from a graph derived from the regular partition.

- Although the notion of "quasirandomness" of functions defined by the Gowers uniformity norm [15, 16] is incomparable with the computer science notions of pseudorandomness and indistinguishability, the recently proved *Gowers inverse conjectures* can be seen as asserting that two functions are close in Gowers norm if and only if they are indistinguishable by a certain set of adversaries.

- Several "decomposition" results in additive combinatorics [20, 46, 24, 23], as well as various versions of the Szemerédi regularity lemma [35, 36, 8, 2, 39] can be generalized to "regularity lemmas for functions", which, stated in computer science language, assert that every high entropy distribution is indistinguishable from a high-entropy efficiently computable (and samplable) distribution. A "weak regularity lemma for functions" in which all complexity parameters are polynomial can be proven using computer science techniques [43].

- The Green–Tao theorem [20] that the primes contain arbitrarily long arithmetic progressions is proved by constructing a "dense model" of the prime numbers, showing that the primes and the dense model are "indistinguishable" in a sense that has implications for the counting of arithmetic progressions, and using the fact that the dense model must have many long arithmetic progressions from Szemerédi's theorem. A more abstract "dense-model theorem" can be proven with computer science techniques and fully polynomial parameters and has applications to cryptography [32, 7, 27, 30].

Overview of the paper. In Sections 2 and 3 we define the computer science notions of indistinguishability, pseudorandomness and simulability of distributions, and in Section 4 we define the combinatorial notion of quasirandom graph and state three versions of the Szemerédi regularity lemma.

In Section 5 we give a first simple example of "translation" to the computer science definitions, by showing how to formulate the notion of quasirandomness of graph as a notion of pseudorandomness of a distribution, and how to interpret regularity lemmas as simulability results. We define the Gowers uniformity norms in Section 6, and state the recently established Gowers inverse conjectures; we view the Gowers inverse conjectures as characterizing the Gowers uniformity in terms of indistinguishability. We tie several strands together in Section 7, in which we state various "regularity lemmas for functions", which arise as natural generalizations of the regularity lemmas for graphs (when the latter are formulated in terms of indistinguishability); pairing such lemmas with the Gowers inverse conjectures one obtains (weak versions of) decomposition results of Green and Tao and of Gowers and Wolf. Finally, in Section 9 we show how the transference result in the proof of the Green–Tao theorem that the primes contain arbitrarily long arithmetic progressions, can also be seen in terms of indistinguishability, and it has a computational analog with applications to cryptography.

Further reading. While the main goal of this paper is to illustrate via several examples a "dictionary" between additive combinatorics and computer science, we hope that we will also inspire the computer scientist reader to learn more about additive combinatorics and the mathematician reader to learn more about computer science. The former is referred to the book of Tao and Vu [44] and to the extraordinary blog maintained by Terry Tao at `terrytao.wordpress.com`. For the latter, a great starting point is a monograph by Oded Goldreich [13], which introduces the reader to pseudorandomness and the foundations of cryptography. A broad treatment of all areas of computational complexity is given in the recent book of Arora and Barak [1].

2. Indistinguishability and Pseudorandomness

Indistinguishability and *pseudorandomness* are two fundamental notions in modern cryptography; their formalization in the work of Blum, Goldwasser, Micali and Yao [3, 10, 47] marked the start of the rigorous approach to definitions of security in the early 1980s.

Definition 1 (Indistinguishability). If X and Y are two distributions over a finite set Ω, and $\mathcal{F}$ is a class of boolean functions $f : \Omega \to \{0,1\}$, then we say that *X and Y are ε-indistinguishable by $\mathcal{F}$* if

$$\forall f \in \mathcal{F}. \quad \left| \mathop{\mathbb{P}}_{x \sim X} \left[f(x) = 1 \right] - \mathop{\mathbb{P}}_{y \sim Y} \left[f(y) = 1 \right] \right| \leq \varepsilon$$

or, equivalently

$$\forall f \in \mathcal{F}. \quad \left| \mathop{\mathbb{E}}_{x \sim X} \left[f(x) \right] - \mathop{\mathbb{E}}_{y \sim Y} \left[f(y) \right] \right| \leq \varepsilon.$$

When $\mathcal{F}$ is defined as the (appropriately quantified) class of efficiently computable algorithms, and ε is small, then when two distributions are indistinguishable we have that, "from the point of view" of algorithms from $\mathcal{F}$, there is almost no observable difference between the distributions. If we think of the algorithms from $\mathcal{F}$ as "statistical tests", then they detect almost no statistical difference between the distributions. Indeed, if we take $\mathcal{F}$ to be the class of *all* boolean functions $f : \Omega \to \{0,1\}$, then X and Y are ε-indistinguishable by $\mathcal{F}$ if and only if their *total variation distance* is at most ε.

In a computational setting, one usually studies an asymptotic setting in which there is a collections (or *ensembles*) of distributions $\mathbf{X} = X_1, \ldots, X_n, \ldots$, indexed by a parameter n which usually denotes the length of the bit strings in the support of X_n (that is, X_n is a distribution over the sample space $\{0,1\}^n$). Then two ensembles $\mathbf{X}$ and $\mathbf{Y}$ are *computationally indistinguishable* if for every two polynomials p, q and for every sufficiently large n we have that X_n and Y_n are $1/q(n)$-indistinguishable by $\mathrm{SIZE}\left(p(n)\right)$, the set of functions computable by circuits of size at most $p(n)$. In this paper, we will work with the "concrete" notion involving only single distributions as in Definition 1.

The notion of indistinguishability originated with the work of Goldwasser and Micali [10] on the proper formal definition of security for encryption. In one of their two definitions (which they prove to be equivalent), an encryption scheme is secure if (simplifying a bit), for every two messages m_1 and m_2, the ensemble of distributions of possible encryptions (the definition requires encryption to be a probabilistic process) of m_1 is computationally indistinguishable from the distribution of possible encryptions of m_2, where the ensembles are indexed by the length of the secret key.

Note that, in an encryption scheme that satisfies the Goldwasser–Micali definition of security, one has pairs of distributions that are ε-indistinguishable for very small ε even though *they have disjoint support*, and thus their total variation distance is maximal.

Definition 2 (Pseudorandomness). If a distribution X over a sample space Ω is ε-indistinguishable by $\mathcal{F}$ from the *uniform distribution* over Ω, then we say that X is *ε-pseudorandom* for $\mathcal{F}$.

In the computational setting, an ensemble $\mathbf{X} = X_1, \ldots, X_n, \ldots$ of distributions is *pseudorandom* if for every two polynomials p, q and for every sufficiently large n we have that X_n is $1/q(n)$-pseudorandom for $\mathrm{SIZE}\left(p(n)\right)$, the set of functions computable by circuits of size at most $p(n)$.

A deterministic length-increasing function $G : \{0,1\}^\ell \to \{0,1\}^n$ is an (S, ε) *pseudorandom generator* if the distribution of outputs of $G(x)$ given a random input "seed" x is ε-pseudorandom for the set of functions computable by circuits of size at most S. This notion is usually studied in an asymptotic setting: a collection of functions $G_n : \{0,1\}^{\ell(n)} \to \{0,1\}^n$ is a pseudorandom generator if the ensemble $G_1\left(U_{\ell(1)}\right), \ldots, G_n\left(U_{\ell(n)}\right), \ldots$ is pseudorandom, where U_ℓ denotes the uniform distribution over $\{0,1\}^\ell$.

This definition of pseudorandom generator was given by Yao [47], who proved it equivalent to an earlier definition given by Blum and Micali [3]. Yao also observed that a pseudorandom generators can be used to *derandomize* randomized algorithms, that is, to reduce the amount of random bits needed to run them. (This is a direction that has evolved very successfully in the 1990s, see for example the monograph of Goldreich [13] or the expository paper [42].)

The three papers of Blum and Micali, of Goldwasser and Micali and Yao, which appeared within months of each other in 1982, ushered in a revolution in the foundations of cryptography, by providing a formal framework to argue rigorously about tasks that are *impossible* from an information-theoretic perspective.

Shannon [34], for example, had considered in the 1940s the issue of security for encryption, and he had formulated a definition of security based on his newly developed information theory. Shannon's notion can be proved to be equivalent to the "message-indistinguishability" notion of Goldwasser and Micali, but using total variation distance instead of indistinguishability. Shannon was able to show that, according to his definition of security, secure secret-key encryption is possible only if the secret key is at least as long as the concatenation of all the messages that one wishes to send (a bound achieved by the one-time pad). While public-key cryptography had not been conceived yet, it is easy to see that it cannot achieve Shannon's security property. Similarly, if one would like a "pseudorandom generator" to be such that the output is close in total variation distance to the uniform distribution, then the definition is simply impossible to achieve, except in the trivial setting in which the input seed is at least as long as the output. (The identity function would then work.) On this matter, von Neumann famously said that "anyone who considers arithmetical methods of producing random digits is, of course, in a state of sin."

Using the computational indistinguishability definition, however, it is possible (under the standard assumption that one-way functions exist – see [13]) to have secure secret-key encryption schemes in which the messages are longer than the secret key, it is possible to have length-increasing pseudorandom generators and (under the assumption that trapdoor permutations exist) it is possible to have public-key encryption.

3. Simulability

How can we formalize the notion that a probability distribution X over a sample space Ω is "known" to an efficient[1] algorithm?

Consider the following settings:

1. *The cumulative distribution function is efficiently computable,* that is, there is an efficient algorithm that given z computes $\mathbb{P}_{x\sim X}[x \leq z]$, where $\leq$ is an efficiently computable total order on Ω;

2. *The pointwise distribution function is efficiently computable,* that is, there is an efficient algorithm that given z computes $X(z) = \mathbb{P}_{x\sim X}[x = z]$;

3. *X is efficiently samplable,* that is, there is an efficient randomized algorithm (called a "sampler") whose output has the same distribution as X;

[1]For the sake of this informal discussion we will not give a rigorous definition of what we mean by "efficient." In a rigorous treatment, we would work with an ensemble $\mathbf{X} = (X_1, \ldots, X_n, \ldots)$ of distribution, and we would require algorithms for X_n to run in time polynomial in n.

4. *X is efficiently simulable*, that is, there is an efficient randomized algorithm (called a "simulator") whose output distribution is indistinguishable from X.

It is easy to see that (1) $\Rightarrow$ (2) and that (1) $\Rightarrow$ (3) $\Rightarrow$ (4), and it is possible to show that the reverse implications are not always true.

For example, if $G : \{0,1\}^\ell \to \{0,1\}^n$ is an efficiently computable length-increasing pseudorandom generator, then the output distribution $G(w)$ for a random $w \sim \{0,1\}^\ell$ is efficiently samplable (the algorithm $G()$ being the sampler), but such a distribution cannot satisfy (2) (and hence cannot satisfy (1)), because otherwise the property of being a possible output of $G()$ would be an efficiently computable property, but it would be true with very different probability under the uniform distribution and under the distribution of outputs of $G()$, thus contradicting the pseudorandomness of $G()$.

To see that (4) does not necessarily imply (3), suppose that f : $\{0,1\}^n \to \{0,1\}$ is a random function (or, say, a function whose truth-table has Kolmogorov complexity close to 2^n). Then the distribution $\big(x, f(x)\big)$ for a random $x \sim \{0,1\}^n$ is impossible to sample, because otherwise the sampler would give a compact representation of the function f. The uniform distribution over $n+1$ bits, however, that is very easy to sample, is indistinguishable from it. This example is interesting because it shows that even a distribution that is very strongly not computable can still be simulable. Are there non-simulable distributions? We will return to this question later.

An important application of such notions arises when one wants to define security in cryptographic protocols in which a malicious party may eavesdrop on the communication between legitimate parties, or might even participate in the protocol, posing as a legitimate party. In the definition of security, we would like to capture the idea that the protocol is secure as long as the malicious party *gains no information* from his eavesdropping or participation in the protocol. In turn, the notion of "gaining no information" can be formalized by proving that the distribution of what the malicious party receives from his eavesdropping/participation was *already known* beforehand, and is *independent* of the secrets held by the legitimate parties.

The strongest way in which this can be formalized is that such a distribution is computable (pointwise or as cumulative distribution function) or samplable, but even if the distribution is simulable, it is still satisfactory for a definition of security.

The simulable case is the more interesting, because, in a typical instantiation, we have a simulator which samples a distribution known before-hand and independent of the secrets of legitimate parties, and, for every actual values of the secrets of the legitimate parties, the distribution of views of the malicious party is indistinguishable from the fixed samplable distribution. In such a case, even if the views of the malicious party is correlated to the secrets, the indistinguishability from a fixed known samplable distribution can be used to show that the malicious party has no way of learning any information about the secrets. (Otherwise, the procedure used to learn such information could be used to distinguish the view of the malicious party from the samplable distribution.)

Such an approach to the definition of security is implicit in the work of of Goldwasser and Micali [10] on security of encryption and explicit in the definition of *zero knowledge* by Goldwasser, Micali and Rackoff [11], and in the treatment of general distributed protocols of Goldreich, Micali and Wigderson [12].

3.1. An Aside on Entropy, Pseudoentropy and Samplability

We remark that, although property (2) does not imply samplability in general, it does for *high min-entropy* distributions.

Definition 3 (Min-Entropy). The min-entropy of a distribution X over a sample space Ω, denoted $H^\infty(X)$, is defined as

$$H^\infty(X) := \log_2 \min_{x\in\Omega} \frac{1}{X(x)}.$$

That is,

$$H^\infty(X) \geq k \Leftrightarrow \forall x \in \Omega. \quad X(x) \leq 2^{-k}.$$

Suppose that X is a distribution of min-entropy at least $\log_2 \delta|\Omega|$, for a not-too-small $\delta > 0$, and that, given z, the probability $X(z)$ is efficiently computable. Then X is efficiently samplable via the following algorithm:

1. Pick a random $z \in \Omega$ uniformly at random
2. With probability $\delta \cdot |\Omega| \cdot X(z)$, output z and halt; with probability $1 - \delta|\Omega| \cdot X(z)$ go to step 1

In each iteration of the algorithm, it either halts and produces an output, or it moves to the next iteration; conditioned on producing an output, the output of the algorithm has the distribution X. The probability of halting in a given iteration is δ, and so the average number of iterations is $1/\delta$.

The final definition of this section is that of *pseudoentropy.*

Definition 4 (Pseudoentropy). Let X be a distribution over a sample space Ω and $\mathcal{F}$ be a family of boolean functions $f : \Omega \to \{0,1\}$.

We say that X has $(\mathcal{F}, \varepsilon)$-pseudoentropy at least k if there is a distribution Y such that

- X and Y are ε-indistinguishable by $\mathcal{F}$;
- Y has min-entropy at least k.

The notion of pseudoentropy was introduced by Håstad et al. [25] in the context of constructing pseudorandom generators. Barak et al. [4] consider the relationships between variants of the above definition.

4. The Szemerédi Regularity Lemma

Chung, Graham and Wilson [6] and Thomason [41] give a definition of *quasirandomness* for graphs. In their definition, a quasirandom graph is a fixed graph which satisfies several properties which are true with high probability for a random graph. The following is an equivalent characterization.

Definition 5 (Quasirandom graph). An undirected graph $G = (V, E)$ is ε-quasirandom if, for every two disjoint subset of vertices $A, B \subseteq V$,

$$\left| \text{edges}_G(A,B) - |A| \cdot |B| \cdot \frac{|E|}{\binom{V}{2}} \right| \le \varepsilon \cdot \binom{V}{2}$$

where $\text{edges}_G(A,B)$ is the number of edges in G that have one endpoint in A and one endpoint in B.

Intuitively, the definition is saying that the number of edges between A and B is the same number that we would expect on average in a random graph with $|E|$ edges.

The Szemerédi regularity lemma (and its variants) states that every dense graph G is well approximated by the edge-disjoint union of a small number of "quasirandom bipartite graphs." We need a couple of definition in order to state the lemma.

Definition 6 (Regular Pair). In a graph $G = (V, E)$, two disjoint subsets of vertices U_1, U_2 of the same size are an "ε-regular pair" if for every subsets $A_1 \subseteq U_1$ and $A_2 \subseteq U_2$ the number of edges with one endpoint in A_1 and one endpoint in A_2 is what we would expect, that is the number of edges between U_1 and U_2 multiplied by $\frac{|A_1|\cdot|A_2|}{|U_1|\cdot|U_2|}$, plus or minus $\varepsilon \cdot |U_1| \cdot |U_2|$.

Note that the notion of regular pair is a natural generalization of the notion of quasirandom graph to the case of bipartite graphs.

Definition 7 (Regular Partition). A partition $(U_1, \ldots, U_k)$ of the vertices of a graph $G = (V, E)$ is ε-regular if all the sets have the same size (except possibly U_k, which has then size smaller than the others) and for at least a $(1 - \varepsilon)$ fraction of the pairs (i, j) we have that (U_i, U_j) is an ε-regular pair.

Lemma 8 (Szemerédi Regularity Lemma [35, 36]). *For every ε and ℓ there is an $k = k(\ell, \varepsilon)$ such that for every graph $G = (V, E)$ there is an ε-regular partition with at least ℓ and at most k sets.*

Given a graph G and a regular partition $(U_1, \ldots, U_k)$ for it, consider the weighted graph G' that has the same vertex set as G, and such that the edge (u, v) has weight 0 if u, v belong to the same set U_i in the partition, and weight $e(U_i, U_j)/\big(|U_i| \cdot |U_j|\big)$ otherwise, where $u \in U_i$, $v \in U_j$, and $e(U_i, U_j)$ is the number of edges between U_i and U_j in G.

The graph G' is the edge-disjoint union of $\binom{k}{2}$ complete bipartite weighted graphs and, given the partition, it is completely specified by $\binom{k}{2}$ numbers.

The existence of G' leads to several very useful consequence. Roughly speaking, many properties of G are, at least approximately, determined by the structure of G', and so are verifiable with "constant complexity" because of the very simple structure of G'.

A concrete instantiation of this approach is the proof the Triangle Removal Lemma [31], which states that if a graph $G = (V, E)$ has $o\big(|V|^3\big)$ triangles, then there is a set of $o\big(|V|^2\big)$ edges whose removal makes the graph triangle-free; a more precise formulation follows.

Theorem 9 (Triangle Removal Lemma [31]). *For every $\varepsilon > 0$ there is a $\delta = \delta(\varepsilon) > 0$ such that, in a graph $G = (V, E)$ that has $\leq \delta|V|^3$ triangles, there is a set of $\leq \varepsilon|V|^2$ edges whose removal makes the graph triangle-free.*

The proof proceeds by constructing a Szemeredi regular partition of G, removing all edges between pairs of sets in the partition that are not regular, and between pairs that have small density of edges between them. If, in the residual graph, there are three blocks of the partition with edges between them, it can be argued that the graph has $\Omega_\varepsilon(|V|^3)$ triangles. If not, the removal of the above-mentioned edges, which can be set to be at most $\varepsilon|V|^2$, makes the graph triangle-free.

The Triangle Removal Lemma gives a simple combinatorial proof of Roth's theorem, that every set of integers of positive density contains length-3 artihmetic progressions. It is also a prototype of several algorithms in the area of *property testing*, an algorithmic model introduced by Goldreich, Goldwasser and Ron [9].

Unfortunately, the "complexity parameter" k in the statement of the Regularity Lemma grows like a *tower of exponentials* of height $\varepsilon^{-O(1)}$, and such a dependency affects its applications. For example, the dependency between $1/\delta$ and $1/\varepsilon$ in the only known proof of the Triangle Removal Lemma is also of tower-of-exponentials type. Gowers has proved that such tower-of-exponentials dependency of k on ε is necessary [14] in the Regularity Lemma.

Frieze and Kannan [8] prove that a much better dependency is possible if one works with a weaker notion of regularity. To introduce the Frieze–Kannan notion of *weak regularity*, it is helpful to consider the following useful property of regular partitions: suppose that $(B_1, \ldots, B_k)$ is an ε-regular partition, that $d_{i,j}$ is the density of the edges between B_i and B_j in the graph, and that A and B are two disjoint sets of vertices. Then the number of edges between A and B in the approximating graph G' is

$$\sum_{i,j} |A \cap B_i| \cdot |B \cap B_j| \cdot d_{ij}$$

and the number of edges in G with one endpoint in A and one endpoint in B deviates from the above formula by at most $2\varepsilon|V|^2$, with a contribution of at most $\varepsilon|V|^2$ coming from the pairs that are not regular, and another $\varepsilon|V|^2$ coming from the regular pairs.

Frieze and Kannan define a partition to be weakly regular if it satisfies just such a property.

Definition 10. A partition $(B_1, \ldots, B_k)$ of the set of vertices of a graph $G = (V, E)$ is weak ε-regular if for every two disjoint sets of vertices A, B we have

$$\left| \text{edges}_G(A, B) - \sum_{i,j} |A \cap B_i| \cdot |B \cap B_j| \cdot d_{ij} \right| \leq \varepsilon |V|^2$$

where $d_{ij} := \text{edges}_G(B_i, B_j)/\big(|V_i| \cdot |V_j|\big)$ is the density of the set of edges between B_i and B_j in G.

Theorem 11 (Weak Regularity Lemma [8]). *For every ε there is a $k = k(\varepsilon) \leq 2^{\varepsilon^{-O(1)}}$ such that for every graph $G = (V, E)$ there is a weak ε-regular partition with at most k sets.*

Interesting results, in particular efficient approximation algorithms for dense graphs, can be derived from the weak regularity lemma. In particular, if $G = (V, E)$ is a graph, $(B_1, \ldots, B_k)$ is a weak ε-regular partition, and G' is the weighted graph obtained from G as before by having an edge of weight $d_{i,j}$ between every vertex of B_i and every vertex of B_j, then we have that, for example, an optimal solution to the Maximum Cut problem in G' is, without loss of generality, a solution that "respects" the partition (meaning such that each block of the partition is entirely contained on one side of the cut), and there are only $\exp\big(\exp\big(\varepsilon^{-O(1)}\big)\big)$ such partitions; furthermore, the cost of a Max Cut solution in G or in G' is approximately the same, up to an additive term $\varepsilon|V|^2$, which means that in time $\exp\big(\exp\big(\varepsilon^{-O(1)}\big)\big)$ we can solve Max Cut in G up to an additive error $2\varepsilon|V|^2$, which is a good result if the graph is dense and ε is small. (This is true provided that the partition itself can be found efficiently, which Frieze and Kannan proved to be the case. Frieze and Kannan also show how to improve the running time to $\exp\big(\varepsilon^{-O(1)}\big)$ via a refined formulation of the Weak Regularity Lemma and a more sophisticated algorithm; we will not discuss this improvement here.)

Unfortunately the existence of a weak regular partition is not sufficient to prove the Triangle Removal Lemma. Indeed, in the proof of the triangle removal lemma one argues that either G' (after removing small weight edges) is triangle free, and hence G is close to triangle free, or else the existence of triangles in G' implies the existence of $\Omega\big(|V|^3\big)$ triangles in G. The latter statement is not true in a weak regular partition.

Even the original Regularity Lemma, indeed, is not sufficient to prove other "global versus local" results that are useful in property testing and additive combinatorics. For example, it is possible to show by other techniques (and with good quantitative parameters, see the foundational paper of Goldreich, Goldwasser and Ron [9] which introduced the model of property testing) that either a graph is ε-close to being bipartite (meaning that it is sufficient to delete $\leq \varepsilon|V|^2$ edges to make the graph bipartite) or a random sample of $\varepsilon^{-O(1)}$ vertices hash high probability to induce a subgraph containing an odd cycles. Although any proof based on regularity would yield much, much, worse bounds, one would like to derive such a proof from a regularity statement by arguing that either G' (after removing low weight edges) is bipartite, and hence G is ε-close to bipartite, or else the existence of odd cycles in G' implies that odd cycles in G can be found by sampling $O_\varepsilon(1)$ vertices.

If one tries to adapt the proof of the Triangle Removal Lemma to this setting, however, one runs into troubles that, roughly speaking, come from the fact that the odd cycle in G' might be as long as the number of sets k in the regular partition. Thus, in order to make sure that an odd cycle in G' corresponds to many odd cycles in G one would need the approximation parameter ε to be much smaller than $1/k$. We know, however, that k is necessarily more than exponential in $1/\varepsilon$.

Alon, Fischer, Krivelevich and Szegedy [2] and Tao [39] get around this seeming impossibility by showing that every graph is *close* to a graph admitting a regular partition in which the regularity parameter ε is arbitrarily smaller than the number of sets k in the partition.

Lemma 12 ("Strong" Szemeredi Regularity Lemma). *For every ε and every function $F(\cdot,\cdot)$ there is a $k = k(\varepsilon, F)$ such that for every graph $G = (V, E)$ there is a graph $\tilde{G}$ that differs from G in at most $\varepsilon|V|^2$ edges and such that $\tilde{G}$ admits a partition $(B_1, \ldots, B_k)$ in which every pair is $1/F(k,\varepsilon)$-regular.*

5. Quasirandomness and Regularity as Indistinguishability

In this section we begin to describe a "dictionary" between notions in additive combinatorics and notions in computer science by reformulating the definition of quasirandom graph and regular partition in terms of pseudorandomness and indistinguishability.

Given a graph $G = (V, E)$, let $X := \binom{V}{2}$ be the set of pairs of vertices in V, and associate to it the distribution μ_G over the set of pairs in X by letting μ_G be uniform over E. Consider the set of boolean function

$$\mathcal{F} := \{ f_{A,B} \mid A, B \subseteq V,\ |A \cap B| = \emptyset \} \tag{1}$$

such that for every two disjoint sets of vertices A, B the function $f_{A,B} : X \to \{0,1\}$ outputs one given a pair $(u, v) \in X$ with one endpoint in A and one endpoint in B and zero otherwise.

Our first observation is that a graph G is ε-quasirandom in the definition of Chung, Graham, Wilson and Thomason if and only μ_G is ε-pseudorandom for $\mathcal{F}$.

Now, if we apply the Weak Regularity Lemma to G with parameter ε, and we let G' be the resulting approximating graph, then we see that the weak regularity property of the partition defining G' is equivalent to the fact that μ_G is indistinguishable from $\mu_{G'}$ by $\mathcal{F}$. Recall also that G' is a disjoint union of bipartite complete graphs and, if one thinks about this for a minute, we can write

$$\mu_{G'}(u, v) := h\big(f_{A_1,B_1}(u, v), \cdots, f_{A_k,B_k}(u, v)\big)$$

for a proper combining function h. From the proof of the weak regularity lemma, it actually follows that k can be chosen to be polynomial in $1/\varepsilon$. This means that not only is $\mu_{G'}$ indistinguishable from μ_G by the "cuts" in $\mathcal{F}$, but the indistinguishable approximation itself is constructed out of cuts.

Something similar is true for the "strong" regularity lemma.

Theorem 13 (Strong Regularity Lemma – Indistinguishability Formulation)**.** *For every function F and ε there is a k such that for every graph G there is a graph $\tilde{G}$ that differs from G in only an ε fraction of edges and such that there is an approximating graph G' such that μ_G and $\mu_{G'}$ are $1/F(k, \varepsilon)$-indistinguishable by $\mathcal{F}$, and*

$$\mu_{G'}(u, v) := h\big(f_1(u, v), \cdots, f_k(u, v)\big)$$

for functions $f_1, \ldots, f_k \in \mathcal{F}$.

If we think of the functions in $\mathcal{F}$ as "efficient algorithms", then the Lemma is saying that every distribution μ_G is indistinguishable from a distribution $\mu_{G'}$ that is "efficiently computable."

As we shall see in Section 7, an analogous statement can be formulated in a much more general setting, with no reference to graphs.

6. Norms and Local Checkability

The *Gowers uniformity norms* [15, 16] have several applications in additive combinatorics and are powerful measures of "quasirandomness" for functions.

Definition 14 (Gowers Uniformity Norms). If G is a finite abelian group and k is an integer parameter, then the k-th Gowers norm of a function $g : G \to \mathbb{R}$ is defined as

$$\|g\|_{U^k} := \left(\mathop{\mathbb{E}}_{x,y_1,\ldots,y_k} \prod_{S \subseteq [k]} g\left(x + \sum_{i \in S} y_i\right)\right)^{1/2^k}.$$

For complex-valued functions $g : G \to \mathbb{C}$, the definition is analogous but every occurrence of $g\big(x + \sum_{i\in S} y_i\big)$ is replaced by the conjugate

$$\overline{g\left(x + \sum_{i \in S} y_i\right)}$$

when $|S|$ is odd.

Gowers introduced the uniformity norms in his work [15, 16] on an analytic proof of Szemerédi's theorem that every set of integers of positive density contains arbitrarily long arithmetic progressions.

The connection comes from the following fact, which is provable using several applications of Cauchy–Schwarz.

Definition 15. For a function $g : G \to \mathbb{R}$ and a positive integer k, let

$$\lambda_k(g) := \mathop{\mathbb{E}}_{x,y} g(x) \cdot g(x+y) \cdot g(x+2y) \cdots g(x+(k-1)y)$$

where ty stands for $y + y + \cdots + y$ added t times.

If $A \subseteq \mathbb{Z}/N\mathbb{Z}$, then the number of length-k arithmetic progressions in A is $N^2 \cdot \lambda_k(1_A)$. (There are a couple of caveats: we are counting arithmetic progressions mod N, and we are counting the trivial progressions in which the increment is zero.)

Theorem 16 [16]. [2] *For every two bounded functions $g, h : G \to [0,1]$ we have*

$$\left|\lambda_k(g) - \lambda_k(h)\right| \leq 2^k \|g - h\|_{U^{k-1}}.$$

The above theorem is interesting in the following special case: suppose $A \subseteq G$ is a subset of G of density δ, and suppose that $\|1_A - \delta\|_{U^{k-1}}$ is small (compared to $\delta^k/2^k$); then the number of length-k arithmetic progressions in A, which is $|G|^2\lambda_k(1_A)$, is close to $|G|^2\lambda_k(\delta) = |G|^2 \cdot \delta^k$, which is the number of length-k arithmetic progressions that one expects to find in a random set of density δ. In particular, A does contain arithmetic progressions.

This notion is used in Gowers's proof of Szemeredi's theorem within an iterative argument. Given a sufficiently large subset $A \subseteq \mathbb{Z}/N\mathbb{Z}$, if $\left\|1_A - |A|/N\right\|_{U^{k-1}}$ is small, then there is nothing to prove. If $\left\|1_A - |A|/N\right\|_{U^{k-1}}$ is large, then one shows that finding a length-k progression in A reduces to finding a length-k progression in $A' \subseteq \mathbb{Z}/N'\mathbb{Z}$, where $N' = N^{\Omega(1)}$ and A' has density $\delta' \geq \delta + \delta^{O(1)}$. This step is by far the hardest one in the paper.

If two sets $A, B \subseteq G$ are such that $\|1_A - 1_B\|_{U^k}$ is small, then the two sets have approximately the same number of length-$(k+1)$ arithmetic progressions, which is, in turn, a form of "indistinguishability", against "adversaries" that count the number of arithmetic progressions in a set.

However, neither the property of having approximately the same number of arithmetic progressions, nor the stronger property of being close in Gowers norm, is, seemingly, a property that can be formulated in terms of indistinguishability, as defined in Section 2. Recall that two sets are indistinguishable (for a class of adversaries) if a random sample from a set is indistinguishable from a random sample from the other set (by the adversaries in the class). The number of arithmetic progressions in a set can certainly not be estimated by looking at one sample, so it does not seem that there could be a class of adversaries such that indistinguishability under the class captures closeness under Gowers norms.

The existence of such a set of adversaries, however, is the subject of the *Gowers inverse conjectures*, which were formulated in [18, 33], and whose proof has recently been announced [5, 45, 21, 22], except for certain cases of the original conjecture that have been shown to be false [19, 28].

[2] Gowers proves the special case of the above theorem in which $h := 0$ as [16, Theorem 3.2], and derives the special case $h := \mathbb{E}\, g$ as [16, Corollary 3.3]. The above slightly more general statement has a proof similar to the one of [16, Corollary 3.3].

Theorem 17 (Gowers Inverse Theorem for Finite Fields [5, 45]). *Let $g : \mathbb{F}_p^n \to \mathbb{C}$ be such that $|g(x)| \leq 1$ for every x, and let $k < p$. Then if*

$$\|g\|_{U^k} \geq \varepsilon$$

there is a polynomial q of degree $\leq k-1$ over $\mathbb{F}_p$ such that

$$|\langle g, \omega^q \rangle| := \left| \mathop{\mathbb{E}}_x g(x) \cdot \omega^{q(x)} \right| \geq \varepsilon'$$

where ω is a primitive p-th root of unity and ε' depends only on p, ε, k and is independent of n and of g.

Theorem 18 (Gowers Inverse Theorem in Cyclic Groups [21, 22]). *Let $g : \mathbb{Z}/N\mathbb{Z} \to \mathbb{C}$ be such that $|g(x)| \leq 1$ for every x, and let $k < p$. Then if*

$$\|g\|_{U^k} \geq \varepsilon$$

there is a degree-$(k-1)$ nilsequence of bounded complexity $h : \mathbb{Z}/N\mathbb{Z} \to \mathbb{C}$ such that

$$|\langle g, h \rangle| := \left| \mathop{\mathbb{E}}_x g(x) \cdot h(x) \right| \geq \varepsilon'$$

where ε' depends only on p, ε, k and is independent of n and of g.

It would take us too far to define the notion of bounded-degree and bounded-complexity nilsequence. We state the theorem to illustrate how it claims the existence of a class of "adversaries" such that if g_1, g_2 are indistinguishable by such adversaries then g_1, g_2 are close in Gowers uniformity norm.

7. Decomposition Results

Several results in additive combinatorics are based on "decomposition" results, showing that every bounded function can be written as a "sum" of a "structured part" which enjoys a nice algebraic structure, and a "uniform part" which only gives a negligible contribution to certain quantitative properties of the function that we are trying to bound.

As a simple example, suppose $g : \mathbb{F}_2^n \to [-1, 1]$ is any bounded function, and that, for every $r \in \mathbb{F}_2^n$, the function $\chi_r : \mathbb{F}_n^2 \to \{-1, 1\}$ is the character $\chi_r(x) := (-1)^{\langle r, x \rangle}$. Then we have the Fourier expansion

$$g(x) = \sum_r \hat{g}(r) \chi_r(x)$$

where

$$\hat{g}_r(x) := \langle g, \chi_r \rangle = \mathop{\mathbb{E}}_{x \in \mathbb{F}_2^n} g(x)\chi_r(x).$$

After we fix an approximation parameter ε, we can write

$$g(x) = \left(\sum_{r : |\hat{g}(r)| \geq \varepsilon} g(r)\chi_r(x) \right) + \left(\sum_{r : |\hat{g}(r)| < \varepsilon} g(r)\chi_r(x) \right)$$

where the first term has the property of being a weighted sum of at most ε^{-2} characters, and the second term is a function all whose Fourier coefficients are at most ε, and has small second Gowers norm.

This has a natural generalization, whose proof follows an argument of Frieze and Kannan [8], and which I have learned in this setting from a lecture of Tao [40]:

Theorem 19 ("Very Weak" Regularity Lemma for Functions). *Let $\mathcal{F}$ be any set of bounded functions $f : X \to [-1, 1]$ and $g : X \to \mathbb{R}$ be a function such that $\mathbb{E}_{x \sim X}\, g^2(x) \leq 1$. Then we can write*

$$g := g_s + g_u$$

where g_s is a structured *function such that*

$$g_s(x) := \sum_{i=1}^{k} c_i f_i(x)$$

where $k \leq \varepsilon^{-2}$, $f_i \in \mathcal{F}$, $c_i \leq 1$, and $\sum_i |c_i| \leq \varepsilon^{-1}$; and g_u is a uniform *function such that*

$$\forall f \in \mathcal{F}. \quad \left| \mathop{\mathbb{E}}_{x \in X} g_u(x) f(x) \right| \leq \varepsilon.$$

Note that the theorem specializes to the trivial Fourier decomposition when the functions in $\mathcal{F}$ are the characters of $\mathbb{F}_2^n$, but that the proof makes no assumptions whatsoever about the functions in $\mathcal{F}$.

Proof. We give an iterative construction of a sequence of "candidates" for g_s, and we use a potential function argument to show that the process must quickly converge to a function satisfying the conclusions of the theorem.

We assume without loss of generality that if $f \in \mathcal{F}$ then $-f \in \mathcal{F}$. (Otherwise we will work with the "closure of $\mathcal{F}$ under negation" and then switch the signs of some of the coefficients c_i if needed.)

We initialize $g_0(x) := 0$ for all x. At step i, if g_i is such that

$$\forall f \in \mathcal{F} \quad \left| \mathop{\mathbb{E}}_{x \in X} \left(g(x) - g_i(x) \right) \cdot f(x) \right| \leq \varepsilon$$

then we output $g_s := g_i$ and $g_u := g - g_i$. Otherwise, let $f_{i+1} \in \mathcal{F}$ be a function such that

$$\mathop{\mathbb{E}}_{x \in X} \left(g(x) - g_i(x) \right) \cdot f_{i+1}(x) \geq \varepsilon$$

then define

$$g_{i+1} := g_i + \varepsilon f_{i+1}.$$

We want to show that the algorithm terminates within ε^{-2} steps. To see this, consider the "energy"

$$E_i := \mathop{\mathbb{E}}_{x \in X} \left(g(x) - g_i(x) \right)^2$$

Initially, $E_0 := \mathbb{E}_{x \in X}\, g^2(x) \leq 1$. We claim that, at every step, the energy decreases by at least ε^2. To see this:

$$\begin{aligned}
&E_i - E_{i+1} \\
&= \mathbb{E} \left(g(x) - g^i(x) \right)^2 - \left(g(x) - g^i(x) - \varepsilon f_{i+1}(x) \right)^2 \\
&= 2\, \mathbb{E} \left(g(x) - g^i(x) \right) \varepsilon f_{i+1}(x) - \varepsilon^2 f_{i+1}^2(x) \\
&\geq \varepsilon^2. \qquad \blacksquare
\end{aligned}$$

Together with the Gowers inverse theorems, it gives the following corollary (cf. [24, 23]).

Corollary 20. *Let q be a prime, $g : \mathbb{F}_q^n \to \mathbb{R}$ be a function such that $\mathbb{E}_{x \sim X}\, g^2(x) \leq 1$, $\varepsilon > 0$ be an approximation parameter and $2 \leq k < q$ be an integer. Then there is a $C = C(q, k, \varepsilon)$ such that we can write*

$$g := g_s + g_u$$

where g_s is a structured *function such that*

$$g_s(x) := \sum_{i=1}^{k} c_i w^{p_i(x)}$$

where ω *is a* q*-th primitive root of unity,* $k \leq C^2$, p_i *is a degree-*$(k-1)$ *polynomial over* $\mathbb{F}_q$, $c_i \leq 1$, *and* $\sum_i |c_i| \leq C$; *and* g_u *is a* uniform *function such that*

$$\|g_u\|_{U^k} \leq \varepsilon.$$

Theorem 19 does not imply the weak regularity lemma for graphs, but it shows that for every graph G there is a graph G' which is a *weighted sum* of $O(\varepsilon^{-2})$ bipartite complete graphs and such that G' is an ε-approximation to G. This also implies that there is a partition of V into at most $2^{\varepsilon^{-2}}$ sets such that the weight of an edge in G' is entirely determined by the blocks of the partition to which its endpoints belong.[3] It does not follow, however, that G' is a weighted graph in which all weights are between 0 and 1, so we do not get a result in terms of indistinguishability.

The following theorem, which is folklore, is the general analog.

Theorem 21 (Weak Regularity Lemma for Functions). *Let* $\mathcal{F}$ *be any set of boolean functions* $f : X \to \{-1,1\}$ *and* $g : X \to [-1,1]$ *be a bounded function. Then we can write*

$$g := g_s + g_u$$

where g_s *is a* structured *function such that*

$$g_s(x) := h\big(f_1(x), \ldots, f_k(x)\big)$$

where $k \leq \varepsilon^{-O(1)}$ $h : \{-1,1\}^k \to [-1,1]$ *is a combining function; and* g_u *is a* uniform *function such that*

$$\forall f \in \mathcal{F}. \quad \Big|\mathop{\mathbb{E}}_{x \in X} g_u(x) f(x)\Big| \leq \varepsilon.$$

One way to prove Theorem 21 is to follow rather closely the proof of the weak regularity lemma. Thus one obtains $k = O\big(\varepsilon^{-2}\big)$, and h has the property that

$$h(b_1, \ldots, b_k) = \mathbb{E}\left[g(x) \mid f_1(x) = b_1 \wedge \cdots \wedge f_k(x) = b_k\right].$$

This theorem has the advantage that h is bounded, so it gives the Weak Regularity Lemma as stated in Section 4. It also abstracts more general

[3]Frieze and Kannan prove such a result for graph, and indeed the proof of Theorem 19 is an adaptation of their argument.

statements, such as the analytic statement of the Weak Regularity Lemma proved by Lovász and Szegedy [29].

Since Theorem 21 works with any collection $\mathcal{F}$ of functions, it is interesting to see what it implies if one chooses it to be a set of efficiently computable functions, for example the set of all functions computable by circuits of size $\leq s$. Then, if D is a probability distribution over $\{0,1\}^n$ of min-entropy at least $n-\ell$ define $g := 2^{n-\ell} \cdot D$ and apply the theorem. We see that there exists a function g' which is the composition of $O(\varepsilon^{-2})$ circuits of size $\leq s$, and such that the distribution $D' := g'/2^{n-\ell}$ is indistinguishable from D up to $\varepsilon 2^{\ell}$. This gives an efficiently computable (and, because of the high min-entropy, efficiently samplable) distribution D' that is indistinguishable from D, leading to the conclusion that every high min-entropy distribution is indistinguishable from a samplable high min-entropy distribution. Unfortunately, the combining function h in Theorem 21 might have complexity exponential in $1/\varepsilon$, and so the theorem cannot be applied to the interesting setting in which ε is smaller than $1/n$.

In [43], we give a new proof of Theorem 21 in which the combining function is computable in linear time.

Theorem 22 (Weak Regularity Lemma for Functions – Polynomial Efficiency). *Let $\mathcal{F}$ be any set of boolean functions $f : X \to \{-1,1\}$ and $g : X \to [-1,1]$ be a bounded function. Then we can write*

$$g := g_s + g_u$$

where g_s is a structured *function such that*

$$g_s(x) := \min\left\{1, \max\left\{-1, \sum_i c_i f_i(x)\right\}\right\}$$

where $k \leq \varepsilon^{-O(1)}$, $f_i \in \mathcal{F}$, $c_i \leq 1$; and g_u is a uniform *function such that*

$$\forall f \in \mathcal{F}. \quad \left| \mathop{\mathbb{E}}_{x \in X} g_u(x) f(x) \right| \leq \varepsilon.$$

This theorem gives

Corollary 23. *Let D be any distribution over $\{0,1\}^n$ of min-entropy at least $n-t$. Then there is a distribution D', also of min-entropy at least $n-t$ such that D and D' are ε-indistinguishable by circuits of size $\leq s$ and $D'(x)$ is computable by a circuit of size $O(2^{2t}\varepsilon^{-2}s)$.*

In particular, D' is samplable by a small circuits, and hence D is simulable.

It would be desirable if D' was computable and samplable in less complexity than the distinguisher, but this is easily seen to be impossible.

It is possible, however, to give the following.

Theorem 24 (Strong Regularity Lemma for Functions). *Let $\mathcal{F}$ be any set of boolean functions $f : X \to \{-1, 1\}$ F be any function, ε an approximation parameter.*

Then there is a $k = k(\varepsilon, F)$ such that if $g : X \to [-1, 1]$ is a bounded function, we can write

$$g := g_s + g_u + g_e$$

where g_s is a structured *function such that*

$$g_s(x) := h\big(f_1(x), \ldots, f_k(x)\big)$$

g_u is a uniform *function such that*

$$\forall f \in \mathcal{F}. \quad \Big| \mathop{\mathbb{E}}_{x \in X} g_u(x) f(x) \Big| \leq 1/F(k, \varepsilon)$$

and g_e is bounded in ℓ_2

$$\mathbb{E}\, g_e^2(x) \leq \varepsilon.$$

This can be seen to yield the strong regularity lemma, taking X to be $\binom{V}{2}$, $g := 1_E$ to be the characteristic function of the edges of G, and then taking the weights of edges of $\tilde{G}$ to be $g_s + g_u$ and the regular approximation G' of $\tilde{G}$ to be g_s. It also implies the strong regularity lemma in [29].

While not explicitly stated in this form, Theorem 24 can be found in [38, 37]. Proposition 10.36 in [44] is the special case in which $X = \mathbb{F}_2^n$ and $\mathcal{F}$ is the set of Fourier characters. (The proof is easily adapted to the general case.)

8. Indistinguishability as a Norm

In this section we take the opposite perspective as the rest of the paper, and we show how to view indistinguishability as a norm.

8.1. Indistinguishability of Distributions

Recall that two distributions X, Y over Ω are ε-indistinguishable by $\mathcal{F}$ if for every $f \in \mathcal{F}$ we have

$$\left| \mathop{\mathbb{E}}_{z \sim X} f(z) - \mathop{\mathbb{E}}_{z \sim Y} f(z) \right| \leq \varepsilon$$

that is

$$\left| \sum_{z \in \Omega} X(z) f(z) - Y(z) f(z) \right| \leq \varepsilon$$

which is equivalent to

$$|\Omega| \cdot \left| \mathop{\mathbb{E}}_{z \in \Omega} \left(X(z) - Y(z) \right) \cdot f(z) \right| = |\Omega| \cdot \left| \langle X - Y, f \rangle \right| \leq \varepsilon.$$

Now, notice that if $\mathcal{F}$ is any family of functions $f : \Omega \to \mathbb{R}$, then

$$\|g\|_{\mathcal{F}} := \max_{f \in \mathcal{F}} \left| \langle g, f \rangle \right|$$

is a (semi-) norm over functions $g : \Omega \to \mathbb{R}$. Thus, we see that two distributions X, Y are ε-indistinguishable by $\mathcal{F}$ if and only if the functions X, Y have distance at most $\varepsilon / |\Omega|$ according to the norm $\| \cdot \|_{\mathcal{F}}$.

It will be convenient to say that two functions g_1, g_2 are ε-indistinguishable by $\mathcal{F}$ if $\|g_1 - g_2\|_{\mathcal{F}} \leq \varepsilon$, that is, if $\left| \langle g_1 - g_2, f \rangle \right| \leq \varepsilon$ for all $f \in \mathcal{F}$.

8.2. Indistinguishability of Sets

If A is a subset of a finite set Ω, then the distribution U_A is the uniform distribution over A, that is the distribution such that $U_A(x) = 1_A(x)/|A|$.

Definition 25 (Indistinguishability of Sets). We say that two sets $A, B \subseteq \Omega$ are ε-indistinguishable by $\mathcal{F}$ if the distributions U_A, U_B are ε-indistinguishable by $\mathcal{F}$.

Note that if A, B are indistinguishable, then, $\forall f \in \mathcal{F}$ we have

$$\left| \sum_{x \in \Omega} \frac{1_A(x)}{|A|} f(x) - \frac{1_B(x)}{|B|} f(x) \right| \leq \varepsilon$$

which is equivalent to

$$\forall f \in \mathcal{F}. \quad \left|\left\langle \frac{|\Omega|}{|A|} 1_A - \frac{|\Omega|}{|B|} 1_B, f \right\rangle\right| \leq \varepsilon.$$

9. Dense Models

In this section we give an overview of the proof of the Green–Tao theorem [20], state a key combinatorial step in the proof, and discuss a computational version of such a step. We will gloss over substantial parts of the proof, and we will be imprecise about certain technical issues (for example we will refer to certain functions as taking 0/1 values while in the proof they take fractional values).

A very high-level view of the proof is as follows: restricting to an interval $\{1,\ldots,N\}$, one shows that $P :=$ PRIMES $\cap \{1,\ldots,N\}$ is "modeled" by a set M of size $\Omega(N)$. By (a standard extension of) Szemeredi's theorem, M has $\Omega(N^2)$ length-k progressions, and then so does P because of the technical sense in which M is a "model" of P.

The proof is based on considering the set R of "almost primes" in $\mathbb{Z}/N\mathbb{Z}$, integers with few, large, prime factors, and exploiting the fact that much is known about the pseudorandomness of R. The idea is that if R is "indistinguishable" from $\mathbb{Z}/N\mathbb{Z}$, and P is a subset containing a constant factor of the elements of R, then there should be a subset $M \subseteq \mathbb{Z}/N\mathbb{Z}$ which is indistinguishable from P and that has about the same density in $\mathbb{Z}/N\mathbb{Z}$ as P does in R.[4]

A first difficulty in realizing this program is how to define the notion of indistinguishability between M and P such that the existence of $\Omega(N^2)$ length-k arithmetic progressions in M implies the existence of length-k arithmetic progressions in P. A simple choice would be, in order to deal with progressions of length k, to require $\|1_M - 1_P\|_{U^{k-1}}$ to be small, but this is impossible: it is easy to see that if two sets have different densities,

[4] Here is where we are going to be imprecise: from now on we refer to the characteristic functions 1_P and 1_R, which are boolean, while Green and Tao deal with fractional-valued functions which are related to, but different from, 1_P and 1_R.

then their characteristic functions are always far in uniformity norm. We could, however, hope for the "normalized" distance in U^{k-1} norm, that is

$$\left\| 1_M - \frac{|M|}{|P|} \cdot 1_P \right\|_{U^{k-1}} \tag{2}$$

to be small. This is indeed the approach taken by Green and Tao. An immediate difficulty is that the quantity in (2) being small does not immediately imply that if M has $\Omega(N^2)$ length-k arithmetic progressions then P has $\Omega\big(N^2 \cdot |P|^2/|M|^2\big) = \Omega\big(|P|^2\big)$ arithmetic progressions.

The problem is that the relation between $\lambda_k(f) - \lambda_k(g)$ and $\|f - g\|_{U^{k-1}}$ holds only when f, g are bounded functions, while $\frac{|M|}{|P|} \cdot 1_P$ takes values which are around $\log N$ in magnitude. Fortunately, based on the fact that $1_P \leq 1_R$ and that 1_R is known to satisfy certain strong pseudorandomness properties, it is possible to prove that if $\left\|1_M - \frac{|M|}{|P|} \cdot 1_P\right\|_{U^{k-1}}$ is small then $\lambda_k(1_M)$ is close to $\lambda_k\big((|M|/|P|) \cdot 1_P\big)$.

With these preliminaries, the goal is to find a set M of size $\Omega(N)$ such that $\left\|1_M - \frac{|M|}{|P|} 1_P\right\|_{U^{k-1}}$ is small. The next step[5] is to reduce this problem to finding a set M of size $\Omega(N)$ such that the set M and P are *indistinguishable* for a certain class of adversaries.

The challenge, here, is to find a collection $\mathcal{F}$ of bounded functions such that if

$$\left| \left\langle 1_M - \frac{|M|}{|P|} 1_P, f \right\rangle \right|$$

is small for every $f \in \mathcal{F}$, then $\left\|\langle 1_M - \frac{|M|}{|P|} 1_P\right\|_{U^{k-1}}$ is small.

One approach would be via the Gowers inverse conjecture, taking $\mathcal{F}$ to be the set of degree-$(k-2)$ nilsequences, but this runs into two problems: one, which is not hard to overcome, is that the Gowers inverse conjecture holds for bounded functions, while $\frac{|M|}{|P|} 1_P$ takes very large values. More importantly, the Gowers inverse conjecture was still open at the time of the work of Green and Tao, and so they had to find a different sufficient condition that implies a small norm.

Consider the following definition: if $g : G \to \mathbb{R}$ is a function, define its *k-th dual* $\tilde{g}_k : G \to \mathbb{R}$ to be

$$\tilde{g}_k := (x) \mathop{\mathbb{E}}_{y_1,\dots,y_k} \prod_{S \subseteq [k],\ S \neq \emptyset} g(x + \sum_{i \in S} y_i).$$

[5]The order of the presentation in this section is not the order of presentation in [20], nor probably the order in which these ideas where developed.

Now we notice that

$$\|g\|_{U^k}^{2^k} = \mathop{\mathbb{E}}_x g(x) \cdot \tilde{g}_k(x) = \langle g, \tilde{g}_k \rangle.$$

This implies that if, for example, we take $\mathcal{F}$ to be *the set of k-th dual function $\tilde{g}_k$ of all bounded functions $g : G \to [-1,1]$* we get that for every two bounded functions $g, h : G \to [0,1]$, the "indistinguishability" condition that

$$\forall f \in \mathcal{F}. \big| \langle g - h, f \rangle \big| \leq \varepsilon$$

implies that Gowers uniformity closeness condition that

$$\|g - h\|_{U^k} \leq \varepsilon^{1/2^k}.$$

In our setting, $1_M - \big(|M|/|P|\big) \cdot 1_P$ is not a bounded function. We know, however, that $1_P \leq 1_R$, because P is a subset of R, and we are looking for a construction in which $|M|/|N| = |P|/|R|$, so at least we know that

$$\big|1_M - \big(|M|/|P|\big) \cdot 1_P\big| \leq 1 + \frac{|N|}{|R|} \cdot 1_R.$$

Green and Tao then proceed by defining $\mathcal{F}$ as the set of all dual functions $\tilde{g}_k$ such that $g : G \to \mathbb{R}$ is a function such that $\big|g(x)\big| \leq 1 + \frac{|N|}{|R|} \cdot 1_R$. If we find a set M of size $\Omega(N)$ such that

$$\forall f \in \mathcal{F}. \quad \big|\langle 1_M - \big(|N|/|R|\big) \cdot 1_P \rangle\big|$$

is small, then we are done.

This approach might seem hopeless, because the set of dual functions of essentially arbitrary functions seems itself to be an essentially arbitrary set. It turns out, however, that the set does contain some structure. For starters, Green and Tao are able to prove that all functions in $\mathcal{F}$ take value that are upper bounded by an absolute constant. (Even though they are dual functions of functions that can take values as large as about $\log N$.)

Having reduced to a problem about indistinguishability, the next step is to prove the following general theorem.

Theorem 26 (Dense Model Theorem [20, 46]). *Let X be a finite set of size N, $\mathcal{F}$ a set of bounded functions $f : X \to [0,1]$, $\varepsilon > 0$ be an approximation parameter $\nu : X \to \mathbb{R}_{\geq 0}$ be a measure such that $\sum_x \nu(x) = |X|$, $d : X \to \mathbb{R}_{\geq 0}$ be such that $d(x) \leq \nu(x)$ for all X and $\sum_x d(x) = \delta N$.*

Then (at least) one of the following two conditions holds;

- ν is not pseudorandom, *that is, there are functions* $f_1, \ldots, f_k \in \mathcal{F}$ *such that*

$$\left|\left\langle \nu - 1, \prod_{i=1}^{k} f_i \right\rangle\right| \geq \frac{1}{\exp\left(\varepsilon^{-O(1)}\delta^{-O(1)}\right)}.$$

- d has a dense model, *that is, there is a bounded measure* $m : X \to [0,1]$ *such that* $\sum_x m(x) \geq \delta N/2$ *and such that*

$$\forall f \in \mathcal{F}. \quad \left|\langle m - d, f\rangle\right| \leq \varepsilon.$$

The Theorem is applied in the following way: we set $\nu := \left(|N|/|R|\right) \cdot 1_R$, $d := \left(|N|/|R|\right) \cdot 1_P$, and $\mathcal{F}$ as above. Then one needs to prove that for every $f_1, \ldots, f_k \in \mathcal{F}$, the inner product $\left|\langle \nu - 1, \prod_i f_i\rangle\right.$ is small, which is the same as saying that the set R is indistinguishable from $\mathbb{Z}/N\mathbb{Z}$ by small products of functions from $\mathcal{F}$. This is the number-theoretic part of [20].

This implies that the second condition is true, that is there is $m : X \to [0,1]$ such that m and d are indistinguishable by $\mathcal{F}$. This means that $\|m - d\|_{U^k}$ is small, and so $\lambda_{k+1}(m)$ and $\lambda_{k+1}(d)$ are close. By Szemeredi's theorem, $\lambda_{k+1}(m) \geq \Omega(1)$, and so $\lambda_{k+1}\left(\left(|N|/|R|\right)1_P\right) \geq \Omega(1)$, and so P has $\Omega\left(N^2/\left(|N|/|R|\right)^{k+1}\right)$ length-$(k+1)$ progressions, that is $\Omega\left(N^2/(\log N)^{k+1}\right)$.

If one prefers to talk about *distributions* instead of measures, then the Dense Model Theorem has an equivalent restatement as follows.

Say that a distribution A is δ-*dense* in a distribution B if for every x we have $A(x) \leq B(x)/\delta$. The motivation for the definition is that A is the uniform distribution over a set S and B is the uniform distirbution over a set T, then A is δ-dense in B if and only if S is a subset of T of density δ.

Theorem 27 (Dense Model Theorem for Distributions). *Let X be a finite set, $\mathcal{F}$ a set of bounded functions $f : X \to [0,1]$, $\varepsilon > 0$ be an approximation parameter R be a distribution over X, D be another distribution over X that is δ-dense in R.*

Then (at least) one of the following two conditions holds;

- R is not pseudorandom, *that is, there are functions* $f_1, \ldots, f_k \in \mathcal{F}$, $k = \varepsilon^{-O(1)} \cdot \delta^{-O(1)}$, *such that*

$$\left|\mathop{\mathbb{E}}_{x\sim X} \prod_i f_i(x) - \mathop{\mathbb{E}}_{x\sim R} \prod_i f_i(x)\right| \geq \frac{1}{\exp\left(\varepsilon^{-O(1)}\delta^{-O(1)}\right)}.$$

- D has a dense model, *that is, there is a distribution M over X which is $\delta/2$-dense in the uniform distribution over X such that M and D are ε-indistinguishable by $\mathcal{F}$.*

Note that we may apply the theorem to $\mathcal{F}$ being a class of bounded-size circuits, or any other complexity class, and so derive that if R is pseudorandom and D is dense in R then D is indistinguishable from a high-entropy distribution. There is, however, the undesirable exponential loss in the pseudorandomness parameter of R versus the indistinguishability parameter of R.

Looking into the proof of Green, Tao and Ziegler, one sees that the distinguishing probability in the first case can be made polynomial in $1/\varepsilon$ and $1/\delta$ if one allows the distinguisher to be not just a product of functions from $\mathcal{F}$, but a function of the form $h\big(f_1(x),\dots,f_k(x)\big)$, where the functions f_i come from $\mathcal{F}$. Unfortunately, however, the combining function h has complexity exponential in k, so this just transfers the exponential dependency from the approximation parameter to the complexity parameter.

Using the Hanh–Banach theorem, instead of the iterative partitioning technique of Green, Tao and Ziegler, it is possible to provide an alternative proof which works better in the computational setting.

Theorem 28 (Dense Model Theorem – Efficient Version). *Let X be a finite set, $\mathcal{F}$ a set of bounded functions $f : X \to [0,1]$, $\varepsilon > 0$ be an approximation parameter R be a distribution over X, D be another distribution over X that is δ-dense in R.*

Then (at least) one of the following two conditions holds;

- R is not pseudorandom, *that is, there are functions $f_1,\dots,f_k \in \mathcal{F}$, $k = \varepsilon^{-O(1)} \cdot \delta^{-O(1)}$, and a function h which is a combination of sums and thresholds such that*

$$\Big| \mathop{\mathbb{E}}_{x\sim X} h\big(f_1(x),\dots,f_k(x)\big) - \mathop{\mathbb{E}}_{x\sim R} h\big(f_1(x),\dots,f_k(x)\big) \Big| \geq \frac{1}{\varepsilon^{-O(1)}\delta^{-O(1)}}.$$

- d has a dense model, *that is, there is a distribution M over X which is δ-dense in the uniform distribution over X such that M and D are ε-indistinguishable by $\mathcal{F}$.*

In particular, if $\mathcal{F}$ is the set of functions computable by circuits of size s, we get that for every distribution R that is ε-pseudorandom for circuits of size $s \cdot \varepsilon^{-O(1)}\delta^{-O(1)}$ and for every distribution D that is δ-dense in R, we can find a model distribution M that is δ-dense in the uniform distribution and which is ε-indistinguishable from D by circuits of size s.

The proof was discovered independently by Gowers [17] and by Reingold, Tulsiani, Trevisan and Vadhan [32]. The computational statement was also formulated independently by Dziemboksi and Pietrzak [7], although there is a gap in their published proof.[6] The proof in [32] is motivated by the proof of a result of Impagliazzo [26], the Impagliazzo Hardcore Set Lemma. Impagliazzo [27] proves that, in fact, Theorem 28 can be derived from the *statement* of his Hardcore Set Lemma, and that one can prove a stronger version in which either D has a dense model or it fails a certain "pseudodensity" condition which is a stronger statement than saying that is not dense in a pseudorandom distribution.

Theorem 28 has the following equivalent formulation: suppose we have a pseudorandom generator, that is a length-increasing procedure $G : \{0,1\}^\ell \to \{0,1\}^n$ such that $G(U_\ell)$ is ε-pseudorandom for circuits of size s. Suppose that B is a distribution of entropy $\ell - t$; then $G(B)$ is indistinguishable from a model distribution of entropy $n - t$. This means that, for example, if an adversary is able to bias (or know) part of the distribution of seeds used by the generator, the output of the generator is not completely compromised, but remains indistinguishable from a high-entropy distribution.

This is the starting point for the Dziemboksi and Pietrzak [7] construction of "leakage-resilient" cryptosystems, which remain secure even if parts of the keys become known to an adversary. The dense model theorem has also been used in the study of computational definitions of privacy [30].

References

[1] Sanjeev Arora and Boaz Barak, *Computational Complexity: A Modern Approach,* Cambridge University Press, 2009.

[2] Noga Alon, Eldar Fischer, Michael Krivelevich and Mario Szegedy, Efficient testing of large graphs, *Combinatorica,* **20(4)** (2000), 451–476.

[6]They attribute a result to [4] which is stronger than the result appearing there, and then derive the dense model theorem from the stronger statement. The gap between the statement in [4] and the stronger one which implies the dense model theorem can be filled with the techniques in Gowers [17] and Reingold, Tulsiani, Trevisan, Vadhan [32].

[3] Manuel Blum and Silvio Micali, How to generate cryptographically strong sequences of pseudorandom bits, *SIAM Journal on Computing,* **13(4)** (1984), 850–864. Preliminary version in *Proc. of FOCS'82.*

[4] Boaz Barak, Ronen Shaltiel and Avi Wigderson, Computational analogues of entropy, in: *Proceedings of RANDOM'03,* 2003, pages 200–215.

[5] Vitaly Bergelson, Terence Tao and Tamar Ziegler, An inverse theorem for the uniformity seminorms associated with the action of f^{ω}, arXiv:0901.2602, 2009.

[6] Fan R. K. Chung, Ronald L. Graham and Richard M. Wilson, Quasi-random graphs, *Combinatorica,* **9(4)** (1989), 345–362.

[7] Stefan Dziembowski and Krzysztof Pietrzak, Leakage-resilient cryptography, in: *Proceedings of the 49th IEEE Symposium on Foundations of Computer Science,* 2008, pages 293–302.

[8] Alan M. Frieze and Ravi Kannan, Quick approximation to matrices and applications, *Combinatorica,* **19(2)** (1999), 175–220.

[9] Oded Goldreich, Shafi Goldwasser and Dana Ron, Property testing and its connection to learning and approximation, *Journal of the ACM,* **45(4)** (1998), 653–750.

[10] Shafi Goldwasser and Silvio Micali, Probabilistic encryption, *Journal of Computer and System Sciences,* **28(2)** (1984), 270–299. Preliminary Version in *Proc. of STOC'82.*

[11] Shafi Goldwasser, Silvio Micali and Charles Rackof, The knowledge complexity of interactive proof systems, *SIAM Journal on Computing,* **18(1)** (1989), 186–208. Preliminary version in *Proc of STOC'85.*

[12] Oded Goldreich, S. Micali and A. Wigderson, How to play any metal game, in: *Proceedings of 19th Symposium on Theory of Computing,* 1987, pages 218–229.

[13] Oded Goldreich, *Modern Cryptography, Probabilistic Proofs and Pseudorandomness,* Springer-Verlag, 1999.

[14] Timothy Gowers, Lower bounds of tower type for Szemerédi's uniformity lemma, *Geometric and Functional Analysis,* **7(2)** (1997), 322–337.

[15] Timothy Gowers, A new proof of Szeméredi's theorem for progressions of length four, *Geometric and Functional Analysis,* **8(3)** (1998), 529–551.

[16] Timothy Gowers, A new proof of Szeméredi's theorem, *Geometric and Functional Analysis,* **11(3)** (2001), 465–588.

[17] Timothy Gowers, Decompositions, approximate structure, transference, and the Hahn–Banach theorem, arXiv:0811.3103, 2008.

[18] Ben Green and Terence Tao, An inverse theorem for the Gowers U^3 norm, math.NT/0503014, 2005.

[19] Ben Green and Terence Tao, The distribution of polynomials over finite fields, with applications to the Gowers norms, arXiv:0711.3191, 2007.

[20] Ben Green and Terence Tao, The primes contain arbitrarily long arithmetic progressions, *Annals of Mathematics,* **167** (2008), 481–547.

[21] Ben Green, Terence Tao and Tamar Ziegler, An inverse theorem for the Gowers U^4 norm, arXiv:0911.5681, 2009.

[22] Ben Green, Terence Tao and Tamar Ziegler, An inverse theorem for the Gowers U^k norm. Preprint, 2009.

[23] Timothy Gowers and Julia Wolf, Linear forms and higher-degree uniformity for functions on $\mathbb{F}_p^n$, arXiv:1002.2208, 2010.

[24] Timothy Gowers and Julia Wolf, Linear forms and quadratic uniformity for functions on $\mathbb{F}_p^n$, arXiv:1002.2209, 2010.

[25] Johan Håstad, Russell Impagliazzo, Leonid Levin and Michael Luby, A pseudorandom generator from any one-way function, *SIAM Journal on Computing,* **28(4)** (1999), 1364–1396.

[26] Russell Impagliazzo, Hard-core distributions for somewhat hard problems, in: *Proceedings of the 36th IEEE Symposium on Foundations of Computer Science,* 1995, pages 538–545.

[27] Russell Impagliazzo, Personal Communication, 2008.

[28] Shachar Lovett, Roy Meshulam and Alex Samorodnitsky, Inverse conjecture for the Gowers norm is false, in: *Proceedings of the 40th ACM Symposium on Theory of Computing,* 2008, pages 547–556.

[29] László Lovász and Balázs Szegedy, Szemerédis lemma for the analyst, *Geometric And Functional Analysis,* **17** (2007), 252–270.

[30] Ilya Mironov, Omkant Pandey, Omer Reingold and Salil P. Vadhan, Computational differential privacy, in: *Proceedings of CRYPTO'09,* 2009, pages 126–142.

[31] Imre Ruzsa and Endre Szemerédi, Triple systems with no six points carrying three triangles, in: *Proceedings of the Fifth Hungarian Colloquium on Combinatorics,* 1976, pages 939–945. Volume II.

[32] Omer Reingold, Luca Trevisan, Madhur Tulsiani and Salil Vadhan, Dense subsets of pseudorandom sets, in: *Proceedings of the 49th IEEE Symposium on Foundations of Computer Science,* 2008, pages 76–85.

[33] Alex Samorodnitsky, Low-degree tests at large distances, in: *Proceedings of the 39th ACM Symposium on Theory of Computing,* 2007, pages 506–515.

[34] Claude Shannon, Communication theory of secrecy systems, *Bell System Technical Journal,* **28** (1949), 656–715.

[35] Endre Szemerédi, On sets of integers containing no four elements in arithmetic progression, *Acta Math. Acad. Sci. Hung.,* **20** (1969), 89–104.

[36] Endre Szemerédi, On sets of integers containing no k elements in arithmetic progression, *Acta Arithmetica,* **27** (1975), 199–245.

[37] Terence Tao, The ergodic and combinatorial approaches to Szemerédi's theorem, arXiv:math.CO/0604456, 2006.

[38] Terence Tao, A quantitative ergodic theory proof of Szemerédi's theorem, *Electronic Journal of Combinatorics,* **13(4)** (2006).

[39] Terence Tao, Szemerédi's regularity lemma revisited, *Contributions to Discrete Mathematics,* **1** (2006), 8–28.

[40] Terence Tao, Structure and randomness in combinatorics, in: *Proceedings of the 48th IEEE Symposium on Foundations of Computer Science,* 2007, pages 3–18.

[41] A. G. Thomason, Pseudo-random graphs, *Annals of Discrete Mathematics,* **33** (1985), 307–331.

[42] Luca Trevisan, Pseudorandomness and combinatorial constructions, in: *Proceedings of the International Congress of Mathematicians,* 2006.

[43] Luca Trevisan, Madhur Tulsiani and Salil Vadhan, Regularity, boosting, and efficiently simulating every high-entropy distribution, in: *Proceedings of the 24th IEEE Conference on Computational Complexity,* 2009.

[44] Terence Tao and Van Vu, *Additive Combinatorics,* Cambridge University Press, 2006.

[45] Terence Tao and Tamar Ziegler, The inverse conjecture for the Gowers norm over finite fields via the correspondence principle, arXiv:0810.5527, 2008.

[46] Terence Tao and Tamar Ziegler, The primes contain arbitrarily long polynomial progressions. *Acta Mathematica,* **201** (2008), 213–305.

[47] Andrew C. Yao, Theory and applications of trapdoor functions, in: *Proceedings of the 23th IEEE Symposium on Foundations of Computer Science,* 1982, pages 80–91.

Luca Trevisan

U.C. Berkeley
Computer Science Division

e-mail: luca@cs.berkeley.edu

BOLYAI SOCIETY
MATHEMATICAL STUDIES, 21

An Irregular Mind
(Szemerédi is 70)
pp. 651–657.

Introduction to the Polymath Project and "Density Hales–Jewett and Moser Numbers"

MICHAEL A. NIELSEN

At first appearance, the paper which follows this essay [7] appears to be a typical mathematical paper. It poses and partially answers several combinatorial questions, and follows the standard forms of mathematical discourse, with theorems, proofs, conjectures, and so on. Appearances are deceiving, however, for the paper has an unusual origin, a clue to which is in the name of the author, one D. H. J. Polymath. Behind this unusual name is a bold experiment in how mathematics is done. This experiment was initiated in January of 2009 by W. Timothy Gowers [5], and was an experiment in what Gowers termed "massively collaborative mathematics". The idea, in brief, was to attempt to solve a mathematical research problem working entirely in the open, using Gowers's blog as a medium for mathematical collaboration. The hope was that a large number of mathematicians would contribute, and that their collective intelligence would make easy work of what would ordinarily be a difficult problem. Gowers dubbed the project the "Polymath Project". In this essay I describe how the Polymath Project proceeded, and reflect on similarities to online collaborations in the open source and open science communities. Although I followed the Polymath Project closely, my background is in theoretical physics, not combinatorics, and so I did not participate directly in the mathematical discussions. The perspective is that of an interested outsider, one whose main creative interests are in open science and collective intelligence.

Gowers began the Polymath Project with posts on his blog describing the problem to be attacked (see below), a list of rules of collaboration, and a list of 38 brief observations he'd made about the problem, intended to serve as starting inspiration for discussion. At that point, on February 1, 2009, other people were invited to contribute their thoughts on the problem. Anyone with an interest and an internet connection could follow along and, if they wished, contribute their ideas in the comment section of Gowers's blog. In just the first 24 hours after Gowers opened his blog up for discussion, six people offered 24 comments. In a sign of things to come, those contributors came from four countries on three continents, and included a high-school teacher, a graduate student, and four professors of mathematics. A collaboration was underway, a collaboration which expanded in the weeks that followed to involve more than twenty people.

The problem originally posed by Gowers was to investigate a new approach to a special case of the density Hales–Jewett theorem (DHJ). Let me briefly describe the statement of the theorem, before describing the special case Gowers proposed to attack. Let $[k]^n$ be the set of all length n strings over the alphabet $1, 2, \ldots, k$. A *combinatorial line* is a set of k points in $[k]^n$, formed by taking a string with one or more wildcards ("x") in it, e.g., $14x1xx3$, and replacing those wildcards by $1, 2, \ldots, k$, respectively. In the example I've given, the resulting combinatorial line is: $\{1411113, 1421223, \ldots, 14k1kk3\}$. The density Hales–Jewett theorem says that as n becomes large, even very low density subsets of $[k]^n$ must contain a combinatorial line. More precisely, let us define the density Hales–Jewett number $c_{n,k}$ to be the size of the largest subset of $[k]^n$ which does not contain a combinatorial line. Then the density Hales–Jewett theorem may be stated as:

Theorem (DHJ). $\lim_{n\to\infty} c_{n,k}/k^n = 0$.

DHJ was originally proved in 1991 by Furstenberg and Katznelson [4], using techniques from ergodic theory. Gowers proposed to find a combinatorial proof of the $k = 3$ case of the theorem, using a strategy that he outlined on his blog. As the Polymath Project progressed, that goal gradually evolved. Four days after Gowers opened his blog up for discussion, Terence Tao used his blog to start a discussion [8] aimed at understanding the behaviour of $c_{n,3}$ for small n. This discussion rapidly gained momentum, and the Polymath Project split into two subprojects, largely carried out, respectively, on Gowers's blog and Tao's blog. The first subproject pursued and eventually found an elementary combinatorial proof of the full DHJ theorem. (An outline of the proof may be found in a paper by Gowers which appears later in this volume [6]). The results of the second subproject are described in the paper which follows this one, "Density Hales–Jewett and Moser Numbers". As mentioned, this second subproject began with the goal of understanding the behaviour of $c_{n,3}$ for small n. It gradually broadened to consider several related questions, including the behaviour of $c_{n,k}$ for small n and k, as well as the behaviour of the *Moser numbers*, $c'_{n,k}$, defined to be the size of the largest subset of $[k]^n$ which contains no *geometric line.* As for a combinatorial line, a geometric line is defined by taking a strinq in $[k]^n$ with one or more wildcard characters present. But unlike a combinatorial line, there are two distinct types of wildcards allowed ("x" and "$\overline{x}$"), with x taken to vary over the range $1, \ldots, k$, and $\overline{x} = k+1-x$. So, for example, $13x\overline{x}2$ generates the geometric line $\{131k2, 132(k-1)2, \ldots, 13k12\}$.

Both subprojects of the Polymath Project progressed quickly. On March 10, Gowers announced that he was confident that the polymaths had found a new combinatorial proof of DHJ. Just 37 days had passed since the collaboration began, and 27 people had contributed approximately 800 mathematical comments, containing 170,000 words. Much work remained to be done, but the original goal had already been surpassed, and this was a major milestone for the first subproject. By contrast, the goals of the second subproject were more open-ended, and no similarly decisive announcement was possible. Work on both continued for months thereafter, gradually shifting to focus on the writeup of results for publication.

Although the Polymath Project is unusual from the perspective of current practice in mathematics, there is another perspective from which it does not appear so unusual. That is the tradition of open source software development in the computer programming community. Perhaps the best known example of open source software is the Linux operating system. Begun by Linus Torvalds in 1991 as a hobby project, Linux has since grown to become one of the world's most popular operating systems. Although not as widely used in the consumer market as Microsoft Windows, Linux is used on giant computer clusters at companies such as Google and Yahoo!, and also dominates in markets such as the movie industry, where it plays a major role at companies such as Dreamworks and Pixar.

A key feature of Linux is that, unlike proprietary software such as Microsoft Windows, the original source code for the operating system is freely available to be downloaded and modified. In his original message announcing Linux in 1991, Torvalds commented that "I've enjouyed [*sic*] doing it, and somebody might enjoy looking at it and even modifying it for their own needs. It is still small enough to understand, use and modify, and I'm looking forward to any comments you might have." Because he had made the code publicly available, other people could add features if they desired. People began emailing code to Torvalds, who incorporated the changes he liked best into the main Linux code base. A Linux kernel discussion group was set up to co-ordinate work, and the number of people contributing code to Linux gradually increased. By 1994, 80 people were named in the Linux credits file as contributors.

Today, nearly twenty years later, Linux has grown enormously. The kernel of Linux contains 13 million lines of code. On an average day in 2007 and 2008, Linux developers added 4,300 lines of code, deleted 1,800 lines, and modified 1,500 lines. The social processes and tools used to create Linux have also changed enormously. In its early days, Linux used off-the-shelf tools and *ad hoc* social processes to manage development. But as Linux and the broader open source community have grown, that community has developed increasingly powerful tools to share and integrate code, and to manage discussion of development. They have also evolved increasingly sophisticated social structures to govern the process of large-scale open source development. None of this was anticipated at the outset by Torvalds – in 2003 he told *Business Week* "If someone had told me 12 years ago what would happen, I'd have been flabbergasted" – but instead happened organically.

Linux is just one project in a much broader ecosystem of open source projects. Deshpande and Riehle [3] have conservatively estimated that more than a billion lines of open source software have been written, and more than 300 million lines are being added each year. Many of these are single-person projects, often abandoned soon after being initiated. But there are hundreds and perhaps thousands of projects with many active developers.

Although it began in the programming community, the open source collaboration process can in principle be applied to any digital artifact. It's possible, for example, for a synthetic biologist to do open source biology, by freely sharing their DNA designs for living things in an online repository such as the Registry of Standard Biological Parts (see `http://partsregistry.org`), and then allowing others to contribute back changes that improve upon those designs. It's possible to write an open source encyclopedia, by freely sharing the text of articles, and making it possible for others to contribute back changes. That's how Wikipedia was written: Wikipedia is an open source project.

The Polymath Project is a natural extension of open source collaboration to mathematics. At first glance it appears to differ in one major way, for in programming the open source process aims to produce an artifact, the source code for the desired software. Similarly, in synthetic biology and the writing of an encyclopedia the desired end is an artifact of some sort. At least in the early stages of the Polymath Project there was no obviously analogous artifact. It's tempting to conclude that the two papers produced by the polymaths play this role, but I don't think that's quite right. In mathematics, the desired end isn't an artifact, it's mathematical understanding. And the Polymath process was a way of sharing that understanding openly, and gradually improving it through the contributions of many people.

The Polymath Project's open approach to collaboration is part of a broader movement toward open science. Other prominent examples include the human genome project and the Sloan Digital Sky Survey, which use the internet to openly share data with the entire scientific community. This enables other scientists to find ingenius ways of reusing that data, often posing and answering questions radically different to those that motivated the people who originally took the data. An example which gives the flavour of this reuse is the recent work by Boroson and Lauer [1], who used a computer algorithm to search through the spectra of 17,000 quasars from the Sloan Digital Sky Survey, looking for a subtle signature that they believed would indicate a pair of orbiting black holes. The result was the discovery of a candidate quasar containing a pair of supermassive black holes, 20 million and 800 million times the mass of the sun, respectively, and a third of a light year apart, orbiting one another roughly once every 100 years. This is just one of more than 3,000 papers to have cited the Sloan data, most of those papers coming from outside the Sloan collaboration. This open data approach has been carried still further by people practicing open notebook science [2], sharing their entire laboratory record in real time. The Polymath Project and the open data and open notebook projects are all examples of scientists sharing information which, historically, has not been openly available, whether it be raw experimental data, observations made in a laboratory notebook, or ideas for the solution of a mathematical problem. Opening information up in this way creates opportunities for other people to build upon that information in unanticipated ways, taking advantage of the diversity of expertise and perspectives within the scientific community.

There are some striking similarities between this situation and events in the early days of modern science. For example, when Galileo first observed what would later be recognized as Saturn's rings, he sent an anagram to the astronomer Kepler so that if Kepler (or anyone else) later made the same discovery, Galileo could disclose the anagram and claim the credit. Such secretive behaviour was common at the time, and other scientists such as Huygens and Hooke also used devices such as anagrams to "publish" their discoveries. Many scientists waited decades before genuine publication, if they published at all. What changed this situation – the first open science revolution – was the gradual establishment of a link between the act of publishing a scientific discovery and the scientist's prospects for employment. This establishment of scientific papers as a reputational currency gave scientists an incentive to share their knowledge. Today, we take this reputational currency for granted, yet it was painstakingly developed over a period of many decades in the 17th and 18th centuries. During that time community norms around authorship, citation, and attribution were slowly worked out by the scientific community.

A similar process is beginning today. Will pseudonyms such as D. H. J. Polymath become a commonplace? How should young scientists report their role in such collaborations, for purposes of job and grant applications? How should new types of scientific contribution – contributions such as data or blog comments or lab notebook entries – be valued by other scientists? All these questions and many more will need answers, if we are to take full advantage of the potential of new ways of working together to generate knowledge.

References

[1] Todd A. Boroson and Tod R. Lauer, A candidate sub-parsec supermassive binary black hole system, *Nature,* **458** (2009), 53–55.

[2] Jean-Claude Bradley, *Open notebook science,* 2006. `http://drexel-coas-elearning.blogspot.com/2006/09/open-notebook-science.html`.

[3] Amit Deshpande and Dierk Riehle, The total growth of open source, in: *Proceedings of the Fourth Conference on Open Source Systems,* 2008.

[4] Hillel Furstenberg and Yitzhak Katznelson, A density version of the Hales–Jewett theorem, *Journal d'Analyse Mathematique,* **57** (1991), 64–119.

[5] W. Timothy Gowers, *Is massively collaborative mathematics possible?,* 2009. `http://gowers.wordpress.com/2009/01/27/is-massively-collaborative-mathematics-possible/`.

[6] W. Timothy Gowers, Polymath and the density Hales–Jewett theorem, in: *This volume,* 2010.

[7] D. H. J. Polymath, Density Hales–Jewett and Moser numbers, in: *This volume,* 2010.

[8] Terence Tao, Upper and lower bounds for the density Hales–Jewett problem. http://terrytao.wordpress.com/2009/02/05/upper-and-lower-bounds-for-the-density-hales-jewett-problem/.

Michael A. Nielsen

2-120 Christie Street
Toronto, Ontario M6G 3B2
Canada

e-mail: mn@michaelnielsen.org

BOLYAI SOCIETY
MATHEMATICAL STUDIES, 21

An Irregular Mind
(Szemerédi is 70)
pp. 659–687.

Polymath and the Density Hales–Jewett Theorem

W. T. GOWERS*

Van der Waerden's theorem has two well-known and very different generalizations. One is the Hales–Jewett theorem, one of the cornerstones of Ramsey theory. The other is Endre Szemerédi's famous density version of the theorem, which has played a pivotal role in the recent growth of additive combinatorics. In 1991 Furstenberg and Katznelson proved the density Hales–Jewett theorem, a result that has the same relationship to the Hales–Jewett theorem that Szemerédi's theorem has to van der Waerden's theorem. Furstenberg and Katznelson used a development of the ergodic-theoretic machinery introduced by Furstenberg. Very recently, a new and much more elementary proof was discovered in a rather unusual way – by a collaborative process carried out in the open with the help of blogs and a wiki. In this informal paper, we briefly discuss this discovery process and then give a detailed sketch of the new proof.

1. Introduction

Let us begin by stating four major theorems in Ramsey theory and extremal combinatorics. The first is van der Waerden's theorem [14]. Throughout this paper we shall use the notation $[n]$ to stand for the set $\{1, 2, \dots, n\}$.

Theorem 1.1. *For any two positive integers k and r there exists a positive integer N such that if the elements of $[N]$ are coloured with r colours, then there must be an arithmetic progression of length k with all its elements of the same colour.*

*The author is a Royal Society 2010 Anniversary Research Professor and gratefully acknowledges the support of the Royal Society.

If the elements of a set X are coloured, and Y is a subset of X with all its elements of the same colour, then one usually says that Y is *monochromatic.* So van der Waerden's theorem tells us that if we colour a sufficiently large set $[N]$ with r colours, then we can find a monochromatic arithmetic progression of length k.

The next theorem on our list is the Hales–Jewett theorem [7]. This can be thought of as an abstract version of van der Waerden's theorem. It implies van der Waerden's theorem and arises naturally when one examines the proof of van der Waeden's theorem. To state it, we need some definitions.

The Hales–Jewett theorem is another colouring theorem. In this case, the structure that we colour is the set $[k]^n$ for some pair of positive integers k and n. That is, it consists of all sequences of length n whose elements belong to the set $\{1, 2, \ldots, k\}$. This structure can also be thought of as an n-dimensional discrete cube with sidelength k. For instance, when $k = 3$ and $n = 2$ it is a noughts-and-crosses (or tic-tac-toe) board.

In the game of noughts and crosses, each player is trying to make a line. What is a line, mathematically speaking? In terms of coordinates, it is a triple of points $\big((x_1, y_1), (x_2, y_2), (x_3, y_3)\big)$ such that (x_1, x_2, x_3) and (y_1, y_2, y_3) are both either constant sequences or one of the sequences $(1, 2, 3)$ or $(3, 2, 1)$, and, to avoid degeneracy, they are not both constant sequences. For example, $\big((1, 3), (2, 3), (3, 3)\big)$ and $\big((1, 3), (2, 2), (3, 1)\big)$ are two lines. This definition extends easily to larger boards and higher dimensions.

Definition 1.2. A *geometric line* in $[k]^n$ is a sequence $\left(x^{(1)}, \ldots, x^{(k)}\right)$ of elements of $[k]^n$ such that for every i, the sequence $\left(x_1^{(i)}, \ldots, x_k^{(i)}\right)$ is either constant or equal to one of the two sequences $(1, 2, \ldots, k)$ or $\big(k, (k-1), \ldots, 1\big)$, and for at least one i the sequence $\left(x_1^{(i)}, \ldots, x_k^{(i)}\right)$ is not constant.

The Hales–Jewett theorem concerns a restricted class of geometric lines called combinatorial lines. These are lines where all the varying coordinates vary in the same direction. Let us spell this out.

Definition 1.3. A *combinatorial line* in $[k]^n$ is a sequence $\left(x^{(1)}, \ldots, x^{(k)}\right)$ of elements of $[k]^n$ such that for every i, the sequence $\left(x_1^{(i)}, \ldots, x_k^{(i)}\right)$ is either constant or equal to the sequence $(1, 2, \ldots, k)$, and for at least one i the sequence $\left(x_1^{(i)}, \ldots, x_k^{(i)}\right)$ is not constant.

A useful way of representing a combinatorial line is as an element of the set $\big([k] \cup \{*\}\big)^n$, where the element $*$ is thought of as indicating the presence of a variable coordinate, or "wildcard". This is best illustrated with an example: if $k = 3$ and $n = 9$, then the sequence $(*, 1, 3, *, *, 2, 2, *, 1)$ corresponds to the line whose three points are

$$(\mathbf{1}, 1, 3, \mathbf{1}, \mathbf{1}, 2, 2, \mathbf{1}, 1),$$

$$(\mathbf{2}, 1, 3, \mathbf{2}, \mathbf{2}, 2, 2, \mathbf{2}, 1)$$

and

$$(\mathbf{3}, 1, 3, \mathbf{3}, \mathbf{3}, 2, 2, \mathbf{3}, 1),$$

where for clarity we have marked the variable coordinates in bold.

This representation shows that there is almost a one-to-one correspondence between combinatorial lines in $[k]^n$ and points in $[k+1]^n$. The only qualification is that a combinatorial line must have at least one wildcard (since one does not wish to allow degenerate combinatorial lines).

It is not hard to see that the Hales–Jewett theorem implies van der Waerden's theorem. To prove this, it is sufficient to find a map from $[k]^n$ to $\mathbb{N}$ that takes combinatorial lines to arithmetic progressions of length k. If this maps $[k]^n$ into $[N]$, then any colouring of $[N]$ can be pulled back to $[k]^n$, and a monochromatic combinatorial line in $[k]^n$ then maps to a monochromatic arithmetic progression of length k in $[N]$. Two obvious examples of such maps are the map that takes each element of $[k]^n$ to the sum of its coordinates, and the map that interprets each element of $[k]^n$ as the base-k representation of some integer (except that to do this it is easier to use the set $\{0, 1, \ldots, k-1\}^n$ instead).

We briefly mention here the notion of an r-dimensional *combinatorial subspace* of $[k]^n$, since we shall need it later. It is similar to a combinatorial line, but now instead of having just one wildcard set one has r such sets. Within each wildcard set all the coordinates have to be equal, but the coordinates in different wildcard sets can vary independently. The importance of this definition to us is that, as is easy to see, there is an isomorphism between $[k]^r$ and an r-dimensional combinatorial subspace of $[k]^n$. (We do not bother to define what "isomorphism" means here, but the important property of isomorphisms is that they preserve combinatorial lines and combinatorial subspaces.)

Let us move on to our third theorem. It is scarcely necessary to state Szemerédi's theorem in this volume, but for completeness we shall do so. Let X be a finite set and let Y be a subset of X. We define the *density* of Y (in X) to be $|Y|/|X|$.

Theorem 1.4. *For every $\delta > 0$ and every positive integer k there exists a positive integer N such that every subset A of $[N]$ of density at least δ contains an arithmetic progression of length k.*

This is the *density version* of van der Waerden's theorem, proved in 1975 [13]. It trivially implies van der Waerden's theorem, since if you colour $[N]$ with r colours, then there must be a monochromatic subset of density at least $1/r$. As this makes clear, one can formulate a density version for pretty well any Ramsey-type theorem. Often this results in a false statement: it is not true, for instance, that if n is sufficiently large, then every subgraph of the complete graph K_n with at least $\binom{n}{2}/10$ edges contains a triangle. It is also not true that every subset of $[n]$ of density at least $1/10$ contains elements x, y and z such that $x+y=z$. Thus, Ramsey's theorem and Schur's theorem do not have density versions.

The simplicity of the statement of Szemerédi's theorem might lead one to expect that it is either fairly simple to prove, or false. However, neither of these is the case: although a number of proofs are now known, none of them is truly simple. In particular, they are all significantly harder than the proof of van der Waerden's theorem. However, these proofs have led to the discovery of a remarkable number of ideas that have gone on to be applied elsewhere: a notable example being Szemerédi's regularity lemma. Another notable example is the ergodic-theoretic machinery of Furstenberg, which he used to give a second proof of Szemerédi's theorem in 1977 [4], and which he and others used to prove many remarkable extensions of the theorem.

The fourth theorem, which is one of these remarkable extensions, "completes the square". It simultaneously generalizes the Hales–Jewett theorem and Szemerédi's theorem in the way that Szemerédi's theorem and the Hales–Jewett theorem, respectively, generalize van der Waerden's theorem. It is not hard to guess the statement.

Theorem 1.5. *For every $\delta > 0$ and every k there exists n such that every subset of $[k]^n$ of density at least δ contains a combinatorial line.*

This, the density Hales–Jewett theorem, was proved by Furstenberg and Katznelson in 1991 [6]. The proof required a significant extension of Furstenberg's techniques, and involved infinitary Ramsey theorems such as Carlson's theorem, making it one of the most difficult results in extremal combinatorics.

For several years, the only known way of proving a number of density theorems was to use Furstenberg's methods. This was true, for example, of the multidimensional version of Szemerédi's theorem [5], the polynomial version of Bergelson and Leibman [3], and the density Hales–Jewett theorem of Furstenberg and Katznelson. Recently this has begun to change. There is now a proof of the multidimensional Szemerédi theorem using a hypergraph analogue of Szemerédi's regularity lemma, and a natural next target was to try to find a finitary proof of the density Hales–Jewett theorem.

How might one go about this? There are two obvious approaches. One is to look very carefully at the proof of Furstenberg and Katznelson and try to discretize it. The other is to try to find an entirely new proof, perhaps drawing inspiration from the methods that go into the proof of the multidimensional Szemerédi theorem (since this is a strong statement and also a notable consequence of the density Hales–Jewett theorem).

This is the mathematical background to an experiment that I began in early 2009. Let me now explain the extra-mathematical background.

2. Polymath and Open Mathematics

The traditional model of research in mathematics is that mathematicians work very privately, either alone or in small groups. If at some point they obtain some results that are in a reasonably final form, it is only then that they make their thoughts public. This process has served us well and is likely to continue to do so, but there are various ways in which it is inefficient.

One is that papers produced in this way can be very hard for other mathematicians to learn from. Sometimes this is for the simple reason that they are badly written. But even a clear and well-written paper can leave its reader thinking, "I understood that but I don't see how the author(s) thought of the argument, so I don't know how I would ever think of that kind of argument for myself."

Another becomes clear if you think about what it takes for certain proofs to be discovered. Imagine an unsolved mathematics problem with a (not yet discovered) solution that needs a couple of clever tricks. Suppose that the solution is reasonably easy, though not trivial, to work out if you are given the hint that those tricks should be used. Suppose moreover that the tricks are quite well known. What, then, is holding up the discovery? The answer is that the two tricks are somewhat different in style (though not necessarily in subject matter). As a consequence, while there are people who are likely to think of using one trick or the other, there is nobody who is likely to think of both. Even this is not necessarily a problem, since mathematicians like doing joint work. But how does the right team assemble itself? Traditionally, the process is rather haphazard: mathematicians collaborate with people they know or happen to hear about, and hope that the pooling of their expertise will be fruitful. This may seem to work well, but perhaps that is because we are blissfully unaware of all the beautiful proofs that we would have discovered if we assembled our teams more systematically. And even if proofs that need more than one idea do eventually get discovered, perhaps the whole process could be much quicker.

A third inefficiency is closely related to both the first two. Although a proof, when written out, is a fairly linear object, starting from the premises and taking a direct route to the conclusion, the discovery of a proof is far from linear. It is more like a tree with many branches; but when you finally discover the branch that leads to the subbranch that leads to the twig that has at the end of it the solitary fruit that is the conclusion you were looking for, you throw away the rest of the tree. A single mathematician may take a very long time to explore this tree, but one of the benefits of increased mathematical experience is that one learns to recognise certain types of branches that do not have fruit at the end. Now different mathematicians will have different expertise of this sort, so if several mathematicians are exploring the tree together, they may well end up with a far more linear discovery process (since when mathematician A says, "How about trying this?" and "this" is not in fact a good approach, there may well be a mathematician B who knows why it is not a good approach).

A fourth inefficiency is like the third, but in the opposite direction. Another important aspect of exploring the tree is finding the branches to look at in the first place. A single mathematician may well only think of certain kinds of branches, but a large group is likely to think of any kind of branch that is not too strange, and maybe even some that are.

At some point in 2008 it occurred to me that many of these inefficiencies could in principle be removed if research took place publicly rather than privately. Moreover, with the advent of the internet, public research was in principle possible: there were now several media that could be used, of which an obvious one was the blog, especially because Wordpress supports LaTeX. One might think that, despite these technical aids, public research would never work in practice because nobody would be prepared to give away their ideas until they had fully exploited them. But the second thought that occurred to me was that this was not necessarily a problem: if a research project was taking place in public, then it would be difficult to steal its ideas and get away with it, and it would also be difficult to exploit those ideas if you were working alone and were in competition with a big team. In fact, the pressure might well be the exact opposite of the usual pressure: if you had an idea, you would have a strong incentive to share it as soon as possible, since otherwise somebody else might think of it and mention it first.

Having had these thoughts, I filed them away in the part of my brain marked "Something I'd like to try one day" and got on with other things. I then chanced on a blog post by Michael Nielsen entitled "Doing science online," which mentioned that a mathematician called Garrett Lisi was using a mathematical wiki to develop his thinking online. I looked into this and found that it was not quite what I had had in mind, but it was close enough to make it clear that such ideas were in the air. This was the spur I needed to propose an online collaboration on my blog with whoever felt like joining in. (Incidentally, Nielsen's essay is highly recommended, and discusses at much greater length than I have above the kinds of efficiency gains that could in principle come with open science.)

What about the choice of problem to work on? I needed a question that would be sufficiently appealing to enough other mathematicians that there would be a realistic chance of assembling a suitable team of collaborators. Ideally the problem should be difficult enough to be well worth solving, but not so difficult that it would put people off even trying. It should also be a problem where I had at least some ideas about what to do – simply proposing a problem in a vacuum did not seem a good way of persuading others to participate – and these ideas should be ones that I was prepared to make public.

To find an elementary proof of the density Hales–Jewett theorem seemed just right. The first difficult case, when $k = 3$, felt as though it was not *hopelessly* difficult, and in a conversation several years earlier Jozsef Solymosi and I had once come up with a programme for solving it by imitating a proof of the "corners theorem" (the first case of the multidimensional Szemerédi theorem, which will be discussed in detail later in this paper). It can be seen that the density Hales–Jewett theorem follows from an adaptation of the so-called triangle-removal lemma, which itself follows from Szemerédi's regularity lemma, so the idea would be to try to find an appropriate adaptation of the regularity lemma, prove the modified triangle-removal lemma and deduce the density Hales–Jewett theorem for $k = 3$. Of course, there were difficulties with this approach, so after I had announced my intention to carry out such an experiment, my initial mathematical posts described the problem and described the approach I wanted to try, up to the point where I had got stuck.

I suggested various rules, some firm and others just guidelines, for how the collaboration should work. A particularly firm rule was that if anything publishable was to result from the collaboration, then it would be published under a pseudonym, and there would be a link from the published paper to the blog discussion. The pseudonym I chose was Polymath. An important guideline was that comments should be kept reasonably short and (if possible) not too technical. In particular, one should try to avoid the temptation to think hard about ideas offline and present them in polished form. Better (and more efficient) to express thoughts that were not yet fully worked out and let them be developed collectively.

To cut a long story short, the result of the experiment was that in under six weeks the problem was solved. In addition to this remarkably quick outcome, there were a number of other interesting aspects to what happened, which I shall briefly list.

(i) The eventual proof was modelled not on the triangle-removal argument of Ruzsa and Szemerédi [10] (later applied to the corners problem by Solymosi [12]) but on a different proof of the corners theorem due to Ajtai and Szemerédi [1]. (It will surely not have escaped the reader's notice how often Endre's name comes up in this discussion.)

(ii) The proof was strictly simpler than that of Furstenberg and Katznelson, because it did not use Carlson's theorem or anything like it. In other words, it was not just a translation of their ideas into a different language.

(iii) Once we had a proof for the case $k = 3$, it was almost immediately clear that it could be generalized to a proof for every k. This was a major surprise, since in all other proofs of Szemerédi's theorem the case $k = 4$ is substantially harder.

(iv) It is arguable (and I myself would argue) that the proof gave the simplest known proof of Szemerédi's theorem.

(v) The set of people who worked seriously on the project eventually settled down to a core (with a non-sharp boundary) of around half a dozen people, who were all people one might have expected in advance to be interested. However, it would have been impossible in advance to predict who these people were going to be.

(vi) The progress of the project was followed by a large number of people, and was commented on in many places around the internet, at one point even making the front page of the Slashdot aggregator.

In addition, one of our lemmas contained an idea that helped Tim Austin to complete a project of finding a simpler ergodic proof of the density Hales–Jewett theorem [2], and Terence Tao ran on his blog a project to investigate upper and lower bounds for the theorem in small dimensions, the results of which have now been written up and can be found in this volume [9].

3. The Corners Theorem of Ajtai and Szemerédi

The rest of this paper is devoted to a discussion of Polymath's proof of the density Hales–Jewett theorem. We do not give full details, but we aim to give a sufficiently detailed account that the interested reader can easily reconstruct a complete proof (and we secretly hope that some readers will find it easier to do this than to read the full details). The full proof appears in a paper of D. H. J. Polymath [8].

As mentioned above, the proof was modelled on a proof of Ajtai and Szemerédi of the following result, which is the first genuinely multidimensional case of the multidimensional Szemerédi theorem [1].

Theorem 3.1. *For every $\delta > 0$ there exists N such that for every subset $A \subset [N]^2$ of density at least δ there exist (x, y) and $d > 0$ such that (x, y), $(x + d, y)$ and $(x, y + d)$ all belong to A.*

The three points that belong to A form the vertices of an isosceles right-angled triangle with two of its sides horizontal and vertical and the right angle pointing downwards to the left. It has become common to call these configurations *corners*. Thus, the theorem of Ajtai and Szemerédi asserts that a density-δ subset of a sufficiently large grid must contain a corner.

Here is a sketch of a proof of Theorem 3.1. It is essentially the same as the proof of Ajtai and Szemerédi, but the steps have been slightly rearranged in a way that was crucial when it came to adapting the argument to prove the density Hales–Jewett theorem.

Let us assume that A is a subset of $[N]^2$ of density at least δ that contains no corners.

The first step is to find a "dense diagonal": that is, to find some z such that there are at least cn pairs $(x, y) \in A$ with $x + y = z$. This is easily done with an averaging argument. There are $2N - 1$ sets of the form $D_z = \{(x, y) \in [N]^2 : x + y = z\}$, and they partition $[N]^2$. Therefore, if A has size at least δN^2, at least one of these sets must contain at least $\delta N/2$ points of A. That gives us our dense diagonal.

The second step is to find two large sets $X, Y \subset [N]$ such that $X \times Y$ is disjoint from A. This is where we use the assumption that A contains no corners. Let the dense diagonal be D_z and suppose that (x_1, y_1) and (x_2, y_2) are two points in $D_z \cap A$. Then we know that $x_1 + y_1 = x_2 + y_2 = z$. Without loss of generality $x_1 < x_2$. Then the point (x_1, y_2) cannot belong to A, since otherwise the three points (x_1, y_2), (x_1, y_1) and (x_2, y_2) would form a corner. (The right angle would be at (x_1, y_2) and d would equal $x_2 - x_1 = y_1 - y_2$.) Let us partition $A \cap D_z$ into two sets E_1, E_2 of approximately equal size, with the x-coordinate of every point in E_1 less than the x-coordinate of every point in E_2. Let X be the set of x-coordinates of points in E_1 and let Y be the set of y-coordinates of points in E_2. Then the argument just given shows that no point in $X \times Y$ can belong to A. Moreover, we can choose the partition so that both X and Y have at least $\delta N/4$ points. (For simplicity I am ignoring the need to take integer parts.)

The third step is to find two large sets $U, V \subset [N]$ such that the density of A inside $U \times V$ is at least $\delta + \eta$, where η is a positive constant that depends on δ only. (By "the density of A inside $U \times V$" I mean the ratio $|A|/|U|\,|V|$.) This follows from a very easy averaging argument: we can partition $[N]^2$ into the four sets $X \times Y$, $X \times Y^c$, $X^c \times Y$ and $X^c \times Y^c$. Since $X \times Y$ is reasonably dense and disjoint from A, the density of A inside one of the other three sets must be noticeably bigger than δ. In fact we can take η to be $c\delta^2$ for an absolute constant $c > 0$.

Define a *grid* to be a subset of $[N]^2$ of the form $P \times Q$, where P and Q are arithmetic progressions with the same common difference. We shall try to find a grid $P \times Q$ in which the density of A is at least $\delta + \eta$ (for some possibly different η). If we can always do this, then a standard argument shows that we are done. Indeed, we can restrict our attention to the grid $P + Q$ and run the argument again. Once again, if there is no corner then we can find a further subgrid in which the density of A is yet larger. Provided that η is an increasing function of δ, this process cannot continue for ever, since the density will reach 1 in a finite number of steps. Therefore, there must be a corner in A.

Although a Cartesian product $U \times V$ is not a grid, it is a useful step towards a grid, because it somehow reduces the problem to a one-dimensional one. What we shall do to complete the proof is show, with the help of Szemerédi's theorem, that it is possible to partition almost all of $U \times V$ into grids. If we can do this, then an averaging argument shows that if A has density $\delta + \eta$ in $U \times V$, then it must have density almost $\delta + \eta$ (and certainly at least $\delta + \eta/2$, say) in at least one of the grids.

The fourth step is to partition almost all of U into long arithmetic progressions. Since U is a dense subset of $[N]$, Szemerédi's theorem tells us that it contains an arithmetic progression P_1 with length tending to infinity. If $U \setminus P$ is still of non-negligible density, then $U \setminus P_1$ contains an arithmetic progression P_2 with length tending to infinity. Continuing this process, we can partition U into a union $P_1 \cup \cdots \cup P_k \cup W$, where all the P_i are arithmetic progressions of length tending to infinity and W is a set of negligible size. It will be useful later if we also insist that the common difference of each P_i is at most $\sqrt{N}$, which we clearly can (since if a set is dense in $[N]$ then by averaging it will be dense in some interval of width $\sqrt{N}$).

The fifth step is to partition almost all of each set $P_i \times [N]$ into grids. This can be done as follows. If P_i has common difference d and length k, partition almost all of $[N]$ into intervals of length dk and then partition each interval into arithmetic progressions of length k and common difference d. Thus, we have partitioned almost all of $U \times [N]$ into grids $P_i \times Q_{ij}$.

The sixth step is to repeat the argument, but reversing the roles of the two coordinates. Let us restrict attention to one of the grids $P_i \times Q_{ij}$. The intersection of U with $P_i \times Q_{ij}$ is of the form $P_i \times U_{ij}$. Precisely the same argument that we gave in steps 4 and 5 can be used to prove that this set can be partitioned into grids plus a set of very small density.

We have therefore partitioned almost all of $U \times V$ into large grids. This, together with the standard density-increment strategy explained above, is sufficient to complete the proof.

4. From Corners to Combinatorial Lines

How can we model a proof of the density Hales–Jewett theorem (for $k = 3$) on the proof just sketched of the corners theorem? Ideally we would have a dictionary that would allow us to carry the steps over in a straightforward way. The reality is that we do indeed have a very helpful dictionary, but that when we use it to carry over the proof, we have to overcome certain difficulties. In this section we shall explain the dictionary, and in the next section what the difficulties are.

Before we start, it will be helpful to reformulate the corners statement in a more symmetrical way, so that it is about equilateral triangles rather than isosceles right angled triangles. To do this, we observe first that the corners theorem is equivalent to the statement that for sufficiently large N, every subset of $T_N = \{(x,y) \in \mathbb{Z}^2 : x,y \geq 0,\ x+y \leq N\}$ contains a corner. This is because if A is a dense subset of $[N]^2$, then we can cover $[N]^2$ by translates of T_N and apply an averaging argument to find a translate inside which A is dense, and in the other direction if A is dense in T_N then $A \cap [N]^2$ is dense in $[N]^2$. However, T_N is a more natural space for the corners problem because it has three privileged directions (rather than just two) and they are the same as the three directions one can form by joining two points of a corner with a line.

To reformulate this in terms of equilateral triangles, we can simply apply an affine transformation that turns all aligned isosceles right angled triangles into equilateral ones. However, a nicer way to do this is to embed $\mathbb{Z}^2$ into $\mathbb{Z}^3$, by sending the point (x,y) to the point $(x,y,N-y)$. Then T_N maps to the set $\{(x,y,z) : x,y,z \geq 0,\ x+y+z=N\}$. A corner will map to a triple of the form $(x+d,y,z)$, $(x,y+d,z)$, $(x,y,z+d)$, with $d>0$ and $x+y+z+d=N$. Thus, we have a reformulation of the corners theorem that treats all three directions in the same way.

How can we think of a combinatorial line in $[3]^n$ as "similar to a corner"? To begin with, we want to represent points in our space $[3]^n$ as triples somehow. But there is a very natural way of doing that. A point in $[3]^n$ is determined by a partition of $[n]$ into three sets: the set of coordinates where it takes the value 1, the set where it takes the value 2, and the set where it takes the value 3. Thus, we could regard the density Hales–Jewett theorem for $k=3$ as a theorem about such partitions.

Let us do so: we shall think of $[3]^n$ as the set of all triples (X,Y,Z) such that X, Y and Z are disjoint and $X \cup Y \cup Z = [n]$. This already suggests that in our dictionary we should translate addition into disjoint union. That in turn suggests that the analogue of a corner should be a triple of triples of the form $(X \cup D, Y, Z)$, $(X, Y \cup D, Z)$ and $(X, Y, Z \cup D)$, where the unions are disjoint unions. The obvious translation of the condition that $d>0$ is that D should be non-empty. And if we think about this for a moment we see that we have just defined the combinatorial line that is 1 on X, 2 on Y and 3 on Z and has wildcard set D.

An important role in the corners proof was played by Cartesian products. What sorts of subsets of $[3]^n$ play the role that is played in the corners problem by subsets of $[N]^2$ of the form $U \times V$? Well, $U \times V = \{(x,y) \in [N]^2 : u \in U,\ v \in V\}$. In terms of our more symmetric coordinate system, we might describe $(U \times V) \cap T_n$ as $\{(u,v,w) : u \in U,\ v \in V,\ u+v \leq N,\ u+v+w = N\}$. Now let $\mathcal{U}$ and $\mathcal{V}$ be collections of subsets of $[n]$. Since we are turning sums into disjoint unions, the natural analogue $\mathcal{U} \boxtimes \mathcal{V}$ is the set of all (U,V,W) such that $U \in \mathcal{U}$, $V \in \mathcal{V}$ and U, V and W partition $[n]$. Equivalently, it is the set of all triples $\big(U, V, [n] \setminus (U \cup V)\big)$ such that $U \in \mathcal{U}$, $V \in \mathcal{V}$ and U and V are disjoint. We shall call $\mathcal{U} \boxtimes \mathcal{V}$ the *disjoint product* of $\mathcal{U}$ and $\mathcal{V}$.

5. Two Apparently Serious Difficulties

This all looks very promising, so where do we run into trouble? There are two observations that might lead one to think that a translation of the corners proof was not after all possible.

The first is that measure concentration has undesirable effects when one is trying to prove the density Hales–Jewett theorem. This difficulty manifests itself in several ways, so let us just pick one: a typical point in a typical line is not a typical point.

To be more explicit, recall that a combinatorial line in $[3]^n$ can be thought of as a point in $[4]^n$ that takes the value 4 at least once (or, in our new language, as a quadruple of sets (X,Y,Z,D) that partition $[n]$, with the condition that $D \neq \emptyset$). Now if you choose a random partition of $[n]$ into four sets, then with extremely high probability each set will have size very close to $n/4$. This implies that a typical point in a random line has about $n/2$ of its coordinates equal to one value and about $n/4$ of its coordinates equal to each of the other two values. This is very unlike a typical point, which takes all three values about $n/3$ times.

Something similar is true in the corners problem. If you choose a random corner and then a random point in that random corner, the resulting distribution is not uniform over all points. However, it is close enough to uniform for many arguments to work. (The main property one needs is that a set that has small probability in one distribution has small probability in the other.) For example, it allows one to show that a dense set has not just one corner but cN^3 corners for some constant $c > 0$ that depends just on the density. That is, if A is a set of density δ, then a random corner has a positive probability $c \geq c(\delta) > 0$ of belonging to A.

The analogous statement for the density Hales–Jewett theorem is easily seen to be false, as a result of the phenomenon just described. Indeed, let A be the set of all points (X, Y, Z) such that each of X, Y and Z has size approximately $n/3$. Then the density of A is almost 1, but the probability that a random combinatorial line has all its points in A (or even just one of its points in A) is almost 0. This is potentially a serious problem, since in a standard density argument one tends to be able to prove a density increment from the assumption that a set contains *few* configurations of the kind one is looking for: it is not necessary to assume that it has none at all.

The second difficulty looks even more serious. As commented in the introduction, the density Hales–Jewett theorem implies Szemerédi's theorem, and it seems as though in some sense the strength of Szemerédi's theorem for progressions of length k is comparable to the strength of the density Hales–Jewett theorem in $[k]^n$. But the proof of the corners theorem used Szemerédi's theorem for all k. If we are going to translate this proof into a proof of the density Hales–Jewett theorem in $[3]^n$, then what will play the role of Szemerédi's theorem? Obviously it cannot be the general case of the density Hales–Jewett theorem!

In the next two sections we shall explain how we deal with these two difficulties.

6. Sperner's Theorem and Equal-Slices Measure

The short answer to the question of how we deal with the problems associated with the uniform measure on $[3]^n$ is that we switch to a different measure.

However, the density Hales–Jewett theorem concerns the uniform measure. How can it be sufficient to prove the result for sets that are dense in a different measure? We must address this question before we start thinking about what the different measure is.

The answer is suggested by the following very simple lemma.

Lemma 6.1. *Let μ and $\nu_1, \ldots, \nu_M$ be probability measures on a finite set S and let ν be a convex combination of $\nu_1, \ldots, \nu_M$. Suppose that $|\mu(A) - \nu(A)| \leq \eta$ for every subset A of S. Then for every subset $A \subset S$ with $\mu(A) \geq \delta$ there exists i such that $\nu_i(A) \geq \delta - \eta$.*

The proof is trivial: we know that $\nu(A) \geq \delta - \eta$ and therefore by averaging this must be true for at least one of the ν_i as well.

Now suppose that we have a way of defining a probability measure on $[3]^m$ for each m, and can prove that sets that are dense for this new measure must contain combinatorial lines. We attempt to apply Lemma 6.1 as follows. Let μ be the uniform measure on $[3]^n$, choose m-dimensional combinatorial subspaces $S_1, \ldots, S_M$ of $[3]^m$, and for each i let ν_i be the new measure applied to S_i. (The new measure is defined on $[3]^m$ but can be transferred to S_i since the two spaces are isomorphic.) If we can do this in such a way that the uniform measure is well approximated by a convex combination of the ν_i, then for every set $A \subset [3]^n$ that is dense in the uniform measure Lemma 6.1 gives us a subspace S_i such that $A \cap S_i$ is dense in the new measure. We then apply our result for the new measure to obtain a combinatorial line and we are done.

That still leaves open the question of what measure to take. To answer this question, we turn to a theorem that will play an important role in our argument: Sperner's theorem.

Sperner's theorem answers the following question: how many subsets of an n-element set can you choose before one of your sets must be a proper subset of another?

Theorem 6.2. *Let $\mathcal{A}$ be a collection of subsets of $[n]$ of size greater than $\binom{n}{\lfloor n/2 \rfloor}$. Then there exist distinct sets $A, B \in \mathcal{A}$ such that $A \subset B$.*

This result is clearly best possible, since the set of all sets of size $\lfloor n/2 \rfloor$ has size $\binom{n}{\lfloor n/2 \rfloor}$ and no one of them is a subset of any other.

A very nice proof of Sperner's theorem, due to Lubell, goes as follows. Suppose that $\mathcal{A}$ does not contain two sets such that one is a proper subset of the other. Now consider the following way of choosing a subset of $[n]$ at random. You first choose a random permutation π of $[n]$, then you choose a random integer $r \in \{0, 1, \ldots, n\}$, and finally you let A be the set $\{\pi(1), \ldots, \pi(r)\}$.

The probability that A belongs to $\mathcal{A}$ if you choose it like this is trivially at most $1/(n+1)$, or there would be at least one permutation π for which there were two or more values of r with $\{\pi(1), \ldots, \pi(r)\}$ belonging to $\mathcal{A}$. This would give us two sets in $\mathcal{A}$, one of which was a proper subset of another, contradicting our assumption.

Thus, this simple averaging argument has given us a probability measure ν with respect to which $\mathcal{A}$ cannot have measure greater than $1/(n+1)$. However, we were interested in the largest possible *uniform* measure of $\mathcal{A}$, so now we have to work out how big the uniform measure can be if its ν-measure is at most $1/(n+1)$.

Now an equivalent way of choosing a set A ν-randomly is as follows. You choose the cardinality r of A uniformly from the set $\{0, 1, \ldots, n\}$, and then you choose A uniformly from the set of all subsets of $[n]$ of size r. Thus, $\nu(A)$ is equal to $\frac{1}{n+1}\binom{n}{|A|}^{-1}$. If we want $\mathcal{A}$ to consist of as many sets as possible while making sure that $\nu(\mathcal{A}) \leq 1/(n+1)$, we therefore want to choose the sets $A \in \mathcal{A}$ in such a way that $\binom{n}{|A|}^{-1}$ is as small as possible. It is therefore obvious that we cannot do better than choosing all sets of size $\lfloor n/2 \rfloor$.

Why should Sperner's theorem be important to us? The answer is simple: it proves the density Hales–Jewett theorem in the case $k = 2$. To see this, one just has to think about combinatorial lines in terms of partitions of $[n]$. A typical point in $[2]^n$ can be thought of as a pair (X, Y) that partitions $[n]$, and a combinatorial line is a pair of points $(X \cup D, Y)$ and $(X, Y \cup D)$. There is a one-to-one correspondence between subsets of $[n]$ and points in $[2]^n$, namely the function that takes the set X to the pair $(X, [n] \setminus X)$. Thus, avoiding combinatorial lines is equivalent to avoiding having two sets X and $X \cup D$ such that D is non-empty and disjoint from X. But that is the same as avoiding having one set that is a proper subset of another.

This is a respect in which the density Hales–Jewett theorem is genuinely different from Szemerédi's theorem: the case $k = 2$, though fairly easy, is not trivial. Strangely, this is a big advantage to us, since it allows us to think at an earlier stage about how the argument might be generalized, whereas the case $k = 2$ of Szemerédi's theorem is too trivial to give one any useful clues about how to proceed. (This seems to be connected to the reason that the $k = 3$ case of the density Hales–Jewett theorem led so rapidly to the general case.)

Note that in order to prove Sperner's theorem, which concerns the uniform measure, we proved a result for a different measure instead, the measure where first you choose the cardinality of a set randomly and then you choose the set randomly given the cardinality. This is quite suggestive, since not only did the new measure arise naturally, but also it does not have the awkward property of being concentrated on sets with roughly the same cardinality. It is therefore reasonable to hope that a generalization of this measure to $[3]^n$ will help us prove the density Hales–Jewett theorem with $k = 3$.

The obvious way of generalizing the measure is this. To pick a random triple (X, Y, Z) of sets that partition $[n]$, you first choose a triple of non-negative integers (r, s, t) such that $r + s + t = n$, choosing it uniformly at random from the set of all such triples, and then you randomly partition $[n]$ into three sets, the first of size r, the second of size s and the third of size t. For each triple (r, s, t) we shall call the corresponding set of tripartitions a *slice*, and we shall call the probability measure we have just defined the *equal-slices measure* on $[3]^n$ (since it gives equal measure to each slice). One could equivalently talk in terms of coordinates. In that case to choose a random point, you again choose a random triple (r, s, t) with $r+s+t$ and then you choose a random point in $[3]^n$ with r 1s, s 2s and t 3s.

Before we finish this section, we should mention a new difficulty, which is that the uniform measure has a very useful property that equal-slices measure lacks. That property is that if you write $[3]^n$ as $[3]^m \times [3]^{n-m}$, as it is sometimes useful to do, and if $B \subset [3]^m$ and $C \subset [3]^{n-m}$, then the uniform measure of $B \times C$ is the product of the uniform measures of B and C. In short, the uniform measure is a product measure whenever you want it to be. Unfortunately, the equal-slices measure is not.

What this means is that we have to apply Lemma 6.1 in both directions. We start with a set that is dense in the uniform measure, apply Lemma 6.1 to obtain a set that is dense in the equal-slices measure on some subspace, apply a further argument to obtain a density increment on a certain kind of subset, apply Lemma 6.1 to move back to uniform measure, and finally complete the argument making repeated use of the product property of the uniform measure.

7. The Multidimensional Density Hales–Jewett Theorem

As an illustration of the usefulness of product measures, let us begin this section with a sketch proof that the density Hales–Jewett theorem implies a multidimensional version in which the object one is trying to find is an r-dimensional combinatorial subspace rather than just a combinatorial line.

Theorem 7.1. *For every $\delta > 0$ and for every pair of positive integers k and r there exists n such that every subset of $[k]^n$ of density at least δ contains an r-dimensional combinatorial subspace.*

The proof is by induction on the dimension r, and assumes the 1-dimensional theorem. So let us suppose that we know the result for $r-1$ and for all densities. (That is, in order to prove the result for r and density δ we shall need to know the result for $r-1$ and some much smaller density.)

Let n be some very large integer, to be chosen later, and write $[k]^n$ as $[k]^m \times [k]^{n-m}$ where m too is to be chosen later, but should be thought of as much smaller than n. We shall write a typical element of $[k]^n$ as (x, y), where it is to be understood that $x \in [k]^m$ and $y \in [k]^{n-m}$.

Let A be a subset of $[k]^n$ of density at least δ. For each $y \in [k]^{n-m}$, let $A_y = \{x \in [k]^m : (x, y) \in A\}$. Let Y be the set of all y such that A_y has density at least $\delta/2$ in $[k]^m$. Then Y has density at least $\delta/2$ in $[k]^{n-m}$, since otherwise a simple argument would show that A had density less than δ. By the 1-dimensional theorem, if m is sufficiently large, A_y contains a combinatorial line for every $y \in Y$.

Since there are at most $(k+1)^m$ combinatorial lines in $[k]^m$, the pigeonhole principle gives us a combinatorial line L in $[k]^m$ and a set $Y' \subset Y$ of density at least $(k+1)^{-m}\delta/2$ in $[k]^{n-m}$ such that $L \times Y' \subset A$. And now if n is sufficiently large the inductive hypothesis implies that Y' contains an $(r-1)$-dimensional combinatorial subspace S. Since $L \times S$ is an r-dimensional combinatorial subspace, the proof is complete.

Note that to make this proof work, we have to iterate the bound for the 1-dimensional theorem r times. As we shall see, this has serious consequences for the bound we shall eventually obtain for the density Hales–Jewett theorem.

That last remark gives a big clue about the answer to a question that we asked earlier: if we are imitating the Ajtai–Szemerédi proof of the corners theorem, then what will play the role that is played in that proof by Szemerédi's theorem? The answer is the multidimensional version of Sperner's theorem. It is not at all obvious in advance that this will be enough, but it turns out that it is. And that, in a sense, is why the theorem is straightforward to prove once one has proved it for $k = 3$: once we have worked out how to use the multidimensional Sperner theorem, we find that more or less the same argument allows us to use the multidimensional density Hales–Jewett theorem for k to prove the 1-dimensional density Hales–Jewett theorem for $k+1$.

Although it is a slight surprise that the multidimensional Sperner theorem, which is a much easier result than Szemerédi's theorem, is enough for the proof, it is worth stressing that it was very natural to try using it. This is partly because one would hope that the density Hales–Jewett theorem could be proved inductively, and, given that hope, it is natural to start with the strongest result one can deduce from Sperner's theorem. It is also natural in the light of known inductive proofs of the colouring Hales–Jewett theorem. Finally, it presents itself in a fairly natural way as the necessary tool when one tries to modify the Ajtai–Szemerédi proof, as we shall see.

8. A Local Density Increment on a Disjoint Product

The first few steps of the Ajtai–Szemerédi proof yielded a Cartesian product $U \times V$ inside which our corner-free dense set A had increased density. Our dictionary suggests that if A is a subset of $[3]^n$ that contains no combinatorial line, then we should try to find large set systems $\mathcal{U}$ and $\mathcal{V}$ such that A has increased density inside the disjoint product $\mathcal{U} \boxtimes \mathcal{V}$, which we defined to be $\{(X,Y,Z) : X \in \mathcal{U}, Y \in \mathcal{V}\}$, it being implicit in this notation that X and Y are disjoint subsets of $[n]$ and $Z = [n] \setminus (X \cup Y)$. It even tells us how we should go about doing this, so let us make a start.

The first step was to find a diagonal that contained many points of A. What was a diagonal? It was the set of all points (x,y,z) with some fixed value for z. So a diagonal in $[3]^n$ should be a set of the form $D(Z_0) = \{(X,Y,Z) : Z = Z_0\}$. If A has density at least δ, then A must have density δ in $D(Z_0)$ for some Z_0. We will also want the set $D(Z_0)$ itself not to be too small, but let us pursue the argument a little further before we think about that.

The next observation in the Ajtai–Szemerédi argument was that if (x,y,z) and (x',y',z) are two points in $A \cap D_z$ such that $x < x'$, then the point $(x,y',z-x-y')$ does not belong to A (assuming that A has no corners). We have thought about how to translate almost all the concepts involved in this statement: the one exception is the notion of "less than". But there is an obvious set-theoretic analogue of this, namely "is a proper subset of", so let us try that.

And we do indeed find that if (X,Y,Z) and (X',Y',Z) are two points in $A\cap D(Z)$, and if X is a proper subset of X', then the point $\big(X,Y', [n]\setminus(X\cup Y')\big)$ cannot belong to A. That is because if we set $D=X'\setminus X$, then the three points (X,Y,Z), (X',Y',Z) and $(X,Y',[n]\setminus(X\cup Y')$ can be rewritten as $(X,Y'\cup D,Z)$, $(X\cup D,Y',Z)$ and $(X,Y',Z\cup D)$, which manifestly form a combinatorial line.

Let us set $\mathcal{U}=\big\{X:\ (X,[n]\setminus(X\cup Z),Z)\in A\big\}$ and $\mathcal{V}=\big\{Y:\ ([n]\setminus(Y\cup Z),Y,Z)\in A\big\}$. Then the above argument is telling us that $\mathcal{U}\boxtimes\mathcal{V}$ is disjoint from A. Indeed, let $\big(X,Y',[n]\setminus(X\cup Y')\big)\in\mathcal{U}\boxtimes\mathcal{V}$. Then both X and Y' must be disjoint from Z, so $Z\subset[n]\setminus(X\cup Y')$. Setting $Y=[n]\setminus(X\cup Z)$ and $X'=[n]\setminus(Y'\cup Z)$, we have that the points (X,Y,Z) and (X',Y',Z) both belong to $A\cap D(Z)$, and that X is a proper subset of X'. The resulting combinatorial line has fixed sets X, Y' and Z and wildcard set $[n]\setminus(X\cup Y'\cup Z)$.

It may look as though we have done exactly what we wanted, but as soon as we try to continue the argument a serious problem arises. Just as with the corners problem, we can partition $[3]^n$ into four sets of the form $\mathcal{W}\boxtimes\mathcal{Z}$, where $\mathcal{W}$ is either $\mathcal{U}$ or $\mathcal{U}^c$ and $\mathcal{Z}$ is either $\mathcal{V}$ or $\mathcal{V}^c$. But the set $\mathcal{U}\boxtimes\mathcal{V}$ has no reason to be dense in $[3]^n$. In fact, it will usually have a density that is an exponentially small function of n. This means that we cannot get a useful density increment on one of the sets $\mathcal{W}\boxtimes\mathcal{Z}$ from the knowledge that A is disjoint from $\mathcal{U}\boxtimes\mathcal{V}$.

To get round this problem we need two ideas. The first is the observation that $\mathcal{U}\boxtimes\mathcal{V}$ is contained in a combinatorial subspace of $[3]^n$, the space of all points (X',Y',Z') such that $Z\subset Z'$ (or equivalently the space of all points whose coordinates are equal to 3 everywhere on the set Z). Therefore, even if $\mathcal{U}\boxtimes\mathcal{V}$ is not dense in $[3]^n$, it might perhaps be dense in this subspace, which typically has a much smaller cardinality than 3^n.

On its own, this thought is not enough, because of the measure-concentration problem discussed earlier. For almost every diagonal $D(Z)$ the set Z has cardinality roughly $n/3$, and for almost every point (X,Y,Z) in $D(Z)$ the sets X and Y also have cardinality roughly $n/3$, so when we restrict to the subspace just mentioned, which has dimension roughly $2n/3$, we may well find that all the points in A have about $n/3$ 1s, $n/3$ 2s and just a tiny number of 3s.

However, if we combine the idea of passing to a subspace with the use of equal-slices measure, it turns out that we can obtain a density increment of the kind we want. Here is a sketch of how the argument works.

We first want to be sure that when we pass to a subspace the restriction of A to that subspace will still have density at least δ, or perhaps $\delta - \varepsilon$ for some very small ε. This can be achieved by an averaging argument: if there is any subspace (from a given class of subspaces that covers $[3]^n$ evenly) in which A has density at least $\delta + \eta$, then we can forget all about the corners argument and simply pass to that subspace, and we have obtained our density increment; otherwise, the proportion of subspaces inside which A has density at most $\delta - \varepsilon$ is at most η/ε. Thus, after choosing ε and η appropriately, we may safely assume that in almost every subspace of the given kind the density of A is scarcely any less than δ. For this argument to be useful to us, the very slight sacrifice we are prepared to make to the density must be more than compensated for by a later density increase, but that turns out to be easy to arrange.

The argument is not quite as straightforward as that description makes it sound, because we start with the assumption that the *uniform* density of A is at least δ, and we find a subspace in which the *equal-slices* density of A is at least $\delta - \varepsilon$. For this we need to know that the average of the equal-slices densities on all the subspaces we consider is approximately equal to the uniform distribution on $[3]^n$.

At this point, let us simply state what works. For any set $J \subset [n]$ and any point $y \in [3]^{[n]\setminus J}$, let $S_{J,y}$ denote the $|J|$-dimensional combinatorial subspace of all points $z \in [3]^n$ such that the restriction of z to $[n] \setminus J$ is equal to y. Now choose a smallish m (it turns out that $m = n^{1/4}$ will do), and let J be a random set of size m. We make two claims.

(i) If you choose the pair (J, y) uniformly at random (with $|J| = m$) and then choose a random point from $S_{J,y}$ according to equal-slices measure, then the distribution of the random point will be approximately uniform in $[3]^n$.

(ii) If you choose (J, y) randomly as above and then choose values for the coordinates in J using the equal-slices measure on $[2]^J$ rather than $[3]^J$, then the distribution of the resulting point is still approximately uniform on $[3]^n$.

The idea behind (i) and (ii) is that if m is small, then the fact that the coordinates in J are chosen according to a highly non-uniform distribution does not disturb the measure all that much. (In fact, it is not hard to prove that if you make the coordinates in J all equal to 1 then the distribution is still approximately uniform.) Actually proving these claims is a straightforward technical exercise.

Claim (i) is what we need in order to be able to assume that the equal-slices density of A inside the subspace $S_{J,y}$ is at least $\delta - \varepsilon$. Let us write A_y for $A \cap S_{J,y}$. Claim (ii) implies that for a reasonable proportion of pairs (J, y), the equal-slices density of the set D_y of points $x \in [2]^J$ such that $(x, y) \in A$ is at least $\delta/2$. (Here we are writing (x, y) for the point whose restriction to J is x and whose restriction to $[n] \setminus J$ is y.)

Note that the set D_y is very like a diagonal. In particular, both it and a diagonal $D(Z)$ are isomorphic to $[2]^m$ for some m, and this turns out to be what matters to us. If we restrict to the subspace $S_{J,y}$, then inside that subspace D_y *is* a diagonal: it is the diagonal $D(\emptyset)$. Therefore, if we change our ground set to J and replace A by A_y, then we can find, just as we did before, a pair of set-systems $\mathcal{U}$ and $\mathcal{V}$, each consisting of subsets of J, such that $\mathcal{U} \boxtimes \mathcal{V}$ is disjoint from A_y. Moreover it can be shown that if the equal-slices density of D_y in $[2]^J$ is at least $\delta/2$, then the equal-slices density of $\mathcal{U} \boxtimes \mathcal{V}$ inside $[3]^J$ is bounded below by a constant that depends on δ only.

Briefly, the reason for this last assertion is as follows. For simplicity let us assume that $J = [m]$. Our argument at the beginning of this section relied on the fact that if (X, Y, Z) and (X', Y', Z) are two points in the diagonal $D(Z)$ that also belong to A, and if X is a proper subset of X', then (X, Y', Z) cannot belong to A. Each point (X, Y, Z) in $D(Z)$ is determined by X, since Z is fixed and $Y = [n] \setminus (X \cup Z)$. In our case, $Z = \emptyset$, so $A \cap D(Z)$ can be thought of as a collection B of subsets of $[m]$. Therefore, the number of points that this argument shows us that A cannot contain is equal to the number of pairs X, X' in B such that X is a proper subset of X'. Sperner's theorem told us that if B is bigger than $\binom{m}{\lfloor m/2 \rfloor}$ then there will be at least one such pair. But precisely the same method of proof also tells us that if B is large, then the number of such pairs will be large, where large means large *in equal-slices measure.*

Earlier we commented that it was not the case that a dense subset of $[k]^n$ has to contain a dense collection of combinatorial lines. Here we see the great advantage of equal-slices measure: if we reinterpret "dense" to mean "equal-slices dense", then it *is* the case that a dense subset of $[k]^n$ contains a dense collection of combinatorial lines. As one might expect, this follows from the density Hales–Jewett theorem and an averaging argument.

So now, after passing to the subspace $S_{J,y}$ and switching to equal-slices measure, we have the analogue of the statement in the corners theorem that

A is disjoint from a large Cartesian product. We have shown that A has equal-slices density at least $\delta - \varepsilon$ on $S_{J,y}$ and is disjoint from an equal-slices-large set of the form $\mathcal{U} \boxtimes \mathcal{V}$.

The next step is easy. By changing at least one of $\mathcal{U}$ and $\mathcal{V}$ to its complement, we obtain, just as we did in the corners proof, an equal-slices-dense set of the form $\mathcal{W} \boxtimes \mathcal{Z}$ inside which the equal-slices density of A is at least $\delta + \gamma$ for some constant γ that depends on δ only.

To complete this stage of the proof, we then switch back to the uniform measure. It is possible to approximate equal-slices measure by an average of uniform measures on subspaces, and restricting a set of the form $\mathcal{W} \boxtimes \mathcal{Z}$ to a subspace gives another set of the same form. This allows us to use an averaging argument to pass to a subspace in which the conclusions we have just drawn for equal-slices measure apply with the uniform measure instead.

To conclude, we have managed to obtain a density increment on a dense set of the form $\mathcal{W} \boxtimes \mathcal{Z}$ but we had to pass to a subspace to do it.

9. Partitioning 23-Insensitive Sets into Combinatorial Subspaces

In the proof of the corners theorem, once we found a set $X \times Y$ inside which A had density slightly greater than δ, it was sufficient to partition almost all of $X \times Y$ into grids of size tending to infinity. The analogous statement here is that it is sufficient to partition almost all of a set of the form $\mathcal{W} \boxtimes \mathcal{Z}$ into combinatorial subspaces with dimension tending to infinity.

In the proof of the corners theorem, we could simplify the task further: it was sufficient to partition X into long arithmetic progressions. Once we had done that, we used a very simple argument to show that a set of the form $P \times [N]$ could be partitioned into grids. This allowed us to partition $X \times [N]$ into grids, and then we partitioned the intersection of each of these grids with $[N] \times Y$ into yet smaller grids.

The first two steps taken together give us the following statement: any subset of $[N]^2$ of the form $X \times [N]$ can be partitioned into grids (plus a small error set – for simplicity I am sometimes not bothering to mention this).

The analogue in $[3]^n$ of a set of the form $X \times [N]$ is what we call a *23-insensitive set.* The characteristic property of a set of the form $X \times [N]$ is that whether or not a point (x, y) belongs to the set depends entirely on x. The natural analogue of this if we use three coordinates is to say that whether or not (x, y, z) belongs to the set depends only on x. In that sense, the set is "insensitive" to changes to the second and third coordinates. Thus, it makes sense to define a subset of $[3]^n$ to be 23-insensitive if whether or not a point (X, Y, Z) belongs to the subset depends only on X. Equivalently, it is a set of the form $\{(X, Y, Z) : X \in \mathcal{U}\}$ for some set-system $\mathcal{U}$.

For the next part of the proof, it will be more convenient to use the language of sequences with values in $\{1, 2, 3\}$. In these terms, a set A of sequences is 23-insensitive if changing 2s to 3s or vice versa in some sequence $x \in [3]^n$ makes no difference to whether x belongs to A.

After these remarks, it should not be surprising that the proof will be essentially finished if we can show that every 23-insensitive set can be almost entirely partitioned into combinatorial subspaces with dimensions tending to infinity.

Note that it was at the corresponding stage of the corners proof that we used Szemerédi's theorem, so this is the point where we must get away with something weaker. Why might we expect the multidimensional Sperner theorem to be helpful (apart from the fact that it's basically all we've got)? The reason is that if a 23-insensitive set contains a copy of $[2]^m$ (meaning that you can fix some coordinates and take m wildcard sets and set the values in any wildcard set to be 1 or 2), then it automatically contains a copy of $[3]^m$ (because if you change the values of the wildcards from 2 to 3 then it will not change whether you belong to the set, and when they take the value 2 you do belong to the set). Those familiar with Shelah's proof of the Hales–Jewett theorem [11] will recognise an echo of it here.

With the Ajtai–Szemerédi proof in mind, one might expect the argument to be something like this. Let A be a 23-insensitive set. Use the multidimensional Sperner theorem to find in A a copy S of $[2]^m$ for some m that tends to infinity. This 2-subspace plays the role played by the arithmetic progression P in the Ajtai–Szemerédi argument. Using the 23-insensitivity of A, find a 23-insensitive set that contains S and is also contained in A. This set plays the role of $P \times [N]$. Partition it into combinatorial subspaces, remove it from A and repeat.

Unfortunately, this argument runs into difficulties. The problem is that the smallest 23-insensitive set that contains S contains too many sequences.

For example, if E is a wildcard set of S and x is a sequence in S that takes the value 1 on E, then any 23-insensitive set that contains S will have to contain all sequences that can be formed by changing some of the 1s in E to 2s. But if a sequence is not constant on E, then it is not clear how we can include it in a disjoint union of combinatorial subspaces. If we try to get round this difficulty by letting the role of $P \times [N]$ be played by just the combinatorial subspace that has the same fixed coordinates and wildcard sets as S (but is allowed to take the value 3 on the wildcard sets), then we no longer have a 23-insensitive set, so when we remove it from A we cannot repeat the argument.

This is a genuine difference between our proof and the Ajtai–Szemerédi proof. Because of this difficulty, we are forced to use a more complicated argument, though ultimately it does still boil down to the fact that the 23-insensitivity of A allows us to convert copies of $[2]^m$ into copies of $[3]^m$.

Here is how it goes. Let A be a 23-insensitive subset of $[3]^n$. Let m and n be positive integers with m much much smaller than n (but still tending to infinity with n). Let J be a random subset of $[n]$ of size m. A simple technical lemma shows that if you choose a random sequence in $[2]^J$ and randomly extend it to a sequence in $[3]^n$, then, provided m is reasonably small, which it is, the resulting distribution is approximately uniform on $[3]^n$ (despite not being allowed to take the value 3 in J). Let us denote a typical point of $[3]^n$ by (x, y), where $x \in [3]^J$ and $y \in [3]^{[n]-J}$. For each (J, y) let $A_{J,y}$ be $\{x \in [2]^J : (x, y) \in A\}$. Let B_J be the set of y such that $A_{J,y}$ has density at least $\delta/2$ in $[3]^m$. Then the average density of B_J is at least $\delta/3$ in $[3]^{n-m}$, by the technical lemma and an averaging argument.

Choose J such that B_J has density at least $\delta/3$. Without loss of generality, $J = [m]$. Now write B for B_J and A_y for $A_{J,y}$. So we are given that B has density $\delta/3$ and that for every $y \in B$ the set $\{x \in [2]^m : (x, y) \in A\}$ has density at least $\delta/3$.

The multidimensional Sperner theorem tells us that for every $y \in B$ the set A_y contains an r-dimensional combinatorial subspace S_y (over the set [2]), for some r that tends to infinity with m. By the pigeonhole principle, there is a subset $C \subset B$ of density at least $\gamma = (r+2)^{-m}(\delta/3)$ and an r-dimensional combinatorial subspace S such that $S_y = S$ for every $y \in B'$. (We are using here the simple result that the number of r-dimensional combinatorial subspaces of $[2]^m$ is at most $(r+2)^m$.) We may assume that B' is maximal with this property.

The set $S \times B'$ is a subset of A. Let T be the combinatorial subspace of $[3]^m$ that has the same fixed coordinates and wildcard sets as S. By the 23-insensitivity of A, $T \times B'$ is also a subset of A. Note that it has density at least $3^{-m}\gamma$ and is a union of r-dimensional combinatorial subspaces.

We now remove $T \times B'$ from A. Since $T \times B'$ is not 23-insensitive, there is no reason to suppose that $A \backslash T \times B'$ is 23-insensitive either. However, at this point we exploit a crucial property of $T \times B'$ that we have not mentioned, which explains why we chose B' to be maximal: *the 23-sensitivity of $T \times B'$ affects only the first m coordinates.* Indeed, y belongs to B' if and only if $S \times \{y\} \subset A$ (the "if" following from the maximality of B'), so the 23-insensitivity of A implies that B' is also 23-insensitive.

In order to exploit this, we simply partition $A_1 = A \setminus (T \times B')$ into (at most) 3^m sets according to what the first m coordinates are. Writing $A_1(x)$ for $\{y \in [3]^{n-m} : (x, y) \in A_1\}$, we see that for every x the set $A_1(x)$ is a 23-insensitive subset of $[3]^{n-m}$. By induction, either $A_1(x)$ is extremely sparse or it can be almost entirely partitioned into combinatorial subspaces. Taking the union over all x, we find that A_1 can be almost entirely partitioned into r-dimensional combinatorial subspaces.

Some care is needed to ensure (i) that the size of the resulting error set is not too large and (ii) that the size of the ground set remains large enough for the argument to continue to be valid as the induction proceeds. But neither presents any serious problem. Provided each contribution to the error set is a sparse subset of some combinatorial subspace and all those subspaces are disjoint, the combined density of all the contributions cannot be large. And in order to ensure that the ground set remains large enough, we need to make sure that if θ is the density of the union of combinatorial subspaces that we remove from A at each stage of the induction, then m is at most $\theta n/2$, say. That way, since the induction cannot continue for more than θ^{-1} steps, the ground set always has size at least $n - \theta^{-1}m \geq n/2$.

Since θ is some constant that is derived from the multidimensional Sperner theorem, and later from lower cases of the multidimensional density Hales–Jewett theorem, and since the multidimensional density Hales–Jewett theorem is proved by iteratively applying the density Hales–Jewett theorem, m has to be very small indeed. For this reason, the bound that we eventually obtain is of Ackermann type.

We have not quite finished the proof, but the rest is easy. Recall that what we actually wanted to do was partition almost all of a set of the form $\mathcal{W} \boxtimes \mathcal{Z}$ into combinatorial subspaces. We are splitting this task into two by noting that $\mathcal{W} \boxtimes \mathcal{Z} = \big(\mathcal{W} \boxtimes \mathbb{P}([n])\big) \cap \big(\mathbb{P}([n]) \boxtimes \mathcal{Z}\big)$. That is, we are expressing it as the intersection of a 23-insensitive set and a 13-insensitive set. We have just proved that $\mathcal{W} \boxtimes \mathbb{P}([n])$, the 23-insensitive set, can be almost entirely partitioned into combinatorial subspaces. To complete the proof, we note that the intersection of $\mathbb{P}([n]) \boxtimes \mathcal{Z}$ with any one of these subspaces remains 13-insensitive (inside that subspace). We can therefore repeat the argument with 1 and 2 exchanged and end up with the desired partition of almost all of $\mathcal{W} \boxtimes \mathcal{Z}$.

10. Conclusion and Open Problems

We have shown that it is possible, though not entirely straightforward, to mimic Ajtai and Szemerédi's proof of the corners theorem to prove the density Hales–Jewett theorem in the case $k = 3$. In order to prove the result for general k it is helpful to start with the following exercise: assuming the $(k-1)$-dimensional Szemerédi theorem, prove the k-dimensional corners theorem. The rough idea is this. Let A be a dense subset of $[N]^k$. By averaging, find a hyperplane H of the form $\{(x_1, \ldots, x_k) : x_1 + \cdots + x_k = z\}$ inside which A is dense. For each i let X_i be the projection of $A \cap H$ in direction i and let Y_i be the "$(i, k-1)$-insensitive set" of all points $(x_1, \ldots, x_k)$ such that $(x_1, \ldots, x_{i-1}, x_{i+1}, \ldots, x_k) \in X_i$. (We can think of Y_i as a Cartesian product of X_i with $[N]$, but $[N]$ has to occupy the ith coordinate.) Then the intersection $X_1 \cap \cdots \cap X_k$ must be disjoint from A or A will contain a k-dimensional corner. (For the purposes of this discussion I am glossing over the question of ensuring that $d > 0$, but this can be done by a similar method to the one we used in the 2-dimensional case.)

Using the $(k-1)$-dimensional Szemerédi theorem, we can partition the set Y_1 almost entirely into k-dimensional grids, just as we did when $k = 2$. And then we can pass to each of those grids and partition its intersection with Y_2, and so on.

With that argument as a model, it is now routine to generalize the proof of the density Hales–Jewett theorem from the case $k = 3$ to the general case, with the role of the multidimensional Sperner theorem being played by the multidimensional density Hales–Jewett theorem for $k - 1$.

We end this paper with two open problems. First is the obvious problem about bounds. As mentioned earlier, the proof in this paper gives rise to a bound of Ackermann type, mainly because of our use of an iteration to prove the multidimensional theorem. It seems that to obtain a "reasonable" bound would require a radically new idea. However, what if we set ourselves the more modest aim of obtaining a primitive recursive (but possibly quite bad) bound? There are enough similarities between our proof and the proof of the Hales–Jewett theorem for this to feel like a more realistic target. Perhaps the proof can be reorganized à la Shelah, so that instead of using the multidimensional theorem for $k-1$ to prove the result for k, we do the induction "the other way round" and use the much less expensive multidimensional Sperner theorem.

The other problem is more specialized. Recall that the first main step of the proof was to find a large disjoint product $\mathcal{U} \boxtimes \mathcal{V}$ inside which A had increased density, but that to do this we had to use equal-slices measure and pass to a subspace. The use of equal-slices measure is very natural, but do we really have to pass to a subspace? We do not know of any examples that show that it is necessary. Accordingly, we ask the following rough question. Let A be an equal-slices dense subset of $[3]^n$ that contains no combinatorial lines. Does this imply that there is an equal-slices dense disjoint product $\mathcal{U} \boxtimes \mathcal{V}$ inside which A has increased density?

Clearly, the answer, "Yes, since the density Hales–Jewett theorem shows that no such A exists," is not satisfactory. To formulate a precise question that avoids this problem, one can either ask for much better bounds than the density Hales–Jewett theorem would imply (that is, one can ask the question when n is not necessarily a vast function of the density of A), or one can ask for a direct proof, or one can change the question slightly and assume merely that A contains few combinatorial lines, in some suitable sense of the word "few".

A slightly less precise question might also be worth thinking about. Suppose that A is an equal-slices dense subset of $[3]^n$, and suppose that the number of combinatorial lines in A is significantly different from the number that you would expect in a random set of the same density. Must A have increased density on some dense "structured" subset of $[3]^n$? For a positive answer to be interesting, a minimum requirement of the notion of "structure" would be that a structured set could be almost entirely partitioned into combinatorial subspaces.

REFERENCES

[1] M. Ajtai and E. Szemerédi, Sets of lattice points that form no squares, *Stud. Sci. Math. Hungar.*, **9** (1974), 9–11.

[2] T. Austin, *Deducing the density Hales–Jewett theorem from an infinitary removal lemma,* `http://arxiv.org/abs/0903.1633, 2009.`

[3] V. Bergelson and A. Leibman, Polynomial extensions of van der Waerden's and Szemerédi's theorems, *Jour. AMS,* **9** (1996), 725–753.

[4] H. Furstenberg, Ergodic behavior of diagonal measures and a theorem of Szemerédi on arithmetic progressions, *J. d'Analyse Math.*, **31** (1977), 204–256.

[5] H. Furstenberg and Y. Katznelson, An ergodic Szemerédi theorem for commuting transformations, *J. d'Analyse Math.,* **34** (1978), 275–291.

[6] H. Furstenberg and Y. Katznelson, A density version of the Hales–Jewett Theorem, *J. d'Analyse Math.*, **57** (1991), 64–119.

[7] A. W. Hales and R. I. Jewett, Regularity and positional games, *Trans. AMS,* **106(2)** (1963), 222–229.

[8] D. H. J. Polymath, *A new proof of the density Hales–Jewett theorem,* `http://arxiv.org/abs/0910.3926`

[9] D. H. J. Polymath, *Density Hales–Jewett and Moser numbers,* this volume.

[10] I. Z. Ruzsa and E. Szemerédi, Triple systems with no six points carrying three triangles, *Colloq. Math. Soc. J. Bolyai,* **18** (1978), 939–945.

[11] S. Shelah, Primitive recursive bounds for van der Waerden numbers, *Jour. AMS,* 1:683-697, 1988.

[12] J. Solymosi, A note on a question of Erdős and Graham, *Combin. Probab. Comput.*, **13(2)** (2004), 263–267.

[13] E. Szemerédi, On sets of integers containing no k elements in arithmetic progression, *Acta Arith.*, **27** (1975), 199–245.

[14] B. L. van der Waerden, Beweis einer Baudetschen Vermutung, *Nieuw Archief voor Wiskunde,* **15** (1927), 212–216.

W. T. Gowers

University of Cambridge
Department of Pure Mathematics and
Mathematical Statistics
Wilberforce Road
Cambridge CB3 0WB
UK.

`e-mail: w.t.gowers@dpmms.cam.ac.uk`

BOLYAI SOCIETY
MATHEMATICAL STUDIES, 21

An Irregular Mind
(Szemerédi is 70)
pp. 689–753.

Density Hales–Jewett and Moser numbers

D. H. J. POLYMATH

For any $n \geq 0$ and $k \geq 1$, the *density Hales–Jewett number* $c_{n,k}$ is defined as the size of the largest subset of the cube $[k]^n := \{1,\ldots,k\}^n$ which contains no combinatorial line; similarly, the Moser number $c'_{n,k}$ is the largest subset of the cube $[k]^n$ which contains no geometric line. A deep theorem of Furstenberg and Katznelson [11], [12], [19] shows that $c_{n,k} = o(k^n)$ as $n \to \infty$ (which implies a similar claim for $c'_{n,k}$); this is already non-trivial for $k = 3$. Several new proofs of this result have also been recently established [23], [2].

Using both human and computer-assisted arguments, we compute several values of $c_{n,k}$ and $c'_{n,k}$ for small n, k. For instance the sequence $c_{n,3}$ for $n = 0,\ldots,6$ is 1, 2, 6, 18, 52, 150, 450, while the sequence $c'_{n,3}$ for $n = 0,\ldots,6$ is 1, 2, 6, 16, 43, 124, 353. We also prove some results for higher k, showing for instance that an analogue of the LYM inequality (which relates to the $k = 2$ case) does not hold for higher k, and also establishing the asymptotic lower bound $c_{n,k} \geq k^n \exp\left(-O(\sqrt[\ell]{\log n}\,)\right)$ where ℓ is the largest integer such that $2k > 2^\ell$.

1. Introduction

For any integers $k \geq 1$ and $n \geq 0$, let $[k] := \{1,\ldots,k\}$, and define $[k]^n$ to be the cube of words of length n with alphabet in $[k]$. Thus for instance $[3]^2 = \{11,12,13,21,22,23,31,32,33\}$.

We define a *combinatorial line* in $[k]^n$ to be a set of the form $\{w(i) : i = 1,\ldots,k\} \subset [k]^n$, where $w \in \left([k] \cup \{x\}\right)^n \backslash [k]^n$ is a word of length n with alphabet in $[k]$ together with a "wildcard" letter x which appears at least once, and $w(i) \in [k]^n$ is the word obtained from w by replacing x by i; we often abuse notation and identify w with the combinatorial line $\{w(i) : i = 1,\ldots,k\}$ it generates. Thus for instance, in $[3]^2$ we have $x2 = \{12,22,32\}$ and $xx = \{11,22,33\}$ as typical examples of combinatorial lines. In general, $[k]^n$ has k^n words and $(k+1)^n - k^n$ lines.

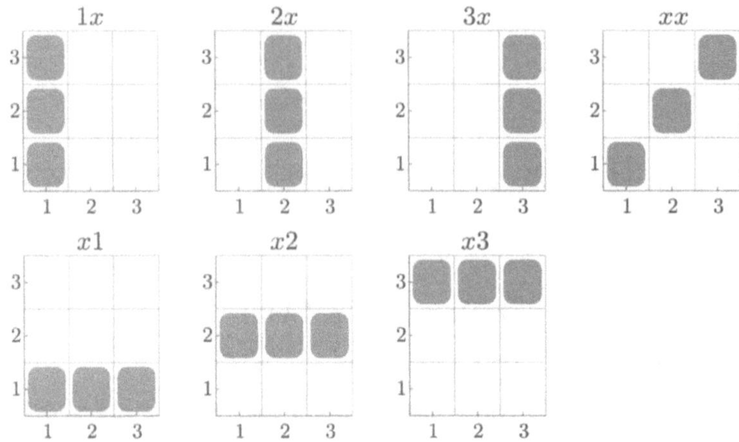

Fig. 1. Combinatorial lines in $[3]^2$.

A set $A \subset [k]^n$ is said to be *line-free* if it contains no combinatorial lines. Define the (n,k) *density Hales–Jewett number* $c_{n,k}$ to be the maximum cardinality $|A|$ of a line-free subset of $[k]^n$. Clearly, one has the trivial bound $c_{n,k} \leq k^n$. A deep theorem of Furstenberg and Katznelson [11], [12] asserts that this bound can be asymptotically improved:

Theorem 1.1 (Density Hales–Jewett theorem). *For any fixed $k \geq 2$, one has $\lim_{n\to\infty} c_{n,k}/k^n = 0$.*

Remark 1.2. The difficulty of this theorem increases with k. For $k = 1$, one clearly has $c_{n,1} = 1$. For $k = 2$, a classical theorem of Sperner [28] asserts, in our language, that $c_{n,2} = \binom{n}{\lfloor n/2 \rfloor}$. The case $k = 3$ is already non-trivial (for instance, it implies Roth's theorem [26] on arithmetic progressions of length three) and was first established in [11] (see also [19]). The case of general k was first established in [12] and has a number of implications, in particular implying Szemerédi's theorem [29] on arithmetic progressions of arbitrary length.

The Furstenberg–Katznelson proof of Theorem 1.1 relies on ergodic-theory techniques and does not give an explicit decay rate for $c_{n,k}$. Recently, two further proofs of this theorem have appeared, by Austin [2] and by the sister Polymath project to this one [23]. The proof of [2] also uses ergodic theory, but the proof in [23] is combinatorial and gave effective bounds for $c_{n,k}$ in the limit $n \to \infty$. For example, if n can be written as an exponential tower $2 \uparrow 2 \uparrow 2 \uparrow \ldots \uparrow 2$ with m 2s, then $c_{n,3} \ll 3^n m^{-1/2}$. However, these bounds are not believed to be sharp, and in any case are only non-trivial in the asymptotic regime when n is sufficiently large depending on k.

Our first result is the following asymptotic lower bound. The construction is based on the recent refinements [9, 14, 20] of a well-known construction of Behrend [4] and Rankin [25]. The proof of Theorem 1.3 is in Section 2. Let $r_k(n)$ be the maximum size of a subset of $[n]$ that does not contain a k-term arithmetic progression.

Theorem 1.3 (Asymptotic lower bound for $c_{n,k}$). *For each $k \geq 3$, there is an absolute constant $C > 0$ such that*

$$c_{n,k} \geq Ck^n \left(\frac{r_k(\sqrt{n})}{\sqrt{n}} \right)^{k-1} = k^n \exp\big(- O(\sqrt[\ell]{\log n}\,)\big),$$

where ℓ is the largest integer satisfying $2k > 2^\ell$. More specifically,

$$c_{n,k} \geq Ck^{n-\alpha(k)\sqrt[\ell]{\log n}+\beta(k)\log\log n},$$

where all logarithms are base-k, and $\alpha(k) = (\log 2)^{1-1/\ell}\ell 2^{(\ell-1)/2-1/\ell}$ and $\beta(k) = (k-1)/(2\ell)$.

In the case of small n, we focus primarily on the first non-trivial case $k = 3$. We have computed the following explicit values of $c_{n,3}$ (entered in the OEIS [21] as A156762):

Theorem 1.4 (Explicit values of $c_{n,3}$). *We have* $c_{0,3} = 1$, $c_{1,3} = 2$, $c_{2,3} = 6$, $c_{3,3} = 18$, $c_{4,3} = 52$, $c_{5,3} = 150$, *and* $c_{6,3} = 450$.

This result is established in Sections 2, 3. Initially these results were established by an integer program, but we provide completely computer-free proofs here. The constructions used in Section 2 give reasonably efficient constructions for larger values of n; for instance, they show that $3^{99} \leq c_{100,3} \leq 2 \times 3^{99}$. See Section 2 for further discussion.

A variant of the density Hales–Jewett theorem has also been studied in the literature. Define a *geometric line* in $[k]^n$ to be any set of the form $\{a + ir : i = 1, \dots, k\}$ in $[k]^n$, where we identify $[k]^n$ with a subset of $\mathbb{Z}^n$, and $a, r \in \mathbb{Z}^n$ with $r \neq 0$. Equivalently, a geometric line takes the form $\{w(i, k+1-i) : i = 1, \dots, k\}$, where $w \in \big([k] \cup \{x, \overline{x}\}\big)^n \backslash [k]^n$ is a word of length n using the numbers in $[k]$ and two wildcards $x, \overline{x}$ as the alphabet, with at least one wildcard occurring in w, and $w(i, j) \in [k]^n$ is the word formed by substituting i, j for $x, \overline{x}$ respectively. Figure 2 shows the eight

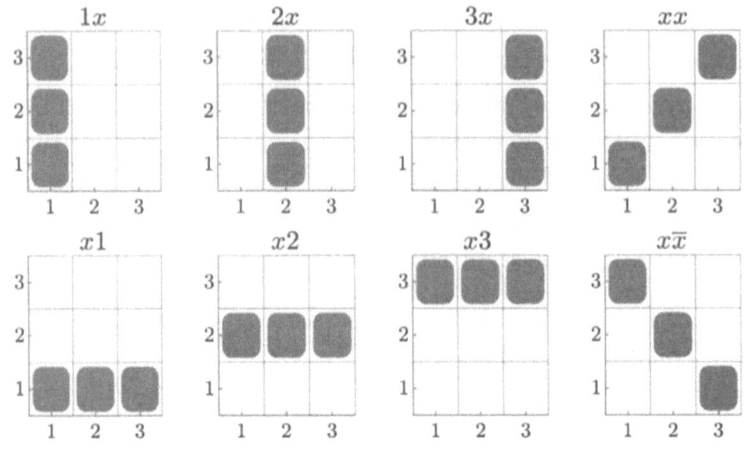

Fig. 2. Geometric lines in $[3]^2$.

geometric lines in $[3]^2$. Clearly every combinatorial line is a geometric line, but not conversely. In general, $[k]^n$ has $\left((k+2)^n - k^n\right)/2$ geometric lines.

Define a *Moser set* in $[k]^n$ to be a subset of $[k]^n$ that contains no geometric lines, and let $c'_{n,k}$ be the maximum cardinality $|A|$ of a Moser set in $[k]^n$. Clearly one has $c'_{n,k} \leq c_{n,k}$, so in particular from Theorem 1.1 one has $c'_{n,k}/k^n \to 0$ as $n \to \infty$. (Interestingly, there is no known proof of this fact that does not go through Theorem 1.1, even for $k = 3$.) Again, $k = 3$ is the first non-trivial case: it is clear that $c'_{n,1} = 0$ and $c'_{n,2} = 1$ for all n.

The question of computing $c'_{n,3}$ was first posed by Moser [18]. Prior to our work, the values

$$c'_{0,3} = 1; \quad c'_{1,3} = 2; \quad c'_{2,3} = 6; \quad c'_{3,3} = 16; \quad c'_{4,3} = 43$$

were known [8], [6] (this is Sequence A003142 in the OEIS [21]). We extend this sequence slightly:

Theorem 1.5 (Values of $c'_{n,3}$ for small n). *We have* $c'_{0,3} = 1$, $c'_{1,3} = 2$, $c'_{2,3} = 6$, $c'_{3,3} = 16$, $c'_{4,3} = 43$, $c'_{5,3} = 124$, *and* $c'_{6,3} = 353$.

This result is established in Sections 4, 5. The arguments given here are computer-assisted; however, we have found alternate (but lengthier) computer-free proofs for the above claims with the the exception of the proof of $c'_{6,3} = 353$, which requires one non-trivial computation (Lemma 5.13). These alternate proofs are not given in this paper to save space, but can be found at [24].

We establish a lower bound for this problem of $(2+o(1))\binom{n}{i}2^i \le c'_{n,3}$, which is maximized for i near $2n/3$. This bound is around one-third better than the previous literature [18], [7]. We also give methods to improve on this construction.

Earlier lower bounds were known. Indeed, let $A(n,d)$ denote the size of the largest binary code of length n and minimal distance d. Then

$$c'_{n,3} \ge \max_k \left(\sum_{j=0}^{k} \binom{n}{j} A(n-j, k-j+1) \right). \tag{1.1}$$

which, with $A(n,1) = 2^n$ and $A(n,2) = 2^{n-1}$, implies in particular that

$$c'_{n,3} \ge \binom{n+1}{\lfloor \frac{2n+1}{3} \rfloor} 2^{\lfloor \frac{2n+1}{3} \rfloor - 1} \tag{1.2}$$

for $n \ge 2$. This bound is not quite optimal; for instance, it gives a lower bound of $c'_{6,3} \ge 344$.

Remark 1.6. Let $c''_{n,3}$ be the size of the largest subset of $\mathbb{F}_3^n$ which contains no lines $x, x+r, x+2r$ with $x, r \in \mathbb{F}_3^n$ and $r \neq 0$, where $\mathbb{F}_3$ is the field of three elements. Clearly one has $c''_{n,3} \le c'_{n,3} \le c_{n,3}$. It is known that

$$c''_{0,3} = 1; \quad c''_{1,3} = 2; \quad c''_{2,3} = 4; \quad c'_{3,3} = 9; \quad c'_{4,3} = 20; \quad c''_{5,3} = 45; \quad c''_{6,3} = 112;$$

see [22].

As mentioned earlier, the sharp bound on $c_{n,2}$ comes from Sperner's theorem. It is known that Sperner's theorem can be refined to the *Lubell–Yamamoto–Meshalkin (LYM) inequality,* which in our language asserts that

$$\sum_{a_1,a_2 \ge 0; a_1+a_2=n} \frac{|A \cap \Gamma_{a_1,a_2}|}{|\Gamma_{a_1,a_2}|} \le 1$$

for any line-free subset $A \subset [2]^n$, where the *cell* $\Gamma_{a_1,\dots,a_k} \subset [k]^n$ is the set of words in $[k]^n$ which contain exactly a_i i's for each $i = 1, \dots, k$. It is natural to ask whether this inequality can be extended to higher k. Let $\Delta_{n,k}$ denote the set of all tuples $(a_1, \dots, a_k)$ of non-negative integers summing to n, define a *simplex* to be a set of k points in $\Delta_{n,k}$ of the form $(a_1+r, a_2, \dots, a_k)$, $(a_1, a_2+r, \dots, a_k)$, $\dots$, $(a_1, a_2, \dots, a_k+r)$ for some $0 < r \le n$ and $a_1, \dots, a_k$

summing to $n-r$, and define a *Fujimura set*[1] to be a subset $B \subset \Delta_{n,k}$ which contains no simplices. Observe that if w is a combinatorial line in $[k]^n$, then

$$w(1) \in \Gamma_{a_1+r,a_2,\ldots,a_k}, w(2) \in \Gamma_{a_1,a_2+r,\ldots,a_k}, \ldots, w(k) \in \Gamma_{a_1,a_2,\ldots,a_k+r}$$

for some simplex $(a_1 + r, a_2, \ldots, a_k)$, $(a_1, a_2 + r, \ldots, a_k)$, $\ldots$, $(a_1, a_2, \ldots, a_k + r)$. Thus, if B is a Fujimura set, then $A := \bigcup_{\vec{a}\in B} \Gamma_{\vec{a}}$ is line-free. Note also that

$$\sum_{\vec{a}\in\Delta_{n,k}} \frac{|A \cap \Gamma_{\vec{a}}|}{|\Gamma_{\vec{a}}|} = |B|.$$

This motivates a "hyper-optimistic" conjecture:

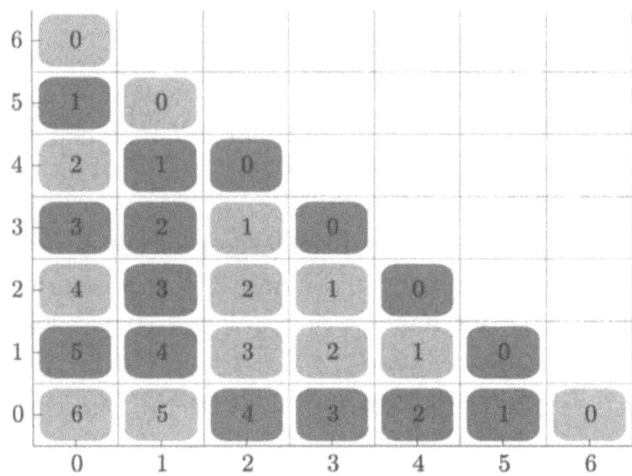

Fig. 3. A Fujimura set in $\Delta_{6,3}$, displayed in "rectangular" coordinates. The point (a, b, c) is represented by a square at (a, b) labeled with c. The Fujimura set is shown in red; its complement in $\Delta_{6,3}$ is shown in gray.

Conjecture 1.7. For any $k \geq 1$ and $n \geq 0$, and any line-free subset A of $[k]^n$, one has

$$\sum_{\vec{a}\in\Delta_{n,k}} \frac{|A \cap \Gamma_{\vec{a}}|}{|\Gamma_{\vec{a}}|} \leq c^{\mu}_{n,k},$$

where $c^{\mu}_{n,k}$ is the maximal size of a Fujimura set in $\Delta_{n,k}$.

[1]Fujimura actually proposed the related problem of finding the largest subset of $\Delta_{n,k}$ that contained no equilateral triangles; see [10]. Our results for Fujimura sets can be found at the page `Fujimura's problem` at [24].

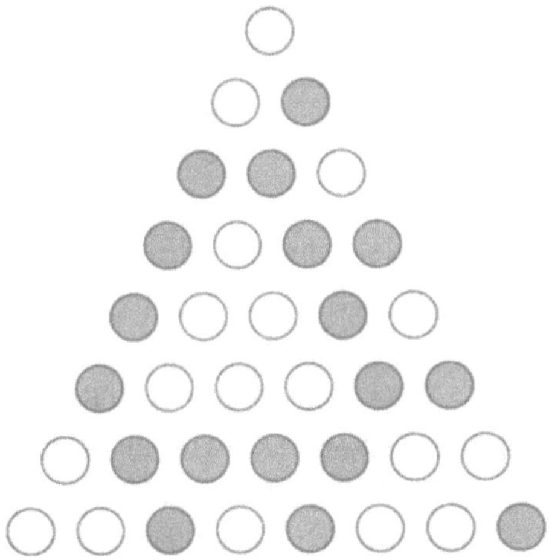

Fig. 4. A Fujimura set in $\Delta_{7,3}$, expressed in "triangular" coordinates.

One can show that this conjecture for a fixed value of k would imply Theorem 1.1 for the same value of k, in much the same way that the LYM inequality is known to imply Sperner's theorem. The LYM inequality asserts that Conjecture 1.7 is true for $k \leq 2$. As far as we know, this conjecture could hold in $k = 3$. However, we found a simple counterexample for $k = 4$ and $n = 2$, given by the line-free set

$$A := \{(1,1),(1,2),(1,3),(2,1),(2,3),(2,4),(3,2),(3,3),\\ (3,4),(4,1),(4,2),(4,4)\}$$

together with the computation that $c^{\mu}_{4,2} = 7$. It is in fact likely that this conjecture fails for all higher k also.

1.1. Notation

There are several subsets of $[k]^n$ which will be useful in our analysis. We have already introduced combinatorial lines, geometric lines, and cells. One can generalise the notion of a combinatorial line to that of a *combinatorial subspace* in $[k]^n$ of dimension d, which is indexed by a word w in $\big([k] \cup \{x_1,\ldots,x_d\}\big)^n$ containing at least one of each wildcard $x_1,\ldots,x_d$, and which forms the set $\{w(i_1,\ldots,i_d) : i_1,\ldots,i_d \in [k]\}$, where $w(i_1,\ldots,i_d) \in [k]^d$ is the word formed by replacing $x_1,\ldots,x_d$ with $i_1,\ldots,i_d$ respectively. Thus for instance, in $[3]^3$, we have the two-dimensional combinatorial subspace

$xxy = \{111, 112, 113, 221, 222, 223, 331, 332, 333\}$. We similarly have the notion of a *geometric subspace* in $[k]^n$ of dimension d, which is defined similarly but with d wildcards $x_1, \ldots, x_d, \overline{x_1}, \ldots, \overline{x_d}$, with at least one of either x_i or $\overline{x_i}$ appearing in the word w for each $1 \leq i \leq d$, and the space taking the form $\{w(i_1, \ldots, i_d, k+1-i_1, \ldots, k+1-i_d) : i_1, \ldots, i_d \in [k]\}$. Thus for instance $[3]^3$ contains the two-dimensional geometric subspace $x\overline{x}y = \{131, 132, 133, 221, 222, 223, 311, 312, 313\}$.

An important class of combinatorial subspaces in $[k]^n$ will be the *slices* consisting of $n-1$ distinct wildcards and one fixed coordinate. We will denote the distinct wildcards here by asterisks, thus for instance in $[3]^3$ we have $2 * * = \{211, 212, 213, 221, 222, 223, 231, 232, 233\}$. Two slices are *parallel* if their fixed coordinate are in the same position, thus for instance $1 * *$ and $2 * *$ are parallel, and one can subdivide $[k]^n$ into k parallel slices, each of which is isomorphic to $[k]^{n-1}$. In the analysis of Moser slices with $k = 3$, we will make a distinction between *centre slices*, whose fixed coordinate is equal to 2, and *side slices*, in which the fixed coordinate is either 1 or 3, thus $[3]^n$ can be partitioned into one centre slice and two side slices.

Another important set in the study of $k = 3$ Moser sets are the *spheres* $S_{i,n} \subset [3]^n$, defined as those words in $[3]^n$ with exactly $n - i$ 2's (and hence i letters that are 1 or 3). Thus for instance $S_{1,3} = \{122, 322, 212, 232, 221, 223\}$. Observe that $[3]^n = \bigcup_{i=0}^{n} S_{i,n}$, and each $S_{i,n}$ has cardinality $|S_{i,n}| = \binom{n}{i} 2^i$.

It is also convenient to subdivide each sphere $S_{i,n}$ into two components $S_{i,n} = S^o_{i,n} \cup S^e_{i,n}$, where $S^o_{i,n}$ are the words in $S_{i,n}$ with an odd number of 1's, and $S^e_{i,n}$ are the words with an even number of 1's. Thus for instance $S^o_{1,3} = \{122, 212, 221\}$ and $S^e_{1,3} = \{322, 232, 223\}$. Observe that for $i > 0$, $S^o_{i,n}$ and $S^e_{i,n}$ both have cardinality $\binom{n}{i} 2^{i-1}$.

The *Hamming distance* between two words w, w' is the number of coordinates in which w, w' differ, e.g. the Hamming distance between 123 and 321 is two. Note that $S_{i,n}$ is nothing more than the set of words whose Hamming distance from $2 \ldots 2$ is i, which justifies the terminology "sphere".

In the density Hales–Jewett problem, there are two types of symmetries on $[k]^n$ which map combinatorial lines to combinatorial lines (and hence line-free sets to line-free sets). The first is a permutation of the alphabet $[k]$; the second is a permutation of the n coordinates. Together, this gives a symmetry group of order $k!n!$ on the cube $[k]^n$, which we refer to as the *combinatorial symmetry group* of the cube $[k]^n$. Two sets which are related by an element of this symmetry group will be called (combinatorially) *equivalent*, thus for instance any two slices are combinatorially equivalent.

For the analysis of Moser sets in $[k]^n$, the symmetries are a bit different. One can still permute the n coordinates, but one is no longer free to permute the alphabet $[k]$. Instead, one can *reflect* an individual coordinate, for instance sending each word $x_1 \dots x_n$ to its reflection $x_1 \dots x_{i-1}(k+1-x_i)x_{i+1} \dots x_n$. Together, this gives a symmetry group of order $2^n n!$ on the cube $[k]^n$, which we refer to as the *geometric symmetry group* of the cube $[k]^n$; this group maps geometric lines to geometric lines, and thus maps Moser sets to Moser sets. Two Moser sets which are related by an element of this symmetry group will be called (geometrically) *equivalent*. For instance, a sphere $S_{i,n}$ is equivalent only to itself, and $S^o_{i,n}$, $S^e_{i,n}$ are equivalent only to each other.

1.2. About this project

This paper is part of the *Polymath project*, which was launched by Timothy Gowers in February 2009 as an experiment to see if research mathematics could be conducted by a massive online collaboration. The first project in this series, *Polymath1*, was focused on understanding the density Hales–Jewett numbers $c_{n,k}$, and was split up into two sub-projects, namely an (ultimately successful) attack on the density Hales–Jewett theorem $c_{n,k} = o(k^n)$ (resulting in the paper [23]), and a collaborative project on computing $c_{n,k}$ and related quantities (such as $c'_{n,k}$) for various small values of n and k. This project (which was administered by Terence Tao) resulted in this current paper.

Being such a collaborative project, many independent aspects of the problem were studied, with varying degrees of success. For reasons of space (and also due to the partial nature of some of the results), this paper does not encompass the entire collection of observations and achievements made during the research phase of the project (which lasted for approximately three months). In particular, alternate proofs of some of the results here have been omitted, as well as some auxiliary results on related numbers, such as coloring Hales–Jewett numbers. However, these results can be accessed from the web site of this project at [24]. We are indebted to Michael Nielsen for hosting this web site, which performed a crucial role in the project. A list of contributors to the project (and the grants that supported these individuals) can also be found at this site.

2. Lower Bounds for the Density Hales–Jewett Problem

The purpose of this section is to establish various lower bounds for $c_{n,3}$, in particular establishing Theorem 1.3 and the lower bound component of Theorem 1.4.

As observed in the introduction, if $B \subset \Delta_{n,3}$ is a Fujimura set (i.e. a subset of $\Delta_{n,3} = \{(a,b,c) \in \mathbb{N}^3 : a+b+c = n\}$ which contains no upward equilateral triangles $(a+r,b,c)$, $(a,b+r,c)$, $(a,b,c+r)$), then the set $A_B := \bigcup_{\vec{a}\in B} \Gamma_{a,b,c}$ is a line-free subset of $[3]^n$, which gives the lower bound

$$c_{n,3} \geq |A_B| = \sum_{(a,b,c)\in B} \frac{n!}{a!b!c!}. \tag{2.1}$$

All of the lower bounds for $c_{n,3}$ in this paper will be constructed via this device. (Indeed, one may conjecture that for every n there exists a Fujimura set B for which (2.1) is attained with equality; we know of no counterexamples to this conjecture.)

In order to use (2.1), one of course needs to build Fujimura sets B which are "large" in the sense that the right-hand side of (2.1) is large. A fruitful starting point for this goal is the sets

$$B_{j,n} := \{(a,b,c) \in \Delta_{n,3} : a+2b \neq j \bmod 3\}$$

for $j = 0,1,2$. Observe that in order for a triangle $(a+r,b,c)$, $(a,b+r,c)$, $(a,b,c+r)$ to lie in $B_{j,n}$, the length r of the triangle must be a multiple of 3. This already makes $B_{j,n}$ a Fujimura set for $n < 3$ (and $B_{0,n}$ a Fujimura set for $n = 3$).

When n is not a multiple of 3, the $B_{j,n}$ are all rotations of each other and give equivalent sets (of size $2 \times 3^{n-1}$). When n is a multiple of 3, the sets $B_{1,n}$ and $B_{2,n}$ are reflections of each other, but $B_{0,n}$ is not equivalent to the other two sets (in particular, it omits all three corners of $\Delta_{n,3}$); the associated set $A_{B_{0,n}}$ is slightly larger than $A_{B_{1,n}}$ and $A_{B_{2,n}}$ and thus is slightly better for constructing line-free sets.

As mentioned already, $B_{0,n}$ is a Fujimura set for $n \leq 3$, and hence $A_{B_{0,n}}$ is line-free for $n \leq 3$. Applying (2.1) one obtains the lower bounds

$$c_{0,3} \geq 1; \quad c_{1,3} \geq 2; \quad c_{2,3} \geq 6; \quad c_{3,3} \geq 18.$$

For $n > 3$, $B_{0,n}$ contains some triangles $(a+r, b, c)$, $(a, b+r, c)$, $(a, b, c+r)$ and so is not a Fujimura set, but one can remove points from this set to recover the Fujimura property. For instance, for $n \leq 6$, the only triangles in $B_{0,n}$ have side length $r = 3$. One can "delete" these triangles by removing one vertex from each; in order to optimise the bound (2.1) it is preferable to delete vertices near the corners of $\Delta_{n,3}$ rather than near the centre. These considerations lead to the Fujimura sets

$$B_{0,4} \backslash \{(0,0,4), (0,4,0), (4,0,0)\}$$

$$B_{0,5} \backslash \{(0,4,1), (0,5,0), (4,0,1), (5,0,0)\}$$

$$B_{0,6} \backslash \{(0,1,5), (0,5,1), (1,0,5), (0,1,5), (1,5,0), (5,1,0)\}$$

which by (2.1) gives the lower bounds

$$c_{4,3} \geq 52; \quad c_{5,3} \geq 150; \quad c_{6,3} \geq 450.$$

Thus we have established all the lower bounds needed for Theorem 1.4.

One can of course continue this process by hand, for instance the set

$$B_{0,7} \backslash \{(0,1,6), (1,0,6), (0,5,2), (5,0,2), (1,5,1), (5,1,1), (1,6,0), (6,1,0)\}$$

gives the lower bound $c_{7,3} \geq 1302$, which we tentatively conjecture to be the correct bound.

A simplification was found when n is a multiple of 3. Observe that for $n = 6$, the sets excluded from $B_{0,6}$ are all permutations of $(0,1,5)$. So the remaining sets are all the permutations of $(1,2,3)$ and $(0,2,4)$. In the same way, sets for $n = 9$, 12 and 15 can be described as:

- $n = 9$: $(2,3,4)$, $(1,3,5)$, $(0,4,5)$ and permutations;
- $n = 12$: $(3,4,5)$, $(2,4,6)$, $(1,5,6)$, $(0,2,10)$, $(0,5,7)$ and permutations;
- $n = 15$: $(4,5,6)$, $(3,5,7)$, $(2,6,7)$, $(1,3,11)$, $(1,6,8)$, $(0,4,11)$, $(0,7,8)$ and permutations.

When n is not a multiple of 3, say $n = 3m-1$ or $n = 3m-2$, one first finds a solution for $n = 3m$. Then for $n = 3m-1$, one restricts the first digit of the $3m$ sequence to equal 1. This leaves exactly one-third as many points for $3m-1$ as for $3m$. For $n = 3m-1$, one restricts the first two digits of the $3m$ sequence to be 12. This leaves roughly one-ninth as many points for $3m-2$ as for $3m$.

An integer program[2] was solved to obtain the maximum lower bound one could establish from (2.1). The results for $1 \leq n \leq 20$ are displayed in Figure 5. More complete data, including the list of optimisers, can be found at [17].

n	lower bound	n	lower bound
1	2	11	96338
2	6	12	287892
3	18	13	854139
4	52	14	2537821
5	150	15	7528835
6	450	16	22517082
7	1302	17	66944301
8	3780	18	198629224
9	11340	19	593911730
10	32864	20	1766894722

Fig. 5. Lower bounds for c_n obtained by the A_B construction.

For medium values of n, in particular for integers $21 \leq n \leq 999$ that are a multiple of 3, $n = 3m$, the best general lower bound for $c_{n,3}$ was found by applying (2.1) to the following Fujimura set construction. It is convenient to write $[a, b, c]$ for the point $(m+a, m+b, m+c)$, together with its permutations, with the convention that $[a, b, c]$ is empty if these points do not lie in $\Delta_{n,3}$. Then a Fujimura set can be constructed by taking the following groups of points:

1. The thirteen groups of points

$$[-7,-3,+10],\ [-7,0,+7],\ [-7,+3,+4],\ [-6,-4,+10],$$
$$[-6,-1,+7],\ [-6,+2,+4],\ [-5,-1,+6],\ [-5,+2,+3],$$
$$[-4,-2,+6],\ [-4,+1,+3],\ [-3,+1,+2],\ [-2,0,+2],\ [-1,0,+1];$$

2. The four families of groups

$$[-8-y-2x,-6+y-2x,14+4x],\ [-8-y-2x,-3+y-2x,11+4x],$$
$$[-8-y-2x, x+y, 8+x],\ [-8-2x, 3+x, 5+x]$$

for $x \geq 0$ and $y = 0, 1$.

[2]Details of the integer programming used in this paper can be found at the page `Integer.tex` at [24].

Numerical computation shows that this construction gives a line-free set in $[3]^n$ of density approximately $2.7\sqrt{\frac{\log n}{n}}$ for $n \leq 1000$; for instance, when $n = 99$, it gives a line-free set of density at least $1/3$. Some additional constructions of this type can be found at the page `Upper and lower bounds` at [24].

However, the bounds in Theorem 1.3, which we now prove, are asymptotically superior to these constructions.

Proof of Theorem 1.3. Let M be the circulant matrix with first row $(1, 2, \ldots, k-1)$, second row $(k-1, 1, 2, \ldots, k-2)$, and so on. Note that M has nonzero determinant by well-known properties[3] of circulant matrices, see e.g. [13, Theorem 3].

Let S be a subset of the interval $[-\sqrt{n}/2, \sqrt{n}/2)$ that contains no nonconstant arithmetic progressions of length k, and let $B \subset \Delta_{n,k}$ be the set

$$B := \left\{ \left(n - \sum_{i=1}^{k-1} a_i, a_1, a_2, \ldots, a_{k-1} \right) : \right.$$

$$\left. (a_1, \ldots, a_{k-1}) = c + \det(M) M^{-1} s,\ s \in S^{k-1} \right\},$$

where c is the $k-1$-dimensional vector, all of whose entries are equal to $\lfloor n/k \rfloor$. The map $(m, a_1, \ldots, a_{k-1}) \mapsto M(a_1, \ldots, a_{k-1})$ takes simplices in $\Delta_{n,k}$ to nonconstant arithmetic progressions in $\mathbb{Z}^{k-1}$, and takes B to $\{ Mc + \det(M)\, s : s \in S^{k-1} \}$, which is a set containing no nonconstant arithmetic progressions. Thus, B is a Fujimura set and so does not contain any combinatorial lines.

If all of $a_1, \ldots, a_k$ are within $C_1\sqrt{n}$ of n/k, then $|\Gamma_{\vec{a}}| \geq Ck^n/n^{(k-1)/2}$ (where C depends on C_1) by the central limit theorem. By our choice of S and applying (2.1) (or more precisely, the obvious generalisation of (2.1) to other values of k), we obtain

$$c_{n,k} \geq Ck^n/n^{(k-1)/2}|S|^{k-1} = Ck^n \left(\frac{|S|}{\sqrt{n}} \right)^{k-1}.$$

[3]For instance, if we let A_i denote the ith row, we see that $(A_1 - A_2) + (A_{i+1} - A_i)$ is of the form $(0, \ldots, 0, -k+1, 0, \ldots, 0, k-1)$, and so the row space spans all the vectors whose coordinates sum to zero; but the first row has a non-zero coordinate sum, so the rows in fact span the whole space.

One can take S to have cardinality $r_k(\sqrt{n})$, which from the results of O'Bryant [20] satisfies (for all sufficiently large n, some $C > 0$, and ℓ the largest integer satisfying $k > 2^{\ell-1}$)

$$\frac{r_k(\sqrt{n})}{\sqrt{n}} \geq C(\log n)^{1/(2\ell)} \exp_2\left(- \ell 2^{(\ell-1)/2-1/\ell} \sqrt[\ell]{\log_2 n} \right),$$

which completes the proof. ■

3. Upper Bounds for the $k = 3$ Density Hales–Jewett Problem

To finish the proof of Theorem 1.4 we need to supply the indicated upper bounds for $c_{n,3}$ for $n = 0, \ldots, 6$.

It is clear that $c_{0,3} = 1$ and $c_{1,3} = 2$. By subdividing a line-free set into three parallel slices we obtain the bound

$$c_{n+1,3} \leq 3c_{n,3}$$

for all n. This is already enough to get the correct upper bounds $c_{2,3} \leq 6$ and $c_{3,3} \leq 18$, and also allows us to deduce the upper bound $c_{6,3} \leq 450$ from $c_{5,3} \leq 150$. So the remaining tasks are to establish the upper bounds

$$c_{4,3} \leq 52 \tag{3.1}$$

and

$$c_{5,3} \leq 150. \tag{3.2}$$

In order to establish (3.2), we will rely on (3.1), together with a classification of those line-free sets in $[3]^4$ of size close to the maximal number 52. Similarly, to establish (3.1), we will need the bound $c_{3,3} \leq 18$, together with a classification of those line-sets in $[3]^3$ of size close to the maximal number 18. Finally, to achieve the latter aim one needs to classify the line-free subsets of $[3]^2$ with exactly $c_{2,3} = 6$ elements.

3.1. $n = 2$

We begin with the $n = 2$ theory.

Lemma 3.1 ($n = 2$ extremals). *There are exactly four line-free subsets of* $[3]^2$ *of cardinality* 6*:*

- *The set* $x := A_{B_{2,2}} = \{12, 13, 21, 22, 31, 33\}$;
- *The set* $y := A_{B_{2,1}} = \{11, 12, 21, 23, 32, 33\}$;
- *The set* $z := A_{B_{2,0}} = \{11, 13, 22, 23, 31, 32\}$;
- *The set* $w := \{12, 13, 21, 23, 31, 32\}$.

Proof. A line-free subset of $[3]^2$ must have exactly two elements in every row and column. The claim then follows by brute force search. ■

3.2. $n = 3$

Now we turn to the $n = 3$ theory. We can slice $[3]^3$ as the union of three slices $1 * *$, $2 * *$, $3 * *$, each of which are identified with $[3]^2$ in the obvious manner. Thus every subset A in $[3]^3$ can be viewed as three subsets A_1, A_2, A_3 of $[3]^2$ stacked together; if A is line-free then A_1, A_2, A_3 are necessarily line-free, but the converse is not true. We write $A = A_1A_2A_3$, thus for instance xyz is the set

$$xyz = \{112, 113, 121, 122, 131, 133\} \cup \{211, 212, 221, 223, 232, 233\} \cup \{311, 313, 322, 323, 331, 332\}.$$

Observe that

$$A_{B_{0,3}} = xyz; \quad A_{B_{1,3}} = yzx; \quad A_{B_{2,3}} = zxy.$$

Lemma 3.2 ($n = 3$ extremals). *The only* 18*-element line-free subset of* $[3]^3$ *is* xyz*. The only* 17*-element line-free subsets of* $[3]^3$ *are formed by removing a point from* xyz*, or by removing either* 111, 222*, or* 333 *from* yzx *or* zxy*.*

Proof. We prove the second claim. As $17 = 6 + 6 + 5$, and $c_{2,3} = 6$, at least two of the slices of a 17-element line-free set must be from x, y, z, w, with the third slice having 5 points. If two of the slices are identical, the last slice must lie in the complement and thus has at most 3 points, a contradiction. If one of the slices is a w, then the 5-point slice consists of the complement of the other two slices and thus contains a diagonal, contradiction. By symmetry we may now assume that two of the slices are x and y, which force the last slice to be z with one point removed. Now one sees that the slices must be in the order xyz, yzx, or zxy, because any other combination has too many lines that need to be removed. The sets yzx, zxy contain the diagonal $\{111, 222, 333\}$ and so one additional point needs to be removed.

The first claim follows by a similar argument to the second. ∎

3.3. $n = 4$

Now we turn to the $n = 4$ theory.

Lemma 3.3. $c_{4,3} \leq 52$.

Proof. Let A be a line-free set in $[3]^4$, and split $A = A_1A_2A_3$ for $A_1, A_2, A_3 \in [3]^3$ as in the $n = 3$ theory. If at least two of the slices A_1, A_2, A_3 are of cardinality 18, then by Lemma 3.2 they are of the form xyz, and so the third slice then lies in the complement and has at most six points, leading to an inferior bound of $18+18+6 = 42$. Thus at most one of the slices can have cardinality 18, leading to the bound $18 + 17 + 17 = 52$.

Now we classify extremisers. Observe that we have the following (equivalent) 52-point line-free sets, which were implicitly constructed in the previous section;

- $E_0 := A_{B_{0,4}} \backslash \{1111, 2222\}$;
- $E_1 := A_{B_{1,4}} \backslash \{2222, 3333\}$;
- $E_2 := A_{B_{2,4}} \backslash \{1111, 3333\}$.

Lemma 3.4.

- *The only 52-element line-free sets in $[3]^4$ are E_0, E_1, E_2.*

- *The only 51-element line-free sets in* $[3]^4$ *are formed by removing a point from* E_0, E_1 *or* E_2.

- *The only 50-element line-free sets in* $[3]^4$ *are formed by removing two points from* E_0, E_1 *or* E_2 *OR are equal to one of the three permutations of the set* $X := \Gamma_{3,1,0} \cup \Gamma_{3,0,1} \cup \Gamma_{2,2,0} \cup \Gamma_{2,0,2} \cup \Gamma_{1,1,2} \cup \Gamma_{1,2,1} \cup \Gamma_{0,2,2}$.

Proof. We will just prove the third claim, which is the hardest; the first two claims follow from the same argument (and can in fact be deduced directly from the third claim).

It suffices to show that every 50-point line-free set is either contained in the 54-point set $A_{B_{j,4}}$ for some $j = 0, 1, 2$, or is some permutation of the set X. Indeed, if a 50-point line-free set is contained in, say, $A_{B_{0,4}}$, then it cannot contain 2222, since otherwise it must omit one point from each of the four pairs formed from $\{2333, 2111\}$ by permuting the indices, and must also omit one of $\{1111, 1222, 1333\}$, leading to at most 49 points in all; similarly, it cannot contain 1111, and so omits the entire diagonal $\{1111, 2222, 3333\}$, with two more points to be omitted. By symmetry we see the same argument works when $A_{B_{0,4}}$ is replaced by one of the other $A_{B_{j,4}}$.

Next, observe that every three-dimensional slice of a line-free set can have at most $c_{3,3} = 18$ points; thus when one partitions a 50-point line-free set into three such slices, it must divide either as $18+16+16$, $18+17+15$, $17+17+16$, or some permutation of these. Suppose that we can slice the set into two slices of 17 points and one slice of 16 points. By the various symmetries, we may assume that the $1***$ slice and $2***$ slices have 17 points, and the $3***$ slice has 16 points. By Lemma 3.2, the 1-slice is $\{1\} \times D_{3,j}$ with one point removed, and the 2-slice is $\{2\} \times D_{3,k}$ with one point removed, for some $j, k \in \{0, 1, 2\}$. If $j = k$, then the 1-slice and 2-slice have at least 15 points in common, so the 3-slice can have at most $27 - 15 = 12$ points, a contradiction. If $jk = 01$, 12, or 20, then observe that from Lemma 3.2 the $*1**$, $*2**$, $*3**$ slices cannot equal a 17-point or 18-point line-free set, so each have at most 16 points, leading to only 48 points in all, a contradiction. Thus we must have $jk = 10$, 21, or 02.

First suppose that $jk = 02$. Then by Lemma 3.2, the $2***$ slice contains the nine points formed from $\{2211, 2322, 2331\}$ and permuting the last three indices, while the $1***$ slice contains at least eight of the nine points formed from $\{1211, 1322, 1311\}$ and permuting the last three indices. Thus the $3***$ slice can contain at most one of the nine points formed from

$\{3211, 3322, 3311\}$ and permuting the last three indices. If it does contain one of these points, say 3211, then it must omit one point from each of the four pairs $\{3222, 3233\}$, $\{3212, 3213\}$, $\{3221, 3231\}$, $\{3111, 3311\}$, leading to at most 15 points on this slice, a contradiction. So the $3{*}{*}{*}$ slice must omit all nine points, and is therefore contained in $\{3\} \times D_{3,1}$, and so the 50-point set is contained in $D_{4,1}$, and we are done by the discussion at the beginning of the proof.

The case $jk = 10$ is similar to the $jk = 02$ case (indeed one can get from one case to the other by swapping the 1 and 2 indices). Now suppose instead that $jk = 12$. Then by Lemma 3.2, the $1{*}{*}{*}$ slice contains the six points from permuting the last three indices of 1123, and similarly the $2{*}{*}{*}$ slice contains the six points from permuting the last three indices of 2123. Thus the $3{*}{*}{*}$ slice must avoid all six points formed by permuting the last three indices of 3123. Similarly, as 1133 lies in the $1{*}{*}{*}$ slice and 2233 lies in the $2{*}{*}{*}$ slice, 3333 must be avoided in the $3{*}{*}{*}$ slice.

Now we claim that 3111 must be avoided also; for if 3111 was in the set, then one point from each of the six pairs formed from $\{3311, 3211\}$, $\{3331, 3221\}$ and permuting the last three indices must lie outside the $3{*}{*}{*}$ slice, which reduces the size of that slice to at most $27 - 6 - 1 - 6 = 14$, which is too small. Similarly, 3222 must be avoided, which puts the $3{*}{*}{*}$ slice inside $\{3\} \times D_3$ and then places the 50-point set inside D_4, and we are done by the discussion at the beginning of the proof.

We have handled the case in which at least one of the slicings of the 50-point set is of the form $50 = 17 + 17 + 16$. The only remaining case is when all slicings of the 50-point set are of the form $18 + 16 + 16$ or $18 + 17 + 15$ (or a permutation thereof). So each slicing includes an 18-point slice. By the symmetries of the situation, we may assume that the $1{*}{*}{*}$ slice has 18 points, and thus by Lemma 3.2 takes the form $\{1\} \times D_3$. Inspecting the ${*}1{*}{*}$, ${*}2{*}{*}$, ${*}3{*}{*}$ slices, we then see (from Lemma 3.2) that only the ${*}1{*}{*}$ slice can have 18 points; since we are assuming that this slicing is some permutation of $18 + 17 + 15$ or $18 + 16 + 16$, we conclude that the ${*}1{*}{*}$ slice must have exactly 18 points, and is thus described precisely by Lemma 3.2. Similarly for the ${*}{*}1{*}$ and ${*}{*}{*}1$ slices. Indeed, by Lemma 3.2, we see that the 50-point set must agree exactly with $D_{4,1}$ on any of these slices. In particular, there are exactly six points of the 50-point set in the remaining portion $\{2, 3\}^4$ of the cube.

Suppose that 3333 was in the set; then since all permutations of 3311, 3331 are known to lie in the set, then 3322, 3332 must lie outside the set. Also, as 1222 lies in the set, at least one of 2222, 3222 lie outside the set.

This leaves only 5 points in $\{2,3\}^4$, a contradiction. Thus 3333 lies outside the set; similarly 2222 lies outside the set.

Let a be the number of points in the 50-point set which are some permutation of 2233, thus $0 \leq a \leq 6$. If $a = 0$ then the set lies in $D_{4,1}$ and we are done. If $a = 6$ then the set is exactly X and we are done. Now suppose $a = 1$. By symmetry we may assume that 2233 lies in the set. Then (since 2133, 1233, 2231, 2213 are known to lie in the set) 2333, 3233, 2223, 2232 lie outside the set, which leaves at most 5 points inside $\{2,3\}^4$, a contradiction. A similar argument holds if $a = 2, 3$.

The remaining case is when $a = 4, 5$. Then one of the three pairs $\{2233, 3322\}$, $\{2323, 3232\}$, $\{2332, 3223\}$ lie in the set. By symmetry we may assume that $\{2233, 3322\}$ lie in the set. Then by arguing as before we see that all eight points formed by permuting 2333 or 3222 lie outside the set, leading to at most 5 points inside $\{2,3\}^4$, a contradiction. ∎

3.4. $n = 5$

Finally, we turn to the $n = 5$ theory. Our goal is to show that $c_{5,3} \leq 150$. Accordingly, suppose for contradiction that we can find a line-free subset A of $[3]^5$ of cardinality $|A| = 151$. We will now prove a series of facts about A which will eventually give the desired contradiction.

Lemma 3.5. *A is not contained inside $A_{B_{j,5}}$ for any $j = 0, 1, 2$.*

Proof. Suppose for contradiction that $A \subset A_{B_{j,5}}$ for some j. By symmetry we may take $j = 0$. The set $A_{B_{0,5}}$ has 162 points. By looking at the triplets $\{10000, 11110, 12220\}$ and cyclic permutations we must lose 5 points; similarly from the triplets $\{20000, 22220, 21110\}$ and cyclic permutations. Finally from $\{11000, 11111, 11222\}$ and $\{22000, 22222, 22111\}$ we lose two more points. Since $162 - 5 - 5 - 2 = 150$, we obtain the desired contradiction. ∎

Observe that every slice of A contains at most $c_{4,3} = 52$ points, and hence every slice of A contains at least $151 - 52 - 52 = 47$ points.

Lemma 3.6. *A cannot have two parallel $[3]^4$ slices, each of which contain at least 51 points.*

Proof. Suppose not that A has two parallel $[3]^4$ slices. By symmetry, we may assume that the $1****$ and $2****$ slices have at least 51 points. Meanwhile, the $3****$ slice has at least 47 points as discussed above.

By Lemma 3.4, the $1****$ slice takes the form $\{1\} \times D_{4,j}$ for some $j = 0, 1, 2$ with the diagonal $\{11111, 12222, 13333\}$ and possibly one more point removed, and similarly the $2****$ slice takes the form $\{2\} \times D_{4,k}$ for some $k = 0, 1, 2$ with the diagonal $\{21111, 22222, 23333\}$ and possibly one more point removed.

Suppose first that $j = k$. Then the 1-slice and 2-slice have at least 50 points in common, leaving at most 31 points for the 3-slice, a contradiction. Next, suppose that $jk = 01$. Then observe that the $*i***$ slice cannot look like any of the configurations in Lemma 3.4 and so must have at most 50 points for $i = 1, 2, 3$, leading to 150 points in all, a contradiction. Similarly if $jk = 12$ or 20. Thus we must have jk equal to 10, 21, or 02.

Let's suppose first that $jk = 10$. The first slice then is equal to $\{1\} \times D_{4,1}$ with the diagonal and possibly one more point removed, while the second slice is equal to $\{2\} \times D_{4,0}$ with the diagonal and possibly one more point removed. Superimposing these slices, we thus see that the third slice is contained in $\{3\} \times D_{4,2}$ except possibly for two additional points, together with the one point 32222 of the diagonal that lies outside of $\{3\} \times D_{4,2}$.

The lines $x12xx, x13xx$ (plus permutations of the last four digits) must each contain one point outside the set. The first two slices can only absorb two of these, and so at least 14 of the 16 points formed by permuting the last four digits of 31233, 31333 must lie outside the set. These points all lie in $\{3\} \times D_{4,2}$, and so the $3****$ slice can have at most $|D_{4,2}| - 14 + 3 = 43$ points, a contradiction.

The case $jk = 02$ is similar to the case $jk = 10$ (indeed one can obtain one from the other by swapping 1 and 2). Now we turn to the case $jk = 21$. Arguing as before we see that the third slice is contained in $\{3\} \times D_4$ except possibly for two points, together with 33333.

If 33333 was in the set, then each of the lines $xx333$, $xxx33$ (and permutations of the last four digits) must have a point missing from the first two slices, which cannot be absorbed by the two points we are permitted to remove; thus 33333 is not in the set. For similar reasons, 33331 is not in the set, as can be seen by looking at $xxx31$ and permutations of the last four digits. Indeed, any string containing four threes does not lie in the set; this means that at least 8 points are missing from $\{3\} \times D_4$, leaving only at most 46 points inside that set. Furthermore, any point in the $3****$

slice outside of $\{3\} \times D_4$ can only be created by removing a point from the first two slices, so the total cardinality is at most $46 + 52 + 52 = 150$, a contradiction. ■

Remark 3.7. This already gives the bound $c_{5,3} \leq 52 + 50 + 50 = 152$, but of course we wish to do better than this.

Lemma 3.8. *A has a slice $j * * * *$ with $j = 1, 2, 3$ that has at most 49 points.*

Proof. Suppose not, thus all three slices of A has at least 50 points. Using earlier notation, we split subsets of $[3]^4$ into nine subsets of $[3]^2$. So we think of x, y, z, a, b and c as subsets of a square. By Lemma 3.4, each slice is one of the following:

- $E_0 = y'zx, zx'y, xyz$ (with one or two points removed)
- $E_1 = xyz, yz'x, zxy'$ (with one or two points removed)
- $E_2 = z'xy, xyz, yzx'$ (with one or two points removed)
- $X = xyz, ybw, zwc$
- $Y = axw, xyz, wzc$
- $Z = awx, wby, xyz$

where a, b and c have four points each: $a = \{2,3\}^2$, $b = \{1,3\}^2$ and $c = \{1,2\}^2$. x', y' and z' are subsets of x, y and z respectively, and have five points each.

Suppose all three slices are subsets of E_{j_1}, E_{j_2}, E_{j_3} respectively for some $j_1, j_2, j_3 \in \{0, 1, 2\}$, E_1, or E_2. We can remove at most five points from the full set $E_{j_1} \uplus E_{j_2} \uplus E_{j_3}$. Consider columns 2, 3, 4, 6, 7, 8. At most two of these columns contain xyz, so one point must be removed from the other four. This uses up all but one of the removals. So the slices must be E_2, E_1, E_0 or a cyclic permutation of that. Then the cube, which contains the first square of slice 1; the fifth square of slice 2; and the ninth square of slice 3, contains three copies of the same square. It takes more than one point removed to remove all lines from that cube. So we can't have all three slices subsets of E_j.

Suppose one slice is X, Y or Z, and two others are subsets of E_j. We can remove at most three points from the two E_j. By symmetry, suppose one slice is X. Consider columns 2, 3, 4 and 7. They must be cyclic permutations of x, y, z, and two of them are not xyz, so must lose a point. Columns 6 and 8 must both lose a point, and we only have 150 points left. So if one slice is X, Y or Z, the full set contains a line.

Suppose two slices are from X, Y and Z, and the other is a subset of E_j. By symmetry, suppose two slices are X and Y. Columns 3, 6, 7 and 8 all contain w, and therefore at most 16 points each. Columns 1, 5 and 9 contain a, b, or c, and therefore at most 16 points. So the total number of points is at most $7 \times 16 + 2 \times 18 = 148 < 151$, a contradiction. ■

This, combined with Lemma 3.6, gives

Corollary 3.9. *Any three parallel slices of A must have cardinality* 52, 50, 49 *(or a permutation thereof).*

Note that this argument already gives the bound $c_{5,3} \leq 151$.

Lemma 3.10. *No slice $j****$ of A is of the form X, where X was defined in Lemma* 3.4.

Proof. Suppose one slice is X; then by the previous discussion one of the parallel slices has 52 points and is thus of the form E_j for some $j = 0, 1, 2$, by Lemma 3.4.

Suppose that X is the first slice $1****$. We have $X = xyz\ ybw\ zwc$. Label the other rows with letters from the alphabet, thus

$$A = \begin{pmatrix} xyz & ybw & zwc \\ mno & pqr & stu \\ def & ghi & jkl \end{pmatrix}$$

Reslice the array into a left nine, middle nine and right nine. One of these squares contains 52 points, and it can only be the left nine. One of its three columns contains 18 points, and it can only be its left-hand column, xmd. So $m = y$ and $d = z$. But none of the E_j begins with y or z, which is a contradiction. So X is not in the first row.

So X is in the second or third row. By symmetry, suppose it is in the second row, so that A has the following shape:

$$A = \begin{pmatrix} def & ghi & jkl \\ xyz & ybw & zwc \\ mno & pqr & stu \end{pmatrix}$$

Again, the left-hand nine must contain 52 points, so it is E_2. Now, to get 52 points in any row, the first row must be E_2. Then the only way to have 50 points in the middle or right-hand nine is if the middle nine is X:

$$A = \begin{pmatrix} z'xy & xyz & yzx' \\ xyz & ybw & zwc \\ yzx' & zwc & stu \end{pmatrix}$$

In the seventh column, s contains 5 points and in the eighth column, t contains 4 points. The final row can now contain at most 48 points, contradicting Corollary 3.9.

A similar argument is possible if X is in the third row; or if X is replaced by Y or Z. Thus, given any decomposition of A into three parallel slices, one slice is a 52-point set E_j and another slice is 50 points contained in E_k. ■

Now we can obtain the desired contradiction:

Lemma 3.11. *There is no* 151*-point line-free set* $A \subset [3]^5$.

Proof. Assume by symmetry that the first row contains 52 points and the second row contains 50. If E_1 is in the first row, then the second row must be contained in E_0:

$$A = \begin{pmatrix} xyz & yz'x & zxy' \\ y'zx & zx'y & xyz \\ def & ghi & jkl \end{pmatrix}$$

But then none of the left nine, middle nine or right nine can contain 52 points, which contradicts Corollary 3.9. Suppose the first row is E_0. Then the second row is contained in E_2, otherwise the cubes formed from the nine columns of the diagram would need to remove too many points:

$$A = \begin{pmatrix} y'zx & zx'y & xyz \\ z'xy & xyz & yzx' \\ def & ghi & jkl \end{pmatrix}.$$

But then neither the left nine, middle nine nor right nine contain 52 points. So the first row contains E_2, and the second row is contained in E_1. Two points may be removed from the second row of this diagram:

$$A = \begin{pmatrix} z'xy & xyz & yzx' \\ xyz & yz'x & zxy' \\ def & ghi & jkl \end{pmatrix}.$$

Slice it into the left nine, middle nine and right nine. Two of them are contained in E_j so at least two of def, ghi, and jkl are contained in the corresponding slice of E_0. Slice along a different axis, and at least two of dgj, ehk, fil are contained in the corresponding slice of E_0. So eight of the nine squares in the bottom row are contained in the corresponding square of E_0. Indeed, slice along other axes, and all points except one are contained within E_0. This point is the intersection of all the 49-point slices. So, if there is a 151-point solution, then after removal of the specified point, there is a 150-point solution, within $D_{5,j}$, whose slices in each direction are $52+50+48$.

$$A = \begin{pmatrix} z'xy & xyz & yzx' \\ xyz & yz'x & zxy' \\ y'zx & zx'y & xyz \end{pmatrix}$$

One point must be lost from columns 3, 6, 7 and 8, and four more from the major diagonal $z'z'z$. That leaves 148 points instead of 150. So the 150-point solution does not exist with $52+50+48$ slices; so the 151 point solution does not exist. ■

This establishes that $c_{5,3} \leq 150$, and thus $c_{6,3} \leq 3c_{5,3} \leq 450$.

4. Lower Bounds for the Moser Problem

Just as for the density Hales–Jewett problem, we found that Gamma sets $\Gamma_{a,b,c}$ were useful in providing large lower bounds for the Moser problem. This is despite the fact that the symmetries of the cube do not respect Gamma sets.

Observe that if $B \subset \Delta_n$, then the set $A_B := \bigcup_{\vec{a}\in B} \Gamma_{a,b,c}$ is a Moser set as long as B does not contain any "isosceles triangles" $(a+r, b, c+s)$, $(a+s, b, c+r)$, $(a, b+r+s, c)$ for any $r, s \geq 0$ not both zero; in particular, B cannot contain any "vertical line segments" $(a+r, b, c+r)$, $(a, b+2r, c)$. An example of such a set is provided by selecting $0 \leq i \leq n-3$ and letting B consist of the triples $(a, n-i, i-a)$ when $a \neq 3 \bmod 3$, $(a, n-i-1, i+1-a)$ when $a \neq 1 \bmod 3$, $(a, n-i-2, i+2-a)$ when $a = 0 \bmod 3$, and $(a, n-i-3, i+3-a)$ when $a = 2 \bmod 3$. Asymptotically, this set includes about two thirds of the spheres $S_{n,i}$, $S_{n,i+1}$ and one third of the spheres $S_{n,i+2}$, $S_{n,i+3}$ and (setting i close to $n/3$) gives a lower bound

$$c'_{n,3} \geq (C - o(1))\, 3^n/\sqrt{n} \tag{4.1}$$

with $C = 2\times\sqrt{\frac{9}{4\pi}}$. This lower bound is the asymptotic limit of our methods; see Proposition 4.1 below.

An integer program was solved to obtain the optimal lower bounds achievable by the A_B construction (using (2.1), of course). The results for $1 \leq n \leq 20$ are displayed in Figure 6. More complete data, including the list of optimisers, can be found at [17].

n	lower bound	n	lower bound
1	2	11	71766
2	6	12	212423
3	16	13	614875
4	43	14	1794212
5	122	15	5321796
6	353	16	15455256
7	1017	17	45345052
8	2902	18	134438520
9	8622	19	391796798
10	24786	20	1153402148

Fig. 6. Lower bounds for $c'_{n,3}$ obtained by the A_B construction.

Unfortunately, any method based purely on the A_B construction cannot do asymptotically better than the previous constructions:

Proposition 4.1. *Let $B \subset \Delta_n$ be such that A_B is a Moser set. Then* $|A_B| \leq \left(2\sqrt{\frac{9}{4\pi}} + o(1)\right)\frac{3^n}{\sqrt{n}}$.

Proof. By the previous discussion, B cannot contain any pair of the form $(a, b+2r, c)$, $(a+r, b, c+r)$ with $r > 0$. In other words, for any $-n \leq h \leq n$, B can contain at most one triple (a,b,c) with $c - a = h$. From this and (2.1), we see that

$$|A_B| \leq \sum_{h=-n}^{n} \max_{(a,b,c)\in\Delta_n : c-a=h} \frac{n!}{a!b!c!}.$$

From the Chernoff inequality (or the Stirling formula computation below) we see that $\frac{n!}{a!b!c!} \leq \frac{1}{n^{10}}3^n$ unless $a, b, c = n/3 + O\big(n^{1/2}\log^{1/2} n\big)$, so we may restrict to this regime, which also forces $h = O\big(n^{1/2}\log^{1/2} n\big)$. If we

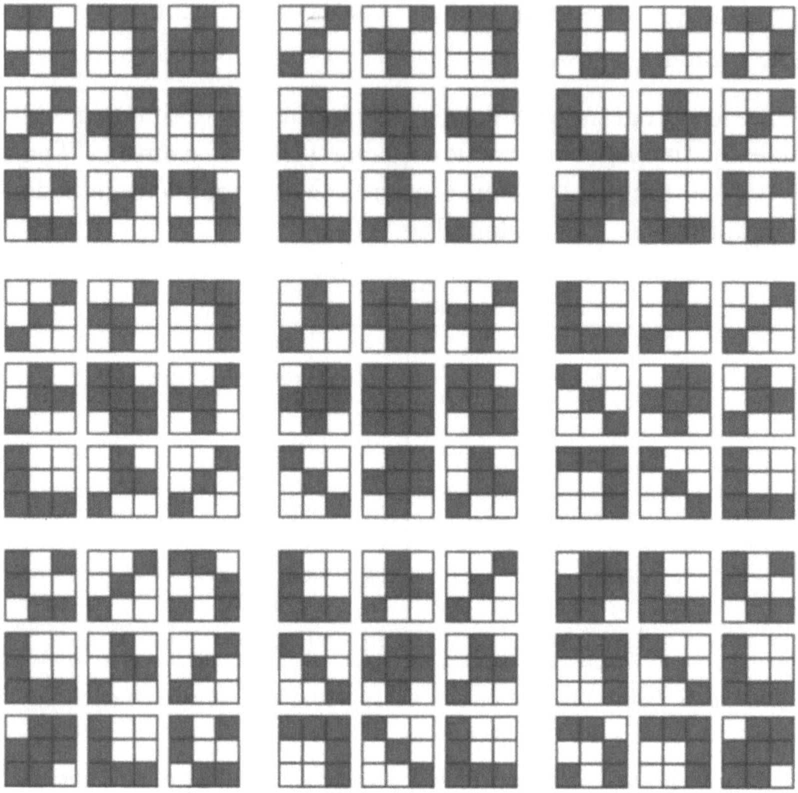

Fig. 7. One of the examples of 353-point sets in $[3]^6$ (elements of the set being indicated by white squares). This example was generated by a genetic algorithm.

write $a = n/3 + \alpha$, $b = n/3 + \beta$, $c = n/3 + \gamma$ and apply Stirling's formula $n! = \big(1 + o(1)\big)\sqrt{2\pi n}n^n e^{-n}$, we obtain

$$\frac{n!}{a!b!c!} = \big(1 + o(1)\big)\frac{3^{3/2}}{2\pi n}3^n \exp\left(-\left(\frac{n}{3} + \alpha\right)\log\left(1 + \frac{3\alpha}{n}\right)\right.$$
$$\left. - \left(\frac{n}{3} + \beta\right)\log\left(1 + \frac{3\beta}{n}\right) - \left(\frac{n}{3} + \gamma\right)\log\left(1 + \frac{3\gamma}{n}\right)\right).$$

From Taylor expansion one has

$$-\left(\frac{n}{3} + \alpha\right)\log\left(1 + \frac{3\alpha}{n}\right) = -\alpha - \frac{3}{2}\frac{\alpha^2}{n} + o(1)$$

and similarly for β, γ; since $\alpha+\beta+\gamma=0$, we conclude that

$$\frac{n!}{a!b!c!} = \big(1+o(1)\big)\frac{3^{3/2}}{2\pi n}3^n \exp\left(-\frac{3}{2n}(\alpha^2+\beta^2+\gamma^2)\right).$$

If $c-a=h$, then $\alpha^2+\beta^2+\gamma^2 = \frac{3\beta^2}{2}+\frac{h^2}{2}$. Thus we see that

$$\max_{(a,b,c)\in\Delta_n : c-a=h} \frac{n!}{a!b!c!} \leq \big(1+o(1)\big)\frac{3^{3/2}}{2\pi n}3^n \exp\left(-\frac{3}{4n}h^2\right).$$

Using the integral test, we thus have

$$|A_B| \leq \big(1+o(1)\big)\frac{3^{3/2}}{2\pi n}3^n \int_{\mathbb{R}} \exp\left(-\frac{3}{4n}x^2\right)\,dx.$$

Since $\int_{\mathbb{R}} \exp\big(-\frac{3}{4n}x^2\big)\,dx = \sqrt{\frac{4\pi n}{3}}$, we obtain the claim. ∎

Actually it is possible to improve upon these bounds by a slight amount. Observe that if B is a maximiser for the right-hand side of (2.1) (subject to B not containing isosceles triangles), then any triple (a,b,c) not in B must be the vertex of a (possibly degenerate) isosceles triangle with the other vertices in B. If this triangle is non-degenerate, or if (a,b,c) is the upper vertex of a degenerate isosceles triangle, then no point from $\Gamma_{a,b,c}$ can be added to A_B without creating a geometric line. However, if $(a,b,c)=(a'+r,b',c'+r)$ is only the lower vertex of a degenerate isosceles triangle $(a'+r,b',c'+r)$, $(a',b'+2r,c')$, then one can add any subset of $\Gamma_{a,b,c}$ to A_B and still have a Moser set as long as no pair of elements in that subset is separated by Hamming distance $2r$. For instance, in the $n=5$ case, we can start with the 122-point set built from

$$B = \big\{(005),(023),(113),(122),(221),(311),(320),(500))\big\}$$

and add a point each from (104) and (401). This gives an example of the maximal, 124-point solution. Again, in the $n=10$ case, the set

$$\begin{aligned}B = \big\{&(0010),(028),(037),(046),(145),(217),(235),(325),(334),(343),\\ &(442),(514),(532),(622),(631),(640),(811),(901),(910)\big\}\end{aligned}$$

generates the lower bound $c'_{10,3} \geq 24786$ given above (and, up to reflection $a \leftrightarrow c$, is the only such set that does so); but by adding the following twelve elements from $\Gamma_{5,0,5}$ one can increase the lower bound slightly to 24798: 1111133333, 1111313333, 1113113333, 1133331113, 1133331131, 1133331311, 3311333111, 3313133111, 3313313111, 3331111133, 3331111313, 3331111331.

A more general form goes with the B set described at the start of this section. Include points from $\Gamma_{(a,n-i-4,i+4-a)}$ when $a = 1 (\mathrm{mod}\ 3)$, subject to no two points being included if they differ by the interchange of a 1 and a 3. Each of these Gamma sets is the feet of a degenerate isosceles triangle with vertex $\Gamma_{(a-1,n-i-2,a+3-a)}$.

Lemma 4.2. *If A is a subset of $\Gamma_{(a,b,c)}$ such that no two points of A differ by the interchange of a 1 and a 3, then $|A| \leq |\Gamma_{a,b,c}|/\bigl(1+\max(a,c)\bigr)$.*

Proof. Say that two points of $\Gamma_{a,b,c}$ are neighbours if they differ by the exchange of a 1 and a 3. Each point of A has ac neighbours, none of which are in A. Each point of $\Gamma_{(a,b,c)} \backslash A$ has ac neighbours, but only $\min(a,c)$ of them may be in A. So for each point of A there are on average $ac/\min(a,c) = \max(a,c)$ points not in A. So the proportion of points of $\Gamma_{(a,b,c)}$ that are in A is at most one in $1+\max(a,c)$. ∎

The proportion of extra points for each of the cells $\Gamma_{(a,n-i-4,i+4-a)}$ is no more than $2/(i+6)$. Only one cell in three is included from the $b = n-i-4$ layer, so we expect no more than $\binom{n}{i+4}2^{i+5}/(3i+18)$ new points, all from $S_{n,i+4}$. One can also find extra points from $S_{n,i+5}$ and higher spheres.

Earlier solutions may also give insight into the problem. Clearly we have $c'_{0,3} = 1$ and $c'_{1,3} = 2$, so we focus on the case $n \geq 2$. The first lower bounds may be due to Komlós [16], who observed that the sphere $S_{i,n}$ of elements with exactly $n-i$ 2 entries (see Section 1.1 for definition), is a Moser set, so that

$$c'_{n,3} \geq |S_{i,n}| \tag{4.2}$$

holds for all i. Choosing $i = \left\lfloor \frac{2n}{3} \right\rfloor$ and applying Stirling's formula, we see that this lower bound takes the form (4.1) with $C := \sqrt{\frac{9}{4\pi}}$. In particular $c'_{3,3} \geq 12, c'_{4,3} \geq 24, c'_{5,3} \geq 80, c'_{6,3} \geq 240$.

These values can be improved by studying combinations of several spheres or semispheres or applying elementary results from coding theory.

Observe that if $\{w(1), w(2), w(3)\}$ is a geometric line in $[3]^n$, then $w(1), w(3)$ both lie in the same sphere $S_{i,n}$, and that $w(2)$ lies in a lower sphere $S_{i-r,n}$ for some $1 \leq r \leq i \leq n$. Furthermore, $w(1)$ and $w(3)$ are separated by Hamming distance r.

As a consequence, we see that $S_{i-1,n} \cup S^e_{i,n}$ (or $S_{i-1,n} \cup S^o_{i,n}$) is a Moser set for any $1 \leq i \leq n$, since any two distinct elements $S^e_{i,n}$ are separated by a Hamming distance of at least two. (Recall Section 1.1 for definitions), This leads to the lower bound

$$c'_{n,3} \geq \binom{n}{i-1} 2^{i-1} + \binom{n}{i} 2^{i-1} = \binom{n+1}{i} 2^{i-1}. \tag{4.3}$$

It is not hard to see that $\binom{n+1}{i+1}2^i > \binom{n+1}{i}2^{i-1}$ if and only if $3i < 2n+1$, and so this lower bound is maximised when $i = \left\lfloor \frac{2n+1}{3} \right\rfloor$ for $n \geq 2$, giving the formula (1.2). This leads to the lower bounds

$$c'_{2,3} \geq 6; \quad c'_{3,3} \geq 16; \quad c'_{4,3} \geq 40; \quad c'_{5,3} \geq 120; \quad c'_{6,3} \geq 336$$

which gives the right lower bounds for $n = 2, 3$, but is slightly off for $n = 4, 5$. Asymptotically, Stirling's formula and (4.3) then give the lower bound (4.1) with $C = \frac{3}{2} \times \sqrt{\frac{9}{4\pi}}$, which is asymptotically 50% better than the bound (4.2).

The work of Chvátal [7] already contained a refinement of this idea which we here translate into the usual notation of coding theory: Let $A(n,d)$ denote the size of the largest binary code of length n and minimal distance d.

Then

$$c'_{n,3} \geq \max_k \left(\sum_{j=0}^{k} \binom{n}{j} A(n-j, k-j+1) \right). \tag{4.4}$$

The following values of $A(n,d)$ for small n, d are known, see [5]:

$A(1,1)=2$							
$A(2,1)=4$	$A(2,2)=2$						
$A(3,1)=8$	$A(3,2)=4$	$A(3,3)=2$					
$A(4,1)=16$	$A(4,2)=8$	$A(4,3)=2$	$A(4,4)=2$				
$A(5,1)=32$	$A(5,2)=16$	$A(5,3)=4$	$A(5,4)=2$	$A(5,5)=2$			
$A(6,1)=64$	$A(6,2)=32$	$A(6,3)=8$	$A(6,4)=4$	$A(6,5)=2$	$A(6,6)=2$		
$A(7,1)=128$	$A(7,2)=64$	$A(7,3)=16$	$A(7,4)=8$	$A(7,5)=2$	$A(7,6)=2$	$A(7,7)=2$	
$A(8,1)=256$	$A(8,2)=128$	$A(8,3)=20$	$A(8,4)=16$	$A(8,5)=4$	$A(8,6)=2$	$A(8,7)=2$	$A(8,8)=2$
$A(9,1)=512$	$A(9,2)=256$	$A(9,3)=40$	$A(9,4)=20$	$A(9,5)=6$	$A(9,6)=4$	$A(9,7)=2$	$A(9,8)=2$
$A(10,1)=1024$	$A(10,2)=512$	$A(10,3)=72$	$A(10,4)=40$	$A(10,5)=12$	$A(10,6)=6$	$A(10,7)=2$	$A(10,8)=2$
$A(11,1)=2048$	$A(11,2)=1024$	$A(11,3)=144$	$A(11,4)=72$	$A(11,5)=24$	$A(11,6)=12$	$A(11,7)=2$	$A(11,8)=2$
$A(12,1)=4096$	$A(12,2)=2048$	$A(12,3)=256$	$A(12,4)=144$	$A(12,5)=32$	$A(12,6)=24$	$A(12,7)=4$	$A(12,8)=2$
$A(13,1)=8192$	$A(13,2)=4096$	$A(13,3)=512$	$A(13,4)=256$	$A(13,5)=64$	$A(12,6)=32$	$A(13,7)=8$	$A(13,8)=4$

In addition, one has the general identities $A(n,1) = 2^n$, $A(n,2) = 2^{n-1}$, $A(n-1, 2e-1) = A(n, 2e)$, and $A(n,d) = 2$, if $d > \frac{2n}{3}$.

Inserting this data into (4.4) for $k = 2$ we obtain the lower bounds

$$c'_{4,3} \geq \binom{4}{0}A(4,3) + \binom{4}{1}A(3,2) + \binom{4}{2}A(2,1)$$
$$= 1 \cdot 2 + 4 \cdot 4 + 6 \cdot 4 \qquad = 42$$
$$c'_{5,3} \geq \binom{5}{0}A(5,3) + \binom{5}{1}A(4,2) + \binom{5}{2}A(3,1)$$
$$= 1 \cdot 4 + 5 \cdot 8 + 10 \cdot 8 \qquad = 124$$
$$c'_{6,3} \geq \binom{6}{0}A(6,3) + \binom{6}{1}A(5,2) + \binom{6}{2}A(4,1)$$
$$= 1 \cdot 8 + 6 \cdot 16 + 15 \cdot 16 = 344.$$

Similarly, with $k = 3$ we obtain

$$c'_{7,3} \geq \binom{7}{0}A(7,4) + \binom{7}{1}A(6,3) + \binom{7}{2}A(5,2) + \binom{7}{3}A(4,1) = 960$$
$$c'_{8,3} \geq \binom{8}{0}A(8,4) + \binom{8}{1}A(7,3) + \binom{8}{2}A(6,2) + \binom{8}{3}A(5,1) = 2832$$
$$c'_{9,3} \geq \binom{9}{0}A(9,4) + \binom{9}{1}A(8,3) + \binom{9}{2}A(7,2) + \binom{9}{3}A(6,1) = 7880$$

and for $k = 4$ we have

$$c'_{10,3} \geq \binom{10}{0}A(10,5) + \binom{10}{1}A(9,4) + \binom{10}{2}A(8,3)$$
$$+ \binom{10}{3}A(7,2) + \binom{10}{4}A(6,1) = 22232$$
$$c'_{11,3} \geq \binom{11}{0}A(11,5) + \binom{11}{1}A(10,4) + \binom{11}{2}A(9,3)$$
$$+ \binom{11}{3}A(8,2) + \binom{11}{4}A(7,1) = 66024$$
$$c'_{12,3} \geq \binom{12}{0}A(12,5) + \binom{12}{1}A(11,4) + \binom{12}{2}A(10,3)$$
$$+ \binom{12}{3}A(9,2) + \binom{12}{4}A(8,1) = 188688.$$

and for $k = 5$ we have

$$c'_{13,3} \geq 539168.$$

It should be pointed out that these bounds are even numbers, so that $c'_{4,3} = 43$ shows that one cannot generally expect this lower bound to give the optimum.

The maximum value appears to occur for $k = \lfloor \frac{n+2}{3} \rfloor$, but even after optimising in these parameters and using explicit bounds on $A(n,d)$ we were unable to improve upon the constant $C = 2 \times \sqrt{\frac{9}{4\pi}}$ for (4.1) arising from previously discussed constructions. Using the singleton bound $A(n,d) \leq 2^{n-d+1}$ Chvátal [7] proved that the expression on the right hand side of (4.4) is also $O\big(\frac{3^n}{\sqrt{n}}\big)$, so that the refinement described above gains a constant factor over the initial construction only.

For $n = 4$ the above does not yet give the exact value. The value $c'_{4,3} = 43$ was first proven by Chandra [6]. A uniform way of describing examples for the optimum values of $c'_{4,3} = 43$ and $c'_{5,3} = 124$ is as follows.

Let us consider the sets

$$A := S_{i-1,n} \cup S^e_{i,n} \cup A'$$

where $A' \subset S_{i+1,n}$ has the property that any two elements in A' are separated by a Hamming distance of at least three, or have a Hamming distance of exactly one but their midpoint lies in $S^o_{i,n}$. By the previous discussion we see that this is a Moser set, and we have the lower bound

$$c'_{n,3} \geq \binom{n+1}{i} 2^{i-1} + |A'|. \tag{4.5}$$

This gives some improved lower bounds for $c'_{n,3}$:

- By taking $n = 4$, $i = 3$, and $A' = \{1111, 3331, 3333\}$, we obtain $c'_{4,3} \geq 43$;
- By taking $n = 5$, $i = 4$, and $A' = \{11111, 11333, 33311, 33331\}$, we obtain $c'_{5,3} \geq 124$.
- By taking $n = 6$, $i = 5$, and $A' = \{111111, 111113, 111331, 111333, 331111, 331113\}$, we obtain $c'_{6,3} \geq 342$.

This gives the lower bounds in Theorem 1.5 up to $n = 5$, but the bound for $n = 6$ is inferior to the lower bound $c'_{6,3} \geq 344$ given above.

4.1. Higher k values

We now consider lower bounds for $c'_{n,k}$ for some values of k larger than 3. Here we will see some further connections between the Moser problem and the density Hales–Jewett problem.

For $k = 4$, we have the lower bounds $c'_{n,4} \geq \binom{n}{n/2}2^n$. To see this, observe that the set of points with a 1s, b 2s, c 3s and d 4s, where $a + d$ has the constant value $n/2$, does not form geometric lines because points at the ends of a geometric line have more a or d values than points in the middle of the line.

The following lower bound is asymptotically twice as large. Take all points with a 1s, b 2s, c 3s and d 4s, for which:

- Either $a + d = q$ or $q - 1$, a and b have the same parity; or
- $a + d = q - 2$ or $q - 3$, a and b have opposite parity.

This includes half the points of four adjacent layers, and therefore may include $\left(1 + o(1)\right)\binom{n}{n/2}2^{n+1}$ points.

We also have a lower bound for $c'_{n,5}$ similar to Theorem 1.3, namely $c'_{n,5} = 5^{n-O(\sqrt{\log n})}$. Consider points with a 1s, b 2s, c 3s, d 4s and e 5s. For each point, take the value $a + e + 2(b + d) + 3c$. The first three points in any geometric line give values that form an arithmetic progression of length three.

Select a set of integers with no arithmetic progression of length 3. Select all points whose value belongs to that sequence; there will be no geometric line among those points. By the Behrend construction[4], it is possible to choose these points with density $\exp -O\left(\sqrt{\log n}\right)$.

For $k = 6$, we observe that the asymptotic $c'_{n,6} = o(6^n)$ would imply the $k = 3$ density Hales–Jewett theorem $c_{n,3} = o(3^n)$. Indeed, any $k = 3$ combinatorial line-free set can be "doubled up" into a $k = 6$ geometric line-free set of the same density by pulling back the set from the map that maps $1, 2, 3, 4, 5, 6$ to $1, 2, 3, 3, 2, 1$ respectively; note that this map sends $k = 6$ geometric lines to $k = 3$ combinatorial lines. So $c'_{n,6} \geq 2^n c_{n,3}$, and more generally, $c'_{n,2k} \geq 2^n c_{n,k}$.

5. Upper Bounds for the $k = 3$ Moser Problem in Small Dimensions

In this section we finish the proof of Theorem 1.5 by obtaining the upper bounds on $c'_{n,3}$ for $n \leq 6$.

5.1. Statistics, densities and slices

Our analysis will revolve around various *statistics* of Moser sets $A \subset [3]^n$, their associated *densities*, and the behavior of such statistics and densities with respect to the operation of passing from the cube $[3]^n$ to various *slices* of that cube.

Definition 5.1 (Statistics and densities). Let $A \subset [3]^n$ be a set. For any $0 \leq i \leq n$, set $a_i(A) := |A \cap S_{n-i,n}|$; thus we have

$$0 \leq a_i(A) \leq |S_{n-i,n}| = \binom{n}{i} 2^{n-i}$$

for $0 \leq i \leq n$ and

$$a_0(A) + \ldots + a_n(A) = |A|.$$

We refer to the vector $\big(a_0(A), \ldots, a_n(A)\big)$ as the *statistics* of A. We define the ith *density* $\alpha_i(A)$ to be the quantity

$$\alpha_i(A) := \frac{a_i(A)}{\binom{n}{i} 2^{n-i}},$$

thus $0 \leq \alpha_i(A) \leq 1$ and

$$|A| = \sum_{i=0}^{n} \binom{n}{i} 2^{n-i} \alpha_i(A).$$

Example 5.2. Let $n = 2$ and A be the Moser set $A := \{12, 13, 21, 23, 31, 32\}$. Then the statistics $\big(a_0(A), a_1(A), a_2(A)\big)$ of A are $(2, 4, 0)$, and the densities $\big(\alpha_0(A), \alpha_1(A), \alpha_2(A)\big)$ are $\big(\frac{1}{2}, 1, 0\big)$.

When working with small values of n, it will be convenient to write $a(A)$, $b(A)$, $c(A)$, etc. for $a_0(A)$, $a_1(A)$, $a_2(A)$, etc., and similarly write $\alpha(A), \beta(A), \gamma(A)$, etc. for $\alpha_0(A)$, $\alpha_1(A)$, $\alpha_2(A)$, etc. Thus for instance in Example 5.2 we have $b(A) = 4$ and $\alpha(A) = \frac{1}{2}$.

Definition 5.3 (Subspace statistics and densities). If V is a k-dimensional geometric subspace of $[3]^n$, then we have a map $\phi_V : [3]^k \to [3]^n$ from the k-dimensional cube to the n-dimensional cube. If $A \subset [3]^n$ is a set and $0 \le i \le k$, we write $a_i(V, A)$ for $a_i\big(\phi_V^{-1}(A)\big)$ and $\alpha_i(V, A)$ for $\alpha_i\big(\phi_V^{-1}(A)\big)$. If the set A is clear from context, we abbreviate $a_i(V, A)$ as $a_i(V)$ and $\alpha_i(V, A)$ as $\alpha_i(V)$.

Recall from Section 1.1 that the cube $[3]^n$ can be subdivided into three *slices* in n different ways, and each slice is an $n-1$-dimensional subspace. For instance, $[3]^3$ can be partitioned into $1**$, $2**$, $3**$. We call a slice a *centre slice* if the fixed coordinate is 2 and a *side slice* if it is 1 or 3.

Example 5.4. We continue Example 5.2. Then the statistics of the side slice $1*$ are $\big(a(1*), b(1*)\big) = (1, 1)$, while the statistics of the centre slice $2*$ are $\big(a(2*), b(2*)\big) = (2, 0)$. The corresponding densities are $\big(\alpha(1*), \beta(1*)\big) = (1/2, 1)$ and $\big(\alpha(2*), \beta(2*)\big) = (1, 0)$.

A simple double counting argument gives the following useful identity:

Lemma 5.5 (Double counting identity). *Let $A \subset [3]^n$ and $0 \le i \le n-1$. Then we have*

$$\frac{1}{n-i-1} \sum_{V \text{ a side slice}} a_{i+1}(V) = \frac{1}{i+1} \sum_{W \text{ a centre slice}} a_i(W) = a_{i+1}(A)$$

where V ranges over the $2n$ side slices of $[3]^n$, and W ranges over the n centre slices. In other words, the average value of $\alpha_{i+1}(V)$ for side slices V equals the average value of $\alpha_i(W)$ for centre slices W, which is in turn equal to $\alpha_{i+1}(A)$.

Indeed, this lemma follows from the observation that every string in $A \cap S_{n-i-1,n}$ belongs to $i+1$ centre slices W (and contributes to $a_i(W)$) and to $n-i-1$ side slices V (and contributes to $a_{i+1}(V)$). One can also view this lemma probabilistically, as the assertion that there are three equivalent ways to generate a random string of length n:

- Pick a side slice V at random, and randomly fill in the wildcards in such a way that $i+1$ of the wildcards are 2's (i.e. using an element of $S_{n-i-2,n-1}$).
- Pick a centre slice V at random, and randomly fill in the wildcards in such a way that i of the wildcards are 2's (i.e. using an element of $S_{n-i-1,n-1}$).

- Randomly choose an element of $S_{n-i-1,n}$.

Example 5.6. We continue Example 5.2. The average value of β for side slices is equal to the average value of α for centre slices, which is equal to $\beta(A) = 1$.

Another very useful fact (essentially due to [8]) is that linear inequalities for statistics of Moser sets at one dimension propagate to linear inequalities in higher dimensions:

Lemma 5.7 (Propagation lemma). *Let $n \geq 1$ be an integer. Suppose one has a linear inequality of the form*

$$\sum_{i=0}^{n} v_i \alpha_i(A) \leq s \tag{5.1}$$

for all Moser sets $A \subset [3]^n$ and some real numbers $v_0, \ldots, v_n, s$. Then we also have the linear inequality

$$\sum_{i=0}^{n} v_i \alpha_{qi+r}(A) \leq s$$

whenever $q \geq 1$, $r \geq 0$, $N \geq nq + r$ are integers and $A \subset [3]^N$ is a Moser set.

Proof. We run a probabilistic argument (one could of course also use a double counting argument instead). Let $n, v_0, \ldots, v_n, s, q, r, N, A$ be as in the lemma. Let V be a random n-dimensional geometric subspace of $[3]^N$, created in the following fashion:

- Pick n wildcards $x_1, \ldots, x_n$ to run independently from 1 to 3. We also introduce dual wildcards $\overline{x_1}, \ldots, \overline{x_n}$; each $\overline{x_j}$ will take the value $4 - x_j$.
- We randomly subdivide the N coordinates into n groups of q coordinates, plus a remaining group of $N - nq$ "fixed" coordinates.
- For each coordinate in the jth group of q coordinates for $1 \leq j \leq n$, we randomly assign either a x_j or $\overline{x_j}$.
- For each coordinate in the $N - nq$ fixed coordinates, we randomly assign a digit $1, 2, 3$, but condition on the event that exactly r of the digits are equal to 2 (i.e. we use a random element of $S_{N-nq-r,N-nq}$).

- Let V be the subspace created by allowing $x_1, \ldots, x_n$ to run independently from 1 to 3, and $\overline{x_j}$ to take the value $4 - x_j$.

For instance, if $n = 2, q = 2, r = 1, N = 6$, then a typical subspace V generated in this fashion is

$$2x_1\overline{x_2}3x_2x_1 = \{213311, 212321, 211331, 223312, 222322, 221332, \\ 233313, 232323, 231333\}.$$

Observe from that the following two ways to generate a random element of $[3]^N$ are equivalent:

- Pick V randomly as above, and then assign $(x_1, \ldots, x_n)$ randomly from $S_{n-i,n}$. Assign $4 - x_j$ to $\overline{x_j}$ for all $1 \leq j \leq n$.
- Pick a random string in $S_{N-qi-r,N}$.

Indeed, both random variables are invariant under the symmetries of the cube, and both random variables always pick out strings in $S_{N-qi-r,N}$, and the claim follows. As a consequence, we see that the expectation of $\alpha_i(V)$ (as V ranges over the recipe described above) is equal to $\alpha_{qi+r}(A)$. On the other hand, from (5.1) we have

$$\sum_{i=0}^{n} v_i \alpha_i(V) \leq s$$

for all such V; taking expectations over V, we obtain the claim. ∎

In view of Lemma 5.7, it is of interest to locate linear inequalities relating the densities $\alpha_i(A)$, or (equivalently) the statistics $a_i(A)$. For this, it is convenient to introduce the following notation.

Definition 5.8. Let $n \geq 1$ be an integer.

- A vector $(a_0, \ldots, a_n)$ of non-negative integers is *feasible* if it is the statistics of some Moser set A.
- A feasible vector $(a_0, \ldots, a_n)$ is *Pareto-optimal* if there is no other feasible vector $(b_0, \ldots, b_n) \neq (a_0, \ldots, a_n)$ such that $b_i \geq a_i$ for all $0 \leq i \leq n$.

- A Pareto-optimal vector $(a_0, \ldots, a_n)$ is *extremal* if it is not a non-trivial convex linear combination of other Pareto-optimal vectors.

To establish a linear inequality of the form (5.1) with the v_i non-negative, it suffices to test the inequality against densities associated to extremal vectors of statistics. (There is no point considering linear inequalities with negative coefficients v_i, since one always has the freedom to reduce a density $\alpha_i(A)$ of a Moser set A to zero, simply by removing all elements of A with exactly i 2's.)

We will classify exactly the Pareto-optimal and extremal vectors for $n \leq 3$, which by Lemma 5.7 will lead to useful linear inequalities for $n \geq 4$. Using a computer, we have also located a partial list of Pareto-optimal and extremal vectors for $n = 4$, which are also useful for the $n = 5$ and $n = 6$ theory.

5.2. Up to three dimensions

We now establish Theorem 1.5 for $n \leq 3$, and establish some auxiliary inequalities which will be of use in higher dimensions.

The case $n = 0$ is trivial. When $n = 1$, it is clear that $c'_{1,3} = 2$, and furthermore that the Pareto-optimal statistics are $(2, 0)$ and $(1, 1)$, which are both extremal. This leads to the linear inequality

$$2\alpha(A) + \beta(A) \leq 2$$

for all Moser sets $A \subset [3]^1$, which by Lemma 5.7 implies that

$$2\alpha_r(A) + \alpha_{r+q}(A) \leq 2 \tag{5.2}$$

whenever $r \geq 0$, $q \geq 1$, $n \geq q + r$, and $A \subset [3]^n$ is a Moser set.

For $n = 2$, we see by partitioning $[3]^2$ into three slices that $c'_{2,3} \leq 3c'_{1,3} = 6$, and so (by the lower bounds in the previous section) $c'_{2,3} = 6$. Writing $(a, b, c) = \big(a(A), b(A), c(A)\big) = \big(4\alpha(A), 4\beta(A), \gamma(A)\big)$, the inequalities (5.2) become

$$a + 2c \leq 4; \quad b + 2c \leq 4; \quad 2a + b <= 8. \tag{5.3}$$

Lemma 5.9. *When $n = 2$, the Pareto-optimal statistics are $(4, 0, 0)$, $(3, 2, 0)$, $(2, 4, 0)$, $(2, 2, 1)$. In particular, the extremal statistics are $(4, 0, 0)$, $(2, 4, 0)$, $(2, 2, 1)$.*

Proof. One easily checks that all the statistics listed above are feasible. Consider the statistics (a, b, c) of a Moser set $A \subset [3]^2$. c is either equal to 0 or 1. If $c = 1$, then (5.3) implies that $a, b \leq 2$, so the only Pareto-optimal statistic here is $(2, 2, 1)$. When instead $c = 0$, the inequalities (5.3) can easily imply the Pareto-optimality of $(4, 0, 0)$, $(3, 2, 0)$, $(2, 4, 0)$. ■

From this lemma we see that we obtain a new inequality $2a+b+2c \leq 8$. Converting this back to densities and using Lemma 5.7, we conclude that

$$4\alpha_r(A) + 2\alpha_{r+q}(A) + \alpha_{r+2q} \leq 4 \tag{5.4}$$

whenever $r \geq 0$, $q \geq 1$, $n \geq q + 2r$, and $A \subset [3]^n$ is a Moser set.

The line-free subsets of $[3]^2$ can be easily exhausted by computer search; it turns out that there are 230 such sets.

Now we look at three dimensions. Writing (a, b, c, d) for the statistics of a Moser set $A \subset [3]^n$ (which thus range between $(0, 0, 0, 0)$ and $(8, 12, 6, 1)$), the inequalities (5.2) imply in particular that

$$a + 4d \leq 8; b + 6d \leq 12; c + 3d \leq 6; 3a + 2c \leq 24; b + c \leq 12 \tag{5.5}$$

while (5.4) implies that

$$3a + b + c \leq 24; b + c + 3d \leq 12. \tag{5.6}$$

Summing the inequalities $b + c \leq 12, 3a + b + c \leq 24, b + c + 3d \leq 12$ yields

$$3(a + b + c + d) \leq 48$$

and hence $|A| = a + b + c + d \leq 16$; comparing this with the lower bounds of the preceding section we obtain $c'_{3,3} = 16$ as required. (This argument is essentially identical to the one in [8]).

We have the following useful computation:

Lemma 5.10 (3D Pareto-optimals). *When $n = 3$, the Pareto-optimal statistics are*

$$(3,6,3,1),\ (4,4,3,1),\ (4,6,2,1),\ (2,6,6,0),\ (3,6,5,0),\ (4,4,5,0),$$
$$(3,7,4,0),\ (4,6,4,0),\ (3,9,3,0),\ (4,7,3,0),\ (5,4,3,0),\ (4,9,2,0),$$
$$(5,6,2,0),\ (6,3,2,0),\ (3,10,1,0),\ (5,7,1,0),\ (6,4,1,0),\ (4,12,0,0),$$
$$(5,9,0,0),\ (6,6,0,0),\ (7,3,0,0),\ (8,0,0,0).$$

In particular, the extremal statistics are

$$(3,6,3,1),\ (4,4,3,1),\ (4,6,2,1),\ (2,6,6,0),\ (4,4,5,0),\ (4,6,4,0),$$
$$(4,12,0,0),\ (8,0,0,0).$$

Proof. This can be established by a brute-force search over the $2^{27} \approx 1.3 \times 10^8$ different subsets of $[3]^3$. Actually, one can perform a much faster search than this. Firstly, as noted earlier, there are only 230 line-free subsets of $[3]^2$, so one could search over $230^3 \approx 1.2 \times 10^7$ configurations instead. Secondly, by symmetry we may assume (after enumerating the 230 sets in a suitable fashion) that the first slice $A \cap 1**$ has an index less than or equal to the third $A \cap 3**$, leading to $\binom{231}{2} \times 230 \approx 6 \times 10^6$ configurations instead. Finally, using the first and third slice one can quickly determine which elements of the second slice $2**$ are prohibited from A. There are $2^9 = 512$ possible choices for the prohibited set in $2**$. By crosschecking these against the list of 230 line-free sets one can compute the Pareto-optimal statistics for the second slices inside the prohibited set (the lists of such statistics turns out to length at most 23). Storing these statistics in a lookup table, and then running over all choices of the first and third slice (using symmetry), one now has to perform $O(512 \times 230) + O(\binom{231}{2} \times 23) \approx O(10^6)$ computations, which is quite a feasible computation.

One could in principle reduce the computations even further, by a factor of up to 8, by using the symmetry group D_4 of the square $[3]^2$ to reduce the number of cases one needs to consider, but we did not implement this.

A computer-free proof of this lemma can be found at the page `Human proof of the 3D Pareto-optimal Moser statistics` at [24]. ■

Remark 5.11. A similar computation revealed that the total number of line-free subsets of $[3]^3$ was 3813884. With respect to the $2^3 \times 3! = 48$-element group of geometric symmetries of $[3]^3$, these sets partitioned into 83158 equivalence classes:

$$3813884 = 76066 \times 48 + 6527 \times 24 + 51 \times 16 + 338 \times 12 + 109 \times 8$$
$$+ 41 \times 6 + 13 \times 4 + 5 \times 3 + 3 \times 2 + 5 \times 1.$$

Lemma 5.10 yields the following new inequalities:

$$2a + b + 2c + 4d \leq 22$$

$$3a + 2b + 3c + 6d \leq 36$$
$$7a + 2b + 4c + 8d \leq 56$$
$$6a + 2b + 3c + 6d \leq 48$$
$$a + 2c + 4d \leq 14$$
$$5a + 4c + 8d \leq 40.$$

Applying Lemma 5.7, we obtain new inequalities:

$$8\alpha_r(A) + 6\alpha_{r+q}(A) + 6\alpha_{r+2q}(A) + 2\alpha_{r+3q}(A) \leq 11 \tag{5.7}$$
$$4\alpha_r(A) + 4\alpha_{r+q}(A) + 3\alpha_{r+2q}(A) + \alpha_{r+3q}(A) \leq 6 \tag{5.8}$$
$$7\alpha_r(A) + 3\alpha_{r+q}(A) + 3\alpha_{r+2q}(A) + \alpha_{r+3q}(A) \leq 7$$
$$8\alpha_r(A) + 3\alpha_{r+q}(A) + 3\alpha_{r+2q}(A) + \alpha_{r+3q}(A) \leq 8 \tag{5.9}$$
$$4\alpha_{r+q}(A) + 2\alpha_{r+2q}(A) + \alpha_{r+3q}(A) \leq 4$$
$$4\alpha_r(A) + 6\alpha_{r+2q}(A) + 2\alpha_{r+3q}(A) \leq 7$$
$$5\alpha_r(A) + 3\alpha_{r+2q}(A) + \alpha_{r+3q}(A) \leq 5$$

whenever $r \geq 0$, $q \geq 1$, $n \geq r + 3q$, and Moser sets $A \subset [3]^n$.

We also note some further corollaries of Lemma 5.10:

Corollary 5.12 (Statistics of large 3D Moser sets). *Let (a, b, c, d) be the statistics of a Moser set A in $[3]^3$. Then $|A| = a+b+c+d \leq 16$. Furthermore:*

- *If $|A| = 16$, then $(a, b, c, d) = (4, 12, 0, 0)$.*
- *If $|A| = 15$, then $(a, b, c, d) = (4, 11, 0, 0)$ or $(3, 12, 0, 0)$.*
- *If $|A| \geq 14$, then $b \geq 6$ and $d = 0$.*
- *If $|A| = 13$ and $d = 1$, then $(a, b, c, d) = (4, 6, 2, 1)$ or $(3, 6, 3, 1)$.*

5.3. Four dimensions

Now we establish the bound $c'_{4,3} = 43$. Let A be a Moser set in $[3]^4$, with attendant statistics (a, b, c, d, e), which range between $(0, 0, 0, 0, 0)$ and $(16, 32, 24, 8, 1)$. In view of the lower bounds, our task here is to establish the upper bound $a + b + c + d + e \leq 43$.

The linear inequalities already established just barely fail to achieve this bound, but we can obtain the upper bound $a+b+c+d+e \leq 44$ as follows.

First suppose that $e = 1$; then from the inequalities (5.2) (or by considering lines passing through 2222) we see that $a \leq 8$, $b \leq 16$, $c \leq 12$, $d \leq 4$ and hence $a + b + c + d + e \leq 41$, so we may assume that $e = 0$.

From Lemma 5.5, we see that $a+b+c+d+e$ is now equal to the sum of $a(V)/4+b(V)/3+c(V)/2+d(V)$, where V ranges over all side slices of $[3]^4$. But from Lemma 5.10 we see that $a(V)/4+b(V)/3+c(V)/2+d(V)$ is at most $\frac{11}{4}$, with equality occurring only when $\big(a(V),b(V),c(V),d(V)\big) = (2,6,6,0)$. This gives the upper bound $a + b + c + d + e \leq 44$.

The above argument shows that $a + b + c + d + e = 44$ can only occur if $e = 0$ and if $\big(a(V),b(V),c(V),d(V)\big) = (2,6,6,0)$ for all side slices V. Applying Lemma 5.10 again this implies $(a,b,c,d,e) = (4,16,24,0,0)$. But then A contains all of the sphere $S_{2,4}$, which implies that the four-element set $A \cap S_{4,4}$ cannot contain a pair of strings which differ in exactly two positions (as their midpoint would then lie in $S_{2,4}$, contradicting the hypothesis that A is a Moser set).

Recall that we may partition $S_{4,4} = S^e_{4,4} \cup S^o_{4,4}$, where

$$S^e_{4,4} := \{1111, 1133, 1313, 3113, 1331, 3131, 3311, 3333\}$$

is the strings in $S_{4,4}$ with an even number of 1's, and

$$S^o_{4,4} := \{1113, 1131, 1311, 3111, 1333, 3133, 3313, 3331\}$$

are the strings in $S_{4,4}$ with an odd number. Observe that any two distinct elements in $S^e_{4,4}$ differ in exactly two positions unless they are antipodal. Thus $A \cap S^e_{4,4}$ has size at most two, with equality only when $A \cap S^e_{4,4}$ consists of an antipodal pair. Similarly for $A \cap S^o_{4,4}$. Thus A must consist of two antipodal pairs, one from $S^e_{4,4}$ and one from $S^o_{4,4}$.

By the symmetries of the cube we may assume without loss of generality that these pairs are $\{1111, 3333\}$ and $\{1113, 3331\}$ respectively. But as A is a Moser set, A must now exclude the strings 1112 and 3332. These two strings form two corners of the eight-element set

$$***2 \cap S_{3,4} = \{1112, 1132, 1312, 3112, 1332, 3132, 3312, 3332\}.$$

Any pair of points in this set which are "adjacent" in the sense that they differ by exactly one entry cannot both lie in A, as their midpoint would then lie in $S_{3,4}$, and so A can contain at most four elements from this set, with equality only if A contains all the points in $***2 \cap S_{3,4}$ of the same parity (either all the elements with an even number of 3s, or all the elements with an odd number of 3s). But because the two corners removed from this set have the opposite parity (one has an even number of 1s and one has an odd number), we see in fact that A can contain at most 3 points from this set. Meanwhile, the same arguments give that A contains at most four points from $**2*\cap S_{3,4}$, $*2**\cap S_{3,4}$, and $2***\cap S_{3,4}$. Summing we see that $b = |A \cap S_{3,4}| \leq 3+4+4+4 = 15$, a contradiction. Thus we have $c'_{4,3} = 43$ as claimed.

We have the following four-dimensional version of Lemma 5.10:

Lemma 5.13 (4D Pareto-optimals). *When $n = 4$, the Pareto-optimal statistics are given by the table in Figure* 8.

Proof. This was computed by computer search as follows. First, one observed that if (a, b, c, d, e) was Pareto-optimal, then $a \geq 3$. To see this, it suffices to show that for any Moser set $A \subset [3]^4$ with $a(A) = 0$, it is possible to add three points from $S_{4,4}$ to A and still have a Moser set. To show this, suppose first that A contains a point from $S_{1,4}$, such as 2221. Then A must omit either 2211 or 2231; without loss of generality we may assume that it omits 2211. Similarly we may assume it omits 2121 and 1221. Then we can add 1131, 1311, 3111 to A, as required. Thus we may assume that A contains no points from $S_{1,4}$. Now suppose that A omits a point from $S_{2,4}$, such as 2211. Then one can add 3333, 3111, 1311 to A, as required. Thus we may assume that A contains all of $S_{2,4}$, which forces A to omit 2222, as well as at least one point from $S_{3,4}$, such as 2111. But then 3111, 1111, 3333 can be added to the set, a contradiction.

Thus we only need to search through sets $A \subset [3]^4$ for which $|A \cap S_{4,4}| \geq 3$. A straightforward computer search shows that up to the symmetries of the cube, there are 391 possible choices for $A \cap S_{4,4}$. For each such choice, we looped through all the possible values of the slices $A \cap 1***$ and $A \cap 3***$, i.e. all three-dimensional Moser sets which had the indicated intersection with $S_{3,3}$. (For fixed $A \cap S_{4,4}$, the number of possibilities for $A \cap 1***$ ranges from 1 to 87123, and similarly for $A \cap 3***$). For each pair of slices $A \cap 1***$ and $A \cap 3***$, we computed the lines connecting these two sets to see what subset of $2***$ was excluded from A; there are 2^{27} possible such exclusion sets. We precomputed a lookup table that gave the Pareto-optimal statistics for $A \cap 2***$ for each such choice of exclusion set; using this lookup table for each choice of $A \cap 1***$ and $A \cap 3***$ and collating the results, we obtained the above list. On a Linux cluster, the lookup table took 22 minutes to create, and the loop over the $A \cap 1***$ and $A \cap 3***$ slices took two hours, spread out over 391 machines (one for each choice of

a	(b,c,d,e)
3	(16,24)
4	(14,19,2), (15,24), (16,8,4,1), (16,14,4), (16,23), (17,21), (18,19)
5	(12,12,4,1), (12,13,6), (12,15,5), (12,19,2), (13,10,4,1), (13,14,5), (13,21,1), (15,9,4,1), (15,12,3,1), (15,13,5), (15,18,3), (15,20,1), (15,22), (16,7,4,1), (16,10,3,1), (16,11,5), (16,12,2,1), (16,16,3), (16,19,1), (16,21), (17,12,4), (17,14,3), (17,16,2), (17,18,1), (17,20), (18,13,3), (18,14,2), (20,8,4), (20,10,3), (20,13,2), (20,14,1), (20,18), (21,10,2), (21,15), (22,13)
6	(8,12,8), (10,11,4,1), (11,12,7), (12,10,7), (12,13,5), (12,18,4), (13,16,4), (14,9,4,1), (14,9,7), (14,12,6), (14,16,3), (14,19,1), (14,21) (15,7,4,1), (15,10,3,1), (15,10,6), (15,11,2,1), (15,12,5), (15,15,4), (15,20), (16,7,3,1), (16,8,6), (16,9,2,1), (16,10,5), (16,12,1,1), (16,13,4), (16,14,3), (16,18,2), (16,19), (17,9,5), (17,10,4), (17,13,3), (17,15,2), (17,17,1), (17,18), (18,13,2), (18,16,1), (18,17), (19,9,4), (19,12,3), (19,15,1), (20,7,4), (20,9,3), (20,12,2), (20,13,1), (20,15), (21,8,3), (21,9,2), (21,12,1), (21,14), (22,7,3), (22,8,2), (22,10,1), (23,9,1), (24,7,2), (24,8,1), (24,12), (25,9), (26,7)
7	(8,6,8), (11,9,4,1), (11,12,6), (12,8,4,1), (12,8,6), (12,12,3,1), (12,12,5), (12,13,4), (12,15,3), (12,17,2), (13,7,4,1), (13,10,3,1), (13,11,5), (13,12,2,1), (13,12,4), (13,14,3), (13,16,2), (14,6,4,1), (14,6,7), (14,9,5), (14,10,2,1), (14,12,1,1), (14,17,1), (14,19), (15,7,5), (15,8,3,1), (15,9,2,1), (15,11,1,1), (15,11,4), (15,13,3), (15,16,1), (16,6,3,1), (16,6,6), (16,8,2,1), (16,10,1,1), (16,10,4), (16,12,0,1), (16,12,3), (16,15,2), (16,17), (17,6,5), (17,7,4), (17,11,3), (17,13,2), (17,14,1),(17,16), (18,10,3), (18,13,1), (18,15), (19,9,3), (20,6,4), (20,11,2), (20,12,1), (20,14), (21,8,2), (21,10,1), (21,12), (22,9,1), (22,11), (23,6,3), (23,7,1), (23,10), (24,6,2), (24,9), (25,6,1), (25,8), (26,3,1), (28,6), (29,3), (30,1)
8	(8,0,8), (8,9,7), (8,12,6), (9,9,4,1), (9,10,6), (9,12,3,1), (9,12,5), (9,13,4), (9,15,3), (10,7,4,1), (10,10,3,1), (10,10,5), (10,12,2,1), (10,12,4), (10,13,3), (10,15,2), (11,6,4,1), (11,9,6), (11,10,2,1), (11,11,4), (12,7,6), (12,9,3,1), (12,9,5), (12,10,4), (12,12,1,1), (12,14,2), (12,16,1), (12,18), (13,7,3,1), (13,7,5), (13,9,2,1) ,(13,12,0,1), (13,12,3), (14,0,7), (14,6,6), (14,7,2,1), (14,8,1,1), (14,9,4), (14,11,0,1), (14,11,3), (14,13,2), (14,15,1), (14,17), (15,6,3,1), (15,6,5), (15,7,1,1), (16,0,6), (16,4,3,1), (16,4,5), (16,6,2,1), (16,8,4), (16,9,0,1), (16,10,3), (16,12,2), (16,14,1), (16,16), (17,0,5), (17,3,4), (17,8,3), (17,10,2), (17,12,1), (17,14), (18,9,2), (18,11,1), (18,12), (19,6,3), (19,8,2), (20,0,4), (20,4,3), (20,7,2), (20,9,1), (20,11), (21,4,2), (21,7,1), (22,3,2), (22,6,1), (22,9), (23,0,3), (23,4,1), (24,0,2), (24,3,1), (24,8), (25,1,1), (25,6), (26,0,1), (26,4), (28,3), (32)
9	(8,10,4), (9,9,4), (9,12,3), (10,8,4), (10,10,3), (10,12,2), (10,13,1), (10,15), (11,11,2), (12,7,4), (12,9,3), (12,12,1), (12,14), (13,7,3), (13,10,2), (14,9,2), (14,11,1), (14,13), (15,6,3), (16,0,4), (16,4,3), (16,8,2), (16,10,1), (16,12), (17,3,3), (17,6,2), (17,8,1), (17,10), (18,2,3) (18,4,2), (18,7,1), (18,9), (19,0,3), (19,3,2), (19,6,1), (20,1,2), (20,5,1), (20,8), (21,4,1), (21,6), (22,1,1), (22,5), (24,4), (25,2), (28)
10	(8,6,4), (8,8,3), (9,7,3), (9,10,2), (9,11,1), (9,13), (10,5,4), (10,9,2), (10,12), (11,6,3), (12,4,4), (12,5,3), (12,7,2), (12,10,1), (12,11), (13,6,2), (13,8,1), (13,10), (14,3,3), (14,5,2), (14,9), (15,2,3), (15,7,1), (16,4,2), (16,6,1), (16,8), (17,4,1), (17,6), (18,2,1), (18,5), (20,4), (21,2), (22,1), (24)
11	(4,6,4), (6,5,4), (7,6,3), (8,4,4), (8,5,3), (9,6,2), (9,8,1), (9,10), (10,3,3), (10,5,2), (10,9), (11,2,3), (11,7,1), (12,4,2), (12,6,1), (12,8), (13,4,1), (13,6), (14,2,1), (14,5), (16,4), (17,2), (18,1), (20)
12	(4,3,3), (6,2,3), (6,5,2), (6,7,1), (6,9), (8,4,2), (8,6,1), (8,8), (9,4,1), (9,6), (10,2,1), (10,5), (12,4) (13,2), (14,1), (16)
13	(6,5), (8,4), (9,2), (10,1), (12)
14	(4,3), (5,2), (6,1), (8)
15	(4)
16	(0)

Fig. 8. The Pareto-optimal statistics (a,b,c,d,e) of Moser sets in $[3]^4$. To save space, all statistics with the same value of a have been collected in a single row; also, trailing zeroes for (b,c,d,e) have been dropped, thus for instance (b,c) is short for $(b,c,0,0)$. This table can also be found at
`http://spreadsheets.google.com/ccc?key=rwXB_Rn3Q1Zf5yaeMQL-RDw`.

$A \cap S_{4,4}$). Further details (including source code) can be found at the page `4D Moser brute force search` of [24]. ∎

As a consequence of this data, we have the following facts about the statistics of large Moser sets:

Proposition 5.14. *Let $A \subset [3]^4$ be a Moser set with statistics (a,b,c,d,e).*

(i) *If $|A| \geq 40$, then $e = 0$.*

(ii) *If $|A| \geq 43$, then $d = 0$.*

(iii) *If* $|A| \geq 42$, *then* $d \leq 2$.

(iv) *If* $|A| \geq 41$, *then* $d \leq 3$.

(v) *If* $|A| \geq 40$, *then* $d \leq 6$.

(vi) *If* $|A| \geq 43$, *then* $c \geq 18$.

(vii) *If* $|A| \geq 42$, *then* $c \geq 12$.

(viii) *If* $|A| \geq 43$, *then* $b \geq 15$.

Remark 5.15. This proposition was first established by an integer program, see the file `integer.tex` at [24]. A computer-free proof can be found at

`terrytao.files.wordpress.com/2009/06/polymath2.pdf.`

5.4. Five dimensions

Now we establish the bound $c'_{5,3} = 124$. In view of the lower bounds, it suffices to show that there does not exist a Moser set $A \subset [3]^5$ with $|A| = 125$.

We argue by contradiction. Let A be as above, and let $\big(a(A), \dots, f(A)\big)$ be the statistics of A.

Lemma 5.16. $f(A) = 0$.

Proof. If $f(A)$ is non-zero, then A contains 22222, then each of the $\frac{3^5-1}{2} = 121$ antipodal pairs in $[3]^5$ can have at most one point in A, leading to only 122 points. ■

Let us slice $[3]^5$ into three parallel slices, e.g. $1{*}{*}{*}{*}$, $2{*}{*}{*}{*}$, $3{*}{*}{*}{*}$. The intersection of A with each of these slices has size at most 43. In particular, this implies that

$$|A \cap 1****| + |A \cap 3****| = 125 - |A \cap 2****| \geq 82. \tag{5.10}$$

Thus at least one of $A \cap 1****$, $A \cap 3****$ has cardinality at least 41; by Proposition 5.14(iv) we conclude that

$$\min\big(d(1****), d(3****)\big) \leq 3. \tag{5.11}$$

Furthermore, equality can only hold in (5.11) if $A \cap 1****$, $A \cap 3****$ both have cardinality exactly 41, in which case from Proposition 5.14(iv) again we must have

$$d(1****) = d(3****) = 3. \tag{5.12}$$

Of course, we have a similar result for permutations.

Now we improve the bound $|A \cap 2****| \leq 43$:

Lemma 5.17. $|A \cap 2****| \leq 41$.

Proof. Suppose first that $|A \cap 2****| = 43$. Let $A' \subset [3]^4$ be the subset of $[3]^4$ corresponding to $A \cap 2****$, thus A' is a Moser set of cardinality 43. By Proposition 5.14(vi), $c(A') \geq 18$. By Lemma 5.5, the sum of the $c(V)$, where V ranges over the eight side slices of $[3]^4$, is therefore at least 36. By the pigeonhole principle, we may thus find two opposing side slices, say $1***$ and $3***$, with $c(1***) + c(3****) \geq 9$. Since $c(1***), c(3***)$ cannot exceed 6, we thus have $c(1***), c(3***) \geq 3$, with at least one of $c(1***), c(3***)$ being at least 5. Passing back to A, this implies that $d(*1***), d(*3***) \geq 3$, with at least one of $d(*1***), d(*3***)$ being at least 5. But this contradicts (5.11) together with the refinement (5.12).

We have just shown that $|A \cap 2****| \leq 42$; we can thus improve (5.10) to

$$|A \cap 1****| + |A \cap 3****| \geq 83.$$

Combining this with Proposition 5.14(ii)–(v) we see that

$$d(1****) + d(3****) \leq 6 \tag{5.13}$$

with equality only if $|A \cap 2****| = 42$, and similarly for permutations.

Now let A' be defined as before. Then we have

$$c(1***) + c(3***) \leq 6$$

and similarly for permutations. Applying Lemma 5.5, this implies that $c(2****) = c(A') \leq 12$.

Now suppose for contradiction that $|A'| = |A \cap 2****| = 42$. Then by Proposition 5.14(vii) we have

$$c(2****) = 12; \tag{5.14}$$

applying Lemma 5.5 again, this forces $c(1{*}{*}{*})+c(3{*}{*}{*}) = 6$ and similarly for permutations, which then implies that

$$(5.15)\qquad d({*}1{*}{*}{*}) + d({*}3{*}{*}{*}) = d({*}{*}1{*}{*}) + d({*}{*}3{*}{*})$$
$$= d({*}{*}{*}1{*}) + d({*}{*}{*}3{*}) = d({*}{*}{*}{*}1) + d({*}{*}{*}{*}3) = 6$$

and hence

$$|A \cap {*}2{*}{*}{*}| = |A \cap {*}{*}2{*}{*}| = |A \cap {*}{*}{*}2{*}| = |A \cap {*}{*}{*}{*}2| = 42$$

and thus

$$(5.16)\qquad c({*}2{*}{*}{*}) = c({*}{*}2{*}{*}) = c({*}{*}{*}2{*}) = c({*}{*}{*}{*}2) = 12.$$

Combining (5.14), (5.15), (5.16) we conclude that

$$d(1{*}{*}{*}{*}) + d(3{*}{*}{*}{*}) = 16,$$

contradicting (5.13). ∎

With this proposition, the bound (5.10) now improves to

$$(5.17)\qquad |A \cap 1{*}{*}{*}{*}| + |A \cap 3{*}{*}{*}{*}| \geq 84$$

and in particular

$$(5.18)\qquad |A \cap 1{*}{*}{*}{*}|, |A \cap 3{*}{*}{*}{*}| \geq 41.$$

from this and Proposition 5.14(ii)–(iv) we now have

$$(5.19)\qquad d(1{*}{*}{*}{*}) + d(3{*}{*}{*}{*}) \leq 4$$

and similarly for permutations.

Lemma 5.18. $e(A) = 0$.

Proof. From (5.17), the intersection of A with any side slice has cardinality at least 41, and thus by Proposition 5.14(i) such a side slice has an e-statistic of zero. The claim then follows from Lemma 5.5. ∎

We need a technical lemma:

Lemma 5.19. *Let $B \subset S_{5,5}$. Then there exist at least $|B| - 4$ pairs of strings in B which differ in exactly two positions.*

Proof. The first non-vacuous case is $|B| = 5$. It suffices to establish this case, as the higher cases then follow by induction (locating a pair of the desired form, then deleting one element of that pair from B).

Suppose for contradiction that one can find a 5-element set $B \subset S_{5,5}$ such that no two strings in B differ in exactly two positions. Recall that we may split $S_{5,5} = S^e_{5,5} \cup S^o_{5,5}$, where $S^e_{5,5}$ are those strings with an even number of 1's, and $S^o_{5,5}$ are those strings with an odd number of 1's. By the pigeonhole principle and symmetry we may assume B has at least three elements in $S^o_{5,5}$. Without loss of generality, we can take one of them to be 11111, thus excluding all elements in $S^o_{5,5}$ with exactly two 3s, leaving only the elements with exactly four 3s. But any two of them differ in exactly two positions, a contradiction. ■

We can now improve the trivial bound $c(A) \leq 80$:

Corollary 5.20 (Non-maximal c). *$c(A) \leq 79$. If $a(A) \geq 7$, then $c(A) \leq 78$.*

Proof. If $c(A) = 80$, then A contains all of $S_{3,5}$, which then implies that no two elements in $A \cap S_{5,5}$ can differ in exactly two places. It also implies (from (5.2)) that $d(A)$ must vanish, and that $b(A)$ is at most 40. By Lemma 5.19, we also have that $a(A) = |A \cap S_{5,5}|$ is at most 4. Thus $|A| \leq 4 + 40 + 80 + 0 + 0 = 124$, a contradiction.

Now suppose that $a(A) \geq 7$. Then by Lemma 5.19 there are at least three pairs in $A \cap S_{5,5}$ that differ in exactly two places. Each such pair eliminates one point from $A \cap S_{3,5}$; but each point in $S_{3,5}$ can be eliminated by at most two such pairs, and so we have at least two points eliminated from $A \cap S_{3,5}$, i.e. $c(A) \leq 78$ as required. ■

Next, we rewrite the quantity $125 = |A|$ in terms of side slices. From Lemmas 5.16, 5.18 we have

$$a(A) + b(A) + c(A) + d(A) = 125$$

and hence by Lemma 5.5, the quantity

$$s(V) := a(V) + \frac{5}{4}b(V) + \frac{5}{3}c(V) + \frac{5}{2}d(V) - \frac{125}{2},$$

where V ranges over side slices, has an average value of zero.

Proposition 5.21 (Large values of $s(V)$). *For all side slices, we have $s(V) \leq 1/2$. Furthermore, we have $s(V) < -1/2$ unless the statistics $\big(a(V), b(V), c(V), d(V), e(V)\big)$ are of one of the following four cases:*

- *(Type 1)* $\big(a(V), b(V), c(V), d(V), e(V)\big) = (2, 16, 24, 0, 0)$ *(and $s(V) = -1/2$ and $|A \cap V| = 42$);*
- *(Type 2)* $\big(a(V), b(V), c(V), d(V), e(V)\big) = (4, 16, 23, 0, 0)$ *(and $s(V) = -1/6$ and $|A \cap V| = 43$);*
- *(Type 3)* $\big(a(V), b(V), c(V), d(V), e(V)\big) = (4, 15, 24, 0, 0)$ *(and $s(V) = 1/4$ and $|A \cap V| = 43$);*
- *(Type 4)* $\big(a(V), b(V), c(V), d(V), e(V)\big) = (3, 16, 24, 0, 0)$ *(and $s(V) = 1/2$ and $|A \cap V| = 43$);*

Proof. Let V be a side slice. From (5.18) we have

$$41 \leq a(V) + b(V) + c(V) + d(V) = |A \cap V| \leq 43.$$

First suppose that $|A \cap V| = 43$, then from Proposition 5.14(ii), (viii), $d(V) = 0$ and $b(V) \geq 15$. Also, we have the trivial bound $c(V) \leq 24$, together with the inequality

$$3b(V) + 2c(V) \leq 96$$

from (5.2). To exploit these facts, we rewrite $s(V)$ as

$$s(V) = \frac{1}{2} - \frac{1}{2}\big(24 - c(V)\big) - \frac{1}{12}\big(96 - 3b(V) - 2c(V)\big).$$

Thus $s(V) \leq 1/2$ in this case. If $s(V) \geq -1/2$, then

$$6\big(24 - c(V)\big) + \big(96 - 3b(V) - 2c(V)\big) \leq 12,$$

which together with the inequalities $b(V) \leq 15$, $c(V) \leq 24$, $3b(V) + 2c(V) \leq 96$ we conclude that $\big(b(V), c(V)\big)$ must be one of $(16, 24)$, $(15, 24)$, $(16, 23)$, $(15, 23)$. The first three possibilities lead to Types 4,3,2 respectively. The fourth type would lead to $\big(a(V), b(V), c(V), d(V), e(V)\big) = (5, 15, 23, 0, 0)$, but this contradicts (5.7).

Next, suppose $|A \cap V| = 42$, so by Proposition 5.14(iii) we have $d(V) \leq 2$. From (5.2) we have

$$2c(V) + 3d(V) \leq 48 \tag{5.20}$$

while from (5.4) we have

$$3b(V) + 2c(V) + 3d(V) \leq 96 \tag{5.21}$$

and so we can rewrite $s(V)$ as

$$\begin{aligned} s(V) = -\frac{1}{2} - \frac{1}{4}(48 - 2c(V) - 3d(V)) \\ -\frac{1}{12}(96 - 3b(V) - 2c(V) - 3d(V)) + \frac{1}{2}d(V). \end{aligned} \tag{5.22}$$

This already gives $s(V) \leq 1/2$. If $d(V) = 0$, then $s(V) \leq -1/2$, with equality only in Type 1. If $d(V) = 1$, then the set $A' \subset [3]^4$ corresponding to $A \cap V$ contains a point in $S_{3,4}$, which without loss of generality we can take to be 2221. Considering the three lines $*221$, $2*21$, $22*1$, we see that at least three points in $S_{2,4}$ must be missing from A', thus $c(V) \leq 21$. This forces $48 - 2c(V) - 3d(V) \geq 3$, and so $s(V) < -3/4$. Finally, if $d(V) = 2$, then A' contains two points in $S_{3,4}$. If they are antipodal (e.g. 2221 and 2223), the same argument as above shows that at least six points in $S_{2,4}$ are missing from A'; if they are not antipodal (e.g. 2221 and 2212) then by considering the lines $*221$, $2*21$, $22*1$, $*212$, $2*12$ we see that five points are missing. Thus we have $c(V) \leq 19$, which forces $48 - 2c(V) - 3d(V) \geq 4$. This forces $s(V) \leq -1/2$, with equality only when $c(V) = 19$ and $3b(V) + 2c(V) + 3d(V) = 96$, but this forces $b(V)$ to be the non-integer $52/3$, a contradiction, which concludes the treatment of the $|A \cap V| = 42$ case.

Finally, suppose $|A \cap V| = 41$. Using (5.20), (5.21) as before we have

$$\begin{aligned} s(V) = -\frac{3}{2} - \frac{1}{4}(48 - 2c(V) - 3d(V)) \\ -\frac{1}{12}(96 - 3b(V) - 2c(V) - 3d(V)) + \frac{1}{2}d(V), \end{aligned} \tag{5.23}$$

while from Proposition 5.14(vi) we have $d(V) \leq 3$. This already gives $s(V) \leq 0$, and $s(V) \leq -1$ when $d(V) = 1$. In order to have $s(V) \geq -1/2$, we must then have $d(V) = 2$ or $d(V) = 3$. But then the arguments of the preceding paragraph give $48 - 2c(V) - 3d(V) \geq 4$, and so $s(V) \leq -1$ in this case. ∎

Since the $s(V)$ average to zero, by the pigeonhole principle we may find two opposing side slices (e.g. $1****$ and $3****$), whose total s-value is non-negative. Actually we can do a little better:

Lemma 5.22. *There exists two opposing side slices whose total s-value is strictly positive.*

Proof. If this is not the case, then we must have $s(1****)+s(3****)=0$ and similarly for permutations. Using Proposition 5.21 we thus see that for every opposing pair of side slices, one is Type 1 and one is Type 4. In particular $c(V)=24$ for all side slices V. But then by Lemma 5.5 we have $c(A)=80$, contradicting Lemma 5.20. ∎

Let V, V' be the side slices in Lemma 5.22 By Proposition 5.21, the V, V' slices must then be either Type 2, Type 3, or Type 4, and they cannot both be Type 2. Since $a(A)=a(V)+a(V')$, we conclude

$$6 \leq a(A) \leq 8. \tag{5.24}$$

In a similar spirit, we have

$$c(V)+c(V') \leq 23+24.$$

On the other hand, by considering the 24 lines connecting c-points of V, V' to c-points of the centre slice W between V and V', each of which contains at most two points in A, we have

$$c(V)+c(W)+c(V') \leq 24 \times 2.$$

Thus $c(W) \leq 1$; since

$$d(A)=d(V)+d(V')+c(W)$$

we conclude from Proposition 5.21 that $d(A) \leq 1$. Actually we can do better:

Lemma 5.23. $d(A)=0$.

Proof. Suppose for contradiction that $d(A)=1$; without loss of generality we may take $11222 \in A$. This implies that $d(1****)=d(*1***)=1$. Also, by the above discussion, $c(**1**)$ and $c(**3**)$ cannot both be 24, so by Proposition 5.21, $s(**1**)+s(**3**) \leq 1/3$; similarly $s(***1*)+s(***3*) \leq 1/3$ and $s(****1)+s(****3) \leq 1/3$. Since the s average to zero, we see from the pigeonhole principle that either $s(1****)+s(3****) \geq -1/2$ or $s(*1***)+s(*3***) \geq -1/2$. We may assume by symmetry that

$$s(1****)+s(3****) \geq -1/2. \tag{5.25}$$

Since $s(3****) \leq 1/2$ by Proposition 5.21, we conclude that

$$s(1****) \geq -1. \tag{5.26}$$

If $|A \cap 1****| = 41$, then by (5.23) we have

$$s(1****) = -1 - \frac{1}{4}(48 - 2c(1****) - 3d(1****))$$
$$- \frac{1}{12}(96 - 3b(1****) - 2c(1****) - 3d(1****))$$

but the arguments in Proposition 5.21 give $48 - 2c(1****) - 3d(1****) \geq 3$ and $96 - 3b(1****) - 2c(1****) - 3d(1****) \geq 0$, a contradiction. So we must have $|A \cap 1****| = 42$ (by Proposition 5.14(ii) and (5.18)). In that case, from (5.22) we have

$$s(1****) = \frac{1}{4}(48 - 2c(1****) - 3d(1****))$$
$$- \frac{1}{12}(96 - 3b(1****) - 2c(1****) - 3d(1****))$$

while also having $48 - 2c(1****) - 3d(1****) \geq 3$ and $96 - 3b(1****) - 2c(1****) - 3d(1****) \geq 0$. Since $s(1****) \geq -1$ and $d(1****) = 1$, we soon see that we must have $48 - 2c(1****) - 3d(1****) = 3$ and $96 - 3b(1****) - 2c(1****) - 3d(1****) \leq 3$, which forces $c(1****) = 21$ and $b(1****) = 16$ or $b(1****) = 17$; thus the statistics of $1****$ are either $(4, 16, 21, 1, 0)$ or $(3, 17, 21, 1, 0)$.

We first eliminate the $(3, 17, 21, 1, 0)$ case. In this case $s(1****)$ is exactly -1. Inspecting the proof of (5.26), we conclude that $s(3****)$ must be $+1/2$ and that $s(**1**) + s(**3**) = 1/3$. From the former fact and Proposition 5.21 we see that $a(A) = a(1****) + a(3****) = 3 + 3 = 6$; on the other hand, from the latter fact and Proposition 5.21 we have $a(A) = a(**1**) + a(**3**) = 4 + 3 = 7$, a contradiction.

So $1****$ has statistics $(4, 16, 21, 1, 0)$, which implies that $s(1****) = -3/4$ and $|A \cap 1****| = 42$. By (5.25) we conclude

$$s(3****) \geq 1/4, \tag{5.27}$$

which by Proposition 5.21 implies that $|A \cap 3****| = 43$, and hence $|A \cap 2****| = 40$. On the other hand, since $e(A) = f(A) = 0$ and $d(A) = 1$, with the latter being caused by 11222, we see that $c(2****) =$

$d(2****) = e(2****) = 0$. From (5.2) we have $4a(2****)+b(2****) \le 64$, and we also have the trivial inequality $b(2****) \le 32$; these inequalities are only compatible if $2****$ has statistics $(8,32,0,0,0)$, thus $A \cap 2****$ contains $S_{2,5} \cap 2****$.

If $a(3****) = 4$, then $a(A) = a(1****) + a(3****) = 8$, which by Proposition 5.21 implies that $s(**1**) + s(**3**)$ cannot exceed $1/12$, and similarly for permutations. On the other hand, from Proposition 5.21 $s(**1**) + s(**3**)$ cannot exceed $-3/4 + 1/4 = -1/2$, and so the average value of s cannot be zero, a contradiction. Thus $a(3****) \neq 4$, which by (5.27) and Proposition 5.21 implies that $**3**$ has statistics $(3,16,24,0,0)$.

In particular, A contains 16 points from $3**** \cap S_{1,5}$ and all of $3**** \cap S_{2,5}$. As a consequence, no pair of the 16 points in $A \cap 3**** \cap S_{1,5}$ can differ in only one coordinate; partitioning the 32-point set $3**** \cap S_{1,5}$ into 16 such pairs, we conclude that every such pair contains exactly one element of A. We conclude that $A \cap 3**** \cap S_{1,5}$ is equal to either $3**** \cap S^e_{1,5}$ or $3**** \cap S^o_{1,5}$.

On the other hand, A contains all of $2**** \cap S_{2,5}$, and exactly sixteen points from $1**** \cap S_{1,5}$. Considering the vertical lines $*xyzw$ where $xyzw \in S_{1,4}$, we conclude that $A \cap 1**** \cap S_{1,5}$ is either equal to $1**** \cap S^o_{1,5}$ or $1**** \cap S^e_{1,5}$. But either case is incompatible with the fact that A contains 11222 (consider either the line $11xx2$ or $11x\overline{x}2$, where $x = 1,2,3$ and $\overline{x} = 4 - x$), obtaining the required contradiction. ■

We can now eliminate all but three cases for the statistics of A:

Proposition 5.24 (Statistics of A). *The statistics $\big(a(A), b(A), c(A), d(A), e(A), f(A)\big)$ of A must be one of the following three tuples:*

- *(Case 1)* $(6,40,79,0,0)$;
- *(Case 2)* $(7,40,78,0,0)$;
- *(Case 3)* $(8,39,78,0,0)$.

Proof. Since $d(A) = e(A) = f(A) = 0$, we have

$$c(2****) = d(2****) = e(2****) = 0.$$

On the other hand, from (5.2) we have $4a(2****) + b(2****) \le 64$ as well as the trivial inequality $b(2****) \le 24$, and also we have

$$|A \cap 2****| = 125 - |A \cap 1****| - |A \cap 3****| \ge 125 - 43 - 43 = 39.$$

Putting all this together, we see that the only possible statistics for $2{*}{*}{*}{*}$ are $(8,32,0,0,0)$, $(7,32,0,0,0)$, or $(8,31,0,0,0)$. In particular, $7 \leq a(2{*}{*}{*}{*}) \leq 8$ and $31 \leq b(2****) \leq 32$, and similarly for permutations. Applying Lemma 5.5 we conclude that

$$35 \leq b(A) \leq 40$$

and

$$77.5 \leq c(A) \leq 80.$$

Combining this with the first part of Corollary 5.20 we conclude that $c(A)$ is either 78 or 79. From this and (5.24) we see that the only cases that remain to be eliminated are $(7,39,79,0,0)$ and $(8,38,79,0,0)$, but these cases are incompatible with the second part of Corollary 5.20. ■

We now eliminate each of the three remaining cases in turn.

5.5. Elimination of $(6,40,79,0,0)$

Here $A \cap S_{5,5}$ has six points. By Lemma 5.19, there are at least two pairs in this set which differ in two positions. Their midpoints are eliminated from $A \cap S_{3,5}$. But A omits exactly one point from $S_{3,5}$, so these midpoints must be the same. By symmetry, we may then assume that these two pairs are $(11111, 11133)$ and $(11113, 11131)$. Thus the eliminated point in $S_{3,5}$ is 11122, i.e. A contains $S_{3,5} \backslash \{11122\}$. Also, A contains $\{11111, 11133, 11113, 11131\}$ and thus must omit $\{11121, 11123, 11112, 11132\}$.

Since $11322 \in A$, at most one of $11312, 11332$ lie in A. By symmetry we may assume $11312 \notin A$, thus there is a pair $(xy1z2, xy3z2)$ with $x, y, z = 1, 3$ that is totally omitted from A, namely $(11112, 11312)$. On the other hand, every other pair of this form can have at most one point in the A, thus there are at most seven points in A of the form $xyzw2$ with $x, y, z, w = 1, 3$. Similarly there are at most 8 points of the form $xyz2w$, or of $xy2zw$, $x2yzw$, $2xyzw$, leading to $b(A) \leq 7+8+8+8+8 = 39$, contradicting the statistic $b(A) = 40$.

5.6. Elimination of $(7, 40, 78, 0, 0)$

Here $A \cap S_{5,5}$ has seven points. By Lemma 5.19, there are at least three pairs in this set which differ in two positions. As we can only eliminate two points from $S_{3,5}$, two of the midpoints of these pairs must be the same; thus, as in the previous section, we may assume that A contains $\{11111, 11133, 11113, 11131\}$ and omits $\{11121, 11123, 11112, 11132\}$ and 11122.

Now consider the 160 lines ℓ connecting two points in $S_{4,5}$ to one point in $S_{3,5}$ (i.e. $*2xyz$ and permutations, where $x, y, z = 1, 3$). By double counting, the total sum of $|\ell \cap A|$ over all 160 lines is $4b(A) + 2c(A) = 316 = 158 \times 2$. On the other hand, each of these lines contain at most two points in A, but two of them (namely $1112*$ and $1112*$) contain no points. Thus we must have $|\ell \cap A| = 2$ for the remaining 158 lines ℓ.

Since A omits $1112x$ and $111x2$ for $x = 1, 3$, we thus conclude (by considering the lines $11 * 2x$ and $11 * x2$) that A must contain $1132x$, $113x2$, $1312x$, and $131x2$. Taking midpoints, we conclude that A omits 11322 and 13122. But together with 11122 this implies that at least three points are missing from $A \cap S_{3,5}$, contradicting the hypothesis $c(A) = 78$.

5.7. Elimination of $(8, 39, 78, 0, 0)$

Now $A \cap S_{5,5}$ has eight points. By Lemma 5.19, there are at least three pairs in this set which differ in two positions. As we can only eliminate two points from $S_{3,5}$, two of these pairs (a, b), (c, d) must have the same midpoint p, and two other pairs (a', b'), (c', d') must have the same midpoint p', and A contains $S_{3,5} \backslash \{p, p'\}$. As p, p' are distinct, the plane containing a, b, c, d is distinct from the plane containing a', b', c', d'.

Again consider the 160 lines ℓ from the previous section. This time, the sum of the $|\ell \cap A|$ is $4b(A) + 2c(A) = 312 = 156 \times 2$. But the two lines in the plane of a, b, c, d passing through p, and the two lines in the plane of a', b', c', d' passing through p', have no points; thus we must have $|\ell \cap A| = 2$ for the remaining 156 lines ℓ.

Without loss of generality we have $(a, b) = (11111, 11133)$, $(c, d) = (11113, 11131)$, thus $p = 11122$. By permuting the first three indices, we may assume that p' is not of the form $x2y2z, x2yz2, xy22z, xy2z2$ for any $x, y, z = 1, 3$. Then we have $1112x \notin A$ and $1122x \in A$ for every $x = 1, 3$, so by the preceding paragraph we have $1132x \in A$; similarly for $113x2$, $1312x$, $131x2$. Taking midpoints, this implies that $13122, 11322 \notin A$, but this (together with 11122) shows that at least three points are missing from $A \cap S_{3,5}$, contradicting the hypothesis $c(A) = 78$.

5.8. Six dimensions

Now we establish the bound $c'_{6,3} = 353$. In view of the lower bounds, it suffices to show that there does not exist a Moser set $A \subset [3]^5$ with $|A| = 354$.

We argue by contradiction. Let A be as above, and let $\big(a(A), \ldots, g(A)\big)$ be the statistics of A.

Lemma 5.25. $g(A) = 0$.

Proof. For any four-dimensional slice V of A, define

$$S(V) := 15a(V) + 5b(V) + 5c(V)/2 + 3d(V)/2 + e(V).$$

From Lemma 5.5 we see that $|A|$ is equal to $a(A) + b(A)$ plus the average of $S(V)$ where V ranges over the twenty slices which are some permutation of the center slice $22 * * * *$.

If $g(A) = 1$, then $a(A) \leq 32$ and $b(A) \leq 96$ by (5.2). Meanwhile, $e(V) = g(A) = 1$ for every center slice V, so from Lemma 5.13, one can show that $S(V) \leq 223.5$ for every such slice. We conclude that $|A| \leq 351.5$, a contradiction. ∎

For any four-dimensional slice V of A, define the *defects* to be

$$D(V) := 356 - \left[4a(V) + 6b(V) + 10c(V) + 20d(V) + 60e(V)\right].$$

Define a *corner slice* to be one of the permutations or reflections of $11****$, thus there are 60 corner slices. From Lemma 5.5 we see that $356 - |A| + f(A) = 2 + f(A)$ is the average of the defects of all the 60 corner slices. On the other hand, from Lemma 5.13 and a straightforward computation, one concludes

Lemma 5.26. *Let A be a four-dimensional Moser set. Then $D(A) \geq 0$. Furthermore:*

- *If A has statistics $(6, 12, 18, 4, 0)$, then $D(A) = 0$.*

- *If A has statistics* $(5,12,18,4,0)$, $(5,12,12,4,1)$, *or* $(6,8,12,8,0)$, *then* $D(A)=4$.
- *For all other A,* $D(A)\geq 6$.
- *If* $a(A)=4$, *then* $D(A)\geq 8$.
- *If* $a(A)\geq 7$, *then* $D(A)\geq 16$ *(with equality iff A has statistics* $(7,11,12,6,0)$*).*
- *If* $a(A)\geq 8$, *then* $D(A)\geq 30$.
- *If* $a(A)\geq 9$, *then* $D(A)\geq 86$.

Define a *family* to be a set of four parallel corner slices, thus there are 15 families, which are all a permutation of $\{11****, 13****, 31****, 33****\}$. We refer to the family $\{11****, 13****, 31****, 33****\}$ as $ab****$, and similarly define the family $a*b***$, etc.

Lemma 5.27. $f(A)=0$.

Proof. For any four-dimensional slice V of A, define

$$s(V):=12a(V)+15b(V)/2+20c(V)/3+15d(V)/2+12e(V),$$

and define an *edge slice* to be one of the 60 permutations or reflections of $12****$. From double counting we see that $|A|-a(A)$ is equal to the average of the 60 values of $s(V)$ as V ranges over edge slices.

From Lemma 5.13 one can verify that $s(V)\leq 336$, and that $s(V)\leq 296=336-40$ if $e(V)=1$. The number of edge slices V for which $e(V)=1$ is equal to $5f(A)$, and so the average value of the $s(V)$ is at most $336-\frac{40\times 5}{60}f(A)$, and so

$$|A|-a(A)\leq 336-\frac{40\times 5}{60}f(A)$$

which we can rearrange (using $|A|=354$) as

$$a(A)\geq 18+\frac{10}{3}f(A).$$

Suppose first that $f(A)\geq 3$; then $a(A)\geq 28$. Then in any given family, there is a corner slice with an a value at least 9, or four slices with a value at least 7, or two slices with a value at least 8, or one slice with a value 8 and two of a value at least 7, leading to a total defect of at least 60 by Lemma 5.26. Thus the average defect is at least 15; on the other hand, the average defect is $2 + f(A) \leq 2 + 12$, a contradiction.

Now suppose that $f(A) = 2$; then $a(A) \geq 25$. This means that in any given family, one of the four corner slices has an a value of at least 7, and thus by Lemma 5.26 has a defect of at least 16. Thus the average defect is at least 4; on the other hand, the average defect is $2 + f(A) = 4$. From Lemma 5.26, this implies that in any given family, three of the corner slices have statistics $(6, 12, 18, 4, 0)$ and the last one has statistics $(7, 11, 12, 6, 0)$. But this forces $b(A) = 70.5$ by double counting, which is absurd.

The remaining case is when $f(A) = 1$. Here we need a different argument. Without loss of generality we may take $122222 \in A$. The average defect among all 60 slices is $2 + f(A) = 3$. Equivalently, the average defect among all 15 families is 12.

First suppose that $a(A) \neq 24$. Then in every family, at least one of the corner slices needs to have an a value distinct from six, and so the average defect in each family is at least 4. Thus the five families $ab{*}{*}{*}{*}$, $a{*}b{*}{*}{*}$, $a{*}{*}b{*}{*}$, $a{*}{*}{*}b{*}$, $a{*}{*}{*}{*}b$ have an average defect of at most 28, which implies that the ten corner slices beginning with 1 (or equivalently, adjacent to an edge slice containing 122222) is at most 14. In other words, if (a, b, c, d, e) is the average of the statistics of these ten corner slices, then

$$4a + 6b + 10c + 20d + 60e \geq 342.$$

On the other hand, (a, b, c, d, e) must lie in the convex hull of the statistics of four-dimensional Moser sets, which are described by Lemma 5.10. Also, as A contains 122222, one has $c/24, d/8, e \leq 1/2$ by considering lines with centre 122222. Finally, from (5.8) and double-counting one has $7c/24 + 3d/8 + 3e + 1 \leq 6$. Inserting these facts into a standard linear program yields a contradiction; indeed, the maximal value of $4a + 6b + 10c + 20d + 60e$ with these constraints is $338\frac{2}{3}$, attained when $(a, b, c, d, e) = \left(\frac{17}{3}, 16, 12, 4, \frac{1}{3}\right)$.

Finally, we consider the case when $f(A) = 1$ and $a(A) = 24$. The preceding arguments allow the average defect of the ten corner slices beginning with 1 to be as large as 18, which implies that $4a+6b+10c+20d+60e \geq 338$. Linear programming shows that this is not possible if $a \geq 6$, thus $a < 6$. But this forces one of the corner slices beginning with 3 to have an a value of at least 7, and thus to have a defect of at least 16 by Lemma 5.10. Repeating the preceding arguments, this increases the lower bound for $4a + 6b + 10c + 20d + 60e$ by $\frac{16}{10}$, to 339.6; but this is now inconsistent with the upper bound of $338\frac{2}{3}$ from linear programming. ■

As a consequence of the above lemma, we see that the average defect of all corner slices is 2, or equivalently that the total defect of these slices is 120.

Call a corner slice *good* if it has statistics $(6, 12, 18, 4, 0)$, and *bad* otherwise. Thus good slices have zero defect, and bad slices have defect at least four. Since the average defect of the 60 corner slices is 2, there are at least 30 good slices.

One can describe the structure of the good slices completely:

Lemma 5.28. *The subset of* $[3]^4$ *consisting of the strings* 1111, 1113, 3333, 1332, 1322, 1222, 3322 *and permutations is a Moser set with statistics* $(6, 12, 18, 4, 0)$. *Conversely, every Moser set with statistics* $(6, 12, 18, 4, 0)$ *is of this form up to the symmetries of the cube* $[3]^4$.

Proof. This can be verified by computer. By symmetry, one assumes 1222, 2122, 2212 and 2221 are in the set. Then 18 of the 24 'c' points with two 2s must be included; it is quick to check that 1122 and permutations must be the six excluded. Next, one checks that the only possible set of six 'a' points with no 2s is 1111, 1113, 3333 and permutations. Lastly, in a rather longer computation, one finds there is only possible set of twelve 'b' points, that is points with one 2. A computer-free proof can be found at the page `Classification of` $(6, 12, 18, 4, 0)$ `sets` at [24]. ■

As a consequence of this lemma, given any $x, y, z, w \in \{1, 3\}$, there is a unique good Moser set in $[3]^4$ set whose intersection with $S_{1,4}$ is $\{x222, 2y22, 22z2, 222w\}$, and these are the only 16 possibilities. Call this set the *good set of type* $xyzw$. It consists of

- The four points $x222, 2y22, 22z2, 222w$ in $S_{1,4}$;
- All 24 elements of $S_{2,4}$ except for $xy22$, $x2z2$, $x22w$, $2yz2$, $2y2w$, $22zw$;
- The twelve points $xYZ2$, $xY2W$, $x2ZW$, $XyZ2$, $Xy2W$, $2yZW$, $XYz2$, $X2zW$, $2YzW$, $XY2w$, $X2Zw$, $2YZw$ in $S_{3,4}$, where $X = 4 - x$, $Y = 4 - y$, $Z = 4 - z$, $W = 4 - w$;
- The six points $xyzw$, $xyzW$, $xyZw$, $xYzw$, $Xyzw$, $XYZW$ in $S_{4,4}$.

We can use this to constrain the types of two intersecting good slices:

Lemma 5.29. *Suppose that the* $pq****$ *slice is of type* $xyzw$, *and the* $p*r***$ *slice is of type* $x'y'z'w'$, *where* p, q, r, x, y, z, w, x', y', z', w' *are in* $\{1,3\}$. *Then* $x' = x$ *iff* $q = r$, *and* $y'z'w'$ *is equal to either* yzw *or* YZW. *If* $x = r$ *(or equivalently if* $x' = q$*), then* $y'z'w' = yzw$.

Proof. By reflection symmetry we can take $p = q = r = 1$. Observe that the $11****$ slice contains 111222 iff $x = 1$, and the $1*1***$ slice similarly contains 111222 iff $x' = 1$. This shows that $x = x'$.

Suppose now that $x = x' = 1$. Then the $111***$ slice contains the three elements $111y22$, $1112z2$, $11122w$, and excludes $111Y22$, $1112Z2$, $11122W$, and similarly with the primes, which forces $yzw = y'z'w'$ as claimed.

Now suppose that $x = x' = 3$. Then the $111***$ slice contains the two elements $111yzw$, $111YZW$, but does not contain any of the other six points in $S_{6,6} \cap 111***$, and similarly for the primes. Thus $y'z'w'$ is equal to either yzw or YZW as claimed. ∎

Now we look at two adjacent parallel good slices, such as $11****$ and $13****$. The following lemma asserts that such slices either have opposite type, or else will create a huge amount of defect in other slices:

Lemma 5.30. *Suppose that the* $11****$ *and* $13****$ *slices are good with types* $xyzw$ *and* $x'y'z'w'$ *respectively. If* $x = x'$, *then the* $1*x***$ *slice has defect at least* 30, *and the* $1*X***$ *slice has defect at least* 8. *Also, the* $1**1**$, $1**3**$, $1***1*$, $1***3*$, $1****1$, $1****3$ *slices have defect at least* 6. *In particular, the total defect of slices beginning with* $1*$ *is at least* 74.

Proof. Observe from the $11****$, $13****$ hypotheses that $a(1*x***) = 9$ and $a(1*X***) = 4$, which gives the first two claims by Lemma 5.26. For the other claims, one sees from Lemma 5.29 that the other six slices cannot be good; also, they have an a-value of 6 and a d-value of at most 7, and the claims then follow from Lemma 5.26. ∎

Now we look at two diagonally opposite parallel good slices, such as $11****$ and $33****$.

Lemma 5.31. *The* $11****$ *and* $33****$ *slices cannot both be good and of the same type.*

Proof. Suppose not. By symmetry we may assume that $11****$ and $33****$ are of type 1111. This excludes a lot of points from $22****$. Indeed, by connecting lines between the $11****$ and $33****$ slices, we see that the only points that can still survive in $22****$ are 221133, 221333, 221132, 223332, and permutations of the last four indices. Double counting the lines $22133*$ and permutations we see that there are at most 12 points one can place in the permutations of 221133, 221333, 221132, and so the $22****$ slice has at most 16 points. Meanwhile, the two five-dimensional slices $1*****$, $3*****$ have at most $c'_{5,3} = 124$ points, and the other two four-dimensional slices $21****$, $23****$ have at most $c'_{4,3} = 43$ points, leading to at most $16 + 124 * 2 + 43 * 2 = 350$ points in all, a contradiction. ∎

Lemma 5.32. *It is not possible for all four slices in a family to be good.*

Proof. Suppose not. By symmetry we may assume that $11****$, $13****$, $31****$, $33****$ are good. By Lemma 5.31, the $11****$ and $33****$ slices cannot be of the same type, and so they cannot both be of the opposite type to either $13****$ or $31****$. If $13****$ is not of the opposite type to $11****$, then by (a permutation of) Lemma 5.30, the total defect of slices beginning with $1*$ is at least 74; otherwise, if $13****$ is not of the opposite type to $33****$, then by (a permutation and reflection of) Lemma 5.30, the total defect of slices beginning with $*3$ is at least 74. Similarly, the total defect of slices beginning with $3*$ or $*1$ is at least 74, leading to a total defect of at least 148. But the total defect of all the corner slices is $2 \times 60 = 120$, a contradiction. ∎

Corollary 5.33. *At most one family can have a total defect of at least* 38.

Proof. Suppose there are two families with defect at least 38. The remaining thirteen families have defect at least 4 by Lemma 5.32 and Lemma 5.26, leading to a total defect of at least $38*2+13*4 = 128$. But the total defect is $2 \times 60 = 120$, a contradiction. ∎

Actually we can refine this:

Proposition 5.34. *No family can have a total defect of at least* 38.

Proof. Suppose for contradiction that the $ab****$ family (say) had a total defect of at least 38, then by Corollary 5.33 no other families have total defect at least 38.

We claim that the $**ab**$ family can have at most two good slices. Indeed, suppose the $**ab**$ has three good slices, say $**11**$, $**13**$, $**33**$. By Lemma 5.31, the $**11**$ and $**33**$ slices cannot be of the same type, and so cannot both be of opposite type to $**13**$. Suppose $**11**$ and $**13**$ are not of opposite type. Then by (a permutation of) Lemma 5.30, one of the families $a*b***$, $*ab***$, $**b*a*$, $**b**a$ has a net defect of at least 38, contradicting the normalisation.

Thus each of the six families $**ab**$, $**a*b*$, $**a**b$, $***ab*$, $***a*b$, $****ab$ have at least two bad slices. Meanwhile, the eight families $a*b***$, $a**b**$, $a***b*$, $a****b$, $*ab***$, $*a*b**$, $*a**b*$, $*a***b$ have at least one bad slice by Corollary 5.32, leading to twenty bad slices in addition to the defect of at least 38 arising from the $ab****$ slice. To add up to a total defect of 120, we conclude from Lemma 5.26 that all bad slices outside of the $ab****$ family have a defect of four, with at most one exception; but then by Lemma 5.30 this shows that (for instance) the $1*1***$ and $1*3***$ slices cannot be good unless they are of opposite type. The previous argument then shows that the $a*b***$ slice cannot have three good slices, which increases the number of bad slices outside of $ab****$ to at least twenty-one, and now there is no way to add up to 120, a contradiction. ∎

Corollary 5.35. *Every family can have at most two good slices.*

Proof. If for instance $11****$, $13****$, $33****$ are all good, then by Lemma 5.31 at least one of $11****$, $33****$ is not of the opposite type to $13****$, which by Lemma 5.30 implies that there is a family with a total defect of at least 38, contradicting the previous proposition. ∎

From this corollary and Lemma 5.26, we see that every family has a defect of at least 8. Since there are 15 families, and 8×15 is exactly equal to 120, we conclude

Corollary 5.36. *Every family has* exactly *two good slices, and the remaining two slices have defect 4. In particular, by Lemma 5.26, the bad slices must have statistics* $(5, 12, 18, 4, 0)$, $(5, 12, 12, 4, 1)$, *or* $(6, 8, 12, 8, 0)$.

We now limit how these slices can interact with good slices.

Lemma 5.37. *Suppose that* $1*1***$ *is a good slice.*

(i) *The* $11****$ *slice cannot have statistics* $(6, 8, 12, 8, 0)$.

(ii) *The* $11****$ *slice cannot have statistics* $(5, 12, 12, 4, 1)$.

(iii) *If the* $11****$ *slice has statistics* $(5, 12, 18, 4, 0)$, *then the* $112***$ *slice has statistics* $(3, 9, 3, 0)$.

Proof. This can be verified through computer search; there are 16 possible configurations for the good slices, and one can calculate that there are 27520 configurations for the $(5, 12, 12, 4, 1)$ slices, 4368 configurations for the $(5, 12, 18, 4, 0)$ slices, and 80000 configurations for the $(6, 8, 12, 8, 0)$ slices. It is then a routine matter to inspect by computer all the potential counterexamples to the above lemma. ∎

Corollary 5.38. *The* $111***$ *slice has statistics* $(4, 3, 3, 1)$, $(2, 6, 6, 0)$, $(3, 3, 3, 1)$, *or* $(1, 6, 6, 0)$.

Proof. From Corollary 5.36, we know that at least one of the slices $13****$, $31****$, $11****$ are good. If $11****$ or $1*1***$ is good, then the slice $111***$ has statistics $(4, 3, 3, 1)$ or $(2, 6, 6, 0)$, by Lemma 5.28. By symmetry we may thus reduce to the case where $13****$ is good and $1*1***$ is bad. Then by Lemma 5.37, the $1*1***$ slice has statistics $(5, 12, 18, 4, 0)$ and the $121***$ slice has statistics $(3, 9, 3, 0)$. Since the $131***$ slice, as a side slice of the good $13****$ slice, has statistics $(4, 3, 3, 1)$ or $(2, 6, 6, 0)$, we conclude that the $111***$ slice has statistics $(1, 6, 6, 0)$ or $(3, 3, 3, 1)$, and the claim follows. ∎

Corollary 5.39. *All corner slices have statistics* $(6, 12, 18, 4, 0)$ *or* $(5, 12, 18, 4, 0)$.

Proof. Suppose first that a corner slice, say $11****$ has statistic $(6, 8, 12, 8, 0)$. Then $111***$ and $113***$ contain one "*d*" point each, and have six "*a*" points between them, so by Corollary 5.38, they both have statistic $(3, 3, 3, 1)$. This forces the $1*1***$, $1*3***$ slices to be bad, which by Corollary 5.36 forces the $3*1***$, $3*3***$ slices to be good. This forces the $311***$, $313***$ slices to have statistics either $(2, 6, 6, 0)$ or $(4, 3, 3, 1)$. But the $311***$ slice (say) cannot have statistic $(4, 3, 3, 1)$, since when combined with the $(3, 3, 3, 1)$ statistics of $111***$ would give $a(*11***) = 7$, which contradicts Corollary 5.36; thus the $311***$ slice has statistic $(2, 6, 6, 0)$, and similarly for $331***$. But then $a(3*1***) = 4$, which again contradicts Corollary 5.36.

Thus no corner slice has statistic $(6, 8, 12, 8, 0)$. Now suppose that a corner slice, say $11****$ has statistic $(5, 12, 12, 4, 1)$. By Lemma 5.37, the

$1*1***$, $1*3***$ slices are bad, so by repeating the preceding arguments we conclude that the $311***$, $313***$ slices have statistics $(2,6,6,0)$ or $(4,3,3,1)$; in particular, their a-value is even. However, the $*11**$ and $*13**$ slices are bad by Lemma 5.37, and thus have an a-value of 5; thus the $111***$ and $113***$ slices have an odd a-value. Thus forces $a(11****)$ to be even; but it is equal to 5, a contradiction. ■

From this and Lemma 5.5, we see that A has statistics $(22, 72, 180, 80, 0, 0, 0)$. In particular, we have $2\alpha_2(A)+\alpha_3(A) = 2$, which by double counting (cf. (5.2)) shows that for every line of the form $11122*$ (or a reflection or permutation thereof) intersects A in exactly two points. Note that such lines connect a "*d*" point to two "*c*" points.

Also, we observe that two adjacent "*d*" points, such as 111222 and 113222, cannot both lie in A; for this would force the $*13***$ and $*11***$ slices to have statistics $(4,3,3,1)$ or $(3,3,3,1)$ by Corollary 5.38, which forces $a(*1****) = 6$, and thus $*1****$ must be good by Corollary 5.39; but this contradicts Lemma 5.28. Since $\alpha_3(A) = 1/2$, we conclude that given any two adjacent "*d*" points, exactly one of them lies in A. In particular, the d points of the form $***222$ consist either of those strings with an even number of 1s, or those with an odd number of 1s.

Let's say it's the former, thus the set contains 111222, 133222, and permutations of the first three coordinates, but omits 113222, 333222 and permutations of the first three coordinates. Since the "*d*" points 113222, 333222 are omitted, we conclude that the "*c*" points 113122, 113322, 333122, 333322 must lie in the set, and similarly for permutations of the first three and last three coordinates. But this gives at least 15 of the 16 "*c*" points ending in 22; by symmetry this leads to 225 c-points in all; but $c(A) = 180$, contradiction. This (finally!) completes the proof that $c'_{6,3} = 353$.

References

[1] M. Ajtai, E. Szemerédi, Sets of lattice points that form no squares, *Studia Scientiarum Mathematicarum Hungarica,* **9** (1974-1975), 9–11.

[2] T. Austin, *Deducing the density Hales–Jewett theorem from an infinitary removal lemma,* preprint, available at `arxiv.org/abs/0903.1633`.

[3] J. Beck, *Combinatorial Games: Tic-Tac-Toe Theory.* Cambridge University Press, 2008, Cambridge.

[4] F. Behrend, On the sets of integers which contain no three in arithmetic progression, *Proceedings of the National Academy of Sciences,* **23** (1946), 331-332.

[5] A. Brower, www.win.tue.nl/~aeb/codes/binary-1.html.

[6] A. Chandra, On the solution of Moser's problem in four dimensions, *Canad. Math. Bull.,* **16** (1973), 507–511.

[7] V. Chvátal, Remarks on a problem of Moser, *Canad. Math. Bull.,* **15** (1972) 19–21.

[8] V. Chvátal, Edmonds polytopes and a hierarchy of combinatorial problems, *Discrete Math.,* **4** (1973) 305–337.

[9] M. Elkin, *An Improved Construction of Progression-Free Sets,* preprint.

[10] K. Fujimura, www.puzzles.com/PuzzlePlayground/CoinsAndTriangles/CoinsAndTriangles.htm

[11] H. Furstenberg and Y. Katznelson, *A density version of the Hales–Jewett theorem for $k = 3$,* Graph Theory and Combinatorics (Cambridge, 1988). *Discrete Math.,* **75** (1989), 227-241.

[12] H. Furstenberg and Y. Katznelson, A density version of the Hales–Jewett theorem, *J. Anal. Math.,* **57** (1991), 64-119.

[13] D. Geller, I. Kra, S. Popescu and S. Simanca, *On circulant matrices,* www.math.sunysb.edu/~sorin/eprints/circulant.pdf

[14] B. Green and J. Wolf, *A note on Elkin's improvement of Behrend's construction,* preprint, available at arxiv.org/abs/0810.0732.

[15] M. Heule, presentation at www.st.ewi.tudelft.nl/sat/slides/waerden.pdf

[16] J. Komlós, solution to problem P.170 by Leo Moser, *Canad. Math. Bull.,* **15** (1972), 312–313, 1970.

[17] K. Markström, abel.math.umu.se/~klasm/Data/HJ/

[18] L. Moser, Problem P.170 in *Canad. Math. Bull.,* **13** (1970), 268.

[19] R. McCutcheon, *The conclusion of the proof of the density Hales–Jewett theorem for $k = 3$,* unpublished.

[20] K. O'Bryant, *Sets of integers that do not contain long arithmetic progressions,* preprint, available at arxiv.org/abs/0811.3057.

[21] N. J. A. Sloane, Ed. (2008), The On-Line Encyclopedia of Integer Sequences, www.research.att.com/~njas/sequences/

[22] A. Potechin, Maximal caps in $AG(6,3)$, *Des. Codes Cryptogr.,* **46** (2008), 243–259.

[23] D. H. J. Polymath, *A new proof of the density Hales–Jewett theorem,* preprint, available at arxiv.org/abs/0910.3926.

[24] D. H. J. Polymath, michaelnielsen.org/polymath1/index.php?title=Polymath1

[25] R. A. Rankin, Sets of integers containing not more than a given number of terms in arithmetical progression, *Proc. Roy. Soc. Edinburgh Sect. A,* **65** (1960/1961), 332-344.

[26] K. Roth, On certain sets of integers, I, *J. Lond. Math. Soc.,* **28** (1953), 104-109.

[27] S. Shelah, Primitive recursive bounds for van der Warden numbers, *J. Amer. Math. Soc.,* **28** (1988), 683-697.

[28] E. Sperner, Ein Satz über Untermengen einer endlichen Menge, *Mathematische Zeitschrift,* **27** (1928), 544-548.

[29] E. Szemerédi, On sets of integers containing no k elements in arithmetic progression, *Acta Arithmetica,* **27** (1975), 199-245.

D. H. J. Polymath

Address:

`e-mail:`

`http://michaelnielsen.org/polymath1/index.php`

BOLYAI SOCIETY
MATHEMATICAL STUDIES, 21

An Irregular Mind
(Szemerédi is 70)
pp. 755–758.

My Early Encounters with Szemerédi

ANDRÁS HAJNAL

I first met Endre in Szigliget, a place on Lake Balaton, about forty five years ago, in the summer of 1965. We were staying as guests in the Retreat House of the Writer's union with Erdős and his mother. The idea was to have a quiet place to work with Paul on one of our lengthy projects. Among the numerous visitors of Paul was Endre, a shy student of twenty five, attired in an old jeans quite long for him that looked preowned. He hitchiked from Budapest and complained that the police had stopped him several times objecting to his outfit (a novelty in those years in Hungary) He claimed to have told them that the *farmer* was that much taller and complained that the police would not understand it. (Jeans are called farmers in Hungarian.)

I knew that everybody referred to him as "Srác" (pronounced Shratz), meaning "Kid", a habit I soon accepted and still practice. He was not like our other stars, who could solve all the problems of the Schweitzer competitions by the second day, but he had original ideas and a deep insight in the Erdős type combinatorics. In fact after graduating from high school he attended medical school for a year and switched to math because he could not stand autopsies. Paul soon recognized his exceptional talents and expected great things from him. Later they became good friends and bantered a lot. Paul called him a mediocre old tennis player if he played too much and Endre pulled his leg by claiming that Jean-Claude Killy (the ski champ) was a greater man then any mathematician.

Soon, in 67 or early 68, Endre lived up to Paul's expectations. He proved his first famous result that a sequence of integers of positive upper density contains a four element arithmetic progression. The result for three element arithmetic progressions was proved by F. K. Roth, a Fields medal winner in 1958, and was considered as one of his major achievements. The problem of writing down the result was solved, as Halberstamm invited him to Nottingham for two months where Endre lectured in a seminar and a manuscript materialized. Witnesses say that Srác, who then hardly spoke any English, left his room only for the seminars and to buy some beer.

I happened to be in England about the same time, I visited Richard Rado in Reading for six weeks, to discuss our joint book with Paul. Because of shortage of funds I had stayed with the Rados. It goes without saying that we were both happy when we arrived to a prearranged meeting to the Piccadilly. I found him soon among the other two hundred youngsters sitting around the fountain. We had a pleasant dinner in a French restaurant in the Soho. This involved a lengthy discussion with the waitress explaining his position that even if the vegetables contained green peas, these unwanted objects should not touch the beef. During our long friendship this conversation was repeated many times, in many languages, in Hungarian, in Russian, and even in Polish which neither of us could speak. Endre mentioned that he never saw striptease before, so I suggested we go to a striptease bar. We were of course cheated with the obligatory consumption. We had to wait in a very hot small theatre quite long for the first girl to come out, and as soon she started her production I fainted. As many times later in our lives Srác had to save me. He dragged me to the street, where I soon recovered. I just had nicotine poisoning from the cheap cigars I smoked to save money on cigarettes. I have no information if Endre has ever seen striptease in a bar again.

The next year found us both in Moscow, in the Soviet Union. In his quest to learn modern abstract mathematics, Endre came to work with Israel Moiseevich Gelfand as a graduate student for three years. I was on an exchange program of the Academies in the Steklov Institute for six months. I did not speak Russian, and there was no one in the logic department interested in my work. None of them accepted the Axiom of Choice. I was hoping that this visit would help me with 'joe' to get an exit visa to Canada next year. I had an invitation to Calgary, then recently taken over by the British mathematicians. Still, this was quite a good deal. We stayed in a three bedroom apartment with my wife and three year old son, and I had lots of time to work.

Endre had to stay in a student hostel, and being finicky about food he was practically starving. His interest in functional analysis rapidly diminished. Now he wanted to learn forcing. Here he made some progress, but soon we started to talk about Paul's conjecture about the uniform decomposition of graphs into the union of $D+1$ independent sets, where D is the maximal degree. I started to admire his way of thinking and learned to translate or formalize his thoughts. He claimed that he can only think while walking, and on rainy days when we needed several sub lemmas he got totally wet. We finally proved the conjecture and this became one our most quoted papers. His Moscow adventure ended well. Gelfand was sensible enough to accept his famous result on four term arithmetical progressions as a PhD-thesis.

Moreover his extracurricular activity to boost the commerce between the population of Hungary and the Soviet Union ended with substantial profit. Even I got involved in it, at one point my living room held three huge boat motors to be exported to Hungary by younger students.

I spent the academic year 1970–71 at the University of Calgary. In the fall semester they organized a Number Theory Year in honor of Erdős. They invited his old friends from the UK, Mordell and Heilbronn, and several of his followers from the US. Szemerédi was also invited for six weeks. He arrived from Budapest practically with no luggage. He had in one hand a black container with an oil painting to be forwarded to the Grätzers in Winnipeg, and a shopping bag with a change of brief and a bottle containing a large baked goose liver. The only explanation we ever got was that originally he also had a salami which was of course confiscated at the border, but they did not think of the impertinence of the goose liver. Endre was staying with us. We used the time to write another not so well known tour de force paper with him and with Eric Milner. Let me copy here its review.

"Consider the following situation: There are n people, each knowing an item of information not known to any of the others. They communicate by telephone and whenever two people talk to one another during a call, each informs the other of all the information known at the time. The problem is to determine the minimum number $f(n)$ of calls needed for everyone to know all the information. In this note the authors elegantly prove the surprisingly stubborn result that $f(n) = 2n - 4$ for $n \geq 4$."

Eric's wife Estelle donated a suitcase to Endre for the trip back, but for reasons unknown, he refused to buy anything though he had enough money. The suitcase held only a large dartboard he got from us for Christmas.

In 1973 Endre announced that he generalized his old theorem for arbitrary arithmetic progressions. Paul was very excited, first because this was a problem he and Turán had raised about thirty years ago, and also because he promised one thousand dollars for it.

He was in the US and could not come to Budapest. He quarelled with 'joe' because he did not let some Israeli mathemathicians attend the conference held in Hungary for his sixtieth birthday. Endre was telling me the proof and I was writing it down as we went along. It took unusually long because we checked every detail. We were about halfway when Paul called from the US asking if there was a proof. I told him that I do not know, since I did not yet see the end of it, but I would buy it for $500. This convinced him that the matter is serious. Endre submitted the manuscript and it appeared quite fast in the Acta Arithmetica in 1975. It helped that Ron Graham kindly read it, but it appeared basically as it was submitted. Endre was lecturing about it at the ICM held in Vancouver. We were the only invited speakers there from Hungary, he in the Discrete Mathematics section and me in the Topology section.

I intend to stop here the play by play account of our youth. Though we remained close friends, built a house together, worked ten years the same time at Rutgers, visit each others family frequently, wrote a number of joint papers, and we vote together on the meetings of the Hungarian Academy, this is a good place to stop.

Endre has grown to be a great mathematician, one of the best our country has ever given to the world. That is what this conference and this book will be about.

To finish I want to quote something very characteristic Endre told me in a conversation. I complained about some idiot who got an undeservedly well paying job back in Hungary. He looked at me and asked: What do you want? Would you give up, even for a million dollars, the pleasure that you are able to think?!

MIX
Papier aus verantwortungsvollen Quellen
Paper from responsible sources
FSC® C105338

FSC
www.fsc.org

If you have any concerns about our products,
you can contact us on
ProductSafety@springernature.com

In case Publisher is established outside the EU,
the EU authorized representative is:
Springer Nature Customer Service Center GmbH
Europaplatz 3, 69115 Heidelberg, Germany

Printed by Libri Plureos GmbH
in Hamburg, Germany